Probability and Mathematical Statistics (C

PURI, VILAPLANA, and WERTZ
Applied Statistics
RANDLES and WOLFE • Introduction
Statistics
RAO • Linear Statistical Inference and Its Applications, *Second Edition*
RAO • Real and Stochastic Analysis
RAO and SEDRANSK • W.G. Cochran's Impact on Statistics
RAO • Asymptotic Theory of Statistical Inference
ROHATGI • An Introduction to Probability Theory and Mathematical Statistics
ROHATGI • Statistical Inference
ROSS • Stochastic Processes
RUBINSTEIN • Simulation and The Monte Carlo Method
SCHEFFE • The Analysis of Variance
SEBER • Linear Regression Analysis
SEBER • Multivariate Observations
SEN • Sequential Nonparametrics: Invariance Principles and Statistical Inference
SERFLING • Approximation Theorems of Mathematical Statistics
SHORACK and WELLNER • Empirical Processes with Applications to Statistics
TJUR • Probability Based on Radon Measures

Applied Probability and Statistics

ABRAHAM and LEDOLTER • Statistical Methods for Forecasting
AGRESTI • Analysis of Ordinal Categorical Data
AICKIN • Linear Statistical Analysis of Discrete Data
ANDERSON, AUQUIER, HAUCK, OAKES, VANDAELE, and WEISBERG • Statistical Methods for Comparative Studies
ARTHANARI and DODGE • Mathematical Programming in Statistics
ASMUSSEN • Applied Probability and Queues
BAILEY • The Elements of Stochastic Processes with Applications to the Natural Sciences
BAILEY • Mathematics, Statistics and Systems for Health
BARNETT • Interpreting Multivariate Data
BARNETT and LEWIS • Outliers in Statistical Data, *Second Edition*
BARTHOLOMEW • Stochastic Models for Social Processes, *Third Edition*
BARTHOLOMEW and FORBES • Statistical Techniques for Manpower Planning
BECK and ARNOLD • Parameter Estimation in Engineering and Science
BELSLEY, KUH, and WELSCH • Regression Diagnostics: Identifying Influential Data and Sources of Collinearity
BHAT • Elements of Applied Stochastic Processes, *Second Edition*
BLOOMFIELD • Fourier Analysis of Time Series: An Introduction
BOX • R. A. Fisher, The Life of a Scientist
BOX and DRAPER • Empirical Model-Building and Response Surfaces
BOX and DRAPER • Evolutionary Operation: A Statistical Method for Process Improvement
BOX, HUNTER, and HUNTER • Statistics for Experimenters: An Introduction to Design, Data Analysis, and Model Building
BROWN and HOLLANDER • Statistics: A Biomedical Introduction
BUNKE and BUNKE • Statistical Inference in Linear Models, Volume I
CHAMBERS • Computational Methods for Data Analysis
CHATTERJEE and PRICE • Regression Analysis by Example
CHOW • Econometric Analysis by Control Methods
CLARKE and DISNEY • Probability and Random Processes: A First Course with Applications, *Second Edition*
COCHRAN • Sampling Techniques, *Third Edition*
COCHRAN and COX • Experimental Designs, *Second Edition*
CONOVER • Practical Nonparametric Statistics, *Second Edition*
CONOVER and IMAN • Introduction to Modern Business Statistics
CORNELL • Experiments with Mixtures: Designs, Models and The Analysis of Mixture Data

Applied Probability and Statistics (Continued)

COX • Planning of Experiments
COX • A Handbook of Introductory Statistical Methods
DANIEL • Biostatistics: A Foundation for Analysis in the Health Sciences, *Fourth Edition*
DANIEL • Applications of Statistics to Industrial Experimentation
DANIEL and WOOD • Fitting Equations to Data: Computer Analysis of Multifactor Data, *Second Edition*
DAVID • Order Statistics, *Second Edition*
DAVISON • Multidimensional Scaling
DEGROOT, FIENBERG and KADANE • Statistics and the Law
DEMING • Sample Design in Business Research
DILLON and GOLDSTEIN • Multivariate Analysis: Methods and Applications
DODGE • Analysis of Experiments with Missing Data
DODGE and ROMIG • Sampling Inspection Tables, *Second Edition*
DOWDY and WEARDEN • Statistics for Research
DRAPER and SMITH • Applied Regression Analysis, *Second Edition*
DUNN • Basic Statistics: A Primer for the Biomedical Sciences, *Second Edition*
DUNN and CLARK • Applied Statistics: Analysis of Variance and Regression, *Second Edition*
ELANDT-JOHNSON and JOHNSON • Survival Models and Data Analysis
FLEISS • Statistical Methods for Rates and Proportions, *Second Edition*
FLEISS • The Design and Analysis of Clinical Experiments
FOX • Linear Statistical Models and Related Methods
FRANKEN, KÖNIG, ARNDT, and SCHMIDT • Queues and Point Processes
GALLANT • Nonlinear Statistical Models
GIBBONS, OLKIN, and SOBEL • Selecting and Ordering Populations: A New Statistical Methodology
GNANADESIKAN • Methods for Statistical Data Analysis of Multivariate Observations
GREENBERG and WEBSTER • Advanced Econometrics: A Bridge to the Literature
GROSS and HARRIS • Fundamentals of Queueing Theory, *Second Edition*
GUPTA and PANCHAPAKESAN • Multiple Decision Procedures: Theory and Methodology of Selecting and Ranking Populations
GUTTMAN, WILKS, and HUNTER • Introductory Engineering Statistics, *Third Edition*
HAHN and SHAPIRO • Statistical Models in Engineering
HALD • Statistical Theory with Engineering Applications
HAND • Discrimination and Classification
HOAGLIN, MOSTELLER and TUKEY • Exploring Data Tables, Trends and Shapes
HOAGLIN, MOSTELLER, and TUKEY • Understanding Robust and Exploratory Data Analysis
HOCHBERG and TAMHANE • Multiple Comparison Procedures
HOEL • Elementary Statistics, *Fourth Edition*
HOEL and JESSEN • Basic Statistics for Business and Economics, *Third Edition*
HOGG and KLUGMAN • Loss Distributions
HOLLANDER and WOLFE • Nonparametric Statistical Methods
IMAN and CONOVER • Modern Business Statistics
JESSEN • Statistical Survey Techniques
JOHNSON • Multivariate Statistical Simulation
JOHNSON and KOTZ • Distributions in Statistics
Discrete Distributions
Continuous Univariate Distributions—1
Continuous Univariate Distributions—2
Continuous Multivariate Distributions

(*continued on back*)

Linear Models for Unbalanced Data

About the author

Shayle R. Searle is Professor of Biological Statistics in the Biometrics Unit, New York State College of Agriculture and Life Sciences at Cornell University. After obtaining his MA degree in Mathematics at the University of New Zealand he received the Diploma of Mathematical Statistics from the University of Cambridge, and later the PhD degree from Cornell University. He is a Fellow of the Royal Statistical Society and the American Statistical Association. Professor Searle's other books are *Matrix Algebra for the Biological Sciences* (Wiley, 1966), *Matrix Algebra for Business and Economics* (with W.H. Hausman, Wiley, 1970), the very successful *Linear Models* (Wiley, 1971), and *Matrix Algebra Useful for Statistics* (Wiley, 1982).

Linear Models for Unbalanced Data

SHAYLE R. SEARLE

Professor of Biological Statistics
Biometrics Unit
New York State College of Agriculture and Life Sciences
Cornell University
Ithaca, New York

JOHN WILEY & SONS

New York · Chichester · Brisbane · Toronto · Singapore

To my wife Helen, for her elegant patience.

Library of Congress Cataloging in Publication Data:

Searle, S. R. (Shayle R.), 1928–
Linear models for unbalanced data.

(Wiley series in probability and mathematical statistics. Applied probability and statistics, ISSN 0271-6356)
Bibliography: p.
Includes index.
1. Analysis of variance. 2. Linear models (Statistics) I. Title. II. Title: Unbalanced data. III. Series.

QA279.S42 1987 519.5′352 87-2113
ISBN 0-471-84096-3

Printed in the United States of America

10 9 8 7 6 5 4 3 2 1

PREFACE

This book is designed for those who want to learn the fundamentals of linear statistical models: both college students and working statisticians can use it. Students who want one course to provide them with a basic knowledge for using linear models can get that knowledge from this book; and students who want a first course in linear models that can be the foundation for more advanced study can begin with this book. It is also designed for trained statisticians who want to broaden their understanding of linear models in order to better interpret their data from analyses that come from statistical computing packages.

The three salient features of the book are (i) its total devotion to unbalanced data—data having unequal numbers of observations in the subclasses, (ii) its emphatic dependence on the up-to-date cell means model approach to linear models for unbalanced data and (iii) its organization, in that the first half of the book requires no knowledge of matrices, whereas the second half does. The intent is first to present a basic corpus of linear model methodology for unbalanced data in a manner that requires no matrix algebra, and then to show the matrix description of linear models and some of the more applicable results that stem therefrom. In doing this there is no thought of the book being all-embracing mathematically, and accordingly the literature references may seem to some readers to be a little scanty. So be it: the object is to emphasize features of a practicable nature.

Unbalanced data are the focal point of the book because (a) they are harder to analyze and understand than balanced data; (b) unbalanced data are becoming increasingly prevalent through computer storage of data; and (c) since the staggering amount of arithmetic often required for analyzing unbalanced data can be done so quickly and cheaply by today's statistical computing packages, analyses of data are increasingly being done by people who are not fully knowledgeable about the statistical implications. There is therefore an urgent need for spreading the understanding of these analyses to such people. This need is especially catered to by the first six chapters of this book.

Confusion has reigned for many years about the interpretation of linear models analyses of unbalanced data—almost entirely as a result of unnecessary adherence to overparametrized models. But the difficulties that they can incur are now widely appreciated. Hence the time is ripe for learning about linear models, not in terms of these difficulties but through using the easier and more informative (for unbalanced data) cell means models. These models merit far wider adoption than they currently receive: when some-cells-empty data are to be analyzed in the presence of interactions they are almost the only form of linear model that may offer some help. Moreover, they have the advantage that they can be studied, used and understood without requiring matrix and vector notation. Such notation is necessary for certain aspects of linear models, but these can be dealt with after the fundamentals have been assimilated on the basis of cell means models.

Such is the plan of this book. Chapter 1 is introductory. Chapters 2–6, based entirely on the cell means approach and using no matrices whatever, deal with analyzing unbalanced data from the 1-way, 2-way nested and 2-way crossed classifications, and with the 1-way classification with a covariate. Chapter 8 has a general matrix description of linear model theory, with some of the prerequisite matrix algebra and results on quadratic forms (being generalizations of mean squares) contained in Chapter 7. Following Chapter 8 come topics that are extensions of Chapters 2–6, but for which dependence on the matrix description of linear models in Chapter 8 is found useful. Examples of such topics are overparameterized models (Chapter 9), restricted models (Section 9.4), multi-factor models (Section 10.1), cell means models omitting some interactions (Section 10.3), mixed models (briefly, in Chapter 13) and analysis of covariance, the general case (Section 11.2) and some particular cases (Section 11.4). There is also a brief discussion in Chapter 12 of some computing-package output and selected topics that are intimately connected with computer output such as Σ-restrictions, a faulty dummy-variable algorithm, and SAS Type IV sums of squares.

The book can be used in college courses in at least three ways: (1) Chapters 1–6 for an introductory course, without matrix algebra; (2) Chapters 1–13 for a fast-moving course which, for Chapters 7–13, has matrix algebra as a firm prerequisite; and (3) Chapters 1–13 for a two-course sequence, requiring matrix algebra for the second of those courses. The prerequisite for using this book is course work in statistical methods and/or mathematical statistics sufficient only to be well at ease with practical usage of the normal, χ^2-, F-, and t-distributions, and with such basic ideas as expectation, variance and covariance, functions of independent variables being independent, correlation coefficients, uncorrelated normally distributed variables being independent, and so on.

Readers do need to be thoroughly conversant with what is involved in using an F-statistic for testing a hypothesis. Although testing hypotheses is not necessarily the most important outcome of using a linear model, so many sums of squares and F-statistics are today put out by computing packages that users must be able to understand what their meaning is. The book therefore devotes considerable space to describing a variety of sums of squares, in order to inform readers what the various possibilities are. For each sum of squares, detailed description and illustration are given of the hypothesis tested by the corresponding F-statistic, because this hypothesis is what the usefulness of a sum of squares depends on. Some readers will undoubtedly feel that undue attention is paid to these topics. But computing packages are here to stay; and, rightly or wrongly, those packages produce an array of F-statistics for unbalanced data. Since many of those F-statistics are, in fact, of little or no use in general, it is only by providing explicit (and perhaps emphatic) description of the hypotheses being tested by all of them that we can be in a position to decide which are useful and which are not. Without such description the risk that many of those statistics will be wrongly used will continue unabated.

The book deals with the fixed effects model (save for Chapter 13, which is a summary review of the mixed model). Regression is not considered, because that topic now has its own vast literature. Nor is variance component estimation dealt with (except in summary manner, in Chapter 13)—that deserves its own books, and one day may get some. No real data are analyzed; it is assumed that readers know how to use an F-statistic garnered from their own data, once they know the form of its associated hypothesis. The book therefore directs attention to, and illustrates, these hypotheses for a variety of F-statistics available from unbalanced data. The many numerical illustrations and exercises use hypothetical data, designed with easy arithmetic in mind. This is because I agree with C. C. Li (1964), who points out that we do not learn to solve quadratic equations by working with something like

$$683125x^2 + 1268.4071x - 213.69825 = 0$$

just because it occurs in real life. Learning to first solve $x^2 + 3x + 2 = 0$ is far more instructive. Whereas real-life examples are certainly motivating, they usually involve arithmetic that becomes as cumbersome and as difficult to follow as is the algebra it is meant to illustrate. Furthermore, if one is going to use real-life examples, they must come from a variety of sources in order to appeal to a wide audience, but the changing from one example to another as succeeding points of analysis are developed and illustrated brings an inevitable loss of continuity. No apology is made, therefore, for the artificiality of the numerical examples used, nor for repeated use of the same example in many places. The attributes of continuity and of relatively easy arithmetic more than compensate for the lack of reality by ensuring that examples achieve their purpose of illustrating the algebra.

Computer output, *per se*, is neither given nor discussed in detail. Instead, following extensive discussion and numerical illustration of available calculations in Chapters 2–6 and 9–11, tables in Chapter 12 show which calculations are produced by a few of the widely-used statistical packages. Even this information will soon become outdated by continuing up-dates that the packages have; and to provide more detail would not only aggravate the outdating problem, but would be prohibitively expensive in book form. The Annotated Computer Output (ACO) that has been available from the Biometrics Unit at Cornell University for several years is the appropriate vehicle for this extensive detail.

Sections within chapters are numbered in the form 1.1, 1.2, 1.3, ... ; e.g., Section 1.3 is Section 3 of Chapter 1. These numbers are also shown in the running head of each page: e.g., [1.3] is found on page 11. Equations are numbered (1), (2), ... throughout each chapter. Equation references across chapters are few, but include explicit mention of the chapter concerned; otherwise "equation (4)" or just "(4)" means the equation numbered (4) in the chapter concerned. Exercises are in the final section of each chapter (except Chapters 1 and 12), with running heads such as [E5] meaning exercises of Chapter 5. Reference to exercise 2 of Chapter 5, for example, is then in the form E5.2.

I am greatly indebted to the Alexander von Humboldt Foundation of Bonn, Federal Republic of Germany, for awarding me a U.S. Senior Scientist Award in 1984, which enabled me to do most of the writing of this book. Grateful thanks also go to my hosts, Friedrich Pukelsheim and his colleagues, at the Mathematisches Institut of Universität Augsburg, where that writing was done. I also thank Harold V. Henderson, Stephen R. Lowry, Charles E. McCulloch, William H. Swallow and Adalbert Wilhelm for exceedingly helpful detailed comments on early drafts of the manuscript. Students in my Cornell courses and in a variety of short courses both on and off campus have also contributed many useful ideas. Special thanks go to Helen Seamon and Norma Phalen, the one for converting handwritten scrawl to the word processor with supreme care and accuracy; and the other for patiently and efficiently dealing with almost endless revisions for finalizing the manuscript: such helpful support is greatly appreciated. Thanks also go to Beatrice Shube of John Wiley & Sons for her endless patience and encouragement, which were of enormous help.

SHAYLE R. SEARLE

Ithaca, New York
June 1987

LIST OF CHAPTERS

CONTENTS

CHAPTER 1

AN UP-DATED VIEWPOINT: CELL MEANS MODELS

1.1. STATISTICS AND COMPUTERS

The age of the computer is upon us—all too obviously—and it includes statisticians. Statistical computing packages available today do our arithmetic for us in a way that was totally unthinkable thirty years ago. The capacity of today's computers for voluminous arithmetic, the great speed with which it is accomplished, and the low operating cost per unit of arithmetic—these characteristics are such as were totally unimaginable to most statisticians in the late 1950s. Solving equations for a 40-variable regression analysis could take six working weeks, using (electric) mechanical desk calculators. No wonder that regression analyses then seldom involved many variables. Today that arithmetic takes no more than ten seconds, an 86,400-fold reduction in time.

Statisticians' early uses of stored-program computers often involved calculating the analysis of variance of repetitive experiments such as, for example, those regularly used by plant breeders in their annual field experiments, e.g., split plots, latin squares and balanced incomplete blocks. Computing analyses of variance of data from these experiments was at first done from what would nowadays be considered a simple stand-alone computer program. Nevertheless, in those early days of using computers it was considered almost spectacular that once a program for, say, split plots was available it could in five minutes produce the sums of squares for such an experiment, and five minutes later those for another experiment, and so on, for a long line of experiments; each five minutes of computer time was replacing possibly three to four hours of desk calculator work. From this beginning, programmers soon found that computer code designed for calculating sums of squares for randomized complete blocks experiments could, by small adaptations, be made into a program for data from 3-factor experiments; and more adaptations produced one for 4-factor experiments. Next came the realization that programs could be written for any k-factor factorial experiment where k was limited, by the then current machine capacity, to some number such as 8 or 10, say. Modest though that limit seems today, it was far from modest in terms of the alternative desk calculator time required for the same arithmetic. Just try using one of today's pocket computers solely for accumulating a single sum of squares, without using any stored-program facilities whatever: with that as one's only computing aid, calculate the sums of squares of a 6-factor factorial experiment with all interactions; and remember, all arithmetic has to be checked by doing it twice. As a contrast to that effort, even the earliest of stored-program computers were soon recognized by statisticians as being enormously useful.

Program adaptations followed one another in a long succession, paralleling the development of increasingly larger, faster and cheaper (to use, per unit of arithmetic) computers. Programs went from calculating just sums of squares for equal-subclass-numbers data to also doing it for unequal-subclass-numbers data; extensions were incorporated for using covariables, for data editing and description (e.g., means, ranges, percentiles), for calculating contrasts and their confidence intervals, and for calculating mean squares, F-ratios and, ultimately, even their P-values; and on and on. Thus were born through some thirty years gestation, the extensive computing packages that we have today, with their numerous and large capacity routines. They provide us, for example, with the ability (if we so wish it) to relatively easily calculate the analysis of variance for a 20-factor factorial with all interactions, using 100,000 observations on 15 variables, with 10 covariables. But the all-important question would then be: Does such an analysis make sense?

Thinking about such a question is essential to sane usage of statistical computing packages. Indeed, a more fundamental question prior to doing an intended analysis is "Is it sensible to do this analysis?". Consider how the environment in which we contemplate this question has changed as a result of the existence of today's packages. Prior to having high-speed computing, the six weeks that it took for solving the least squares equations for a 40-variable regression analysis had a very salutory effect on planning the analysis. One did not embark on such a task lightly; much forethought would first be given as to whether such voluminous arithmetic would likely be worthwhile or not. Questions about which variables to use would be argued at length: are all forty necessary, or could fewer suffice, and if so, which ones? Thought-provoking questions of this nature were not lightly dismissed. Once the six-week task were to be settled on and begun, there would be no going back; at least not without totally wasting effort up to that point. Inconceivable was any notion of "try these 40 variables, and then a different set of maybe 10, 15 or 20 variables". Yet this is an attitude that can be taken today, because computing facilities (machines and programs) enable the arithmetic to be done in minutes, not weeks, and at very small cost compared to six weeks of human labor. Further, and this is the flash-point for embarking on thoughtless analyses, these computing facilities can be initiated with barely a thought either for the subject-matter of the data being analyzed or for that all-important question "Is this a sensible analysis?"

This is the danger to the discipline and profession of statistics, having such easily accessible, cheap and fast computing facilities as are available nowadays. The easy access and low cost mean that very minimal (maybe zero) statistical knowledge is needed for getting what can be voluminous and sophisticated arithmetic easily accomplished. But that same minimal knowledge may be woefully inadequate for understanding the computer output, for knowing what it means and how to use it. Nowhere is this more true than with linear model analyses. Several large computing packages will now easily carry out extensive linear model calculations on large, very large, data sets—including those with unequal numbers of observations in the subclasses (unbalanced data). Interpreting computing package output for such data simply by analogy with that from data coming from well-designed and well-executed experiments can lead to many wrong interpretations.

The purpose of this book is to provide information about linear model analyses for unbalanced data that will, hopefully, be of assistance to a wide group of readers: students who want just a single course in linear models as part of their general training in statistics; those who want a first course in linear models as a prelude to more advanced courses, and practitioners who want to know enough about linear models in order to easily and correctly use and understand the output from statistical computing packages.

The book puts the understanding of linear models first. Based on that understanding, readers should then be able to make correct use of computing packages. Thus the bulk of the book is on linear models—only in Chapter 12 is computer output specifically addressed. Output available from a few of the larger, more widely distributed computing packages is related to the linear model analyses discussed in the preceding chapters.

1.2. BALANCED AND UNBALANCED DATA

a. Factors, levels, effects and cells

Analysis of variance is concerned with attributing the variability that is evident in data to the various classifications by which the sources of data can be categorized. For example, consider a clinical trial where three different tranquilizer drugs are used on both men and women, some of whom are married and some not. The resulting data could be arrayed in the tabular form indicated by Table 1.1.

TABLE 1.1. A FORMAT FOR SUMMARIZING DATA

	Marital Status					
	Married			Not Married		
	Drug			Drug		
Sex	A	B	C	A	B	C
Male						
Female						

The three classifications, sex, drug and marital status, that identify the source of each datum are called *factors*. The individual classes of a classification are the *levels* of the factor; e.g., the three different drugs are the three levels of the factor "drug"; and male and female are the two levels of the factor "sex". The subset of data occurring at the "intersection" of one level of every factor being considered is said to be in a *cell* of the data. Thus with the three factors, sex (2 levels), drug (3 levels) and marital status (2 levels), there are $2 \times 3 \times 2 = 12$ cells.

In classifying data in terms of factors and their levels, the feature of interest is the extent to which different levels of a factor affect the variable of interest. We refer to this as the *effect* of a level of a factor on that variable. Effects are of two kinds. First are *fixed effects*, which are the effects attributable to a finite set of levels of a factor that occur in the data and which are there because we are interested in them; e.g., the effect on crop yield of three levels of a factor called fertilizer could correspond to the three different fertilizer regimes used in an agricultural experiment. They would be three regimes of particular interest, the effects of which we would want to quantify from the data to be collected from the experiment. Kempthorne (1975) contains, among other things, a good discussion of fixed effects.

The second kind of effects are *random effects*. These are attributable to a (usually) infinite set of levels of a factor, of which only a random sample are deemed to occur in the data. For example, four loaves of bread are taken from each of six batches of bread baked at three different temperatures. Whereas the effects due to temperature would be considered fixed effects (presumably we are interested in the particular temperatures used), the effects due to batches would be considered random effects because the batches chosen would be deemed to be random samples of batches from some hypothetical, infinite population of batches. Since there is definite interest in the particular baking temperatures used, the statistical concern is to estimate those temperature effects; they are fixed effects. In contrast, there is no particular interest in the individual batches, because those that occur in the data are considered as just a random sample of batches, and so batch effects are random effects. There is therefore no interest in quantifying individual batch effects—instead there is usually great interest in estimating the variance of those effects. Thus such data are considered as having two sources of random variation: batch variance and, as usual, error variance. These two variances are known as *variance components.*

Models in which the only effects are fixed effects are called *fixed effects models*, or sometimes just *fixed models*. Models that contain both fixed effects and random effects are called *mixed models*. And those having (apart from a single, general mean common to all observations) only random effects are called *random effects models* or, more simply, *random models*.

This book deals mostly with fixed effects models. Chapter 13 considers mixed models just briefly. They and random models will be treated elsewhere.

b. Balanced data

Data can be usefully characterized in several ways that depend on whether or not each cell contains the same number of observations. When these numbers are the same, the data shall be described as *balanced data*; they typically come from designed factorial experiments that have been executed as planned.

A formal, rigorous, mathematical definition of balanced data is elusive. Definitions in terms of Kronecker products of matrices are implicit in Smith and Hocking (1978), Seifert (1979), Searle and Henderson (1979), and Anderson *et al.* (1984); and an explicit definition of a very broad class of balanced data is given in Searle (1987). These definitions are beyond the scope of this book.

The analysis of balanced data, whether the models are fixed, mixed or random, and whether there are interactions or not, is relatively easy and is certainly well known. It is recorded in numerous texts on the design and analysis of experiments; since it is to unbalanced data that this book is directed, little more will be said about balanced data. The analyses of them are labeled "standard" in the left-hand part of Figure 1.1.

c. Special cases of unbalanced data

In a general sense all data that are not balanced are, quite clearly, unbalanced. Nevertheless, there are at least two special cases of that broad class of unbalanced data which need to be identified. They can then be dispensed with because their analyses come within the purview of the standard (so-called) analyses of balanced data.

The first of these two cases is what can be called *planned unbalancedness*. This is usually when there are no observations on certain, carefully planned, combinations of levels of the factors involved in an experiment. An example of this is shown in Table 1.2. It is, in fact, a particular one-third of a 3-factor experiment (of rows, columns and treatments with three levels of each), which is known as a latin square of order 3, as shown in Table 1.3.

TABLE 1.2. AN EXAMPLE OF PLANNED UNBALANCEDNESS: A LATIN SQUARE (SEE TABLE 1.3)

	Number of Observations								
	Treatment								
	A			B			C		
	Column			Column			Column		
Row	1	2	3	1	2	3	1	2	3
1	1	0	0	0	1	0	0	0	1
2	0	1	0	0	0	1	1	0	0
3	0	0	1	1	0	0	0	1	0

TABLE 1.3. A LATIN SQUARE OF ORDER 3 (TREATMENTS A, B, C)

	Column		
Row	1	2	3
1	A	B	C
2	C	A	B
3	B	C	A

Another example of planned unbalancedness is an experiment involving 3 fertilizer treatments A, B and C, say, used on three blocks of land in which treatment pairs A and C, A and B, and B and C are used. This can be represented as a two-factor situation, each factor having three levels, with certain cells empty, as shown in Table 1.4.

TABLE 1.4. NUMBER OF OBSERVATIONS IN A BALANCED INCOMPLETE BLOCKS EXPERIMENT

	Block		
Treatment	1	2	3
A	1	1	0
B	0	1	1
C	1	0	1

This is a simple example of what is known as a balanced incomplete blocks experiment. In a manner more general than either of the two preceding examples, planned unbalancedness need not require that a planned subset of cells be empty; it could be that subsets of cells are just used unequally; e.g., Table 1.4 with every 0 and 1 being a 1 and 2, respectively, would still represent planned unbalancedness.

Analyses of variance of data exhibiting planned unbalancedness of the nature just illustrated have well-known and relatively straightforward analyses that are often to be found in the same places as those describing the analysis of variance of balanced data. We therefore consider these analyses to be "standard" as indicated in Figure 1.1—and consider them no further.

The second special case of unbalanced data is when the number of observations in every cell is the same, except that in a very few cells (one, two or three, say) the number of observations is just one or two less than all the other cells. This usually occurs when a very few intended observations have inadvertently been lost or gone missing somehow, possibly due to misadventure during the course of an experiment. Maybe in a laboratory experiment, equipment got broken or animals died; or in an agricultural experiment, farm animals broke fences and ate some experimental plots. Under these circumstances there are well-known techniques for *estimating* such *missing observations* [e.g., Steel and Torrie (1980), pp. 209, 227, and 388], following which one then uses the standard analyses for balanced data as indicated in Figure 1.1. We therefore give no further consideration to the case of missing observations.

d. Unbalanced data

After defining balanced data and excluding from all other data those that can be described as exhibiting planned unbalancedness or involving just a few missing observations, we are left with what shall be called unbalanced data. This is data where the numbers of observations in the cells (defined by one level of each factor) are not all equal, and may in fact be quite unequal. This can include some cells having no data, but, in contrast to planned unbalancedness, with those cells occurring in an unplanned manner. Survey data are often like this, where data are sometimes collected simply because they exist and so the numbers of observations in the cells are just those that are available. Records of many human activities are of this nature; e.g., yearly income for people classified by age, sex, education, education of each parent, and so on. This is the kind of data that shall be called *unbalanced data*.

Within the class of unbalanced data we make two divisions. One is for data in which all cells contain data; none are empty. We call these *all-cells-filled* data. Complementary to this are *some-cells-empty* data, wherein there are some cells that have no data. This division is vitally useful when we come to consider whether or not analyses shall pay heed to possibly interactions. This, too, is indicated in Figure 1.1.

All-cells-filled data can be analyzed using with-interaction models by means of the well-known (Yates, 1934) weighted-squares-of-means analysis. This analysis cannot be used on some-cells-empty data, although it can be described in terms of the cells means analysis that is virtually mandatory for some-cells-empty data. Both kinds of data can be analyzed on the basis of no-interaction models by using the main-effects-only analysis.

e. A summary

Figure 1.1 is a schematic summary of the four classifications of data just discussed: balanced, planned unbalanced, missing observations and unbalanced. The figure also shows, for unbalanced data, how the use of models either with or without interactions affects the preferred kinds of analyses to be undertaken, and in which sections of this book these analyses are discussed. (The broken line indicates an analysis that may not always be required.)

Two features of Figure 1.1 are important. First, the analyses suggested as useful for unbalanced data are quite separate from those shown as standard for balanced data and data exhibiting planned unbalancedness or involving just missing observations. True it is, that when analysis procedures for unbalanced data are applied to balanced data they do simplify to the standard analyses. But too often, and for too many years, have the analyses of unbalanced data been described and taught as a natural (and maybe even called a simple) extension of the standard analyses of balanced data. This is a fragile way to view the situation. It is preferable by far to think of the analysis of unbalanced data as quite separate from that of balanced data. To connect the two at the start of learning about unbalanced data is misleading and has led to a whole array of misunderstandings of unbal-

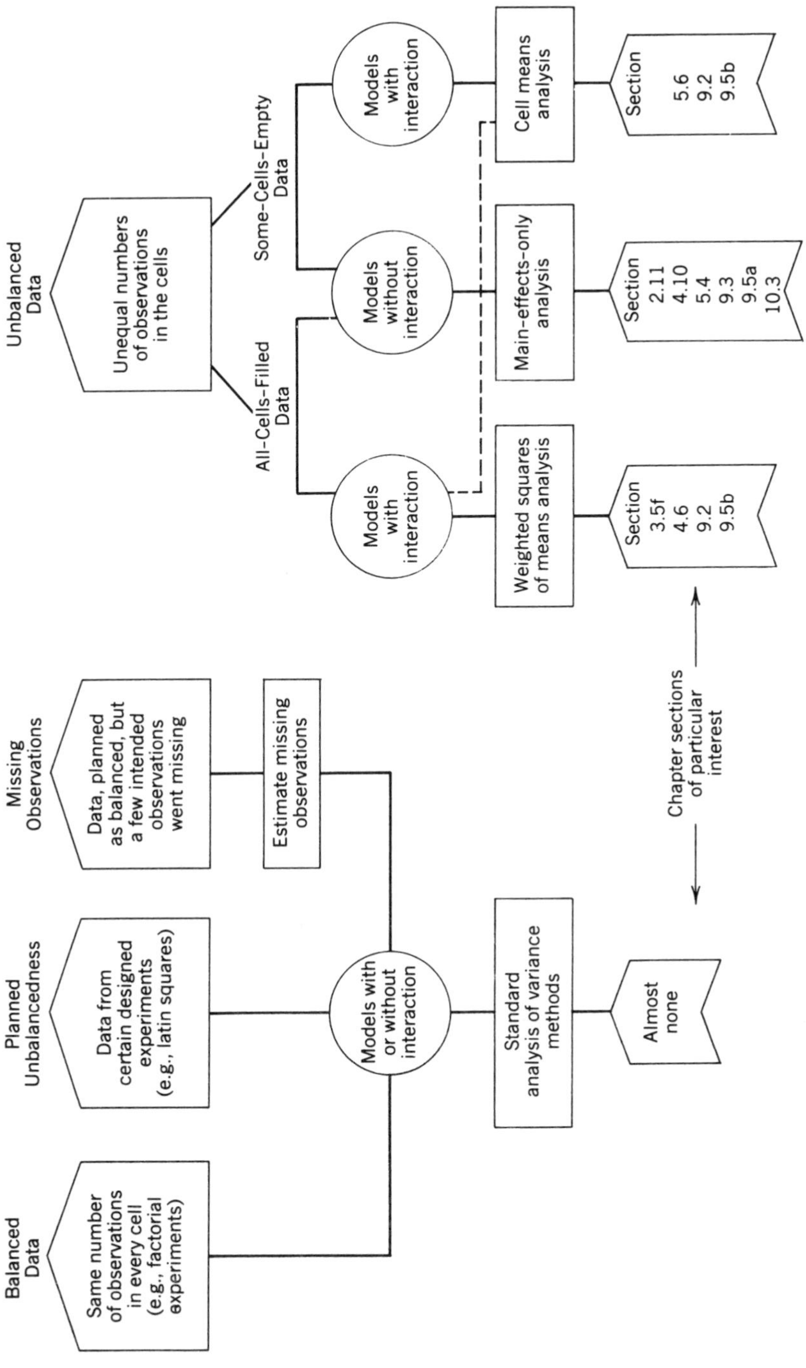

Figure 1.1. A characterization of data, models and analyses.

anced data. The useful connection comes at the end of learning about unbalanced data, for then one easily appreciates that balanced data are just a (very) special case of unbalanced data.

A particularly important feature of Figure 1.1 is that for unbalanced data the cell means model is the mainstay of most chapters of this book. To a large extent, the cell means model is the vehicle that makes the analysis of unbalanced data much more rational and easier to use and understand than does any attempt at expanding the traditional analysis procedures. Figure 1.1 shows the cell means model as a feature of analyzing with-interaction models, useful as secondary analysis (broken line) for all-cells-filled data and as primary analysis for some-cells-empty data. For the latter it is virtually mandatory for making any sense out of such data and, as shown in subsequent chapters, it is a very useful model in a variety of other ways. This, of course, is not seen in the figure, so demonstrating that any attempt to diagram all possible features of data and their connections to models and analyses simply cannot succeed. Other characteristics not appearing in Figure 1.1 are fixed and random effects; neither do the possible connections between missing-observations data and analyzing them by the methods indicated for unbalanced data. No doubt, readers will see other omissions. So be it. At least as a broad sweep, Figure 1.1 is useful for a general impression of data, models and analyses as we shall be concerned with them.

1.3. CELL MEANS MODELS

A customary practice of the last 20–30 years has been that of writing a model equation as a vehicle for describing many analysis of variance procedures. For example, for data classified by two factors that shall be called (quite generally) rows and columns, suppose there are a rows and b columns. Let y_{ijk} be the kth observation in row i and column j for $i = 1, 2, \ldots, a$ and $j = 1, 2, \ldots, b$, with there being n observations in every combination of row and column, so that $k = 1, 2, \ldots, n$. Without going into any great detail, which is reserved for Chapter 9, a customary model equation for y_{ijk} is

$$y_{ijk} = \mu + \alpha_i + \beta_j + \gamma_{ij} + e_{ijk}, \tag{1}$$

where μ is a general mean, α_i is the effect due to the ith row, β_j is the effect due to the jth column, γ_{ij} is the effect due to the interaction of the ith row and jth column, and e_{ijk} is a random residual error term.

One advantage of (1) is that the parameters (μ, the αs, βs and γs) in that equation make it easy to be specific about what we might like to estimate.

For example, concerning row effects, we might be interested in estimating a difference such as $\alpha_1 - \alpha_2$. However, a counteracting difficulty is that (1) involves more parameters than there are observed cell means to estimate them from. There are $1 + a + b + ab$ parameters but only ab cell means

$$\bar{y}_{ij\cdot} = \sum_{k=1}^{n} y_{ijk}/n, \tag{2}$$

for $i = 1, \ldots, a$ and $j = 1, \ldots, b$. Hence there are too many parameters for us to be able to estimate them all as linear functions of the observed $\bar{y}_{ij\cdot}$ cell means defined by (2). And the row and column means and the grand mean (the mean of all the observations), respectively

$$\bar{y}_{i\cdot\cdot} = \sum_{j=1}^{b} \bar{y}_{ij\cdot}/b, \qquad \bar{y}_{\cdot j\cdot} = \sum_{i=1}^{a} \bar{y}_{ij\cdot}/a \quad \text{and} \quad \bar{y}_{\cdot\cdot\cdot} = \sum_{i=1}^{a} \sum_{j=1}^{b} \bar{y}_{ij\cdot}/ab, \tag{3}$$

are of no help in this regard, additional to the cell means because, as is clear in (3), they are only linear combinations of those cell means.

This feature of having more parameters in a model than there are observed cell means to estimate them from is well known as *overparameterization* and leads to such a model being described as an *overparameterized model*. To circumvent this situation we usually invoke one of two procedures: either we use estimable functions (Section 8.7), which has us confine attention to only certain functions of the parameters that can be estimated satisfactorily from the data; or we use reparameterization (e.g., Sections 9.2c and 9.3c), wherein we define relationships among parameters of an overparameterized model, which has the implicit effect of rewriting the model in terms of no more than the maximum number of what may be termed "new" parameters as can be estimated from the data. Each of these procedures is easily applied and easily interpreted with balanced data. But the difficulties brought about by using overparameterized models are not always as easily circumvented for unbalanced data; estimable functions or reparameterization do not necessarily simplify the situation as they do with balanced data.

This is where the cell means model becomes so useful. Instead of the model equation (1) we use

$$y_{ijk} = \mu_{ij} + e_{ijk}, \tag{4}$$

where μ_{ij} is defined as the population mean of cell i, j, from which the observations y_{ijk} are deemed to be a random sample. Details of how this kind of model is used and of why it is such a help for unbalanced data are given at length in succeeding chapters. But even at this stage the reader can see how much simpler (4) is than (1). It is the keystone of this book, beginning in Chapter 2.

Using cell means as the basis of a model, as in (4), is in keeping with the early development of analysis of variance by R. A. Fisher as an analysis of differences among observed means. Thus for (3) and (4) he noted the obvious identity

$$y_{ijk} - \bar{y}_{...} \equiv (\bar{y}_{i..} - \bar{y}_{...}) + (\bar{y}_{.j.} - \bar{y}_{...}) + (\bar{y}_{ij.} - \bar{y}_{i..} - \bar{y}_{.j.} + \bar{y}_{...}) + (y_{ijk} - \bar{y}_{ij.}) \qquad (5)$$

and then, with an interest in sums of squares of the form $\Sigma_{t=1}^{r}(x_t - \bar{x})^2$ for $\bar{x} = \Sigma_{t=1}^{r}x_t/r$, established the further identity that

$$\Sigma\Sigma\Sigma(y_{ijk} - \bar{y}_{...})^2 \equiv \Sigma\Sigma\Sigma(\bar{y}_{i..} - \bar{y}_{...})^2 + \Sigma\Sigma\Sigma(\bar{y}_{.j.} - \bar{y}_{...})^2 + \Sigma\Sigma\Sigma(\bar{y}_{ij.} - \bar{y}_{i..} - \bar{y}_{.j.} + \bar{y}_{...})^2 + \Sigma\Sigma\Sigma(y_{ijk} - \bar{y}_{ij.})^2, \qquad (6)$$

where, in each case, the triple summation is $\Sigma_{i=1}^{a}\Sigma_{j=1}^{b}\Sigma_{k=1}^{n}$.

It seems that at no point did Fisher use a model like (1) or (4). He simply looked at cell means—in his case, observed cell means. The cell means model (4) is addressed to cell means too—population cell means, but, as shall be seen, this leads in many cases to estimates that are observed cell means.

The ultimate development of an analysis of variance table from an identity such as (6) starts with just summarizing that identity. Then, on assuming normality with homoscedastic variances, each sum of squares on the right-hand side of (6) is distributed proportional to a χ^2-variable, independently of the others; and from this come the familiar F-statistics. These and (6) are then summarized in tabular form as an analysis of variance table.

Fisher has an interesting comment on such a table. In a letter dated 6/Jan/'34 (on display at the 50th Anniversary Conference of the Statistics Department at Iowa State University, June, 1983), Fisher writes to

Snedecor that

> "the analysis of variance is (not a mathematical theorem but) a simple method of arranging arithmetical facts so as to isolate and display the essential features of a body of data with the utmost simplicity."

That the analysis of variance table is indeed, as Fisher says, no more than "a simple method of arranging arithmetical facts" is worth emphasizing in these days of computer-generated tables, which too many computer package users are inclined to uncritically treat as sacrosanct. What is important about this, though, is that whilst computers are efficient at doing arithmetic, the intervention of human thinking is always required for making valid interpretation.

1.4. STATISTICAL COMPUTING PACKAGES

Statistical computing packages for analysis of variance that can handle unbalanced data can, of course, also deal with balanced data since they represent just a special case. Yet an important distinction exists concerning package output from those two kinds of data. What we are calling the standard analyses of variance for balanced data are generally well known, open to little question and described in many books; and for many experiment designs the particular analysis is essentially unique. As a result, for a given experiment and resulting data set, most statistical packages produce essentially the same analysis, namely that which is well known, widely documented and ubiquitously accepted. Moreover, in the resulting output, the labeling of the sums of squares is usually sufficiently descriptive that, through knowing the well-established standard analysis for balanced data, one's interpretation of the computer output is unambiguous. For example, a sum of squares labeled "treatments" from the latin square of Table 1.2 would be well known as the numerator sum of squares for an F-statistic suitable for testing the hypothesis that treatment effects are all equal.

This feature of balanced data, namely that the general availability and acceptance of the standard analysis of many a particular case makes for straightforward interpretation of computer output, unfortunately does not carry over to unbalanced data. First, for unbalanced data there is often no unique, unambiguous method of analysis. Instead, several methods are usually available, methods that are often not as easily interpreted as methods for balanced data; nor are they as well known, nor so widely documented. Second, as a result of having several methods of analysis, not all statistical computing packages necessarily do the same analysis on any given set of unbalanced data. Consequently, in the context of hypothesis testing, or of arraying sums of squares in an analysis of variance format, there is often, for the one data set, a variety of sums of squares available from computing packages. The problem is to identify those that are useful.

This directing of attention to sums of squares must not be taken as implying that sums of squares are the only important calculations in linear model work; far from it. But, whether we like it or not, the real-world usage of linear model analysis is coming to depend on calculations available from statistical computing packages. And reasonably so, too, since the calculations can be voluminous and tedious. Therefore, since a large (but by no means total) part of the output from computer packages consists of sums of squares, it behooves users of packages to know how to interpret those sums of squares.

It is true, of course, that computer output includes a label for almost every calculated value shown therein. Unfortunately, though, the same label in different places can sometimes mean different things. For example, in analyzing unbalanced data from a completely randomized design with a covariate, Searle and Hudson (1982) report the label "sum of squares due to the mean" being attributable to five different values. This illustrates how the labeling of computed values does not always adequately describe underlying calculations and their meaning. Users of computing packages therefore face problems: for a given computed sum of squares, what is the underlying calculation, what is its meaning and what is its use? To answer these questions the user must first know what the procedures for analyzing unbalanced data are. They are the basis on which computer output can be understood and from which appropriate use of that output can be made. And also, of course, from which improved packages can be designed.

1.5. HYPOTHESIS TESTING

Although hypothesis testing is certainly one very useful feature of linear models analysis, it is by no means the only useful aspect of that analysis. Interval estimation (confidence intervals) may well be more useful in many situations. Nevertheless, the tradition of arraying sums of squares and resulting F-statistics in an analysis of variance table is so firmly entrenched in the analysis of balanced data that, through the unhappy extension of that analysis to the analysis of unbalanced data, the tradition has been perpetuated in many computing packages that handle unbalanced data.

One's first (and often most lasting) acquaintance with F-statistics is from calculating them in analysis of variance tables for data from carefully designed and executed experiments such as randomized complete blocks, latin squares, factorial experiments and so on. In most of these the hypotheses being tested by the F-statistics available in the resulting analysis of variance tables are quite clear and straightforward, and usually useful. But such is not often the case with unbalanced data. For these data, the many different sums of squares available do, if used in the numerators of F-statistics, provide tests of a wide variety of hypotheses, some of which are decidedly more useful than others, including some of which are of no general use whatever. Faced as we are with numerous sums of squares available from unbalanced data, the assessment of the usefulness of each is therefore tied to the question "What hypothesis is it testing?".

The mere existence of the preceding question emphasizes the topsy-turvy nature of the situation that prompts it. We all know that the proper logic for hypothesis testing is to set up a hypothesis of interest in the context of the scientific investigation at hand, construct a test for it and then collect data to carry out that test. Unfortunately, in the presence of today's statistical computing packages, this logic can all too easily not be followed: data are often first collected, then fed to a computer, and only then are hypotheses formulated corresponding, perhaps, to just the important-looking (does one dare say "significant"?) F-statistics. Data can so readily be subjected to computer processing without sufficient forethought as to what hypotheses are to be tested that computing power gets used to calculate values whose need has not necessarily been carefully planned in accord with the investigation at hand. Data are simply fed to a package and the (ofttimes voluminous) output that ensues immediately prompts the question "What does this output mean?". Deciding from numerical output alone whether or not a ratio of mean squares, when compared to tabulated values of the F-distribution, is significant or not is easy. However, this can be useful only if one is certain that that ratio of mean squares posing as an F-statistic does indeed (under the usual normality conditions) have an F-distribution. For example, some ratios of mean squares which in fixed effects models have F-distributions do not do so if some of those effects are random effects, that is, in mixed models. It is for this kind of reason that one needs to know how to ascertain whether or not any particular ratio of mean squares does indeed have an F-distribution. Only when it does, does comparing a computed ratio to tabulated Fs make sense. That is the first thing. The second is to then ascertain what hypothesis is being tested by each such statistic. For example, even with a model as simple as $E(y_{ij}) = \mu_i$ it is essential to appreciate that $\Sigma_i \Sigma_j (\bar{y}_{i\cdot} - \bar{y}_{\cdot\cdot})^2$ is a sum of squares that tests not H: $\mu_i = \mu$ *for all* i, for some pre-assigned value μ, but that it tests

H: μ_i *all equal*. Deriving precisely what hypothesis is being tested by an F is not always as easy as this simple example might suggest. Particularly is this so for unbalanced data, where labels attached to computed values are often not sufficient for a user to be certain of knowing what hypothesis is being tested. Therefore, since reliance is undoubtedly coming to be placed on computing packages for doing the arithmetic of linear model analysis, it becomes essential that we know what possible sums of squares can be computed, that we know which ratios of mean squares are F-statistics, that we know how to derive from a sum of squares the hypothesis tested when it is used as the numerator of an F-statistic, and that we know what hypotheses are tested by the sums of squares produced by the computing packages. Only then can wise use be made of those packages, and of the F-statistics they produce. Computing package output cannot be usefully employed without intervention from *homo sapiens*—the "sapiens" of our species must be utilized, for otherwise a computer is no better than *homo insipiens*.

CHAPTER 2

BASIC RESULTS FOR CELL MEANS MODELS: THE 1-WAY CLASSIFICATION

We begin with the simplest case of a cell means model, but one which at the same time provides all the basic results for more complicated cases. Chapters 3–6, on nested classifications, the 2-way crossed classification and covariance analyses, are based directly on the 1-way classification of this chapter—all without matrix notation. Not until Chapter 7 do we introduce matrices and in Chapter 8 give the general description of linear model theory that their use facilitates; and then Chapters 9–11 deal with more difficult topics stemming from Chapters 2–6.

2.1. AN EXAMPLE

Numerical examples are used somewhat for motivation but primarily to illustrate algebra. All of them involve trite hypothetical data, using numbers that have been chosen with an eye towards simplifying arithmetic. The implausibility of such numbers as real data is more than compensated for by improved simplicity of the ensuing arithmetic. Since the *raison d'être* of that arithmetic is to illustrate techniques, simplicity seems more to be desired than mimicry of real life. The latter inevitably involves numbers that become as difficult to follow as the algebra they purport to illustrate.

Suppose seven tomato plants are grown with three different fertilizer treatments: three plants receive one treatment, two receive another and two get the third. Table 2.1 shows the height of the seven plants, eight weeks after being planted out as seedlings all of the same age.

TABLE 2.1. HEIGHT OF TOMATO PLANTS (IN INCHES)

	Treatment			
	1	2	3	
	74	76	87	
	68	80	91	
	77			
Total	219	156	178	553 = grand total
Number	(3)	(2)	(2)	(7)
Mean	73	78	89	79 = grand mean

The data of Table 2.1 are classified by just one factor: fertilizer treatment, which has three levels. The data are therefore described as coming from a 1-way classification. Interest lies in estimating the effect of treatment on growth, in testing hypotheses about treatments and in estimating variability among plants.

In treatment i, for $i = 1, 2$ or 3, let y_{ij} denote the jth observation, where for $i = 1$ we have $j = 1, 2$ or 3; and for both $i = 2$ and $i = 3$, $j = 1$ or 2. The plants in treatment i are thought of as being a random sample from an infinite population of plants receiving that treatment, having a mean μ_i. Thus, using E to represent expectation over repeated sampling we have $E(y_{ij}) = \mu_i$. This is our first statement about a linear model; yet it turns out to be the foundation of almost everything else we do.

Notation. Discourses on mathematical statistics usually distinguish between random variables and realizations of them by using, for example, Y for a random variable and y for a realization thereof. We dispense with this distinguishing notation and make lowercase letters do double duty: occasionally as random variables, as in $E(y_{ij}) = \mu_i$, but more often as realizations of random variables.

2.2. THE MODEL AND MODEL EQUATIONS

a. The model equations

Generalizing the example to the case of a classes with n_i observations in the ith level for $i = 1, 2, \ldots, a$ gives y_{ij} as the jth observation in the ith level with $j = 1, 2, \ldots, n_i$. Then μ_i is defined as the population mean of the ith class and

$$E(y_{ij}) = \mu_i \, . \tag{1}$$

Since there is only a single factor, fertilizer treatment, individual levels of that factor (the classes) are themselves the cells of the data. Thus the μ_is are the (population) *cell means*, and we have what we call a *cell means model.* Although it is the simplest example of such a model, it turns out to be basic to a great deal of what is done with other cell means models.

Of course y_{ij} and $E(y_{ij})$ are not, in general, equal. Their difference is defined as

$$e_{ij} = y_{ij} - E(y_{ij}) \tag{2}$$

and is assumed to be a random variable. It accounts for everything that contributes to y_{ij} being different from its expectation $E(y_{ij})$. And, quite obviously, (1) and (2) give

$$y_{ij} = \mu_i + e_{ij} \ . \tag{3}$$

Equation (3) is the *model equation* for observation y_{ij}. One such equation exists for every observation.

Example (continued). The model equations for the data of Table 2.1 are

$$\begin{array}{lll} y_{11} = \mu_1 + e_{11} & y_{21} = \mu_2 + e_{21} & y_{31} = \mu_3 + e_{31} \\ y_{12} = \mu_1 + e_{12} & y_{22} = \mu_2 + e_{22} & y_{32} = \mu_3 + e_{32} \ . \\ y_{13} = \mu_1 + e_{13} & & \end{array} \tag{4}$$

b. The model

A model is more than just its model equations. Nevertheless, for ease of reference, a model is often referred to solely through its equation; e.g., the "model" $E(y_{ij}) = \mu_i$. But, more accurately, a model is its equation plus statements that describe the terms in the equation. Describing μ_i of (3) has already been done: it is the population mean of the ith class. And since e_{ij} is $y_{ij} - E(y_{ij})$ it is the extent to which, in a rough sense, $E(y_{ij})$ is in "error" as representing y_{ij}. Thus it is called the *error term*, or *residual error term*, and since it is assumed to be a random variable we need to attribute to it some probability distribution properties. Clearly, its expectation is zero:

$$E(e_{ij}) = E\left[y_{ij} - E(y_{ij})\right] = E(y_{ij}) - E(y_{ij}) = 0; \tag{5}$$

and its variance, to be denoted $v(e_{ij})$, is assumed to be the same for all e_{ij}s, namely σ^2:

$$v(e_{ij}) = E\left[e_{ij} - E(e_{ij})\right]^2 = E(e_{ij} - 0)^2 = E\left(e_{ij}^2\right) = \sigma^2 . \tag{6}$$

And we also assume that the covariance between every pair of (different) e_{ij}s is zero, i.e., unless $h = i$ and $k = j$, whereupon (6) applies,

$$\text{cov}(e_{ij}, e_{hk}) = 0, \tag{7}$$

$$\begin{aligned} E\left[e_{ij} - E(e_{ij})\right]\left[e_{hk} - E(e_{hk})\right] &= E\ (e_{ij} - 0)(e_{hk} - 0) \\ &= E(e_{ij}e_{hk}) = 0 . \end{aligned} \tag{8}$$

Notation. Equations (1) and (8) illustrate two ways of using E for expectation. In (1), a variable whose expectation is being considered is enclosed in parentheses—and in (5) it is in square brackets. The right-hand side of (8) has no such enclosing of the variable but, to avoid excessive parentheses, a space follows the E.

We can now make a complete statement of the model: for μ_i being the population mean of the ith class

$$y_{ij} = \mu_i + e_{ij}, \tag{9}$$

with

$$E(e_{ij}) = 0 \tag{10}$$

and

$$v(e_{ij}) = \sigma^2 \quad \forall\ i \text{ and } j; \tag{11}$$

and, for all i, j, h and k, except $i = h$ and $j = k$ together,

$$\text{cov}(e_{ij}, e_{hk}) = 0 . \tag{12}$$

Note that at this point the only distributional properties attributed to the error terms are those concerning first and second moments. Nothing has been said about what form the underlying distribution is, only that it has zero mean and finite variance σ^2. In point of fact, normality is the form of distribution usually attributed to the error term, but the need for this does not arise until after considering estimation. (See Section 2.10.)

Since, in contrast to e_{ij} being a random variable, μ_i is a constant, albeit unknown, we now have from (10), (11) and (12)

$$E(y_{ij}) = \mu_i, \qquad v(y_{ij}) = \sigma^2 \qquad \forall\ i \text{ and } j \tag{13}$$

and, for all i, j, h and k except $i = h$ and $j = k$ together,

$$\text{cov}(y_{ij}, y_{hk}) = 0 \ . \tag{14}$$

The preceding definitions of variances and covariances are widely used in linear models work, and are broadly adopted in this book. Nevertheless, they are by no means the only definitions that can be used. For example, it is quite feasible to define a separate variance for each class, such as $v(e_{ij}) = \sigma_i^2$ for $i = 1, \ldots, a$ in place of (6); and also to give some structure to $\text{cov}(e_{ij}e_{hk})$ other than all covariances zero: e.g., in combination with (6), $\text{cov}(e_{ij}, e_{hk}) = \rho\sigma^2$ for all pairs of $i \neq h$ for which $i - h = \pm 1$. The handling of such generalities is briefly considered in its broadest terms in Section 8.10.

2.3. ESTIMATION

The μ_is in (9) are usually estimated by the method of least squares. This involves momentarily viewing the μ_is as mathematical variables and minimizing

$$S = \sum_{i=1}^{a} \sum_{j=1}^{n_i} (y_{ij} - \mu_i)^2 \tag{15}$$

with respect to the μ_is. This can be achieved by applying elementary calculus to S, which leads to equating $-2\Sigma_{j=1}^{n_i}(y_{ij} - \mu_i)$ to zero. Denoting the μ_i that results from this by $\hat{\mu}_i$, which will be our estimator of μ_i, gives a set of a equations

$$\sum_{j=1}^{n_i} \hat{\mu}_i = \sum_{j=1}^{n_i} y_{ij} \quad \text{for } i = 1, \ldots, a \ . \tag{16}$$

Notation. In line with (3) of Section 1.3, we use $y_{i\cdot}$ as the total of the observations in class i:

$$y_{i\cdot} = \sum_{j=1}^{n_i} y_{ij}, \tag{17}$$

and $\bar{y}_{i\cdot}$ as the corresponding observed mean for that class:

$$\bar{y}_{i\cdot} = y_{i\cdot}/n_i = \sum_{j=1}^{n_i} y_{ij}/n_i \ . \tag{18}$$

This is the familiar dot notation for sums, and dot and bar notation for means. It extends directly to the grand total and grand mean,

$$y_{..} = \sum_{i=1}^{a} \sum_{j=1}^{n_i} y_{ij} = \sum_{i=1}^{a} n_i \bar{y}_{i.} \tag{19}$$

and

$$\bar{y}_{..} = y_{..}/N = \sum_{i=1}^{a} n_i \bar{y}_{i.} \bigg/ \sum_{i=1}^{a} n_i, \tag{20}$$

where N and $n.$ are used interchangeably,

$$N = n. = \sum_{i=1}^{a} n_i \tag{21}$$

for the total number of observations. To simplify appearances, limits of summation are omitted except when deemed essential. Thus for the remainder of this chapter $\Sigma_{i=1}^{a}$ appears as Σ_i and $\Sigma_{j=1}^{n_i}$ as Σ_j.

Using (17) in (16) gives $n_i\hat{\mu}_i = y_{i.}$, so that with (18) the estimator for μ_i is, for $i = 1, \ldots, a$,

$$\hat{\mu}_i = \bar{y}_{i.} \; . \tag{22}$$

Equation (16) is a simple example of what are generally called *normal equations*, the general form of which is described in Section 8.2.

The simple calculus for deriving (22) is that equating $\partial S/\partial\mu_i = -2\Sigma_j(y_{ij} - \mu_i)$ to zero with $\hat{\mu}_i$ in place of μ_i gives (16) and thence (22); and since $\partial^2 S/\partial\mu_i^2 = 2n_i > 0$ the solution of (16) for $\hat{\mu}_i$, namely (22), *is* minimizing S.

The least squares estimator $\hat{\mu}_i = \bar{y}_{i.}$ of (22) can also be derived (especially with benefit of hindsight) from (15) without using calculus. Observe that

$$\begin{aligned} S &= \Sigma_i\Sigma_j(y_{ij} - \mu_i)^2 = \Sigma_i\Sigma_j\left[y_{ij} - \bar{y}_{i.} - (\mu_i - \bar{y}_{i.})\right]^2 \\ &= \Sigma_i\Sigma_j(y_{ij} - \bar{y}_{i.})^2 + \Sigma_i\Sigma_j(\mu_i - \bar{y}_{i.})^2 \\ &\quad -2\Sigma_i\left[\Sigma_j(y_{ij} - \bar{y}_{i.})(\mu_i - \bar{y}_{i.})\right] \; . \end{aligned} \tag{23}$$

Then, because $\mu_i - \bar{y}_{i.}$ within the square brackets plays no part in the summation over j and because $\Sigma_j(y_{ij} - \bar{y}_{i.}) = 0$, the term in those square brackets is zero. Hence

$$S = \Sigma_i\Sigma_j(y_{ij} - \bar{y}_{i\cdot})^2 + \Sigma_i n_i(\mu_i - \bar{y}_{i\cdot})^2 \ .$$

Since both terms in S are sums of squares of real numbers, no negatives occur in S and so the only way to minimize S with respect to the μ_is is to make each squared term in the second summation in S zero; i.e., make $\mu_i - \bar{y}_{i\cdot}$ zero. This gives $\hat{\mu}_i = \bar{y}_{i\cdot}$ of (22).

Attention has been confined to estimation by the widely-accepted method of least squares. Three other methods that are briefly discussed in Section 8.10 are weighted least squares, best linear unbiased estimation, and maximum likelihood under normality. It so happens that for the model (9)–(12), all three of these methods give the same estimator as does least squares, namely $\hat{\mu}_i = \bar{y}_{i\cdot}$. Nevertheless, in more complicated situations these methods of estimation do not always lead to the same estimators, whereupon distinctions among them become important. But, at this point there need be no concern for such differences.

Example (continued). The means of the observations in the three classes of Table 2.1 are shown there as 73, 78 and 89. Hence

$$\hat{\mu}_1 = \bar{y}_{1\cdot} = 73, \qquad \hat{\mu}_2 = \bar{y}_{2\cdot} = 78 \quad \text{and} \quad \hat{\mu}_3 = \bar{y}_{3\cdot} = 89 \ .$$

2.4. EXPECTED VALUES AND SAMPLING VARIANCES

The expected value of $\hat{\mu}_i$ is μ_i:

$$E(\hat{\mu}_i) = \mu_i \ . \tag{24}$$

This is so because

$$E\hat{\mu}_i = E\bar{y}_{i\cdot} = E\,\Sigma_j y_{ij}/n_i = \Sigma_j E(y_{ij})/n_i = n_i\mu_i/n_i,$$

which gives (24). Thus the estimator $\hat{\mu}_i = \bar{y}_{i\cdot}$ is unbiased.

The variance of any random variable x is denoted by $v(x)$. We then have

$$v(\hat{\mu}_i) = \sigma^2/n_i \ . \tag{25}$$

This comes from (13) and (14):

$$v(\hat{\mu}_i) = v(\bar{y}_{i\cdot}) = v(\Sigma_j y_{ij})/n_i^2 = n_i\sigma^2/n_i^2 = \sigma^2/n_i \ .$$

And from (14) the covariance between any two different $\hat{\mu}_i$s is zero, i.e.,

$$\text{cov}(\hat{\mu}_i, \hat{\mu}_h) = 0 \quad \text{for } i \neq h \; . \tag{26}$$

A consequence of (25) and (26) is that the variance of any weighted sum of the $\hat{\mu}_i$s is very easily obtained. For $\Sigma_{i=1}^a k_i \hat{\mu}_i$ it is

$$v(\Sigma_i k_i \hat{\mu}_i) = v(\Sigma_i k_i \bar{y}_{i\cdot}) = \sigma^2 \Sigma_i \big(k_i^2/n_i\big) \; . \tag{27}$$

Example (continued). From the numbers of observations shown in Table 2.1, $n_1 = 3$, $n_2 = 2$ and $n_3 = 2$, so that $v(\hat{\mu}_1) = \frac{1}{3}\sigma^2$, $v(\hat{\mu}_2) = \frac{1}{2}\sigma^2$ and $v(\hat{\mu}_3) = \frac{1}{2}\sigma^2$. And $v(2\hat{\mu}_1 + 3\hat{\mu}_2) = \sigma^2(4/3 + 9/2) = 35\sigma^2/6$.

2.5. ESTIMATING $E(y_{ij})$ AND FUTURE OBSERVATIONS

The expected value of y_{ij} is $E(y_{ij}) = \mu_i$, from (13). Therefore an unbiased estimator of $E(y_{ij})$, which shall be denoted $\hat{y}_{ij}$, is

$$\hat{y}_{ij} = \hat{\mu}_i = \bar{y}_{i\cdot} \; . \tag{28}$$

The idea of estimating the expected value $E(y_{ij})$ is particularly easy here, because $E(y_{ij}) = \mu_i$ and the μ_is are the parameters we estimate. In more complicated models, where estimation is directed at parameters that have to be combined in some fashion in order to define a cell mean, the concept embodied in $\hat{y}_{ij}$ is more useful; for instance, it corresponds to what are often (in regression) called predicted y-values. Nevertheless, it does have an important use in considering the estimate of a future observation.

For each datum y_{ij} we already have $E(y_{ij}) = \mu_i$ estimated by $\hat{y}_{ij} = \hat{\mu}_i = \bar{y}_{i\cdot}$. In contrast, consider a future observation y_{if} from class i. By the model, $y_{if} = \mu_i + e_{if}$ where e_{if} is a random error term that can be neither observed nor estimated. Hence the best available prediction for y_{if}, which we shall call $\tilde{y}_{if}$, is $\tilde{y}_{if} = \hat{\mu}_i$. Thus $\hat{\mu}_i$ can be used both as a prediction of a future observation in class i as well as for its more customary use, that of estimating the expected value $E(y_{if})$. The first of these uses prompts inquiring how some future observation varies about its prediction, $\tilde{y}_{if} = \hat{\mu}_i$. To do this consider the deviation of y_{if} from $\tilde{y}_{if}$:

$$y_{if} - \tilde{y}_{if} = y_{if} - \hat{\mu}_i = \mu_i - \hat{\mu}_i + e_{if} \; .$$

The variance of this deviation is derived by noting, since y_{if} is conceived of as a future observation obtained independently of the observations used in

$\hat{\mu}_i$, that $\hat{\mu}_i$ and e_{if} have zero covariance, and so

$$v(y_{if} - \tilde{y}_{if}) = v(\mu_i - \hat{\mu}_i) + v(e_i) = \sigma^2(1/n_i + 1) \ . \tag{29}$$

Thus the estimated expected value of y_{ij} is $\hat{y}_{ij} = \hat{\mu}_i$ with variance σ^2/n_i; and the predicted value of a future observation in class i is the same value, $\tilde{y}_{if} = \hat{\mu}_i$, with the variance of deviations from this prediction being $\sigma^2(1/n_i + 1)$. These results are true for each class in the data. The variance of y_{if} itself is, of course, σ^2 at all times.

2.6. ESTIMATING THE ERROR VARIANCE

The error term in the model (3) is $e_{ij} = y_{ij} - \mu_i$, and so

$$\hat{e}_{ij} = y_{ij} - \hat{\mu}_i$$

is often called the *estimated residual*. It is the extent to which the observation y_{ij} differs from its estimated expected value $\hat{y}_{ij} = \hat{\mu}_i$. The sum of squares of these estimated residuals is known as the *error* (or *residual*) *sum of squares* for which the symbol SSE shall be used:

$$\text{SSE} = \Sigma_i\Sigma_j(y_{ij} - \hat{y}_{ij})^2 \tag{30}$$

$$= \Sigma_i\Sigma_j(y_{ij} - \bar{y}_{i\cdot})^2 \tag{31}$$

$$= \Sigma_i\left(\Sigma_j y_{ij}^2 - n_i\bar{y}_{i\cdot}^2\right) = \Sigma_i\Sigma_j y_{ij}^2 - \Sigma_i n_i\bar{y}_{i\cdot}^2 \ . \tag{32}$$

From SSE comes an unbiased estimator of σ^2. The model equation is

$$y_{ij} = \mu_i + e_{ij},$$

with $$\bar{y}_{i\cdot} = \frac{1}{n_i}\Sigma_j y_{ij} = \frac{1}{n_i}\Sigma_j(\mu_i + e_{ij}) = \mu_i + \frac{1}{n_i}\Sigma_j e_{ij} \ .$$

Therefore for $\bar{e}_{i\cdot} = (1/n_i)\Sigma_j e_{ij}$

$$y_{ij} - \bar{y}_{i\cdot} = e_{ij} - \bar{e}_{i\cdot} \ . \tag{33}$$

Using (33) in (31) gives

$$\text{SSE} = \Sigma_i\Sigma_j(e_{ij} - \bar{e}_{i\cdot})^2 = \Sigma_i\Sigma_j e_{ij}^2 - \Sigma_i n_i \bar{e}_{i\cdot}^2 \ .$$

Hence

$$E(\text{SSE}) = \Sigma_i\Sigma_j E(e_{ij}^2) - \Sigma_i n_i E(\bar{e}_{i\cdot}^2)$$
$$= \Sigma_i\Sigma_j v(e_{ij}) - \Sigma_i n_i v(\bar{e}_{i\cdot})$$

because $E(e_{ij}) = 0$ as in (10). Hence

$$E(\text{SSE}) = \Sigma_i\Sigma_j\sigma^2 - \Sigma n_i\sigma^2/n_i = \sigma^2(N - a),$$

for $N = n_\cdot$ defined in (21). Thus an unbiased estimator of $\hat{\sigma}^2$ is

$$\hat{\sigma}^2 = \frac{\text{SSE}}{N - a} = \frac{\Sigma_i\Sigma_j(y_{ij} - \bar{y}_{i\cdot})^2}{N - a} \ . \tag{34}$$

Example (continued). Table 2.1 shows values of y_{ij} and $\bar{y}_{i\cdot}$ so that (34) is

$$\hat{\sigma}^2 = \frac{\left[1^2 + (-5)^2 + 4^2\right] + \left[(-2)^2 + 2^2\right] + \left[(-2)^2 + 2^2\right]}{7 - 3} = \frac{58}{4} = 14\tfrac{1}{2} \ . \tag{35}$$

The numerator, SSE = 58, can also be calculated using (32):

$$\begin{aligned} \text{SSE} &= \Sigma_i\Sigma_j y_{ij}^2 - \Sigma_i n_i \bar{y}_{i\cdot}^2 \\ &= 74^2 + 68^2 + 77^2 + 76^2 + 80^2 + 87^2 + 91^2 \\ &\qquad - \left[3(73^2) + 2(78^2) + 2(89^2)\right] \\ &= 44055 - 43997 = 58 \ . \end{aligned} \tag{36}$$

2.7. REDUCTIONS IN SUMS OF SQUARES: THE $R(\cdot)$ NOTATION

The total sum of squares of each observation is

$$\text{SST} = \Sigma_i\Sigma_j y_{ij}^2 \ . \tag{37}$$

The error sum of squares after fitting the model $E(y_{ij}) = \mu_i$ is, from (31),

$$\text{SSE} = \Sigma_i\Sigma_j(y_{ij} - \bar{y}_{i\cdot})^2 \ . \tag{38}$$

The difference between these two is

$$\text{SST} - \text{SSE} = \Sigma_i\Sigma_j y_{ij}^2 - \Sigma_i\Sigma_j(y_{ij} - \bar{y}_{i\cdot})^2 = \Sigma_i n_i \bar{y}_{i\cdot}^2 \ . \tag{39}$$

Since SST is the sum of squares of the data, it is effectively the error sum of squares after fitting no model at all; whereas SSE is the sum of squares after fitting $E(y_{ij}) = \mu_i$. Therefore SST − SSE can be viewed as the reduction in sum of squares due to fitting the model $E(y_{ij}) = \mu_i$. It is the amount by which the error sum of squares for no model is reduced when the model $E(y_{ij}) = \mu_i$ is used. We denote it by $R(\mu_i)$—the reduction in sum of squares due to fitting the model $E(y_{ij}) = \mu_i$:

$$R(\mu_i) = \text{SST} - \text{SSE} = \Sigma_i n_i \bar{y}_{i\cdot}^2 \ . \tag{40}$$

The mnemonic nature of the symbol $R(\mu_i)$ is apparent: R is for "reduction" (in sum of squares) and the μ_i in parentheses indicates the model involved, $E(y_{ij}) = \mu_i$.

The $R(\cdot)$ notation is useful for its mnemonic value for comparing different linear models in terms of the extent to which fitting each of them accounts for a different reduction in the sum of squares. Each reduction is measured, as in (40), as SST minus the error sum of squares after fitting the model, i.e., as SST − SSE where SSE is, in turn, the error sum of squares due to fitting each model that is of interest. An example follows.

Consider fitting the simplest of all linear models

$$E(y_{ij}) = \mu \ . \tag{41}$$

Then, similar to (2), defining e_{ij} as $e_{ij} = y_{ij} - E(y_{ij})$ which for (41) is $e_{ij} = y_{ij} - \mu$, gives

$$y_{ij} = \mu + e_{ij} \ . \tag{42}$$

And on using least squares and minimizing $\Sigma_i\Sigma_j(y_{ij} - \mu)^2$ to estimate μ, the normal equation comparable to (22) would be $N\hat{\mu} = y_{\cdot\cdot}$ with solution

$$\hat{\mu} = \bar{y}_{\cdot\cdot} \ . \tag{43}$$

Hence for the model (41) the estimate of $E(y_{ij})$ is $\hat{y}_{ij} = \hat{\mu} = \bar{y}_{\cdot\cdot}$ and the error sum of squares, (30), is what shall be momentarily called

$$\text{SSE}(\mu) = \Sigma_i\Sigma_j(y_{ij} - \bar{y}_{\cdot\cdot})^2 \ .$$

Then SST − SSE(μ) shall be denoted as $R(\mu)$, because the model being used is $E(y_{ij}) = \mu$:

$$\begin{aligned} R(\mu) &= \text{SST} - \text{SSE}(\mu) \\ &= \Sigma_i\Sigma_j y_{ij}^2 - \Sigma_i\Sigma_j(y_{ij} - \bar{y}_{..})^2 \\ &= n_.\bar{y}_{..}^2 \ . \end{aligned} \tag{44}$$

This is the reduction in sum of squares due to fitting the model $E(y_{ij}) = \mu$.

The two models $E(y_{ij}) = \mu_i$ and $E(y_{ij}) = \mu$ can now be compared in terms of their respective reductions in sums of squares. $R(\mu_i) - R(\mu)$ is the extent to which fitting $E(y_{ij}) = \mu_i$ brings about a greater reduction in sum of squares than does fitting $E(y_{ij}) = \mu$. That difference is therefore of interest, and the $R(\cdot)$ notation is extended in a natural way to label it

$$\mathscr{R}(\mu_i \mid \mu) = R(\mu_i) - R(\mu) \ . \tag{45}$$

Thus $\mathscr{R}(\mu_i \mid \mu)$ is the additional reduction in sum of squares that will be achieved by fitting $E(y_{ij}) = \mu_i$ over and above fitting $E(y_{ij}) = \mu$. In this sense $\mathscr{R}(\mu_i \mid \mu)$ is then referred to as the sum of squares due to the μ_is, *adjusted for* μ; i.e., due to fitting $E(y_{ij}) = \mu_i$ adjusted for fitting (or after having fitted) $E(y_{ij}) = \mu$.

Notation. Readers familiar with the $R(\cdot)$ notation in overparameterized models [e.g., Searle (1971), p. 246] should notice the important distinction between $R(\cdot \mid \cdot)$ and $\mathscr{R}(\cdot \mid \cdot)$ that follows from (45). For two sets of parameters denoted by θ and Δ we define, in line with (45),

$$\mathscr{R}(\theta \mid \Delta) = R(\theta) - R(\Delta) \ . \tag{46}$$

In the earlier notation the definition of $R(\theta \mid \Delta)$ is

$$R(\theta \mid \Delta) = R(\theta, \Delta) - R(\Delta) \ .$$

This shall be retained, with the relationships

$$R(\theta \mid \Delta) = \mathscr{R}(\theta, \Delta \mid \Delta) \quad \text{and} \quad R(\theta \mid \Delta) \neq \mathscr{R}(\theta \mid \Delta) \ .$$

With the cell means models we find that $\mathscr{R}(\theta \mid \Delta)$ is a more useful notation than $R(\theta \mid \Delta)$. The latter implies that θ and Δ are both in a model together, whereas $\mathscr{R}(\theta \mid \Delta)$ does not necessarily imply this.

Example. Readers accustomed to $R(\alpha\,|\,\mu) = R(\mu, \alpha) - R(\mu)$ will now see it as $\mathscr{R}(\mu, \alpha\,|\,\mu) = R(\mu, \alpha) - R(\mu) = R(\alpha\,|\,\mu)$.

2.8. PARTITIONING THE TOTAL SUM OF SQUARES

a. For the analysis of variance

It is now very easy to partition the total sum of squares, SST, into terms that arise in the analysis of variance. From the underlying identity

$$\text{SST} = R(\mu) + [R(\mu_i) - R(\mu)] + [\text{SST} - R(\mu_i)]$$

we have, using (45) and (40) for the last two terms, respectively, that

$$\text{SST} = R(\mu) + \mathscr{R}(\mu_i\,|\,\mu) + \text{SSE} .$$

This is summarized in Table 2.2, where the familiar summation formulae are shown; $R(\mu)$ and SSE come from (44) and (38), respectively; and from (45), (40) and (44) we obtain

$$\mathscr{R}(\mu_i\,|\,\mu) = R(\mu_i) - R(\mu) = \Sigma_i n_i \bar{y}_{i\cdot}^2 - n_{\cdot}\bar{y}_{\cdot\cdot}^2$$

$$= \Sigma_i n_i(\bar{y}_{i\cdot} - \bar{y}_{\cdot\cdot})^2 = \Sigma_i\Sigma_j(\bar{y}_{i\cdot} - \bar{y}_{\cdot\cdot})^2 . \tag{47}$$

From these last two expressions we see that $\mathscr{R}(\mu_i\,|\,\mu)$ is a sum of squares. Note in general, though, that $\mathscr{R}(\theta\,|\,\Delta)$ as defined in (46) is not necessarily a sum of squares, since it is simply the difference between two reductions in sums of squares. Nevertheless, on almost all occasions that we use such a difference it does turn out to be a sum of squares.

TABLE 2.2. PARTITIONING THE TOTAL SUM OF SQUARES

Source of Variation	Sum of Squares			
Mean	$R(\mu)$		$= N\bar{y}_{\cdot\cdot}^2$	$=$ SSM
Model (a.f.m.)[1]	$\mathscr{R}(\mu_i\,|\,\mu)$	$= R(\mu_i) - R(\mu)$	$= \Sigma_i n_i(\bar{y}_{i\cdot} - \bar{y}_{\cdot\cdot})^2$	$=$ SSR_m
Residual	SSE	$= \text{SST} - R(\mu_i)$	$= \Sigma_i\Sigma_j(y_{ij} - \bar{y}_{i\cdot})^2$	
Total	SST		$= \Sigma_i\Sigma_j y_{ij}^2$	

[1] a.f.m. = adjusted for mean, in this and subsequent tables.

Along with the $R(\cdot)$ and $\mathscr{R}(\cdot)$ labels in Table 2.2 are

$$\text{SSM} = N\bar{y}_{..}^2 = R(\mu), \tag{48}$$

which is familiarly known as the *sum of squares due to the mean*, or as the *correction factor for the mean*; and also

$$\text{SSR}_m = \Sigma_i n_i(\bar{y}_{i\cdot} - \bar{y}_{..})^2 = \mathscr{R}(\mu_i \mid \mu), \tag{49}$$

the reduction in sum of squares due to fitting the model $E(y_{ij}) = \mu_i$, adjusted for the mean, as in (45).

Table 2.2 shows a partitioning of the total sum of squares SST that provides sums of squares $R(\mu)$ due to fitting $E(y_{ij}) = \mu$, and $\mathscr{R}(\mu_i \mid \mu)$ due to fitting $E(y_{ij}) = \mu_i$ adjusted for $E(y_{ij}) = \mu$. An alternative is Table 2.3, which explicitly shows $R(\mu_i)$ for fitting $E(y_{ij}) = \mu_i$ and then its partitioning into $R(\mu)$ and $\mathscr{R}(\mu_i \mid \mu)$. This table also shows the values of these sums of squares for the example of Table 2.1—which the reader should confirm.

A second alternative to Table 2.2 is Table 2.4, which is a partitioning of

$$\text{SST}_m = \text{SST} - R(\mu) = \Sigma_i\Sigma_j y_{ij}^2 - \bar{N}y_{..}^2 = \Sigma_i\Sigma_j(y_{ij} - \bar{y}_{..})^2 \,. \tag{50}$$

This is simply Table 2.2 with its first line removed from the body of the table so that the total is SST_m rather than SST.

Having developed Tables 2.2–2.4 it would seem natural to immediately proceed to discussing analysis of variance, F-statistics and tests of hypotheses. Those topics utilize the distribution properties of normality, which has not yet been introduced. Before making that introduction we deal with two

TABLE 2.3. PARTITIONING THE TOTAL SUM OF SQUARES WITH A SUB-PARTITIONING OF $R(\mu_i)$

Source of Variation	Sum of Squares: General Case		Example
Model	$\text{SSR} = R(\mu_i)$	$= \Sigma_i n_i \bar{y}_{i\cdot}^2$	43997
Mean	$\text{SSM} = R(\mu)$	$= N\bar{y}_{..}^2$	43687
Model, a.f.m.	$\text{SSR}_m = \mathscr{R}(\mu_i \mid \mu)$	$= \Sigma_i n_i(\bar{y}_{i\cdot} - \bar{y}_{..})^2$	310
Residual	$\text{SSE} = \text{SST} - R(\mu_i)$	$= \Sigma_i\Sigma_j(y_{ij} - \bar{y}_{i\cdot})^2$	58
Total	$\text{SST} = \Sigma_i\Sigma_j y_{ij}^2$		44055

TABLE 2.4. PARTITIONING THE TOTAL SUM OF SQUARES ADJUSTED FOR THE MEAN

Source of Variation	Sum of Squares: General Case	Example
Model, a.f.m.	$\text{SSR}_m = \mathscr{R}(\mu_i \mid \mu) \quad = \Sigma_i n_i(\bar{y}_{i\cdot} - \bar{y}_{\cdot\cdot})^2$	310
Residual	$\text{SSE} \quad = \text{SST} - R(\mu_i) = \Sigma_i\Sigma_j(y_{ij} - \bar{y}_{i\cdot})^2$	58
Total, a.f.m.	$\text{SST}_m = \text{SST} - R(\mu) \quad = \Sigma_i\Sigma_j(y_{ij} - \bar{y}_{\cdot\cdot})^2$	368

topics that also make no demand on normality: one to finish this section and one in the succeeding section. We return to these tables in Section 2.11 when discussing analysis of variance.

b. The coefficient of determination

The square of the product–moment correlation between the y_{ij}s and the $\hat{y}_{ij}$s is called the *coefficient of determination*:

$$R^2 = \frac{\left[\Sigma_i\Sigma_j(y_{ij} - \bar{y}_{\cdot\cdot})\left(\hat{y}_{ij} - \bar{\hat{y}}_{\cdot\cdot}\right)\right]^2}{\Sigma_i\Sigma_j(y_{ij} - \bar{y}_{\cdot\cdot})^2\Sigma_i\Sigma_j\left(\hat{y}_{ij} - \bar{\hat{y}}_{\cdot\cdot}\right)^2} . \tag{51}$$

Since $\hat{y}_{ij} = \bar{y}_{i\cdot}$, and $\bar{\hat{y}}_{ij} = \Sigma_i\Sigma_j\hat{y}_{ij}/N = \Sigma_i\Sigma_j\bar{y}_{i\cdot}/N = \Sigma_i n_i\bar{y}_{i\cdot}/N = \bar{y}_{\cdot\cdot}$,

$$R^2 = \frac{\left[\Sigma_i\Sigma_j(y_{ij} - \bar{y}_{\cdot\cdot})(\bar{y}_{i\cdot} - \bar{y}_{\cdot\cdot})\right]^2}{\Sigma_i\Sigma_j(y_{ij} - \bar{y}_{\cdot\cdot})^2\Sigma_i\Sigma_j(\bar{y}_{i\cdot} - \bar{y}_{\cdot\cdot})^2}$$

$$= \frac{\left[\Sigma_i\Sigma_j(\bar{y}_{i\cdot} - \bar{y}_{\cdot\cdot})^2\right]^2}{\Sigma_i\Sigma_j(y_{ij} - \bar{y}_{\cdot\cdot})^2\Sigma_i n_i(\bar{y}_{i\cdot} - \bar{y}_{\cdot\cdot})^2} = \frac{(\text{SSR}_m)^2}{\text{SST}_m\text{SSR}_m} = \frac{\text{SSR}_m}{\text{SST}_m} . \tag{52}$$

The reader should confirm that $\text{SSR}_m/\text{SST}_m = 310/368$, and that the same value is obtained by calculating R^2 from its definitional form, (51), using the data of Table 2.1.

In passing, we note that the result $R^2 = \text{SSR}_m/\text{SST}_m$ of (52) applies quite generally, as is shown in Section 8.4c, with SSR_m and SST_m being, respectively, the sums of squares for fitting a linear model and the total sum of squares, each adjusted for the mean. From its ratio form, $R^2 = \text{SSR}_m/\text{SST}_m$, we see that as well as being the square of a correlation coefficient, R^2 is also the fraction of SST_m that is contained in SSR_m. Thus R^2 represents the fraction of the total sum of squares (adjusted for the mean) that is accounted for by fitting the model, adjusted for the mean.

2.9. THE BEST LINEAR UNBIASED ESTIMATOR (BLUE)

Near the end of Section 2.3, mention was made of three methods of estimation additional to least squares. One of these is best linear unbiased estimation (BLUE), and although its general development is not given until Section 8.7c, we show here that $\hat{\mu}_i$ has the properties of, and therefore is, the BLUE of μ_i.

a. Linear functions

The graph of the simple function $y = bx$ in Cartesian co-ordinates x, y is a straight line. In this context y is said to be a *linear function* of x. The algebraic extension of this is that for variables $x_1, x_2, \ldots, x_t$ and any constants $k_1, k_2, \ldots, k_t$, the sum of the xs weighted by the ks, namely $\sum_{m=1}^{t} k_m x_m$, is said to be a *linear function* of the xs (or, equivalently, a *linear combination* of the xs). This definition is used extensively in what follows.

b. Definition of BLUE

The best linear unbiased estimator of a parameter is that linear function of the observations which is unbiased for the parameter and which, among all such linear unbiased functions, has smallest variance.

c. Verification: $\hat{\mu}_i$ is the BLUE of μ_i

Clearly, $\hat{\mu}_i = \bar{y}_{i\cdot} = \sum_j y_{ij}/n_i$ is a linear function of observations (even though all y_{hj} for which $h \neq i$ have zero coefficients). And, as in (24), $\hat{\mu}_i$ is unbiased for μ_i. Notice, though, that $\frac{1}{2}(y_{i1} + y_{i2})$ is also a linear function of observations which is unbiased for μ_i: and $\frac{1}{2}(y_{i1} + y_{i2})$ is different from $\bar{y}_{i\cdot}$ provided $n_i > 2$. Thus, in general, there are many linear unbiased estimators of μ_i. We now show that of all such estimators, $\hat{\mu}_i = \bar{y}_{i\cdot}$ has the smallest variance.

For some constants λ_{rj} for $r = 1, \ldots, a$ and $j = 1, \ldots, n_r$, consider the linear combination of observations

$$y_\lambda = \sum_{r=1}^{a} \sum_{j=1}^{n_r} \lambda_{rj} y_{rj} \,. \tag{53}$$

For y_λ to be an unbiased estimator of μ_i we must have

$$E(y_\lambda) = \mu_i; \quad \text{i.e.,} \quad E\sum_{r=1}^{a}\sum_{j=1}^{n_r}\lambda_{rj}y_{rj} = \mu_i \ .$$

Therefore the λs must satisfy

$$\sum_{r=1}^{a}\left(\sum_{j=1}^{n_r}\lambda_{rj}\right)\mu_r = \mu_i \ .$$

Hence, by equating coefficients, they must satisfy

$$\sum_{j=1}^{n_r}\lambda_{rj} = 0 \quad \text{for } r \neq i, \quad \text{and} \quad \sum_{j=1}^{n_i}\lambda_{ij} = 1 \ . \tag{54}$$

Thus y_λ of (53) with λs satisfying (54) is a linear unbiased estimator of μ_i.

Now consider the variance of y_λ. Using $v(y_{ij}) = \sigma^2$ and the zero covariances among the ys, as in (13) and (14), the variance of (53) is

$$v(y_\lambda) = \sigma^2\sum_{r=1}^{a}\sum_{j=1}^{n_r}\lambda_{rj}^2 = \sigma^2\left(\sum_{j=1}^{n_i}\lambda_{ij}^2 + \sum_{r\neq i}^{a}\sum_{j=1}^{n_r}\lambda_{rj}^2\right) \tag{55}$$

$$\geq \sigma^2\sum_{j=1}^{n_i}\lambda_{ij}^2 \ . \tag{56}$$

Observe, in general, that $$\sum_{m=1}^{t} x_m^2 \geq \left(\sum_{m=1}^{t} x_m\right)^2\Big/ t \tag{57}$$

because $$\sum_{m=1}^{t} x_m^2 - \left(\sum_{m=1}^{t} x_m\right)^2\Big/ t = \sum_{m=1}^{t}\left(x_m - \sum_{m=1}^{t} x_m/t\right)^2 \geq 0 \ . \tag{58}$$

Applying (57) to (56) gives $v(y_\lambda) \geq \sigma^2(\Sigma_{j=1}^{n_i}\lambda_{ij})^2/n_i$, which, because $\Sigma_j\lambda_{ij} = 1$ of (54), produces

$$v(y_\lambda) \geq \sigma^2/n_i; \tag{59}$$

i.e., $$v(y_\lambda) \geq v(\hat{\mu}_i) \quad \text{for} \quad \hat{\mu}_i = \bar{y}_{i\cdot} \ .$$

This shows that every linear function of the observations that is unbiased for μ_i has variance larger than (or equal to) that of $\hat{\mu}_i$: hence $\hat{\mu}_i$ is the BLUE of μ_i.

d. Derivation of $\hat{\mu}_i$ as the BLUE

Although $\hat{\mu}_i = \bar{y}_{i\cdot}$ was derived by using least squares, and we have just shown that it satisfies all conditions to be the BLUE of μ_i, and therefore is the BLUE, we can derive it directly as the BLUE by the following argument. Recall that a sum of squared real numbers is zero only if every one of those numbers is zero. Thus equality with 0 in (58) holds only when all xs are equal, so that this is also the case for the equality part of (57). Therefore $\sigma^2 \Sigma_j \lambda_{ij}^2$ of (56) can equal its minimum, σ^2/n_i of (59), only when λ_{ij} for $j = 1, 2, \ldots, n_i$ are all the same. This means, in conjunction with $\Sigma_j \lambda_{ij} = 1$ of (54), that $\lambda_{ij} = 1/n_i$ for $j = 1, 2, \ldots, n_i$. Furthermore, the equality part of (56) is, from (55), true only for $\Sigma_{r=1}^{a} \Sigma_{j=1}^{n_r} \lambda_{rj}^2 = 0$ for $r \neq i$, which means $\lambda_{rj} = 0$ for $r \neq i$ and $j = 1, 2, \ldots, n_r$. These conditions on the λs reduce y_λ to be $\bar{y}_{i\cdot}$, i.e., $\hat{\mu}_i = \bar{y}_{i\cdot}$.

e. A linear function of BLUEs is a BLUE

Consider a linear function of BLUEs: for simplicity's sake, of just two BLUEs, $\hat{\mu}_1$ and $\hat{\mu}_2$, say:

$$\hat{\omega} = k_1\hat{\mu}_1 + k_2\hat{\mu}_2 \quad \text{with} \quad v(\hat{\omega}) = \sigma^2\left(k_1^2/n_1 + k_2^2/n_2\right) . \tag{60}$$

First, since $\hat{\mu}_1$ and $\hat{\mu}_2$ are linear functions of the observations, so is $\hat{\omega}$. Second, on defining $\omega = k_1\mu_1 + k_2\mu_2$ it is clear that $E(\hat{\omega}) = \omega$, i.e., $\hat{\omega}$ is unbiased for ω. Suppose y_λ of (53) is considered as a possible alternative to $\hat{\omega}$. The argument that yielded (59) easily extends to showing that y_λ as an unbiased estimator of ω never has variance smaller than that of $\hat{\omega}$ given in (60). For suppose $E(y_\lambda) = \omega$. Then like (54),

$$\sum_{j=1}^{n_r} \lambda_{rj} = 0 \quad \text{for} \quad r \neq 1, 2, \qquad \sum_{j=1}^{n_1} \lambda_{1j} = k_1 \quad \text{and} \quad \sum_{j=1}^{n_2} \lambda_{2j} = k_2 . \tag{61}$$

Hence

$$\begin{aligned} v(y_\lambda) &= \sigma^2\left(\sum_{j=1}^{n_1} \lambda_{1j}^2 + \sum_{j=1}^{n_2} \lambda_{2j}^2 + \sum_{r=3}^{a} \sum_{j=1}^{n_r} \lambda_{rj}^2\right) \\ &\geq \sigma^2\left(\sum_{j=1}^{n_1} \lambda_{1j}^2 + \sum_{j=1}^{n_2} \lambda_{2j}^2\right) \\ &\geq \sigma^2\left[\left(\sum_{j=1}^{n_1} \lambda_{1j}\right)^2 \Big/ n_1 + \left(\sum_{j=1}^{n_2} \lambda_{2j}\right)^2 \Big/ n_2\right] \\ &\geq \sigma^2\left(k_1^2/n_1 + k_2^2/n_2\right) \end{aligned}$$

because of (61). Hence (60) gives $v(y_\lambda) \geq v(\hat{\omega})$ and so $\hat{\omega}$ is the BLUE of ω.

This result for the BLUE of a linear combination of two μ_is extends very naturally to the general result that the BLUE of any linear combination of μ_is is that same linear combination of the BLUE of the μ_is, i.e., of the $\hat{\mu}_i$s. We return to this result more generally in Section 8.7.

2.10. NORMALITY ASSUMPTIONS

Up till now, no particular form of probability distribution has been assumed for the e_{ij}s, only that they have zero mean, nonzero variance and zero covariances. Within these conditions we now assume normality, i.e., that the e_{ij}s are normally distributed. Then, because every e_{ij} has the same mean and variance, and because the mean and variance of a normal distribution completely determine the form of that distribution, all the e_{ij}s have the same normal distribution, that with zero mean and variance σ^2. We denote this by $\mathcal{N}(0, \sigma^2)$, and describe the e_{ij}s as being identically distributed $\mathcal{N}(0, \sigma^2)$. Furthermore, recall that a zero covariance between two normally distributed random variables means that they are independent. Hence, on assuming that the e_{ij}s are normally distributed, and because of the zero covariances among them, they are also independent. Thus we write

$$e_{ij} \sim \text{i.i.d. } \mathcal{N}(0, \sigma^2)$$

for $i = 1, \ldots, a$ and $j = 1, 2, \ldots, n_i$, meaning that the e_{ij}s are identically and independently distributed $\mathcal{N}(0, \sigma^2)$; and we often use the phrase "is normal" as abbreviation of "is normally distributed".

a. The y_{ij}s are normal

Because $e_{ij} \sim$ i.i.d. $\mathcal{N}(0, \sigma^2)$, properties (13) and (14) mean that for each $i = 1, \ldots, a$,

$$y_{ij} \sim \text{i.i.d. } \mathcal{N}\left(\mu_i, \sigma^2\right) \quad \text{for } j = 1, \ldots, n_i \ . \tag{62}$$

Also, because observations in class i have zero covariance with those in class $h \neq i$, the independence stated in (62) also applies from one class to another.

b. The BLUE, $\hat{\mu}_i$, is normal

It is a property of normal variables that linear functions of them are normal. Thus, with $E(\hat{\mu}_i) = \mu_i$ and $v(\hat{\mu}_i) = \sigma^2/n_i$ of (24) and (25)

$$\hat{\mu}_i = \bar{y}_{i\cdot} \sim \mathcal{N}\left(\mu_i, \sigma^2/n_i\right) \ . \tag{63}$$

Also, since for $h \neq i$, $\bar{y}_{i\cdot}$ and $\bar{y}_{h\cdot}$ are independent, so are $\hat{\mu}_i$ and $\hat{\mu}_h$.

c. $\hat{\mu}_i$ and $\hat{\sigma}^2$ are independent

From (34)

$$(N-a)\hat{\sigma}^2 = \sum_{h=1}^{a}\sum_{j=1}^{n_h}(y_{hj}-\bar{y}_{h\cdot})^2$$

and so for some particular i

$$(N-a)\hat{\sigma}^2 = \sum_{h\neq i}^{a}\sum_{j=1}^{n_h}(y_{hj}-\bar{y}_{h\cdot})^2 + \sum_{j=1}^{n_i}(y_{ij}-\bar{y}_{i\cdot})^2 \,. \tag{64}$$

But $\hat{\mu}_i = \bar{y}_{i\cdot}$; and for $h \neq i$ every y_{hj} has zero covariance with $\bar{y}_{i\cdot}$. Hence for the first term in (64), the covariance of $\bar{y}_{i\cdot}$ with $y_{hj} - \bar{y}_{h\cdot}$ is zero: and for the second term

$$\operatorname{cov}\left[\bar{y}_{i\cdot}, (y_{ij}-\bar{y}_{i\cdot})\right] = \sigma^2(1/n_i - 1/n_i) = 0 \,.$$

Thus, because of joint normality and these zero covariances, $\bar{y}_{i\cdot}$ is independent of every $y_{hj} - \bar{y}_{h\cdot}$ for $h = 1, \ldots, a$, including $h = i$. Further, using the important theorem about independent random variables (that functions of them are also independent), $\bar{y}_{i\cdot} = \hat{\mu}_i$ is therefore independent of every squared term in (64), and hence of $\hat{\sigma}^2$ itself.

d. SSR_m / σ^2 is distributed as χ^2_{a-1}

Recall the definition of the χ^2-distribution on t degrees of freedom χ^2_t:

$$\text{when} \quad u_i \sim \text{i.i.d. } \mathcal{N}(0,1) \quad \text{then} \quad \sum_{i=1}^{t} u_i^2 \sim \chi^2_t \,. \tag{65}$$

In terms of the model $E(y_{ij}) = \mu_i$, it is shown in a variety of texts that on assuming

$$\mu_1 = \mu_2 = \cdots = \mu_a, \tag{66}$$

then for SSR_m of (49),

$$\text{SSR}_m/\sigma^2 = \Sigma_i n_i(\bar{y}_{i\cdot} - \bar{y}_{\cdot\cdot})^2/\sigma^2 \sim \chi^2_{a-1} \,. \tag{67}$$

Derivation is usually based on results in matrix algebra and multivariate normal distribution theory. (See, for example, Sections 7.5b and 8.6b.) An alternative derivation based solely on induction and elementary properties of the univariate normal and taken from Searle and Pukelsheim (1985) is given in the Appendix (Section 2.14a). The proof is important because it also applies to a more general result used in Section 4.5.

e. SSE / σ^2 is distributed as χ^2_{N-a}

Write SSE $= \Sigma_i\Sigma_j(y_{ij} - \bar{y}_{i\cdot})^2 = \Sigma_i W_i$ for $W_i = \Sigma_j(y_{ij} - \bar{y}_{i\cdot})^2$. It is then a standard result in normal theory that $W_i/\sigma^2 \sim \chi^2_{n_i-1}$. [This is just a special case of (67): let Σ_i, n_i and a be Σ_j, 1, and n_i, respectively, and (67) becomes $W_i/\sigma^2 \sim \chi^2_{n_i-1}$.] Then, because $y_{ij} - \bar{y}_{i\cdot}$ is independent of $y_{hj} - \bar{y}_{h\cdot}$ for $h \neq i$, we have the W_is being independent. Therefore the standard result that a sum of independent χ^2-variables is a χ^2-variable leads to

$$\text{SSE}/\sigma^2 = \Sigma W_i/\sigma^2 \sim \chi^2_{\Sigma_i(n_i-1)} \sim \chi^2_{N-a} . \tag{68}$$

This is just a special case of the general result developed in (103) of Section 8.6a.

f. SSR_m and SSE are independent

Since $\text{SSR}_m = \Sigma_i n_i(\bar{y}_{i\cdot} - \bar{y}_{\cdot\cdot})^2$ and SSE $= \Sigma_i\Sigma_j(y_{ij} - \bar{y}_{i\cdot})^2$, consider the covariance of terms that are squared, one in SSR_m and one in SSE: using (13) and (14)

$$\text{cov}\left[(\bar{y}_{i\cdot} - \bar{y}_{\cdot\cdot}), (y_{ij} - \bar{y}_{i\cdot})\right] = \sigma^2(1/n_i - 1/n_i - 1/N + n_i/Nn_i) = 0$$

and, for $h \neq i$,

$$\text{cov}\left[(\bar{y}_{i\cdot} - \bar{y}_{\cdot\cdot}), (y_{hj} - \bar{y}_{h\cdot})\right] = \sigma^2(0 - 0 - 1/N + n_h/Nn_h) = 0 .$$

Since the ys are normally distributed, so are linear combinations of them; and since jointly normal variables that have zero covariance are independent, the two preceding zero covariances mean that every squared term in SSR_m is independent of every squared term in SSE. Therefore SSR_m and SSE are independent, this being a particular case of results developed in Section 8.6b-iii.

g. *F*-statistics

The basic definition of a variable that has an F-distribution is as follows: when $u \sim \chi^2_{n_1}$ and, independently, $v \sim \chi^2_{n_2}$ then

$$F = \frac{u/n_1}{v/n_2} \sim \mathscr{F}_{n_1 n_2}, \tag{69}$$

the F-distribution on n_1 and n_2 degrees of freedom. Based directly on this definition, we see from the immediately preceding subsections d, e and f that, provided we assume $\mu_1 = \mu_2 = \cdots = \mu_a$ from (66), then

$$F = \frac{\text{SSR}_m/(a-1)}{\text{SSE}/(N-a)} \sim \mathscr{F}_{a-1,\, N-a} \,. \tag{70}$$

Defining mean squares as

$$\text{MSR}_m = \text{SSR}_m/(a-1) \quad \text{and} \quad \text{MSE} = \text{SSE}/(N-a) \tag{71}$$

gives

$$F = \frac{\text{MSR}_m}{\text{MSE}} \sim \mathscr{F}_{a-1,\, N-a}; \tag{72}$$

or, on using $\text{MSE} = \text{SSE}/(N-a) = \hat{\sigma}^2$ of (34) an equivalent form of F is, using (49),

$$F = \frac{\text{SSR}_m}{(a-1)\hat{\sigma}^2} = \frac{\mathscr{R}(\mu_i \mid \mu)}{(a-1)\hat{\sigma}^2} \,. \tag{73}$$

2.11. THE ANALYSIS OF VARIANCE TABLE

a. A summary of arithmetic

The sums of squares involved in (70) are tabulated in Tables 2.2–2.4. To each of these the simple calculation of the mean squares in (71) and F in (72) can be added. We use the form of Table 2.4.

In Table 2.5, we see that an analysis of variance table is just what Fisher wrote that it is: "... a simple method of arranging arithmetical facts" (see

TABLE 2.5. ANALYSIS OF VARIANCE FOR THE 1-WAY CLASSIFICATION CELL MEANS MODEL

Source of Variation	d.f.[1]	Sum of Squares	Mean Square	F-statistic
Model	$a-1$	$\text{SSR}_m = \Sigma_i n_i(\bar{y}_{i\cdot} - \bar{y}_{\cdot\cdot})^2$	$\text{MSR}_m = \dfrac{\text{SSR}_m}{a-1}$	$\dfrac{\text{MSR}_m}{\text{MSE}}$
Residual	$N-a$	$\text{SSE} = \Sigma_i\Sigma_j(y_{ij} - \bar{y}_{i\cdot})^2$	$\text{MSE} = \dfrac{\text{SSE}}{N-a}$	
Total	$N-1$	$\text{SST}_m = \Sigma_i\Sigma_j(y_{ij} - \bar{y}_{\cdot\cdot})^2$		

[1]d.f. = degrees of freedom.

the end of Section 1.3) and, he went on, "to isolate and display the essential features of a body of data ...". We have done the arithmetic; we now describe those "essential features".

b. Tests of hypotheses

In starting off with the model $E(y_{ij}) = \mu_i$ there is no thought of assuming equality of the μ_is. Nevertheless, to derive $\text{SSR}_m/\sigma^2 \sim \chi^2_{a-1}$ of (67) and hence $F \sim \mathscr{F}_{a-1,\,N-a}$ of (72), for $F = \text{MSR}_m/\text{MSE}$ of (72) and Table 2.5, the assumption $\mu_1 = \mu_2 = \cdots = \mu_a$ was used. Through appealing to the basic properties of hypothesis testing using F-statistics (familiarity with which is presumed of the reader), using the assumption $\mu_1 = \mu_2 = \cdots = \mu_a$ makes F suitable for testing the hypothesis that the assumption (made initially for considering distributional properties of F) is true of the model. Thus $F = \text{MSR}_m/\text{MSE}$ of (72) and Table 2.5 can be used as a test statistic for testing the hypothesis

$$\text{H: } \mu_1 = \mu_2 = \cdots = \mu_a\,. \tag{74}$$

This is the first example of a feature of this book that occurs repeatedly: a hypothesis tested by an F-statistic. It is the simplest case of the hard, practical problem faced by a user of statistical computing packages discussed in Chapter 1: faced with not one but maybe several F-statistics in computer output, users must decide what applicability those F-statistics have. That decision is usually more difficult than in Table 2.5 not only because there may be several F-values, but also because most computer output shows numerical values of F-statistics with brief labeling, but not their background derivation as is given in Table 2.5. Addressing this problem of interpretation is an important part of this book, not simply by explaining the various computer outputs, but by giving the reader the foundation on which those explanations must necessarily be built.

The case in Table 2.5 is straightforward:

$$F = \text{MSR}_m/\text{MSE} \text{ tests } \quad \text{H: } \mu_1 = \mu_2 = \cdots = \mu_a\,. \tag{75}$$

This is the nature of the explanations of F-statistics that will be used: a statement of the hypothesis tested by an F. Armed therewith, one can then describe the usefulness of computer-calculated Fs to whomsoever's data are being analyzed, so that decisions can be made against the context of the data as to which hypotheses are useful and which are not. But more than just statements like (75) are needed for those computer-calculated Fs: the material between the model in (1) and the hypothesis in (75) is the foundation for (75). Indeed, it is also foundation to the numerous more complicated applications of cell means models described in subsequent chapters.

We see in $F = \text{MSR}_m/\text{MSE} = \text{SSR}_m/(a-1)\hat{\sigma}^2$ that its denominator is $\text{MSE} = \hat{\sigma}^2$. This is the case with all fixed effects models where the only randomness in the data is assumed to arise from error terms (denoted e_{ij} here) that are i.i.d. $\mathcal{N}(0, \sigma^2)$. These are the models dealt with throughout most of this book (except in Chapter 13, where mixed models are briefly considered).

Let us generalize $\text{SSR}_m/(a-1)\hat{\sigma}^2$ slightly, so as to consider the computer user's problem a little more broadly. Suppose Q is a sum of squares such that $Q/\sigma^2 \sim \chi^2_f$ independently of $\hat{\sigma}^2$. Then $F = Q/f\hat{\sigma}^2 \sim \mathscr{F}_{f,N-a}$, and to decide on the usefulness of F we need to know what hypothesis it tests. Since Q is always used in F in the form $Q/f\hat{\sigma}^2$, we might just as well ask "What is the hypothesis tested by Q?" rather than the longer, but of course more accurate, question "When Q is converted to a mean square to be used as the numerator of an F-statistic, with denominator $\hat{\sigma}^2$, what hypothesis is that F-statistic testing?". Just as the first form of the question is shorter, so is its answer "Q tests H:", and this is the form that shall be used. We call H the hypothesis associated with Q.

Definition. When $Q/f\hat{\sigma}^2$ is an F-statistic, the hypothesis that it tests is called the Associated Hypothesis for Q.

2.12. LINEAR COMBINATIONS OF CELL MEANS

A simple extension of (60) is the general linear function of μ_is

$$\omega = \Sigma_i k_i \mu_i \tag{76}$$

with BLUE

$$\hat{\omega} = \Sigma_i k_i \bar{y}_{i\cdot} \quad \text{and} \quad v(\hat{\omega}) = \sigma^2 \Sigma_i k_i^2/n_i \ . \tag{77}$$

We consider properties of ω and its BLUE. Since $\omega = \mu_i$ by choosing $k_i = 1$ and $k_r = 0$ for all $r \neq i$, dealing with ω also includes being able to deal with each μ_i individually.

a. Confidence intervals

Applying the normality assumptions of (62) to $\hat{\omega}$ gives

$$\frac{\hat{\omega} - \omega}{\sqrt{v(\hat{\omega})}} \sim \mathcal{N}(0,1) \ . \tag{78}$$

Replacing σ^2 in (77) by $\hat{\sigma}^2$ based on $N - a$ degrees of freedom, so that

$$\hat{v}(\hat{\omega}) = \hat{\sigma}^2 \Sigma_i k_i^2 / n_I, \quad \text{gives} \quad \frac{\hat{\omega} - \omega}{\sqrt{\hat{v}(\omega)}} \sim \mathscr{T}_{N-a}, \tag{79}$$

where $\mathscr{T}_{N-a}$ represents the t-distribution on $N - a$ degrees of freedom.

Define $t_{L, N-a, \alpha}$ and $t_{U, N-a, \alpha}$ as a pair of lower and upper points, respectively, of the $\mathscr{T}_{N-a}$ distribution such that for $t \sim \mathscr{T}_{N-a}$,

$$\Pr\{t \le t_{L, N-a, \alpha}\} + \Pr\{t \ge t_{U, N-a, \alpha}\} = \alpha \tag{80}$$

and so
$$\Pr\{t_{L, N-a, \alpha} \le t \le t_{U, N-a, \alpha}\} = 1 - \alpha \ .$$

Then from (79)

$$\Pr\left\{ t_{L, N-a, \alpha} \le \frac{\hat{\omega} - \omega}{\sqrt{\hat{v}(\hat{\omega})}} \le t_{U, N-a, \alpha} \right\} = 1 - \alpha$$

and rearrangement of this probability statement in the form

$$\Pr\left\{ \hat{\omega} - t_{U, N-a, \alpha}\sqrt{\hat{v}(\hat{\omega})} \le \omega \le \hat{\omega} - t_{L, N-a, \alpha}\sqrt{\hat{v}(\hat{\omega})} \right\} = 1 - \alpha$$

provides
$$\hat{\omega} - t_{U, N-a, \alpha}\sqrt{\hat{v}(\hat{\omega})} \quad \text{to} \quad \hat{\omega} - t_{L, N-a, \alpha}\sqrt{\hat{v}(\hat{\omega})} \tag{81}$$

as a $100(1 - \alpha)\%$ confidence interval for ω. This interval will be symmetric with respect to $\hat{\omega}$, as is often required, when

$$-t_{L, N-a, \alpha} = t_{U, N-a, \alpha} = t_{N-a, \frac{1}{2}\alpha} \quad \text{where } \Pr\{t \ge t_{N-a, \frac{1}{2}\alpha}\} = \tfrac{1}{2}\alpha,$$

and the interval then becomes

$$\hat{\omega} \pm t_{N-a, \frac{1}{2}\alpha}\sqrt{\hat{v}(\hat{\omega})} \ . \tag{82}$$

Of all intervals (81), i.e., of all different possible lower and upper points on the $\mathscr{T}_{N-a}$ distribution as defined by (80), the interval (82) is the one of minimum width, that width being $2t_{N-a, \frac{1}{2}\alpha}\sqrt{\hat{v}(\hat{\omega})}$.

When degrees of freedom are large ($N - a > 100$, say), the distribution in (79) is approximately normal $\mathscr{N}(0, 1)$ and on defining $z_{L, \alpha}$ and $z_{U, \alpha}$ such that

$$\text{for } z \sim \mathscr{N}(0, 1), \qquad \Pr\{z_{L, \alpha} \le z \le z_{U, \alpha}\} = 1 - \alpha$$

then $z_{L,\alpha}$ and $z_{U,\alpha}$ can be used in (81) in place of $t_{L,N-a,\alpha}$ and $t_{U,N-a,\alpha}$, respectively. In particular, for a symmetric confidence interval, we need

$$z_{L,\alpha} = -z_{U,\alpha} = z_{\frac{1}{2}\alpha} \quad \text{for } (2\pi)^{-\frac{1}{2}}\int_{z_{\frac{1}{2}\alpha}}^{\infty} e^{-\frac{1}{2}x^2}\,dx = \tfrac{1}{2}\alpha$$

and the interval is then
$$\hat{\omega} \pm z_{\frac{1}{2}\alpha}\sqrt{\hat{v}(\hat{\omega})}\ . \tag{83}$$

Example (continued). Consider $\omega = 2\mu_1 + \frac{1}{3}\mu_2 + \frac{2}{3}\mu_3$. Then using the cell means $\hat{\mu}_i = \bar{y}_{i\cdot}$ from Table 2.1, $\hat{\omega} = 2(73) + \frac{1}{3}(78) + \frac{2}{3}(89) = 231\frac{1}{3}$; and from (35) and (79), $\hat{v}(\hat{\omega}) = (14\frac{1}{2})[4/3 + (1/9)/2 + (4/9)/2] = (29/6)^2$. Thus the symmetric interval (82) is $231\frac{1}{3} \pm (29/6)t_{4,\frac{1}{2}\alpha}$; and for a 5% confidence interval this is $231\frac{1}{3} \pm (29/6)(2.78) = 231\frac{1}{3} \pm 13.44 = (217.89, 244.77)$.

b. Hypothesis tests

-i. One-part hypotheses. We consider what shall be called a *one-part hypothesis*, namely that a linear combination of the μ_is equals some pre-assigned constant m: i.e.,

$$\text{H: } \Sigma_i k_i \mu_i = m\ .$$

In keeping with the usual likelihood ratio methods of hypothesis testing [e.g., Searle (1971), Section 3.7a] we derive the error sum of squares under this hypothesis, to be denoted SSE_H, and the F-statistic for testing H will be $(\text{SSE}_H - \text{SSE})/\hat{\sigma}^2$. To do this, proceed as follows.

Take the model as

$$E(y_{ij}) = \mu_i \quad \text{and} \quad \Sigma_i k_i \mu_i = m\ .$$

From this model, estimate the μ_i by least squares through minimizing

$$\Sigma_i \Sigma_j (y_{ij} - \mu_i)^2 + 2\lambda(\Sigma_i k_i \mu_i - m)$$

with respect to the μ_is and the Lagrange multiplier, λ. Denoting the resulting estimator of μ_i by $\hat{\mu}_{i,H}$ to distinguish it from $\hat{\mu}_i$ leads to equations

$$\hat{\mu}_{i,H} = \bar{y}_{i\cdot} - \hat{\lambda}k_i/n_i \quad \text{and} \quad \Sigma k_i \hat{\mu}_{i,H} = m\ . \tag{84}$$

Using the first of these in the second gives

$$\Sigma_i k_i \bar{y}_{i\cdot} - \hat{\lambda}\Sigma_i k_i^2/n_i = m$$

and hence

$$\hat{\lambda} = \frac{\Sigma_i k_i \bar{y}_{i\cdot} - m}{\Sigma_i (k_i^2/n_i)} \, . \tag{85}$$

Then SSE_H, which is defined as

$$\text{SSE}_\text{H} = \Sigma_i \Sigma_j (y_{ij} - \hat{\mu}_{i,\text{H}})^2,$$

is, from (84),

$$\begin{aligned} \text{SSE}_\text{H} &= \Sigma_i \Sigma_j \left(y_{ij} - \bar{y}_{i\cdot} + \hat{\lambda} k_i/n_i \right)^2 \\ &= \Sigma_i \Sigma_j \left(y_{ij} - \bar{y}_{i\cdot} \right)^2 + \hat{\lambda}^2 \Sigma k_i^2/n_i - 2\hat{\lambda}\Sigma_i\Sigma_j (y_{ij} - \bar{y}_{i\cdot}) k_i/n_i \\ &= \text{SSE} + \frac{(\Sigma_i k_i \bar{y}_{i\cdot} - m)^2}{\Sigma_i k_i^2/n_i}, \end{aligned} \tag{86}$$

on using (85) and $\Sigma_j(y_{ij} - \bar{y}_{i\cdot}) = 0$. Hence

$$F = (\text{SSE}_\text{H} - \text{SSE})/\hat{\sigma}^2 = \frac{(\Sigma_i k_i \bar{y}_{i\cdot} - m)^2}{\hat{\sigma}^2 \Sigma_i k_i^2/n_i} \sim \mathscr{F}_{1, N-a} \tag{87}$$

is the F-statistic that tests H: $\Sigma_i k_i \mu_i = m$.

The following features of F of (87) are worth noting. First, its denominator, $\hat{\sigma}^2\Sigma_i k_i^2/n_i$, is the estimated variance of the term in its numerator that is squared, namely $\Sigma_i k_i \bar{y}_{i\cdot} - m$. Second, under the normality assumptions the numerator of (87) is independent of $\hat{\sigma}^2$. Third, under those same assumptions, F is the square of a variable that has a $\mathscr{T}$-distribution on $N - a$ degrees of freedom. It is left to the reader to verify these three facts (see E2.4).

-ii. ***Two-part hypotheses.*** A hypothesis that each of two linear functions of the μ_is simultaneously equal some pre-assigned constants shall be called a *two-part hypothesis*. For example,

$$\text{H: } \begin{cases} \Sigma_i k_i \mu_i = m_1 \\ \Sigma_i l_i \mu_i = m_2 \end{cases} \tag{88}$$

is a two-part hypothesis. Discussion of hypotheses of more than two parts is deferred until Section 8.8, wherein the use of matrix notation permits convenient description. In the meantime we simply display a procedure that yields the F-statistic for testing (88). Calculate

$$v_1 = \Sigma_i k_i^2/n_i, \qquad v_2 = \Sigma_i l_i^2/n_i, \qquad c = \Sigma_i k_i l_i/n_i,$$

$$d_1 = \Sigma_i k_i \bar{y}_{i\cdot} - m_1 \quad \text{and} \quad d_2 = \Sigma_i l_i \bar{y}_{i\cdot} - m_2 \ . \tag{89}$$

Then

$$F = \frac{\left(v_2 d_1^2 + v_1 d_2^2 - 2cd_1 d_2\right)/\left(v_1 v_2 - c^2\right)}{2\hat{\sigma}^2} \sim \mathscr{F}_{2,\, N-a} \tag{90}$$

can be used for testing H of (88). The expression in (90) is a special case of (146) in Section 8.8b.

Connections between F of (90) and the Fs for testing the two one-part hypotheses

$$\text{H}_1\text{: } \Sigma_i k_i \mu_i = m_1 \quad \text{and} \quad \text{H}_2\text{: } \Sigma_i l_i \mu_i = m_2 \tag{91}$$

are worth noting. First, the Fs for (91) are, from (98),

$$F_1 = \left(d_1^2/v_1\right)/\hat{\sigma}^2 \quad \text{and} \quad F_2 = \left(d_2^2/v_2\right)/\hat{\sigma}^2 \ . \tag{92}$$

Second, the numerator of F for the two-part hypothesis is not the sum of the numerators of the Fs for the two one-part hypotheses that make up the two-part hypothesis; i.e., the numerator of (90) does not equal the sum of the numerators in (92). Equality does occur if c of (89) is zero; i.e., if

$$\Sigma_i k_i l_i/n_i = 0 \ . \tag{93}$$

This is a generalization of the idea of orthogonality of two linear combinations of μ_is, an idea that is very familiar in the context of analyzing balanced data (see subsection f, which follows). There though, the orthogonality concept relies on a simplified form of (93) that comes from effectively having the n_i all equal in the designed experiment context, whereupon (93) reduces to the familiar $\Sigma_i k_i l_i = 0$. That is for balanced data; but with unbalanced data that definition of orthogonality has to be modified to be (93).

c. Fitting the mean

Consider a particular form of $\omega = \Sigma_i k_i \mu_i$ with $k_i = n_i/N$ and define it as ρ', a weighted mean of the μ_is, weighted in proportion to the n_is:

$$\rho' = \Sigma_i n_i \mu_i / N \ . \tag{94}$$

Then
$$\hat{\rho}' = \Sigma_i n_i \bar{y}_{i\cdot}/N = \bar{y}_{\cdot\cdot} \quad \text{and} \quad v(\hat{\rho}') = \sigma^2/N \ . \tag{95}$$

From (87)
$$F = \frac{(\bar{y}_{\cdot\cdot} - m)^2}{\hat{\sigma}^2(1/N)} \quad \text{tests} \quad \text{H: } \rho' = m, \tag{96}$$

so that for m being zero

$$F = \frac{N\bar{y}_{\cdot\cdot}^2}{\hat{\sigma}^2} \quad \text{tests} \quad \text{H: } \Sigma_i n_i \mu_i = 0 \ . \tag{97}$$

Thus we see that the sum of squares $R(\mu) = \text{SSM} = N\bar{y}_{\cdot\cdot}^2$ in Table 2.3 can be used to test H: $\rho' = 0$, i.e., H: $\Sigma n_i \mu_i = 0$. This is summarized in Table 2.6, where the F-statistics are denoted by $F(M)$ and $F(R_m)$, where the hypothesis tested by each F is described as the associated hypothesis and where the sums of squares are calculated as in Table 2.3.

TABLE 2.6. ANALYSIS OF VARIANCE FOR THE 1-WAY CLASSIFICATION CELL MEANS MODEL

Source of Variation	d.f.	Sum of Squares	Mean Square	F-statistic	Associated Hypothesis
Mean	1	SSSM	$\text{MSM} = \frac{\text{SSM}}{1}$	$F(M) = \frac{\text{MSM}}{\hat{\sigma}^2}$	H: $\Sigma_i n_i \mu_i = 0$
Classes	$a - 1$	SSR_m	$\text{MSR} = \frac{\text{SSR}_m}{a-1}$	$F(R_m) = \frac{\text{MSR}_m}{\hat{\sigma}^2}$	H: μ_i *all equal*
Residual	$N - a$	SSE	$\text{MSE} = \frac{\text{SSE}}{N-a} = \hat{\sigma}^2$		
Total	N	SST			

TABLE 2.7. ANALYSIS OF VARIANCE FOR THE 1-WAY CLASSIFICATION (SUMS OF SQUARES AND ASSOCIATED HYPOTHESES)

Source of Variation	d.f.	Sum of Squares	Associated Hypothesis
Mean	1	$\text{SSM} = N\bar{y}_{..}^2$	$H: \Sigma_i n_i\mu_i = 0$
Classes	$a-1$	$\text{SSR}_m = \Sigma_i n_i(\bar{y}_{i.} - \bar{y}_{..})^2$	$H: \mu_i$ *all equal*
Residual	$N-a$	$\text{SSE} = \Sigma_i\Sigma_j(y_{ij} - \bar{y}_{i.})^2$	
Total	N	$\text{SST} = \Sigma_i\Sigma_j y_{ij}^2$	

A curtailed form of Table 2.6 is Table 2.7. It contains all the salient information on sums of squares and associated hypotheses and omits the standard details of calculating mean squares and F-statistics in the familiar, generic form $\text{MS} = \text{SS}/df$ and $F = \text{MS}/\hat{\sigma}^2$, respectively .

This is the format for most analysis of variance tables in this book. It shows just sums of squares and their associated hypotheses.

d. Other forms of a grand mean

$\rho' = \Sigma_i n_i\mu_i/N$ of (94) is a weighted mean of the μ_is using the n_is as weights. Thus, by its very definition, ρ' is governed by the numbers of observations in the samples drawn from each class. In contrast to this is the equally-weighted average

$$\rho = \Sigma_i\mu_i/a \tag{98}$$

with

$$\hat{\rho} = \Sigma_i\bar{y}_{i.}/a \quad \text{and} \quad v(\hat{\rho}) = \sigma^2\Sigma_i(1/n)/a^2 \ . \tag{99}$$

This is just the mean of the means, regardless of the number of observations in the classes. For this reason, it might appeal as a definition of an overall mean more than does ρ'. Nevertheless, its BLUE has a larger variance than does that of ρ', as is now shown.

We begin with the well-known Cauchy–Schwarz inequality, that for any a pairs of real numbers p_i and q_i for $i = 1, 2, \ldots, t$

$$\Sigma_i p_i^2 \Sigma_i q_i^2 \geq (\Sigma_i p_i q_i)^2 \ . \tag{100}$$

Proof.

$$\Sigma_i\left(q_i - \frac{\Sigma_i p_i q_i}{\Sigma_i p_i^2}p_i\right)^2 \geq 0 \text{ gives } \Sigma_i q_i^2 - \frac{(\Sigma_i p_i q_i)^2}{\Sigma_i p_i^2} \geq 0, \text{equivalent to (100)} \ .$$

Using (100) with $p_i^2 = n_i$ and $q_i^2 = 1/n_i$ leads to

$$(\Sigma_i 1/n_i)/a^2 \geq 1/\Sigma n_i; \quad \text{hence} \quad v(\hat{\rho}) \geq v(\hat{\rho}') \ . \tag{101}$$

Of course, with every $n_i = n$ in balanced data, $\rho = \rho'$, their BLUEs are equal and so are their variances.

An even more general form of weighted mean is

$$\rho'' = \Sigma_i w_i \mu_i / \Sigma_i w_i \tag{102}$$

with BLUE

$$\hat{\rho}'' = \Sigma w_i \bar{y}_{i\cdot} / \Sigma w_i \quad \text{and} \quad v(\hat{\rho}'') = \hat{\sigma}^2 \big(\Sigma w_i^2/n_i\big)/(\Sigma_i w_i)^2, \tag{103}$$

where the w_is can be any weights (real numbers) at all. In the same way that (101) was derived, another application of (100) with $p_i^2 = n_i$ and $q_i^2 = w_i^2/n_i$ leads to $v(\hat{\rho}'') \geq v(\hat{\rho}')$. Note, of course, that ρ'_i and ρ are special cases of ρ'': with $w_i = n_i$ we have $\rho'' = \rho'$, and $w_i = 1$ gives $\rho \equiv \rho$.

e. Contrasts

The difference $\mu_i - \mu_h$ for $i \neq h$ is obviously a form of linear function of special interest. A generalization of that difference is any linear combination of such differences. It is called a *contrast*, and it has the property that its coefficients of the μ_is sum to zero; i.e., its general form is

$$\gamma = \Sigma_i k_i \mu_i \quad \text{with} \quad \Sigma_i k_i = 0 \ . \tag{104}$$

Examples are $\mu_1 + \mu_2 - 2\mu_3$ and $2\mu_1 + 3\mu_2 - 5\mu_3$. Properties of γ and of its BLUE $\hat{\gamma}$ are precisely those of ω and $\hat{\omega}$ in (76) and (77) but with the additional property that $\Sigma_i k_i = 0$ of (104).

f. Orthogonal contrasts for balanced data

Although this book is directed towards unbalanced data, a few features of balanced data warrant description. One such is orthogonal contrasts.

Two contrasts

$$\gamma_1 = \Sigma_i k_{1i} \mu_i \quad \text{and} \quad \gamma_2 = \Sigma_i k_{2i} \mu_i \quad \text{with} \quad \Sigma_i k_{1i} = 0 = \Sigma_i k_{2i} \tag{105}$$

are, in the context of balanced data, said to be orthogonal when

$$\Sigma_i k_{1i} k_{2i} = 0 \ . \tag{106}$$

Their BLUEs are

$$\hat{\gamma}_1 = \Sigma_i k_{1i} \bar{y}_{i\cdot} \quad \text{and} \quad \hat{\gamma}_2 = \Sigma_i k_{2i} \bar{y}_{i\cdot} \tag{107}$$

and, since with balanced data every $\bar{y}_{i\cdot}$ has the same number of observations, n say, the covariance of the BLUEs in (107) is

$$\mathrm{cov}(\hat{\gamma}_1, \hat{\gamma}_2) = (\sigma^2/n)\Sigma_i k_{1i}k_{2i} = 0$$

from (106). Hence, under normality, the definition in (106) for orthogonality of contrasts (indeed of any pair of linear functions of μ_is) implies independence of their BLUEs.

A special case of particularly useful orthogonal contrasts for balanced data, for the case of $a = 4$ by way of example, is

$$\begin{aligned} \gamma_2 &= \mu_1 - \mu_2 \\ \gamma_3 &= \mu_1 + \mu_2 - 2\mu_3 \\ \gamma_4 &= \mu_1 + \mu_2 + \mu_3 - 3\mu_4 . \end{aligned} \tag{108}$$

Extension to the general case is clear:

$$\gamma_i = \sum_{h=1}^{i-1} \mu_h - (i-1)\mu_i \quad \text{for } i = 2, \ldots, a, \tag{109}$$

Since γ_i can also be written as

$$\gamma_i = -(i-1)\left[\mu_i - \frac{\sum_{h=1}^{i-1} \mu_h}{i-1}\right]$$

and as

$$\gamma_i = \sum_{h=1}^{i-1} (\mu_h - \mu_i)$$

we see that each γ_i provides opportunity for comparing μ_i against the average of $(i-1)$ other μs; or, equivalently, that γ_i is a sum of differences of μ_i from $(i-1)$ other μs. Thus the hypothesis

$$\mathrm{H}_i\colon\ \gamma_i = 0 \tag{110}$$

is equivalent to

$$\mathrm{H}_i\colon\ \mu_i = \sum_{h=1}^{i-1} \mu_h/(i-1), \tag{111}$$

i.e., the hypothesis that μ_i equals the mean of $\mu_1, \mu_2, \ldots, \mu_{i-1}$.

Denote by q_i the sum of squares for which H_i of (110) is the associated hypothesis, i.e.,

$$F_i = q_i/\hat{\sigma}^2 \quad \text{tests} \quad \text{H}_i\text{: } \gamma_i = 0 \ . \tag{112}$$

Then

$$q_i = \hat{\gamma}_i^2 / [v(\hat{\gamma}_i)/\sigma^2], \tag{113}$$

with

$$\hat{\gamma}_i = \sum_{h=1}^{i-1} \bar{y}_{h\cdot} - (i-1)\bar{y}_{i\cdot} \quad \text{and} \quad v(\hat{\gamma}_i) = i(i-1)\sigma^2/n \ . \tag{114}$$

Furthermore, the $\hat{\gamma}_i$s are, under normality, independent, and hence so are the q_is; and the sum of the q_is is SSR_m, as can be shown from (113) and (114) using $n_i = n$ (for balanced data):

$$\sum_{i=2}^{a} q_i = \text{SSR}_m = n\sum_{i=1}^{a} (\bar{y}_{i\cdot} - \bar{y}_{\cdot\cdot})^2 \ . \tag{115}$$

Thus, in (115), we see that the sum of squares SSR_m on $a - 1$ degrees of freedom used for testing H: *all* μ_i *equal* is equal to the sum of the $a - 1$ independent sums of squares q_i used for testing H_i: $\gamma_i = 0$ *for* $i = 2, 3, \ldots, a$. This provides opportunity for comparing the μ_is when the hypothesis H: μ_i *all equal* is rejected.

Derivation of (113), (114) and (115) is left to the reader (E2.5), although these results are just special cases of the orthogonal contrasts for unbalanced data described in the next subsection.

Example. Suppose there are four sample means, $\bar{y}_{1\cdot} = 12$, $\bar{y}_{2\cdot} = 10$, $\bar{y}_{3\cdot} = 13$ and $\bar{y}_{4\cdot} = 9$, each based on seven observations. Then $\bar{y}_{\cdot\cdot} = (12 + 10 + 13 + 9)/4 = 11$, and so $\text{SSR}_m = 7\Sigma(\bar{y}_{i\cdot} - \bar{y}_{\cdot\cdot})^2 = 7[1^2 + (-1)^2 + 2^2 + (-2)^2] = 70$. From (114) we obtain the values in Table 2.8. Using those

TABLE 2.8. CONTRASTS AMONG FOUR MEANS, 12, 10, 13 AND 9, EACH OF SEVEN OBSERVATIONS—USING (114)

i	$\hat{\gamma}_i$	$v(\hat{\gamma}_i)/\sigma^2$
2	$\bar{y}_{1\cdot} - \bar{y}_{2\cdot} = 12 - 10 = 2$	2/7
3	$\bar{y}_{1\cdot} + \bar{y}_{2\cdot} - 2\bar{y}_{3\cdot} = 12 + 10 - 26 = -4$	6/7
4	$\bar{y}_{1\cdot} + \bar{y}_{2\cdot} + \bar{y}_{3\cdot} - 3\bar{y}_{4\cdot} = 12 + 10 + 13 - 27 = 8$	12/7

values in (113) gives

$$\sum_{i=2}^{4} q_i = 2^2/(2/7) + 4^2/(6/7) + 8^2/(12/7)$$

$$= 7(2 + 8/3 + 16/3) = 70 = \mathrm{SSR}_m,$$

so demonstrating (115).

g. Orthogonal contrasts for unbalanced data

The contrasts in (105) with the BLUEs of (107) have, in the case of unbalanced data,

$$\mathrm{cov}(\hat{\gamma}_1, \hat{\gamma}_2) = (\Sigma_i k_{1i}k_{2i}/n_i)\sigma^2 \ . \tag{116}$$

On defining orthogonality with unbalanced data as being

$$\Sigma_i \frac{k_{1i}k_{2i}}{n_i} = 0, \tag{117}$$

the $\mathrm{cov}(\hat{\gamma}_1, \hat{\gamma}_2)$ in (116) is zero and, under normality, $\hat{\gamma}_1$ and $\hat{\gamma}_2$ are independent.

Obviously, when all n_is are equal, (116) reduces to (106), the definition of contrasts being orthogonal for balanced data. In that case it is the independence of the $\hat{\gamma}_i$s that is useful and consistent with $\Sigma_{i=2}^{a} q_i = \mathrm{SSR}_m$ of (115), and that independence implies $\Sigma_i k_{1i}k_{2i} = 0$ of (106). But (106) is also the definition that the set of numbers $k_{11}, \ldots, k_{1a}$ are orthogonal to the set $k_{21}, \ldots, k_{2a}$ (more correctly, from vector algebra, that two vectors are "perpendicular", namely "orthogonal")—and, presumably, this is how the name "orthogonal contrasts" arose. If we think of this name as meaning "independent estimated contrasts" then the same name for unbalanced data can be attached to contrasts for which $\Sigma_i k_{1i}k_{2i}/n_i = 0$ of (117) is true; it is the condition for $\hat{\gamma}_1$ and $\hat{\gamma}_2$ to be independent. (It is also the condition that the two sets of numbers $k_{11}/\sqrt{n_1}, \ldots, k_{1a}/\sqrt{n_a}$ and $k_{21}/\sqrt{n_1}, \ldots, k_{2a}/\sqrt{n_a}$ are orthogonal.) And with that condition a set of orthogonal contrasts analogous to the γ_is of balanced data can be derived such that they, too, have an equality similar to $\Sigma_{i=2}^{a} q_i = \mathrm{SSR}_m$ of (115). These contrasts are now given.

First, define

$$s_i = \sum_{h=1}^{i} n_h \quad \text{so that} \quad s_{i+1} = s_i + n_{i+1} \tag{118}$$

and

$$s_1 = n_1 \quad \text{and} \quad s_a = n_{.} = N \ . \tag{119}$$

Then the contrasts corresponding to (108) for $a = 4$ but for unbalanced data are

$$\delta_2 = \sqrt{n_2}(n_1\mu_1 - s_1\mu_2)$$

$$\delta_3 = \sqrt{n_3}(n_1\mu_1 + n_2\mu_2 - s_2\mu_3)$$

$$\delta_4 = \sqrt{n_4}(n_1\mu_1 + n_2\mu_2 + n_3\mu_3 - s_3\mu_4) \ . \tag{120}$$

The general form is

$$\delta_i = \sqrt{n_i}\left(\sum_{h=1}^{i-1} n_h\mu_h - s_{i-1}\mu_i\right) \ . \tag{121}$$

It is easily seen that δ_i is a contrast, and for balanced data with $n_i = n \ \forall \ i$, that $\delta_i = n\sqrt{n}\,\gamma_i$ for γ_i of (109).

From (121) it is easy to rewrite δ_i as

$$\delta_i = s_{i-1}\sqrt{n_i}\left(\frac{\sum_{h=1}^{i-1} n_h\mu_h}{\sum_{h=1}^{i-1} n_h} - \mu_i\right)$$

and as

$$\delta_i = \sqrt{n_i}\sum_{h=1}^{i-1} n_h(\mu_h - \mu_i) \ .$$

From the first of these we see that comparable to H_i: $\gamma_i = 0$ of (110) for balanced data, we here have

$$H_i\colon \ \delta_i = 0 \tag{122}$$

equivalent to

$$H_i\colon \ \mu_i = \frac{\sum_{h=1}^{i-1} n_h\mu_h}{\sum_{h=1}^{i-1} n_h} \tag{123}$$

which is just the hypothesis that μ_i equals the weighted mean of $\mu_1, \mu_2, \ldots, \mu_{i-1}$, using the ns as weights. Thus (123) is just the weighted mean counterpart of the average of means in (111).

The analogy of δ_i with γ_i is that the sums of squares for the hypotheses H_i: $\delta_i = 0$ of (122), summed over $i = 2, \ldots, a$, add to SSR_m, just as the q_i sum to SSR_m in (115). Thus in (121), replacing h by k and μ_i by $\hat{\mu}_i = \bar{y}_{i\cdot}$ gives

$$\hat{\delta}_i = \sqrt{n_i}\left(\sum_{k=1}^{i-1} n_k \bar{y}_{k\cdot} - s_{i-1}\bar{y}_{i\cdot}\right). \tag{124}$$

Then, as shown in the Appendix (Section 2.14b), we find that

$$v(\hat{\delta}_i) = s_{i-1}s_i\sigma^2, \tag{125}$$

and under normality the $\hat{\delta}_i$s are independent. Also, for

$$u_i = \hat{\delta}_i^2/[v(\hat{\delta}_i)/\sigma^2], \tag{126}$$

$$F_i = u_i/\hat{\sigma}^2 \quad \text{tests} \quad H_i\text{: } \delta_i = 0, \tag{127}$$

TABLE 2.9. CONTRASTS FOR UNBALANCED DATA, WITH MEANS 12, 10, 13 AND 9, BASED ON 4, 2, 3 AND 4 OBSERVATIONS

Term	$i = 1$	$i = 2$	$i = 3$	$i = 4$	Total
$\bar{y}_{i\cdot}$	12	10	13	9	
n_i	4	2	3	4	$13 = n.$
$n_i\bar{y}_{i\cdot} = y_{i\cdot}$	48	20	39	36	$143 = y_{\cdot\cdot}$ $(\bar{y}_{\cdot\cdot} = 11)$
$n_i(\bar{y}_{i\cdot} - \bar{y}_{\cdot\cdot})^2$	$4(12-11)^2 = 4$	$2(10-11)^2 = 2$	$3(13-11)^2 = 12$	$4(9-11)^2 = 16$	$34 = SSR_m$
From (118) $s_i = \sum_{h=1}^{i} n_h$	4	6	9	13	
From (125) $v(\hat{\delta}_i)/\sigma^2 = s_i s_{i-1}$		24	54	117	
$\hat{\delta}_i$ from (124)					
	$\hat{\delta}_2 = \sqrt{2}[4(12) - 4(10)]$			$= 8\sqrt{2}$	
	$\hat{\delta}_3 = \sqrt{3}[4(12) + 2(10) - 6(13)]$			$= -10\sqrt{3}$	
	$\hat{\delta}_4 = \sqrt{4}[4(12) + 2(10) + 3(13) - 9(9)]$			$= 26\sqrt{4}$	
$u_i = \dfrac{\hat{\delta}_i^2}{s_i s_{i-1}}$		$\dfrac{2(8^2)}{24}$	$\dfrac{3(10^2)}{54}$	$\dfrac{4(26^2)}{117}$	$34 = SSR_m$

and
$$\sum_{i=2}^{a} u_i = \text{SSR}_m = \sum_{i=1}^{a} n_i(\bar{y}_{i\cdot} - \bar{y}_{\cdot\cdot})^2 \ . \tag{128}$$

Example. Suppose the means $\bar{y}_{1\cdot} = 12$, $\bar{y}_{2\cdot} = 10$, $\bar{y}_{3\cdot} = 13$ and $\bar{y}_{4\cdot} = 9$ of the preceding example are based on 4, 2, 3 and 4 observations, respectively. Then Table 2.9 shows the calculations that illustrate (128).

2.13. THE OVERPARAMETERIZED MODEL

One objective of this book is to show how much easier it is to use cell means models than overparameterized models—and this is done not so much by detailed comparison but by concentrating on cell means models and leaving it to readers who know something of overparameterized models to judge for themselves. (Moreover, not only are the cell means models easier, they are often more useful.) Nevertheless, it seems appropriate to occasionally give brief consideration to overparameterized models.

In the 1-way classification the customary overparameterized model is

$$E(y_{ij}) = \mu + \alpha_i \tag{129}$$

where we describe μ as a general mean and α_i as the effect on y_{ij} due to its being in class i. Immediately we see that there is one μ and a α_is to be estimated; i.e., $1 + a$ parameters to be estimated and only a cell means to estimate them from. This is what is meant by overparameterization: too many parameters to estimate. It leads to the problem that although the BLUE of $\mu + \alpha_i$ is

$$\text{BLUE}(\mu + \alpha_i) = \bar{y}_{i\cdot} \quad \text{for } i = 1, \ldots, a, \tag{130}$$

one cannot, from (129), obtain the BLUE of just μ on its own, nor of just α_i on its own. The $\mu + \alpha_i$ for $i = 1, \ldots, a$ are declared to be estimable functions (of the parameters) with available BLUEs, and any linear combination of them has its BLUE. Thus is the problem of overparameterization overcome through the concept of estimable functions, as mentioned just before equation (4) in Section 1.3. Clearly, in this case,

$$\mu + \alpha_i = \mu_i \tag{131}$$

for μ_i of the cell means model—and so the overparameterization seems totally unnecessary.

An alternative to using estimable functions is that of including restrictions on the parameters of the model to reduce the number of effective parameters to be the same as the number of cells. This, too, is mentioned in Section 1.3. The most usual restrictions for this purpose are those known as the Σ-restrictions. To distinguish the parameters in the model that have restrictions from those of (129) without restrictions we use the symbols $\dot{\mu}$ and $\dot{\alpha}$. Then what shall be called the Σ-restricted model is

$$E(y_{ij}) = \dot{\mu} + \dot{\alpha}_i \quad \text{and} \quad \Sigma_i \dot{\alpha}_i = 0\,. \tag{132}$$

It is the second equation in (132) that is the Σ-restriction.

A relationship between parameters of the restricted model (132) and those of the unrestricted model (129) is

$$\dot{\mu} = \mu + \Sigma\alpha_i/a \quad \text{and} \quad \dot{\alpha}_i = \alpha_i - \Sigma\alpha_i/a,$$

from which it is easily seen from $\mu_i = \mu + \alpha_i$ of (131) that

$$\dot{\mu} = \Sigma\mu_i/a = \rho \quad \text{and} \quad \dot{\alpha}_i = \mu_i - \Sigma_i\mu_i/a = \mu_i - \rho,$$

for ρ of (98). Thus using the Σ-restrictions in the 1-way classification is simply equivalent to defining a general mean as ρ and defining each class effect as $\mu_i - \rho$. It is clear that little is gained by the use of the Σ-restrictions. Their use for 2-way classifications is considered in Section 9.4.

2.14. APPENDIX

a. Proof that $\mathrm{SSR}_m/\sigma^2 \sim \chi^2_{a-1}$ (Section 2.10d)

Theorem. When $y_{ij} \sim$ i.i.d. $\mathcal{N}(\mu_i, \sigma^2)$ for $j = 1, \ldots, n_i$, and independently for $i = 1, \ldots, a$, then if $\mu_1 = \mu_2 = \cdots = \mu_a$, the distribution of $\mathrm{SSR}_m/\sigma^2 = \Sigma_{i=1}^a n_i(\bar{y}_{i\cdot} - \bar{y}_{\cdot\cdot})^2/\sigma^2$ is χ^2_{a-1}, where $\bar{y}_{\cdot\cdot} = \Sigma_i n_i \bar{y}_{i\cdot}/\Sigma_i n_i$.

Notation. For simplicity, drop the dot subscript from $\bar{y}_{i\cdot}$ so as to write it as $\bar{y}_i$ and have $\bar{y}_i \sim \mathcal{N}(\mu_i, \sigma^2/n_i)$. Then, using $s_r = \Sigma_{i=1}^r n_i$ of (118), similarly define the mean of all observations in classes $1, 2, \ldots, r$ as m_r:

$$m_r = \frac{\sum_{i=1}^{r} n_i \bar{y}_i}{\sum_{i=1}^{r} n_i} = \frac{\sum_{i=1}^{r-1} n_i \bar{y}_i + n_r \bar{y}_r}{\sum_{i=1}^{r-1} n_i + n_r} = \frac{s_{r-1} m_{r-1} + n_r \bar{y}_r}{s_{r-1} + n_r}\,. \tag{133}$$

Then for r classes, SSR_m will be denoted B_r and we have

$$B_r = \sum_{i=1}^{r} n_i(\bar{y}_{i\cdot} - m_r)^2 = \sum_{i=1}^{r} n_i \bar{y}_i^2 - s_r m_r^2 . \tag{134}$$

Distributional results. The following standard results of normal and χ^2-distributions are utilized:

D(i) Two normally distributed random variables that have zero covariance are independent.

D(ii) A linear function of normal variables is itself normally distributed.

D(iii) When, for $i = 1, \ldots, t$, $x_i \sim$ i.i.d. $\mathcal{N}(0, 1)$, then $\Sigma_{i=1}^{t} x_i^2 \sim \chi_t^2$.

D(iv) The sum of independently distributed χ^2-variables is distributed as a χ^2-variable.

Proof of theorem. By induction on a we show that, when $\mu_1 = \mu_2 = \cdots = \mu_a$, then $B_a/\sigma^2 \sim \chi_{a-1}^2$.

First consider B_2:

$$B_2 = n_1\bar{y}_1^2 + n_2\bar{y}_2^2 - \frac{(n_1\bar{y}_1 + n_2\bar{y}_2)^2}{n_1 + n_2} = \frac{n_1 n_2}{n_1 + n_2}(\bar{y}_1 - \bar{y}_2)^2 . \tag{135}$$

By D(ii), when $\mu_1 = \mu_2$,

$$\bar{y}_1 - \bar{y}_2 \sim \mathcal{N}\left[0, \sigma^2\left(\frac{1}{n_1} + \frac{1}{n_2}\right)\right] \sim \mathcal{N}\left(0, \frac{n_1 + n_2}{n_1 n_2}\sigma^2\right) .$$

Therefore, by D(iii),

$$\frac{B_2}{\sigma^2} = \frac{n_1 n_2}{n_1 + n_2}\frac{(\bar{y}_1 - \bar{y}_2)^2}{\sigma^2} = \left(\frac{\bar{y}_1 - \bar{y}_2}{\sqrt{v(\bar{y}_1 - \bar{y}_2)}}\right)^2 \sim \chi_1^2 . \tag{136}$$

Thus $B_a/\sigma^2 \sim \chi_{a-1}^2$ holds for $a = 2$.

Now consider B_a for some value of a, say r, with $r > 2$. First we develop a recurrence relationship between B_{r+1} and B_r: from (133) and (134)

$$\begin{aligned} B_{r+1} &= \sum_{i=1}^{r+1} n_i \bar{y}_i^2 - s_{r+1} m_{r+1}^2 \\ &= \sum_{i=1}^{r} n_i \bar{y}_i^2 + n_{r+1}\bar{y}_{r+1}^2 - (s_r m_r + n_{r+1}\bar{y}_{r+1})^2 / s_{r+1} \\ &= B_r + s_r m_r^2 + n_{r+1}\bar{y}_{r+1}^2 - (s_r m_r + n_{r+1}\bar{y}_{r+1})^2 / s_{r+1} \\ &= B_r + s_r m_r^2 (1 - s_r/s_{r+1}) + n_{r+1}\bar{y}_{r+1}^2 (1 - n_{r+1}/s_{r+1}) \\ &\qquad - 2 s_r m_r n_{r+1} \bar{y}_{r+1} / s_{r+1} \end{aligned}$$

and on using $s_{r+1} = s_r + n_{r+1}$ this reduces to

$$B_{r+1} = B_r + \frac{n_{r+1}s_r}{s_{r+1}}(m_r - \bar{y}_{r+1})^2 = B_r + \Delta \tag{137}$$

for

$$\Delta = \frac{n_{r+1}s_r}{s_{r+1}}(m_r - \bar{y}_{r+1})^2 . \tag{138}$$

Now by D(ii), when $\mu_1 = \mu_2 = \cdots = \mu_r$,

$$m_r - \bar{y}_{r+1} \sim \mathcal{N}\left[0, \sigma^2\left(\frac{1}{s_r} + \frac{1}{n_{r+1}}\right)\right] \sim \mathcal{N}\left(0, \frac{s_{r+1}}{n_{r+1}s_r}\right) .$$

Therefore for Δ of (138), by D(iii)

$$\Delta/\sigma^2 \sim \chi_1^2 . \tag{139}$$

In (137), consider the covariance of $m_r - \bar{y}_{r+1}$ with each term that is squared in $B_r = \Sigma_{i=1}^r n_i(\bar{y}_i - m_r)^2$. Because $\text{cov}(\bar{y}_i, \bar{y}_h) = 0$ for all $i \neq h$, $\text{cov}(\bar{y}_{r+1}, \bar{y}_i) = 0$ for $i \leq r$, and so $\text{cov}(\bar{y}_{r+1}, m_r) = 0$. Therefore with $i \leq r$

$$\text{cov}(m_r - \bar{y}_{r+1}, \bar{y}_i - m_r) = \text{cov}(m_r, \bar{y}_i - m_r) = \sigma^2(n_i/s_r n_i - 1/s_r) = 0 .$$

Therefore by D(i), $m_r - \bar{y}_{r+1}$ is independent of every squared term in B_r, and hence of B_r itself. Therefore B_r and Δ in (137) are independent. This independence, along with $\Delta/\sigma^2 \sim \chi_1^2$ from (139), indicates from D(iv) that on assuming $B_r/\sigma^2 \sim \chi^2_{r-1}$ we have $B_{r+1}/\sigma^2 \sim \chi^2_r$; i.e., on assuming $B_r/\sigma^2 \sim \chi^2_{r-1}$ we have shown $B_{r+1}/\sigma^2 \sim \chi^2_r$. But from (136) we know that $B_2/\sigma^2 \sim \chi^2_1$. Thus by induction $B_a/\sigma^2 \sim \chi^2_{a-1}$ for all a, and so the theorem is proved. **Q.E.D.**

This proof is taken from Searle and Pukelsheim (1985), based on Stigler (1984). It holds more generally: suppose $\bar{y}_i$ is of some form such that $v(\bar{y}_i) = \sigma^2/w_i$ for some w_i. Then in the theorem and proof replace n_i by w_i and $\bar{y}_{..}$ by $\Sigma_i w_i \bar{y}_i/\Sigma_i w_i$. This form of the theorem has an important use in Section 4.5a.

b. Properties of $\hat{\delta}_i$ (Section 2.12g)

Using $v(\bar{y}_{i\cdot}) = \sigma^2/n_i$ and $\text{cov}(\bar{y}_{i\cdot}, \bar{y}_{h\cdot}) = 0$ for $i \neq h$, the variance of

$$\hat{\delta}_i = \sqrt{n_i}\left(\sum_{k=1}^{i-1} n_k \bar{y}_{k\cdot} - s_{i-1}\bar{y}_{i\cdot}\right)$$

of (124) is

$$v(\hat{\delta}_i) = n_i\left(\sum_{k=1}^{i-1} n_k^2\sigma^2/n_k + s_{i-1}^2\sigma^2/n_i\right) = n_i\left(s_{i-1} + s_{i-1}^2/n_i\right)\sigma^2$$

$$= n_i s_{i-1}(n_i + s_{i-1})\sigma^2/n_i = s_{i-1}s_i\sigma^2 .$$

To consider the independence of the $\hat{\delta}_i$s we determine the covariance of $\hat{\delta}_i$ and $\hat{\delta}_h$ for $h > i$:

$$\text{cov}(\hat{\delta}_i, \hat{\delta}_h) = \sqrt{n_i}\sqrt{n_h}\,\text{cov}\left(\sum_{k=1}^{i-1} n_k\bar{y}_{k\cdot} - s_{i-1}\bar{y}_{i\cdot},\ \sum_{k=1}^{i} n_k\bar{y}_{k\cdot}\right)$$

$$= \sqrt{n_i n_h}\left(\sum_{k=1}^{i-1} n_k^2\sigma^2/n_k - s_{i-1}n_i\sigma^2/n_i\right)$$

$$= \sqrt{n_i n_h}\,(s_{i-1} - s_{i-1})\sigma^2 = 0 .$$

Hence, for every $i \neq h$, under normality, $\hat{\delta}_i$ and $\hat{\delta}_h$ are independent.

To show that $\Sigma_{i=2}^a u_i = \text{SSR}_m$ of (128), begin from (124), (125) and (126) by expressing u_i as

$$u_i = n_i\left(\sum_{k=1}^{i-1} n_k\bar{y}_k - s_{i-1}\bar{y}_i\right)^2 \Bigg/ s_{i-1}s_i,$$

so that on using m_r of (133)

$$u_i = n_i(s_{i-1}m_{i-1} - s_{i-1}\bar{y}_i)^2/s_{i-1}s_i = (n_i s_{i-1}/s_i)(m_{i-1} - \bar{y}_i)^2 \ . \quad (140)$$

Again we use induction, to show that $\Sigma_{i=2}^a u_i = B_a$. First, for $a = 2$ observe that

$$\sum_{i=2}^{a} u_i = u_2 = \frac{n_1 n_2}{n_1 + n_2}(\bar{y}_1 - \bar{y}_2)^2 \quad \text{from (140)} = B_2 \quad \text{from (135)} \ .$$

Therefore $\Sigma_{i=2}^a u_i = B_a$ is certainly true for $a = 2$. Suppose it is true for $a = r$. Then

$$\sum_{i=2}^{r+1} u_i = \sum_{i=2}^{r} u_i + u_{r+1} = B_r + \frac{n_{r+1}s_r}{s_{r+1}}(m_r - \bar{y}_{r+1})^2 = B_{r+1} \quad \text{from (137)} \ .$$

Thus the inductive argument shows that $\Sigma_{i=2}^a u_i = B_a$ for any a.

2.15. SUMMARY

a. Model and estimation

$$E(y_{ij}) = \mu_i \quad \text{for} \quad i = 1, \ldots, a \quad \text{and} \quad j = 1, \ldots, n_i \ . \qquad (1)$$

$$y_{ij} = \mu_i + e_{ij} \quad \text{for} \quad e_{ij} \sim \text{i.i.d. } \mathcal{N}(0, \sigma^2) \ . \qquad (9), (62)$$

$$\hat{\mu}_i = \bar{y}_{i\cdot} \quad \text{with} \quad \hat{\mu}_i \sim \mathcal{N}(\mu_i, \sigma^2/n_i) \ . \qquad (22), (63)$$

$$\tilde{y}_{if} = \hat{\mu}_i \quad \text{and} \quad v(y_{if} - \tilde{y}_i) = \sigma^2(1/n_i + 1) \ . \qquad (29)$$

$$\hat{\sigma}^2 = \Sigma_i\Sigma_j(y_{ij} - \bar{y}_{i\cdot})^2/(n_\cdot - a) \ . \qquad (34)$$

b. Sums of squares and *F*-statistics

$$\text{SST}_m = \Sigma_i\Sigma_j(y_{ij} - \bar{y}_{\cdot\cdot})^2 \ . \qquad (50)$$

$$\text{SSR}_m = \Sigma_i n_i(\bar{y}_{i\cdot} - \bar{y}_{\cdot\cdot})^2 = \mathcal{R}(\mu_i \mid \mu) \ . \qquad (49)$$

$$R^2 = \text{SSR}_m/\text{SST}_m \ . \qquad (52)$$

$$\text{SSR}_m/(a-1)\hat{\sigma}^2 \quad \text{tests} \quad \text{H: } \mu_i \textit{ all equal} \ . \qquad (75)$$

$$N\bar{y}_{\cdot\cdot}^2/\hat{\sigma}^2 \quad \text{tests} \quad \text{H: } \Sigma_i n_i\mu_i = 0 \ . \qquad (97)$$

2.16. EXERCISES

E2.1. Data sets A through F each represent data from a 1-way classification with four classes (columns of data). For each data set

(a) Write out the model equations,

and then calculate

(b) BLUEs of the population cell means,

(c) Analysis of variance Table 2.6,

(d) ρ', ρ, their BLUEs, variances of their BLUEs and estimates thereof,

(e) Coefficient of determination, and

(f) F-statistics for testing the three hypotheses:

$$H_1:\ \mu_1 - \mu_2 = 7$$

$$H_2:\ \mu_1 - \mu_3 = 3$$

$$H_3:\ \mu_1 - \mu_2 = 7 \quad \text{and} \quad \mu_1 - \mu_3 = 3.$$

Data A

14	12	7	1
10	4	1	5
13	14		
7			

Data B

11	8	17	9
5	18	11	5
	16	14	
		18	

Data C

5	8	2	8
13	2	6	11
15			14
			15

Data D

30	12	22	4
10	4	30	16
26		28	
		16	

Data E

19	31	13	7
7	25	29	15
	33	33	
	19		

Data F

9	13	13	7
1	27	1	27
	25		23
	19		

E2.2. The following data are from a 1-way classification with three classes:

8	4	8
6	12	18
5	8	
2		
9		

(a) Calculate Table 2.6.

(b) Calculate F-statistics for testing each of the hypotheses

$$\mathrm{H}_1: \begin{array}{l} \mu_1 - \mu_2 = 0 \\ \mu_1 + \mu_2 - 2\mu_3 = 0 \end{array} \qquad \mathrm{H}_2: \begin{array}{l} \mu_1 - \mu_3 = 0 \\ 3\mu_1 + 2\mu_2 - 5\mu_3 = 0 \end{array}$$

(c) For your answers in (b), discuss their relationship to each other and to the answer in (a).

E2.3. Under the normality assumptions of Section 2.10, show that SSM is independent of SSR_m and of SSE.

E2.4. Verify the three features of F of (87) that are described following equation (87).

E2.5. For γ_i of (109), show that for balanced data with $n_i = n \;\forall\; i$

(a) γ_i is a contrast,

(b) the BLUE of γ_i has variance $\sigma^2 i(i-1)/n$,

(c) the BLUEs of γ_i and γ_h for $i \neq h$ are uncorrelated,

(d) equation (115) is true for $a = 2$, and

(e) on the basis of part (d) and using induction on a, equation (115) is true for any a.

E2.6. Show that (128) for unbalanced data reduces to (115) when $n_i = n \;\forall\; i$.

E2.7. From (130) show that it is impossible to derive a BLUE of μ or of α_i.

E2.8. Prove (128) directly, not by induction.

CHAPTER 3

NESTED CLASSIFICATIONS

Nested (or hierarchical) classifications are usually analyzed using mixed models. Nevertheless, the analysis of a 2-way nested classification using a fixed effects model is a convenient stepping stone from the analysis of a 1-way classification in Chapter 2 to that of a 2-way crossed classification in Chapter 4. The fixed effects model applied to a 2-way nested classification is therefore considered in some detail.

Nested classifications of more than two factors are dealt with briefly in Section 3.6; briefly, because they are of minor practical importance insofar as fixed effects models are concerned. (Section 3.6 can therefore be easily omitted at a first reading.) Of more use is the application of mixed models to nested classifications: these are considered in Chapter 13. This is, therefore, a short chapter.

3.1. NOTATION

It is helpful to start with an example.

Example. Table 3.1 shows partial data of results of a student opinion poll of teachers' classroom use of computers in courses in English and Geology. These are data from a 2-way nested classification. (Sections of a course are presumed taught by different teachers, and English teachers are presumed different from Geology teachers.)

Let y_{ijk} be the kth observation in the jth level of (what shall be called) the secondary factor nested within the ith level of (what shall be called) the primary factor. In Table 3.1, course is the primary factor and section within course is the secondary factor. In general, let the number of levels of the

TABLE 3.1. DATA FROM STUDENT OPINION POLL OF TEACHERS' CLASSROOM USE OF COMPUTERS

Course	Section of Course	Observations: Individual		Total	Number[1]	Mean
English	1	5		5	(1)	5
	2	8, 10, 9		27	(3)	9
			Total:	32	(4)	8
Geology	1	8, 10		18	(2)	9
	2	6, 2, 1, 3		12	(4)	3
	3	3, 7		10	(2)	5
			Total:	40	(8)	5
			Grand total:	72	(12)	6

[1]For clarity, number of observations are in parentheses.

primary factor (courses, in the example) be a, so that i takes values $i = 1, \ldots, a$; and $a = 2$ in Table 3.1. Also, let b_i denote the number of levels of the secondary factor nested within the ith level of the primary factor. Let n_{ij} represent the number of observations in cell i, j, i.e., the cell defined by the jth level of the secondary factor within the ith level of the primary factor (sections within courses in the example). Thus for y_{ijk}, $j = 1, \ldots, b_i$, and $k = 1, \ldots, n_{ij}$; in the example, $b_1 = 2$ and $b_2 = 3$, $n_{11} = 1$, $n_{12} = 3$, $n_{21} = 2$, $n_{22} = 4$ and $n_{23} = 2$.

The number of observations in the ith level of the primary factor and the total number over all levels are, respectively,

$$n_{i\cdot} = \sum_{j=1}^{b_i} n_{ij} \quad \text{and} \quad n_{\cdot\cdot} = \sum_{i=1}^{a} n_{i\cdot} = N, \quad \text{say} \ .$$

Similarly, the total of the observations in cell i, j is

$$y_{ij\cdot} = \sum_{k=1}^{n_{ij}} y_{ijk} \quad \text{with mean} \quad \bar{y}_{ij\cdot} = y_{ij\cdot}/n_{ij} \ . \tag{1}$$

And for the ith level of the primary factor the total and mean are, respectively,

$$y_{i\cdot\cdot} = \sum_{j=1}^{b_i} y_{ij\cdot} = \sum_{j=1}^{b_i} \sum_{k=1}^{n_{ij}} y_{ijk} = \sum_{j=1}^{b_i} n_{ij}\bar{y}_{ij\cdot} \quad \text{and} \quad \bar{y}_{i\cdot\cdot} = y_{i\cdot\cdot}/n_{i\cdot} \ .$$

Finally, the grand total and mean are

$$y_{\cdot\cdot\cdot} = \sum_{i=1}^{a} y_{i\cdot\cdot} = \sum_{i=1}^{a} \sum_{j=1}^{b_i} \sum_{k=1}^{n_{ij}} y_{ijk} \quad \text{and} \quad \bar{y}_{\cdot\cdot\cdot} = y_{\cdot\cdot\cdot}/n_{\cdot\cdot} \ .$$

Values of these numbers of observations, totals, and means are shown in Table 3.1; for example, $y_{1\cdot\cdot} = 32$, $n_{1\cdot\cdot} = 4$ and $\bar{y}_{1\cdot\cdot} = 8$.

3.2. THE MODEL

The cell means model for data of the nature just described (exemplified in Table 3.1) is based on

$$E(y_{ijk}) = \mu_{ij} \ . \tag{2}$$

μ_{ij} is deemed to be the mean of a conceptual population in cell i, j, from which the available data are considered to be a random sample. E represents expectation over repeated sampling.

Associated with the observation y_{ijk} is a random error term defined, just as in (2) of Section 2.2, to be

$$e_{ijk} = y_{ijk} - E(y_{ijk}),$$

so giving

$$y_{ijk} = \mu_{ij} + e_{ijk} \ . \tag{3}$$

By its definition, e_{ijk} has expected value zero. Further, to each e_{ijk} we attribute a common variance σ^2 and a zero covariance between every pair of (different) e_{ijk}s. To these assumptions that of normality is added later, just as in Section 2.10.

3.3. ESTIMATION

The model equation (3) has precisely the same form as the cell means model $y_{ij} = \mu_i + e_{ij}$ of the 1-way classification. The only difference is notation: i of y_{ij} in the 1-way classification has become ij in (3), and j of the 1-way case has become k of (3). Hence estimating μ_{ij} proceeds exactly as in Chapter 2, and the $\hat{\mu}_i = \bar{y}_{i\cdot}$ of (22) in that chapter becomes

$$\hat{\mu}_{ij} = \bar{y}_{ij\cdot} \tag{4}$$

from (3), for $\bar{y}_{ij\cdot}$ of (1). Furthermore, exactly as in Chapter 2, $\hat{\mu}_{ij}$ is the BLUE of μ_{ij}:

$$\hat{\mu}_{ij} = \text{BLUE of } \mu_{ij} \text{ is } \hat{\mu}_{ij} = \bar{y}_{ij\cdot}$$

with
$$E(\hat{\mu}_{ij}) = \mu_{ij} \quad \text{and} \quad v(\hat{\mu}_{ij}) = \sigma^2/n_{ij},$$

and with the covariance between every pair of (different) $\hat{\mu}_{ij}$s being zero.

Example (continued). From Table 3.1:

$$\hat{\mu}_{11} = 5, \quad v(\hat{\mu}_{11}) = \sigma^2/1 \qquad \hat{\mu}_{21} = 9, \quad v(\hat{\mu}_{21}) = \sigma^2/2$$

$$\hat{\mu}_{12} = 9, \quad v(\hat{\mu}_{12}) = \sigma^2/3 \qquad \hat{\mu}_{22} = 3, \quad v(\hat{\mu}_{22}) = \sigma^2/4$$

$$\hat{\mu}_{23} = 5, \quad v(\hat{\mu}_{23}) = \sigma^2/2.$$

There is absolutely no new concept in these results. They follow directly and easily from Chapter 2: the BLUE of the population mean of a cell is the observed mean of that cell. And other estimation results follow similarly.

More generally, consider any linear combination of μ_{ij}s,

$$\omega = \sum_{i=1}^{a} \sum_{j=1}^{b_i} k_{ij}\mu_{ij}, \tag{5}$$

a direct extension of (76) in Section 2.12. The BLUE of ω is

$$\hat{\omega} = \sum_{i=1}^{a} \sum_{j=1}^{b_i} k_{ij}\hat{\mu}_{ij} = \sum_{i=1}^{a} \sum_{j=1}^{b_i} k_{ij}\bar{y}_{ij\cdot}. \tag{6}$$

with
$$v(\hat{\omega}) = \sigma^2 \sum_{i=1}^{a} \sum_{j=1}^{b_i} k_{ij}^2/n_{ij}, \tag{7}$$

just as in (77) of Section 2.12.

Lastly, so far as estimation is concerned, we have the usual unbiased estimator of $\hat{\sigma}^2$,

$$\hat{\sigma}^2 = \frac{\sum_{i=1}^{a} \sum_{j=1}^{b_i} \sum_{k=1}^{n_{ij}} (y_{ijk} - \bar{y}_{ij\cdot})^2}{n_{\cdot\cdot} - b.}, \tag{8}$$

which is a straightforward extension of (34) of Section 2.6.

3.4. CONFIDENCE INTERVALS, AND ONE-PART HYPOTHESES

On adopting normality assumptions, that $e_{ijk} \sim$ i.i.d. $\mathcal{N}(0, \sigma^2)$, confidence intervals on ω are precisely those of (81)–(83) of Section 2.12a, with appropriate degrees of freedom for t; thus the non-symmetric $100\alpha\%$ interval is

$$\hat{\omega} - t_{U,f,\alpha}\sqrt{\hat{v}(\hat{\omega})} \quad \text{to} \quad \hat{\omega} - t_{L,f,\alpha}\sqrt{\hat{v}(\hat{\omega})} \tag{9}$$

based on the $\mathcal{T}$-distribution with $f = n_{..} - b.$ degrees of freedom. The symmetric form of (9) is

$$\hat{\omega} \pm t_{f,\frac{1}{2}\alpha}\sqrt{\hat{v}(\hat{\omega})}\,; \tag{10}$$

and, when f is large (e.g., exceeds 100, say), this can be replaced by

$$\hat{\omega} \pm z_{\frac{1}{2}\alpha}\sqrt{\hat{v}(\hat{\omega})}\ . \tag{11}$$

In all of these intervals, $\hat{\omega}$ comes from (6), and $\hat{v}(\hat{\omega})$ is $v(\hat{\omega})$ of (7) with σ^2 replaced by $\hat{\sigma}^2$ of (8).

Also, similar to (87) of Section 2.12b(i), the one-part hypothesis

$$\text{H:}\ \ \omega = m \quad \text{is tested by} \quad \frac{(\hat{\omega} - m)^2}{\hat{v}(\hat{\omega})} \sim \mathcal{F}_{1,f}\ . \tag{12}$$

The extension to a two-part hypothesis involving two linear functions of the μ_{ij}s is a simple extension of (90) of Chapter 2 and consideration of multi-part hypotheses in general is left to Chapter 8. And (12) is, of course, the square of the t-statistic $(\hat{\omega} - m)/\sqrt{\hat{v}(\hat{\omega})}$.

Although the results (9)–(12) are in terms of the general linear function of μ_{ij}s defined in (5), they are easily applied to any individual μ_{rs}. This is done simply by using $\hat{\mu}_{rs} = \bar{y}_{rs\cdot}$ as $\hat{\omega}$, and $\hat{\sigma}^2/n_{rs}$ as $\hat{v}(\hat{\omega})$, this being so because if every k_{ij} in (5) is given the value zero except $k_{rs} = 1$, then $\omega = \mu_{rs}$.

3.5. ANALYSIS OF VARIANCE

a. For the cell means

The numerator of $\hat{\sigma}^2$ in (8) is the residual sum of squares

$$\text{SSE} = \sum_{i=1}^{a}\sum_{j=1}^{b_i}\sum_{k=1}^{n_{ij}} (y_{ijk} - \bar{y}_{ij\cdot})^2 \tag{13}$$

TABLE 3.2. ANALYSIS OF VARIANCE FOR THE 2-WAY NESTED MODEL

Source of Variation	d.f.	Sum of Squares	Associated Hypothesis
Mean	1	$\text{SSM} = N\bar{y}^2_{\cdot\cdot\cdot}$	H: $\sum_{i=1}^{a}\sum_{j=1}^{b_i} n_{ij}\mu_{ij} = 0$
Model	$b. - 1$	$\text{SSR}_m = \sum_{i=1}^{a}\sum_{j=1}^{b_i} n_{ij}(\bar{y}_{ij\cdot} - \bar{y}_{\cdot\cdot\cdot})^2$	H: μ_{ij} *all equal*
Residual	$N - b.$	$\text{SSE} = \sum_{i=1}^{a}\sum_{j=1}^{b_i}\sum_{k=1}^{n_{ij}} (y_{ijk} - \bar{y}_{ij\cdot})^2$	
Total	N	$\text{SST} = \sum_{i=1}^{a}\sum_{j=1}^{b_i}\sum_{k=1}^{n_{ij}} y^2_{ijk}$	

after fitting the model $E(y_{ijk}) = \mu_{ij}$ of (2). This is just like

$$\text{SSE} = \sum_{i=1}^{a}\sum_{j=1}^{n_i} (y_{ij} - \bar{y}_{i\cdot})^2$$

of Table 2.7 (Section 2.12c) for the 1-way classification. Indeed, that table extends very easily to be Table 3.2 for the 2-way nested classification.

The analysis of variance shown in Table 3.2 treats the data as if they were just a 1-way classification of $b.$ classes. That is why the hypothesis associated with SSR_m is H: μ_{ij} *all equal*; it is just a simple extension of H: μ_i *all equal* in Table 2.7 for the 1-way classification model $E(y_{ij}) = \mu_i$.

There is, however, an important practical difference between H: μ_i *all equal*, of the 1-way classification and H: μ_{ij} *all equal*, of the 2-way case. The former, in the context of the 1-way classification, is useful; it is a hypothesis that the class means are equal. But H: μ_{ij} *all equal* is seldom of any interest. It is a hypothesis that the mean of every cell in every level of the primary factor is the same. The example illustrates why this is not, generally speaking, an interesting or useful hypothesis.

Example (continued). For the example of Table 3.1, the hypothesis associated with SSR_m of Table 3.2 is

$$\text{H: } \mu_{11} = \mu_{12} = \mu_{21} = \mu_{22} = \mu_{23} \ .$$

Clearly this is of little interest, a hypothesis that section means within

English are not only equal to each other but are also equal to section means within Geology, which are also equal to each other. But what might be of interest would be to test the hypothesis that section means within English are all equal and (either separately, or simultaneously) that section means within Geology are all equal, but without their being equal to those within English. Thus we would be interested in testing

$$\mathrm{H}_1\colon\ \mu_{11} = \mu_{12} \quad \text{and} \quad \mathrm{H}_2\colon\ \mu_{21} = \mu_{22} = \mu_{23}\ . \tag{14}$$

This can easily be done.

b. Cell means within levels of the primary factor

Generalization of (14) is

$$\mathrm{H}_i\colon\ \mu_{i1} = \mu_{i2} = \cdots = \mu_{ib_i}\ \forall\ i = 1, \ldots, a\ . \tag{15}$$

This is a many-part hypothesis that within each level of the primary factor the cell means for the different levels of the secondary factor are equal. To test one of the H_i of (15) we use Table 2.7 (Section 2.12c) again, and also the $\mathscr{R}(\mu_i | \mu)$ notation of Table 2.4 (Section 2.8). From them we find that in the model $E(y_{ij}) = \mu_i$

$$\mathscr{R}(\mu_i | \mu) = \sum_{i=1}^{a} n_i(\bar{y}_{i\cdot} - \bar{y}_{\cdot\cdot})^2 \ \text{tests}\quad \mathrm{H}\colon\ \mu_i\ \textit{equal}\ \forall\ i = 1, 2, \ldots, a\ . \tag{16}$$

Therefore, on confining attention to the data of just the ith level of the primary factor in our 2-way nested classification, we have by analogy with (16), on defining

$$\mathscr{R}_i(\mu_{ij} | \mu_i) = \sum_{j=1}^{b_i} n_{ij}(\bar{y}_{ij\cdot} - \bar{y}_{i\cdot\cdot})^2, \tag{17}$$

that $\qquad \mathscr{R}_i(\mu_{ij} | \mu_i)$ tests $\mathrm{H}_i\colon\ \mu_{ij}\ \textit{equal}\ \forall\ j = 1, 2, \ldots, b_i\ .$ (18)

Within the ith level of the primary factor, μ_{ij}, μ_i, b_i, n_{ij}, $\bar{y}_{ij\cdot}$ and $\bar{y}_{i\cdot\cdot}$ of the nested classification are playing the same roles in (18) as are μ_i, μ, a, n_i, $\bar{y}_{i\cdot}$ and $\bar{y}_{\cdot\cdot}$ of the 1-way classification in (16). Thus (18) provides a test of each hypothesis H_i in (15).

Suppose (15) is written as $\mathrm{H}_i\colon\ \mu_{ij} = \mu_i\ \forall\ j$. To test all the H_i simultaneously we write

$$\mathrm{H}\colon\ \{\mu_{ij} = \mu_i\,\forall\ j\}\ \forall\ i, \tag{19}$$

and then, similar to F in (87) of Section 2.12b, we calculate

$$F = \frac{\text{SSE}_\text{H} - \text{SSE}}{f\hat{\sigma}^2},$$

where f is degrees of freedom, SSE_H is the residual sum of squares under the hypothesis (19), and $\hat{\sigma}^2$ and SSE are as in (8) and (13), respectively. Under H of (19) the model equation is

$$E(y_{ijk}) = \mu_i \ .$$

Based on (38) of Section 2.7, we then have

$$\text{SSE}_\text{H} = \Sigma_i\Sigma_j\Sigma_k(y_{ijk} - \bar{y}_{i..})^2 = \Sigma_i\Sigma_j\Sigma_k y_{ijk}^2 - \Sigma_i n_{i.}\bar{y}_{i..}^2 \ .$$

And from (13)

$$\text{SSE} = \Sigma_i\Sigma_j\Sigma_k(y_{ijk} - \bar{y}_{ij.})^2 = \Sigma_i\Sigma_j\Sigma_k y_{ijk}^2 - \Sigma_i\Sigma_j n_{ij}\bar{y}_{ij.}^2 \ .$$

Hence

$$\begin{aligned} \text{SSE}_\text{H} - \text{SSE} &= \Sigma_i\Sigma_j n_{ij}\bar{y}_{ij.}^2 - \Sigma_i n_{i.}\bar{y}_{i..}^2 \\ &= \Sigma_i\Sigma_j n_{ij}(\bar{y}_{ij.} - \bar{y}_{i..})^2 \\ &= \Sigma_i\mathscr{R}_i(\mu_{ij}|\mu_i) = \mathscr{R}(\mu_{ij}|\mu_i) \end{aligned} \tag{20}$$

for $\mathscr{R}_i(\mu_{ij}|\mu_i)$ of (18). Furthermore, the degrees of freedom are $b. - a$. Therefore

$$F = \frac{\Sigma_i\Sigma_j n_{ij}(\bar{y}_{ij.} - \bar{y}_{i..})^2}{(b. - a)\hat{\sigma}^2} \quad \text{tests} \quad \{\text{H}_i\text{: } \mu_{ij} \textit{ equal } \forall\ j\}\ \forall\ i\ . \tag{21}$$

c. Means for the levels of the primary factor

Chapter 2 deals with the 1-way classification having cell means $\mu_1, \mu_2, \ldots, \mu_a$. As a result of Sections 2.10d, e and f we can say from (73) of Section 2.10g (and as summarized in Table 2.6) that

$$F = \frac{\Sigma_i n_i(\bar{y}_{i.} - \bar{y}_{..})^2}{(a-1)\hat{\sigma}^2} \quad \text{tests} \quad \text{H: } \mu_i \textit{ equal } \forall\ i\ . \tag{22}$$

Part of the derivation of this result is the theorem in the Appendix of

Chapter 2, which starts with $y_{ij} \sim$ i.i.d. $\mathcal{N}(\mu_i, \sigma^2)$ for $j = 1, \ldots, n_i$, independently for $i = 1, \ldots, a$. This is used to establish

$$\bar{y}_{i\cdot} \sim \mathcal{N}\left(\mu_i, \quad \sigma^2/n_i\right)$$

independently for $i = 1, \ldots, a$. But for the 2-way nested classification of this chapter,

$$\bar{y}_{i\cdot\cdot} \sim \mathcal{N}\left(\rho'_i, \quad \sigma^2/n_{i\cdot}\right) \quad \text{with } \rho'_i = \frac{\Sigma_j n_{ij}\mu_{ij}}{n_{i\cdot}}, \tag{23}$$

independently for $i = 1, \ldots, a$. Hence the theorem of the Appendix of Chapter 2 applies to (23), and so from (22)

$$F = \frac{\Sigma_i n_{i\cdot}(\bar{y}_{i\cdot\cdot} - \bar{y}_{\cdot\cdot\cdot})^2}{(a-1)\hat{\sigma}^2} \quad \text{tests} \quad \text{H: } \rho'_i \textit{ equal } \forall\ i\ . \tag{24}$$

Moreover, since $R(\mu_i) = \Sigma_i n_{i\cdot}\bar{y}^2_{i\cdot\cdot}$ and $R(\mu) = n_{\cdot\cdot}\bar{y}^2_{\cdot\cdot\cdot}$, we have for (24)

$$\Sigma_i n_{i\cdot}(\bar{y}_{i\cdot\cdot} - \bar{y}_{\cdot\cdot\cdot})^2 = R(\mu_i) - R(\mu) = \mathcal{R}(\mu_i|\mu)\ .$$

Hence, from (24), in terms of sums of squares,

$$\mathcal{R}(\mu_i|\mu) = \Sigma_i n_{i\cdot}(\bar{y}_{i\cdot\cdot} - \bar{y}_{\cdot\cdot\cdot})^2 \quad \text{tests} \quad \text{H: } \rho'_i \textit{ equal } \forall\ i \quad \text{for } \rho'_i = \frac{\Sigma_i n_{ij}\mu_{ij}}{n_{i\cdot}}\ . \tag{25}$$

A more direct derivation of (24) from (22) is simply to replace in (22) the terms n_i, $\bar{y}_{i\cdot}$, $\bar{y}_{\cdot\cdot}$ and μ_i by $n_{i\cdot}$, $\bar{y}_{i\cdot\cdot}$, $\bar{y}_{\cdot\cdot\cdot}$ and ρ'_i of (23), respectively: (24) follows at once. And ρ'_i is just the same form as ρ' is in (94) of Section 2.12c.

For the example, the hypothesis in (25) is

$$\text{H: } \frac{\mu_{11} + 3\mu_{12}}{4}, \text{ for English, } = \frac{2\mu_{21} + 4\mu_{22} + 2\mu_{23}}{8}, \text{ for Geology}\ .$$

Clearly, this is unlikely to be useful. And in general the form of the hypothesis given in (25) is not very appealing. For one thing, it is a hypothesis that is dependent on the data; not on the observations themselves, but on the numbers of them that occur in the various cells of the classifications. Since hypotheses are statements about populations (in linear models, usually about means of populations) it does not seem altogether reasonable to be interested in hypotheses that are determined by how much data has been collected in the different cells.

An all-important aspect of the hypothesis in (25) is that we understand exactly what that hypothesis is; for, armed with that understanding, one can then decide, for whatever data are being analyzed, whether or not that hypothesis is of interest.

A final comment: each of (21) and (25) is a particular case of the general result given in (144) and (146) of Section 8.8b.

d. An overall mean

The first line of Table 2.7 shows $N\bar{y}_{..}^2$ for the model $E(y_{ij}) = \mu_i$ as having for its associated hypothesis H: $\Sigma_i n_i \mu_i = 0$. The extension to the 2-way nested model is straightforward:

$$R(\mu) = n_{..}\bar{y}_{...}^2 \quad \text{tests} \quad \text{H: } \sum_{i=1}^{a} \sum_{j=1}^{b_i} n_{ij}\mu_{ij} = 0 \ .$$

e. Summary

The arithmetic involved in the preceding sums of squares is as follows, where Σ_i, Σ_j and Σ_k represent, respectively, summations over $i = 1, \ldots, a$, over $j = 1, \ldots, b_i$ and over $k = 1, \ldots, n_{ij}$.

$$R(\mu) = n_{..}\bar{y}_{...}^2$$

$$\mathscr{R}(\mu_i \mid \mu) = \Sigma_i n_{i.}(\bar{y}_{i..} - \bar{y}_{...})^2 = \Sigma_i n_i \bar{y}_{i..}^2 - N\bar{y}_{...}^2$$

$$\mathscr{R}(\mu_{ij} \mid \mu_i) = \Sigma_i \Sigma_j n_{ij}(\bar{y}_{ij.} - \bar{y}_{i..})^2 = \Sigma_i \Sigma_j n_{ij}\bar{y}_{ij.}^2 - \Sigma_i n_{i.}\bar{y}_{i..}^2$$

$$\text{SSE} = \Sigma_i \Sigma_j \Sigma_k (y_{ijk} - \bar{y}_{ij.})^2 = \Sigma_i \Sigma_j \Sigma_k y_{ijk}^2 - \Sigma_i \Sigma_j n_{ij}\bar{y}_{ij.}^2$$

$$\text{SST} = \Sigma_i \Sigma_j \Sigma_k y_{ijk}^2 .$$

With these calculations we summarize the sums of squares and their associated hypotheses in the analysis of variance table of Table 3.3.

The reader should confirm the necessary arithmetic to verify that Table 3.3 for the example of Table 3.1 is as shown in Table 3.4.

TABLE 3.3. ANALYSIS OF VARIANCE TABLE FOR A 2-WAY NESTED CLASSIFICATION USING THE MODEL $E(y_{ijk}) = \mu_{ij}$

Source of Variation	d.f.	Sum of Squares	Associated Hypothesis
Mean	1	$R(\mu)$	H: $\Sigma_i\Sigma_j n_{ij}\mu_{ij} = 0$
Primary factor	$a - 1$	$\mathscr{R}(\mu_i \mid \mu)$	H: $\dfrac{\Sigma_j n_{ij}\mu_{ij}}{n_{i\cdot}}$ *equal* $\forall\ i$
Secondary factor within levels of the primary factor	$b_{\cdot} - a$	$\mathscr{R}(\mu_{ij} \mid \mu_i)$	H: $\{\text{H}_i\text{: } \mu_{ij}$ *equal* $\forall\ j\}\ \forall\ i$
Residual	$N - b_{\cdot}$	SSE	
Total	N	SST	

TABLE 3.4. ANALYSIS OF VARIANCE (TABLE 3.3) FOR THE DATA OF TABLE 3.1

Source of Variation	d.f.	Sum of Squares	Associated Hypothesis
Mean	1	$R(\mu) = 432$	H: $\mu_{11} + 3\mu_{12} + 2\mu_{21} + 4\mu_{22} + 2\mu_{23} = 0$
Primary factor	1	$\mathscr{R}(\mu_i \mid \mu) = 24$	H: $\dfrac{\mu_{11} + 3\mu_{12}}{4} = \dfrac{2\mu_{21} + 4\mu_{22} + 2\mu_{23}}{8}$
Secondary factor, within levels of the primary factor	3	$\mathscr{R}(\mu_{ij} \mid \mu_i) = 60$	H: $\begin{cases} \mu_{11} = \mu_{12} \\ \mu_{21} = \mu_{22} = \mu_{23} \end{cases}$
Residual	7	SSE = 26	
Total	12	SST = 542	

f. Using the hypotheses in the analysis of variance

A useful way to use the hypotheses of Table 3.3 is to first of all consider the hypothesis associated with $\mathscr{R}(\mu_{ij} \mid \mu_i)$. If we accept (fail to reject) that hypothesis, then we are accepting the hypothesis that within the ith level of the primary factor $\mu_{i1} = \mu_{i2} = \cdots = \mu_{ib_i}$, and that this is so for each $i = 1, 2, \ldots, a$. That being so, let $\mu_{i1} = \mu_{i2} = \mu_{ib_i} = \mu_i$; then the hypothesis associated with $\mathscr{R}(\mu_i \mid \mu)$ becomes H: μ_i *equal* $\forall\ i$. This is often of real interest; it is the hypothesis that levels of the primary factor all have the same mean.

Of course, one is not confined to the hypothesis H: μ_i *equal* $\forall\ i$. Using $\hat{\mu}_i = \bar{y}_{i\cdot\cdot}$ for $\hat{\omega}$ in (6) one can test any linear hypothesis about the μ_i.

Example (continued). From Table 3.1

$$\hat{\mu}_1 = 32/4 = 8 \quad \text{and} \quad \hat{\mu}_2 = 40/8 = 5;$$

and using $\hat{\sigma}^2 = 26/7$ from Table 3.4 the F-statistic for testing H: $\mu_1 - \mu_2 = 0$ is, from (12) and (7),

$$F = \frac{(8-5)^2}{(26/7)(1/4 + 1/8)} = \frac{84}{13}\ .$$

Similarly, (9)–(11) can be used for establishing confidence intervals.

Suppose, though, that we reject the hypothesis associated with $\mathscr{R}(\mu_{ij}|\mu_i)$ in Table 3.3. Then the hypothesis associated with $\mathscr{R}(\mu_i|\mu)$ in that table is as shown there; it can still be interpreted in the same manner as when accepting the hypothesis associated with $\mathscr{R}(\mu_{ij}|\mu_i)$, namely as H: μ_i *equal* $\forall\ i$, except that μ_i must be interpreted as $\rho'_i = \Sigma_j n_{ij}\mu_{ij}/n_{i\cdot}$; and this in turn becomes $\rho_i = \Sigma_j \mu_{ij}/b_i$ when $n_{i1} = n_{i2} = \cdots = n_{ib_i}$. Of course, equal numbers of observations in the cells within each level of the primary factor is not a necessary condition for considering $\rho_i = \Sigma_j\mu_{ij}/b_i$. Indeed, the hypothesis

$$\text{H: } \rho_i \ \textit{equal}\ \forall\ i, \quad \text{for } \rho_i = \Sigma_j\mu_{ij}/b_i \tag{26}$$

can be tested using

$$\Sigma_i w_i(\tilde{y}_{i\cdot\cdot} - \tilde{y}_{\cdot\cdot\cdot})^2 \tag{27}$$

for $$\frac{1}{w_i} = \frac{1}{b_i^2}\Sigma_j\frac{1}{n_{ij}}, \qquad \tilde{y}_{i\cdot\cdot} = \frac{\Sigma_j\bar{y}_{ij\cdot}}{b_i} \quad \text{and} \quad \tilde{y}_{\cdot\cdot\cdot} = \frac{\Sigma_i w_i\tilde{y}_{i\cdot\cdot}}{\Sigma w_i}\ .$$

This is established by replacing in (22) the terms n_i, $\bar{y}_{i\cdot}$, $\bar{y}_{\cdot\cdot}$ and μ_i by w_i, $\tilde{y}_{i\cdot\cdot}$, $\tilde{y}_{\cdot\cdot\cdot}$ and ρ_i, respectively. (See E3.7.)

By its very nature, (26) is likely to be a more generally useful hypothesis than hypothesis (25), because formulation of the latter depends on the number of observations in each cell. On the other hand, (26) is a hypothesis that, over all levels of the primary factor, the simple within-level averages of the cell means are equal. Despite the more general appeal of this hypothesis (uncluttered by the numbers of observations in each cell), readers should be aware that default procedures of computer packages do not always yield what is of greatest use to an investigator. Intervention by *Homo sapiens* is often required in order to get test statistics for useful hypotheses: that is where the "sapiens" applies.

3.6. MORE THAN TWO FACTORS

Details for 3-way, 4-way, ..., r-way nested classifications can easily be derived as extensions of those already given for the 1-way classification in Chapter 2, and for the 2-way nested classification in the preceding sections of this chapter. Those details are voluminous and tedious, and are of little practical use for fixed effects models, especially with unbalanced data. Instead of presenting the details we therefore briefly describe the principles by which extensions from the 2-way nested classification can be made.

a. Models

The model equations for the 1-way and 2-way cases are, respectively,

$$y_{ij} = \mu_i + e_{ij} \quad \text{for } i = 1, \ldots, a \text{ and } j = 1, \ldots, n_i; \tag{28}$$

and

$$y_{ijk} = \mu_{ij} + e_{ijk} \quad \text{for } i = 1, \ldots, a, \ j = 1, \ldots, b_i \text{ and } k = 1, \ldots, n_{ij}. \tag{29}$$

Similarly, that for a 3-way nested classification could be taken as

$$y_{ijkl} = \mu_{ijk} + e_{ijkl} \quad \text{for } i = 1, \ldots, a, \ j = 1, \ldots, b_i \quad k = 1, \ldots, c_{ij} \text{ and } l = 1, \ldots, n_{ijk}. \tag{30}$$

The extension is clear: for an r-way nested classification we can write

$$y_{i_1 i_2 \ldots i_r i_{r+1}} = \mu_{i_1 i_2 \ldots i_r} + e_{i_1 i_2 \ldots i_r i_{r+1}} \tag{31}$$

for $i_1 = 1, \ldots, a_1$, $i_2 = 1, \ldots, a_{2, i_1}$, $i_3 = 1, \ldots, a_{3, i_1 i_2}$, and for $i_t = 1, \ldots, a_{t, i_1 i_2 \ldots i_{t-1}}$ for $t = 2, \ldots, r+1$, where, for example, a_1 is the number of levels of the primary factor and $a_{r+1, i_1 i_2 \ldots i_r}$ is defined as the number of observations in the i_rth level of the rth factor, nested within the (i_{r-1})th level of the $(r-1)$th factor,... and so on,... within the i_1th level of the primary factor. Thus, for example, for $r = 3$, $a_1 = a$, $a_{2,i_1} = b_i$, $a_{3,i_1i_2} = c_{ij}$ and $a_{4,i_1i_2i_3} = n_{ijk}$ of (30). Then, from the progression that is evident in (28) through (31), we can conclude that the model equation for the $(r+1)$-way nested classification model differs from that for the r-way model, namely (31), in only three respects:

(i) One further subscript, i_{r+2}, must be put on y and e.
(ii) One further subscript, i_{r+1}, must be put on μ.
(iii) The limits of i_{r+2} are $1, 2, \ldots, a_{r+2, i_1 i_2 \ldots i_{r+1}}$.

b. Sums of squares

The sums of squares in the analysis of variance Tables 2.7 and 3.4 are set alongside each other in Table 3.5. For simplicity, limits of summation have been omitted, and in each case the summations are over all subscripts rather than, where possible, replacing a summation by an n with appropriate subscripts. For example, preceding Table 3.3 we have

$$\mathscr{R}(\mu_i|\mu) = \Sigma_i n_i(\bar{y}_{i\cdot} - \bar{y}_{\cdot\cdot})^2$$

but in Table 3.5 have used its equivalent form

$$\mathscr{R}(\mu_i|\mu) = \Sigma_i\Sigma_j(\bar{y}_{i\cdot} - \bar{y}_{\cdot\cdot})^2 .$$

This makes for easier discussion of the general form of the sums of squares in Table 3.5. Scrutiny of that table reveals that notationally the sums of

TABLE 3.5. SUMS OF SQUARES IN THE ANALYSIS OF VARIANCE OF THE 1-WAY CLASSIFICATION (TABLE 2.7) AND THE 2-WAY NESTED CLASSIFICATION (TABLE 3.3)

1-way $E(y_{ij}) = \mu_i$	2-way nested $E(y_{ijk}) = \mu_{ij}$
$R(\mu) = n_{\cdot}\bar{y}_{\cdot\cdot}^2$	$R(\mu) = n_{\cdot\cdot}\bar{y}_{\cdot\cdot\cdot}^2$
$\mathscr{R}(\mu_i\|\mu) = \Sigma_i\Sigma_j(\bar{y}_{i\cdot} - \bar{y}_{\cdot\cdot})^2$	$\mathscr{R}(\mu_i\|\mu) = \Sigma_i\Sigma_j\Sigma_k(\bar{y}_{i\cdot\cdot} - \bar{y}_{\cdot\cdot\cdot})^2$
$\text{SSE} = \Sigma_i\Sigma_j(y_{ij} - \bar{y}_{i\cdot})^2$	$\mathscr{R}(\mu_{ij}\|\mu_i) = \Sigma_i\Sigma_j\Sigma_k(\bar{y}_{ij\cdot} - \bar{y}_{i\cdot\cdot})^2$
	$\text{SSE} = \Sigma_i\Sigma_j\Sigma_k(y_{ijk} - \bar{y}_{ij\cdot})^2$
$\text{SST} = \Sigma_i\Sigma_j y_{ij}^2$	$\text{SST} = \Sigma_i\Sigma_j\Sigma_k y_{ijk}^2$

squares for the 2-way nested case are just those of the 1-way case but with

(i) each y having an additional dot in its subscript;
(ii) each term involving one more summation;
(iii) y_{ij} becoming $\bar{y}_{ij\cdot}$;
(iv) SSE becoming $\mathscr{R}(\mu_{ij} | \mu_i)$; and
(v) inclusion of a new term, SSE $= \Sigma_i \Sigma_j \Sigma_k (y_{ijk} - \bar{y}_{ij\cdot})^2$.

These changes apply quite generally: the sums of squares for the $(r + 1)$-way nested classification are those of the r-way case with (i) and (ii) applying as is; and with (iii), (iv) and (v) taking the form

(iii)′ $y_{i_1 i_2 \ldots i_{r+1}}$ becoming $\bar{y}_{i_1 i_2 \ldots i_{r+1}\cdot}$;
(iv)′ SSE becoming $\mathscr{R}(\mu_{i_1 i_2 \ldots i_{r+1}} | \mu_{i_1 i_2 \ldots i_r})$; and
(v)′ inclusion of a new term,

$$\text{SSE} = \Sigma_{i_1} \Sigma_{i_2} \ldots \Sigma_{i_{r+2}} \left(y_{i_1 i_2 \ldots i_{r+1} i_{r+2}} - \bar{y}_{i_1 i_2 \ldots i_{r+1}\cdot} \right)^2 .$$

With this as an algorithm one can, if desired, build up from the sums of squares for the 2-way nested classification of Table 3.3 those for any r-way nested classification ($r > 2$).

c. Hypotheses tested by analysis of variance sums of squares in the fixed effects model

Using $N\bar{y}^2$ to generically denote either of the $R(\mu)$ forms in Table 3.5, we always have the result

$$R(\mu) = N\bar{y}^2 \quad \text{tests} \quad \text{H: } E(\bar{y}) = 0 . \tag{32}$$

This applies quite generally.

Other than that, observe from Table 2.7 that for the 1-way classification

$$\mathscr{R}(\mu_i | \mu) = \Sigma_i \Sigma_j (\bar{y}_{i\cdot} - \bar{y}_{\cdot\cdot})^2$$

tests

$$\text{H: } \mu_i \textit{ equal } \forall\ i, \quad \text{equivalent to} \quad \text{H: } E(\bar{y}_{i\cdot}) \textit{ equal } \forall\ i . \tag{33}$$

And from Table 3.3 for the 2-way case

$$\mathscr{R}(\mu_i | \mu) = \Sigma_i \Sigma_j \Sigma_k (\bar{y}_{i\cdot\cdot} - \bar{y}_{\cdot\cdot\cdot})^2 \quad \text{tests} \quad \text{H: } \frac{\Sigma_j n_{ij} \mu_{ij}}{n_{i\cdot}} \textit{ equal } \forall\ i .$$

This can be rewritten, by observing that $E(\bar{y}_{i\cdot\cdot}) = \Sigma_j n_{ij}\mu_{ij}/n_{i\cdot}$, as

$$\mathscr{R}(\mu_i \mid \mu) \quad \text{tests} \quad \text{H: } E(\bar{y}_{i\cdot\cdot}) \textit{ equal } \forall \ i \ . \tag{34}$$

And (34) is a simple extension of (33). Indeed, in the general r-way nested model of (31)

$$\mathscr{R}(\mu_{i_1} \mid \mu) \quad \text{tests} \quad \text{H: } E(\bar{y}_{i_1 *}) \textit{ equal } \forall \ i_1 \tag{35}$$

where the subscript $*$ indicates averaging over all subscripts not otherwise shown.

The principle suggested in (35) extends to all other sums of squares in nested classifications with fixed effects models. First, as an example, from Table 3.3,

$$\mathscr{R}(\mu_{ij} \mid \mu_i) = \Sigma_i \Sigma_j \Sigma_k (\bar{y}_{ij\cdot} - \bar{y}_{i\cdot\cdot})^2$$

we have $\quad \mathscr{R}(\mu_{ij} \mid \mu_i) \quad \text{tests} \quad \text{H: } \mu_{ij} \textit{ equal } \forall \ j, \textit{ within each } i$

equivalent to $\quad \text{H: } E(\bar{y}_{ij\cdot}) \textit{ equal } \forall \ j, \textit{ within each } i \ . \qquad (36)$

This principle applies in general. For notational clarity replace one or more dots in the subscript of a mean by an asterisk; thus $\bar{y}_{ij\cdot}$ and $\bar{y}_{i\cdot\cdot}$ become $\bar{y}_{ij*}$ and $\bar{y}_{i*}$, respectively. Then the generalization of (36) for the r-way nested model is

$$\Sigma_{i_1} \ldots \Sigma_{i_{r+1}} \left(\bar{y}_{i_1 i_2 \ldots i_{k*}} - \bar{y}_{i_1 i_2 \ldots i_{k-1*}} \right)^2$$

$$\text{tests} \quad \text{H: } E\left(\bar{y}_{i_1 i_2 \ldots i_{k*}}\right) = 0 \ \forall \ i_k, \textit{ within each } i_1 i_2 \ldots i_{k-1} \ . \tag{37}$$

This does, in fact, cover all of the cases (32) through (36). Thus (32) is (37) with k not existing, and so the sum of squares is $N\bar{y}^2$; (33), (34) and (35) are cases where k is just the first subscript of the model; and (36) is where k of (37) is the second subscript of the model.

In practice, of course, nested classifications are seldom used with fixed effects models; they arise mostly in situations concerned with sampling, whereupon the nested factors are random effects factors for which the parameters of interest are variance components. This is discussed briefly in Chapter 13.

3.7. EXERCISES

E3.1. Suppose the data of a student opinion poll, similar to that of Table 3.1, are as shown below. (Each column is a section of a course.)

English			Geology		Chemistry			
2	7	8	2	10	8	6	1	8
5	9	4	6	8		2	3	6
2		3		9		3	2	
		6				1		
		4						

(a) Calculate an analysis of variance table like that of Table 3.4.

(b) Write out the hypothesis associated with the sums of squares of your analysis.

E3.2. Repeat (a) and (b) of E3.1 for the following similar kind of data.

History		Math		
17	24	13	10	7
	16	7	12	
			6	
			8	

E3.3. Suppose that the data of E3.1 are from one school and those of E3.2 are from another. Write the model and calculate the analysis of variance for combining all the data into a 3-way nested classification, sections within courses within schools.

E3.4. Explain why $R(\mu) = N\bar{y}^2$ always provides a test of $E(\bar{y}) = 0$.

E3.5. Explain why, in Table 3.3, $R(\mu)$, $\mathcal{R}(\mu_i|\mu)$ and $\mathcal{R}(\mu_{ij}|\mu_i)$ are, under normality,

(i) independent of one another;

(ii) distributed in proportion to χ^2-variables;

(iii) independent of SSE.

E3.6. Describe why the properties (i), (ii) and (iii) of E3.5 apply generally to the sums of squares of any r-way nested classification using a fixed effects model.

E3.7. Using the methods of Chapter 2 (particularly Sections 2.10 and 2.14a), confirm that $\Sigma_i w_i(\tilde{y}_{i\cdot\cdot} - \tilde{y}_{\cdot\cdot\cdot})^2/\sigma^2$ of (27) in Section 3.5 has a χ^2_{a-1}-distribution.

CHAPTER 4

THE 2-WAY CROSSED CLASSIFICATION WITH ALL-CELLS-FILLED DATA: CELL MEANS MODELS

Literature on the 2-way crossed classification is extensive. This is so because the 2-way crossed classification is the simplest situation that displays most of the difficulties associated with using overparameterized models with unbalanced data. Many of these difficulties can be circumvented by using cell means models, and those that cannot be avoided can at least be more easily understood by considering them in terms of cell means models. Nevertheless, there are some features of the 2-way crossed classification that cannot be thoroughly described in terms of cell means models as already propounded in Chapters 2 and 3, so complete presentation of the 2-way crossed classification encompasses three chapters, this and Chapters 5 and 9. This chapter deals with the case of all-cells-filled data, the next considers some-cells-empty data and Chapter 9 is largely concerned with over-parameterized models and depends on the general matrix algebra description of linear model theory given in Chapter 8. Nevertheless, without such dependence, this chapter explains most of the useful results for the 2-way crossed classification, with only an occasional forward reference to, and use of, results developed in Chapter 9. This dividing of a topic between chapters, before and after the general theory, is also done for the analysis of covariance—in Chapters 6 and 11.

4.1. NOTATION

As usual, we begin with a numerical example.

Example. Suppose we are interested in the effect of two different kinds of potting soil on the number of days to germination of three varieties of

TABLE 4.1. NUMBER OF DAYS TO FIRST GERMINATION OF THREE VARIETIES OF CARROT SEED GROWN IN TWO DIFFERENT POTTING SOILS

	Variety		
Soil	1	2	3
1	6	13	14
	10	15	22
	11		
	27(3)9	28(2)14	36(2)18
2	12	31	18
	15		9
	19		12
	18		
	64(4)16	31(1)31	39(3)13

carrot seed. Table 4.1 might be data from an experiment designed to consider this situation, in which a large amount of the intended data got lost as a result of seed pots being accidentally knocked off greenhouse shelves.

Let y_{ijk} be the kth observation on soil i for variety j, where $i = 1, 2$, $j = 1, 2, 3$ and $k = 1, 2, \ldots, n_{ij}$, with n_{ij} being the number of observations on soil i for variety j. For the sake of generality we refer to soils as rows and varieties as columns; and the data on row i and column j are said to be in cell i, j, with

$$y_{ij\cdot} = \sum_{k=1}^{n_{ij}} y_{ijk} \quad \text{and} \quad \bar{y}_{ij\cdot}/n_{ij} \tag{1}$$

being the total and mean, respectively, of the data in cell i, j. Table 4.1 shows individual data values for each cell, and the triplet of numbers for each cell is $y_{ij\cdot}, (n_{ij}), \bar{y}_{ij\cdot}$; e.g., in the 1, 1 cell, $y_{11\cdot} = 27$, $n_{11} = 3$ and $\bar{y}_{11\cdot} = 9$.

In general we take the number of rows to be a; and the number of columns to be b. Then $i = 1, \ldots, a$ and $j = 1, \ldots, b$; and in the example of Table 4.1, $a = 2$ and $b = 3$.

Inevitably there is need for the numbers of observations in row i and column j, denoted respectively by

$$n_{i\cdot} = \sum_{j=1}^{b} n_{ij} \quad \text{and} \quad n_{\cdot j} = \sum_{i=1}^{a} n_{ij}; \tag{2}$$

and for the total of the observations in row i and column j, namely

$$y_{i\cdot\cdot} = \sum_{j=1}^{b} \sum_{k=1}^{n_{ij}} y_{ijk} \quad \text{and} \quad y_{\cdot j\cdot} = \sum_{i=1}^{a} \sum_{k=1}^{n_{ij}} y_{ijk} \tag{3}$$

and the corresponding row and column means

$$\begin{aligned} \bar{y}_{i\cdot\cdot} &= y_{i\cdot\cdot}/n_{i\cdot} = \sum_{j=1}^{b} n_{ij}\bar{y}_{ij\cdot}/n_{i} \\ \bar{y}_{\cdot j\cdot} &= y_{\cdot j\cdot}/n_{\cdot j} = \sum_{i=1}^{a} n_{ij}\bar{y}_{ij\cdot}/n_{\cdot j}, \end{aligned} \tag{4}$$

all of these, of course, applying for each $i = 1, \ldots, a$ and $j = 1, \ldots, b$. Although the example of Table 4.1 has all cells filled, i.e., every $n_{ij} \neq 0$ in that table, the notational definitions in (1)–(4) also apply in cases where some cells are empty; i.e., have no data, so that $n_{ij} = 0$ for such cells. In those cases, of course, $\bar{y}_{ij\cdot}$ of (1) does not exist for an empty cell: such cases are pursued in Chapter 5.

Finally, we have the total number of observations, usually denoted by N:

$$N = n_{\cdot\cdot} = \sum_{i=1}^{a} \sum_{j=1}^{b} n_{ij} = \sum_{i=1}^{a} n_{i\cdot} = \sum_{j=1}^{b} n_{\cdot j} \tag{5}$$

and the total of all observations

$$y_{\cdot\cdot\cdot} = \sum_{i=1}^{a} y_{i\cdot\cdot} = \sum_{j=1}^{b} y_{\cdot j\cdot} = \sum_{i=1}^{a} \sum_{j=1}^{b} y_{ij\cdot} = \sum_{i=1}^{a} \sum_{j=1}^{b} \sum_{k=1}^{n_{ij}} y_{ijk} \tag{6}$$

and the corresponding grand mean

$$\begin{aligned} \bar{y}_{\cdot\cdot\cdot} = y_{\cdot\cdot\cdot}/N &= \sum_{i=1}^{a} n_{i\cdot}\bar{y}_{i\cdot\cdot}/N = \sum_{j=1}^{b} n_{\cdot j}\bar{y}_{\cdot j\cdot}/N \\ &= \sum_{i=1}^{a} \sum_{j=1}^{b} n_{ij}\bar{y}_{ij\cdot}/N = \sum_{i=1}^{a} \sum_{j=1}^{b} \sum_{k=1}^{n_{ij}} y_{ijk}/N\,. \end{aligned} \tag{7}$$

These alternative forms get used repeatedly in whatever context they appear useful; and, for simplicity of appearance, we henceforth omit limits of summation unless the context demands it. Thus we use equivalently

$$\Sigma_i \text{ for } \sum_{i=1}^{a}, \quad \Sigma_j \text{ for } \sum_{j=1}^{b}, \quad \text{and } \Sigma_k \text{ for } \sum_{k=1}^{n_{ij}} \tag{8}$$

4.2. THE MODEL

The cell means model for data such as those of Table 4.1 is just like that of Section 3.2. Its description begins with

$$E(y_{ijk}) = \mu_{ij}, \tag{9}$$

where μ_{ij} is deemed to be the mean of a conceptual population corresponding to cell i, j, from which the available data are considered to be a random sample. E in (9) represents expectation over repeated sampling, and associated with each observation y_{ijk} is a random error term defined as

$$e_{ijk} = y_{ijk} - E(y_{ijk}), \tag{10}$$

so that

$$y_{ijk} = \mu_{ij} + e_{ijk} \,. \tag{11}$$

By its definition, $E(e_{ijk}) = 0$. Further, to each e_{ijk} we attribute a common variance σ^2 and a zero covariance between every pair of (different) e_{ijk}s. To these assumptions, that of normality is later added.

4.3. ESTIMATION

The model equation (11) is indistinguishable from (3) of Section 3.2 and so the BLUE is the same as (4) in Section 3.3, just a simple extension of (22) in Chapter 2. Thus

$$\text{BLUE of } \mu_{ij} \text{ is } \hat{\mu}_{ij} = \bar{y}_{ij\cdot}. \tag{12}$$

with

$$E(\hat{\mu}_{ij}) = \mu_{ij} \quad \text{and} \quad v(\hat{\mu}_{ij}) = \sigma^2/n_{ij} \tag{13}$$

and a zero covariance between every pair of (different) $\hat{\mu}_{ij}$s. Hence we have the straightforward result for all-cells-filled data that the BLUE of a cell mean is the corresponding observed cell mean, $\bar{y}_{ij\cdot}$.

Consider a linear combination of the μ_{ij}s, similar to (60) and (76) of Sections 2.9e and 2.12, respectively,

$$\omega = \Sigma_i \Sigma_j k_{ij} \mu_{ij} \ . \tag{14}$$

Then the BLUE of ω is $\hat{\omega} = \Sigma_i \Sigma_j k_{ij} \hat{\mu}_{ij} = \Sigma_i \Sigma_j k_{ij} \bar{y}_{ij\cdot}$. (15)

with

$$v(\hat{\omega}) = \sigma^2 \Sigma_i \Sigma_j k_{ij}^2 / n_{ij} \ . \tag{16}$$

Results (15) and (16) are of course based on (77) of Section 2.12, and are the same as (6) and (7) of Section 3.3.

Lastly, so far as estimation is concerned, we have the usual unbiased estimator of the residual error variance,

$$\hat{\sigma}^2 = \frac{\text{SSE}}{N - s} \quad \text{for SSE} = \Sigma_i \Sigma_j \Sigma_k (y_{ijk} - \bar{y}_{ij\cdot})^2 \tag{17}$$

where

$$s = \text{number of filled cells,}$$

$$= ab, \quad \text{when all cells are filled} \ . \tag{18}$$

Although this chapter is concerned solely with the all-cells-filled case, in which case it is unnecessary to have s as defined above (because s is always ab for all cells filled), having s available provides a convenient connection with the some-cells-empty case of Chapter 5.

4.4. CONFIDENCE INTERVALS, AND ONE-PART HYPOTHESES

We now adopt normality assumptions that each $e_{ijk} \sim \mathcal{N}(0, \sigma^2)$, independently of all other e_{ijk}s. Then confidence intervals on ω are precisely those of (9)–(11) of Section 3.4, based on (81)–(83) of Section 2.12a. Using the $\mathcal{T}$-distribution, the non-symmetric $100\alpha\%$ interval is

$$\hat{\omega} - t_{U, N-s, \alpha} \sqrt{\hat{v}(\hat{\omega})} \quad \text{to} \quad \hat{\omega} - t_{L, N-s, \alpha} \sqrt{\hat{v}(\hat{\omega})} \ , \tag{19}$$

the symmetric interval is

$$\hat{\omega} \pm t_{N-s, \frac{1}{2}\alpha} \sqrt{\hat{v}(\hat{\omega})} \tag{20}$$

and, when $N - s$ is large (e.g., exceeds 100, say) this can be replaced by

$$\hat{\omega} \pm z_{\frac{1}{2}\alpha} \sqrt{\hat{v}(\hat{\omega})} \ . \tag{21}$$

In all cases $\hat{\omega}$ comes from (15), and $\hat{v}(\hat{\omega})$ is $v(\hat{\omega})$ of (16) with $\hat{\sigma}^2$ of (17) replacing σ^2.

Similar to (87) of Section 2.12b-i the one-part hypothesis

$$\text{H: } \omega = m \text{ is tested by } \frac{(\hat{\omega} - m)^2}{\hat{v}(\hat{\omega})} \sim \mathscr{F}_{1, N-s}, \tag{22}$$

similar to (12) of Section 3.4. As there, so here, the statistic in (22) is the square of a t-statistic. Consideration of more general hypotheses is left to Chapter 8, and examples thereof for the 2-way classification to Chapter 9.

Again, as at the end of Section 3.4, the results in (19)–(22) are for a general linear function of μ_{ij}s, as defined in (13) and (14), but they can be applied to any single μ_{ij}, say μ_{rs} simply by using $\hat{\mu}_{rs} = \bar{y}_{rs\cdot}$ as $\hat{\omega}$, and $\hat{\sigma}^2/n_{rs}$ as $\hat{v}(\hat{\omega})$. This is because putting $k_{ij} = 0$ in (13), except for $k_{rs} = 1$, gives $\omega = \mu_{rs}$.

Example. From Table 4.1

$$\begin{array}{lll} \hat{\mu}_{11} = 9 & \hat{\mu}_{12} = 14 & \hat{\mu}_{13} = 18 \\ \hat{\mu}_{21} = 16 & \hat{\mu}_{22} = 31 & \hat{\mu}_{23} = 13. \end{array}$$

It is then not difficult to calculate

$$\text{SSE} = \Sigma_i \Sigma_j \Sigma_k (y_{ijk} - \bar{y}_{ij\cdot})^2 \quad \text{as} \quad \text{SSE} = 120,$$

whereupon (17) gives $\hat{\sigma}^2 = 120/(15 - 6) = 13\frac{1}{3}$. (23)

The BLUE of

$$\omega = \mu_{11} + 4\mu_{12} - 5\mu_{13}$$

is

$$\hat{\omega} = 9 + 4(14) - 5(18) = -25$$

and with $n_{11} = 3$, $n_{12} = 2$ and $n_{13} = 2$,

$$\hat{v}(\hat{\omega}) = (1/3 + 16/2 + 25/2)13\tfrac{1}{3} = 2500/9 = \left(16\tfrac{2}{3}\right)^2 .$$

Therefore a 95% symmetric confidence interval based on (20) is

$$\begin{aligned} \hat{\omega} \pm t_{9, \frac{1}{2}\alpha}\sqrt{\hat{v}(\hat{\omega})} &= -25 \pm t_{9,.025} 16\tfrac{2}{3} \\ &= -25 \pm 2.26(50/3) \\ &= -25 \pm 37.7 = (-62.7, 12.7) . \end{aligned}$$

TABLE 4.2. ANALYSIS OF VARIANCE OF ALL-CELLS-FILLED DATA FROM A 2-WAY CROSSED CLASSIFICATION

Source of Variation	d.f.	Sum of Squares	Associated Hypothesis
Mean	1	SSM $= N\bar{y}_{\cdots}^2$	H: $\Sigma_i\Sigma_j n_{ij}\mu_{ij} = 0$
Model	$ab - 1$	$\text{SSR}_m = \Sigma_i\Sigma_j n_{ij}(\bar{y}_{ij\cdot} - \bar{y}_{\cdots})^2$	H: μ_{ij} *all equal*
Residual	$N - ab$	SSE $= \Sigma_i\Sigma_j\Sigma_k(y_{ijk} - \bar{y}_{ij\cdot})^2$	
Total	N	SST $= \Sigma_i\Sigma_j\Sigma_k y_{ijk}^2$	

And from (22) the F-statistic for testing H: $\omega = 0$ is

$$F_{1,9} = (-25)^2/(50/3)^2 = 2.25 \ .$$

4.5. ANALYSIS OF VARIANCE

a. A standard analysis

The residual sum of squares for fitting $E(y_{ijk}) = \mu_{ij}$ of (9) used in $\hat{\sigma}^2$ of (17) is

$$\text{SSE} = \Sigma_i\Sigma_j\Sigma_k(y_{ijk} - \bar{y}_{ij\cdot})^2,$$

just like in (13) of Section 3.5, both results stemming from SSE of Table 2.7. Thus an analysis of variance table can be developed just like Table 3.2, and is shown as Table 4.2.

Example (continued). The reader should calculate Table 4.2 for the data of Table 4.1 and confirm the values shown in Table 4.3.

The analysis of variance in Table 4.2 is for the data treated as if they came from a 1-way classification of *ab* classes. That is why the hypothesis associated with SSR_m is

$$\text{H: } \mu_{ij} \textit{ equal } \forall \ i, j \ . \tag{24}$$

It is just a simple extension of H: μ_i *all equal* of Table 2.7 for the 1-way classification model $E(y_{ij}) = \mu_i$.

But, as with the 2-way nested classification, following Table 3.2, there is an important distinction between H: μ_i *all equal* of the 1-way classification and (24) of Table 4.2. It is as follows. In the 1-way classification, H: μ_i *all equal* is a hypothesis about the population means of the classes by which

TABLE 4.3. ANALYSIS OF VARIANCE (TABLE 4.2) FOR THE DATA OF TABLE 4.1, USING $E(y_{ijk}) = \mu_{ij}$

Source of Variation	d.f.	Sum of Squares	Associated Hypothesis
Mean	1	3375	H: $3\mu_{11} + 2\mu_{12} + 2\mu_{13} + 4\mu_{21} + \mu_{22} + 3\mu_{23} = 0$
Model	5	400	H: $\mu_{11} = \mu_{12} = \mu_{13} = \mu_{21} = \mu_{22} = \mu_{23}$
Residual	9	120	
Total	15	3895	

the data are classified. Thus it is a hypothesis of some interest. But (24) is a hypothesis of equality of the population means of *all* the cells defined by levels of the row and column factors. These means are therefore not means corresponding to levels of the factors by which the data are classified. Thus (24) is seldom, if ever, of much practical interest. Presumably the factors used in classifying the data *are* of interest, for otherwise the data would not be so classified. To test a hypothesis that essentially ignores those factors does not, therefore, seem very appropriate.

Furthermore, whether the test of (24) provided by SSR_m in Table 4.2 leads to rejection or to non-rejection of (24) we will always be interested in the two factors that constitute the 2-way classification. Rejection of (24) would indicate that differences may exist among the μ_{ij}s; and investigation of those differences in terms of row (and/or column) averages of the μ_{ij}s may well be of interest. Also, although non-rejection of (24) suggests evidence that (24) is not violated, it is no proof that all μ_{ij}s *are* equal; we can still be interested in investigating differences between row (and/or column) averages of the μ_{ij}s. This we proceed to do.

b. An alternative approach

Fisher's comment about the analysis of variance table in his 1934 letter to Snedecor (quoted at the end of Section 1.3) warrants mentioning again here, namely that the "analysis of variance is ... a simple method ... to ... display the essential features of a body of data ...". The implication is that deciding on what are the "essential features of a body of data" is done first, and then those features are summarized in the analysis of variance table. The first step is easy, of course, with balanced data; it depends on identities such as

$$y_{ijk} - \bar{y}_{\cdots} = (\bar{y}_{i\cdot\cdot} - \bar{y}_{\cdots}) + (\bar{y}_{\cdot j\cdot} - \bar{y}_{\cdots}) + (\bar{y}_{ij\cdot} - \bar{y}_{i\cdot\cdot} - \bar{y}_{\cdot j\cdot} + \bar{y}_{\cdots}) + (y_{ijk} - \bar{y}_{ij\cdot}) \qquad (25)$$

and

$$\Sigma_i\Sigma_j\Sigma_k(y_{ijk} - \bar{y}_{...})^2$$

$$= \Sigma_i\Sigma_j\Sigma_k(\bar{y}_{i..} - \bar{y}_{...})^2 + \Sigma_i\Sigma_j\Sigma_k(\bar{y}_{.j.} - \bar{y}_{...})^2 \tag{26}$$

$$+\Sigma_i\Sigma_j\Sigma_k(\bar{y}_{ij.} - \bar{y}_{i..} - \bar{y}_{.j.} + \bar{y}_{...})^2 + \Sigma_i\Sigma_j\Sigma_k(y_{ijk} - \bar{y}_{ij.})^2$$

which are discussed in Section 1.3. For balanced data, the utility of (26) is that, along with the usual normality assumptions and consequent χ^2, independence and F-distribution properties, the sums of squares therein result in tests of useful hypotheses. They are also hypotheses that are easily described in terms of over-parameterized models, such as (1) of Chapter 1.

For unbalanced data, most of these niceties do not arise: certainly (25) is still true, but (26) is not. [The reader should confirm for the data of the example in Table 4.1 that the left-hand side of (26) is 3895 − 3375 = 520, from Table 4.3, but that the right-hand side of (26) is 495; see Exercise E4.1.] Therefore, even without going into details, it is easily appreciated that analysis of variance calculations of unbalanced data that simply mimic those of balanced data are not necessarily going to "display the essential features" of the data (in Fisher's words), nor yield F-statistics that test hypotheses of interest. Indeed such is precisely the case, as is explained in detail in Section 4.9h. It is for this reason that adherence to treating the analysis of unbalanced data as if it were simply an extension of analyzing balanced data is no longer recommended. As numerous writers have pointed out, with varying degrees of emphasis, much of what has been done in terms of this extension, using over-parameterized models, can more easily be understood in terms of cell means models. So we may as well take that approach to begin with, and not just use cell means models to explain results from over-parameterized models, but positively derive results in terms of cell means models. This is now done.

4.6. UNWEIGHTED MEANS

The essence of what we do is to abandon the analysis of variance table with the sanctimonious aura that it has engendered through decades of successful use with balanced data, and go back to the basic cause of its existence: testing hypotheses of interest. We begin with the underlying basis of the cell means, the definition $E(y_{ijk}) = \mu_{ij}$ of (9), and consider several different means of the μ_{ij}s.

a. Definitions

The row mean of the cell means μ_{ij} in row i shall be denoted ρ_i. It is a straightforward extension of ρ in (98) of Section 2.12d:

$$\rho_i = \frac{1}{b}\Sigma_j\mu_{ij} \ . \tag{27}$$

Its BLUE is $$\hat{\rho}_i = \frac{1}{b}\Sigma_j\bar{y}_{ij\cdot} \quad \text{with} \quad v(\hat{\rho}_i) = \frac{1}{b^2}\Sigma_j\frac{1}{n_{ij}}\sigma^2 \tag{28}$$

and $\text{cov}(\hat{\rho}_i, \hat{\rho}_h) = 0$ for $i \neq h$. For notational convenience we define

$$\tilde{y}_{i\cdot\cdot} = \frac{1}{b}\Sigma_j\bar{y}_{ij\cdot} \quad \text{and} \quad \frac{1}{w_i} = \frac{1}{b^2}\Sigma_j\frac{1}{n_{ij}} = v(\hat{\rho}_i)/\sigma^2 \tag{29}$$

Thus $$\hat{\rho}_i = \tilde{y}_{i\cdot\cdot} \quad \text{and} \quad v(\hat{\rho}_i) = \sigma^2/w_i \ . \tag{30}$$

Note in passing that $\tilde{y}_{i\cdot\cdot}$ is not generally the same as $\bar{y}_{i\cdot\cdot}$ of (4). They are always the same when $n_{i1} = n_{i2} = \cdots = n_{ib}$ (e.g., balanced data) and they can be the same for certain values of $\bar{y}_{ij\cdot}$ and n_{ij} with unbalanced data. An example of the latter is $\bar{y}_{11\cdot} = 4$, $\bar{y}_{12\cdot} = 9$ and $\bar{y}_{13\cdot} = 11$, with $n_{11} = 5$, $n_{12} = 2$ and $n_{13} = 6$. The reader will find that $\bar{y}_{1\cdot\cdot} = 8 = \tilde{y}_{1\cdot\cdot}$.

Example (continued). The observed cell means $\bar{y}_{ij\cdot}$ and cell frequencies n_{ij} of Table 4.1 are summarized in Table 4.4. Then from (29)

$$\tilde{y}_{1\cdot\cdot} = \tfrac{1}{3}(9 + 14 + 18) = \tfrac{14}{3} \quad \text{and} \quad \frac{1}{w_1} = \tfrac{1}{9}\left(\tfrac{1}{3} + \tfrac{1}{2} + \tfrac{1}{2}\right) = \tfrac{4}{27} \tag{31}$$

$$\tilde{y}_{2\cdot\cdot} = \tfrac{1}{3}(16 + 31 + 13) = 20 \quad \text{and} \quad \frac{1}{w_2} = \tfrac{1}{9}\left(\tfrac{1}{4} + \tfrac{1}{1} + \tfrac{1}{3}\right) = \tfrac{19}{108} \ . \tag{32}$$

TABLE 4.4. CELL MEANS $\bar{y}_{ij\cdot}$, AND CELL FREQUENCIES FROM TABLE 4.1

$\bar{y}_{ij\cdot}$			n_{ij}		
9	14	18	3	2	2
16	31	13	4	1	3

b. Hypothesis: equality of row means

In the case of having only two rows, as in the example, it seems reasonable that a useful hypothesis of equality of row means might be

$$\text{H: } \rho_1 = \rho_2 \ . \tag{33}$$

Under the usual assumptions of normality,

$$\hat{\rho}_1 - \hat{\rho}_2 - (\rho_1 - \rho_2) \sim \mathcal{N}\left[0, \sigma^2(1/w_1 + 1/w_2)\right] \ . \tag{34}$$

Hence, when (33) is true,

$$t_{N-s} = \frac{\hat{\rho}_1 - \hat{\rho}_2}{\sqrt{\hat{\sigma}^2(1/w_1 + 1/w_2)}} = \frac{\tilde{y}_{1..} - \tilde{y}_{2..}}{\sqrt{\hat{\sigma}^2(1/w_1 + 1/w_2)}} \tag{35}$$

is a t-statistic on $N - s$ degrees of freedom suitable for testing the hypothesis (33). The degrees of freedom are those used in estimating σ^2 as $\hat{\sigma}^2 = \text{SSE}/(N - s)$ from (17) and the corresponding F-statistic on one and $N - s$ degrees of freedom is

$$F_{1, N-s} = t^2 = \frac{Q}{1\hat{\sigma}^2}, \tag{36}$$

for

$$Q = \frac{(\tilde{y}_{1..} - \tilde{y}_{2..})^2}{1/w_1 + 1/w_2} \ . \tag{37}$$

Thus from (31), (32) and $\hat{\sigma}^2 = \text{MSE} = \frac{120}{9} = \frac{40}{3}$ from Table 4.3,

$$F = \frac{\left(\frac{41}{3} - 20\right)^2}{\frac{40}{3}\left(\frac{4}{27} + \frac{19}{108}\right)} = 9\tfrac{99}{350} \ . \tag{38}$$

c. Numerator sums of squares

Q, the numerator of F in (36), is a sum of squares (in this case just a single squared term). It is a sum of squares associated with H: $\rho_1 = \rho_2$, so that we write

$$\text{H: } \rho_1 = \rho_2 \quad \text{can be tested with} \quad F = Q/1\hat{\sigma}^2 \ . \tag{39}$$

The denominator of F is shown as $1\hat{\sigma}^2$ to emphasize in the calculation of F the occurrence of the degrees of freedom corresponding to Q. We do this because in Section 8.8b, for testing a quite general linear hypothesis, we develop a general expression for the F-statistic in the form $F = Q/r\hat{\sigma}^2$, where r is the degrees of freedom corresponding to Q. Thus $F = Q/1\hat{\sigma}^2$ of (39) is just a first, special case of the general result, (146) in Chapter 8.

Notice the line of reasoning here. We began in (33) with a hypothesis that seems interesting, reasonable, useful—call it what you may—and have come to Q of (37) and (39) as the sum of squares for testing that hypothesis. Nowhere has an analysis of variance been laid out and a sum of squares therein been considered in terms of its associated hypothesis. Rather, we have followed the correct logic for hypothesis testing: set up a hypothesis and develop a test statistic for it. A general procedure for doing this is given in Section 8.8b. It yields an F-statistic in the form $F = Q/r\hat{\sigma}^2$, for Q being a sum of squares and r its associated degrees of freedom. Bearing in mind that an analysis of variance table is just a convenient summary of useful arithmetic we could, if it seemed appropriate, summarize our calculations of Q and $\hat{\sigma}^2$ in some tabular form; but such a tabulation is really not very appropriate, at least not in the sense of having some of the properties that a customary analysis of variance table of balanced data has. For example, our sums of squares would not add up to the total sum of squares $\Sigma_i\Sigma_j\Sigma_k(y_{ijk} - \bar{y}_{...})^2$. But this property of "adding up", which we have grown accustomed to in analysis of variance of balanced data, is not essential to the testing of hypotheses. It is nice, in some sense of representing a partitioning of the total sum of squares, but it is not necessary for using an F-statistic. And, it is much more important to calculate F-statistics that test useful hypotheses than it is to calculate sums of squares that "add up" but which test worthless hypotheses. So attention is concentrated on hypotheses of interest, on the numerator sums of squares Q for testing each of them and, of course, on the independence of each Q from $\hat{\sigma}^2$. And, where feasible, dependence or independence of one Q from another is also considered.

d. A general hypothesis: row means equal

We proceed to extend H: $\rho_1 = \rho_2$ of the example to the general case

$$\text{H: } \rho_1 = \rho_2 = \cdots = \rho_a, \tag{40}$$

i.e.,

$$\text{H: } \frac{1}{b}\Sigma_j\mu_{1j} = \frac{1}{b}\Sigma_j\mu_{2j} = \cdots = \frac{1}{b}\Sigma_j\mu_{aj}\,. \tag{41}$$

And here, with benefit of hindsight, rather than develop the numerator sum of squares for testing (41), we simply quote it and then establish its properties. It is an extension of (37):

$$\mathrm{SSA}_w = \sum_{i=1}^{a} w_i \left(\tilde{y}_{i\cdot\cdot} - \sum_{i=1}^{a} w_i \tilde{y}_{i\cdot\cdot} \Big/ \sum_{i=1}^{a} w_i \right)^2 . \tag{42}$$

It is denoted by the symbol SSA_w as mnemonic for the sum of squares for the A-factor in the weighted squares of means analysis propounded by Yates (1934). This is because (42) is indeed a weighted sum of squares of the $\tilde{y}_{i\cdot\cdot}$s using the w_is as weights [i.e., weighting in inverse proportion to $v(\tilde{y}_{i\cdot\cdot})$] and the weighted mean $\Sigma_i w_i \tilde{y}_{i\cdot\cdot}/\Sigma_i w_i$. It remains to show that SSA_w has a χ^2-distribution under the hypothesis (40); and that it is distributed independently of SSE—and hence of $\hat{\sigma}^2$.

First, because $\tilde{y}_{i\cdot\cdot}$ is the BLUE of ρ_i, note that $E(\tilde{y}_{i\cdot\cdot}) = \rho_i$, and, under normality

$$\tilde{y}_{i\cdot\cdot} \sim N(\rho_i, \quad \sigma^2/w_i),$$

with $\tilde{y}_{i\cdot\cdot}$ and $\tilde{y}_{h\cdot\cdot}$ being independent for all $i \neq h$. Under these conditions a proof that SSA_w/σ^2 for (42) is, under H: ρ_i *all equal*, distributed as a χ^2_{a-1}-variable is exactly the same as the proof given in Section 2.14a that $\mathrm{SSR}_m/\sigma^2 = \Sigma_i n_i(\bar{y}_{i\cdot} - \bar{y}_{\cdot\cdot})^2/\sigma^2$ has a χ^2_{a-1}-distribution. In that proof one simply uses $\tilde{y}_{i\cdot\cdot}$ and w_i of SSA_w in place of $\bar{y}_{i\cdot}$ and n_i (of SSR_m), respectively, and the result is that under H: ρ_i *all equal*, $\mathrm{SSA}_w/\sigma^2 \sim \chi^2_{a-1}$.

The independence of SSA_w and SSE is also easily developed. Typical terms that are squared in SSE and in SSA_w are, respectively, $y_{ijk} - \bar{y}_{ij\cdot}$ and $\tilde{y}_{h\cdot\cdot} - \Sigma_i w_i \tilde{y}_{i\cdot\cdot}/\Sigma_i w_i$. Showing that these two terms have zero covariance is straightforward, both for $i = h$ and $i \neq h$; thus under normality those two differences are independent, and so therefore are SSE and SSA_w. Hence $\hat{\sigma}^2$ and SSA_w are independent and so under H: ρ_i *all equal* the statistic $F = \mathrm{SSA}_w/(a-1)\hat{\sigma}^2$ has an $\mathscr{F}$-distribution. Therefore

$$F = \frac{\mathrm{SSA}_w}{(a-1)\hat{\sigma}^2} \quad \text{is a test statistic for} \quad \text{H: } \rho_i \textit{ equal } \forall\ i; \tag{43}$$

i.e., for

$$\text{H: } \frac{1}{b} \sum_{j=1}^{b} \mu_{ij} \textit{ equal } \forall\ i .$$

Example (continued). From (31) and (32)

$$\Sigma_i w_i \tilde{y}_{i\cdot\cdot}/\Sigma_i w_i = \left[\tfrac{27}{4}\left(\tfrac{41}{3}\right) + \tfrac{108}{19}(20)\right]\Big/\left(\tfrac{27}{4} + \tfrac{108}{19}\right) = \tfrac{1739}{105} .$$

Then in (42)

$$\text{SSA}_w = \tfrac{27}{4}\left(\tfrac{41}{3} - \tfrac{1739}{105}\right)^2 + \tfrac{108}{19}\left(20 - \tfrac{1739}{105}\right)^2 = 123\tfrac{27}{35} .$$

This is the same as $F\hat{\sigma}^2 = t^2\hat{\sigma}^2$ in (36), because $t^2\hat{\sigma}^2$ from (35) is $(\tilde{y}_{1\cdot\cdot} - \tilde{y}_{2\cdot\cdot})^2 w_1 w_2/(w_1 + w_2)$, which is what SSA_w of (42) reduces to for $a = 2$ (see E4.4). And the associated hypothesis is, by (43),

$$\text{H: } \tfrac{1}{3}(\mu_{11} + \mu_{12} + \mu_{13}) = \tfrac{1}{3}(\mu_{21} + \mu_{22} + \mu_{23}) .$$

e. A similar result for columns

Whatever choice is made in a two-way crossed classification for defining one factor as rows and other as columns can equally as well be made the other way around. Thus any result such as (43) that is in terms of rows can immediately be expressed in terms of columns. Thus for

$$\tilde{y}_{\cdot j\cdot} = \frac{1}{a}\sum_{i=1}^{a} \bar{y}_{ij\cdot} \quad \text{and} \quad \frac{1}{v_j} = \frac{1}{a^2}\sum_{i=1}^{a}\frac{1}{n_{ij}},$$

and

$$\text{SSB}_w = \Sigma_j v_j\left(\tilde{y}_{\cdot j\cdot} - \Sigma_j v_j \tilde{y}_{\cdot j\cdot}/\Sigma_j v_j\right)^2,$$

$$F = \frac{\text{SSB}_w}{(b-1)\hat{\sigma}^2} \quad \text{tests} \quad \text{H: } \frac{1}{a}\Sigma_i \mu_{ij} \textit{ equal } \forall\ j . \tag{44}$$

f. A caution

It is to be emphasized that the results (43) and (44) very definitely depend upon all cells having data in them. Even though, for example, $\rho_i = \Sigma_{j=1}^{b}\mu_{ij}/b$ is well defined whether all cells are filled or not, it can be estimated as $\hat{\rho}_i = \tilde{y}_{i\cdot\cdot}$ only if all cells in row i are filled; and it is with this condition that $v(\tilde{y}_{i\cdot\cdot}) = \sigma^2/w_i$, which in turn leads to the χ^2 and independence properties of SSA_w. These properties do not arise when some cells are empty. That case is considered in Chapter 5.

g. Least squares means

The row mean $\rho_i = \Sigma_i \mu_{ij}/b$ is sometimes called the *least squares mean* (LSM) for row i; or the *population marginal mean* (PMM) for row i [see Searle, Speed and Milliken (1980)]. Unfortunately, the name least squares mean is often used not just for ρ_i but also for an estimate of it. This is particularly confusing in the case of some-cells-empty data wherein in some circumstances certain ρ_is cannot be estimated (see Sections 5.2a and 5.4a). But for all-cells-filled data

$$\hat{\rho}_i = \Sigma_j \hat{\mu}_{ij}/b = \Sigma_j \bar{y}_{ij\cdot}/b .$$

4.7. WEIGHTED MEANS

a. Weighting by cell frequencies

A different mean of the cell means μ_{ij} in row i that is sometimes thought to be an interesting alternative to $\rho_i = (1/b)\Sigma_j\mu_{ij}$ of the preceding section is

$$\rho_i' = \frac{1}{n_{i\cdot}}\Sigma_j n_{ij}\mu_{ij} . \tag{45}$$

This is simply a ρ' of (94) in Section 2.12c for each row. It has BLUE

$$\hat{\rho}_i' = \frac{1}{n_{i\cdot}}\Sigma_j n_{ij}\bar{y}_{ij\cdot} = \bar{y}_{i\cdot\cdot}$$

with $$v(\rho_i') = \sigma^2/n_{i\cdot} \quad \text{and} \quad \text{cov}(\hat{\rho}_i', \hat{\rho}_h') = 0$$

for $i \neq h$. To test the hypothesis

$$\text{H: } \rho_i' \textit{ equal } \forall \ i$$

we use a simple extension [similar to that described in the paragraph following (26) in Section 3.5f] of the result in Chapter 2 that $\text{SSR}_m/\sigma^2 = \Sigma_i n_i(\bar{y}_{i\cdot} - \bar{y}_{\cdot\cdot})^2/\sigma^2$ there tests H: μ_i *equal* $\forall\ i$. If in that result we replace n_i, $\bar{y}_{i\cdot}$, $\bar{y}_{\cdot\cdot}$ and μ_i by $n_{i\cdot}$, $\bar{y}_{i\cdot\cdot}$, $\bar{y}_{\cdot\cdot\cdot}$ and ρ_i', respectively, then we have at once that

$$\Sigma_i n_{i\cdot}(\bar{y}_{i\cdot\cdot} - \bar{y}_{\cdot\cdot\cdot})^2/\sigma^2 \sim \chi^2_{a-1}; \tag{46}$$

and its independence to $\hat{\sigma}^2$ is readily established.

In the same way as in Chapter 2, it is appropriate to use the symbol $\mathscr{R}(\mu_i|\mu)$ for the sum of squares in (46), i.e.,

$$\mathscr{R}(\mu_i|\mu) = \Sigma_i n_{i\cdot}(\bar{y}_{i\cdot\cdot} - \bar{y}_{\cdot\cdot\cdot})^2 . \tag{47}$$

Then for the model $E(y_{ijk}) = \mu_{ij}$ a consequence of (46) is that

$$F = \frac{\mathscr{R}(\mu_i|\mu)}{(a-1)\hat{\sigma}^2} \quad \text{tests} \quad \text{H: } \rho_i' \textit{ equal } \forall \ i . \tag{48}$$

Example. From Table 4.1

$$\bar{y}_{1\cdot\cdot} = (27 + 28 + 36)/(3 + 2 + 2) = 91/7 = 13,$$

$$\bar{y}_{2\cdot\cdot} = (64 + 31 + 39)/(4 + 1 + 3) = 134/8 = 16\tfrac{3}{4},$$

$$\tilde{y}_{\cdot\cdot\cdot} = (91 + 134)/(7 + 8) = 225/15 = 15.$$

Hence $$\mathscr{R}(\mu_i \,|\, \mu) = 7(13 - 15)^2 + 8(16\tfrac{3}{4} - 15)^2 = 52\tfrac{1}{2} \;.$$

And the associated hypothesis is, by (48),

$$\text{H: } \tfrac{1}{7}(3\mu_{11} + 2\mu_{12} + 2\mu_{13}) = \tfrac{1}{8}(4\mu_{21} + \mu_{22} + 3\mu_{23}) \;.$$

b. Evaluation

It is most important that the two sums of squares SSA_w and $\mathscr{R}(\mu_i \,|\, \mu)$ be distinguished one from the other; and also their respective hypotheses,

$$\text{SSA}_w \text{ of (42) tests H: } \frac{1}{b}\Sigma_j \mu_{ij} \textit{ equal } \forall \; i$$

and $$\mathscr{R}(\mu_i \,|\, \mu) \text{ of (48) tests H: } \frac{1}{n_{i\cdot}}\Sigma_j n_{ij}\mu_{ij} \textit{ equal } \forall \; i \;.$$

They are identical for balanced data, but for unbalanced data they are different; e.g., for the example they are, respectively,

$$\text{H: } \tfrac{1}{3}(\mu_{11} + \mu_{12} + \mu_{13}) = \tfrac{1}{3}(\mu_{21} + \mu_{22} + \mu_{23})$$

and $$\text{H: } \tfrac{1}{7}(3\mu_{11} + 2\mu_{12} + 2\mu_{13}) = \tfrac{1}{8}(4\mu_{21} + \mu_{22} + \mu_{23}) \;.$$

An equally as important feature of these hypotheses is that neither of them is right, and neither is wrong: they are simply different hypotheses, a fact that need not evoke any ethical considerations that one is good and one is bad. The salient feature of the hypotheses is that they are different, and it is through knowing and understanding the differences that we can decide when one may be more appropriate than the other. Furthermore, this decision will seldom (if ever) be made on statistical grounds. Depending on the source of one's data, the reasoning will be to consider which definition of a row mean is more appropriate: $\rho_i = \Sigma_{j=1}^{b}\mu_{ij}/b$ in which every cell mean in a row is weighted the same; or $\rho_i' = \Sigma_{j=1}^{b} n_{ij}\mu_{ij}/n_{i\cdot}$, in which each cell mean in a row is weighted according to the number of observations in the cell. For example, in Table 4.1 are we to define (population) means for the two soil types with seed varieties weighted equally, e.g., $\rho_1 = \frac{1}{3}(\mu_{11} + \mu_{12} + \mu_{13})$, or with seed varieties weighted according to the number of observations in the data, e.g., $\rho_1' = \frac{1}{7}(3\mu_{11} + 2\mu_{12} + \mu_{13})$?

An important difference between ρ_i and ρ_i' is that ρ_i is independent of the data, whereas ρ_i' depends on the data, not through the observations themselves but through the numbers of them in each cell. If this is an appropriate way to define row means for the study at hand, then the test based on $\mathscr{R}(\mu_i \mid \mu)$ should be used; but if equal weighting is more appropriate, then the test based on SSA_w should be used. And if some weighting other than these two possibilities is wanted, then it can be used, as follows.

c. A general weighting

Define

$$\rho_i'' = \Sigma_j t_{ij}\mu_{ij} \quad \text{with} \quad \Sigma_j t_{ij} = 1\ . \tag{49}$$

(There is no loss of generality in having $\Sigma_j t_{ij} = 1$, because any set of values t_{ij}' for which $\Sigma_j t_{ij}' \neq 1$ can be transformed to $t_{ij} = t_{ij}'/\Sigma_j t_{ij}'$ for which $\Sigma_j t_{ij} = 1$.) Also define

$$\bar{y}_i' = \sum_{j=1}^{b} t_{ij}\bar{y}_{ij\cdot} \quad \text{and} \quad 1/m_i = \Sigma_j t_{ij}^2/n_{i\cdot}\ .$$

Then

$$\hat{\rho}_i'' = \bar{y}_i' \quad \text{and} \quad v(\bar{y}_i') = \sigma^2/m_i,$$

and

$$\frac{\Sigma_i m_i\left(\bar{y}_i' - \sum_{i=1}^{a} m_i\bar{y}_i' \Big/ \sum_{i=1}^{a} m_i\right)^2}{(a-1)\hat{\sigma}^2} \quad \text{tests} \quad \text{H: } \rho_i'' \textit{ equal } \forall\ i\ . \tag{50}$$

Two special cases of this have already been considered; $t_{ij} = 1/a$ gives (43), based on SSA_w, and $t_{ij} = n_{ij}/n_{i\cdot}$ gives (48), based on $\mathscr{R}(\mu_i \mid \mu)$.

d. Similar results for columns

The F-statistic analogous to (48) but corresponding to columns is that for

$$\mathscr{R}(\mu_j \mid \mu) = \Sigma_j n_{\cdot j}(\bar{y}_{\cdot j\cdot} - \bar{y}_{\cdot\cdot\cdot})^2 \quad \text{and} \quad \gamma_j' = \Sigma_i n_{ij}\mu_{ij}/n_{\cdot j}$$

$$F = \frac{\mathscr{R}(\mu_j \mid \mu)}{(b-1)\hat{\sigma}^2} \quad \text{tests} \quad \text{H: } \gamma_j' \textit{ equal } \forall\ j\ . \tag{51}$$

Its generalization, comparable to (49) and (50), is that for

$$\gamma_j'' = \sum_{i=1}^{a} f_{ij}\mu_{ij} \quad \text{with} \quad \sum_{i=1}^{a} f_{ij} = 1 \tag{52}$$

and $$\bar{y}_j'' = \Sigma_i f_{ij}\bar{y}_{ij\cdot} \quad \text{and} \quad 1/p_j = \Sigma_i f_{ij}^2/n_{\cdot j},$$

$$\frac{\Sigma_j p_j\left(\bar{y}_j'' - \Sigma_j p_j \bar{y}_j''/\Sigma_j p_j\right)^2}{(b-1)\hat{\sigma}^2} \quad \text{tests} \quad \text{H: } \gamma_j'' \textit{ equal } \forall\ j\ . \tag{53}$$

Example. From Table 4.1, for (51),

$$\bar{y}_{\cdot 1\cdot} = (27 + 64)/(3 + 4) = 91/7 = 13$$

$$\bar{y}_{\cdot 2\cdot} = (28 + 31)/(2 + 1) = 59/3 = 19\tfrac{2}{3}$$

$$\bar{y}_{\cdot 3\cdot} = (36 + 39)/(2 + 3) = 75/5 = 15.$$

Hence, with $\bar{y}_{\cdots} = 15$ calculated at the end of Section 4.7a,

$$\mathscr{R}(\mu_j|\mu) = 7(13 - 15)^2 + 3(19\tfrac{2}{3} - 15)^2 + 5(15 - 15)^2 = 93\tfrac{1}{3}\ .$$

4.8. INTERACTIONS

a. A meaning of interaction

Consider two hypothetical cases where we assume we *know* the μ_{ij}s, as set out in Table 4.5.

Values of the μ_{ij}s in Table 4.5 are shown in Figure 4.1, where the abscissa shows not a continuous variable but just the column numbers 1, 2 and 3. The ordinate in each case is μ_{ij}, and for each row the μ_{ij}s for the three columns are joined by a line.

TABLE 4.5. TWO CASES OF CELL MEANS ASSUMED KNOWN

Case 1			Case 2		
$\mu_{11} = 10$	$\mu_{12} = 15$	$\mu_{13} = 8$	$\mu_{11} = 10$	$\mu_{12} = 15$	$\mu_{13} = 17$
$\mu_{21} = 12$	$\mu_{22} = 8$	$\mu_{23} = 13$	$\mu_{21} = 7$	$\mu_{22} = 12$	$\mu_{23} = 14$

The difference between the two cases in Figure 4.1 is obvious. For case 1 the lines cross, whereas for case 2 they are parallel.

In case 1 the change in μ-values in changing from column 1 to column 2 is $+5$ for row 1 but -4 for row 2; and in changing from column 2 to column 3 the change is -7 for row 1 but $+5$ for row 2. The interesting feature of these changes in μ-values is not their numerical values but the fact that they are different: the change from one column to another is different for different rows. We say that columns affect different rows differently (and similarly, rows affect different columns differently). This is what is meant by interaction: columns (rows) act differently for different rows (columns). Were Table 4.5 to apply to the situation of Table 4.1 where rows are soil types and columns are varieties of carrot, we would say that an interaction between soil types and varieties means that varieties act differently, in regard to germination rate, for different soils.

In contrast to case 1 in Figure 4.1, the lines for case 2 are parallel. This is because in case 2, changing from one column to another has the same effect in all rows: the change from column 1 to column 2 is $+5$ in both rows; and from column 2 to column 3 it is $+2$ in both rows. We say that columns affect all rows similarly (and likewise, rows affect all columns similarly). This is what is meant by no interaction.

b. A measure of interaction

In case 2 of Table 4.5 and Figure 4.1, columns affect all rows the same; e.g., for columns 1 and 2 with rows 1 and 2

$$\mu_{11} - \mu_{12} = 10 - 15 = -5 = 7 - 12 = \mu_{21} - \mu_{22},$$

so that

$$(\mu_{11} - \mu_{12}) - (\mu_{21} - \mu_{22}) = 0 \ . \qquad (54)$$

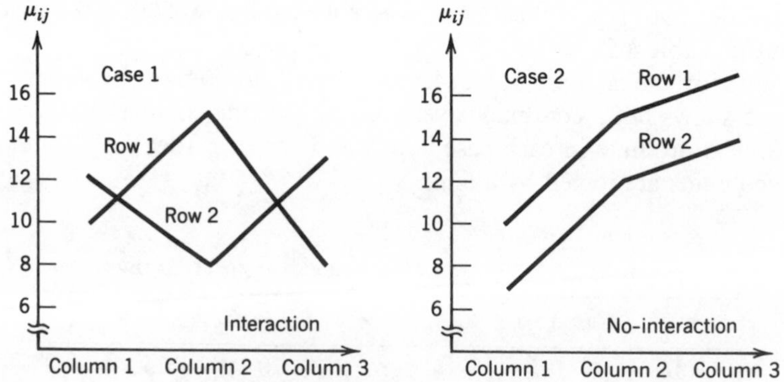

Figure 4.1. Cell means of Table 4.5.

Define

$$\theta_{ij,i'j'} = \mu_{ij} - \mu_{ij'} - \mu_{i'j} + \mu_{i'j'} \quad \text{for } i < i' \text{ and } j < j' . \tag{55}$$

Then (54) is $\theta_{11,22} = 0$. And indeed, for the no-interaction situation of case 2 in Table 4.5 every $\theta_{ij,i'j'}$ that can be defined for that table is zero. That is a direct consequence of what is meant by no interaction.

In contrast, when $\theta_{ij,i'j'}$ is nonzero we take it as being a measure of interaction: specifically, $\theta_{ij,i'j'}$ of (55) is the interaction of rows i and i' with columns j and j'. Written as

$$\theta_{ij,i'j'} = (\mu_{ij} - \mu_{ij'}) - (\mu_{i'j} - \mu_{i'j'})$$

it is seen to be the extent to which the difference between columns j and j' in row i differs from that difference in row i'; likewise, written as

$$\theta_{ij,i'j'} = (\mu_{ij} - \mu_{i'j}) - (\mu_{ij'} - \mu_{i'j'})$$

it is the extent to which the difference between rows i and i' in column j differs from that difference in column j'. More briefly we sometimes refer to $\theta_{ij,i'j'}$ as a *quadrat*.

c. The number of interactions in a grid

Two useful characteristics of the number of possible interactions (θs) in a 2-way classification depend upon the following definitions.

Definitions. The grid of a 2-way crossed classification is defined as the rows-by-columns layout of a rows and b columns of its ab cells.

Mathematical variables are said to be *linearly independent* (LIN) if there is no linear function (see Section 2.9a) of them that is zero; in which case, also, no one of them can be expressed as a linear combination of the others.

We then have the following lemma.

Lemma. In a grid of a rows and b columns:

(i) There are $\frac{1}{4}ab(a-1)(b-1)$ different θs.
(ii) The maximum number of LIN θs is $(a-1)(b-1)$.

Proof. For row i the number of different pairs of columns j and j', with $j < j'$ as in (55), is $\frac{1}{2}b(b-1)$; and so for each i there are $\frac{1}{2}b(b-1)$ different differences, $\delta_{ijj'} = \mu_{ij} - \mu_{ij'}$. And the number of different pairs of rows i and i' with $i < i'$ as in (55), is $\frac{1}{2}a(a-1)$; for each of which, with each pair $j < j'$, there are terms $\delta_{ijj'}$ and $\delta_{i'jj'}$ that can be used in

$$\theta_{ij,i'j'} = \delta_{ijj'} - \delta_{i'jj'} = (\mu_{ij} - \mu_{ij'}) - (\mu_{i'j} - \mu_{i'j'}) \; .$$

Therefore there are $\frac{1}{4}ab(a-1)(b-1)$ different θs.

Although $\frac{1}{2}b(b-1)$ different terms $\delta_{ijj'}$ can be chosen in row i, only $b-1$ of them are linearly independent. This is so because if, for some i, we pick $b-1$ terms $\delta_{ijj'}$ such that none of them can be expressed as a linear combination of the others, then it will be found that the remaining $\frac{1}{2}b(b-1)-(b-1)$ $\delta_{ijj'}$s can be expressed as linear combinations of those $b-1$.

Example. Suppose $b=5$. For any i there are $\frac{1}{2}5(5-1)=10$ different terms $\delta_{ijj'}$. Select as $5-1=4$ of them, none of which is a linear combination of the others, the four

$$\delta_{i12} = \mu_{i1} - \mu_{i2}, \qquad \delta_{i13} = \mu_{i1} - \mu_{i3}, \qquad \delta_{i14} = \mu_{i1} - \mu_{i4}, \quad \text{and} \quad \delta_{i15} = \mu_{i1} - \mu_{i5} \; . \tag{56}$$

Then it is easily verified that each of the other six $\delta_{ijj'}$s can be expressed as a linear combination (a difference, in fact) of two of those in (56); e.g., $\delta_{i24} = \delta_{i14} - \delta_{i12}$.
Q.E.D.

Proof of (ii). The preceding argument, which yields for given i only $b-1$ LIN differences $\delta_{ijj'} = \mu_{ij} - \mu_{ij'}$ for $j<j'$, also yields, in analogous fashion for given $j<j'$, only $a-1$ LIN differences $\theta_{ij,i'j'} = \delta_{ijj'} - \delta_{i'jj'}$ for $i<i'$. Therefore, although there are $\frac{1}{4}ab(a-1)(b-1)$ different θs in a grid of a rows and b columns, the greatest number of LIN θs is $(a-1)(b-1)$. **Q.E.D.**

Example. In a 3×3 grid there are $\frac{1}{4}3(3)2(2) = 9$ different θs, but $2(2)=4$ is the largest number of LIN θs; e.g., $\theta_{11,22}$, $\theta_{11,23}$, $\theta_{11,32}$ and $\theta_{11,33}$ are LIN; and, for example, $\theta_{22,33}$ is a linear function of these:

$$\theta_{22,33} = \theta_{11,22} - \theta_{11,23} - \theta_{11,32} + \theta_{11,33} \; .$$

d. A consequence of interactions

In case 1 of Table 4.5 we find that the row mean $\rho_i = \Sigma_j \mu_{ij}/b$ of (27) has values $\rho_1 = \rho_2$. And yet any test statistic that led us to not reject H: $\rho_1 = \rho_2$ would be misleading to the extent that we interpreted that non-rejection as meaning that the rows are equivalent, not just in some average sense but in each individual column. Clearly, from Figure 4.1, they are not. The plot of cell means for row 1 has a peak in column 2 whereas with row 2 there is a valley. This is an interaction effect, which therefore needs to be taken into account before considering row averages such as ρ_i. It is for this kind of reason that the consideration of interactions is important. Their presence can mislead one in drawing conclusions about ρ_i

(or ρ_i' or ρ_i'') in terms of row effects. It is therefore advisable to assess the presence of interactions before making such conclusions. This is done by testing a hypothesis that the interactions are zero.

4.9. TESTING A HYPOTHESIS OF NO INTERACTIONS

The model $E(y_{ijk}) = \mu_{ij}$ of (9) does not include any requirements that the interactions

$$\theta_{ij,i'j'} = \mu_{ij} - \mu_{ij'} - \mu_{i'j} + \mu_{i'j'}$$

of (55) be zero. Therefore that basic model implicitly includes interactions and we call it either the *with-interaction model* or simply just the *interaction model*.

Information about the overall importance of the magnitude of the terms $\theta_{ij,i'j'}$ in the model $E(y_{ijk}) = \mu_{ij}$ comes from testing the hypothesis in that model that the interactions are zero, i.e., testing

$$\text{H: } \textit{all } \theta_{ij,i'j'} = 0 \ . \tag{57}$$

To do this we define as the *no-interaction model*, the model

$$E(y_{ijk}) = \mu_{ij} \quad \text{and} \quad \text{all } \theta_{ij,i'j'} = 0 \ . \tag{58}$$

Then Q, the numerator sum of squares for testing (57) in the with-interaction model is

$$Q = \text{SSE}_{\text{H}} - \text{SSE},$$

as indicated in the first paragraph of Section 2.12b-i. Hence

$$\begin{aligned} Q = {} & \big[\text{error sum of squares after fitting the model } E(y_{ij}) = \mu_{ij} \\ & \quad \text{that also includes the no-interaction statements of the hypothesis}\big] \\ & - \big[\text{error sum of squares after fitting the model } E(y_{ij}) = \mu_{ij}\big] \\ = {} & (\text{error sum of squares for no-interaction model}) \\ & - (\text{error sum of squares for with-interaction model}) \\ = {} & \Sigma_i\Sigma_j\Sigma_k y_{ijk}^2 - R(\text{no-interaction model}) \\ & - \big[\Sigma_i\Sigma_j\Sigma_k y_{ijk}^2 - R(\text{with-interaction model})\big], \end{aligned}$$

i.e.,

$$Q = R(\text{with-interaction model}) - R(\text{no-interaction model}) \ .$$

In Table 4.2, the error sum of squares is SSE $= \Sigma_i\Sigma_j\Sigma_k(y_{ijk} - \bar{y}_{ij\cdot})^2$, and since this is for the model $E(y_{ijk}) = \mu_{ij}$, which is now being called the with-interaction model, we have

$$\begin{aligned} R(\mu_{ij}) &= R(\text{with-interaction model}) \\ &= \Sigma_i\Sigma_j\Sigma_k y_{ijk}^2 - \Sigma_i\Sigma_j\Sigma_k(y_{ijk} - \bar{y}_{ij\cdot})^2 \\ &= \Sigma_i\Sigma_j n_{ij}\bar{y}_{ij\cdot}^2 \ . \end{aligned} \tag{59}$$

Hence
$$\begin{aligned} Q &= R(\mu_{ij}) - R(\text{no-interaction model}) \\ &= \Sigma_i\Sigma_j n_{ij}\bar{y}_{ij\cdot}^2 - R(\text{no-interaction model}) \ . \end{aligned} \tag{60}$$

Deriving an explicit form for (60) therefore demands considering implications of the no-interaction model (58).

a. A no-interaction model

Suppose we assume that μ_{ij} has the form

$$\mu_{ij} = \mu_i + \tau_j \tag{61}$$

for all $i = 1, \ldots, a$ and $j = 1, \ldots, b$. Then it is easily seen that every $\theta_{ij,i'j'}$ is zero, i.e., all interactions are zero. And incorporating (61) into $E(y_{ijk}) = \mu_{ij}$ yields

$$E(y_{ijk}) = \mu_i + \tau_j, \tag{62}$$

This is equivalent to the no-interaction model of (58) because with (61)

$$\begin{aligned} \theta_{ij,i'j'} &= \mu_{ij} - \mu_{i'j} - \mu_{ij'} + \mu_{i'j'} \\ &= \mu_i + \tau_j - (\mu_{i'} + \tau_j) - (\mu_i + \tau_{j'}) + (\mu_{i'} + \tau_{j'}) \\ &= 0 \ . \end{aligned}$$

An important feature of (61) must now be noted: for any given set of μ_{ij}-values, the set of values of μ_i and τ_j that satisfies (61) is not unique. Consider the no-interaction example of μ_{ij}s in case 2 in Table 4.5. Table 4.6 shows three different ways of writing those μ_{ij}-values in the manner of (61). They are labeled (a), (b) and (c). Each is a different way of writing μ_{ij} of case 2 of Table 4.5 in the form $\mu_{ij} = \mu_i + \tau_j$. Scrutiny reveals a simple

TABLE 4.6. THE NO-INTERACTION VALUES OF μ_{ij} OF CASE 2 IN TABLE 4.5, EXPRESSED IN THREE EQUIVALENT FORMS OF (61)

μ_{ij}

	$j = 1$	$j = 2$	$j = 3$
$i = 1$	$\mu_{11} = 10$	$\mu_{12} = 15$	$\mu_{13} = 17$
$i = 2$	$\mu_{21} = 7$	$\mu_{22} = 12$	$\mu_{23} = 14$

(a)

	$\tau_1' = 6$	$\tau_2' = 11$	$\tau_3' = 13$
$\mu_1' = 4$	$4 + 6$	$4 + 11$	$4 + 13$
$\mu_2' = 1$	$1 + 6$	$1 + 11$	$1 + 13$

(b)

	$\tau_1'' = 1$	$\tau_2'' = 6$	$\tau_3'' = 8$
$\mu_1'' = 9$	$9 + 1$	$9 + 6$	$9 + 8$
$\mu_2'' = 6$	$6 + 1$	$6 + 6$	$6 + 8$

(c)

	$\tau_1 = -7$	$\tau_2 = -2$	$\tau_3 = 0$
$\mu_1 = 17$	$17 - 7$	$17 - 2$	17
$\mu_2 = 14$	$14 - 7$	$14 - 2$	14

relationship between the parts. Where part (a) uses $\mu_i' + \tau_j'$, any other part uses $(\mu_i' + \delta) + (\tau_j' - \delta)$ for some δ; for example, (b) is $\mu_i'' = \mu_i' + 5$ and $\tau_j'' = \tau_j' - 5$. Part (c) is of particular interest because it has $\tau_3 = 0$. This means that the $ab = 2(3) = 6$ values of μ_{ij} are completely determined by $a + b - 1 = 2 + 3 - 1 = 4$ values, $\mu_1 = 17$, $\mu_2 = 14$, $\tau_1 = -7$ and $\tau_2 = -2$. This is true in general, as a result of which we now take our no-interaction model (62) to be

$$\left.\begin{aligned} E(y_{ijk}) &= \mu_i + \tau_j \quad \text{for } j = 1, \ldots, b - 1 \\ E(y_{ijk}) &= \mu_i \qquad\quad \text{for } j = b \end{aligned}\right\} \quad \begin{aligned} &\text{for } i = 1, \ldots, a \\ &\text{and } k = 1, \ldots, n_{ij}. \end{aligned} \tag{63}$$

b. Estimation in the no-interaction model

Least squares estimators of μ_i and τ_j in the no-interaction model (63) are obtained by minimizing the sum of squares

$$\begin{aligned} S &= \Sigma_i \Sigma_j \Sigma_k (y_{ijk} - \mu_{ij})^2 \\ &= \Sigma_i \sum_{j=1}^{b-1} \Sigma_k (y_{ijk} - \mu_i - \tau_j)^2 + \Sigma_i \sum_{k=1}^{n_{ib}} (y_{ibk} - \mu_i)^2 \,. \end{aligned} \tag{64}$$

Notation. In this expression, and subsequently, note the continuance of the abbreviated summation notation of (8) which omits the limits of summation unless needed explicitly. In particular, notice in the first term of (64) that summation over j has an upper limit $b-1$ and not b. This occurs repeatedly in the sequel.

Details of deriving the least squares estimators are shown in equations (118)–(125) of the Appendix in Section 4.11. They turn out to require first solving, for $\hat{\tau}_1, \hat{\tau}_2, \dots, \hat{\tau}_{b-1}$, the $b-1$ equations, for $j = 1, 2, \dots, b-1$

$$c_{jj}\hat{\tau}_j + \sum_{\substack{j'=1 \\ j' \neq j}}^{b-1} c_{jj'}\hat{\tau}_{j'} = r_j \tag{65}$$

where
$$c_{jj} = n_{.j} - \sum_{i=1}^{a} \frac{n_{ij}^2}{n_{i.}} \quad \text{and} \quad c_{jj'} = -\sum_{i=1}^{a} \frac{n_{ij}n_{ij'}}{n_{i.}} \tag{66}$$

and
$$r_j = y_{.j.} - \sum_{i=1}^{a} n_{ij}\bar{y}_{i..} \ . \tag{67}$$

Then, on using the $\hat{\tau}_j$s obtained from (65)

$$\hat{\mu}_i = \bar{y}_{i..} - \frac{1}{n_{i.}} \sum_{j=1}^{b-1} n_{ij}\hat{\tau}_j \quad \text{for } i = 1, \dots, a \ . \tag{68}$$

In contrast to estimates obtained earlier in this chapter, and in Chapters 2 and 3, it is to be noted that the estimator $\hat{\tau}_j$ coming from (65) is not a simple mean as is $\hat{\mu}_{ij} = \bar{y}_{ij.}$ of (12), for example. Indeed, solutions to equations (65) have no simple form. The equations can, of course, be written succinctly in matrix notation and their solution likewise (see Section 9.3j), but except in special cases that solution has no simple form. Neither, of course, does (68).

Example (continued). Table 4.7 shows the cell, row, column and grand means of the data in Table 4.1. Using the entries in Table 4.7 in (66) and (67) gives

$$c_{11} = 7 - \left(3^2/7 + 4^2/8\right) = 208/56, \qquad c_{12} = -6/7 - 4/8 = -76/56,$$

$$c_{22} = 3 - \left(2^2/7 + 1^2/8\right) = 129/56;$$

$$r_1 = 91 - \left[3(13) + 4\left(16\tfrac{3}{4}\right)\right] = -15$$

and

$$r_2 = 59 - \left[2(13) + 1\left(16\tfrac{3}{4}\right)\right] = 16\tfrac{1}{4} \ .$$

TABLE 4.7. MEANS FROM TABLE 4.1

	$y_{ij\cdot}(n_{ij})\bar{y}_{ij\cdot}$			
	$j=1$	$j=2$	$j=3$	$y_{i\cdot\cdot}(n_{i\cdot})\bar{y}_{i\cdot\cdot}$
$i=1$	27(3) 9	28(2)14	36(2)18	91(7)13
$i=2$	64(4)16	31(1)31	39(3)13	134(8)$16\frac{3}{4}$
$y_{\cdot j\cdot}(n_{\cdot j})\bar{y}_{\cdot j\cdot}$	91(7)13	59(3)$19\frac{2}{3}$	75(5)15	$225(15)15 = y_{\cdots}(n_{\cdot\cdot})\bar{y}_{\cdots}$

Hence equations (65) are

$$(208\hat{\tau}_1 - 76\hat{\tau}_2)/56 = -15$$

$$(-76\hat{\tau}_1 + 129\hat{\tau}_2)/56 = 16\tfrac{1}{4} \tag{69}$$

with solutions $\hat{\tau}_1 = -1\frac{81}{94}$ and $\hat{\tau}_2 = 5\frac{90}{94}$. (70)

Then from (68):

$$\begin{aligned} \hat{\mu}_1 &= 13 - \tfrac{1}{7}\left[3\left(-1\tfrac{81}{94}\right) + 2\left(5\tfrac{90}{94}\right)\right] = 12\tfrac{9}{94} \\ \hat{\mu}_2 &= 16\tfrac{3}{4} - \tfrac{1}{8}\left[4\left(-1\tfrac{81}{94}\right) + 1\left(5\tfrac{90}{94}\right)\right] = 16\tfrac{88}{94}\ . \end{aligned} \tag{71}$$

c. Reduction in sum of squares

We denote the reduction in sum of squares for fitting the no-interaction model (63) as $R(\mu_i, \tau_j)$. Then

$$R(\mu_i, \tau_j) = R(\text{no-interaction model})$$

$$= \Sigma_i\Sigma_j\Sigma_k y_{ijk}^2 - \Sigma_i \sum_{j=1}^{b-1} \Sigma_k\left(y_{ijk} - \hat{\mu}_i - \hat{\tau}_j\right)^2 - \Sigma_i \sum_{k=1}^{n_{ib}} \left(y_{ibk} - \hat{\mu}_i\right)^2 .$$

This simplifies, as shown in equations (126)–(128) of the Appendix (Section 4.11), to

$$R(\mu_i, \tau_j) = \Sigma_i n_{i\cdot}\bar{y}_{i\cdot\cdot}^2 + \sum_{j=1}^{b-1} \hat{\tau}_j r_j \tag{72}$$

for $\hat{\tau}_j$ and r_j of (65) and (67), respectively.

Example (continued). With $n_{i\cdot}$ and $\bar{y}_{i\cdot\cdot}$ from Table 4.7, and $\hat{\tau}_j$ and r_j from (69) and (70), we have (72) as

$$R(\mu_i, \tau_j) = 7(13^2) + 8\left(16\tfrac{3}{4}\right)^2 + \left(-1\tfrac{81}{94}\right)(-15) + \left(5\tfrac{90}{94}\right)16\tfrac{1}{4} = 3552\tfrac{11}{47}$$

with
$$\sum_{j=1}^{b-1} \hat{\tau}_j r_j = \left(-1\tfrac{81}{94}\right)(-15) + \left(5\tfrac{90}{94}\right)16\tfrac{1}{4} = 124\tfrac{69}{94}\ . \tag{73}$$

d. The *F*-statistic

As discussed following equation (58),

$$\text{H: } all\ \theta_{ij,i'j'} = 0 \tag{74}$$

of (57) is tested using

$$Q = R(\text{with-interaction model}) - R(\text{no-interaction model})\ .$$

This is
$$Q = R(\mu_{ij}) - R(\mu_i, \tau_j) = \mathscr{R}(\mu_{ij} \mid \mu_i, \tau_j) \tag{75}$$

which, from (60) and (72), is

$$Q = \mathscr{R}(\mu_{ij} \mid \mu_i, \tau_j) = \Sigma_i \Sigma_j n_{ij} \bar{y}_{ij\cdot}^2 - \Sigma_i n_{i\cdot} \bar{y}_{i\cdot\cdot}^2 - \Sigma_j \hat{\tau}_j r_j\ . \tag{76}$$

In order to use Q as the numerator sum of squares of an F-statistic we need to establish that it has the following properties:

(i) Q is a sum of squares.
(ii) Q is distributed independently of $\hat{\sigma}^2$ of (17).
(iii) Under H of (74), the distribution of Q is χ_f^2 for known degrees of freedom f.

In all preceding situations of requiring these properties of a sum of squares we have been able to establish them by showing that the sum of squares of interest was a special case of the 1-way classification sum of squares $\text{SSR}_m = \Sigma_{i=1}^a n_i(\bar{y}_{i\cdot} - \bar{y}_{\cdot\cdot})^2$ for which, in Sections 2.10d and 2.10f, these properties are made evident. Examples of such special cases are SSR_m of Table 3.2 and (17) in Section 3.5b; $\mathscr{R}(\mu_i \mid \mu)$ of (16) in Section 3.5b; and SSA_w of (42) and $\mathscr{R}(\mu_i \mid \mu)$ of (47) of this chapter. However, for $Q = \mathscr{R}(\mu_{ij} \mid \mu_i, \tau_j)$ of (76) there seems to be no readily apparent way of showing that it is a special case of $\Sigma_{i=1}^a n_{i\cdot}(\bar{y}_{i\cdot} - \bar{y}_{\cdot\cdot})^2$. Nevertheless, it certainly does have the requisite properties, derivation of which stems most easily from the general results developed in matrix terminology in Section 8.6. Their application to Q is shown specifically in Section 9.2g.

The degrees of freedom $f = (a-1)(b-1)$ associated with $\mathscr{R}(\mu_{ij}|\mu_i, \tau_j)$ of (76) is established rigorously through application of the same results in Section 8.6 as those by which the χ^2 and independence properties are established. But one can gain an appreciation of that value from considering (75). For $R(\mu_{ij}) = \Sigma_i\Sigma_j n_{ij}\bar{y}_{ij\cdot}^2$, which is a sum of squares due to fitting $s = ab$ cell means μ_{ij}, the degrees of freedom are $s = ab$; and degrees of freedom for $R(\mu_i, \tau_j)$ are $a + b - 1$, the number of μ_is and τ_js in the no-interaction model (63). The difference between these two,

$$s - (a + b - 1) = ab - (a + b - 1) = (a-1)(b-1), \tag{77}$$

is the degrees of freedom for $\mathscr{R}(\mu_{ij}|\mu_i, \tau_j) = R(\mu_{ij}) - R(\mu_i, \tau_j)$. This corresponds, too, to the number of LIN θs in a grid of a rows and b columns, as developed in Section 4.8c. Thus for all-cells-filled data, with $\mathscr{R}(\mu_{ij}|\mu_i, \tau_j)$ of (76),

$$\frac{\mathscr{R}(\mu_{ij}|\mu_i, \tau_j)}{\hat{\sigma}^2(a-1)(b-1)} \quad \text{tests} \quad \text{H: } \textit{all } \theta_{ij,i'j'} = 0\ . \tag{78}$$

Since the maximum number of LIN θs is $(a-1)(b-1)$ this hypothesis can also be written as

$$\text{H: } \textit{any } (a-1)(b-1) \text{ LIN } \theta_{ij,i'j'}\textit{s all zero}\ . \tag{79}$$

e. Consequences of testing for interactions

-i. Estimating means. The two inferences that can be drawn from (78) are (I) reject H and conclude that interactions are important, or (II) do not reject H and conclude that there is no evidence that H is violated. In either case, one can still estimate linear combinations of the μ_{ij}s (including the individual μ_{ij}s, themselves) and calculate confidence intervals and test hypotheses about those combinations—as in Sections 4.4, 4.6 and 4.7. Those procedures are not invalidated by any rejection or non-rejection of the hypothesis in (78) that "interactions are zero"; just their interpretation is affected. For example, when the hypothesis (79) is rejected, $\rho_1 - \rho_2 = \Sigma_j(\mu_{1j} - \mu_{2j})/b$ for ρ_i of (27) in Section 4.6a can be expressed in one of a variety of ways involving θs as, for example

$$\rho_1 - \rho_2 = \mu_{11} - \mu_{21} - \sum_{j=2}^{b} \theta_{11,2j}/b, \tag{80}$$

in this case involving the LIN interactions $\theta_{11,2j}$ for $j = 2, 3, \ldots, b$. It is, of course, easier to think about $\rho_1 - \rho_2$ as

$$\rho_1 - \rho_2 = \Sigma_j \mu_{1j}/b - \Sigma_j \mu_{2j}/b$$

than in the form (80). Nevertheless, (80) emphasizes that interactions, whether large or small, whether judged significant or non-significant on the basis of data, do occur in linear functions of cell means.

When H of (78) is not rejected, one might be prepared to assume that all interactions $\theta_{ij,i'j'}$ are zero and so assume

$$\mu_{ij} = \mu_i + \tau_j$$

as in (61). On the basis of this assumption, ρ_i simplifies to

$$\rho_i = \Sigma_j \mu_{ij}/b = \mu_i + \bar{\tau}_{.},$$

so that $\text{SSA}_w/(a-1)\hat{\sigma}^2$ of (42), which tests

$$\text{H: } \rho_i \textit{ all equal}, \quad \text{is testing} \quad \text{H: } \mu_i \textit{ all equal} . \tag{81}$$

-ii. Estimating the residual variance. When (78) is rejected the estimator of σ^2 in the with-interaction model is, with $s = ab$,

$$\hat{\sigma}^2 = \text{SSE}/(N - s) \quad \text{for} \quad \text{SSE} = \Sigma_i \Sigma_j \Sigma_k (y_{ijk} - \bar{y}_{ij.})^2$$

of (17). But when (78) is not rejected, a question is sometimes raised about using an estimator of σ^2 different from $\hat{\sigma}^2$. It is based on taking non-rejection of (78) as being tantamount to acceptance, which means accepting that there *are* no interactions. This in turn means using the model (63), for which the estimate of σ^2 is

$$\tilde{\sigma}^2 = \frac{\text{SSE}(\mu_i, \tau_j)}{N - (a + b - 1)} . \tag{82}$$

This is the estimator of σ^2 that is suggested as an alternative to $\hat{\sigma}^2$ when the hypothesis of no interactions is not rejected. It has the following computational forms:

$$
\begin{aligned}
(N-a-b+1)\tilde{\sigma}^2 &= \Sigma_i\Sigma_j\Sigma_k y_{ijk}^2 - R(\mu_i, \tau_j) \\
&= \Sigma_i\Sigma_j\Sigma_k y_{ijk}^2 - \Sigma_i n_{i\cdot}\bar{y}_{i\cdot\cdot}^2 - \sum_{j=1}^{b-1} \hat{\tau}_j r_j, \quad \text{using (72)} \\
&= \Sigma_i\Sigma_j\Sigma_k y_{ijk}^2 - \Sigma_i\Sigma_j n_{ij}\bar{y}_{ij\cdot}^2 + \Sigma_i\Sigma_j n_{ij}\bar{y}_{ij\cdot}^2 \\
&\qquad - \Sigma_i n_i \bar{y}_{i\cdot\cdot}^2 - \sum_{j=1}^{b-1} \hat{\tau}_j r_j \\
&= \text{SSE} + \text{Q}, \quad \text{on using (76)} .
\end{aligned}
$$

Thus
$$\tilde{\sigma}^2 = \frac{\text{SSE} + Q}{N-a-b+1} . \tag{83}$$

The reasoning upon which suggesting the use of $\tilde{\sigma}^2$ is based is that when the hypothesis of no interactions is not rejected we ought to use the no-interaction model. But this reasoning has a weakness: non-rejection of H: *interactions all zero* is no definitive proof that there are no interactions. It is only some reasonably firm (but by no means conclusive) evidence that the hypothesis of zero interactions is not violated. To the extent that that evidence may be misleading, it seems appropriate to take no account of it in estimating σ^2 and to continue using $\hat{\sigma}^2 = \text{SSE}/(N-ab)$. An upshot of this is that if the evidence is misleading, then, for example, $\text{SSA}_w/(a-1)\hat{\sigma}^2$ for $\hat{\sigma}^2 = \text{SSE}/(N-ab)$ remains an appropriate F-statistic for testing H: ρ_i *equal* $\forall$ i, rather than $\text{SSA}/(a-1)\tilde{\sigma}^2$. To whatever extent the true model does include interactions, $\tilde{\sigma}^2$ is undoubtedly a wrong estimator for this purpose. My inclination is therefore always to use $\hat{\sigma}^2$.

The same argument applies to (81), of interpreting H: ρ_i *all equal* as H: μ_i *all equal* on the basis of not rejecting H: *all interactions zero* of (78). To the extent that there are interactions in the true model the hypothesis H: ρ_i *all equal* includes them and is, to my mind, always best thought of as H: $\Sigma_j\mu_{ij}/b$ *equal* $\forall$ i.

-iii. Other F-statistics. The existence of $R(\mu_i, \tau_j) = \Sigma_i n_{i\cdot}\bar{y}_{i\cdot\cdot}^2 + \Sigma_{j=1}^{b-1}\hat{\tau}_j r_j$ and of $R(\mu_i) = \Sigma_i n_{i\cdot}\bar{y}_{i\cdot\cdot}^2$ analogous to (52) of Section 2.8 begs the question: Of what use is

$$\mathscr{R}(\mu_i, \tau_j | \mu_i) = R(\mu_i, \tau_j) - R(\mu_i) = \sum_{j=1}^{b-1} \hat{\tau}_j r_k \ ? \tag{84}$$

Developing the answer for the model $E(y_{ijk}) = \mu_{ij}$, i.e., the with-interaction model, is deferred to Section 9.1 just as is the establishment of properties (i), (ii) and (iii) of Q following (76). Nevertheless, the answer itself is as follows: in *the with-interaction model*

$$\frac{\mathscr{R}(\mu_i, \tau_j \mid \mu_i)}{(b-1)\hat{\sigma}^2} \quad \text{tests} \quad \text{H: } \frac{1}{n_{\cdot j}}\Sigma_i n_{ij}\mu_{ij} = \frac{1}{n_{\cdot j}}\Sigma_i n_{ij}\left(\frac{1}{n_{i\cdot}}\Sigma_k n_{ik}\mu_{ik}\right) \ \forall\ j\ . \tag{85a}$$

Using $\rho_i' = (1/n_{i\cdot})\Sigma_j n_{ij}\mu_{ij}$ of (45) and $\gamma_j' = (1/n_{\cdot j})\Sigma_i n_{ij}\mu_{ij}$ of (51), the hypothesis in (85a) can also be written as

$$\text{H: } \gamma_j' = \frac{1}{n_{\cdot j}}\Sigma_i n_{ij}\rho_i' \ \forall\ j, \quad \text{or as} \quad \text{H: } \Sigma_i n_{ij}\mu_{ij} = \Sigma_i n_{ij}\rho_i' \ \forall\ j\ . \tag{85b}$$

Of course, having the $1/n_{\cdot j}$ on each side of the hypothesis as in (85a) is unnecessary except that its presence makes the statement easier to comprehend. It concerns weighted means of cell means, over rows and over columns, using numbers of observations in the cells as weights. γ_j' is such a weighted mean for column j and ρ_i' is such a weighted mean for row i. Then the hypothesis in (85b) is that the weighted mean of column j, namely γ_j', equals the same weighted mean of the ρ_i's. In truth, it is not easy to understand and seems to be of no general interest.

The hypothesis in (85) has $b - 1$ degrees of freedom even though the hypothesis consists of b statements: those b statements are not LIN. They add to an identity; i.e., each side of the b statements in H of (85b) adds to the same expression. Thus, adding the left-hand sides of H in (85b) gives

$$\Sigma_j(\text{L.H.S.}) = \Sigma_j\Sigma_i n_{ij}\mu_{ij} = \Sigma_i\Sigma_j n_{ij}\mu_{ij},$$

and adding the right-hand sides gives

$$\begin{aligned}\Sigma_j(\text{R.H.S.}) &= \Sigma_j\Sigma_i n_{ij}(\Sigma_k n_{ik}\mu_{ik}/n_{i\cdot}) \\ &= \Sigma_i n_{i\cdot}(\Sigma_k n_{ik}\mu_{ik}/n_{i\cdot}) \\ &= \Sigma_i\Sigma_k n_{ik}\mu_{ik} \equiv \Sigma_i\Sigma_j n_{ij}\mu_{ij} = \Sigma_j(\text{L.H.S.})\ .\end{aligned}$$

Similar to (84) and (85) we also have for

$$\mathscr{R}(\mu_i, \tau_j \mid \tau_j) = \Sigma_i n_{i\cdot}\bar{y}_{i\cdot\cdot}^2 + \sum_{j=1}^{b-1}\hat{\tau}_j r_j - \Sigma_j n_{\cdot j}\bar{y}_{\cdot j\cdot}^2, \tag{86}$$

that in the with-interaction model

$$\frac{\mathcal{R}(\mu_i, \tau_j \mid \tau_j)}{(a-1)\hat{\sigma}^2} \quad \text{tests} \quad \text{H: } \rho_i' = \frac{1}{n_{i\cdot}}\Sigma_j n_{ij}\gamma_j' \ \forall \ i \ . \tag{87}$$

Example (continued). The data of Table 4.1 have the following table of n_{ij}s and subtotals thereof:

3	2	2	7
4	1	3	8
7	3	5	15

Therefore the hypothesis tested in (85) uses

$$\rho_1' = \tfrac{1}{7}(3\mu_{11} + 2\mu_{12} + 2\mu_{13}) \quad \text{and} \quad \rho_2' = \tfrac{1}{8}(4\mu_{21} + \mu_{22} + 3\mu_{23});$$

and

$$\gamma_1' = \tfrac{1}{7}(3\mu_{11} + 4\mu_{21}), \qquad \gamma_2' = \tfrac{1}{3}(2\mu_{12} + \mu_{22}) \quad \text{and} \quad \gamma_3' = \tfrac{1}{5}(2\mu_{13} + 3\mu_{23}) \ .$$

Hence the hypothesis in (85) is

$$\text{H: } \begin{cases} \gamma_1' = \tfrac{1}{7}(3\rho_1' + 4\rho_2') \\ \gamma_2' = \tfrac{1}{3}(2\rho_1' + \ \rho_2') \\ \gamma_3' = \tfrac{1}{5}(2\rho_1' + 3\rho_2') \end{cases} .$$

This example vividly illustrates two very unattractive features of the hypothesis in (85): it is determined by the numbers of observations in the cells; this, and the manner in which those numbers occur, makes the hypothesis generally not very useful.

f. Partitioning the total sum of squares

Of the sums of squares developed in preceding sections for testing hypotheses, SSA_w tests what is probably the most generally useful form of hypothesis in the with-interaction model, with all-cells-filled data.

$$\text{SSA}_w \quad \text{tests} \quad \text{H: } \frac{1}{b}\Sigma_j\mu_{ij} \ \textit{equal} \ \forall \ i \ . \tag{88}$$

This, its counterpart for columns,

$$\text{SSB}_w \quad \text{tests} \quad \text{H: } \frac{1}{a}\Sigma_i\mu_{ij} \ \textit{equal} \ \forall \ j, \tag{89}$$

and the interaction test,

$$\mathscr{R}(\mu_{ij}|\mu_i, \tau_j) \quad \text{tests} \quad \text{H: } \textit{all } \theta_{ij,i'j'} = 0 \tag{90}$$

are of very broad applicability. And note that these three sums of squares do not add up to

$$\text{SST} = \Sigma_i\Sigma_j\Sigma_k y_{ijk}^2, \tag{91}$$

nor to
$$\text{SST}_m = \Sigma_i\Sigma_j\Sigma_k(y_{ijk} - \bar{y}_{...})^2 = \text{SST} - R(\mu)$$

for
$$R(\mu) = N\bar{y}_{...}^2 \ . \tag{92}$$

That is of no concern; in (88), (89) and (90) we have the means for testing three generally useful hypotheses when using all-cells-filled data.

Contrary to the sums of squares in (88), (89) and (90) not adding up to SST or SST_m, there is a set of sum of squares that do add to SST, namely $R(\mu)$, $\mathscr{R}(\mu_i|\mu)$, $\mathscr{R}(\mu_i, \tau_j|\mu_i)$, $\mathscr{R}(\mu_{ij}|\mu_i, \tau_j)$ and SSE. Using the definition of $\mathscr{R}(\cdot \mid \cdot)$, together with SSE, SST and $R(\mu)$ of (17), (91) and (92), respectively, we have

$$\begin{aligned}
R(\mu) &+ \mathscr{R}(\mu_i|\mu) + \mathscr{R}(\mu_i, \tau_j|\mu_i) + \mathscr{R}(\mu_{ij}|\mu_i, \tau_j) + \text{SSE} \\
&= R(\mu) + [R(\mu_i) - R(\mu)] + [R(\mu_i, \tau_j) - R(\mu_i)] \\
&\quad + [R(\mu_{ij}) - R(\mu_i, \tau_j)] + [\text{SST} - R(\mu_{ij})] \\
&= \text{SST} \ .
\end{aligned} \tag{93}$$

Thus it is that these sums of squares are a partitioning of SST. They are summarized as such in Table 4.8, which also shows equation numbers in the text where the pertinent formulae have been established. As well as (93), notice also the following identity:

$$\begin{aligned}
\mathscr{R}(\mu_i|\mu) + \mathscr{R}(\mu_i, \tau_j|\mu_i) &= R(\mu_i) - R(\mu) + R(\mu_i, \tau_j) - R(\mu_i) \\
&= R(\tau_j) - R(\mu) + R(\mu_i, \tau_j) - R(\tau_j) \\
&= \mathscr{R}(\tau_j|\mu) + \mathscr{R}(\mu_i, \tau_j|\tau_j) \ .
\end{aligned} \tag{94}$$

The equivalence in (94) is also shown in Table 4.8, as partitioning II. As can

TABLE 4.8. PARTITIONING SST = $\Sigma_i\Sigma_j\Sigma_k y_{ijk}^2$

Partitioning I			Equation
$R(\mu)$	$= N\bar{y}_{...}^2$	$= N\bar{y}_{...}^2$	(92)
[$\mathscr{R}(\mu_i \mid \mu)$	$= \Sigma_i n_{i.}\bar{y}_{i..}^2 - N\bar{y}_{...}^2$	$= \Sigma_i n_{i.}(\bar{y}_{i..} - \bar{y}_{...})^2$	(47)
[$\mathscr{R}(\mu_i, \tau_j \mid \mu_i)$	$= \Sigma_i n_{i.}\bar{y}_{i..}^2 + \Sigma_j \hat{\tau}_j r_j - \Sigma n_{i.}\bar{y}_{i..}^2$	$= \Sigma_j \hat{\tau}_j r_j$	(84)
$\mathscr{R}(\mu_{ij} \mid \mu_i, \tau_j)$	$= \Sigma_i\Sigma_j n_{ij}\bar{y}_{ij.}^2 - \Sigma_i n_{i.}\bar{y}_{i..}^2 - \Sigma_j \hat{\tau}_j r_j$	$= \Sigma_i\Sigma_j n_{ij}(\bar{y}_{ij.} - \bar{y}_{i..})^2 - \Sigma_j \hat{\tau}_j r_j$	(76)
SSE	$= \Sigma_i\Sigma_j\Sigma_k y_{ijk}^2 - \Sigma_i\Sigma_j n_{ij}\bar{y}_{ij.}^2$	$= \Sigma_i\Sigma_j\Sigma_k (y_{ijk} - \bar{y}_{ij.})^2$	(17)
SST	$= \Sigma_i\Sigma_j\Sigma_k y_{ijk}^2$	$= \Sigma_i\Sigma_j\Sigma_k y_{ijk}^2$	(91)
Partitioning II			
[$\mathscr{R}(\tau_j \mid \mu)$	$= \Sigma_j n_{.j}\bar{y}_{.j.}^2 - N\bar{y}_{...}^2$	$= \Sigma_j n_{.j}(\bar{y}_{.j.} - \bar{y}_{...})^2$	(51)
[$\mathscr{R}(\mu_i, \tau_j \mid \tau_j)$	$= \mathscr{R}(\tau_j \mid \mu) + \mathscr{R}(\mu_i, \tau_j \mid \mu_i) - \mathscr{R}(\tau_j \mid \mu)$	$=$[1] $\Sigma_i n_{i.}\bar{y}_{i..}^2 - \Sigma_j n_{.j}\bar{y}_{.j.}^2 + \Sigma_j \hat{\tau}_j r_j$	(86)

[1]An equivalent form is $\Sigma_i n_{i.}(\bar{y}_{i..} - \bar{y}_{...})^2 - \Sigma_j n_{.j}(\bar{y}_{.j.} - \bar{y}_{...})^2 + \Sigma_j \hat{\tau}_j r_j$.

be seen, that partitioning is not a complete partitioning of SST but is simply two sums of squares alternative to those that are bracketed in partitioning I; i.e., the bracketed pair in partitioning I are those on the left-hand side of (94) and the pair in partitioning II are the right-hand side of (94).

g. Analysis of variance

The partitionings of SST in Table 4.8 are shown in Table 4.9 in the format of an analysis of variance table. Whereas Table 4.8 shows computational formulae for the sums of squares, Table 4.9 shows their associated hypotheses—and equation numbers where they have been developed in preceding subsections. The first, based on (97) of Section 2.12c, is

$$R(\mu)/\hat{\sigma}^2 \quad \text{tests} \quad \text{H: } \frac{1}{N}\Sigma_i\Sigma_j n_{ij}\mu_{ij} = 0 \ . \tag{95}$$

In Table 4.9 the labels given under the heading "Source of Variation" are descriptions of the sums of squares. For example, "mean" is the label for $R(\mu)$—because it is the reduction in sum of squares for fitting $E(y_{ijk}) = \mu$, which is a model that has but one, overall, mean effect. Also, since

$$\mathscr{R}(\mu_i \mid \mu) = R(\mu_i) - R(\mu)$$

is the sum of squares

for fitting $E(y_{ijk}) = \mu_i$ minus (adjusted for) that for fitting $E(y_{ijk}) = \mu$,

it is called "rows, adjusted for mean"—and so on.

TABLE 4.9. ANALYSIS OF VARIANCE OF THE 2-WAY CROSSED CLASSIFICATION WITH INTERACTION

Source of Variation	d.f.	Sum of Squares	Associated Hypothesis	Equation
Partitioning I				
Mean	1	$R(\mu)$	H: $\frac{1}{N}\Sigma_i\Sigma_j n_{ij}\mu_{ij} = 0$	(95)
Rows, adjusted for mean	$a - 1$	$\mathscr{R}(\mu_i \mid \mu)$	H: ρ_i' *all equal*	(48)
			for $\rho_i' = \frac{1}{n_{i\cdot}}\Sigma_j n_{ij}\mu_{ij}$	(45)
Columns, adjusted for rows	$b - 1$	$\mathscr{R}(\mu_i, \tau_j \mid \mu_i)$	H: $\gamma_j' = \frac{1}{n_{\cdot j}}\Sigma_i n_{ij}\rho_i' \ \forall\ j$	(85)
			for $\gamma_j' = \frac{1}{n_{\cdot j}}\Sigma_i n_{ij}\mu_{ij}$	(51)
Interaction	$s - a - b + 1$	$\mathscr{R}(\mu_{ij} \mid \mu_i, \tau_j)$	1. **All cells filled** H: $\theta_{ij,i'j'}$ *all zero* 2. **Some cells empty** See (42), Chapter 5	(78)
Residual	$N - s$	SSE		
Total	N	SST		
Partitioning II				
Columns, adjusted for mean	$b - 1$	$\mathscr{R}(\tau_j \mid \mu)$	H: γ_j' *all equal*	(51)
Rows, adjusted for columns	$a - 1$	$\mathscr{R}(\mu_i, \tau_j \mid \tau_j)$	H: $\rho_i' = \frac{1}{n_{i\cdot}}\Sigma_j n_{ij}\gamma_i' \ \forall\ i$	(87)

Although this chapter is confined to all-cells-filled data, Tables 4.8 and 4.9 do apply to some-cells-empty data with only a few simple precautions having to be taken into account—as discussed in Chapter 5. One most important difference from all-cells-filled data is the hypothesis tested by the interaction sum of squares—as noted in Table 4.9 and considered at length in Chapter 5. Important features of interpreting the analysis of such data are also dealt with in that chapter.

Example (continued). Table 4.8 provides computing formulae that can be used when starting an analysis from raw data. Such is not our case here, however, since we can draw on calculations already made, as follows.

$$R(\mu) = 15(15)^2 = 3375 \quad \text{using} \quad N = 15 \text{ and } \bar{y}_{\cdots} = 15 \text{ of Table 4.7.}$$

$$\mathscr{R}(\mu_i \mid \mu) = 52\tfrac{1}{2}, \quad \text{as calculated following (48)} . \tag{96}$$

$$\mathscr{R}(\mu_i, \tau_j \mid \mu_i) = \sum_{j=1}^{b-1} \hat{\tau}_j r_j = 124\tfrac{69}{94}, \quad \text{from (73)} . \tag{97}$$

$$\mathscr{R}(\mu_{ij} \mid \mu_i, \tau_j) = \Sigma_i \Sigma_j n_{ij} \bar{y}_{ij\cdot}^2 - \Sigma_i n_{i\cdot} \bar{y}_{i\cdot\cdot}^2 - \sum_{j=1}^{b-1} \hat{\tau}_j r_j, \quad \text{from Table 4.8,}$$

$$= 3(9^2) + \cdots + 3(13^2) - \left[7(13^2) + 8\left(16\tfrac{3}{4}\right)^2\right] - 124\tfrac{69}{94},$$

using Table 4.7 and (97),

$$= 3775 - 3427\tfrac{1}{2} - 124\tfrac{69}{94} = 222\tfrac{36}{47} .$$

$$\text{SSE} = 120, \quad \text{as used in (23)} .$$

$$\text{SST} = 3895, \quad \text{as in Table 4.3.}$$

$$\mathscr{R}(\tau_j \mid \mu) = 93\tfrac{1}{3}, \quad \text{from the end of Section 4.7d,} \tag{98}$$

and, finally,

$$\mathscr{R}(\mu_i, \tau_j \mid \mu_j) = \mathscr{R}(\mu_i \mid \mu) - \mathscr{R}(\tau_j \mid \mu) + \sum_{j=1}^{b-1} \hat{\tau}_j r_j,$$

from the footnote to Table 4.8, which is equivalent to (94). Hence, using (96), (98) and (97), respectively,

$$\mathscr{R}(\mu_i, \tau_j \mid \tau_j) = 52\tfrac{1}{2} - 93\tfrac{1}{3} + 124\tfrac{69}{94} = 83\tfrac{127}{141} .$$

Using these values, Table 4.9 for this example is shown in Table 4.10.

h. Hypotheses in the analysis of variance table

The most striking feature of Table 4.9 and its example in Table 4.10 is that only one of the hypotheses is simple, clear-cut and generally useful, namely that labeled "Interaction". It is the only hypothesis that is a

TABLE 4.10. ANALYSIS OF VARIANCE (TABLE 4.9) FOR THE EXAMPLE OF TABLE 4.1: WITH-INTERACTION MODEL

Source of Variation	d.f.	Sum of Squares	Associated Hypothesis[1]
Partitioning I			
Mean	1	$R(\mu) = 3375$	H: $3\mu_{11} + 2\mu_{12} + 2\mu_{13} + 4\mu_{21} + \mu_{22} + 3\mu_{23} = 0$
Rows, adjusted for mean	1	$\mathscr{R}(\mu_i \mid \mu) = 52\frac{1}{2}$	H: $\rho_1' = \rho_2'$
Columns, adjusted for rows	2	$\mathscr{R}(\mu_i, \tau_j \mid \mu_i) = 124\frac{69}{94}$	H: $\gamma_1' = \frac{1}{7}(3\rho_1' + 4\rho_2')$, $\gamma_2' = \frac{1}{3}(2\rho_1' + \rho_2')$
Interaction	2	$\mathscr{R}(\mu_{ij} \mid \mu_i, \tau_j) = 222\frac{36}{47}$	H: $\theta_{11,22} = \theta_{11,23} = 0$
Residual	9	SSE $= 120$	
Total	15	SST $= 3895$	
Partitioning II			
Columns, adjusted for mean	2	$\mathscr{R}(\tau_j \mid \mu) = 93\frac{1}{3}$	H: $\gamma_1' = \gamma_2' = \gamma_3'$
Rows, adjusted for columns	1	$\mathscr{R}(\mu_i, \tau_j \mid \tau_j) = 83\frac{127}{141}$	H: $\rho_1' = \frac{1}{7}(3\gamma_1' + 2\gamma_2' + 2\gamma_3')$

[1] $\rho_1' = \frac{1}{7}(3\mu_{11} + 2\mu_{12} + 2\mu_{13})$ $\gamma_1' = \frac{1}{7}(3\mu_{11} + 4\mu_{21})$ $\gamma_3' = \frac{1}{5}(2\mu_{13} + 3\mu_{23})$

$\rho_2' = \frac{1}{8}(4\mu_{21} + \mu_{22} + 3\mu_{23})$ $\gamma_2' = \frac{1}{3}(2\mu_{12} + \mu_{22})$

statement about simple functions of the μ_{ij}s that do not involve the numbers of observations in the cells. All the other hypotheses involve those numbers.

Considering that the sums of squares do, by their derivation and notation, appear to be a rational partitioning of the total sum of squares, the hypotheses evident in Table 4.9 are a sorry lot. For example, $\mathscr{R}(\mu_i \mid \mu) = R(\mu_i) - R(\mu)$ is the sum of squares due to fitting $E(y_{ijk}) = \mu_i$ after having fitted $E(y_{ijk}) = \mu$. One might therefore expect it to be testing a hypothesis about row means of the μ_{ij}s, the terms $\rho_i = \sum_j \mu_{ij}/b$. It does so with balanced data. But not with unbalanced data, as is evident in Table 4.9. Indeed, the hypothesis is not even a statement about just ρ_is; it involves the n_{ij}s. As a result it is a hypothesis that depends on the data; not on the observations *per se*, but on how many of them there are in each cell of the data.

The effects of hypotheses depending on the numbers of observations are particularly evident when one looks at an example, such as Table 4.10. The hypothesis associated with $\mathscr{R}(\mu_i|\mu)$, namely H: ρ' *all equal*, may be of interest if in each row the n_{ij}-values are in proportion to the population sizes of those cells; but in general this is most unlikely to be the case. This situation is even worse for the hypothesis tested by $\mathscr{R}(\mu_i, \tau_j|\tau_j)$, namely H: $\rho_i' = \Sigma_j n_{ij}\gamma_j'/n_{i\cdot}\ \forall\ i$. Symbolically that may look meaningful, but written in terms of the cell means, as

$$\text{H: } \Sigma_j n_{ij}\mu_{ij} = \Sigma_j n_{ij}(\Sigma_h n_{hj}\mu_{hj}/n_{\cdot j})\ \forall\ i,$$

it is not recognizable as being of particular interest. Moreover, this is generally quite different from H: ρ_i' *all equal*, i.e.,

$$\text{H: } \Sigma_j n_{ij}\mu_{ij}/n_{i\cdot}\ \textit{all equal}$$

tested by $\mathscr{R}(\mu_i|\mu)$. In only a few special cases (Burdick, 1979) are the two hypotheses the same.

The preceding hypotheses, depending as they do on the data, are clearly not one of general interest. Nor are they in keeping with the general philosophy of hypothesis testing that a hypothesis is a conjectured statement about parameters of a population. This statement should be formulated before data are collected, and so it should in no way depend on the data. If it does, two people with two different sets of data could, with the same test procedure, be testing different hypotheses. And with both of them depending on data, neither of them may be a credible hypothesis about the population. Hence, all the hypotheses in Table 4.9 except that associated with $\mathscr{R}(\mu_{ij}|\mu_i, \tau_j)$ are to be despised because they depend on the n_{ij}s. They are not hypotheses that one would generally formulate before collecting data.

The emphatic conclusion to be drawn from Table 4.9 and its illustrative example in Table 4.10 is that there is only one hypothesis that may be of general interest when all cells are filled, namely that associated with $\mathscr{R}(\mu_{ij}|\mu_i, \tau_j)$. Otherwise, the hypotheses associated with $R(\mu)$, with $R(\mu_i|\mu)$ and $\mathscr{R}(\mu_i, \tau_j|\mu_i)$, and with $\mathscr{R}(\tau_j|\mu)$ and $\mathscr{R}(\mu_i, \tau_j|\tau_j)$ are, in the with-interaction model with unbalanced data, of no general interest whatever. They are dependent on the data (on the n_{ij}s of the data) and are *not* hypotheses about means of cell means unencumbered by those cell numbers.

In contrast to this rather negative outcome of assembling the analysis of variance table (Table 4.9), it should never be forgotten that SSA_w and SSB_w do test hypotheses that may often be of general interest; so does $\mathscr{R}(\mu_{ij}|\mu_i, \tau_j)$. These are summarized in (88), (89) and (90).

This conclusion, of effectively discarding most of the hypotheses of Table 4.9 as useless does, of course, motivate a variety of questions, such as the following. (i) What is the situation in the no-interaction model? See Section 4.10. (ii) What happens with some-cells-empty data? See Chapter 5. (iii) Are there alternative analyses that are useful? Yes, for all-cells-filled data SSA_w and SSB_w are useful: see (88) and (89). (iv) Are there special cases of particular interest? Yes, balanced incomplete blocks, for example; and also other planned unbalanced cases such as proportional subclass numbers; and simple examples such as having just two rows and two columns (see E4.6). (v) What happens when the data are balanced? The next subsection addresses this.

i. Balanced data

Balanced data have $n_{ij} = n \;\forall\; i, j$. This has a tremendous simplifying effect on the preceding developments and leads to the following partial results—partial because this book is directed toward unbalanced data and because the analysis of balanced data is described in many places.

The most important simplification in Table 4.9 for balanced data is that there are not two partitionings of SST. The sums of squares in that table become as follows.

$$R(\mu) = abn\bar{y}_{\cdots}^2 .$$

$$\mathcal{R}(\mu_i \mid \mu) = \mathcal{R}(\mu_i, \tau_j \mid \tau_j) = \text{SSA}_w = bn\Sigma_i(\bar{y}_{i\cdot\cdot} - \bar{y}_{\cdots})^2$$

$$\mathcal{R}(\tau_j \mid \mu) = \mathcal{R}(\mu_i, \tau_j \mid \mu_i) = \text{SSB}_w = an\Sigma_j(\bar{y}_{\cdot j\cdot} - \bar{y}_{\cdots})^2$$

$$\mathcal{R}(\mu_{ij} \mid \mu_i, \tau_j) = n\Sigma_i\Sigma_j(\bar{y}_{ij\cdot} - \bar{y}_{i\cdot\cdot} - \bar{y}_{\cdot j\cdot} + \bar{y}_{\cdots})^2$$

$$\text{SSE} = \Sigma_i\Sigma_j\Sigma_k(y_{ijk} - \bar{y}_{ij\cdot})^2$$

$$\text{SST} = \Sigma_i\Sigma_j\Sigma_k y_{ijk}^2 .$$

And the hypotheses they test are as follows:

$$R(\mu) \quad \text{tests} \quad \text{H: } \Sigma_i\Sigma_j\mu_{ij} = 0 \ .$$

$$\mathcal{R}(\mu_i \mid \mu) \quad \text{tests} \quad \text{H: } \frac{1}{b}\Sigma_j\mu_{ij} \textit{ equal } \forall\; i, \quad \text{i.e.,} \quad \text{H: } \rho_i \textit{ all equal} \ .$$

$$\mathcal{R}(\tau_j \mid \mu) \quad \text{tests} \quad \text{H: } \frac{1}{a}\Sigma_i\mu_{ij} \textit{ equal } \forall\; j, \quad \text{i.e.,} \quad \text{H: } \gamma_j \textit{ all equal} \ .$$

$$\mathcal{R}(\mu_{ij} \mid \mu_i, \tau_j) \quad \text{tests} \quad \text{H: } \textit{all } \theta_{ij,i'j'} \textit{ zero.}$$

These, as is easily agreed upon, are hypotheses that are likely to be useful in a wide variety of situations.

j. Computer output

One might ask: If the sums of squares in Table 4.9 are of so little value for testing hypotheses, then why be bothered with them? Why even tabulate them as has been done? One reason for doing so, as indicated in general terms in Chapter 1, concerns computer-package output. Sums-of-squares output from many packages is based on Table 4.9 (see Tables 12.2 and 12.3). Package users therefore need to know when and why those sums of squares will be of little use. That they are of great use for balanced data has been relied upon for decades, but this has engendered a widely-held view that they are always of great importance. Although this is indeed so for balanced data, it is decidedly not so for unbalanced data. Successfully impressing this upon computer-package users is often helped by having available, as in Table 4.9, the precise form of associated hypothesis corresponding to each of the sum of squares.

4.10. THE NO-INTERACTION MODEL

It has been said that seldom do we definitively know that there are no interactions between whatever two factors we are studying. From this viewpoint it would therefore seem reasonable to never consider models that have no interactions. Despite this rationale, there are at least two situations where no-interaction models can be useful. One is when there is not more than one observation in a cell: for interaction defined as $\theta_{ij,i'j'}$ the no-interaction model then has to be used (although certain forms of with-interaction model can be used when interaction is defined differently —as mentioned in Section 9.1b). Another is some-cells-empty data for which the no-interaction model does not have to be used but where it does provide useful information that is not then otherwise obtainable (see Section 5.6b). For use in these two cases, and because the no-interaction model can also be used with all-cells-filled data, we here consider such a model.

Details of the no-interaction model, estimation and the reduction in sum of squares due to fitting it are exactly as in the preceding section: (63) is the model, (65)–(68) are the estimation equations and $R(\mu_i, \tau_j)$ of (72) is the reduction in sum of squares due to fitting it: the salient features are

$$c_{jj}\hat{\tau}_j + \sum_{\substack{j'=1 \\ j'\neq j}}^{b-1} c_{jj'}\hat{\tau}_{j'} = r_j \quad \text{for } j = 1, 2, \ldots, b-1$$

from (65), with c_{jj}, $c_{jj'}$ and r_j as in (66) and (67);

$$\hat{\mu}_i = \bar{y}_{i\cdot\cdot} - \frac{1}{n_{i\cdot}} \sum_{j=1}^{b-1} n_{ij}\hat{\tau}_j, \quad \text{for} \quad i = 1, 2, \ldots, a$$

from (68) and, from (72),

$$R(\mu_i, \tau_j) = \Sigma_i n_{i\cdot}\bar{y}_{i\cdot\cdot}^2 + \sum_{j=1}^{b-1} \hat{\tau}_j r_j \; .$$

A numerical example is shown in (69)–(71) and (73).

The only question then is: How do we make use of these results in the no-interaction model? There are two main answers.

a. Estimating means

Up to now we have alluded to $\hat{\mu}_i$ and $\hat{\tau}_j$ as estimators of μ_i and τ_j, respectively. But such is not quite the case, for the following reason. In (63) the no-interaction model is defined to have no τ_b, i.e., effectively with $\tau_b = 0$. In terms of expressing a given set of μ_{ij}-values as $\mu_{ij} = \mu_i + \tau_j$ we illustrated in Table 4.6 that this made no difference. However, it does have an effect on $\hat{\mu}_i$ and $\hat{\tau}_j$ that must be noted.

Instead of using $\tau_b = 0$ (and correspondingly $\hat{\tau}_b = 0$) in defining the no-interaction model as (63), we could just as well have defined the model slightly differently in terms of what shall be denoted as μ'_i and τ'_j, with $\tau'_b = \Delta$ being any arbitrarily chosen numerical value. It will then be found that minimizing the corresponding form of S of (64), namely

$$S' = \Sigma_i \sum_{j=1}^{b-1} \Sigma_k \left(y_{ijk} - \mu'_i - \tau'_j\right)^2 + \Sigma_i \sum_{k=1}^{n_{ib}} \left(y_{ibk} - \mu'_i - \Delta\right)^2, \qquad (99)$$

with respect to μ'_i and τ'_j, leads (see E4.15), through appropriate adaptation of (118)–(121) in the Appendix, to

$$\hat{\mu}'_i = \hat{\mu}_i - \Delta \quad \text{and} \quad \hat{\tau}'_j = \hat{\tau}_j + \Delta \; . \qquad (100)$$

This means that although there is only one set of values $\hat{\mu}_i$ and $\hat{\tau}_j$ with $\hat{\tau}_b = 0$, there is also an infinite number of other sets, $\hat{\mu}'_i$ and $\hat{\tau}'_j$, available from (100) using $\hat{\mu}_i$ and $\hat{\tau}_j$ with $\hat{\tau}_b = 0$ and with any value for Δ. Then $\hat{\tau}'_b = \Delta$. (The values $\hat{\mu}_i$ and $\hat{\tau}_j$ with $\hat{\tau}_b = 0$ can then be seen as $\hat{\mu}'_i$ and $\hat{\tau}'_j$ for the special case of $\Delta = 0$.)

Since there is an infinite number of sets of values $\hat{\mu}'_i$ and $\hat{\tau}'_j$ obtainable from (100), none of them can be thought of as BLU estimators of μ_i and τ_j, respectively. Nevertheless, as is easily seen from (100), for every such set of values

$$\hat{\mu}'_i - \hat{\mu}'_h = \hat{\mu}_i - \hat{\mu}_h \quad \text{and} \quad \hat{\tau}'_j - \hat{\tau}'_k = \hat{\tau}_j - \hat{\tau}_k .$$

Furthermore these differences are the BLUEs:

$$\text{BLUE}(\mu_i - \mu_h) = \hat{\mu}_i - \hat{\mu}_h \tag{101}$$

and similarly

$$\text{BLUE}(\tau_j - \tau_k) = \hat{\tau}_j - \hat{\tau}_k \tag{102}$$

for the $\hat{\mu}_i$ and $\hat{\tau}_j$ obtained in (65)–(68). And now, in (102) we can ignore the idea introduced in (63) that $\hat{\tau}_b$ is zero. Equation (102) applies for any $j \neq k$. Thus a difference $\mu_i - \mu_h$ has its BLUE as the difference between the corresponding $\hat{\mu}_i$s—as in (101); and a difference $\tau_j - \tau_k$ has as its BLUE the difference between the corresponding $\hat{\tau}_j$s—as in (102). And the BLUE of any linear function of differences between μs and between τs is the same linear function of those estimated differences.

Example (continued). In (71), $\hat{\mu}_1 = 12\frac{9}{94}$ and $\hat{\mu}_2 = 16\frac{88}{94}$; hence

$$\text{BLUE}(\mu_1 - \mu_2) = \hat{\mu}_1 - \hat{\mu}_2 = 12\tfrac{9}{94} - 16\tfrac{88}{94} = -4\tfrac{79}{94} .$$

Likewise, $\hat{\tau}_1 = -1\frac{84}{94}$, $\hat{\tau}_2 = 5\frac{90}{94}$ and $\hat{\tau}_3 = 0$. Therefore

$$\text{BLUE}(\tau_2 - \tau_1) = \hat{\tau}_2 - \hat{\tau}_1 = 5\tfrac{90}{94} - \left(-1\tfrac{84}{94}\right) = 7\tfrac{80}{94} .$$

Similarly

$$\text{BLUE}(\tau_3 - \tau_1) = \hat{\tau}_3 - \hat{\tau}_1 = 0 - \left(-1\tfrac{84}{94}\right) = 1\tfrac{84}{94} .$$

Hence, for example,

$$\text{BLUE}[2(\tau_2 - \tau_1) - 5(\tau_3 - \tau_1)] = 2\left(7\tfrac{80}{94}\right) - 5\left(1\tfrac{84}{94}\right) = 5\tfrac{22}{94};$$

i.e.,

$$\text{BLUE}(3\tau_1 + 2\tau_2 - 5\tau_3) = 6\tfrac{22}{94} .$$

Sampling variances of $\hat{\mu}_i$, of $\hat{\tau}_j$ and of the estimators in (101) and (102) are also obtainable. Unfortunately, though, there are no simple algebraic forms for these variances without using matrices—so details are deferred to Section 9.3.

Although, as indicated leading up to (100), there are numerous sets of values of the form (100) that satisfy the least squares requirement, we also find from (100) that $\hat{\mu}'_i + \hat{\tau}'_j = \hat{\mu}_i + \hat{\tau}_j$ regardless of Δ. Thus

$$\hat{\mu}_{ij} = \hat{\mu}_i + \hat{\tau}_j \tag{103}$$

is the BLUE of μ_{ij}. Again, the variance $v(\hat{\mu}_{ij})$ is obtainable but its description must be deferred to Section 9.2k.

b. *F*-statistics

The sum of squares for fitting the no-interaction model is $R(\mu_i, \tau_j)$ of (72). And now, because we have definitively chosen to use a no-interaction model the estimator of σ^2 is

$$\tilde{\sigma}^2 = \frac{\Sigma_i\Sigma_j\Sigma_k y_{ijk}^2 - R(\mu_i, \tau_j)}{N - a - b + 1} \tag{104}$$

as in (83).

We now consider

$$\mathscr{R}(\mu_i, \tau_j \mid \mu_i) = \sum_{j=1}^{b-1} \hat{\tau}_j r_j \tag{105}$$

of (84). Similar to properties (i) and (ii) of Q following (76) one can show, using the methods of Section 8.6, that (105) is a sum of squares distributed independently of $\tilde{\sigma}^2$; and those same methods indicate that $\tilde{\sigma}^2$ has a χ^2-distribution on $N - a - b + 1$ degrees of freedom. As explained following (102), that result gives reason to ignore the putting of τ_b to zero in (63). Continuing to ignore this, we can also use the methods of Section 9.1 to show that under the hypothesis H: τ_j *all equal*, (105) has a χ^2_{b-1}-distribution. Hence, in the no-interaction model, using $\tilde{\sigma}^2$ of (104)

$$\frac{\mathscr{R}(\mu_i, \tau_j \mid \mu_i)}{(b-1)\hat{\sigma}^2} = \frac{\sum_{j=1}^{b-1} \hat{\tau}_j r_j}{(b-1)\tilde{\sigma}^2} \quad \text{tests} \quad \text{H: } \tau_j \textit{ all equal} . \tag{106}$$

It can be shown in similar fashion that with

$$\mathscr{R}(\mu_i, \tau_j \mid \tau_j) = \Sigma_i n_i \bar{y}_{i\cdot\cdot}^2 + \sum_{j=1}^{b-1} \hat{\tau}_j r_j - \Sigma_j n_{\cdot j} \bar{y}_{\cdot j\cdot}^2 .$$

then, in the *no-interaction model*

$$\frac{\mathscr{R}(\mu_i, \tau_j \mid \tau_j)}{(a-1)\tilde{\sigma}^2} \quad \text{tests} \quad \text{H: } \mu_i \textit{ all equal} . \tag{107}$$

These are the useful hypotheses in the no-interaction model. They are part of the summary that is the analysis of variance table of Table 4.11.

TABLE 4.11. ANALYSIS OF VARIANCE OF THE 2-WAY CLASSIFICATION WITHOUT INTERACTION

Source of Variation	d.f.	Sum of Squares	Associated Hypothesis
Partitioning I			
Mean	1	$R(\mu)$	H: $\Sigma_i n_{i.}\mu_i + \Sigma_j n_{.j}\tau_j = 0$
Rows, adjusted for mean	$a-1$	$\mathscr{R}(\mu_i \mid \mu)$	H: $\mu_i + \Sigma_j n_{ij}\tau_j/n_{i.}$ *equal* $\forall\ i$
Columns, adjusted for rows	$b-1$	$\mathscr{R}(\mu_i, \tau_j \mid \mu_i)$	H: τ_j *all equal*
Residual	$N-a-b+1$	$\text{SSE}' = \text{SST} - R(\mu_i, \tau_j)$	
Total	N	SST	
Partitioning II			
Columns, adjusted for mean	$b-1$	$\mathscr{R}(\tau_j \mid \mu)$	H: $\tau_j + \Sigma_i n_{ij}\mu_i/n_{.j}$ *equal* $\forall\ j$
Rows, adjusted for columns	$a-1$	$\mathscr{R}(\mu_i, \tau_j \mid \tau_j)$	H: μ_i *all equal*

Following Table 9.2 is an explanation of why (106) and (107) are F-statistics; and Section 9.3g describes derivation of the hypotheses.

c. Analysis of variance

When using the no-interaction model, the analysis of variance table is that of Table 4.9 with $\mathscr{R}(\mu_{ij} \mid \mu_i, \tau_j)$ and SSE there being added together to give

$$\text{SSE}' = \text{SSE} + \mathscr{R}(\mu_{ij} \mid \mu_i, \tau_j) = \text{SST} - R(\mu_i, \tau_j)\ . \qquad (108)$$

The hypotheses change too. For example, ρ_i' becomes

$$\rho_i' = \Sigma_j n_{ij}(\mu_i + \tau_j)/n_{i.} = \mu_i + \Sigma_j n_{ij}\tau_j/n_{i.} \qquad (109)$$

and so $$\mathscr{R}(\mu_i \mid \mu) \quad \text{tests} \quad \text{H: } \mu_i + \Sigma_j n_{ij}\tau_j/n_i \ \textit{equal}\ \forall\ i\ . \qquad (110)$$

The change for $\mathscr{R}(\mu_i, \tau_j \mid \mu_i)$ is more radical:

$$\mathscr{R}(\mu_i, \tau_j \mid \mu_i) \quad \text{tests} \quad \text{H: } \tau_j\ \textit{equal}\ \forall\ i\ . \qquad (111)$$

With these changes, Tables 4.9 and 4.10 become, for the no-interaction model, Tables 4.11 and 4.12.

Deriving (111) from

$$\text{H: } \gamma_j' = \Sigma_i n_{ij}\rho_j'/n_{.j}\ \forall\ j \qquad (112)$$

TABLE 4.12. ANALYSIS OF VARIANCE (TABLE 4.11) FOR THE EXAMPLE OF TABLE 4.1: NO-INTERACTION MODEL

Source of Variation	d.f.	Sums of Squares	Associated Hypothesis
Partitioning I			
Mean	1	$R(\mu) = 3375$	H: $8\mu_1 + 7\mu_2 + 7\tau_1 + 3\tau_2 + 5\tau_3 = 0$
Rows, adjusted for mean	1	$\mathscr{R}(\mu_i \mid \mu) = 52\frac{1}{2}$	H: $\mu_1 + (3\tau_1 + 2\tau_2 + 2\tau_3)/7 = \mu_2 + (4\tau_1 + \tau_2 + 3\tau_3)/8$
Columns, adjusted for rows	2	$\mathscr{R}(\mu_i, \tau_j \mid \mu_i) = 124\frac{69}{94}$	H: $\tau_1 = \tau_2 = \tau_3$
Residual	11	$\text{SSE}' = 342\frac{36}{47}$	
Total	15	$\text{SST} = 3895$	
Partitioning II			
Columns, adjusted for mean	2	$\mathscr{R}(\tau_j \mid \mu) = 83\frac{1}{3}$	H: $\tau_1 + \dfrac{3\mu_1 + 4\mu_2}{7} = \tau_2 + \dfrac{2\mu_1 + \mu_2}{3} = \tau_3 + \dfrac{2\mu_1 + 3\mu_2}{5}$
Rows, adjusted for columns	1	$\mathscr{R}(\mu_i, \tau_j \mid \tau_j) = 93\frac{127}{141}$	H: $\mu_1 = \mu_2$

of Table 4.9 is a mite devious. Consider just a single statement in (112): with

$$\gamma'_j = \tau_j + \Sigma_i n_{ij}\mu_i/n_{\cdot j}$$

analogous to ρ'_i of (109), the hypothesis statement (112) for $j = 1$ is

$$\tau_j + \Sigma_i n_{i1}\mu_i/n_{\cdot 1} = \Sigma_i n_{i1}(\mu_i + \Sigma_k n_{ik}\tau_k/n_{i\cdot})/n_{\cdot 1}$$

$$= \Sigma_i n_{i1}\mu_i/n_{\cdot 1} + \frac{1}{n_{\cdot 1}}\Sigma_i\Sigma_k \frac{n_{i1}n_{ik}}{n_{i\cdot}}\tau_k,$$

which is also

$$\tau_j - \frac{1}{n_{\cdot 1}}\Sigma_k\left(\Sigma_i \frac{n_{i1}n_{ik}}{n_{i\cdot}}\right)\tau_k = 0\ . \tag{113}$$

In (113), summation over k is over all values of $k = 1, \ldots, b$ and so includes $k = 1$. Hence (113) involves *all* the τ_js, for $j = 1, \ldots, b$. Furthermore, the sum of the coefficients of all the τ_js in (113) is

$$1 - \frac{1}{n_{\cdot 1}}\Sigma_k\Sigma_i \frac{n_{i1}n_{ik}}{n_{i\cdot}} = 1 - \frac{1}{n_{\cdot 1}}\Sigma_i \frac{n_{i1}n_{i\cdot}}{n_{i\cdot}} = 1 - \frac{n_{\cdot 1}}{n_{\cdot 1}} = 0\ . \tag{114}$$

This is a property of (113), which is (112) with $j = 1$. But a similar result is also true for (112) with $j = 2, 3, \ldots, a$. As a result, those hypothesis statements are true when all the τ_js are equal. Thus H of (112) reduces to (111).

Example. With the table of n_{ij}s being

3	2	2	7
4	1	3	8
7	3	5	15

the hypothesis consists of (113) for $j = 1, 2, 3$:

$$\text{H:} \begin{cases} \tau_1 - \frac{1}{7}\left[\left(\frac{9}{7} + \frac{16}{8}\right)\tau_1 + \left(\frac{6}{7} + \frac{4}{8}\right)\tau_2 + \left(\frac{6}{7} + \frac{12}{8}\right)\tau_3\right] = 0 \\ \tau_2 - \frac{1}{3}\left[\left(\frac{6}{7} + \frac{4}{8}\right)\tau_1 + \left(\frac{4}{7} + \frac{1}{8}\right)\tau_2 + \left(\frac{4}{7} + \frac{3}{8}\right)\tau_3\right] = 0 \\ \tau_3 - \frac{1}{5}\left[\left(\frac{6}{7} + \frac{12}{8}\right)\tau_1 + \left(\frac{4}{7} + \frac{3}{8}\right)\tau_2 + \left(\frac{4}{7} + \frac{9}{8}\right)\tau_3\right] = 0. \end{cases}$$

This reduces to
$$\text{H:} \begin{cases} 208\tau_1 - 76\tau_2 - 132\tau_3 = 0 \\ -76\tau_1 + 129\tau_2 - 53\tau_3 = 0 \\ -132\tau_1 - 53\tau_2 + 185\tau_3 = 0. \end{cases}$$

In every statement the coefficients add to zero, as in (114); and it is clear that if that hypothesis is satisfied so is

$$\text{H:}\ \tau_1 = \tau_2 = \tau_3 \tag{115}$$

d. Hypotheses in the analysis of variance table

The most useful feature of Table 4.11 is that for the no-interaction model $E(y_{ijk}) = \mu_i + \tau_j$ it provides a test of the eminently useful hypotheses that the row terms, the μ_is, are equal, and that the column terms, the τ_js, are equal. Thus

$$\frac{\mathscr{R}(\mu_i, \tau_j \mid \tau_j)}{(a-1)\tilde{\sigma}^2} \quad \text{tests} \quad \text{H: } \mu_i \textit{ all equal} \tag{116}$$

$$\frac{\mathscr{R}(\mu_i, \tau_j \mid \mu_i)}{(b-1)\tilde{\sigma}^2} \quad \text{tests} \quad \text{H: } \tau_j \textit{ all equal} \tag{117}$$

for
$$\tilde{\sigma}^2 = \frac{\text{SST} - R(\mu_i, \tau_j)}{N - a - b + 1}.$$

Being able to test these useful hypotheses in the no-interaction model is sharply different from the with-interaction model where, in the analysis of variance table (Table 4.9), no analogous test exists. Of course, with all-cells-filled data, SSA_w and SSB_w provide tests of comparably useful hypotheses, namely

$$\text{H: } \frac{1}{b}\Sigma_j\mu_{ij} \text{ equal } \forall\ i \quad\text{and}\quad \text{H: } \frac{1}{a}\Sigma_i\mu_{ij} \text{ equal } \forall\ j\ .$$

But that is *only* for all-cells-filled data; the great advantage of (116) and (117) is that they apply, in the no-interaction model, whether or not there are empty cells, and even if the filled cells contain only one observation each. The only condition that data must satisfy is that they be connected (see Section 5.3).

4.11. APPENDIX

Starting from the model (63), $E(y_{ijk}) = \mu_i + \tau_j$ for $j = 1,\ldots, b-1$ and $E(y_{ijk}) = \mu_i$ for $j = b$, for $i = 1,\ldots, a$ and $k = 1,\ldots, n_{ij}$, we estimate the μ_i and τ_j by least squares through minimizing S of (64):

$$S = \Sigma_i \sum_{j=1}^{b-1} \Sigma_k (y_{ijk} - \mu_i - \tau_j)^2 + \Sigma_i \sum_{k=1}^{n_{ib}} (y_{ibk} - \mu_i)^2\ .$$

Equating to zero the partial derivations $\partial S/\partial\mu_i$ and $\partial S/\partial\tau_j$ gives equations in the estimators as

$$\sum_{j=1}^{b-1} \Sigma_k (y_{ijk} - \hat{\mu}_i - \hat{\tau}_j) + \sum_{k=1}^{n_{ib}} (y_{ibk} - \hat{\mu}_i) = 0 \quad \text{for } i = 1,\ldots, a$$

and
$$\Sigma_i\Sigma_k (y_{ijk} - \hat{\mu}_i - \hat{\tau}_j) = 0 \quad \text{for } j = 1,\ldots, b-1.$$

These reduce to
$$n_{i\cdot}\hat{\mu}_i + \sum_{j=1}^{b-1} n_{ij}\hat{\tau}_j = y_{i\cdot\cdot} \quad \text{for } i = 1,\ldots, a \tag{118}$$

and
$$\Sigma_i n_{ij}\hat{\mu}_i + n_{\cdot j}\hat{\tau}_j = y_{\cdot j\cdot} \quad \text{for } j = 1,\ldots, b-1\ . \tag{119}$$

Hence
$$\hat{\mu}_i = \bar{y}_{i\cdot\cdot} - \frac{1}{n_{i\cdot}} \sum_{j=1}^{b-1} n_{ij}\hat{\tau}_j \tag{120}$$

and substituting (120) into (119) gives the following equations for the $\hat{\tau}_j$s, for

$j = 1, 2, \ldots, b-1$:

$$n_{\cdot j}\hat{\tau}_j - \Sigma_i n_{ij}\left(\frac{1}{n_{i\cdot}}\sum_{j'=1}^{b-1} n_{ij'}\hat{\tau}_{j'}\right) = y_{\cdot j\cdot} - \sum_{i=1}^{a} n_{ij}\bar{y}_{i\cdot\cdot} \ . \tag{121}$$

These can be rewritten as, for $j = 1, 2, \ldots, b-1$,

$$\left(n_{\cdot j} - \Sigma_i \frac{n_{ij}^2}{n_{i\cdot}}\right)\hat{\tau}_j - \sum_{\substack{j'=1 \\ j' \neq j}}^{b-1}\left(\Sigma_i \frac{n_{ij}n_{ij'}}{n_{i\cdot}}\right)\hat{\tau}_{j'} = y_{\cdot j\cdot} - \Sigma_i n_{ij}\bar{y}_{i\cdot\cdot} \ . \tag{122}$$

In the jth equation of (122) the coefficient of $\hat{\tau}_j$, which shall be denoted c_{jj}, is

$$c_{jj} = n_{\cdot j} - \Sigma_i \frac{n_{ij}^2}{n_{i\cdot}} \tag{123}$$

and the coefficient of $\hat{\tau}_{j'}$ for $j' \neq j$, which shall be denoted $c_{jj'}$, is

$$c_{jj'} = -\Sigma_i \frac{n_{ij}n_{ij'}}{n_{i\cdot}} \quad \text{for } j' = 1, \ldots, b-1 \text{ with } j' \neq j; \tag{124}$$

and the term on the right-hand side of (122) shall be denoted r_j:

$$r_j = y_{\cdot j\cdot} - \Sigma_i n_{ij}\bar{y}_{i\cdot\cdot}, \quad \text{for } j = 1, \ldots, b-1 \ . \tag{125}$$

Although c_{bb}, c_{jb} for $j \neq b$ and r_b are, for purposes of (122), not defined, there is nothing to prevent using (123), (124) and (125) to define them, regardless of (122). It is then easily shown that

$$\sum_{j'=1}^{b} c_{jj'} = 0 \quad \text{for all } j; \quad \text{and} \quad \Sigma_j r_j = 0 \ . \tag{126}$$

After solving (122) numerically, the solutions $\hat{\tau}_j$ can be used in (120) to obtain $\hat{\mu}_i$. The reduction in sum of squares for fitting the no-interaction model, to be denoted $R(\mu_i, \tau_j)$, is

$$\begin{aligned} R(\mu_i, \tau_j) &= R(\text{no-interaction model}) \\ &= \sum_{i=1}^{a}\sum_{j=1}^{b}\sum_{k=1}^{n_{ij}} y_{ijk}^2 - \sum_{i=1}^{a}\sum_{j=1}^{b-1}\sum_{k=1}^{n_{ij}} (y_{ijk} - \hat{\mu}_i - \hat{\tau}_j)^2 \\ &\quad - \sum_{i=1}^{a}\sum_{k=1}^{n_{ib}} (y_{ibk} - \hat{\mu}_i)^2 \end{aligned} \tag{127}$$

which reduces to

$$R(\mu_i, \tau_j)$$

$$= \sum_{i=1}^{a}\sum_{j=1}^{b-1}\sum_{k=1}^{n_{ij}} \left(2\hat{\mu}_i y_{ijk} + 2\hat{\tau}_j y_{ijk} - \hat{\mu}_i^2 - \hat{\tau}_j^2 - 2\hat{\mu}_i\hat{\tau}_j\right) + \sum_{i=1}^{a}\sum_{k=1}^{n_{ib}} \left(2\hat{\mu}_i y_{ibk} - \hat{\mu}_i^2\right)$$

$$= 2\sum_{i=1}^{a} \hat{\mu}_i (y_{i..} - y_{ib.}) + 2\sum_{j=1}^{b-1} \hat{\tau}_j y_{.j.} - \sum_{i=1}^{a} (n_{i.} - n_{ib})\hat{\mu}_i^2$$

$$- \sum_{j=1}^{b-1} n_{.j}\hat{\tau}_j^2 - 2\sum_{i=1}^{a}\sum_{j=1}^{b-1} n_{ij}\hat{\mu}_i\hat{\tau}_j + 2\sum_{i=1}^{a} \hat{\mu}_i y_{ib.} - \sum_{i=1}^{a} n_{ib}\hat{\mu}_i^2$$

$$= 2\sum_{i=1}^{a} \hat{\mu}_i y_{i..} + 2\sum_{j=1}^{b-1} \hat{\tau}_j y_{.j.} - \sum_{i=1}^{a} n_{i.}\hat{\mu}_i^2 - \sum_{j=1}^{b-1} n_{.j}\hat{\tau}_j^2 - 2\sum_{i=1}^{a}\sum_{j=1}^{b-1} n_{ij}\hat{\mu}_i\hat{\tau}_j$$

$$= \sum_{i=1}^{a} \hat{\mu}_i \left(2y_{i..} - n_{i.}\hat{\mu}_i - 2\sum_{j=1}^{b-1} n_{ij}\hat{\tau}_j\right) + \sum_{j=1}^{b-1} \hat{\tau}_j (2y_{.j.} - n_{.j}\hat{\tau}_j)$$

$$= \sum_{i=1}^{a} n_{i.}\hat{\mu}_i^2 + \sum_{j=1}^{b-1} \hat{\tau}_j (2y_{.j.} - n_{.j}\hat{\tau}_j), \quad \text{from (120)}$$

$$= \sum_{i=1}^{a} n_{i.}\left(\bar{y}_{i..} - \frac{1}{n_{i.}}\sum_{j=1}^{b-1} n_{ij}\hat{\tau}_j\right)^2 + \sum_{j=1}^{b-1} \hat{\tau}_j (2y_{.j.} - n_{.j}\hat{\tau}_j)$$

$$= \sum_{i=1}^{a} n_{i.}\bar{y}_{i..}^2 + 2\sum_{j=1}^{b-1} \hat{\tau}_j \left(y_{.j.} - \sum_{i=1}^{a} n_{ij}\bar{y}_{i..}\right) - \sum_{j=1}^{b-1} n_{.j}\hat{\tau}_j^2 + \sum_{i=1}^{a} \frac{1}{n_{i.}}\left(\sum_{j=1}^{b-1} n_{ij}\hat{\tau}_j\right)^2 .$$

On using r_j of (125) this is

$$R(\mu_i, \tau_j) = \sum_{i=1}^{a} n_{i.}\bar{y}_{i..}^2 + 2\sum_{j=1}^{b-1} \hat{\tau}_j r_j$$

$$- \sum_{j=1}^{b-1} \hat{\tau}_j \left[\left(n_{.j} - \sum_{i=1}^{a} \frac{n_{ij}^2}{n_{i.}}\right)\hat{\tau}_j - \sum_{\substack{j'=1 \\ j' \neq j}}^{b-1} \left(\sum_{i=1}^{a} \frac{n_{ij}n_{ij'}}{n_{i.}}\right)\hat{\tau}_{j'}\right]$$

and then on using (122) and also (125) again this is

$$R(\mu_i, \tau_j) = \sum_{i=1}^{a} n_{i\cdot}\bar{y}_{i\cdot\cdot}^2 + 2\sum_{j=1}^{b-1} \hat{\tau}_j r_j - \sum_{j=1}^{b-1} \hat{\tau}_j r_j$$

$$= \sum_{i=1}^{a} n_{i\cdot}\bar{y}_{i\cdot\cdot}^2 + \sum_{j=1}^{b-1} \hat{\tau}_j r_j \ . \tag{128}$$

4.12. SUMMARY

a. Model and estimation

$$E(y_{ijk}) = \mu_{ij};\ i = 1, \ldots, a,\ \ j = 1, \ldots, b \text{ and } k = 1, \ldots, n_{ij} > 0 \ . \tag{9}$$

$$y_{ijk} = \mu_{ij} + e_{ijk}, \quad \text{with the } e_{ijk} \text{ i.i.d. } N(0, \sigma^2) \ . \tag{11}$$

$$\hat{\mu}_{ij} = \bar{y}_{ij\cdot}, \quad \text{with the } \hat{\mu}_{ij} \text{ i.d. } N\left(\mu_{ij}, \sigma^2/n_{ij}\right) \ . \qquad (12), (13)$$

$$\omega = \Sigma_i\Sigma_j k_{ij}\mu_{ij},\ \hat{\omega} = \Sigma_i\Sigma_k k_{ij}\hat{\mu}_{ij} \sim N\left(\ \omega, \Sigma_i\Sigma_j k_{ij}^2\sigma^2/n_i\right) \ . \qquad (15), (16)$$

$$\hat{\sigma}^2 = \Sigma_i\Sigma_j\Sigma_k\left(y_{ijk} - \bar{y}_{ij\cdot}\right)^2/(n_{\cdot\cdot} - ab) \ . \tag{17}$$

b. Means of cell means

$$\rho_i = \Sigma_j\mu_{ij}/b \quad \text{has} \quad \hat{\rho}_i = \tilde{y}_{i\cdot\cdot} = \Sigma_j\bar{y}_{ij\cdot}/b \ . \qquad (27), (28)$$

$$\hat{\rho}_i \sim N\left(\rho, \sigma^2/w_i\right) \quad \text{for} \quad \frac{1}{w_i} = \frac{1}{b^2}\Sigma_j\frac{1}{n_{ij}} \ . \qquad (29), (30)$$

$$\text{SSA}_w = \Sigma_i w_i\left(\tilde{y}_{i\cdot\cdot} - \Sigma_i w_i\tilde{y}_{i\cdot\cdot}/\Sigma w_i\right)^2 \tag{42}$$

$$\text{SSA}_w/(a-1)\hat{\sigma}^2 \quad \text{tests} \quad \text{H: } \rho_i \ \textit{equal} \ \forall \ i \ . \tag{43}$$

Similarly for $\gamma_j = \Sigma_i\mu_{ij}/a$. (44)

c. Interactions

$$\theta_{ij,i'j'} = \mu_{ij} - \mu_{ij'} - \mu_{i'j} + \mu_{i'j'} \ . \tag{55}$$

$\frac{1}{4}ab(a-1)(b-1)$ in an $a \times b$ grid: $(a-1)(b-1)$ are LIN (Sect. 4.8c) .

d. The no-interaction model

$$\left.\begin{array}{ll} E(y_{ijk}) = \mu_i + \tau_j & \text{for } j = 1,\ldots,b-1 \\ E(y_{ijk}) = \mu_i & \text{for } j = b \end{array}\right\} \quad \begin{array}{l}\text{for } i = 1,\ldots,a \\ \text{and } k = 1,\ldots,n_{ij}\end{array} \qquad (63)$$

$$c_{jj}\hat{\tau}_j + \sum_{\substack{j'=1 \\ j'\neq j}}^{b-1} c_{jj'}\hat{\tau}_{j'} = r_j \quad \text{for } j = 1,\ldots,b-1 \qquad (65)$$

$$c_{jj} = n_{.j} - \Sigma_i \frac{n_{ij}^2}{n_{i.}}, \qquad c_{jj'} = -\Sigma_i \frac{n_{ij}n_{ij'}}{n_{i.}} \qquad (66)$$

$$r_j = y_{.j.} - \Sigma_i n_{ij}\bar{y}_{i..} \qquad (67)$$

$$\hat{\mu}_i = \bar{y}_{i..} - \frac{1}{n_{i.}}\sum_{j=1}^{b-1} n_{ij}\hat{\tau}_j \quad \text{for } i = 1,\ldots,a\ . \qquad (68)$$

$$R(\mu_i, \tau_j) = \Sigma_i n_{i.}\bar{y}_{i..}^2 + \sum_{j=1}^{b-1}\hat{\tau}_j r_j\ . \qquad (72)$$

e. Analysis of variance for the with-interaction model

See Tables 4.8 and 4.9.

f. Using the no-interaction model

$$\tilde{\sigma}^2 = \frac{\Sigma_i\Sigma_j\Sigma_k y_{ijk}^2 - R(\mu_i, \tau_j)}{N - a - b + 1}\ . \qquad (82)$$

$$\text{BLUE}(\mu_i - \mu_h) = \hat{\mu}_i - \hat{\mu}_h\ . \qquad (101)$$

$$\text{BLUE}(\tau_j - \tau_k) = \hat{\tau}_j - \hat{\tau}_k\ . \qquad (102)$$

$$\text{BLUE}(\mu_{ij}) = \hat{\mu}_i + \hat{\tau}_j\ . \qquad (103)$$

Analysis of variance: Table 4.11.

4.13. EXERCISES

E4.1.

(a) With $\Sigma\Sigma\Sigma$ representing summation over i, j and k, show that for unbalanced data

$$\Sigma\Sigma\Sigma(y_{ijk} - \bar{y}_{\cdots})^2 - \Big[\Sigma\Sigma\Sigma(\bar{y}_{i\cdot\cdot} - \bar{y}_{\cdots})^2 + \Sigma\Sigma\Sigma(y_{\cdot j\cdot} - \bar{y}_{\cdots})^2$$

$$+\Sigma\Sigma\Sigma(\bar{y}_{ij\cdot} - \bar{y}_{i\cdot\cdot} - \bar{y}_{\cdot j\cdot} + \bar{y}_{\cdots})^2 + \Sigma\Sigma\Sigma(y_{ijk} - \bar{y}_{ij\cdot})^2\Big]$$

$$= -2\Sigma\Sigma\Sigma(\bar{y}_{i\cdot\cdot} - \bar{y}_{\cdots})(\bar{y}_{\cdot j\cdot} - \bar{y}_{\cdots}),$$

and thus (26) does not hold for unbalanced data.

(b) Confirm that each side of the identity in (a) has the value 25 for the example of Table 4.1.

(c) Explain why each side of the identity in (a) is zero for balanced data.

(d) Show also that

$$\Sigma_i\Sigma_j\Sigma_k(\bar{y}_{ij\cdot} - \bar{y}_{i\cdot\cdot} - \bar{y}_{\cdot j\cdot} + \bar{y}_{\cdots})^2$$

$$-\Big[\Sigma_i\Sigma_j n_{ij}\bar{y}_{ij\cdot}^2 - \Sigma_i n_{i\cdot}\bar{y}_{i\cdot\cdot}^2 - \Sigma_j n_{\cdot j}\bar{y}_{\cdot j\cdot}^2 + n_{\cdot\cdot}\bar{y}_{\cdots}^2\Big]$$

$$= +2\Sigma_i\Sigma_j\Sigma_k(\bar{y}_{i\cdot\cdot} - \bar{y}_{\cdots})(\bar{y}_{\cdot j\cdot} - \bar{y}_{\cdots})\ .$$

E4.2. On the basis of normality assumptions, prove that

(a) under H: ρ_i *equal* $\forall\ i$, that $\text{SSA}_w/\sigma^2 \sim \chi^2_{a-1}$.

(b) SSE and SSA_w are independent.

E4.3. Verify for Table 4.1 that $\text{SSB}_w = 192\frac{6}{47}$.

E4.4. Show, for $a = 2$, that $\text{SSA}_w = w_1 w_2(\tilde{y}_{1\cdot\cdot} - \tilde{y}_{2\cdot\cdot})^2/(w_1 + w_2)$.

E4.5. (a) Calculate SSA_w and SSB_w for the following data grid.

108, 240	105, 140, 160
50, 56, 62, 64	16, 18, 19, 21, 26

(b) Also calculate Table 4.9.

E4.6. For the 2×2 case show that (78) is

$$\lambda^2(\Sigma_i\Sigma_j 1/n_{ij})/\hat{\sigma}^2 \quad \text{for } \lambda = \frac{\bar{y}_{11\cdot} - \bar{y}_{12\cdot} - \bar{y}_{21\cdot} + \bar{y}_{22\cdot}}{\dfrac{1}{n_{11}} + \dfrac{1}{n_{12}} + \dfrac{1}{n_{21}} + \dfrac{1}{n_{22}}}.$$

Give an alternative reason for this result.

E4.7. (a) Check the results in E4.6 using one of the following data sets.

P:

4, 8, 12	7, 9, 11
1, 2, 3	1, 1, 2, 4

Q:

4, 8	8, 8, 9, 11
3	16, 18, 6, 12

R:

7, 9	3, 8, 13
14	1, 2, 3

S:

3, 7	9, 5
6, 3, 6	25

T:

10	1, 2, 3
40	6, 8

U:

3, 4, 8	2, 8
9	13, 17

(b) Calculate SSA_w and SSB_w.

(c) Calculate Table 4.9.

E4.8. Calculate SSA_w, SSB_w and Table 4.9 for the following 2×3 grid.

3, 7	9, 5	4, 8, 6
6, 3, 6	8, 5, 8, 7	1, 1

E4.9. Fit the no-interaction model (63) for a 3×2 grid. Define

$$m_i = 1/n_{i1} + 1/n_{i2} \quad \text{and} \quad d_i = \bar{y}_{i1\cdot} - \bar{y}_{i2\cdot}$$

and obtain the estimators as

$$\hat{\mu}_i = \bar{y}_{i\cdot\cdot} - n_{i1}\hat{\tau}_1/n_{i\cdot} \quad \text{for} \quad \hat{\tau}_1 = \frac{\Sigma_i d_i/m_i}{\Sigma_i 1/m_i}.$$

E4.10. Confirm the results of E4.9 for the following data set; and also calculate SSA_w, SSB_w and Tables 4.9 and 4.11.

7, 9	3, 8, 13
14	1, 2, 3
4, 6, 14	12

E4.11. Calculate the F-statistic for the hypothesis of no-interaction using one of the following data sets; also calculate SSA_w and SSB_w.

X:

7, 9	8	6, 6, 12
8, 4	11, 12, 13	8

Y:

X:3, 7	9, 5	4, 8, 6
6, 3, 6	8, 5, 8, 7	1, 1

E4.12. Calculate Tables 4.9 and 4.11 for the following data.

11, 12, 13	6, 8	5, 8, 11	10, 9, 16, 17
3, 11	11, 13	14, 18, 16, 12	24, 38

E4.13. Repeat E4.10 for the following data sets.

A

8, 14	15	9, 9
11	25	21

B

1, 7	8	2, 2
4	18	14

C

1, 2, 3	8	2, 2
6	16	14

D

13, 15	11	17
11	7	3, 7

E

1, 1, 7	9	3, 3
7	17	15

F

1, 1, 3, 7	9	3
8	17	7, 8

G

14	6, 4	12
6	2	3, 9

H

1, 3, 5	18	13
7	6, 12	43

K

2, 6, 7	8, 12	7
8, 12	19	9

E4.14. For the no-interaction 2 × 2 case, write μ_{22} as $\mu_{12} + \mu_{21} - \mu_{11}$. Show that

$$\hat{\mu}_{ij} = \bar{y}_{ij\cdot} - (-1)^{i+j}\lambda/n_{ij}$$

for $i, j = 1, 2$, using λ of E4.6.

E4.15. (a) Verify (100).

(b) Show that $\sum_{j=1}^{b} \hat{\tau}_j' r_j = \sum_{j=1}^{b-1} \hat{\tau}_j r_j$.

CHAPTER 5

THE 2-WAY CLASSIFICATION WITH SOME-CELLS-EMPTY DATA: CELL MEANS MODELS

This chapter relies heavily on Chapter 4. Notation and the model are identical to the all-cells-filled case of Chapter 4, and estimation procedures and formula for certain sums of squares are the same except for taking account of cells that have no data. The big differences concern what can be estimated, what hypotheses can be tested and what hypotheses are tested by the sums of squares developed for all-cells-filled data. The differences that do arise do so solely from the occurrence of empty cells—and, unfortunately, they turn out to be such that many procedures useful for all-cells-filled data are not so useful for some-cells-empty data.

5.1. PRELIMINARIES

a. Model

As in Chapter 4, the model is taken as

$$E(y_{ijk}) = \mu_{ij}, \tag{1}$$

with

$$y_{ijk} = \mu_{ij} + e_{ijk} \tag{2}$$

for

$$e_{ijk} \sim \text{i.i.d. } N(0, \sigma^2) \ . \tag{3}$$

b. Estimation

Least squares estimation leads to the same BLUE of μ_{ij} as previously except for a proviso about empty cells. The sum of squares to be minimized is $S = \Sigma_i \Sigma_j \Sigma_k (y_{ijk} - \mu_{ij})^2$, just as previously, except that for the empty cells there will be no observations and the corresponding terms $(y_{ijk} - \mu_{ij})^2$ will not occur in S. Therefore the resulting $\hat{\mu}_{ij} = \bar{y}_{ij\cdot}$ will arise for only the filled cells; i.e.,

$$\hat{\mu}_{ij} = \bar{y}_{ij\cdot} \; \forall \; i, j \text{ for which } n_{ij} > 0 \; . \tag{4}$$

This is precisely the same as in (12) of Chapter 4, except for the proviso of $n_{ij} > 0$.

Notation. n_{ij} represents the number of observations in cell i, j and summing over those observations is represented by $\Sigma_{k=1}^{n_{ij}}$. For a cell having no data, n_{ij} is formally identified as $n_{ij} = 0$; and $\Sigma_{k=1}^{n_{ij}}$ for that cell is implicitly omitted whenever it would otherwise arise. For example, with $n_{11} = 3$ and $n_{12} = 0$, $\Sigma_{j=1}^{2}\Sigma_{k=1}^{n_{ij}} y_{ijk} = \Sigma_{k=1}^{n_{11}} y_{11k} = y_{11\cdot}$.

Complementary to (4), for cells i, j for which $n_{ij} = 0$ (the empty cells), there is no equation (4); i.e., there is no BLUE of the μ_{ij} for such empty cells. Thus, together with (4) there is the additional and important statement for some-cells-empty data, namely that

$$\text{BLUE}(\mu_{ij}) \text{ does not exist when cell } i, j \text{ is empty} \; . \tag{5}$$

It is to be noted that (5) does not mean that such a BLUE is zero; the fact is that that BLUE does not exist. For example, from data in which $n_{25} = 0$ there is no BLUE of μ_{25}. But for data that do have observations in cell 2, 5 there is a BLUE, $\hat{\mu}_{25} = \bar{y}_{25\cdot}$. The existence or non-existence of the BLUE of μ_{ij} depends entirely on the presence or absence of the corresponding $\bar{y}_{ij\cdot}$ in the data at hand.

c. Linear functions of cell means

The same situation applies to the occurrence of each μ_{ij} in a linear function (similar to that of Section 4.3):

$$\omega = \Sigma_i \Sigma_j k_{ij} \mu_{ij} \; . \tag{6}$$

With all-cells-filled data, each k_{ij} in (6) can be any value whatever; and ω can be estimated as $\hat{\omega} = \Sigma_i \Sigma_j k_{ij} \hat{\mu}_{ij}$. But for some-cells-empty data, whereas ω can always be defined, it can be estimated only when k_{ij} for each empty cell is zero; i.e., on defining

$$\omega' = \Sigma_i \Sigma_j k_{ij} \mu_{ij} \quad \text{with } k_{ij} = 0 \text{ for } n_{ij} = 0 \tag{7}$$

its BLUE is

$$\hat{\omega}' = \Sigma_i\Sigma_j k_{ij}\hat{\mu}_{ij} = \Sigma_i\Sigma_j k_{ij}\bar{y}_{ij\cdot} \quad \text{with } k_{ij} = 0 \text{ for } n_{ij} = 0 \; . \tag{8}$$

When, of course, $n_{ij} = 0$, there will be no $\bar{y}_{ij\cdot}$ value, and $k_{ij} = 0$.

Variance properties of these estimators are precisely as in Chapter 4:

$$v(\hat{\mu}_{ij}) = \sigma^2/n_{ij}, \qquad \text{cov}(\hat{\mu}_{ij}, \hat{\mu}_{hk}) = 0$$

$$v(\hat{\omega}) = \Sigma_i\Sigma_j(k_{ij}^2/n_{ij})\sigma^2,$$

and $$\text{cov}(\Sigma_i\Sigma_j k_{ij}\hat{\mu}_{ij}, \Sigma_i\Sigma_j m_{ij}\hat{\mu}_{ij}) = \Sigma_i\Sigma_j(k_{ij}m_{ij}/n_{ij})\sigma^2 \; .$$

d. Estimating the residual variance

The number of filled cells in a data set having some cells empty is denoted by s:

$$s = \text{number of filled cells} \; . \tag{9}$$

There are then s estimators $\hat{\mu}_{ij} = \bar{y}_{ij\cdot}$ obtainable from the data. Consequently σ^2 is estimated as

$$\hat{\sigma}^2 = \frac{\Sigma_i\Sigma_j\Sigma_k(y_{ijk} - \bar{y}_{ij\cdot})^2}{N - s} \tag{10}$$

just as in (17) of Chapter 4, wherein, for all cells filled, $s = ab$.

e. Confidence intervals

The description of confidence intervals on ω and of testing one-part hypotheses as in Section 4.4 applies precisely to ω of (7) here, taking account of the proviso that $k_{ij} = 0$ for $n_{ij} = 0$.

f. Example 1

Estimators of the cell means are the observed cell means:

$$\hat{\mu}_{11} = 8 \qquad \mu_{12} = 8 \qquad \mu_{13} = 3$$
$$\hat{\mu}_{21} = 6 \qquad \hat{\mu}_{22} = 16 \qquad \mu_{23} \text{ has no BLUE in these data.}$$

As particular cases of (6), both

$$\rho_1 = \tfrac{1}{3}(\mu_{11} + \mu_{12} + \mu_{13}) \quad \text{and} \quad \rho_2 = \tfrac{1}{3}(\mu_{21} + \mu_{22} + \mu_{23}) \tag{11}$$

TABLE 5.1. DATA IN A 2 × 3 GRID WITH ONE EMPTY CELL

7, 9	8	2, 4	
16(2)8	8(1)8	6(2)3	30(5)6
5, 7	14, 15, 19	–	
12(2)6	48(3)16		60(5)12
28(4)7	56(4)14	6(2)3	90(10)9

can be defined but, of the two, only ρ_1 has a BLUE from Table 5.1:

$$\hat{\rho}_1 = \tfrac{1}{3}(\bar{y}_{11\cdot} + \bar{y}_{12\cdot} + \bar{y}_{13\cdot}) = \tfrac{1}{3}(8 + 8 + 3) = 6\tfrac{1}{3}\ .$$

The BLUE of ρ_2 does not exist because μ_{23} has no BLUE available from Table 5.1—since cell 2, 3 is empty. A possible alternative to ρ_2 is

$$\rho_2' = \tfrac{1}{2}(\mu_{21} + \mu_{22}), \tag{12}$$

which has a BLUE because it involves just filled cells:

$$\hat{\rho}_2'' = \tfrac{1}{2}(\bar{y}_{21\cdot} + \bar{y}_{22\cdot}) = \tfrac{1}{2}(6 + 16) = 11\ .$$

From (10) with $N = 10$ and $s = 5$

$$\hat{\sigma}^2 = \frac{[(-1)^2 + 1^2] + [(-1)^2 + 1^2] + [(-1)^2 + 1^2] + [(-2)^2 + (-1)^2 + 3^2]}{10 - 5}$$

$$= \frac{20}{5} = 4. \tag{13}$$

g. Analysis of variance

The standard analysis of variance table of Table 4.2 for all-cells-filled data is almost exactly the same for some-cells-empty data, the only differences being the calculation of degrees of freedom for SSR_m and SSE (s is used in place of ab) and in the associated hypothesis for SSR_m (involving μ_{ij}s for only the filled cells). These differences are to be seen in comparing Tables 4.2 and 5.2.

Example (continued). The reader should confirm that for the data of Table 5.1, the analysis of variance table is as shown in Table 5.3.

TABLE 5.2. ANALYSIS OF VARIANCE OF SOME-CELLS-EMPTY DATA FROM A 2-WAY CROSSED CLASSIFICATION

Source of Variation	d.f.	Sum of Squares	Associated Hypothesis
Mean	1	$\text{SSM} = N\bar{y}^2_{...}$	$H: \Sigma_i\Sigma_j n_{ij}\mu_{ij} = 0$
Model	$s-1$	$\text{SSR}_m = \Sigma_i\Sigma_j n_{ij}(\bar{y}_{ij\cdot} - \bar{y}_{...})^2$	H: μ_{ij} *all equal*, $\forall\ n_{ij} \neq 0$
Residual	$N-s$	$\text{SSE} = \Sigma_i\Sigma_j\Sigma_k(y_{ijk} - y_{ij\cdot})^2$	
Total	N	$\text{SST} = \Sigma_i\Sigma_j\Sigma_k y^2_{ijk}$	

TABLE 5.3. ANALYSIS OF VARIANCE (TABLE 5.2) OF THE DATA IN TABLE 5.1, USING $E(y_{ijk}) = \mu_{ij}$

Source of Variation	d.f.	Sum of Squares	Associated Hypothesis
Mean	1	810	$H: 2\mu_{11} + \mu_{12} + 2\mu_{13} + 2\mu_{21} + 3\mu_{22} = 0$
Model	4	240	$H: \mu_{11} = \mu_{12} = \mu_{13} = \mu_{21} = \mu_{22}$
Residual	5	20	
Total	10	1070	

Just like the associated hypothesis for SSR_m in Tables 4.2 and 4.3, that in Tables 5.2 and 5.3 is of little interest, for the same reasons as are given following Table 4.3. We therefore consider alternative methods of analysis as is done in Sections 4.6–4.9. The rationale for doing this, as described in Section 4.5b, applies even more strongly here for some-cells-empty data than it does there for all-cells-filled data.

5.2. ESTIMABILITY: AN INTRODUCTION

Most of Chapter 4 deals with situations where each cell of a data grid contains data. In considering ω and $\hat{\omega}$ of (7) and (8) and examples in (11) and (12), it has been emphasized that, in the with-interaction model, any linear function of μ_{ij}s has a BLUE only if all the μ_{ij}s in that function are for filled cells, For example, ρ_1 of (11) has a BLUE, but ρ_2 does not; and ρ_2'' of (12) does. We say that ρ_1 and ρ_2'' *can be estimated*, or that they are each an *estimable function*. By this we mean that those functions have BLUEs from the available data—in this case the data of Table 5.1. In contrast, ρ_2 has no BLUE from these data and we say that ρ_2 is *not estimable* or is a *non-estimable function*.

The description of being estimable (or non-estimable) applies in this book only to linear functions of parameters of a model. It is usually presented in terms of a formal mathematical definition, as given in Section 8.7, but basically it is equivalent to the existence of a BLUE. Whether a function is estimable or not is a consequence of both the data available and the model being used. As a concept, estimability does not change the definitions of linear functions of μ_{ij}s that we might be interested in, but it identifies those functions that have a BLUE and those that do not.

a. Row means of cell means

With all-cells-filled data we considered three different row means of cell means:

$$\rho_i = \frac{1}{b}\Sigma_j\mu_{ij}, \qquad \rho_i' = \frac{1}{n_{i\cdot}}\Sigma_i n_{ij}\mu_{ij}, \quad \text{and} \quad \rho_i'' = \Sigma_j t_{ij}\mu_{ij} \tag{14}$$

of (27), (45) and (49), respectively, in Chapter 4. As definitions, those functions are still valid for some-cells-empty data but, as already described with such data, some of those functions are estimable and others are not.

Example 1 (continued). The occurrence of data in Table 5.1 is summarized in Grid 5.1, where the presence of a check mark indicates data and the absence indicates no data:

Grid 5.1

√	√	√
√	√	

Then both

$$\rho_1 = \tfrac{1}{3}(\mu_{11} + \mu_{12} + \mu_{13}) \quad \text{and} \quad \rho_2 = \tfrac{1}{3}(\mu_{21} + \mu_{22} + \mu_{23}) \tag{15}$$

can be defined, as in (11); ρ_1 involves no empty cells and so can be estimated, whereas ρ_2 involves μ_{23} of the empty cell, and so cannot be estimated. Thus, since ρ_i is the least squares mean (Section 4.6g) for row i, Grid 5.1 is an example where a least squares mean for one row, ρ_1, is estimable whereas that for another row, ρ_2, is not. Neither is the least squares mean for column 3 estimable, namely $\gamma_3 = (\mu_{13} + \mu_{23})/2$.

Suppose we want to compare rows in some manner: this cannot be done using ρ_1 and ρ_2 since ρ_2 cannot be estimated. Faced with this we go back to the underlying model $E(y_{ijk}) = \mu_{ij}$ and ask "What can be estimated?" The immediate answer is "μ_{ij} for every filled cell, and every linear combination of μ_{ij}s of the filled cells." This being so, the next question is "Which linear combinations of μ_{ij}s of filled cells might be of interest for comparing rows?" Scanning Grid 5.1 it is clear that

$$\rho_1'' = \tfrac{1}{2}(\mu_{11} + \mu_{12}) \quad \text{and} \quad \rho_2'' = \tfrac{1}{2}(\mu_{21} + \mu_{22}) \tag{16}$$

are both estimable and that the BLUE of $\rho_1'' - \rho_2''$ will provide information about the difference between rows 1 and 2, albeit over only columns 1 and 2. Thus in (15) we have ρ_1 that might be used to define an overall row mean, and in (16) a different definition, ρ_1'', is used for comparing rows 1 and 2. The need for this different definition is forced on us by the paucity of the data, because it is obvious that rows 1 and 2 cannot be compared in column 3 using data from Grid 5.1. We therefore compare rows 1 and 2 using the BLUEs of (16) in the form $\hat{\rho}_1'' - \hat{\rho}_2'' = \frac{1}{2}(\bar{y}_{11\cdot} + \bar{y}_{12\cdot}) - \frac{1}{2}(\bar{y}_{21\cdot} + \bar{y}_{2\cdot})$; and this uses only part of the total available data. This using of subsets of data is dealt with at length in Section 5.6.

b. Interactions

With a rows and b columns there are $\frac{1}{4}ab(a-1)(b-1)$ interactions of which a maximum of only $(a-1)(b-1)$ are LIN; see section 4.8c. This is true for both all-cells-filled data and for some-cells-empty data. But whereas with all-cells-filled data, any set of $(a-1)(b-1)$ LIN interactions $\theta_{ij,i'j'}$ can be estimated (from which all interactions can then be estimated), with some-cells-empty data fewer than $(a-1)(b-1)$ LIN interactions can be estimated. In the example of Grid 5.1, only one interaction is estimable, namely

$$\begin{aligned} &\theta_{11,22} = \mu_{11} - \mu_{12} - \mu_{21} + \mu_{22} \\ &\qquad\qquad \text{with} \quad \hat{\theta}_{11,22} = \bar{y}_{11\cdot} - \bar{y}_{12\cdot} - \bar{y}_{21\cdot} + \bar{y}_{22\cdot}\ . \end{aligned} \tag{17}$$

In contrast, although the other two interactions of the grid,

$$\begin{aligned} &\theta_{11,23} = \mu_{11} - \mu_{13} - \mu_{21} + \mu_{23} \\ &\qquad\qquad \text{and} \quad \theta_{12,23} = \mu_{12} - \mu_{13} - \mu_{22} + \mu_{23}, \end{aligned} \tag{18}$$

can be defined, neither of them is estimable because no BLUE of μ_{23} is available from data of Grid 5.1. In this simple 2×3 case there are only $\frac{1}{4}2(3)1(2) = 3$ interactions and no more than $1(2) = 2$ can constitute a LIN set. Hence the three interactions in (17) and (18) comprise all possible interactions for Grid 5.1; and any two are LIN. Therefore there is a relationship between the three, for example,

$$\theta_{11,22} = \theta_{11,23} - \theta_{12,23} \,. \tag{19}$$

Hence, since $\theta_{11,22}$ is estimable (the only estimable interaction for Grid 5.1) its BLUE can also be considered as the BLUE of the difference between the other two interactions as in (19); but neither of those two is estimable.

5.3. CONNECTED DATA

a. Basic ideas

If the occurrence of filled cells in some-cells-empty data is sufficiently sparse, it can happen that with the no-interaction model, $E(y_{ijk}) = \mu_i + \tau_j$, some differences of the form $\mu_i - \mu_{i'}$ for $i \neq i'$ and $\tau_j - \tau_{j'}$ for $j \neq j'$ cannot be estimated. When every such difference can be estimated, the mean $\mu_{ij} = \mu_i + \tau_j$ can be estimated for every cell i, j whether the cell is filled or empty; i.e., every cell mean can be estimated.

Examples.

In Grid 5.2 the mean of the empty $1, 3$ cell has a BLUE in the no-interaction model because

$$\mu_{13} = \mu_1 + \tau_3 = \mu_1 + \tau_1 + \mu_2 + \tau_3 - (\mu_2 + \tau_1) = \mu_{11} + \mu_{23} - \mu_{21}$$

and μ_{11}, μ_{23} and μ_{21} are means of filled cells and hence have BLUEs. Hence, even though cell $1, 3$ is empty (and so in the with-interaction model

Grid 5.2

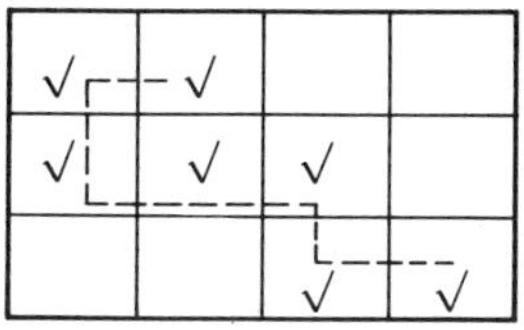

Grid 5.3

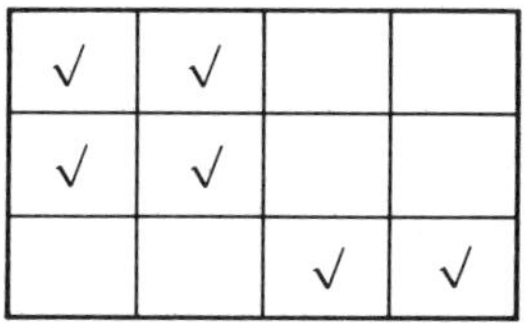

μ_{13} has no BLUE), in the no-interaction model it does have a BLUE:

$$\hat{\mu}_{13} = \hat{\mu}_{11} + \hat{\mu}_{23} - \hat{\mu}_{21} \ .$$

But such an argument fails for Grid 5.3 because in that case it is impossible to express μ_{13} as a linear combination of means of filled cells. A characteristic of the pattern of filled cells in Grid 5.2 which is known as connectedness ensures that with the no-interaction model every cell mean in Grid 5.2 is estimable. Such data are said to be connected. In contrast, the data of Grid 5.3 are not connected; they are said to be disconnected, and not all cell means are estimable.

Data from which all differences $\mu_i - \mu_{i'}$ and $\tau_j - \tau_{j'}$ (for $i \neq i'$ and $j \neq j'$) are estimable provide the opportunity to estimate any contrast among μs and among τs that interest us. Such data are called *connected*. More generally, connected data are data from which, for every main effects factor, all differences between levels of a factor are estimable. Although this gives a name to data that provide this appealing estimability feature, it tells us nothing about the manner in which such data can have empty cells and still be connected. Weeks and Williams (1964) have a characterization that is applicable no matter how many factors there are (see subsection **e** which follows), but its algebraic nature has minimal appeal intuitively. Nevertheless, for the 2-factor case it corresponds to what shall be called *geometric connectedness* (*g-connectedness*), which provides a very straightforward method for ascertaining if data from a 2-way classification are connected, i.e., if every difference $\mu_i - \mu_{i'}$ and $\tau_j - \tau_{j'}$ is estimable. We therefore give the algorithm for data to be *g*-connected in the 2-way classification and discuss its relationship to useful estimability properties. The algorithm is an outcome, for the 2-way crossed classification, of the more general definition of connectedness given by Weeks and Williams (1964).

b. A geometric algorithm

Definition. Data of a 2-way crossed classification are said to be *g-connected* when the filled cells of the grid can be joined by a continuous line, consisting solely of horizontal and vertical segments, that has changes of direction only in filled cells.

Corollary. All-cells-filled data are *g*-connected.

Examples. The requisite kind of line is easily drawn in Grid 5.2 but cannot be drawn in Grid 5.3. Thus data of Grid 5.2 are *g*-connected and those of Grid 5.3 are disconnected. Additional examples are Grids 5.2a and 5.3a:

Grid 5.2a

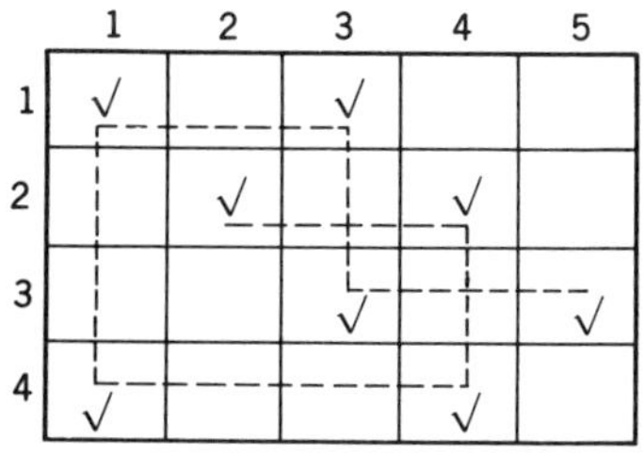

Grid 5.3a

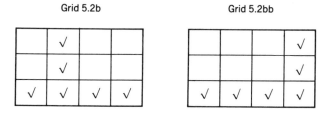

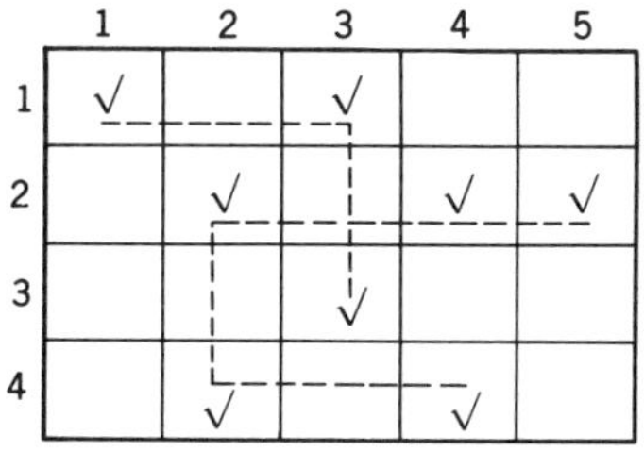

Grid 5.2a data are *g*-connected, but Grid 5.3a data are disconnected; they consist of two sets of *g*-connected data, each of which has its own requisite line, but the lines cannot be made into a single line. They are disconnected.

Re-sequencing rows and/or columns is sometimes required in order to display *g*-connectedness that, by the algorithm, might not be apparent. For example, no single requisite line can be drawn in Grid 5.2b, but it can in Grid 5.2bb, where columns 2 and 4 of Grid 5.2b have been interchanged:

Grid 5.2b

	√		
	√		
√	√	√	√

Grid 5.2bb

			√
			√
√	√	√	√

c. Separation of disconnected data

A major consequence of data not being connected (i.e., being disconnected) is that they separate into two or more disconnected sets of connected data that have to be analyzed separately.

Example 2. Suppose the numbers of observations for data of Grid 5.3 were as shown in Table 5.4. Then the least squares equations (118) and (119) of Section 4.11 for fitting the no-interaction model (63) of Section

TABLE 5.4. NUMBERS OF OBSERVATIONS FOR DATA OF GRID 5.3

	$j = 1$	$j = 2$	$j = 3$	$j = 4$	Total
$i = 1$	2	3	—	—	5
$i = 2$	6	1	—	—	7
$i = 3$	—	—	10	11	21
Total	8	4	10	11	33

4.9a are as follows:

$$\begin{array}{llllllll}
5\hat{\mu}_1 & & & +2\hat{\tau}_1 & +3\hat{\tau}_2 & & = y_{1..} & \text{(i)} \\
& 7\hat{\mu}_2 & & +6\hat{\tau}_1 & +\ \hat{\tau}_2 & & = y_{2..} & \text{(ii)} \\
& & 21\hat{\mu}_3 & & & +10\hat{\mu}_3 & = y_{3..} & \text{(iii)} \\
2\hat{\mu}_1 & +6\hat{\mu}_2 & & +8\hat{\tau}_1 & & & = y_{.1.} & \text{(iv)} \\
3\hat{\mu}_1 & +\ \hat{\mu}_2 & & & +4\hat{\tau}_2 & & = y_{.2.} & \text{(v)} \\
& & 10\hat{\mu}_3 & & & +10\hat{\tau}_3 & = y_{.3.} & \text{(vi)}
\end{array} \qquad (20)$$

It is clear that equations (iii) and (vi) in (20) are the only equations that involve $\hat{\mu}_3$ and $\hat{\tau}_3$; thus the complete set of equations separates into two sets that can be solved quite independently of one another. And these two sets of equations correspond to the two disconnected sets of connected data in Grid 5.3: one is the four filled cells in rows 1 and 2, and the other is the two filled cells in row 3. The consequence of the disconnectedness is that these two data sets have to be analyzed separately.

Further evidence of this separation comes from applying equations (65) of Chapter 4 directly to data having the n_{ij}-values of Table 5.4. We find that those equations are

$$\left[8 - \left(\frac{2^2}{5} + \frac{6^2}{7}\right)\right]\hat{\tau}_1 - \left[\frac{2(3)}{5} + \frac{6(1)}{7}\right]\hat{\tau}_2 = y_{.1.} - (2\bar{y}_{1..} + 6\bar{y}_{2..})$$

$$-\left[\frac{2(3)}{5} + \frac{6(1)}{7}\right]\hat{\tau}_1 + \left[4 - \left(\frac{3^2}{5} + \frac{1^2}{7}\right)\right]\hat{\tau}_2 = y_{.2.} - (3\bar{y}_{1..} + \bar{y}_{2..})$$

$$\left(10 - \frac{10^2}{21}\right)\hat{\tau}_3 = y_{.3.} - 10\bar{y}_{3..}$$

which reduce to

$$72(\hat{\tau}_1 - \hat{\tau}_2)/35 = y_{\cdot 1 \cdot} - 2\bar{y}_{1 \cdot\cdot} - 6\bar{y}_{2 \cdot\cdot}$$

$$-72(\hat{\tau}_1 - \hat{\tau}_2)/35 = y_{\cdot 2 \cdot} - 3\bar{y}_{1 \cdot\cdot} - \bar{y}_{2 \cdot\cdot}$$

$$110\hat{\tau}_3/121 = y_{\cdot 3 \cdot} - 10\bar{y}_{3 \cdot\cdot} \ . \tag{21}$$

Since, by the location of the filled cells in Table 5.4

$$y_{11\cdot} + y_{12\cdot} + y_{21\cdot} + y_{22\cdot} = y_{\cdot 1 \cdot} + y_{\cdot 2 \cdot} = 5\bar{y}_{1 \cdot\cdot} + 7\bar{y}_{2 \cdot\cdot},$$

we see that adding the first two equations in (21) gives $0 = 0$. Thus (21) provides no unique solution for $\hat{\tau}_1$ and $\hat{\tau}_2$; i.e., equations (65) of Chapter 4 do not operate for the disconnected data of Table 5.4.

These results are true in general for disconnected data: they separate into two or more disconnected sets of connected data, to which correspond sets of the least squares equations that can be dealt with independently of one another. At page 157, in Section 5.5c we discuss how such data can be analyzed on a within-set basis, although when the sets themselves have empty cells, little is to be gained by so doing. We therefore confine attention to connected data, and leave it to the reader to ensure that, for whatever data are being analyzed, one is able to draw the continuous line required by the definition of connected data for a 2-way crossed classification so as to be satisfied that one's data are indeed connected.

d. Estimability of contrasts, and of all cell means

Weeks and Williams (1964) give a proof that their characterization of connectedness implies estimability of all contrasts among row terms μ_i and among column terms τ_j. (The proof is not brief.) This means in the 2-way classification with the no-interaction model that data being g-connected is an assurance that all differences $\mu_i - \mu_{i'}$ for $i \neq i'$ and $\tau_j - \tau_{j'}$ for $j \neq j'$ are estimable. Hence in the no-interaction model all contrasts among μ_is and among τ_js are estimable. A corollary to this is that every cell mean is estimable, be the cell filled or empty.

For filled cells, estimability is no problem. But suppose cell i, j is empty. Then, because the data are connected, there is either some cell i, j' in the same row and/or some cell i', j in the same column that is filled. Then, since

$$\mu_{ij} = \mu_i + \tau_j = \mu_i + \tau_{j'} + \tau_j - \tau_{j'} = \mu_{ij'} + \tau_j - \tau_{j'},$$

and because both $\mu_{ij'}$ and $\tau_j - \tau_{j'}$ are estimable, we have μ_{ij} as being estimable.

Thus we have the important result for the 2-way classification that so long as data are g-connected, all cell means in the no-interaction model are estimable, with BLUEs given by (65)–(68) of Section 4.9b. In contrast, in the with-interaction model, cell means of only the filled cells are estimable, the BLUEs being $\hat{\mu}_{ij} = \bar{y}_{ij\cdot}$ for $n_{ij} \neq 0$, as in (4).

e. Connectedness for several factors

The algorithm given earlier for data to be g-connected is an outcome of the Weeks and Williams (1964) formulation of connectedness that can be described as follows. For k main-effect factors, let the array of k subscripts on a sub-most cell mean be $[i_1 \; i_2 \; \cdots \; i_k]$. Then two such i-arrays are said to be *nearly identical* if they differ in only a single element; e.g., [1 3 2 5] and [1 3 4 5] are nearly identical. Connected data are then defined as data for which the i-array of every filled sub-most cell is nearly identical to that of at least one other filled sub-most cell.

For data specified as connected by the preceding definition, the estimability of intra-factor differences between main effects in main-effects-only models is assured. But unfortunately that definition of connectedness is only a sufficient condition for the desired estimability (of all intra-factor differences) and it is not a necessary condition. And of course, in keeping with that definition, the extension of g-connectedness to more than two dimensions is also only a sufficient and not a necessary condition for intra-factor differences to be estimable. This means that for some patterns of filled and empty cells, the data can be not g-connected but will be connected in the sense that every intra-factor difference of main effects is estimable.

Example. Consider the 3-way classification main-effects-only model

$$E(y_{ijk}) = \mu + \alpha_i + \beta_j + \gamma_k$$

for $i, j, k = 1, 2$. Suppose the data are y_{112}, y_{211}, y_{121} and y_{222}. These data do not satisfy the Weeks–Williams definition for being connected nor are they g-connected, as may be seen from Figure 5.1, where the axes represent the levels of the three factors and the four data points are shown as dots. The three-dimensional analog of the two-dimensional algorithm for g-connectedness would be that the data would be g-connected if they could be joined by the line that consisted solely of segments parallel to the three axes in Figure 5.1, and in which changes of direction occur only at data points. Such a line is clearly not feasible in Figure 5.1. Thus the data are not g-connected; and yet intra-factor main effect differences are estimable; e.g.,

$$E\,\tfrac{1}{2}(y_{112} - y_{211} + y_{121} - y_{222}) = \alpha_1 - \alpha_2,$$

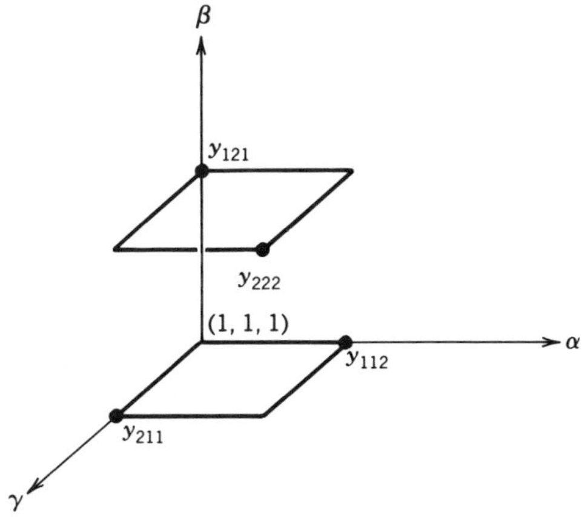

Figure 5.1. Four observations in a 3-factor main-effects-only model.

and thus $\alpha_1 - \alpha_2$ is estimable; so is $\beta_1 - \beta_2$ and $\gamma_1 - \gamma_2$.

5.4. THE NO-INTERACTION MODEL

The no-interaction model to be used for some-cells-empty data is exactly the same as that used with all-cells-filled data of (63) in Section 4.9b:

$$\left.\begin{aligned} E(y_{ijk}) &= \mu_i + \tau_j \quad \text{for } j = 1, \dots, b-1 \\ E(y_{ijk}) &= \mu_i \qquad\quad \text{for } j = b \end{aligned}\right\} \quad \begin{aligned} &\text{for } i = 1, \dots, a \\ &\text{and } k = 1, \dots, n_{ij}\,. \end{aligned} \tag{22}$$

As usual, $n_{ij} = 0$ indicates that cell i, j is empty.

a. Fitting the model

All-cells-filled data are always connected; hence the absence of any discussion of connectedness in Chapter 4. But some-cells-empty data must be connected in order to fit the no-interaction model of (22) to an entire data set. Other than this requirement there is nothing in the derivation of the least squares equations used in Chapter 4 that is invalidated by having empty cells in the data grid. Using $n_{ij} = 0$ for those cells does not affect the equations deleteriously in any way. We therefore use them as they stand, taken from (65)–(68) of Section 4.9b.

$$c_{jj}\hat{\tau}_j + \sum_{\substack{j'=1\\ j'\neq j}}^{b-1} c_{jj'}\hat{\tau}_{j'} = r_j \quad \text{for } j = 1, \dots, b-1 \tag{23}$$

with
$$c_{jj} = n_{.j} - \Sigma_i \frac{n_{ij}^2}{n_{i.}} \quad \text{and} \quad c_{jj'} = -\Sigma_i \frac{n_{ij}n_{ij'}}{n_{i.}}, \tag{24}$$

and
$$r_j = y_{.j.} - \Sigma_i n_{ij}\bar{y}_{i..}; \tag{25}$$

and
$$\hat{\mu}_i = \bar{y}_{i..} - \frac{1}{n_{i.}} \sum_{j=1}^{b-1} n_{ij}\hat{\tau}_j \quad \text{for } i = 1, \ldots, a\,. \tag{26}$$

Then, as in (103) of Section 4.10a, the estimator of μ_{ij} is

$$\hat{\mu}_{ij} = \hat{\mu}_i + \hat{\tau}_j, \tag{27}$$

for $i = 1, \ldots, a$ and for $j = 1, \ldots, b$ with $\hat{\tau}_b = 0$.

Note an important distinction here from the with-interaction model for which $\hat{\mu}_{ij} = \bar{y}_{ij.}$ only for $n_{ij} > 0$. That model provides no $\hat{\mu}_{ij}$ for empty cells. In contrast, (27) provides $\hat{\mu}_{ij}$ for *all* cells, including the empty ones (provided the data are connected). Also, of course,

$$\text{BLUE}(\mu_i - \mu_h) = \hat{\mu}_i - \hat{\mu}_h$$

and

$$\text{BLUE}(\tau_j - \tau_k) = \hat{\tau}_j - \hat{\tau}_k\,.$$

Also, so is *every* row mean, $\rho_i = \Sigma_j \mu_{ij}/b$ (least squares, or population marginal, mean), estimable

$$\hat{\rho}_i = \Sigma_j \hat{\mu}_{ij}/b = \hat{\mu}_i + \Sigma_j \hat{\tau}_j/b\,.$$

This is, of course, for connected data. And a similar result applies for $\hat{\gamma}_j$. Recall that ρ_i is sometimes called a least squares mean or population marginal mean, for row i. It is to be noted that with some-cells-empty data every ρ_i is estimable with the no-interaction model if the data are connected, but with the with-interaction model not every ρ_i is estimable. The same is true of γ_j.

Example 1 (continued). Table 5.5 is the table of totals, numbers of observations and means extracted from Table 5.1.

TABLE 5.5. TOTALS, NUMBERS AND MEANS FROM TABLE 5.1

16(2)8	8(1) 8	6(2)3	30(5) 6
12(2)6	48(3)16	—	60(5)12
28(4)7	56(4)14	6(2)3	90(10)9

From (24):

$$c_{11} = 4 - \left(\frac{2^2}{5} + \frac{2^2}{5}\right) = 2\tfrac{2}{5} \qquad c_{12} = -\left[\frac{2(1)}{5} + \frac{2(3)}{5}\right] = -1\tfrac{3}{5}$$

$$c_{22} = 4 - \left(\frac{1^2}{5} + \frac{3^2}{5}\right) = 2$$

and (25) gives

$$r_1 = 28 - [2(6) + 2(12)] = -8 \qquad r_2 = 56 - [1(6) + 3(12)] = 14\ . \tag{28}$$

Hence (23) is

$$2\tfrac{2}{5}\hat{\tau}_1 - 1\tfrac{3}{5}\hat{\tau}_2 = -8 \quad \text{and} \quad -1\tfrac{3}{5}\hat{\tau}_1 + 2\hat{\tau}_2 = 14$$

with solution $$\hat{\tau}_1 = 2\tfrac{6}{7} \quad \text{and} \quad \hat{\tau}_2 = 9\tfrac{2}{7}\ . \tag{29}$$

Hence (26) is $$\hat{\mu}_1 = 6 - \tfrac{1}{5}\left[2\left(2\tfrac{6}{7}\right) + 1\left(9\tfrac{2}{7}\right)\right] = 3 \tag{30}$$

and $$\hat{\mu}_2 = 12 - \tfrac{1}{5}\left[2\left(2\tfrac{6}{7}\right) + 3\left(9\tfrac{2}{7}\right)\right] = 5\tfrac{2}{7}.$$

Then, using $\hat{\mu}_{ij} = \hat{\mu}_i + \hat{\tau}_j$ of (27), the estimated cell means are

$$\begin{aligned} &\hat{\mu}_{11} = 3 + 2\tfrac{6}{7} = 5\tfrac{6}{7} \qquad \hat{\mu}_{12} = 3 + 9\tfrac{2}{7} = 12\tfrac{2}{7} \qquad \hat{\mu}_{13} = 3 + 0 = 3 \\ &\hat{\mu}_{21} = 5\tfrac{2}{7} + 2\tfrac{6}{7} = 8\tfrac{1}{7} \qquad \hat{\mu}_{22} = 5\tfrac{2}{7} + 9\tfrac{2}{7} + 14\tfrac{4}{7} \qquad \hat{\mu}_{23} = 5\tfrac{2}{7} + 0 = 5\tfrac{2}{7}. \end{aligned} \tag{31}$$

b. The reduction in sum of squares

The reduction in sum of squares for fitting the no-interaction model is, from (72) of Chapter 4,

$$R(\mu_i, \tau_j) = \Sigma_i n_{i\cdot}\bar{y}_{i\cdot\cdot}^2 + \sum_{j=1}^{b-1} \hat{\tau}_j r_j \ . \tag{32}$$

This is also given, equivalently, by

$$R(\mu_i, \tau_j) = \Sigma_i\Sigma_j\Sigma_k y_{ijk}^2 - \Sigma_i\Sigma_j\Sigma_k\left(y_{ijk} - \hat{\mu}_{ij}\right)^2 \ . \tag{33}$$

Example 1 (continued). With $n_{1\cdot} = 5$ and $\bar{y}_{1\cdot\cdot} = 6$ and $n_{2\cdot} = 5$ and $\bar{y}_{2\cdot\cdot} = 12$ from Table 5.5, the first term of (32) is

$$\Sigma_i n_{i\cdot}\bar{y}_{i\cdot\cdot}^2 = 5(6^2) + 5(12^2) = 900; \tag{34}$$

and with $\hat{\tau}_j$ and r_j taken from (28) and (29), the second term of (32) is

$$\sum_{j=1}^{b-1} \hat{\tau}_j r_j = 2\tfrac{6}{7}(-8) + 9\tfrac{2}{7}(14) = 107\tfrac{1}{7} \ . \tag{35}$$

Substituting (34) and (35) into (32) gives

$$R(\mu_i, \tau_j) = 900 + 107\tfrac{1}{7} = 1007\tfrac{1}{7} \ . \tag{36}$$

Similarly (33) is calculated, using $\Sigma_i\Sigma_j\Sigma_k y_{ijk}^2 = 1070$ from Table 5.3 together with the individual y_{ijk}-values from Table 5.1 and the $\hat{\mu}_{ij}$-values of (31), as

$$\begin{aligned} R(\mu_i, \tau_j) = 1070 - \Big[&\left(7 - 5\tfrac{6}{7}\right)^2 + \left(9 - 5\tfrac{6}{7}\right)^2 + \left(8 - 12\tfrac{2}{7}\right)^2 \\ &+(2-3)^2 + (4-3)^2 + \left(5 - 8\tfrac{1}{7}\right)^2 + \left(7 - 8\tfrac{1}{7}\right)^2 \\ &+\left(14 - 14\tfrac{4}{7}\right)^2 + \left(15 - 14\tfrac{4}{7}\right)^2 + \left(19 - 14\tfrac{4}{7}\right)\Big] \\ = 1070 - 62\tfrac{6}{7} &= 1007\tfrac{1}{7} \quad \text{of (36)} \ . \end{aligned}$$

c. Testing the hypothesis of no-interaction

The numerator sum of squares for testing the hypothesis of no interaction is, as discussed at equations (59) and (60) of Section 4.9,

$$\begin{aligned} \mathscr{R}\left(\mu_{ij} \mid \mu_i, \tau_j\right) &= R(\text{with-interaction model}) - R(\text{no-interaction model}) \\ &= \Sigma_i\Sigma_j n_{ij}\bar{y}_{ij\cdot}^2 - R(\mu_i, \tau_j) \end{aligned} \tag{37}$$

for $R(\mu_i, \tau_j)$ of (33), the same as (72) of Section 4.9c. To use this as the numerator of an F-statistic, the same properties concerning χ^2 and independence of $\hat{\sigma}^2$ discussed following (76) in Section 4.9d have to be established; and can be, by the same methods as indicated there and detailed in Section 8.6.

An important feature of $\mathcal{R}(\mu_{ij}|\mu_i, \tau_j)$ is that whereas degrees of freedom for all-cells-filled data are $ab - (a + b - 1) = (a - 1)(b - 1)$, for some-cells-empty data they are

$$s - (a + b - 1) = s - a - b + 1 \ . \tag{38}$$

This is because $\Sigma_i\Sigma_j n_{ij}\bar{y}_{ij\cdot}^2$ has ab degrees of freedom for all-cells-filled data but s for some-cells-empty data: in both cases, the number of filled cells. Hence for some-cells-empty data the F-statistic is

$$F = \frac{\mathcal{R}(\mu_{ij}|\mu_i, \tau_j)}{(s - a - b + 1)\hat{\sigma}^2} = \frac{\Sigma_i\Sigma_j n_{ij}\bar{y}_{ij\cdot}^2 - R(\mu_i, \tau_j)}{(s - a - b + 1)\hat{\sigma}^2} \ . \tag{39}$$

Example 1 (continued). From Table 5.5

$$\Sigma_i\Sigma_j n_{ij}\bar{y}_{ij\cdot}^2 = 2(8^2) + 1(8^2) + 2(3^2) + 2(6^2) + 3(16^2) = 1050$$

as can also be confirmed directly from Table 5.3 in the form

$$\Sigma_i\Sigma_j n_{ij}\bar{y}_{ij\cdot}^2 = \text{SSR}_m + \text{SSM} = 240 + 810 = 1050 \ .$$

Substituting this value and (36) into (37) and calling it Q gives

$$Q = \mathcal{R}(\mu_{ij}|\mu_i, \tau_j) = 1050 - 1007\tfrac{1}{7} = 42\tfrac{6}{7} \tag{40}$$

and so in (39), with $\hat{\sigma}^2 = 4$ from (13),

$$F = \frac{42\frac{6}{7}}{(5 - 2 - 3 + 1)4} = 10\tfrac{5}{7} \ . \tag{41}$$

d. What hypothesis?

With some-cells-empty data that are connected, the all-important question is "What is the hypothesis tested by F of (39)?" For its counterpart with all-cells-filled data in (78) of Section 4.9, the hypothesis is very straightforward: H: *all interactions zero*, where an interaction is represented by $\theta_{ij,i'j'}$. But such is not the case here—if for no other reason than the degrees of freedom associated with Q, the numerator of (39) exemplified in (40), are $s - a - b + 1$ whereas the number of LIN interactions in an $a \times b$ grid is $(a - 1)(b - 1) = ab - a - b + 1$ and, for some-cells-empty data this exceeds $s - a - b + 1$. Therefore the first important feature of

(39) is that with some-cells-empty (connected) data, F of (39) is *not testing* the hypothesis of all interactions zero.

Example 1 (continued). $Q = 42\frac{6}{7}$ in (40) has degrees of freedom, d.f. = 1; but in a 2×3 grid there are 1(2) = 2 LIN interactions, e.g., $\theta_{11,22}$ and $\theta_{11,23}$. Hence $F = 10\frac{5}{7}$ of (41), based on Q, cannot be testing H: $\theta_{11,22} = 0 = \theta_{11,23}$. The question therefore arises: What is it testing? To answer this we look at Grid 5.1 again

Grid 5.1

√	√	√
√	√	

and see that the only estimable interaction is

$$\theta_{11,22} = \mu_{11} - \mu_{12} - \mu_{21} + \mu_{22} .$$

Therefore the t-statistic for testing

$$\text{H: } \theta_{11,22} = 0$$

is

$$t = \frac{\hat{\mu}_{11} - \hat{\mu}_{12} - \hat{\mu}_{21} + \hat{\mu}_{22}}{\sqrt{(1/n_{11} + 1/n_{12} + 1/n_{21} + 1/n_{22})\hat{\sigma}^2}}$$

and the square of this (the F-statistic) is, using the estimates following Table 5.1 from the $E(y_{ijk}) = \mu_{ij}$ model,

$$F = \frac{(8 - 8 - 6 + 16)^2}{(\frac{1}{2} + \frac{1}{1} + \frac{1}{2} + \frac{1}{3})4} = \frac{100}{(2\frac{1}{3})4} = \frac{75}{7} = 10\tfrac{5}{7} .$$

This is precisely the same as the F in (41) based on Q: i.e., for this example,

$$F = Q/(s - a - b + 1)\hat{\sigma}^2 \quad \text{is testing} \quad \text{H: } \theta_{11,12} = 0 .$$

It is not a test that all three interactions in a 2×3 grid are zero, but just a test that the sole interaction that is estimable from Grid 5.1 is zero.

Emphasis to this feature of F of (39) is given by the following example. Suppose the μ_{ij}-values of a 2×3 grid are as shown in Table 5.6. Suppose also that available data were from Grid 5.1 with no observations on cell 2, 3. Then $F = Q/\hat{\sigma}^2$ from such data would be testing H: $\mu_{11} - \mu_{12} - \mu_{21}$

TABLE 5.6. EXAMPLE OF μ_{ij}-VALUES THAT EXHIBIT INTERACTION

$\mu_{11} = 7$	$\mu_{12} = 3$	$\mu_{13} = 2$
$\mu_{21} = 15$	$\mu_{22} = 11$	$\mu_{23} = 38$

$+ \mu_{22} = 0$; and even if the data were such that that hypothesis was not rejected, that would be no evidence whatever of the nonzero value that $\theta_{11,23} = \mu_{11} - \mu_{13} - \mu_{21} + \mu_{23} = 7 - 2 - 15 + 38 = 28$ has in Table 5.6.

This simple demonstration illustrates the following important result that pertains to the F-statistic of (39) of which (41) is but a simple example: using F of (39) for testing a hypothesis about interactions when the data have empty cells is not necessarily testing the hypothesis that all interactions are zero.

e. Specifying the hypothesis

The important aspect of (39) is that whereas its counterpart [(78) of Section 4.9d] for all-cells-filled data tests H: *all interactions zero*, this is *not* the hypothesis tested by (39) for some-cells-empty data. What is tested by (39) in that case is

$$\text{H:}\ \left\{\begin{array}{l}\textit{Any set of } s-a-b+1 \textit{ linearly independent}\\ \textit{functions of } \theta_{ij,i'j'}\textit{s where such functions are either}\\ \textit{estimable } \theta\textit{s or estimable sums or differences of } \theta\textit{s}\end{array}\right\}\ \textit{are all zero}\ . \qquad (42)$$

As with a number of results in this and the preceding chapter, derivation of (42) is deferred until later chapters. [See also Searle (1971), Chapter 7.]

Example 3. For data from Grid 5.4

Grid 5.4

Row	Column 1	2	3	4
1	√	√		√
2	√	√	√	
3			√	√

s, the number of filled cells, is $s = 8$, and $a + b - 1 = 3 + 4 - 1 = 6$. Hence there are only

$$s - (a + b - 1) = 8 - 6 = 2 \tag{43}$$

degrees of freedom for H of (42). The question then is to ascertain from the description of H given in (42) just what the hypothesis is when the pattern of filled cells is as in Grid 5.4.

Clearly one interaction that has a BLUE is

$$\theta_{11,22} = \mu_{11} - \mu_{12} - \mu_{21} + \mu_{22}\,. \tag{44}$$

The cell corresponding to each μ_{ij} in (44) contains data: therefore $\theta_{11,22}$ is estimable. Hence one part of H of (42) can be that $\theta_{11,22}$ is zero.

Since using (43) in (42) indicates that there are just two statements needed for specifying H in this case, we seek a second one. Scrutinizing Grid 5.4 shows that no individual interaction $\theta_{ij,i'j'}$ other than $\theta_{11,22}$ is estimable. There is in Grid 5.4 no quadrat $(ij, i'j')$ other than $(11, 22)$ that has filled cells at all its four corners; i.e., other than μ_{11}, μ_{12}, μ_{21} and μ_{22}, there is no set of four estimable μs (μ_{ij}, $\mu_{ij'}$, $\mu_{i'j}$ and $\mu_{i'j'}$) located at the four corners of any rectangle of cells in Grid 5.4. This means that $\theta_{11,22}$ is the only interaction estimable from data of Grid 5.4.

To complete the hypothesis (42) for Grid 5.4, along with H: $\theta_{11,22} = 0$ being part of it, we must therefore look in Grid 5.4 for patterns of filled cells that provide, in the words of (42), functions that "are estimable sums or differences of θs." These do not have to be sums or differences of estimable θs. They will be sums or differences of non-estimable θs, but such that those sums or differences are themselves estimable. Two examples to be found in Grid 5.4 are

$$\begin{aligned}\theta_{11,23} + \theta_{13,34} &= (\mu_{11} - \mu_{13} - \mu_{21} + \mu_{23}) + (\mu_{13} - \mu_{14} - \mu_{33} + \mu_{34}) \\ &= \mu_{11} - \mu_{21} + \mu_{23} - \mu_{14} - \mu_{33} + \mu_{34}\end{aligned} \tag{45}$$

and

$$\begin{aligned}\theta_{12,24} + \theta_{23,34} &= (\mu_{12} - \mu_{14} - \mu_{22} + \mu_{24}) + (\mu_{23} - \mu_{24} - \mu_{33} + \mu_{34}) \\ &= \mu_{12} - \mu_{14} - \mu_{22} + \mu_{23} - \mu_{33} + \mu_{34}\,.\end{aligned} \tag{46}$$

Each pair of the functions (44), (45) and (46) is a linearly independent (LIN) pair of functions suitable for the hypothesis (42), but all three of them are not LIN: the sum of (44) and (46) is (45). Furthermore, no such

LIN pair of functions is sufficient for deriving all interactions available from a 3×4 grid, of which Grid 5.4 is just one pattern of possible data being considered. Deriving all available interactions requires having not 2 but $(a-1)(b-1) = 2(3) = 6$ LIN θs. Therefore even though (39) can be used to test, using (44) and (45),

$$\text{H:} \begin{cases} \theta_{11,22} = 0 \\ \theta_{11,23} + \theta_{13,34} = 0 \end{cases}, \tag{47}$$

it is not testing H: *all interactions zero.*

f. Using the no-interaction model

Since, as has been described, the F-statistic (39) is, for some-cells-empty data, not testing that all interactions are zero, the procedure of assuming a model that truly has no interactions is even more tenuous here than it is with all-cells-filled data, as discussed in Section 4.10. Nevertheless, using the no-interaction model with some-cells-empty data can provide information that is not otherwise available from such data, namely some overall indication of the importance of differences among row effects (and among column effects) in the *assumed absence* of interactions—and provided, of course, that the some-cells-empty data are connected. It is all too true that if indeed there are interactions between the underlying factors and we use a no-interaction model then any information about rows and columns based on that no-interaction model may be misleading. On the other hand, as seen in the discussion of Grid 5.4 in the preceding subsection, using the with-interaction model with some-cells-empty data can provide very little information about differences among rows (or columns) generally. Therefore it is my feeling that with some-cells-empty data one should not only carry out subset analyses (as briefly described in the preceding subsection and considered at length in Section 5.6) but also, in some circumstances, one should consider using the following results.

With the no-interaction model

$$E(y_{ijk}) = \mu_i + \tau_j \tag{48}$$

the following F-statistics, which are the same as (106) and (107) in Section 4.10, can be very useful:

$$\frac{\mathcal{R}(\mu_i, \tau_j \mid \mu_i)}{(b-1)\hat{\sigma}^2} = \frac{\sum_{j=1}^{b-1} \hat{\tau}_j r_j}{(b-1)\tilde{\sigma}^2} \quad \text{tests} \quad \text{H: } \tau_j \textit{ all equal}; \tag{49}$$

and

$$\frac{\sum_{i=1}^{a} n_{i\cdot}\bar{y}_{i\cdot\cdot}^2 + \sum_{j=1}^{b-1}\hat{\tau}_j r_j - \sum_{j=1}^{b} n_{\cdot j}\bar{y}_{\cdot j\cdot}^2}{(a-1)\tilde{\sigma}^2} \quad \text{tests} \quad \text{H: } \mu_i \textit{ all equal}, \tag{50}$$

where the numerator of (50) is $\mathscr{R}(\mu_i, \tau_j \mid \tau_j)$. In these statistics, $\tilde{\sigma}^2$ is the estimated residual variance for the no-interaction model,

$$\tilde{\sigma}^2 = \frac{\Sigma_i\Sigma_j\Sigma_k y_{ijk}^2 - R(\mu_i, \tau_j)}{N - a - b + 1}$$

$$= \frac{\Sigma_i\Sigma_j\Sigma_k y_{ijk}^2 - \Sigma_i n_{i\cdot}\bar{y}_{i\cdot\cdot}^2 - \sum_{j=1}^{b-1}\hat{\tau}_j r_j}{N - a - b + 1}, \tag{51}$$

the same as in (82) of Section 4.9.

Example 1 (continued). Table 5.3 shows $\Sigma_i\Sigma_j\Sigma_k y_{ijk}^2 = 1070$, and (36) is $R(\mu_i, \tau_j) = 1007\frac{1}{7}$ so that from (51)

$$\tilde{\sigma}^2 = \frac{1070 - 1007\frac{1}{7}}{10 - 2 - 3 + 1} = \frac{62\frac{6}{7}}{6} = 10\tfrac{10}{21},$$

different, of course, from $\hat{\sigma}^2 = 4$ of (13). Also, (35) is $\Sigma_{j=1}^{b-1}\hat{\tau}_j r_j = 107\frac{1}{7}$ so that (49) gives

$$F = \frac{107\frac{1}{7}}{(3-1)10\frac{10}{21}} = 5\tfrac{5}{44} \quad \text{tests} \quad \text{H: } \tau_1 = \tau_2 = \tau_3\,. \tag{52}$$

5.5. ANALYSIS OF VARIANCE

The preceding section emphasizes what the useful procedures are for some-cells-empty data. With the with-interaction model, one considers just linear functions of cell means of filled cells. With the no-interaction model and connected data, contrasts among μ_is and τ_js are estimable, and the analysis of variance of Table 4.11 provides tests of the hypotheses H: μ_i *equal* $\forall\ i$ and H: τ_j *equal* $\forall\ j$. These procedures, which are the very essence of analyzing some-cells-empty data, differ from those useful for all-cells-filled data, which, for analysis of variance, are based on SSA_w and SSB_w and on Table 4.8. Despite this, the question is sometimes asked "What happens if the analysis of variance procedures for all-cells-filled data, particularly Table 4.8, are used for some-cells-empty, connected data?" After all, there is nothing inherent in some-cells-empty data that prevents calculation of Table 4.8; either it will be useful, in which case we should know how useful, or it will be useless, in which case we should know why. We therefore consider these possibilities, despite already knowing from the associated hypotheses of Table 4.9 that even for all-cells-filled data the analysis of variance of Table 4.8 is not generally very useful.

a. Weighted squares of means analysis

This analysis [e.g., SSA_w of (42) in Section 4.6] *cannot be used* for some-cells-empty data. It is suitable to only all-cells-filled data. An obvious reason for this is that SSA_w depends on b^2/w_i defined in (29) of Section 4.6 as $\Sigma_j 1/n_{ij}$, and since some-cells-empty data have some n_{ij}-values that are zero, some or all of the w_i-values are undefined.

b. The all-cells-filled analysis of Table 4.8

Table 4.8 can be calculated for some-cells-empty data, exactly as it stands. Its utility depends upon the associated hypotheses specified in Table 4.9. Notationally those hypotheses are exactly the same for some-cells-empty data as they are in Table 4.9 for all-cells-filled data. There are two major consequences, the first of which is, as noted in Table 4.9, that the sum of squares due to interaction has as its associated hypothesis not H: $\theta_{ij,i'j'}$ *all zero*, but the hypothesis described in (42) of Section 5.4e.

The second consequence of using Table 4.9 for some-cells-empty data is one of great import insofar as usefulness of hypotheses is concerned. For the n_{ij}-values that are zero, the corresponding μ_{ij}-terms get omitted from the ρ_i' and γ_j' that occur in the hypotheses. This makes the hypotheses of even less interest than in the all-cells-filled case. We illustrate with an example.

Example. For Grid 5.4, suppose the numbers of observations were as shown in Table 5.7. Then the ρ_i'- and γ_j'-values for Table 4.9 are as follows.

TABLE 5.7. NUMBERS OF OBSERVATIONS FOR GRID 5.4

	n_{ij}				$n_{i\cdot}$
	2	3	—	5	10
	4	8	7	—	19
	—	—	9	12	21
$n_{\cdot j}$:	6	11	16	17	$50 = n_{\cdot\cdot}$

They are written with spaces in them to emphasize the omission of certain μ_{ij}s corresponding to the empty cells.

$$
\begin{aligned}
\rho_1' &= \tfrac{1}{10}(2\mu_{11} + 3\mu_{12} \qquad\qquad + 5\mu_{14}) \\
\rho_2' &= \tfrac{1}{19}(4\mu_{21} + 8\mu_{22} + 7\mu_{23} \qquad\qquad) \\
\rho_3' &= \tfrac{1}{21}(\qquad\qquad\qquad 9\mu_{33} + 12\mu_{34}) \\
\gamma_1' &= \tfrac{1}{6}(2\mu_{11} + 4\mu_{21} \qquad\qquad) \\
\gamma_2' &= \tfrac{1}{11}(3\mu_{12} + 8\mu_{22} \qquad\qquad) \\
\gamma_3' &= \tfrac{1}{16}(\qquad 7\mu_{23} + 9\mu_{33}) \\
\gamma_4' &= \tfrac{1}{17}(5\mu_{14} \qquad\qquad + 12\mu_{34})\ .
\end{aligned}
\tag{53}
$$

With these definitions, the associated hypotheses of Table 4.9 are as shown in Table 5.8. The effect of empty cells on the form of ρ_i' and γ_j' and on the

TABLE 5.8. ASSOCIATED HYPOTHESES IN THE ANALYSIS OF VARIANCE OF TABLE 4.9 APPLIED TO SOME-CELLS-EMPTY DATA OF TABLE 5.7 USING ρ_i' AND γ_j' OF (53)

d.f.	Sum of Squares	Associated Hypothesis
Partitioning I		
1	$R(\mu)$	H: $\frac{1}{50}(2\mu_{11} + 3\mu_{12} + 5\mu_{14} + 4\mu_{21} + 8\mu_{22} + 7\mu_{23} + 9\mu_{33} + 12\mu_{34}) = 0$
2	$\mathscr{R}(\mu_i \mid \mu)$	H: $\rho_1' = \rho_2' = \rho_3'$
3	$\mathscr{R}(\mu_i, \tau_j \mid \mu_i)$	H: $\begin{cases} \gamma_1' = \frac{1}{6}(2\rho_1' + 4\rho_2' \qquad) \\ \gamma_2' = \frac{1}{11}(3\rho_1' + 8\rho_2' \qquad) \\ \gamma_3' = \frac{1}{16}(\qquad 7\rho_2' + 9\rho_3') \\ \gamma_4' = \frac{1}{17}(5\rho_1' \qquad + 12\rho_3') \end{cases}$
1	$\mathscr{R}(\mu_{ij} \mid \mu_i, \tau_j)$	H: $\mu_{11} - \mu_{12} - \mu_{21} + \mu_{22} = 0$
Partitioning II		
3	$\mathscr{R}(\tau_j \mid \mu)$	H: $\gamma_1' = \gamma_2' = \gamma_3' = \gamma_4'$
2	$\mathscr{R}(\mu_i, \tau_j \mid \tau_j)$	H: $\begin{cases} \rho_1' = \frac{1}{10}(2\gamma_1' + 3\gamma_2' \qquad + 5\gamma_4') \\ \rho_2' = \frac{1}{19}(4\gamma_1' + 8\gamma_2' + 7\gamma_3' \qquad) \\ \rho_3' = \frac{1}{21}(\qquad\qquad 9\gamma_3' + 12\gamma_4') \end{cases}$

hypotheses is clearly evident in (53) and Table 5.8. That effect is simply one of aggravating the uselessness of most of the hypotheses in Table 4.9.

c. Disconnected data

Connectedness of some-cells-empty data has been discussed in Section 5.3 largely in relation to the estimability of contrasts in no-interaction models. But lack of connectedness can also affect the partitioning of sums of squares shown in Table 4.9. Only connected data can be analyzed in the manner of that table. Otherwise, for disconnected data, the disconnected sets of connected data must be analyzed separately and then pooled over sets. If this is not done, blind adherence to Table 4.9 can yield negative degrees of freedom, as in the analysis of Grid 5.5:

Grid 5.5

√	√	
√		
		√

Table 5.9 shows the degrees of freedom for the two different analyses of Grid 5.5. The three connected cells are assumed to have a total of N_1 observations, the other cell N_2 observations.

TABLE 5.9. DEGREES OF FREEDOM IN THE ANALYSIS OF VARIANCE FOR GRID 5.5

	Degrees of Freedom		
	Analyzed wrongly, as one set of data, ignoring disconnectedness	Analyzed as disconnected data	
Sum of Squares		Set I (3 cells)	Set II (1 cell)
$R(\mu)$	1	1	1
$\mathscr{R}(\mu_i \mid \mu)$	2	1	0
$\mathscr{R}(\mu_i, \tau_j \mid \mu_i)$	2	1	0
$\mathscr{R}(\mu_{ij} \mid \mu_i, \tau_j)$	-1	0	0
Error	$N_1 + N_2 - 4$	$N_1 - 3$	$N_2 - 1$
Total	$N_1 + N_2$	N_1	N_2

5.6. SUBSET ANALYSES FOR WITH-INTERACTION MODELS

a. Difficulties

A first difficulty with some-cells-empty data is that there is no test of the hypothesis H: *all interactions zero*, as discussed in Section 5.4. And even if there were such a test, that might not be just cause for using a no-interaction model—for the same reasons as discussed following (83) in Section 4.9e-ii, for all-cells-filled data. A second difficulty is that for the with-interaction model the row means of cell means $\rho_i = (\mu_{i1} + \mu_{i2} + \cdots + \mu_{ib})/b$ cannot be estimated for every row, as illustrated in Section 5.2a. Therefore, as illustrated in (15) and (16), we are driven by the fact of having empty cells to sometimes defining a row mean in different ways for different purposes.

b. What is of interest?

The salient, general question, exemplified just prior to (16) for comparing rows in Grid 5.1, is "What linear functions of μ_{ij}s of filled cells are of interest for whatever purpose we have in mind?"

This is the all-important question for analyzing some-cells-empty data using a with-interaction model. It can be answered in any manner we wish so far as linear functions of μ_{ij}s of filled cells are concerned. All such functions are estimable. Exactly which ones are to be used depends on what is of interest to the person whose data are being analyzed. We call that person the investigator.

c. The investigator's role

It is the investigator's knowledge of the study for which the data were collected, of the data themselves, and of the source of the data that produces what is of interest so far as linear functions of filled-cell μ_{ij}s are concerned. Thus it is, with some-cells-empty data and with-interaction cell means models, that the question of what to estimate (and test hypotheses about) rests squarely on the investigator. Once decisions are made about which linear functions of cell means (of filled cells) are of interest, the mechanics of estimation and hypothesis testing are absolutely straightforward. And making such decisions is not the task of just the statistician—far from it. Investigators familiar with the data, their source and the data-gathering process have to be involved. Not only do these aspects of data have to be taken into account when deciding on what linear functions of cell means (of filled cells) are of interest, but also the purpose for which the linear function is intended must be considered. This means that sometimes a feature of interest has to be defined differently for different purposes. Just as in (15) and (16) for Grid 5.1, to estimate a row mean for row 1 we might use $\rho_1 = \frac{1}{3}(\mu_{11} + \mu_{12} + \mu_{13})$, whereas for comparing rows 1 and 2 we would probably use $\rho_1'' = \frac{1}{2}(\mu_{11} + \mu_{12})$ and be interested in its difference from $\rho_2'' = \frac{1}{2}(\mu_{21} + \mu_{22})$:

$$\omega = \rho_1'' - \rho_2'' = \tfrac{1}{2}(\mu_{11} + \mu_{12}) - \tfrac{1}{2}(\mu_{21} + \mu_{22}), \tag{54}$$

which can be estimated. We could, of course, also use the estimable

$$\omega^* = \rho_1 - \rho_2 = \tfrac{1}{3}(\mu_{11} + \mu_{12} + \mu_{13}) - \tfrac{1}{2}(\mu_{21} + \mu_{22}), \tag{55}$$

realizing in doing so that it is a comparison that, for each column, involves different weights for the two rows. Generally speaking, ω^* does not seem very appealing, compared to ω. Nevertheless, if the situation is such that an investigator finds ω^* to be of interest then, since ω^* is a linear combination of μs of filled cells, it is estimable (it has a BLUE) and hypotheses about it can be tested.

d. A procedure

The first step is for the investigator to decide on what kinds of comparisons among (linear functions of) cell means are of interest. To the extent that these include cell means of empty cells, the linear functions have to be adapted to include cell means of only the filled cells. What has to be done is to look at (the pattern, so to speak, of) which cells are filled and which are not, and from that pattern decide what comparisons among the cell means of filled cells might be of interest in terms of providing at least some information about comparisons of interest to the investigator. For example, if in Grid 5.4 we want to compare rows 2 and 3, the only data that provide a comparison in the same column are those in column 3: i.e., $\mu_{23} - \mu_{33}$ is an estimable function. True, that function provides information that is not very substantive about rows 2 and 3, but the general paucity of filled cells in Grid 5.4 would usually be apparent to any investigator having such data, who would therefore realize the impracticability of comparing rows 2 and 3 over all four columns.

At the heart of this process is the investigator's knowledge of the data; in the presence of empty cells the investigator must contribute knowledge to deciding what combinations of means (of filled cells) are of interest. No longer can the automatic hypotheses like "equality of rows" be tested; considerable thought must be given, under the spotlight of having empty cells, as to what combination of the filled cells are interesting. Knowledge of the data and the pattern of filled cells both have to be utilized. What one would *like* to consider, if all cells were filled, has to be tempered by what *can* be considered in the light of some cells being empty. The consulting statistician and the investigator, working together, must formulate what they think are interesting linear combinations of cell means of cells that contain data. These combinations can then be estimated, the sampling variance of each estimate can be estimated, and tests of hypotheses can be made about them.

e. Subset analyses

It is clear that neither ω of (54) nor ω^* of (55) is using all the data available on rows 1 and 2 in Grid 5.4. Each uses just a subset of the data. We therefore use the name *subset analyses* for analyses of this nature.

Subset analyses of some-cells-empty data are certainly not as informative as either the analysis of variance of balanced data, or the weighted squares of means analysis of unbalanced but all-cells-filled data (see Section 4.6d). But for unbalanced and some-cells-empty data they are much more useful than the analyses of variance of unbalanced data that are an extension of those of balanced data, as described at length in Searle (1971, Chapter 7)—and see Chapter 9 thereof. Furthermore, subset analyses for some-cells-empty data are vastly easier to understand, to interpret and to explain to decision-makers than are the analyses that are extensions from balanced data. Not only are the latter difficult to interpret, but in most cases of some-cells-empty data they are of no real interest.

f. Interactions

The preceding discussion has been largely from the viewpoint of being interested in row (and, of course, column) effects. But it applies even more forcefully to interactions. We have already seen in Section 5.4 that for some-cells-empty data not all interactions are estimable and the traditional sums of squares "for interaction" is not testing the hypothesis of all interactions zero. Despite this, an investigator's desire and/or need to study the occurrence of interactions is often very strong, and whereas the F-statistic of (39) is of little help in this connection when dealing with some-cells-empty data, the use of subset analyses can be very helpful indeed. This is achieved by looking at the data grid to see which sets of filled cells suggest themselves as possibilities for subset analyses that might yield information about interactions. The information so obtained may not be as far-reaching as when all cells are filled, but it will be better than nothing (which is only what a no-interaction model can yield, insofar as interactions are concerned) and it will nearly always be better than the F of (39) or the traditional, overparameterized analysis (see Chapter 9), because that analysis can, in the face of empty cells, be very difficult to interpret.

The crux of the procedure, when wishing to consider interactions with some-cells-empty data, is therefore to view the data from the perspective of a cell means model and to seek subsets of data that can yield information about interactions. Then, as consultant to a client who insists (as do some clients) on considering interactions when data have empty cells, the statistician can offer clarification in the form of helping the client decide which subsets of data might provide analyses of interest.

g. Examples

Suppose data occurred as indicated in Grid 5.6:

Grid 5.6

	1	2	3	4
I	√	√		√
II	√	√		√
III		√	√	√

The F-statistic of (39) would have three degrees of freedom. But for trying both to ascertain which interactions appear to be significant and to understand their occurrence in the data, scrutinizing the pattern of filled cells in the data reveals subsets of the data that may well be more informative and certainly are easier to interpret, namely Grids 5.6a and b:

Grid 5.6a

	1	2	4
I	√	√	√
II	√	√	√

Grid 5.6b

	2	4
I	√	√
II	√	√
III	√	√

True, analyses of these (overlapping) subsets of the data are not independent of one another; but the analyses are simple and interpretation of each is straightforward.

The preceding example is so simple that it may fail to emphasize just how difficult the search for informative subsets of data can be. But consider Grid 5.7, of six rows and eight columns, with 19 of the 48 cells containing data.

Grid 5.7

	1	2	3	4	5	6	7	8
I	√			√		√	√	
II				√	√			√
III				√				√
IV	√	√		√		√	√	√
V								√
VI			√	√				√

As an example, it illustrates how we can find our way towards analyses that may provide more useful interpretation than does analyzing the full data set "warts and all", in this case the warts being the large number of empty cells: 60% are empty. In this regard the very grid itself is useful, because it provides opportunity to scrutinize just which cells have data and whether or not any of them form subsets of the data that may be open to straightforward analysis. Such scrutiny reveals that columns 2, 3 and 5 and row V each have but a single filled cell. Setting these data aside leaves Grid 5.7a:

Grid 5.7a

	1	4	6	7	8
I	√	√	√	√	
II		√			√
III		√			√
IV	√	√	√	√	√
VI		√			√

This is easily seen to divide into two subsets of data: rows I and IV in columns 1, 4, 6 and 7, and rows II, III, IV and VI in columns 4 and 8. These subsets have but one cell in common and account for all six of the degrees of freedom for interaction available in the analysis of the full data set. But now, in directing attention to these two subsets, we see quite clearly what interactions are being considered.

h. Difficulties with subsets

Two difficulties with subset analyses are readily evident. One is that a data set may not always yield subsets that are useful in the way that those of the preceding examples appear to be. For example, consider Grid 5.8:

Grid 5.8

	1	2	3	4
1	√			√
2		√		√
3	√	√	√	

The F-statistic of (39) will have 1 degree of freedom, and will be testing a hypothesis stated, as follows, in three equivalent ways.

$$\text{H:}\ \mu_{11} - \mu_{14} - \mu_{21} + \mu_{24} + (\mu_{21} - \mu_{22} - \mu_{31} + \mu_{32}) = 0$$

$$\text{H:}\ \mu_{11} - \mu_{14} - \mu_{31} + \mu_{34} - (\mu_{22} - \mu_{24} - \mu_{32} + \mu_{34}) = 0$$

$$\text{H:}\ \mu_{11} + \mu_{24} + \mu_{32} - (\mu_{14} + \mu_{22} + \mu_{31}) = 0.$$

The first of these involves the sum of two interactions whereas the second involves the difference between two interactions. Whatever the utility of these may be (if any), scrutiny of Grid 5.8 reveals that no subsets of the data manifest themselves as being candidates for informative analyses. When this kind of situation occurs the statistician can do little more than persuade the investigator that this is so, and fall back on the no-interaction model. Of course with data sets larger than that of Grid 5.8, coming to the conclusion of no useful subsets may not be as easy as it is there, and much resourcefulness might be needed before such a conclusion can be firmly established.

A second and obvious difficulty inherent in subset analyses is that a data set might well be divisible into two or more different subset analyses. For example, Grid 5.9 can easily be divided into two subsets in two different ways: one way consists of rows I and II, and row III; and the other is columns 1 and 4, and columns 2 and 3.

Grid 5.9

	1	2	3	4
I	√	√	√	√
II	√	√	√	√
III	√			√

This situation emphasizes what is so important about analyzing unbalanced data, especially some-cells-empty data; there is seldom just a single, correct way of doing a statistical analysis. Therefore the first responsibility of a consulting statistician to those who have garnered such data is to impress upon them that analyzing those data has no single, easy umbrella of interpretation. Within that umbra, the statistician can certainly provide advice as to what analyses might be helpful; and two different statisticians may well have two different lines of advice for analyzing the same data set. As with lawyers, the advice of statisticians is not necessarily uniform, let alone uniformly right or uniformly wrong.

i. The investigator

Naturally, the investigator must contribute to deciding on possible divisibility of a total data set into subset analyses. In the preceding examples the pattern of empty cells has been the sole criterion for suggesting subsets. That is always a useful criterion because it can lead to analyses that are interpretable. But it must not be the sole criterion; investigators must be urged, from their prior knowledge of similar data and of the context from which the present data have come, to decide what specific levels of the factors (or pooled combinations thereof) are of prime interest, especially in the context of interactions. Indeed, the statistician's advice should be in terms of helping investigators, nay even cajoling them, perhaps, into deciding what specific subset of filled cells might be of real interest. One would hope that when faced with empty cells, the combined efforts of statistician and investigator would usually reveal some data subsets that provide both easy analysis and straightforward interpretation through the use of the cell means model. Examples of doing this in animal science and in agronomy research are to be found, respectively, in Urquhart and Weeks (1978) and Meredith and Cady (1984).

5.7. SUMMARY

a. Model and estimation

$$E(y_{ijk}) = \mu_{ij} \quad \text{for } i = 1, \ldots, a, \; j = 1, \ldots, b \quad \text{and } k = 1, \ldots, n_{ij} \geq 0 \; . \tag{1}$$

$$y_{ijk} = \mu_{ij} + e_{ijk}, \quad \text{with the } e_{ijk} \sim \text{i.i.d. } N(0, \sigma^2) \tag{2), (3}$$

$$\hat{\mu}_{ij} = \bar{y}_{ij\cdot} \quad \text{for } n_{ij} > 0, \; \hat{\mu}_{ij} \text{ i.d. } N\big(\mu_{ij}, \sigma^2/n_{ij}\big) \tag{4}$$

$$\omega' = \Sigma_i \Sigma_j k_{ij} \mu_{ij} \quad \text{with } k_{ij} = 0 \text{ for } n_{ij} = 0 \tag{7}$$

$$s = \text{number of filled cells} \tag{9}$$

$$\hat{\sigma}^2 = \Sigma_i \Sigma_j \Sigma_k (y_{ijk} - \bar{y}_{ij\cdot})^2 / (N - s) \; . \tag{10}$$

b. No-interaction model

Data must be connected . (Section 5.3)

$$\left.\begin{aligned} E(y_{ijk}) &= \mu_i + \tau_j \quad \text{for } j = 1, \ldots, b-1 \\ &= \mu_i \qquad\quad\; \text{for } j = b \end{aligned}\right\} \begin{array}{l}\text{for } i = 1, \ldots, a \\ \text{and } k = 1, \ldots, n_{ij}\end{array} \tag{20}$$

$$c_{jj}\hat{\tau}_j + \sum_{\substack{j'=1 \\ j' \neq j}}^{b-1} c_{jj'}\hat{\tau}_{j'} = r_j \quad \text{for } j = 1, \ldots, b-1 \tag{23}$$

$$c_{jj} = n_{\cdot j} - \Sigma_i \frac{n_{ij}^2}{n_{i\cdot}} \quad \text{and} \quad c_{jj'} = -\Sigma_i \frac{n_{ij}n_{ij'}}{n_{i\cdot}} \tag{24}$$

$$r_j = y_{\cdot j \cdot} - \Sigma_i n_{ij} \bar{y}_{i\cdot\cdot} \tag{25}$$

$$\hat{\mu}_i = \bar{y}_{i\cdot\cdot} - \frac{1}{n_{i\cdot}} \sum_{j=1}^{b-1} n_{ij}\hat{\tau}_j \tag{26}$$

$$\hat{\mu}_{ij} = \hat{\mu}_i + \hat{\tau}_j \tag{27}$$

$$\text{BLUE}(\mu_i - \mu_h) = \hat{\mu}_i - \hat{\mu}_h \quad \text{and} \quad \text{BLUE}(\tau_j - \tau_k) = \hat{\tau}_j - \hat{\tau}_k$$

$$R(\mu_i, \tau_j) = \Sigma_i n_{i\cdot} \bar{y}_{i\cdot\cdot}^2 + \sum_{j=1}^{b-1} \hat{\tau}_j r_j \; . \tag{32}$$

c. Testing for no interactions

An attempted test is

$$\frac{\Sigma_i\Sigma_j n_{ij}\bar{y}_{ij\cdot}^2 - R(\mu_i, \tau_j)}{(s-a-b+1)\hat{\sigma}^2} \quad \text{tests H of (42)}\ . \tag{39}$$

d. Using the no-interaction model

$$\tilde{\sigma}^2 = \frac{\Sigma_i\Sigma_j\Sigma_k y_{ijk}^2 - R(\mu_i, \tau_j)}{N-a-b+1} \tag{51}$$

$$\sum_{j=1}^{b-1} \hat{\tau}_j r_j/(b-1)\tilde{\sigma}^2 \quad \text{tests} \quad \text{H: } \tau_j \textit{ all equal} \tag{49}$$

$$\frac{R(\mu_i, \tau_j) - \Sigma_j n_{\cdot j}\bar{y}_{\cdot j\cdot}^2}{(a-1)\tilde{\sigma}^2} \quad \text{tests} \quad \text{H: } \mu_i \textit{ all equal}\ . \tag{50}$$

5.8. EXERCISES

E5.1. Show that the data patterns of each of the Grids 5.4, 5.6, 5.6a, 5.7, 5.8 and 5.9 are connected.

E5.2. Derive estimators of the cell means for the data grid

n_{11}	n_{12}
n_{21}	—

and use the following data to illustrate your answer:

4, 8	8, 8, 9, 11
3	—

E5.3. Repeat E5.2 for

n_{11}	n_{12}	n_{13}
n_{21}	n_{22}	—

and for

7, 13	6, 10	8, 11, 8
3, 7, 5	65	—

Calculate Tables 4.9 and 4.11.

E5.4. Repeat E5.2 for

n_{11}	—	n_{13}
n_{21}	n_{22}	—

and

7, 13	—	14, 13, 18
3, 8, 4	13	—

E5.5. Analyze the following data on a no-interaction model (a), as is, (b) deleting cell 2, 3, and (c), deleting cells 2, 3 and 1, 2.

7, 9	8	6, 6, 12
8, 4	11, 12, 13	8

E5.6. Calculate Tables 4.9 and 4.11 for the following data:

7, 8, 12	13, 15,	—
13, 14, 19, 18	—	30, 32
—	15, 17	4, 6, 8

E5.7. Repeat E5.6 for the following data:

1, 8, 3	2, 3, 5, 6	12, 14
2, 4	—	18
3, 7	2, 3, 4, 5, 6	2

E5.8. Repeat E5.6 for the following data:

8, 13, 9	—	12	7, 11
6, 12	12, 14	—	—
—	9, 7	14, 16	10, 14, 11, 13

E5.9. Make up a set of data for a 3×4 layout with 4 empty cells and repeat E5.6 for those data.

E5.10. Calculate F of (39) and interpret it for the following data:

5, 6, 6, 7	—	—	—
6, 8, 10	—	—	—
2, 3, 7	6, 8	2, 4, 7, 11	39, 43

Explain why connectedness implies $s > a + b - 1$.

E5.11. Repeat E5.10 for the following data:

3, 9	12, 15, 6	—	—
3, 8, 10, 11, 13, 15	31	—	—
—	—	85	—
—	—	2, 8, 10, 15, 18, 16, 19, 16, 12, 19	5, 8, 8

Repeat the analysis after combining rows 2 and 3.

E5.12. Suppose the data for a 3-way classification are in cells 111, 122, 133, 144, 212, 223 and 234. By drawing a diagram similar to Figure 5.1, show that such data are not connected.

CHAPTER 6

MODELS WITH COVARIABLES (ANALYSIS OF COVARIANCE): THE 1-WAY CLASSIFICATION

There are many occasions when, in studying some response variable, we want to take into account the contribution to variation evident in the response variable that might be attributable to some other (concomitant) variable. For example, in analyzing weight gains in young beef cattle fed different diets one might want to incorporate into the analysis the initial weight of each animal. To do this, instead of using a model equation such as $E(y_{ij}) = \mu_i$ of Chapter 2, we might use $E(y_{ij}) = \mu_i + bz_{ij}$, where y_{ij} and z_{ij} are, respectively, the final and initial weights on animal j fattened on diet i. This is the basic model for what has traditionally been called analysis of covariance. Today's more pertinent description is analysis with *concomitant variables*, also called, interchangeably, *covariables* or *covariates*. This is a description that minimizes any possible confusion with what is sometimes called covariance analysis, which is concerned with the analysis of covariance matrices. Analysis with covariates is accurate because what has been called analysis of covariance is actually analysis of variance in the presence of covariates.

The model $E(y_{ij}) = \mu_i + bz_{ij}$ is a very simple example of an analysis-with-covariates model. It can be extended in many ways, to several factors, to more than one covariate and to the coefficients of those covariates being different for different levels of one or more of the factors. Some of these extensions are considered in Chapter 11, where their presentation makes use of the general matrix description of linear models analysis given in Chapter 8. But even without matrix notation we can describe the analysis of two straightforward models that have many uses. They are two forms of extending the 1-way classification of Chapter 2 to utilize a single covariable. One is

$$E(y_{ij}) = \mu_i + bz_{ij},$$

which shall be called *single slope model*, so named because it contrasts with

$$E(y_{ij}) = \mu_i + b_i z_{ij},$$

the *multiple-slope model*, where there is a separate slope for each class: b_i for class i. This is often called the *intra-class regression model.*

Both of the preceding models are adaptations not only of the 1-way classification of Chapter 2, but also of the simple regression model $E(y_i) = a + bz_i$; indeed, the single slope model is just the no-covariate model $E(y_{ij}) = \mu_i$ with a regression feature bz_{ij} added to it. From this context comes use of "slope" to describe coefficients of covariates; and, although x_{ij} is the usual notation where we here have z_{ij}, the latter is maintained in order to retain X for its very customary use in the general matrix-and-vector form of (linear) model equation $\mathbf{y} = \mathbf{Xb} + \mathbf{e}$ dealt with in Chapter 8.

Example. Suppose tomato plants are grown with three different fertilizers; three in one fertilizer and two each in the other two fertilizers. Table 6.1 shows plant height in inches both at 10 weeks after planting (y_{ij}) and at the time of planting (z_{ij}), and the density of seedlings one week after first germination (w_{ij}, number of seedlings per four-inch square).

Analysis of variance of the response variable, y_{ij} of Table 6.1, could be based on $E(y_{ij}) = \mu_i$ of Chapter 2. Analysis incorporating the covariate z_{ij} could use either the single slope model $E(y_{ij}) = \mu_i + bz_{ij}$ or the multiple slopes model $E(y_{ij}) = \mu_i + b_i z_{ij}$, as just defined. Similar models involving

TABLE 6.1. HEIGHTS OF TOMATO PLANTS AND DENSITY OF GERMINATION

	Fertilizer						
	I Plant			II Plant		III Plant	
	1	2	3	1	2	1	2
Height at 10 weeks, y_{ij}	74	68	77	76	80	87	91
Height at transplant, z_{ij}	3	4	5	2	4	3	7
Density of germination, w_{ij}	22	26	21	21	27	20	24

the density of germination w_{ij} could also be considered:

$$E(y_{ij}) = \mu_i + b'w_{ij} \quad \text{and} \quad E(y_{ij}) = \mu_i + b_i'w_{ij}\ .$$

Analysis of the data in Table 6.1 could also be made using both covariates at once. This can be done in a variety of ways, such as

$$E(y_{ij}) = \mu_i + bz_{ij} + b^+w_{ij},$$

which has a single slope for each covariate; or as either

$$E(y_{ij}) = \mu_i + b_iz_{ij} + b^+w_{ij}$$

or

$$E(y_{ij}) = \mu_i + bz_{ij} + b_i^+w_{ij},$$

each of which is multiple slopes for one covariate and single slope for the other; or as

$$E(y_{ij}) = \mu_i + b_iz_{ij} + b_i^+w_{ij},$$

which is multiple slopes for both covariates. Actually, for Table 6.1, where w_{ij} represents density of germination, it might be more appropriate to use $1/w_{ij}$ in these models in place of w_{ij}.

The procedures for dealing with more than one covariate are best described in terms of matrix notation, and so are deferred until Chapter 11 as an extension of the general matrix description of linear models given in Chapter 8; not only can the four preceding cases be dealt with satisfactorily, but further extensions to more than two covariates and to more than just the 1-way classification can also be taken into account. This chapter, in not utilizing matrix algebra, is confined to the 1-way classification with just one covariate, using the single slope or the multiple slopes model.

6.1. THE SINGLE SLOPE MODEL

a. Model

The model for the 1-way classification with one covariate and a single slope is

$$E(y_{ij}) = \mu_i + bz_{ij}, \tag{1}$$

or, equivalently,

$$y_{ij} = \mu_i + bz_{ij} + e_{ij}, \tag{2}$$

where, as usual, e_{ij} is defined as $e_{ij} = y_{ij} - E(y_{ij})$, and we take $e_{ij} \sim$ i.i.d. $N(0, \sigma^2)$. As in Chapter 2 we let a be the number of classes, so that $i = 1, \ldots, a$; and n_i is the number of observations in the ith class, so that $j = 1, 2, \ldots, n_i$.

b. Estimation of means

-i. Estimation and unbiasedness. Minimizing

$$S = \Sigma_i \Sigma_j (y_{ij} - \mu_i - bz_{ij})^2 \tag{3}$$

with respect to the μ_is and b leads to equations

$$\hat{\mu}_i n_i + \hat{b}\Sigma_j z_{ij} = \Sigma_j y_{ij} \quad \text{for } i = 1, \ldots, a \tag{4}$$

and

$$\Sigma_i \Sigma_j \hat{\mu}_i z_{ij} + \hat{b}\Sigma_i \Sigma_j z_{ij}^2 = \Sigma_i \Sigma_j y_{ij} z_{ij} \tag{5}$$

for the least squares estimators $\hat{\mu}_i$ and $\hat{b}$ of μ_i and b, respectively. As in Chapter 2, so here also: Σ_i denotes $\Sigma_{i=1}^{a}$ and Σ_j denotes $\Sigma_{j=1}^{n_i}$.

From (4), for $i = 1, \ldots, a$,

$$\hat{\mu}_i = \bar{y}_{i\cdot} - \hat{b}\bar{z}_{i\cdot}, \tag{6}$$

and substituting (6) into (5) yields

$$\hat{b} = \frac{\Sigma_i \Sigma_j (y_{ij} - \bar{y}_{i\cdot})(z_{ij} - \bar{z}_{i\cdot})}{\Sigma_i \Sigma_j (z_{ij} - \bar{z}_{i\cdot})^2} . \tag{7}$$

The estimator $\hat{\mu}_i = \bar{y}_{i\cdot} - \hat{b}\bar{z}_{i\cdot}$ in (6) is sometimes called an *adjusted treatment mean*, although that name is more often reserved for

$$\hat{\mu}_i + \hat{b}\bar{z}_{\cdot\cdot} = \bar{y}_{i\cdot} - \hat{b}(\bar{z}_{i\cdot} - \bar{z}_{\cdot\cdot}) .$$

The estimators (6) and (7) are unbiased: using (1) first in (7) gives

$$E(\hat{b}) = \frac{\Sigma_i \Sigma_j b(z_{ij} - \bar{z}_{i\cdot})(z_{ij} - \bar{z}_{i\cdot})}{\Sigma_i \Sigma_j (z_{ij} - \bar{z}_{i\cdot})^2} = b,$$

and then in (6)

$$E(\hat{\mu}_i) = \mu_i + b\bar{z}_{i\cdot} - b\bar{z}_{i\cdot} = \mu_i .$$

-ii. Sampling variances and covariances. Sampling variances and covariances are as follows.

$$v(\hat{b}) = \frac{v[\Sigma_i\Sigma_j(y_{ij} - \bar{y}_{i\cdot})(z_{ij} - \bar{z}_{i\cdot})]}{\left[\Sigma_i\Sigma_j(z_{ij} - \bar{z}_{i\cdot})^2\right]^2}$$

$$= \frac{\left[\Sigma_i\Sigma_j(1 - 1/n_i)(z_{ij} - \bar{z}_{i\cdot})^2 - \Sigma_i(1/n_i)\Sigma_{j' \neq j}\Sigma(z_{ij} - \bar{z}_{i\cdot})(z_{ij'} - \bar{z}_{i\cdot})\right]\sigma^2}{\left[\Sigma_i\Sigma_j(z_{ij} - \bar{z}_{i\cdot})^2\right]^2}$$

$$= \frac{\left\{\Sigma_i\Sigma_j(z_{ij} - \bar{z}_{i\cdot})^2 - \Sigma_i(1/n_i)[\Sigma_j(z_{ij} - \bar{z}_{i\cdot})]^2\right\}\sigma^2}{\left[\Sigma_i\Sigma_j(z_{ij} - \bar{z}_{i\cdot})^2\right]^2}$$

$$= \frac{\sigma^2}{\Sigma_i\Sigma_j(z_{ij} - \bar{z}_{i\cdot})^2} . \tag{8}$$

$$\text{cov}(\bar{y}_{h\cdot}, \hat{b}) = \frac{\Sigma_i\Sigma_j\left[\text{cov}(\bar{y}_{h\cdot}y_{ij}) - \text{cov}(\bar{y}_{h\cdot}\bar{y}_{i\cdot})\right](z_{ij} - \bar{z}_{i\cdot})}{\Sigma_i\Sigma_j(z_{ij} - \bar{z}_{i\cdot})^2}$$

$$= \frac{\Sigma_j(1/n_h - 1/n_h)(z_{ij} - \bar{z}_{i\cdot})\sigma^2}{\Sigma_i\Sigma_j(z_{ij} - \bar{z}_{i\cdot})^2} = 0 .$$

Hence from (6) and (8)

$$v(\hat{\mu}_i) = \frac{\sigma^2}{n_i} + \bar{z}_{i\cdot}^2 v(\hat{b}) = \left(\frac{1}{n_i} + \frac{\bar{z}_{i\cdot}^2}{\Sigma_i\Sigma_j(z_{ij} - \bar{z}_{i\cdot})^2}\right)\sigma^2, \tag{9}$$

$$\text{cov}(\hat{\mu}_i, \hat{\mu}_h) = \bar{z}_{i\cdot}\bar{z}_{h\cdot}v(\hat{b}) = \frac{\bar{z}_{i\cdot}\bar{z}_{h\cdot}}{\Sigma_i\Sigma_j(z_{ij} - \bar{z}_{i\cdot})^2}\sigma^2 \tag{10}$$

and

$$\text{cov}(\hat{\mu}_i, \hat{b}) = -\bar{z}_{i\cdot}v(\hat{b}) = \frac{-\bar{z}_{i\cdot}}{\Sigma_i\Sigma_j(z_{ij} - \bar{z}_{i\cdot})^2}\sigma^2 \tag{11}$$

Example. Table 6.2 shows the data values for y and z from Table 6.1 together with values of expressions needed for the estimation equations (6)–(11).Then, using Table 6.2 gives $\hat{b}$ from (7) as

$$\hat{b} = \frac{3 + 4 + 8}{2 + 2 + 8} = \frac{15}{12} = 1\tfrac{1}{4}; \tag{12}$$

TABLE 6.2. CALCULATIONS FROM THE DATA OF TABLE 6.1

	$i = 1$		$i = 2$		$i = 3$	
Data	y_{ij}	z_{ij}	y_{ij}	z_{ij}	y_{ij}	z_{ij}
	74	3	76	2	87	3
	68	4	80	4	91	7
	77	5				
$y_{i\cdot}$ and $z_{i\cdot}$	219	12	156	6	178	10
n_i	3		2		2	
$\bar{y}_{i\cdot}$ and $\bar{z}_{i\cdot}$	73	4	78	3	89	5
$\Sigma_j(y_{ij} - \bar{y}_{i\cdot})(z_{ij} - \bar{z}_{i\cdot})$	3		4		8	
$\Sigma_j(z_{ij} - \bar{z}_{i\cdot})^2$	2		2		8	
$\Sigma_j(y_{ij} - \bar{y}_{i\cdot})^2$	42		8		8	
$\Sigma_j y_{ij}^2$	16029		12176		15850	
$n_i \bar{y}_{i\cdot}^2$	15987		12168		15842	
$\Sigma_j z_{ij}^2$	50		20		58	
$\Sigma_j z_{ij} y_{ij}$	879		472		898	

(6) gives

$$\hat{\mu}_1 = 73 - 1\tfrac{1}{4}(4) = 68,\ \hat{\mu}_2 = 78 - 1\tfrac{1}{4}(3) = 74\tfrac{1}{4},\ \hat{\mu}_3 = 89 - 1\tfrac{1}{4}(5) = 82\tfrac{3}{4}\ . \tag{13}$$

(8) is

$$v(\hat{b}) = \sigma^2/(2 + 2 + 8) = \sigma^2/12, \tag{14}$$

(9) and (10) give, respectively,

$$\begin{aligned} v(\hat{\mu}_1) &= \left(\frac{1}{3} + \frac{4^2}{12}\right)\sigma^2 = 1\tfrac{2}{3}\sigma^2 & \operatorname{cov}(\hat{\mu}_1, \hat{\mu}_2) &= \frac{4(3)}{12}\sigma^2 = \sigma^2 \\ v(\hat{\mu}_2) &= \left(\frac{1}{2} + \frac{3^2}{12}\right)\sigma^2 = 1\tfrac{1}{4}\sigma^2 \quad \text{and} & \operatorname{cov}(\hat{\mu}_1, \hat{\mu}_3) &= \frac{4(5)}{12}\sigma^2 = 1\tfrac{2}{3}\sigma^2 \\ v(\hat{\mu}_3) &= \left(\frac{1}{2} + \frac{5^2}{12}\right)\sigma^2 = 2\tfrac{7}{12}\sigma^2 & \operatorname{cov}(\hat{\mu}_2, \hat{\mu}_3) &= \frac{3(5)}{12}\sigma^2 = 1\tfrac{1}{4}\sigma^2 \end{aligned} \tag{15}$$

and (11) yields

$$\operatorname{cov}(\hat{\mu}_1, \hat{b}) = -\tfrac{4}{12}\sigma^2 = -\tfrac{1}{2}\sigma^2, \qquad \operatorname{cov}(\hat{\mu}_2, \hat{b}) = -\tfrac{3}{12}\sigma^2 = -\tfrac{1}{4}\sigma^2$$

and

$$\operatorname{cov}(\hat{\mu}_3, \hat{b}) = -\tfrac{5}{12}\sigma^2 \,. \tag{16}$$

c. Estimating residual variance

The residual sum of squares is

$$\text{SSE} = \Sigma_i\Sigma_j(y_{ij} - \hat{y}_{ij})^2 \tag{17}$$

with

$$\hat{y}_{ij} = \hat{\mu}_i + \hat{b}z_{ij} = \bar{y}_{i\cdot} - \hat{b}\bar{z}_{i\cdot} + \hat{b}z_{ij} \tag{18}$$

$$= \bar{y}_{i\cdot} + \hat{b}(z_{ij} - \bar{z}_{i\cdot}) \,. \tag{19}$$

Substituting (19) in (17) gives

$$\begin{aligned} \text{SSE} &= \Sigma_i\Sigma_j\left[(y_{ij} - \bar{y}_{i\cdot}) - \hat{b}(z_{ij} - \bar{z}_{i\cdot})\right]^2 \\ &= \Sigma_i\Sigma_j(y_{ij} - \bar{y}_{i\cdot})^2 + \hat{b}^2\Sigma_i\Sigma_j(z_{ij} - \bar{z}_{i\cdot})^2 \qquad (20) \\ &\quad - 2\hat{b}\Sigma_i\Sigma_j(y_{ij} - \bar{y}_{i\cdot})(z_{ij} - \bar{z}_{i\cdot}) \,. \end{aligned}$$

From (7), for the last term in (20),

$$\hat{b}\Sigma_i\Sigma_j(y_{ij} - \bar{y}_{i\cdot})(z_{ij} - \bar{z}_{i\cdot}) = \hat{b}^2\Sigma_i\Sigma_j(z_{ij} - \bar{z}_{i\cdot})^2 \tag{21}$$

$$= \frac{\left[\Sigma_i\Sigma_j(y_{ij} - \bar{y}_{i\cdot})(z_{ij} - \bar{z}_{i\cdot})\right]^2}{\Sigma_i\Sigma_j(z_{ij} - \bar{z}_{i\cdot})^2} \,. \tag{22}$$

Using (21) in (20) gives

$$\text{SSE} = \Sigma_i\Sigma_j(y_{ij} - \bar{y}_{i\cdot})^2 - \hat{b}^2\Sigma_i\Sigma_j(z_{ij} - \bar{z}_{i\cdot})^2 \,. \tag{23}$$

The equivalence of the three expressions in (21) and (22) is to be noted, because they are used interchangeably to provide expressions for SSE of (23), in (36) which follows.

The expected value of SSE comes from substituting $y_{ij} = \mu_i + bz_{ij} + e_{ij}$ from (2) into (23), and on using W_{zz} for $\Sigma_i\Sigma_j(z_{ij} - \bar{z}_{i\cdot})^2$ this gives

$$\begin{aligned} E(\text{SSE}) &= \Sigma_i\Sigma_j E\left[b(z_{ij} - \bar{z}_{i\cdot}) + e_{ij} - \bar{e}_{i\cdot}\right]^2 - E(\hat{b}^2)W_{zz} \\ &= b^2 W_{zz} + \Sigma_i\Sigma_j E(e_{ij} - \bar{e}_{i\cdot})^2 - \left[v(\hat{b}) + b^2\right]W_{zz} \\ &= \Sigma_i\Sigma_j(1 - 1/n_i)\sigma^2 - (\sigma^2/W_{zz})W_{zz} \\ &= (N - a - 1)\sigma^2 \,. \end{aligned} \tag{24}$$

Hence an unbiased estimator of σ^2 is

$$\hat{\sigma}^2 = \frac{\text{SSE}}{N - a - 1}, \tag{25}$$

with SSE being available from (23). The result (24) can, of course, also be derived by using the general matrix methods of Section 8.7d.

Example (continued). From Table 6.2

$$\Sigma_i\Sigma_j(y_{ij} - \bar{y}_{i\cdot})^2 = 42 + 8 + 8 = 58 \tag{26}$$

and from (12)

$$\hat{b}^2\Sigma_i\Sigma_j(z_{ij} - \bar{z}_{i\cdot})^2 = (1\tfrac{1}{4})^2(2 + 2 + 8) = 18\tfrac{3}{4} \tag{27}$$

so that in (23)

$$\text{SSE} = 58 - 18\tfrac{3}{4} = 39\tfrac{1}{4} \,. \tag{28}$$

Therefore in (25)

$$\hat{\sigma}^2 = \frac{39\frac{1}{4}}{3 + 2 + 2 - 3 - 1} = \frac{39\frac{1}{4}}{3} = 13\tfrac{1}{12} \,. \tag{29}$$

The reader should confirm (28) by calculating it in the form (17).

d. Between- and within-classes sums of squares and products

The sums of squares for the analysis of variance for the 1-way classification using the model $E(y_{ij}) = \mu_i$, as dealt with at length in Chapter 2, are shown in Table 6.3. Included in Table 6.3 are symbols B_{yy}, W_{yy} and T_{yy}, for

TABLE 6.3. SUMS OF SQUARES IN THE ANALYSIS OF VARIANCE OF THE 1-WAY CLASSIFICATION

Source of Variation	d.f.	Sums of Squares
Between classes	$a-1$	$\mathscr{R}(\mu_i \mid \mu) = \Sigma_i n_i(\bar{y}_{i\cdot} - \bar{y}_{\cdot\cdot})^2 = B_{yy}$
Within classes	$N-a$	SSE $= \Sigma_i\Sigma_j(y_{ij} - \bar{y}_{i\cdot})^2 = W_{yy}$
Total (a.f.m.)	$N-1$	$\text{SST}_m = \Sigma_i\Sigma_j(y_{ij} - \bar{y}_{\cdot\cdot})^2 = T_{yy}$

the sum of squares between classes, the sum of squares within classes, and the total sum of squares, respectively. They are introduced as a concise, mnemonic notation for terms that arise in the analysis of variance of the 1-way classification with one covariate, for sums of squares not only of ys, but also of zs, and for sums of products of ys and zs. Thus in addition to the definitions implicit in Table 6.3 we also define

$$\begin{aligned} B_{zz} &= \Sigma_i n_i(\bar{z}_{i\cdot} - \bar{z}_{\cdot\cdot})^2 & B_{yz} &= \Sigma_i n_i(\bar{y}_{i\cdot} - \bar{y}_{\cdot\cdot})(\bar{z}_{i\cdot} - \bar{z}_{\cdot\cdot}) \\ W_{zz} &= \Sigma_i\Sigma_j(z_{ij} - \bar{z}_{i\cdot})^2 \quad \text{and} & W_{yz} &= \Sigma_i\Sigma_j(y_{ij} - \bar{y}_{i\cdot})(z_{ij} - \bar{z}_{i\cdot}) \\ T_{zz} &= \Sigma_i\Sigma_j(z_{ij} - \bar{z}_{\cdot\cdot})^2 & T_{yz} &= \Sigma_i\Sigma_j(y_{ij} - \bar{y}_{\cdot\cdot})(z_{ij} - \bar{z}_{\cdot\cdot}) \end{aligned} \tag{30}$$

and at all times have

$$B_{pq} + W_{pq} = T_{pq} \tag{31}$$

for the subscripts pq being yy, zz or yz.

Example (continued). Table 6.2 yields the following.

$$\begin{aligned} B_{yy} &= 310 & B_{zz} &= 4 & B_{yz} &= 22 \\ W_{yy} &= 58 & W_{zz} &= 12 & W_{yz} &= 15 \\ T_{yy} &= 368 & T_{zz} &= 16 & T_{yz} &= 37 \end{aligned} \tag{32}$$

The reader should confirm these results, and in doing so should note how these values relate to some of the calculated values given in Table 6.2.

The conciseness of this notation, and the numerous opportunities for using it, provide the means for simplifying the appearance of many expressions in the analysis. Already-occurring examples are, from (7) and (8),

$$\hat{b} = W_{yz}/W_{zz} \quad \text{and} \quad v(\hat{b}) = \sigma^2/W_{zz}, \tag{33}$$

and from (9), (10) and (11),

$$v(\hat{\mu}_i) = \left(1/n_i + \bar{z}_{i.}^2/W_{zz}\right)\sigma^2, \qquad \text{cov}(\hat{\mu}_i, \hat{\mu}_h) = \bar{z}_{i.}\bar{z}_{h.}\sigma^2/W_{zz}$$

$$\text{and} \qquad \text{cov}\left(\hat{\mu}_i, \hat{b}\right) = -\bar{z}_{i.}\sigma^2/W_{zz}\ . \tag{34}$$

From (21) and (22), using $\hat{b}$ from (33),

$$\hat{b}W_{yz} = \hat{b}^2 W_{zz} = W_{yz}^2/W_{zz} \tag{35}$$

and (23) is

$$\text{SSE} = W_{yy} - \hat{b}W_{yz} = W_{yy} - \hat{b}^2 W_{zz} = W_{yy} - W_{yz}^2/W_{zz}\ . \tag{36}$$

e. Partitioning the total sum of squares

Up to this point we have dealt with estimating the parameters μ_i, b and σ^2 of the with-covariate model. Now, on using SS as an obvious abbreviation for sum of squares, and on defining

$$R(\mu_i, b) = \text{SS due to fitting } E(y_{ij}) = \mu_i + bz_{ij}, \tag{37}$$

we derive a convenient expression for $R(\mu_i, b)$. With that, and sums of squares due to fitting other models to the 1-way classification, we then establish two useful partitionings of the total sum of squares $\text{SST} = \Sigma_i\Sigma_j y_{ij}^2$, one based on first fitting classes and then the covariate, and the other on fitting the covariate and then classes. In subsection g these partitionings are summarized in analysis of variance tables and used for testing a variety of hypotheses.

Deriving $R(\mu_i, b)$ starts with the basic definition of SSE as $\text{SSE} = \Sigma_i\Sigma_j(y_{ij} - \hat{y}_{ij})^2$ of (17), which leads to (23) and (36). But, given $R(\mu_i, b)$ as defined in (37), SSE is also

$$\text{SSE} = \Sigma_i\Sigma_j y_{ij}^2 - R(\mu_i, b)\ . \tag{38}$$

Equating this to (36) gives

$$\text{SSE} = \Sigma_i\Sigma_j y_{ij}^2 - R(\mu_i, b) = W_{yy} - \hat{b}W_{yz}$$

so that, from the nature of W_{yy} defined in Table 6.3,

$$R(\mu_i, b) = \Sigma_i n_i \bar{y}_{i\cdot}^2 + \hat{b}W_{yz} \ . \tag{39}$$

For the model

$$E(y_{ij}) = \mu_i + \tau_j, \tag{40}$$

the partitioning of SST as

$$\text{SST} = R(\mu) + \mathcal{R}(\mu_i | \mu) + \mathcal{R}(\mu_i, \tau_j | \mu_i) + \text{SSE}(\mu_i, \tau_j) \tag{41}$$

is shown in Table 4.11. The SSE in that table is denoted SSE(μ_i, τ_j) in (41) to distinguish it from SSE(μ_i) $= \Sigma_i\Sigma_j(y_{ij} - \bar{y}_{i\cdot})^2 = W_{yy}$ of Table 6.3 and SSE(μ_i, b) $= W_{yy} - \hat{b}W_{yz}$ of (36). We use (41) to partition SST for the model

$$E(y_{ij}) = \mu_i + bz_{ij} \ . \tag{42}$$

Relate (42) to (40), and with b of (42) then playing the part of τ_j in (40) we immediately have from (41)

$$\text{SST} = R(\mu) + \mathcal{R}(\mu_i | \mu) + \mathcal{R}(\mu_i, b | \mu_i) + \text{SSE}(\mu_i, b) \ . \tag{43}$$

Calculation of the terms in (43) starts with

$$R(\mu_i) = \Sigma_i n_i \bar{y}_{i\cdot}^2, \tag{44}$$

$$R(\mu) = N\bar{y}_{\cdot\cdot}^2, \tag{45}$$

and

$$\mathcal{R}(\mu_i | \mu) = R(\mu_i) - R(\mu) = \Sigma_i n_i(\bar{y}_{i\cdot} - \bar{y}_{\cdot\cdot})^2 = B_{yy}, \tag{46}$$

from (40), (44) and (47), respectively, of Chapter 2. The third term in (43) is

$$\mathcal{R}(\mu_i, b | \mu_i) = R(\mu_i, b) - R(\mu_i) \tag{47}$$

which from (39) and (44) gives

$$\mathcal{R}(\mu_i, b | \mu_i) = \hat{b}W_{yz} \ . \tag{48}$$

And the last term in (43) is (36). In this way the partitioning of (43) as (45), (46), (48) and (36) is shown as the last four lines of part (a) of Table 6.4.

TABLE 6.4. ANALYSIS OF VARIANCE FOR THE 1-WAY CLASSIFICATION, UNBALANCED DATA, WITH ONE COVARIATE WITH A SINGLE SLOPE

Line	Source of Variation	d.f.	Sum of Squares: $R(\cdot)$	Sum of Squares: Calculation	Equation	Associated Hypothesis
	(a) Fitting classes and then covariate					
1.	Classes	a	$R(\mu_i)$	$= \Sigma_i n_i \bar{y}_{i\cdot}^2$	(44)	H: $\mu_i + b\bar{z}_{i\cdot} = 0 \;\forall\; i$
2.	Mean	1	$R(\mu)$	$= N\bar{y}_{\cdot\cdot}^2$	(45)	H: $\Sigma_i n_i \mu_i / N + b\bar{z}_{\cdot\cdot} = 0$
3.	Classes (adjusted for mean)	$a - 1$	$\mathscr{R}(\mu_i \mid \mu)$	$= B_{yy}$	(46)	H: $\mu_i + b\bar{z}_{i\cdot}$ *equal* $\forall\; i$
4.	Covariate (adjusted for classes)	1	$\mathscr{R}(\mu_i, b \mid \mu_i)$	$= \hat{b}W_{yz}$	(48)	H: $b = 0$
5.	Residual	$N - a - 1$	$\text{SSE}(\mu_i, b)$	$= W_{yy} - \hat{b}W_{yz}$	(36)	
6.	Total	N	SST	$= \Sigma_i \Sigma_j y_{ij}^2$		
	(b) Fitting covariate and then classes					
7.	Mean	1	$R(\mu)$	$= N\bar{y}_{\cdot\cdot}^2$	(45)	H: $\Sigma_i n_i \mu_i / N + b\bar{z}_{\cdot\cdot} = 0$
8.	Covariate (adjusted for mean)	1	$\mathscr{R}(\mu, b \mid \mu)$	$= \tilde{b}T_{yz}$	(60)	H: $\Sigma_i k_i \mu_i + b = 0$[1]
9.	Classes (adjusted for covariate)	$a - 1$	$\mathscr{R}(\mu_i, b \mid \mu, b)$	$= B_{yy} + \hat{b}W_{yz} - \tilde{b}T_{yz}$	(62)	H: μ_i *equal* $\forall\; i$
10.	Residual	$N - a - 1$	$\text{SSE}(\mu_i, b)$	$= W_{yy} - \hat{b}W_{yz}$	(36)	
11.	Total	N	SST	$= \Sigma_i \Sigma_j y_{ij}^2$		

[1] $k_i = n_i(\bar{z}_{i\cdot} - \bar{z}_{\cdot\cdot})/T_{zz}$ with $\Sigma_i k_i = 0$.

Example (continued). From Table 6.2, SST = 16029 + 12176 + 15850 = 44055. Then, using values from Table 6.2 and (32) as needed, we have from

$$(33): \qquad \hat{b}_i = 15/12 = 1\tfrac{1}{4}, \quad \text{as in (12)} \tag{49}$$

$$(44): \qquad R(\mu_i) = 15987 + 12168 + 15842 = 43997 \tag{50}$$

$$(45): \qquad R(\mu) = 7[(219 + 156 + 178)/7]^2 = 43687 \tag{51}$$

$$(46): \qquad \mathscr{R}(\mu_i \mid \mu) = B_{yy} = 310 \tag{52}$$

$$(48): \; \mathscr{R}(\mu_i, b \mid \mu_i) = 1\tfrac{1}{4}(15) = 18\tfrac{3}{4} \tag{53}$$

$$(36): \quad \text{SSE}(\mu_i, b) = 58 - 1\tfrac{1}{4}(15) = 39\tfrac{1}{4}, \quad \text{as in (28)} \tag{54}$$

These values in (50)–(54) are shown in Table 6.4E (6.4E indicates Example 6.4).

f. The equal-class-effects sub-model: a single line

For the model (40), equation (41) is the first partitioning of SST in Table 4.11. The second is

$$\text{SST} = R(\mu) + \mathscr{R}(\tau_j \mid \mu) + \mathscr{R}(\mu_i, \tau_j \mid \tau_j) + \text{SSE}(\mu_i, \tau_j)\ . \tag{55}$$

Comparable to (55), the partitioning for $E(y_{ij}) = \mu_i + bz_{ij}$ of (42) is

$$\text{SST} = R(\mu) + \mathscr{R}(\mu, b \mid \mu) + \mathscr{R}(\mu_i, b \mid \mu, b) + \text{SSE}(\mu_i, b)\ . \tag{56}$$

Here, for the two middle terms of (56), we need $R(\mu, b)$, the sum of squares due to fitting

$$E(y_{ij}) = \mu + bz_{ij}\ . \tag{57}$$

The distinction between (42) and (57) is that in (42) each class has its own μ_i whereas in (57) every class has the same μ. This can be further described in terms of graphical representations using Cartesian coordinates $E(y_{ij})$ and z_{ij}. Model (42) is $E(y_{ij}) = \mu_i + bz_{ij}$, which is a series of parallel lines, all with slope b, but intercepts μ_i, one for each class, $i = 1, \ldots, a$. Model (57) is a single line with slope b and intercept μ. It is useful as an alternative to (42) because it provides, in the presence of the covariate, a test of the hypothesis H: μ_i *equal for all i*.

TABLE 6.4E. ANALYSIS OF VARIANCE (TABLE 6.4) FOR y WITH COVARIATE z, OF EXAMPLE IN TABLE 6.1

$E(y_{ij}) = \mu_i + bz_{ij}$

Line	Source of Variation	d.f.	Sum of Squares: $R(\cdot)$	Sum of Squares: Calculation	Sum of Squares: Equation	Associated Hypothesis
	(a) Fitting classes and then covariate					
1.	Classes	3	$R(\mu_i)$	= 43997	(50)	H: $\mu_1 + 4b = 0 = \mu_2 + 3b = \mu_3 + 5b$
2.	Mean	1	$R(\mu)$	= 43687	(51)	H: $\mu + (3\mu_1 + 2\mu_2 + 2\mu_3)/7 + 4b = 0$
3.	Classes (adjusted for mean)	2	$\mathscr{R}(\mu_i \mid \mu)$	= 310	(52)	H: $\mu_1 + 4b = \mu_2 + 3b = \mu_3 + 5b$
4.	Covariate (adjusted for classes)	1	$\mathscr{R}(\mu_i, b \mid \mu_i)$	= $18\frac{3}{4}$	(53)	H: $b = 0$
5.	Residual	3	SSE(μ_i, b)	= $39\frac{1}{4}$	(54)	
6.	Total	7	SST	= 44055		
	(b) Fitting covariate and then classes					
7.	Mean	1	$R(\mu)$	= 43687	(51)	H: $(3\mu_1 + 2\mu_2 + 2\mu_3)/7 + 4b = 0$
8.	Covariate (adjusted for mean)	1	$\mathscr{R}(\mu, b \mid \mu)$	= $85\frac{9}{16}$	(64)	H: $(0\mu_1 - 2\mu_2 + 2\mu_3)/16 + b = 0$[1]
9.	Classes (adjusted for covariate)	2	$\mathscr{R}(\mu_i, b \mid \mu, b)$	= $243\frac{3}{16}$	(65)	H: $\mu_1 = \mu_2 = \mu_3$
10.	Residual	3	SSE(μ_i, b)	= $39\frac{1}{4}$	(54)	
11.	Total	7	SST	= 44055		

[1] $k_1 = 3(4-4)/16 = 0/16$, $k_2 = 2(3-4)/16 = -2/16$, and $k_3 = 2(5-4)/16 = 2/16$.

-i. Fitting the single-line model. Since (57) represents just a single line it is nothing more than a simple regression model for all the data, ignoring the different classes $i = 1, \ldots, a$. It is left as an exercise (E6.4) for the reader to derive least squares estimators $\tilde{\mu}$ and $\tilde{b}$ from minimizing $\Sigma_i\Sigma_j(y_{ij} - \mu - bz_{ij})^2$, and to obtain $R(\mu, b)$. The results are

$$\tilde{b} = T_{yz}/T_{zz} \quad \text{and} \quad \tilde{\mu} = \bar{y}_{..} - \tilde{b}\bar{z}_{..} \tag{58}$$

and

$$R(\mu, b) = N\bar{y}_{..}^2 + \tilde{b}T_{yz} . \tag{59}$$

-ii. Partitioning SST. Individual terms in the partitioning (56) of SST can now be calculated. Starting with $R(\mu) = N\bar{y}_{..}^2$, using (59) gives

$$\mathscr{R}(\mu, b \,|\, \mu) = R(\mu, b) - R(\mu) = \tilde{b}T_{yz} . \tag{60}$$

Then

$$\mathscr{R}(\mu_i, b \,|\, \mu, b) = R(\mu_i, b) - R(\mu, b)$$

$$= \Sigma_i n_i \bar{y}_{i.}^2 + \hat{b}W_{yz} - \left(N\bar{y}_{..}^2 + \tilde{b}T_{yz}\right) \tag{61}$$

from (39) and (59), and so, using B_{yy} of (46)

$$\mathscr{R}(\mu_i, b \,|\, \mu, b) = B_{yy} + \hat{b}W_{yz} - \tilde{b}T_{yz} . \tag{62}$$

As in (43), so in (56): the last term is SSE(μ_i, b) of (36). Thus the sums of squares in part (b) of Table 6.4 are (45), (60), (62) and (36).

Example (continued). Substituting for T_{yz} and T_{zz} from (32) into (58) gives

$$\tilde{b} = \tfrac{37}{16} = 2\tfrac{5}{16} \tag{63}$$

and (60) is

$$\mathscr{R}(\mu, b \,|\, \mu) = 2\tfrac{5}{16}(37) = 85\tfrac{9}{16} . \tag{64}$$

Then (62) is, from (32), (49) and (64)

$$\mathscr{R}(\mu_i, b \,|\, \mu, b) = 310 + 1\tfrac{1}{4}(15) - 85\tfrac{9}{16} = 243\tfrac{3}{16}; \tag{65}$$

and (59) is

$$R(\mu, b) = 7(79)^2 + 2\tfrac{5}{16}(37) = 43772\tfrac{9}{16} . \tag{66}$$

g. Establishing tests of hypotheses

Convenient summarizing of the preceding partitionings of SST is shown in the analysis of variance tables of Table 6.4, analogous to Table 4.11. For using the resulting ratios of mean squares as F-statistics for testing hypotheses we need to assume normality, consider χ^2 and independence properties of the sums of squares, and display their associated hypotheses. Details of doing this can be tedious.

As already indicated in Sections 1.5 and 4.6c, when hypothesis testing is properly planned it is carried out by setting up a hypothesis of interest, H, and deriving a statistic suitable for testing it. With linear hypotheses the suitable statistic is usually of the form $Q/r\hat{\sigma}^2$ (as discussed in Section 4.6c), where Q is a sum of squares and r is its degrees of freedom. Q is calculated from the data, it depends on H and, under H, the distribution of Q/σ^2 must be χ^2 with r degrees of freedom, independent of $\hat{\sigma}^2$. Then $Q/r\hat{\sigma}^2$ is an F-statistic for testing H. Except for a few simple examples in the preceding chapters (e.g., in Sections 1.12b-i and 4.6b), establishing this procedure has been deferred until Chapter 8 and beyond, because in Section 8.8 a method of doing it is described which is both suitable to a very broad class of hypotheses and applicable to any linear model. In matrix notation that method straightforwardly provides, for any legitimate hypothesis stated in a (simple) specified manner, a Q that satisfies all requirements for $Q/r\hat{\sigma}^2$ to be an F-statistic suitable for testing H. It is because of the availability of this method that the preceding chapters contain practically nothing (except Section 2.12b-ii) about setting up hypotheses of interest and testing them.

An alternative aspect of testing hypotheses is simply to use F-statistics provided in the output of statistical computing packages (the easy way out). This is discussed at some length in Sections 1.4 and 1.5. However, for whatever set of sums of squares that are calculated from the data, we need to be assured that what we are provided with as F-statistics are indeed just that; for example, F-statistics based on sums of squares such as those in Table 6.4. Clearly, within the framework of data and arithmetic, calculating $Q/r\hat{\sigma}$ given $\hat{\sigma}^2$ and any Q and its r is easy. But before doing this we must know that Q is such that the ratio $Q/r\hat{\sigma}^2$ *is* an F-statistic. In most computer packages, such ratios are indeed F-statistics. Nevertheless, in the context of the analysis of variance tables of Table 6.4 and displaying the hypotheses tested by the resulting ratios $Q/r\hat{\sigma}^2$, we certainly need to assure ourselves that those ratios *are* F-statistics. This means that each numerator Q must satisfy the following conditions, similar to those in Section 4.9d.

(i) Q must be a sum of squares.
(ii) Q must be distributed independently of $\hat{\sigma}^2$.
(iii) Under some specifiable hypothesis H, we must have $Q/\sigma^2 \sim \chi^2_r$ for known r.

And, although not a condition on Q, (iv), $\hat{\sigma}^2$ must have a χ^2-distribution.

Conditions (i)–(iii) have to be satisfied by each Q that is to be used; and (ii)–(iv) must stem from the normality assumption made about the model equation $y_{ij} = E(y_{ij}) + e_{ij}$, namely

$$e_{ij} \sim \text{i.i.d. } N(0, \sigma^2) \ . \tag{67}$$

In Chapters 2–4 there have been certain sums of squares in simpler situations (e.g., Sections 2.10d, 3.5b and 4.6c) where establishing the preceding conditions has been somewhat straightforward using basic properties of normally distributed variables. But there have also been occasions (e.g., Sections 2.12b-ii, 4.9d, 4.9e-iii, 4.10b and 5.4a) where, because using these principles would be extremely tedious, establishment of those conditions has been deferred until after Chapter 8. There, at the start of Section 8.6 and based on Section 7.6a and in general matrix notation, is a set of theorems that provides a method of straightforwardly ascertaining for any Q stated in a (simple) specified manner whether or not it satisfies conditions (i)–(iii). At the same time, those theorems yield H and r of (ii) and provide opportunity for checking that (iv) is satisfied. And the theorems are very general: they are applicable to any sum of squares from a broad class of linear models.

Many of the sums of squares in Table 6.4 are more complicated than those in earlier chapters—especially when it comes to establishing conditions (ii)–(iv) from (67). The with-covariate feature of the model, bz_{ij}, also adds to the complication. As a result, using the general results from Chapter 8 is *far* easier than applying basic properties of normally distributed random variables for seeing that conditions (i)–(iv) are satisfied, as needed, for the sums of squares of the partitionings of SST shown in Table 6.4. We therefore make no attempt in this chapter to establish that those sums of squares do satisfy the conditions (apart from the two easy exceptions of the following subsection), but leave that to Section 11.3 (in the second chapter on analysis with covariables), which utilizes the general results of Chapter 8.

h. Two examples of χ^2-variables

Just to illustrate how χ^2-distributions are established for the sums of squares of Table 6.4 we consider but two cases. The first is (48):

$$\mathcal{R}(\mu_i, b \mid \mu_i) = \hat{b}W_{yz} = \left[\Sigma_i\Sigma_j(y_{ij} - \bar{y}_{i\cdot})(z_{ij} - \bar{z}_{i\cdot})\right]^2 / W_{zz} \ . \tag{68}$$

Under the hypothesis H: $b = 0$ and based on (67),

$$y_{ij} - \bar{y}_{i\cdot} \sim N\left[0, \sigma^2(1 - 1/n_i)\right]$$

with $$\text{cov}(y_{ij} - \bar{y}_{i\cdot}, y_{hk} - \bar{y}_{h\cdot}) = \begin{cases} 0 & \text{for} \quad i \neq h \\ -\sigma^2/n_i & \text{for} \quad i = h \text{ and } j \neq k \, . \end{cases}$$

Then

$$v\left[\Sigma_j(y_{ij}-\bar{y}_{i\cdot})(z_{ij}-\bar{z}_{i\cdot})\right]$$

$$=\Sigma_j(1-1/n_i)(z_{ij}-\bar{z}_{i\cdot})^2\sigma^2+\Sigma_{j\neq k}(z_{ij}-\bar{z}_{i\cdot})(z_{ik}-\bar{z}_{i\cdot})(-\sigma^2/n_i)$$

$$=\sigma^2\Sigma_j(z_{ij}-\bar{z}_{i\cdot})^2-(\sigma^2/n_i)\left[\Sigma_j(z_{ij}-\bar{z}_{i\cdot})\right]^2$$

$$=\sigma^2\Sigma_j(z_{ij}-\bar{z}_{i\cdot})^2 .$$

Therefore, still under H: $b=0$,

$$\frac{W_{yz}}{\sqrt{W_{zz}}}=\frac{\Sigma_i\Sigma_j(y_{ij}-\bar{y}_{i\cdot})(z_{ij}-\bar{z}_{i\cdot})}{\sqrt{\Sigma_i\Sigma_j(z_{ij}-\bar{z}_{i\cdot})^2}}\sim N(0,\sigma^2) .$$

Hence $$\mathscr{R}(\mu_i, b\,|\,\mu_i)/\sigma^2=\left(W_{yz}^2/W_{zz}\right)/\sigma^2\sim\chi_1^2 .$$

Second, consider

$$\mathscr{B}(\mu_i\,|\,\mu)=\Sigma_i n_{i\cdot}(\bar{y}_{i\cdot}-\bar{y}_{\cdot\cdot})^2$$

$$=\Sigma_i n_i(\mu_i+b\bar{z}_{i\cdot}+\bar{e}_{i\cdot}-\Sigma_i n_i\mu_i/N-b\Sigma n_i\bar{z}_{i\cdot}/N-\bar{e}_{\cdot\cdot})^2 .$$

Under the hypothesis H: $\mu_i+b\bar{z}_{i\cdot}$ *all equal*, this reduces to

$$\mathscr{R}(\mu_i\,|\,\mu)=\Sigma n_i(\bar{e}_{i\cdot}-\bar{e}_{\cdot\cdot})^2,$$

which, on the basis of (67), we know from Chapter 2 has the property

$$\mathscr{R}(\mu_i\,|\,\mu)/\sigma^2\sim\chi_{a-1}^2 .$$

The methods illustrated here for establishing χ^2-distributions utilize basic properties of normally distributed random variables. Extensions for the other sums of squares and for establishing independence properties can become quite involved. For example, in the model $E(y_{ij})=\mu_i$, the distribution of $W_{yy}/\sigma^2=\Sigma_i\Sigma_j(y_{ij}-\bar{y}_{i\cdot})^2/\sigma^2$ is χ_{N-a}^2, and although we have just shown that $\mathscr{R}(\mu_i, b\,|\,\mu_i)/\sigma^2\sim\chi_1^2$, this is no argument for deriving

$$\text{SSE}/\sigma^2=W_{yy}/\sigma^2-\mathscr{R}(\mu_i, b\,|\,\mu_i)/\sigma^2\sim\chi_{N-a-1}^2 .$$

This last statement is correct, but deriving it cannot be achieved as naively as just suggested, using just basic properties of the normally distributed error terms in (67). Yet, from (103) of Section 8.6a, establishing $\text{SSE}/\sigma^2 \sim \chi_{N-a-1}$ becomes a simple example of a standard, universal result. We therefore defer further consideration of developing χ^2 and independence properties of the sums of squares and turn to discussing their consequences in the analysis of variance tables of Table 6.4.

i. Analysis of variance

Table 6.4 shows the analysis of variance tables corresponding to the preceding partitionings of SST in (43) and (56). Both Table 6.4 and Table 6.4E (for the example) contain equation numbers from the text, to help readers locate the various results and calculations. The two tables also show the associated hypothesis for each sum of squares, not because every hypothesis is useful—far from it—but because when computer-package output includes a sum of squares corresponding to a non-useful hypothesis it is convincing to know precisely what the hypothesis is, and so be able to definitively decide on its unworthiness, rather than perhaps being misled by a computer output label into thinking that the hypothesis is useful.

A good example occurs in line 8 of Table 6.4, labeled "Covariate (adjusted for mean)". It is not inconceivable that a user of computer output reading a label of that nature might be led to thinking that that sum of squares, in this case $\mathscr{R}(\mu, b \,|\, \mu)$, was testing $b = 0$ in the model $E(y_{ij}) = \mu_i + bz_{ij}$. In fact, it is testing

$$\text{H: } \Sigma_i k_i \mu_i + b = 0 \quad \text{for} \quad k_i = n_i(\bar{z}_{i\cdot} - \bar{z}_{\cdot\cdot})/T_{zz} \quad \text{with} \quad \Sigma_i k_i = 0 \;.$$

Clearly, this is not a very useful hypothesis and certainly not one that an investigator would set out to test. Furthermore, one well might ask "How has it been derived?" It comes from (123) of Section 8.6c, which provides an expression for the hypothesis tested by any F-statistic after writing the numerator sum of squares in a specified form. For purposes of testing hypotheses of interest, that expression is of little value; for ascertaining hypotheses associated with sums of squares which by their labels might appear to be useful, it is invaluable. It enables one to decide, on the basis of fact, whether or not a computer-calculated F-statistic is useful. (One must, of course, be able to write the numerator sum of squares in the specified form—this is usually not difficult.)

Although the preceding H: $\Sigma_i k_i \mu_i + b = 0$ is of no value in the model $E(y_{ij}) = \mu_i + bz_{ij}$, it is of use if we consider the model $E(y_{ij}) = \mu + bz_{ij}$; and this might well be done if from the test based on $\mathscr{R}(\mu_i, b \,|\, \mu, b)$ we do not reject the hypothesis H: μ_i *equal* $\forall\ i$. If as a consequence we then take

$\mu_i = \mu \quad \forall\ i$, this alters the hypothesis corresponding to $\mathscr{R}(\mu, b \mid \mu)$,

$$\text{H: } \Sigma_i k_i \mu_i + b = 0, \quad \text{to be} \quad \text{H: } \mu \Sigma_i k_i + b = 0, \quad \text{i.e.,} \quad \text{H: } b = 0,$$

because $\Sigma_i k_i = 0$. And this is to be expected, because in the model $E(y_{ij}) = \mu + bz_{ij}$ the numerator sum of squares for testing H: $b = 0$ is $\mathscr{R}(\mu, b \mid \mu) = T_{yz}^2/T_{zz}$.

At all times, what constitutes a useful hypothesis depends on the purpose of the study for which the data at hand are being analyzed. And the logical procedure is for investigators to set up hypotheses that are known to be of interest. In contrast, what we are doing here is precisely the wrong approach to hypothesis testing: looking at F-statistics, contemplating what they test, and asking which is useful. Nevertheless, in this era of prolific computer output ("give 'em everything" says a package developer) it is worthwhile understanding just what the computer output is, even (or perhaps especially) when we can then recognize some parts of that output as having little use. No apologies are needed, therefore, for devoting attention to the utility of all the sums of squares in a table such as 6.4. Discarding what is valueless, based on knowledge, is far to be preferred to using something based only on ignorant belief that it is useful.

j. Tests of intercepts

The three sums of squares in Table 6.4 that are possibly of greatest general use are

$$\mathscr{R}(\mu_i, b \mid \mu_i), \quad \text{for testing} \quad \text{H: } b = 0,$$

a simple hypothesis about the slope coefficient for the covariate; and the sums of squares for testing two different hypotheses about the μ_is:

$$\mathscr{R}(\mu_i \mid \mu) \qquad \text{tests} \quad \text{H: } \mu_i + b\bar{z}_{i\cdot} \textit{ all equal}$$

$$\mathscr{R}(\mu_i, b \mid \mu, b) \quad \text{tests} \quad \text{H: } \mu_i \textit{ all equal}.$$

Yet one other hypothesis concerning the μ_is can be formulated. Suppose that at different predetermined values of the covariate for the different classes it is reasonable to hypothesize equality of the response variable. If $\mathring{z}_i$ denotes the predetermined value of the covariate for class i, then this hypothesis is

$$\text{H: } \mu_i + b\mathring{z}_i \textit{ all equal}\ . \tag{69}$$

This is similar to the hypothesis considered by Urquhart (1982). A further example could be where the concentration of a herbicide is the covariate and the classes are different kinds of weeds. In this case one might use a model of different mortality rates for different weeds and then proceed to test the hypothesis of equal mortality rates for the different weeds at some pre-arranged (perhaps equally spaced) levels of herbicide concentration. The hypothesis in (69) would be appropriate for this.

The numerator sum of squares for testing (69) can be shown to be

$$Q = \Sigma_i n_i (d_i - \bar{d})^2 - \frac{[\Sigma_i n_i (d_i - \bar{d})(g_i - \bar{g})]^2}{\Sigma_i \Sigma_j (z_{ij} - \bar{z}_{i\cdot})^2 + \Sigma_i n_i (g_i - \bar{g})^2} \tag{70}$$

for

$$g_i = \bar{z}_{i\cdot} - \mathring{z}_i \quad \text{and} \quad \bar{g} = \Sigma_i n_i g_i / N$$
$$d_i = \bar{y}_{i\cdot} - \hat{b} g_i \quad \text{and} \quad \bar{d} = \Sigma_i n_i d_i / N \ . \tag{71}$$

Special cases of (69) and (70) are those of $\mathring{z}_i = \bar{z}_{i\cdot}$, which reduces Q to $\mathscr{R}(\mu_i | \mu)$ for testing H: $\mu_i + b\bar{z}_{i\cdot}$ *all equal*; and $\mathring{z}_i = 0$, which gives $Q = \mathscr{R}(\mu_i, b | \mu, b)$ for testing H: μ_i *all equal*. Illustration of the various hypotheses tested by Q is depicted in Figure 6.1 for the case of three classes.

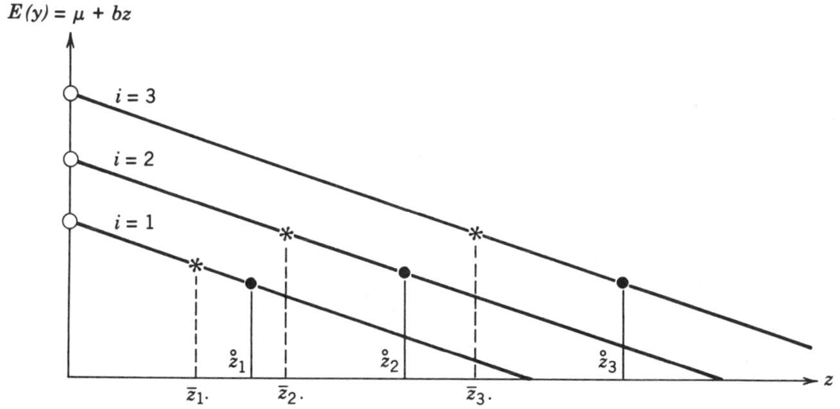

Figure 6.1. Illustration of hypotheses with parallel intra-class regression lines (the single slope model). The hypotheses tested are equality of heights on the $E(y) = \mu + bz$ axis, as follows.

Points Marked	Hypothesis Tested	Sum of Squares
○	H: μ_i *all equal*	$\mathscr{R}(\mu_i, b \mid \mu, b)$
*	H: $\mu_i + b\bar{z}_{i\cdot}$ *all equal*	$\mathscr{R}(\mu_i \mid \mu)$
●	H: $\mu_i + b\mathring{z}_i$ *all equal* for assigned $\mathring{z}_i$	Q of (69)

An important feature of the hypothesis H: μ_i *all equal*, represented in Figure 6.1 by the unfilled circles on the vertical axis, arises from the fact that the model specifies that the slope for the covariate is the same for each class. Therefore the lines in Figure 6.1 are parallel. This means that if, at some level of the covariate, z_0 say, we wish to compare classes adjusted for the covariate, the comparison will, because of the parallelism, be the same for every z_0. Thus the vertical distance between the lines for any particular pair of classes will be the same for every z_0. Hence it is the same for $z_0 = 0$; i.e., on the vertical axis in Figure 6.1. Algebraically, because b is the same for every treatment, $\mu_1 + bz_0 - (\mu_2 + bz_0) = \mu_1 - \mu_2$. Thus, of course,

$$\text{H:}\ \mu_1 + bz_0 = \mu_2 + bz_0 = \cdots = \mu_a + bz_0$$

is the same as

$$\text{H:}\ \mu_1 = \mu_2 = \cdots = \mu_a,$$

which is tested by $\mathscr{R}(\mu_i, b \mid \mu, b)$. This is tantamount to comparing classes adjusted to some uniform value of the covariate, without having to specify that value. The preceding hypothesis is also equivalent to

$$\text{H:}\ \mu_i + b\bar{z}_{..}\ \textit{equal}\ \forall\ i,$$

which is worth noting because, as seen following (7), the BLUE of $\mu_i + b\bar{z}_{..}$ is $\bar{y}_{i.} - \hat{b}(\bar{z}_{i.} - \bar{z}_{..})$, which is often referred to in the literature as an adjusted treatment mean, and one sometimes reads of "testing adjusted treatment means". This, as we see here, is simply the testing of H: μ_i *all equal*.

6.2. THE MULTIPLE SLOPES MODEL: INTRA-CLASS REGRESSION

a. The model

In the preceding development the slope b for the covariate is the same for all classes. We now provide for a different slope for each class, a model that is known as the *intra-class regression* model. Its equation is

$$E(y_{ij}) = \mu_i + b_i z_{ij}, \tag{72}$$

where, as usual, e_{ij} is defined as $e_{ij} = y_{ij} - E(y_{ij})$ and by the normality assumptions we mean $e_{ij} \sim$ i.i.d. $N(0, \sigma^2)$. Also, a represents the number of classes and n_i is the number of observations in the ith class.

b. Estimation of means

-i. Estimation and unbiasedness. Minimizing

$$S = \Sigma_i \Sigma_j (y_{ij} - \mu_i - b_i z_{ij})^2 \tag{73}$$

with respect to the μ_is and b_is leads to a sets of equations, the ith one of which, for $i = 1, \ldots, a$, is (4) together with the ith part of (5):

$$\hat{\mu}_i n_i + \hat{b}_i \Sigma_j z_{ij} = \Sigma_j y_{ij}$$

and

$$\hat{\mu}_i \Sigma_j z_{ij} + \hat{b}_i \Sigma_j z_{ij}^2 = \Sigma_j y_{ij} z_{ij} \tag{74}$$

Hence the estimators are

$$\hat{\mu}_i = \bar{y}_{i\cdot} - \hat{b}_i \bar{z}_{i\cdot} \tag{75}$$

and

$$\hat{b}_i = \frac{\Sigma_j (y_{ij} - \bar{y}_{i\cdot})(z_{ij} - \bar{z}_{i\cdot})}{\Sigma_j (z_{ij} - \bar{z}_{i\cdot})^2} \tag{76}$$

Motivated by (76) we extend the W_{yy}, W_{yz} and W_{zz} notation of Table 6.3 and (30) to have such terms for each class:

$$W_{i,yz} = \Sigma_j (y_{ij} - \bar{y}_{i\cdot})(z_{ij} - \bar{z}_{i\cdot}) \quad \text{and} \quad W_{i,zz} = \Sigma_j (z_{ij} - \bar{z}_{i\cdot})^2 \tag{77}$$

with $W_{i,yy}$ defined similarly. Hence

$$\hat{b}_i = \frac{W_{i,yz}}{W_{i,zz}} . \tag{78}$$

Then, of course,

$$W_{yz} = \Sigma_i W_{i,yz} \quad \text{and} \quad W_{zz} = \Sigma_i W_{i,zz} ; \tag{79}$$

and $\hat{b} = W_{yz}/W_{zz}$ of (7) in the single slope model is the ratio of the summed (pooled) numerators of the $\hat{b}_i$s of (78) to the summed denominators.

The results (75) and (76) are scarcely surprising; they would arise from performing a simple regression of y on z within each class, as can be appreciated by rewriting (72) as

$$\left\{E(y_j) = \mu + bz_j\right\}_i \quad \text{for } i = 1, \ldots, a,$$

where the subscript outside the braces means that every term inside them gets that subscript.

Unbiasedness of these estimators is easily established. Substituting (72) into (76) and using the notation of (77) gives

$$E(\hat{b}_i) = \Sigma_j b_i(z_{ij} - \bar{z}_{i\cdot})(z_{ij} - \bar{z}_{i\cdot})/W_{i,zz} = b_i W_{i,zz}/W_{i,zz} = b_i\,;$$

and then from (75)

$$E(\hat{\mu}_i) = \mu_i + b\bar{z}_{i\cdot} - b\bar{z}_{i\cdot} = \mu_i\ .$$

-ii. Sampling variances and covariances. Derivation of $v(\hat{b}_i)$ is exactly the same as in deriving $v(\hat{b}) = \sigma^2/W_{zz}$ of (8), except that in every step therein the summations over i are dropped. Thus

$$v(\hat{b}_i) = \sigma^2/W_{i,zz}\ . \tag{80}$$

Then

$$v(\hat{\mu}_i) = \left(1/n_i + \bar{z}_{i\cdot}^2/W_{i,zz}\right)\sigma^2 \tag{81}$$

and

$$\text{cov}\left(\hat{\mu}_i, \hat{b}_i\right) = -\bar{z}_{i\cdot}\sigma^2/W_{i,zz}, \tag{82}$$

similar to (9) and (11); and because $\hat{\mu}_i$ and $\hat{b}_i$ are calculated solely from data of the ith class, which are assumed to be independent from those of every other class, covariances of estimators in different classes are zero:

$$\text{cov}(\hat{\mu}_i, \hat{\mu}_h) = 0 = \text{cov}\left(\hat{\mu}_i, \hat{b}_h\right) = \text{cov}\left(\hat{b}_i, \hat{b}_h\right) \quad \text{for } i \neq h\ .$$

Example (continued). For the example of Table 6.1, the requisite details for calculating the $\hat{\mu}_i$ and $\hat{b}_i$ are extracted from Table 6.2 in Table 6.5.

c. Estimating residual variance

The residual sum of squares will be denoted SSE(μ_i, b_i) and, for brevity, SSE$''$; it is

$$\text{SSE}(\mu_i, b_i) = \text{SSE}'' = \Sigma_i\Sigma_j\left(y_{ij} - \hat{y}_{ij}\right)^2 \tag{83}$$

TABLE 6.5. CALCULATING $\hat{\mu}_i$ AND $\hat{b}_i$ FROM (75) AND (76) FOR THE DATA OF TABLE 6.2

Statistic	Class $i = 1$	$i = 2$	$i = 3$
$\bar{y}_{i\cdot}$	73	78	89
$\bar{z}_{i\cdot}$	4	3	5
$W_{i,yz}$	3	4	8
$W_{i,zz}$	2	2	8
$\hat{b}_i$	$\frac{3}{2} = 1\frac{1}{2}$	$\frac{4}{2} = 2$	$\frac{8}{8} = 1$
$\hat{\mu}_i$	$73 - 1\frac{1}{2}(4) = 67$	$78 - 2(3) = 72$	$89 - 1(5) = 84$
$W_{i,yy}$	42	8	8

with, on using (75),

$$\hat{y}_{ij} = \hat{\mu}_i + \hat{b}_i z_{ij} = \bar{y}_{i\cdot} + \hat{b}_i(z_{ij} - \bar{z}_{i\cdot}) \ . \tag{84}$$

Substituting into (83) gives

$$\text{SSE}'' = \Sigma_i\left\{\Sigma_j\left[y_{ij} - \bar{y}_{i\cdot} - \hat{b}_i(z_{ij} - \bar{z}_{i\cdot})\right]^2\right\},$$

which shall be written as

$$\text{SSE}'' = \Sigma_i \text{SSE}_i''$$

for

$$\text{SSE}_i'' = \Sigma_j\left[y_{ij} - \bar{y}_{i\cdot} - \hat{b}_i(z_{ij} - \bar{z}_{i\cdot})\right]^2$$

$$= \Sigma_j(y_{ij} - \bar{y}_{i\cdot})^2 - \hat{b}_i^2\Sigma_j(z_{ij} - \bar{z}_{i\cdot})^2 \tag{85}$$

very similar to the derivation of (23). Hence, using the W-notation of (77) and (79),

$$\text{SSE}_i'' = W_{i,yy} - \hat{b}_i^2 W_{i,zz} \tag{86}$$

and

$$\text{SSE}'' = \Sigma_i \text{SSE}_i'' = \Sigma_i W_{i,yy} - \Sigma_i \hat{b}_i^2 W_{i,zz} \tag{87}$$

$$= W_{yy} - \Sigma_i \hat{b}_i W_{i,yz} \ . \tag{88}$$

The expected value of SSE″ is found by noting that (25) with $a = 1$ and $N = n_i$ gives

$$E(\text{SSE}_i'') = (n_i - 2)\sigma^2 \tag{89}$$

so that from (87)

$$E(\text{SSE}'') = (N - 2a)\sigma^2 . \tag{90}$$

Therefore an unbiased estimator of σ^2 is

$$\hat{\sigma}^2 = \frac{\text{SSE}''}{N-a} = \frac{W_{yy} - \Sigma_i \hat{b}_i W_{i,yz}}{N - 2a} . \tag{91}$$

Example (continued). Entries in Table 6.5 used in (88) give

$$\text{SSE}'' = 42 + 8 + 8 - \left[1\tfrac{1}{2}(3) + 2(4) + 1(8)\right] = 37\tfrac{1}{2}, \tag{92}$$

and so

$$\hat{\sigma}^2 = 37\tfrac{1}{2}/(7-6) = 37\tfrac{1}{2} . \tag{93}$$

d. Partitioning the total sum of squares

Partitioning $\text{SST} = \Sigma_i \Sigma_j y_{ij}^2$ for the multiple slopes model largely involves amending Table 6.4 for the single slope model as is necessary for multiple slopes. We therefore derive appropriate sums of squares using procedures similar to those that led to Table 6.4. As there, so here: we have two cases, first fitting classes and then covariates and, second, fitting covariates and then classes.

For both cases we need $R(\mu_i, b_i)$, the reduction in the sum of squares due to fitting (72). As well as (88) for calculating SSE″ we also have, by the definition of $R(\mu_i, b_i)$ that

$$\text{SSE}'' = \Sigma_i \Sigma_j y_{ij}^2 - R(\mu_i, b_i) . \tag{94}$$

Equating this to (88) gives

$$R(\mu_i, b_i) = \Sigma_i \Sigma_j y_{ij}^2 - W_{yy} + \Sigma_i \hat{b}_i W_{i,yz} . \tag{95}$$

But $W_{yy} = \Sigma_i \Sigma_j (y_{ij} - \bar{y}_{i\cdot})^2$ from Table 6.3 and so

$$R(\mu_i, b_i) = \Sigma_i n_i \bar{y}_{i\cdot}^2 + \Sigma_i \hat{b}_i W_{i,yz} . \tag{96}$$

The starting point for partitioning SST is to use (43) but with the b_i playing the part of b. This gives, using (83),

$$\text{SST} = R(\mu) + \mathscr{R}(\mu_i \mid \mu) + \mathscr{R}(\mu_i, b_i \mid \mu_i) + \text{SSE}(\mu_i, b_i)\ . \qquad (97)$$

The first two terms are as before, in (45) and (46), and the third is

$$\mathscr{R}(\mu_i, b_i \mid \mu_i) = R(\mu_i, b_i) - R(\mu_i) = \Sigma_i \hat{b}_i W_{i,\,yz} \qquad (98)$$

from (96) and (44); and the last term in (97) is SSE″ of (88).

To provide a test of the hypothesis H: b_i *all equal* we make a further partitioning, of $R(\mu_i, b_i \mid \mu_i)$ of (98), doing it in a manner that introduces $R(\mu_i, b)$ of the single slope model:

$$R(\mu_i, b_i \mid \mu_i) = R(\mu_i, b_i) - R(\mu_i)$$

$$= R(\mu_i, b) - R(\mu_i) + R(\mu_i, b_i) - R(\mu_i, b) \qquad (99)$$

$$= \mathscr{R}(\mu_i, b \mid \mu_i) + \mathscr{R}(\mu_i, b_i \mid \mu_i, b) \qquad (100)$$

$$= \hat{b} W_{yz} + \mathscr{R}(\mu_i, b_i \mid \mu_i, b), \qquad (101)$$

from (48). Thus from (98) and (101)

$$\mathscr{R}(\mu_i, b_i \mid \mu_i, b) = \Sigma_i \hat{b}_i W_{i,\,yz} - \hat{b} W_{yz}, \qquad (102)$$

and this provides a test of H: b_i *all equal*. The partitioning of SST based on (44), (45) and (46), as in Table 6.4, and then (98), (48), (102), and (88) is shown in the analysis of variance table of part (a) of Table 6.6.

Example (continued). Many of the values needed for Table 6.6E are available in Table 6.4E. Those not in Table 6.4E are from Table 6.5 and

$$(98):\quad \mathscr{R}(\mu_i, b_i \mid \mu_i) = 1\tfrac{1}{2}(3) + 2(4) + 1(8) = 20\tfrac{1}{2}, \qquad (103)$$

$$(102):\ \mathscr{R}(\mu_i, b_i \mid \mu_i, b) = 20\tfrac{1}{2} - 18\tfrac{3}{4} = 1\tfrac{3}{4},\quad \text{using (103) and (53);} \qquad (104)$$

and, by using (50) and (103), from

$$(96):\qquad R(\mu_i, b_i) = 43997 + \left[1\tfrac{1}{2}(3) + 2(4) + 1(8)\right] = 44017\tfrac{1}{2}\ . \qquad (105)$$

TABLE 6.6. ANALYSIS OF VARIANCE FOR THE 1-WAY CLASSIFICATION, UNBALANCED DATA, WITH ONE COVARIATE, INTRA-CLASS REGRESSION MODEL

$E(y_{ij}) = \mu_i + b_i z_{ij}$

Line	Source of Variation	d.f.	Sum of Squares: $R(\cdot)$	Sum of Squares: Calculation	Equation	Associated Hypothesis
	(a) Fitting classes and then covariate					
1.	Classes	a	$R(\mu_i)$	$= \Sigma_i n_i \bar{y}_{i\cdot}^2$	(44)	
2.	Mean	1	$R(\mu)$	$= N\bar{y}_{\cdot\cdot}^2$	(45)	H: $\Sigma_i n_i(\mu_i + b_i\bar{z}_{i\cdot}) = 0$
3.	Classes (adjusted for mean)	$a - 1$	$\mathscr{R}(\mu_i \mid \mu)$	$= B_{yy}$	(46)	H: $\mu_i + b_i\bar{z}_{i\cdot}$ *equal* $\forall\ i$
4.	Covariate (adjusted for classes)	a	$\mathscr{R}(\mu_i, b_i \mid \mu_i)$	$= \Sigma_i \hat{b}_i W_{i,yz}$	(98)	H: $b_i = 0 \ \forall\ i$
5.	Single slope	1	$\mathscr{R}(\mu_i, b \mid \mu_i)$	$= bW_{yz}$	(48)	H: $\Sigma_i b_i W_{i,zz}/W_{zz} = 0$
6.	Deviations	$a - 1$	$\mathscr{R}(\mu_i, b_i \mid \mu_i, b)$	$= \Sigma_i \hat{b}_i W_{i,yz} - \hat{b}W_{yz}$	(102)	H: b_i *equal* $\forall\ i$
7.	Residual	$N - 2a$	SSE″	$= W_{yy} - \Sigma_i \hat{b}_i W_{i,yz}$	(88)	
8.	Total	N	SST	$= \Sigma_i \Sigma_j y_{ij}^2$		
	(b) Fitting covariate and then classes					
9.	Mean and covariate	$a + 1$	$\mathscr{R}(\mu, b_i)$	$= \tilde{\mu} y_{\cdot\cdot} + \Sigma_i \tilde{b}_i(\Sigma_j y_{ij} z_{ij})$	(115)	
10.	Classes (adjusted for covariate)	$a - 1$	$\mathscr{R}(\mu_i, b_i \mid \mu, b_i)$	$= R(\mu_i) + \mathscr{R}(\mu_i, b_i \mid \mu_i) - R(\mu, b_i)$[1]	(119)	H: μ_i *equal* $\forall\ i$
11.	Residual	$N - 2a$	SSE″	$= W_{yy} - \Sigma_i \hat{b}_i W_{i,yz}$	(88)	
12.	Total	N	SST	$= \Sigma_i \Sigma_j y_{ij}^2$		

[1] This is Line 1 + Line 4 − Line 9.

TABLE 6.6E. ANALYSIS OF VARIANCE (TABLE 6.6) FOR y AND COVARIATE z OF TABLE 6.1, WITH INTRA-CLASS REGRESSION MODEL

Line	Source of Variation	d.f.	Sum of Squares $R(\mu)$	Calculation	Equation	Associated Hypothesis
	(a) Fitting classes and then covariate					
1.	Classes	3	$R(\mu_i)$	= 43997	(50)	
2.	Mean	1	$R(\mu)$	= 43687	(51)	H: $(3\mu_1 + 2\mu_2 + 2\mu_3) + (12b_1 + 6b_2 + 10b_3) = 0$
3.	Classes, adjusted for mean	2	$\mathscr{R}(\mu_i \mid \mu)$	= 310	(52)	H: $\mu_1 + 4b_1 = \mu_2 + 3b_2 = \mu_3 + 5b_3$
4.	Covariate (adjusted for classes)	3	$\mathscr{R}(\mu_i, b_i \mid \mu_i)$	= $20\frac{1}{2}$	(103)	H: $b_1 = b_2 = b_3 = 0$
5.	Single slope	1	$\mathscr{R}(\mu_i, b \mid \mu_i)$	= $18\frac{3}{4}$	(53)	H: $(b_1 + b_2 + 4b_3)/6 = 0$
6.	Deviations	2	$\mathscr{R}(\mu_i, b_i \mid \mu_i, b)$	= $1\frac{3}{4}$	(104)	H: $b_1 = b_2 = b_3$
7.	Residual	1	SSE″	= $37\frac{1}{2}$	(106)	
8.	Total	7	SST_{yy}	= 44055		
	(b) Fitting covariate and then classes					
9.	Mean and covariate	4	$R(\mu, b_i)$	= $43987\frac{5}{48}$	(117)	
10.	Classes (adjusted for covariate)	2	$\mathscr{R}(\mu_i, b_i \mid \mu, b_i)$	= $30\frac{19}{48}$	(120)	H: $\mu_1 = \mu_2 = \mu_3$
11.	Residual	1	SSE″	= $37\frac{1}{2}$	(106)	
12.	Total	7	SST	= 44055		

With this and SST = 44055 we get, from

$$(94): \qquad \text{SSE}'' = 44055 - 44017\tfrac{1}{2} = 37\tfrac{1}{2}\ . \tag{106}$$

These values are shown in part (a) of Table 6.6E.

e. The equal-class-effects sub-model: a pencil of lines

For the multiple slopes model, the preceding section deals with fitting classes before covariate. The reverse sequence is now considered: fitting covariate before classes. In this case, with the multiple slopes model $E(y_{ij}) = \mu_i + b_i z_{ij}$, we use (56) with b_i in place of b and so get the partitioning

$$\text{SST} = R(\mu) + \mathcal{R}(\mu_i, b_i \mid \mu) + \mathcal{R}(\mu_i, b_i \mid \mu, b_i) + \text{SSE}(\mu_i, b_i)\ . \tag{107}$$

Because, as shall be shown, there are two different partitions of the sum of the first two terms in (107), they are combined, and (107) is written as

$$\text{SST} = R(\mu, b_i) + \mathcal{R}(\mu_i, b_i \mid \mu, b_i) + \text{SSE}(\mu_i, b_i)\ . \tag{108}$$

For (108) we need $R(\mu, b_i)$, the sum of squares due to fitting

$$E(y_{ij}) = \mu + b_i z_{ij}\ . \tag{109}$$

The distinction between this and $E(y_{ij}) = \mu + bz_{ij}$ of (57) is that although both have a single μ for all classes, (109) has a different slope for each class whereas (57) has the same slope for all classes. In terms of the Cartesian co-ordinates $E(y_{ij})$ and z_{ij}, (109) represents a pencil of lines, a set of non-parallel lines with common intercept μ. It is useful because it provides a test of the hypothesis H: μ_i *equal* $\forall\ i$ in the multiple slope model $E(y_{ij}) = \mu_i + b_i z_{ij}$.

-i. Fitting the model. Least squares estimators for μ and b_i in (108) come from minimizing

$$S = \Sigma_i \Sigma_j (y_{ij} - \mu - b_i z_{ij})^2,$$

which yields equations

$$N\tilde{\mu} + \Sigma_i \tilde{b}_i z_{i\cdot} = y_{\cdot\cdot} \tag{110}$$

and

$$\tilde{\mu} z_{i\cdot} + \tilde{b}_i \Sigma_i z_{ij}^2 = \Sigma_j y_{ij} z_{ij} \quad \text{for } i = 1, \ldots, a \tag{111}$$

that have solutions

$$\tilde{\mu} = \left[y_{..} - \Sigma_i \frac{z_{i.}(\Sigma_j y_{ij} z_{ij})}{\Sigma_j z_{ij}^2} \right] \Bigg/ \left(N - \Sigma_i \frac{z_{i.}^2}{\Sigma_j z_{ij}^2} \right) \tag{112}$$

and, for $i = 1, \ldots, a$,

$$\tilde{b}_i = (\Sigma_j y_{ij} z_{ij} - \tilde{\mu} z_{i.}) / \Sigma_j z_{ij}^2 \ . \tag{113}$$

The reduction in sum of squares $R(\mu, b_i)$ due to fitting (109) is derived in the same way that (38) led to (39), by equating two equivalent forms of the resulting error sum of squares:

$$\Sigma_i \Sigma_j y_{ij}^2 - R(\mu, b_i) = \Sigma_i \Sigma_j \left(y_{ij} - \tilde{\mu} - \tilde{b}_i z_{ij} \right)^2 \ . \tag{114}$$

Because $\tilde{\mu}$ and $\tilde{b}_i$ of (112) and (113) are not functions of differences between observed means, as are the B, W and T terms of Table 6.3 and (30) for example, equation (114) does not yield $R(\mu, b_i)$ as a function of familiar sums of squares such as those involved earlier in this chapter. Nevertheless, as shown in the Appendix, (114) does simplify, and to an expression that is an example of the useful result developed in (65) of Section 8.4a. Its form here, from (138) of the Appendix, is

$$R(\mu, b_i) = \tilde{\mu} y_{..} + \sum_{i=1}^{a} \tilde{b}_i (\Sigma_j y_{ij} z_{ij}) \ . \tag{115}$$

The important feature of (115) is its being a sum of products: products of each estimator multiplied by the right-hand side of the corresponding equation in the set of least squares (usually called the normal) equations from which the estimators are derived. These are equations (110) and (111). Thus (115) is $\tilde{\mu}$ multiplied by $y_{..}$, the right-hand side of (110), plus the sum of the products of each $\tilde{b}_i$ multiplied by $\Sigma_j y_{ij} z_{ij}$, the right-hand side of (111). This sum of products, of estimators multiplying right-hand sides of least squares equations, is a quite general method of calculating the reduction in sum of squares for any linear model. It has not been utilized in earlier chapters because there has been no real need for doing so until here. More is said about this in Section 8.4a.

Example (continued). Values needed for calculating $\tilde{\mu}$, $\tilde{b}_i$ and $R(\mu, b_i)$ for the example of Table 6.1 are shown in Table 6.2. It is left for the reader to confirm from (112), (113) and (115) that estimates for the example are

$$\tilde{\mu} = 76\tfrac{79}{144} = 76.5486111 \quad \begin{aligned} \tilde{b}_1 &= -\tfrac{19}{24} = -0.7916666 \\ \tilde{b}_2 &= \tfrac{61}{96} = 0.6354166 \\ \tilde{b}_3 &= 2\tfrac{41}{44} = 2.2847222 \end{aligned} \tag{116}$$

and

$$R(\mu, b_i) = 43987\tfrac{5}{48} = 43987.104166 \ . \tag{117}$$

-ii. Partitioning SST. The individual sums of squares in (108) are now easy: the first and last, $R(\mu, b_i)$ and SSE(μ_i, b_i), are (115) and (88), respectively, and the middle term is

$$\mathscr{R}(\mu_i, b_i | \mu, b_i) = R(\mu_i, b_i) - R(\mu, b_i) \ . \tag{118}$$

Although (118) can be expressed in computational terms using (95) and (115), it is more pertinent to write it in terms of other sums of squares that precede it in Table 6.6, namely as

$$\mathscr{R}(\mu_i, b_i | \mu, b_i) = R(\mu_i) + R(\mu_i, b_i | \mu_i) - R(\mu, b_i) \ . \tag{119}$$

This and (115) and (88) are shown in part (b) of Table 6.6.

Example (continued). Using (105) and (117) the value of (118) is

$$\mathscr{R}(\mu_i, b_i | \mu, b_i) = 44017\tfrac{1}{2} - 43987\tfrac{5}{48} = 30\tfrac{19}{48}, \tag{120}$$

shown in line 10 of Table 6.6E.

-iii. Partitioning R(μ, b_i). As indicated earlier, there are two different partitionings of $R(\mu, b_i)$; therefore that term is shown as is, in part (b) of Table 6.6. Both its partitionings are shown in Table 6.7. Neither are of any particular interest insofar as associated hypotheses are concerned, but because parts or all of these partitionings may be produced by computer packages it is important that their details be displayed.

First, there is the straightforward

$$\begin{aligned} R(\mu, b_i) &= R(\mu) + R(\mu, b_i) - R(\mu) \\ &= R(\mu) + \mathscr{R}(\mu, b_i | \mu) \ . \end{aligned} \tag{121}$$

TABLE 6.7. PARTITIONINGS OF $R(\mu, b_i)$ OF LINE 9, TABLE 6.6

Line	Source of Variation	d.f.	Sum of Squares $R(\cdot)$	Sum of Squares Calculation	Equation
			(a) Partitioning to involve $R(\mu)$		
1.	Mean	1	$R(\mu)$	$= N\bar{y}_{..}^2$	(45)
2.	Covariate	a	$\mathscr{R}(\mu, b_i \mid \mu)$	$= R(\mu, b_i) - R(\text{M})$	(121)
3.	Single slope	1	$\mathscr{R}(\mu, b \mid \mu)$	$= \tilde{b}T_{yz}$	(60)
4.	Deviations	$a-1$	$\mathscr{R}(\mu, b_i \mid \mu, b)$	$= \mathscr{R}(\mu, b_i \mid \mu) - \mathscr{R}(\mu, b \mid \mu)$	(123)
5.	Total	$a+1$	$R(\mu, b_i)$	$= \tilde{\mu}y_{..} + \Sigma_i \tilde{b}_i(\Sigma_j y_{ij}z_{ij})$	(115)
			(b) Partitioning to involve $R(b_i)$		
6.	Slopes	a	$R(b_i)$	$= \Sigma_i(\Sigma_j y_{ij}z_{ij})^2/\Sigma_j z_{ij}^2$	(126)
7.	Intercept, adjusted for slopes	1	$\mathscr{R}(\mu, b_i \mid b_i)$	$= R(\mu, b_i) - R(b_i)$	(127)
8.	Total	$a+1$	$R(\mu, b_i)$	$= \tilde{\mu}y_{..} + \Sigma_i \tilde{b}_i(\Sigma_j y_{ij}z_{ij})$	(115)

ASSOCIATED HYPOTHESIS FOR LINES 1, 2 AND 3 UNDER THE MODEL $E(y_{ij}) = \mu_i + b_i z_{ij}$

1.	$\text{H}: \Sigma_i n_i(\mu_i + b_i\bar{z}_{i\cdot}) = 0$
2.	$\text{H}: (\mu_i - \Sigma_i n_i\mu_i/N)z_{i\cdot} + b_i\Sigma_j z_{ij}^2 - z_{i\cdot}\Sigma_i b_i z_{i\cdot}/N = 0 \quad \forall\ i$
3.	H[1]$: \Sigma_i k_i\mu_i + \Sigma_i q_i b_i = 0$

[1] $k_i = n_i(\bar{z}_{i\cdot} - \bar{z}_{..})T_{zz}$ with $\Sigma_i k_i = 0$, and $q_i = W_{i,zz}/W_{zz}$ with $\Sigma_i q_i = 1$.

Then the second term in (121) can be partitioned through replacing μ_i in (100) by μ; and using that gives (121) as

$$R(\mu, b_i) = R(\mu) + \mathscr{R}(\mu, b \mid \mu) + \mathscr{R}(\mu, b_i \mid \mu, b) \ . \tag{122}$$

The first two terms in (122) are in (45) and (60) and the last is

$$\mathscr{R}(\mu, b_i \mid \mu, b) = R(\mu, b_i) - R(\mu, b),$$

for which computational forms are provided by (115) and (59); but a more pertinent form is

$$\mathscr{R}(\mu, b_i \mid \mu, b) = \mathscr{R}(\mu, b_i \mid \mu) - \mathscr{R}(\mu, b \mid \mu), \tag{123}$$

obtainable from (121) and (60).

Example (continued). From (121)

$$\mathscr{R}(\mu, b_i \mid \mu) = R(\mu, b_i) - R(\mu)$$

$$= 43987\tfrac{5}{48} - 43687, \quad \text{from (117) and (53)}$$

$$= 300\tfrac{5}{48}\,; \tag{124}$$

and from using (124) and (64) in (123)

$$\mathscr{R}(\mu, b_i \mid \mu, b) = 300\tfrac{5}{48} - 85\tfrac{9}{16} = 214\tfrac{13}{24}\ . \tag{125}$$

The second partitioning of $R(\mu, b_i)$ is the straightforward

$$\mathscr{R}(\mu, b_i) = R(b_i) + R(\mu, b_i) - R(b_i)$$

$$= R(b_i) + \mathscr{R}(\mu, b_i \mid b_i)\ .$$

$R(b_i)$ is the sum of squares due to fitting $E(y_{ij}) = b_i z_{ij}$, a pencil of lines through the origin. Hence

$$R(b_i) = \Sigma_i \frac{(\Sigma_i y_{ij} z_{ij})^2}{\Sigma_j z_{ij}^2}\ , \tag{126}$$

and then, of course,

$$\mathscr{R}(\mu, b_i \mid b_i) = R(\mu, b_i) - R(b_i), \tag{127}$$

obtainable from (115) and (126).

These two partitionings of $R(\mu, b_i)$, namely (45), (60) and (123) and (126) and (127), are shown in part (b) of Table 6.7.

$R(\mu, b_i)$ not only has the form (115) but, as shown in (139) of the Appendix, can also be expressed as

$$R(\mu, b_i) = \Sigma_i \frac{(\Sigma_j y_{ij} z_{ij})^2}{\Sigma_j z_{ij}^2} + \left(N - \Sigma_i \frac{z_{i\cdot}^2}{\Sigma_j z_{ij}^2} \right) \tilde{\mu}^2$$

$$= R(b_i) + \left(N - \Sigma_i \frac{z_{i\cdot}^2}{\Sigma_j z_{ij}^2} \right) \tilde{\mu}^2 \tag{128}$$

from (126). Hence

$$\mathscr{R}(\mu, b_i \mid b_i) = R(\mu, b_i) - R(b_i) = \left(N - \Sigma_i \frac{z_{i\cdot}^2}{\Sigma_j z_{ij}^2} \right) \tilde{\mu}^2 . \tag{129}$$

And an alternative form for the coefficient of $\tilde{\mu}^2$ in (128) and (129) is

$$N - \Sigma_i \frac{z_{i\cdot}^2}{\Sigma_j z_{ij}^2} = \Sigma_i \frac{n_i W_{i,zz}}{W_{i,zz} + n_i \bar{z}_{i\cdot}^2} .$$

Both of the expressions (115) and (128) can be used for $R(\mu, b_i)$; part (b) of Table 6.6 shows (115).

Example (continued). Using entries in Table 6.2 for (126) gives

$$R(b_i) = \frac{879^2}{50} + \frac{472^2}{20} + \frac{898^2}{58} = 40495\tfrac{779}{1450} \tag{130}$$

and, for (129), with $\tilde{\mu}$ from (116),

$$\mathscr{R}(\mu, b_i \mid b_i) = \left[7 - \left(\frac{12^2}{50} + \frac{6^2}{20} + \frac{10^2}{58} \right) \right] \left(76\tfrac{79}{144} \right)^2 = 3491\tfrac{19729}{34800} . \tag{131}$$

These values are shown in Table 6.7E.

f. Analysis of variance

Table 6.6 shows the analysis of variance tables corresponding to the preceding partitionings of SST, for taking account of $R(\mu_i)$ in part (a) and of $R(\mu, b_i)$ in part (b). Accounting for $R(\mu, b_i)$ on its own, without partitioning it, is the most frequent use of that term. To include the partitionings of $R(\mu, b_i)$ in Table 6.6 would make that table unmanageably large, so they are shown in Table 6.7.

As with Table 6.4, so with 6.6: we here accept that, from the normality assumption on the error terms, all χ^2 and independence properties can be established (as in Section 11.3) to substantiate the necessary distribution properties of the implied F-statistics in Tables 6.6 and 6.7, namely that each sum of squares Q satisfies conditions (i)–(iv) in Section 6.1g, so that $Q/r\hat{\sigma}^2$ is an F-statistic. And that hypotheses shown in Tables 6.6 and 6.7 can be similarly established. Under these conditions, some comments about the associated hypotheses in those two tables are in order.

TABLE 6.7E. TABLE 6.7 FOR THE EXAMPLE

Line	Source of Variation	d.f.	Sum of Squares		Equation
1.	Mean	1	$R(\mu)$	$= 43687$	(53)
2.	Covariate	3	$\mathcal{R}(\mu, b_i \mid \mu)$	$= 300\frac{5}{48}$	(124)
3.	Single slope	1	$\mathcal{R}(\mu, b \mid \mu)$	$= 85\frac{9}{16}$	(64)
4.	Deviations	2	$\mathcal{R}(\mu, b_i \mid \mu, b)$	$= 214\frac{13}{24}$	(125)
5.	Total	4	$R(\mu, b_i)$	$= 43987\frac{5}{48}$	(117)
6.	Slopes	3	$R(b_i)$	$= 40495\frac{779}{1450}$	(130)
7.	Intercept, adjusted for slopes	1	$\mathcal{R}(\mu, b_i \mid b_i)$	$= 3491\frac{19729}{34800}$	(131)
8.	Total	4	$R(\mu, b_i)$	$= 43987\frac{5}{48}$	(117)

ASSOCIATED HYPOTHESIS FOR LINES 1, 2 AND 3 UNDER THE MODEL $E(y_{ij}) = \mu_i + b_i z_{ij}$

1. H: $(3\mu_1 + 2\mu_2 + 2\mu_3)/7 + (12b_1 + 6b_2 + 10b_3)/7 = 0$
2. H: See Table 6.8
3. H: $\mu_3 - \mu_2 + b_1 - 2b_2 + 9b_3 = 0$

First, Table 6.6. The hypotheses in lines 3, 4, 6 and 10 are all of some general interest, being respectively H: $\mu + b_i\bar{z}_{i\cdot}$ *all equal*, H: *all* b_i *zero*, H: *all* b_i *equal*, and H: *all* μ_i *equal*. In contrast, the hypothesis in line 5 is unlikely to be of any interest whatever. Nevertheless, one aspect of it is worth noting. If the hypothesis H: *all* b_i *equal* is not rejected and we take $b_i = b \;\; \forall \; i$, then H: $\Sigma_i b_i W_{i,zz}/W_{zz} = 0$ of line 5 reduces to H: $b = 0$ since $\Sigma_i W_{i,zz}/W_{zz} = 1$. This is as it should be, since the model is then $E(y_{ij}) = \mu_i + bz_{ij}$ and so line 5 is correct for testing H: $b = 0$, as in line 4 of Table 6.4.

Line 9 of Table 6.6 shows no associated hypothesis. It exists, but is not shown because of its complexity and dependence on the z_{ij}s, similar to that seen in the associated hypotheses of Table 6.7. Moreover, it is line 10 that is of more interest than line 9.

Table 6.7 has sums of squares that by their definition do not depend on the μ_is, i.e., not one of the symbols $R(\cdot)$ or $\mathcal{R}(\cdot \mid \cdot)$ in Table 6.7 contains μ_i. But, of course, under the model $E(y_{ij}) = \mu_i + b_i z_{ij}$ the associated hypothesis for each line of Table 6.7 *does* depend on the μ_is—as seen for lines 1, 2 and 3. The hypothesis in line 1 may, on some occasions, be of interest since it is, as usual, H: $E(\bar{y}_{..}) = 0$. But the hypotheses in lines 2 and 3 are of no general interest whatever. Nevertheless, bearing in mind that either part of Table 6.7 can be inserted in Table 6.6 as a partitioning of its line 9, it is worthwhile viewing the hypotheses in Table 6.7 in light of the hypothesis H: *all* μ_i *equal* in line 10 of Table 6.6. Should that not be rejected, and we take $\mu_i = \mu \quad \forall\ i$, then the hypotheses in Table 6.7 become of some interest.

First, with $\mu_i = \mu \quad \forall\ i$, the hypothesis in line 7 will, whatever its form (which is not shown) in the model $E(y_{ij}) = \mu_i + b_i z_{ij}$, reduce for $\mu_i = \mu$ to H: $\mu = 0$. Thus in the pencil of lines represented by the model $E(y_{ij}) = \mu + b_i z_{ij}$, line 7 tests H: $\mu = 0$, namely that the intercept is zero. And, if that hypothesis is not rejected and μ is taken as zero, line 6 for the model $E(y_{ij}) = b_i z_{ij}$ (which represents a pencil of lines through the origin) will test the hypothesis that those lines are all horizontal.

Similarly, with the model $E(y_{ij}) = \mu + bz_{ij}$ (the pencil of lines with common intercept μ), line 4 tests that those are a single (but not necessarily horizontal) line. If that is not rejected and we take $\mu_i = \mu$ and $b_i = b \quad \forall\ i$, then because (as in the footnote to Table 6.7) $\Sigma_i k_i = 0$ and $\Sigma_i q_i = 1$, the hypothesis shown in line 3 becomes H: $b = 0$. And that is correct, because the model $E(y_{ij}) = \mu_i + b_i z_{ij}$ with $\mu_i = \mu$ and $b_i = b \quad \forall\ i$ becomes the model $E(y_{ij}) = \mu + bz_{ij}$ for which $\tilde{b}T_{yz}$ of line 3 certainly is the numerator sum of squares for testing H: $b = 0$.

Finally, if we take $\mu_i = \mu$ but not $b_i = b$, then in line 2 the hypothesis reduces to a set of equations whose only solution is $b_i = 0 \quad \forall\ i$. Hence this hypothesis reduces to H: $b_i = 0 \quad \forall\ i$; this is as one would expect from the nature of the sum of squares $\mathscr{R}(\mu, b_i \mid \mu)$.

Thus, although the associated hypotheses in Table 6.7 for the model $E(y_{ij}) = \mu_i + b_i z_{ij}$ are not of much interest, they do tie together for various sub-models of that model.

Tables 6.6E and 6.7E are for the example of Table 6.1, and Table 6.8 shows calculations needed for the hypotheses in Table 6.7E.

g. Tests of intercepts

Of particular interest in the analyses of variance of Table 6.6 are the two lines for testing different class means:

$$\text{Line 3:} \quad \mathscr{R}(\mu_i \mid \mu) \quad \text{tests} \quad \text{H: } \mu_i + b_i\bar{z}_{i\cdot} \textit{ all equal}, \tag{132}$$

$$\text{Line 10:} \quad \mathscr{R}(\mu_i, b_i \mid \mu, b_i) \quad \text{tests} \quad \text{H: } \mu_i \textit{ all equal}\ . \tag{133}$$

The first of these, (132), is the one familiarly found in most accounts of analysis of covariance, and is a test of equality of the μ_i in the presence of the average observed value of the covariate in that class. In the presence of both different slopes b_i and different means $\bar{z}_{i\cdot}$ in the different classes,

TABLE 6.8. CALCULATION OF TERMS IN TABLES 6.6 AND 6.7 FOR HYPOTHESES IN TABLE 6.7E

Statistic	Class $i=1$	$i=2$	$i=3$	Total
For Table 6.6				
Line 2: n_i	3	2	2	$N=7$
$n_i\bar{z}_{i\cdot}=z_{i\cdot}$	12	6	10	
$\Sigma n_i(\mu_i+b_i\bar{z}_{i\cdot})$	$(3\mu_1+2\mu_2+2\mu_3)+(12b_1+6b_2+10b_3)$			
Line 3: $\bar{z}_{i\cdot}$	4	3	5	
$\mu_i+b_i\bar{z}_{i\cdot}$	μ_1+4b_1	μ_2+3b_2	μ_3+5b_3	
Line 5: $W_{i,zz}$	2	2	8	$W_{zz}=12$
$\Sigma b_iW_{i,zz}/W_{zz}$	$(b_1+b_2+4b_3)/6$			
For Table 6.7				
Line 2: $\rho=\Sigma n_i\mu_i/N$	$\rho=(3\mu_1+2\mu_2+2\mu_3)/7$			
$\Sigma_j z_{ij}^2$	50	20	58	
$\beta=\Sigma b_iz_{i\cdot}/N$	$\beta=(12b_1+6b_2+10b_3)/7$			
$(\mu_i-\rho)z_{i\cdot}$	$12(\mu_1-\rho)$	$6(\mu_2-\rho)$	$10(\mu_3-\rho)$	
$b_i\Sigma_j z_{ij}^2-z_{i\cdot}\beta$	$50b_1-12\beta$	$20b_2-6\beta$	$58b_3-10\beta$	
Line 3: $T_{zz}=16$				
$k_i=\dfrac{n_i(\bar{z}_{i\cdot}-\bar{z}_{\cdot\cdot})}{T_{zz}}$	$\dfrac{3(4-4)}{16}=0$	$\dfrac{2(3-4)}{16}=\dfrac{-1}{8}$	$\dfrac{2(5-4)}{16}=\dfrac{1}{8}$	$\Sigma_i k_i=0$
$\Sigma k_i\mu_i$	$(\mu_3-\mu_2)/8$			
$q_i=\dfrac{\Sigma_j z_{ij}^2-z_{i\cdot}\bar{z}_{\cdot\cdot}}{T_{zz}}$	$\dfrac{2}{16}=\dfrac{1}{8}$	$\dfrac{-4}{16}=\dfrac{-1}{4}$	$\dfrac{18}{16}=1\dfrac{1}{8}$	$\Sigma q_i=1$
Σq_ib_i	$(b_1-2b_2+9b_3)/8$			

this hypothesis may sometimes be of doubtful interest. One situation where this might be useful is where the covariable is an outcome of the treatment represented by the classes. For example,[1] if for hospital patient j who has had surgery of type i the cost is y_{ij} and the length of stay in hospital is z_{ij}, then z_{ij} does, no doubt, depend on i. (Usually one stays in hospital longer for hip replacement surgery than for appendectomy.) In this case the test in (132) could be appropriate.

In (133), using $\mathscr{R}(\mu_i, b_i \mid \mu, b_i)$, the test is one of equality of the intercepts of the individual intra-class regression lines. In the presence of differing slopes for each class, this, too, may often be a test of doubtful interest. Nevertheless, there are situations where it can be of great interest.

[1]Thanks go to C. E. McCulloch for this example.

Economic data, for example, may provide such a situation, if one is interested in class means after taking account of the slopes but without being concerned about the magnitude of the slopes. Another example could be that of testing the efficacy of a drug at different levels of concentration (the covariable) on different classes of patients (different age groups, perhaps). If all patients in the study receive the drug, then $\mathscr{R}(\mu_i, b_i \mid \mu, b_i)$ provides a test of differences among age groups when the drug is not used.

A more general hypothesis, of which (132) and (133) are each a special case, is

$$\text{H: } \mu_i + b_i \mathring{z}_i \textit{ all equal}, \tag{134}$$

where $\mathring{z}_i$ is some pre-assigned value of the covariate for class i. The numerator sum of squares for this hypothesis, with $a - 1$ degrees of freedom, can be shown to be

$$Q = \Sigma_i \frac{1}{v_i}\left(f_i - \Sigma_i \frac{f_i}{v_i} \bigg/ \Sigma_i \frac{1}{v_i}\right)^2 = \Sigma_i \frac{f_i^2}{v_i} - \left(\Sigma_i \frac{f_i}{v_i}\right)^2 \bigg/ \Sigma_i \frac{1}{v_i} \tag{135}$$

for, with similarities to (70),

$$f_i = \bar{y}_{i\cdot} - \hat{b}_i(\bar{z}_{i\cdot} - \mathring{z}_i) \quad \text{and} \quad v_i = \frac{1}{n_i} + \frac{(\bar{z}_{i\cdot} - \mathring{z}_i)^2}{\Sigma_j (z_{ij} - \bar{z}_{i\cdot})^2}. \tag{136}$$

Special cases of this are $\mathring{z}_i = \bar{z}_{i\cdot}$ giving (132), and $\mathring{z}_i = 0$ giving (133). Illustration of the various hypotheses available from (134) is depicted for three classes in Figure 6.2.

Figure 6.2 illustrates a hypothesis that does not occur in Figure 6.1, namely H: $\mu_i + b_i z_0$ *all equal* where z_0 is the same for all i. This is not available in Figure 6.1, where the regression lines are parallel because that parallelism causes H: $\mu_i + b z_0$ *all equal* to be the same as H: μ_i *all equal*. But in Figure 6.2 the lines are not parallel and so differences such as $(\mu_1 + b_1 z_0) - (\mu_2 + b_2 z_0)$ differ for different values of z_0. Of course the points shown on the $E(y)$ axis in Figure 6.2 are simply those for $\mu_i + b_i z_0$ with $z_0 = 0$, and so those points correspond to H: μ_i *all equal* just as they do in Figure 6.1. But, in contrast to Figure 6.1, in the case of different slopes for the covariate for each class (i.e., non-parallel lines in Figure 6.2), this test of H: μ_i *all equal* is not equivalent to comparing adjusted treatment means at *any* level of the covariate. That interpretation holds only when the lines are parallel; i.e., only when all b_is are the same. Hence, in order to have that interpretation one must first test H: $b_i = b \quad \forall\ i$ using

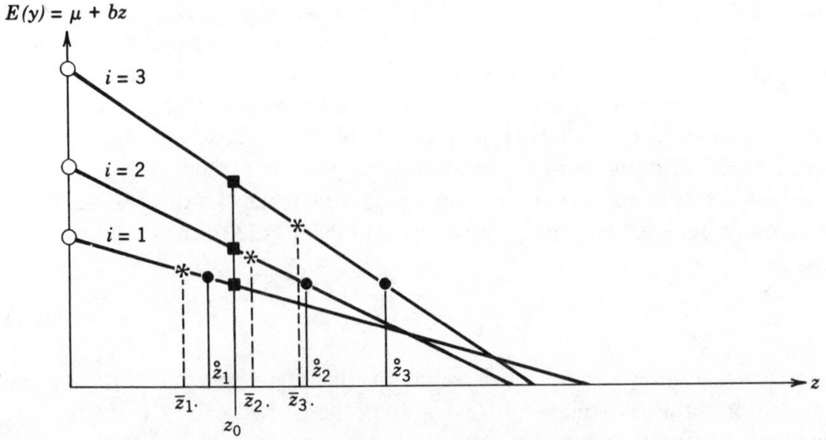

Figure 6.2. Illustration of hypotheses with non-parallel intra-class regression lines (the multiple slope model). The hypotheses tested are equality of heights on the $E(y) = \mu + bz$ axis, as follows.

Points Marked	Hypothesis Tested	Sum of Squares
○	H: μ_i *all equal*	$\mathscr{R}(\mu_i, b_i \mid \mu, b_i)$
∗	H: $\mu_i + b_i\bar{z}_{i\cdot}$ *all equal*	$\mathscr{R}(\mu_i \mid \mu)$
●	H: $\mu_i + b_i\mathring{z}_i$ *all equal* for assigned $\mathring{z}_i$	Q of (135)
■	H: $\mu_i + b_i z_0$ *all equal* for assigned z_0	Q of (135) *with* $\mathring{z}_i = z_0 \quad \forall\ i$

line 6 of Table 6.6. If this hypothesis is not rejected, then apart from the possibility of a type II error at that point, the interpretation of H: μ_i *all equal* being that of equality of adjusted treatment means over all values of the covariate can be made.

Of course, the preceding discussion does not prevent testing H: $\mu_i + b_i z_0$ *all equal* (i.e., equal heights of the points marked ■ in Figure 6.2) for any z_0, but conclusions regarding this hypothesis may well differ for different values of z_0—especially, perhaps, as between two values that occur one on each side of where the lines in Figure 6.2 cross one another. And for $z_0 = \bar{z}_{\cdot\cdot}$ the hypothesis is H: $\mu_i + b_i\bar{z}_{\cdot\cdot}$ *equal* $\forall\ i$, which, by arguments similar to those based on $\hat{\mu} + \hat{b}_i\bar{z}_{\cdot\cdot} = \bar{y}_{i\cdot} - \hat{b}_i(\bar{z}_{i\cdot} - \bar{z}_{\cdot\cdot})$ and following Figure 6.1, can be loosely described as a test of equality of adjusted treatment means.

6.3. Appendix

Derivation of $R(\mu, b_i)$ starts with (114):

$$R(\mu, b_i) = \Sigma_i\Sigma_j y_{ij}^2 - \Sigma_i\Sigma_j\left(y_{ij} - \tilde{\mu} - \tilde{b}_i z_{ij}\right)^2$$

$$= 2\Sigma_i\Sigma_j y_{ij}\left(\tilde{\mu} + \tilde{b}_i z_{ij}\right) - \Sigma_i\Sigma_j\left(\tilde{\mu} + \tilde{b}_i z_{ij}\right)^2 \tag{137}$$

$$= 2\tilde{\mu} y_{..} + 2\Sigma_i\tilde{b}_i\Sigma_j y_{ij}z_{ij} - N\tilde{\mu}^2 - 2\tilde{\mu}\Sigma_i\tilde{b}_i z_{i.} - \Sigma_i b_i^2\Sigma_j z_{ij}^2$$

$$= \tilde{\mu} y_{..} + \tilde{\mu}\left(y_{..} - N\tilde{\mu} - 2\Sigma_i\tilde{b}_i z_{i.}\right) + \Sigma_i\tilde{b}_i\left(2\Sigma_j y_{ij}z_{ij} - \tilde{b}_i\Sigma_j z_{ij}^2\right) .$$

Using $y_{..} - N\tilde{\mu} - \Sigma\tilde{b}_i z_{i.} = 0$ available from (110) and $\tilde{b}_i\Sigma_j z_{ij}^2 = \Sigma_j y_{ij}z_{ij} - \tilde{\mu} z_{i.}$ from (111) gives (137) as

$$R(\mu, b_i) = \tilde{\mu} y_{..} + \tilde{\mu}\left(-\Sigma_i\tilde{b}_i z_{i.}\right) + \Sigma\tilde{b}_i(\Sigma_j y_{ij}z_{ij} + \tilde{\mu} z_{i.})$$

$$= \tilde{\mu} y_{..} + \Sigma_i\tilde{b}_i(\Sigma_j y_{ij}z_{ij}) \tag{138}$$

which is (115). This can also be expressed directly in terms of ys and zs:

$$R(\mu, b_i) = \tilde{\mu} y_{..} + \Sigma_i\frac{\Sigma_j y_{ij}z_{ij} - \tilde{\mu} z_{i.}}{\Sigma_j z_{ij}^2}(\Sigma_j y_{ij}z_{ij})$$

$$= \tilde{\mu}\left(y_{..} - \Sigma_i\frac{z_{i.}\Sigma_j y_{ij}z_{ij}}{\Sigma z_{ij}^2}\right) + \Sigma_i\frac{(\Sigma_i y_{ij}z_{ij})^2}{\Sigma_i z_{ij}^2}$$

$$= \left(N - \Sigma_i\frac{z_{i.}^2}{\Sigma_j z_{ij}^2}\right)\tilde{\mu}^2 + \Sigma_i\frac{(\Sigma_j y_{ij}z_{ij})^2}{\Sigma_j z_{ij}^2} . \tag{139}$$

6.4. Summary

$$\left.\begin{aligned} B_{yz} &= \Sigma_i n_i(\bar{y}_{i.} - \bar{y}_{..})(\bar{z}_{i.} - \bar{z}_{..}) \\ W_{yz} &= \Sigma_i\Sigma_j(y_{ij} - \bar{y}_{i.})(z_{ij} - \bar{z}_{i.}) \\ T_{yz} &= \Sigma_i\Sigma_j(y_{ij} - \bar{y}_{..})(z_{ij} - \bar{z}_{..}) = B_{yz} + W_{yz} \end{aligned}\right\} \text{Also for } yy \text{ and } zz. \tag{30}$$

a. Single slope model: $E(y_{ij}) = \mu_i + bz_{ij}$

$$\hat{b} = W_{yz}/W_{zz}, \qquad v(\hat{b}) = \sigma^2/W_{zz} \tag{33}$$

$$\hat{\mu}_i = \bar{y}_{i\cdot} - \hat{b}\bar{z}_{i\cdot} \tag{6}$$

$$R(\mu_i, b) = \Sigma_i n_i \bar{y}_{i\cdot}^2 + \hat{b}W_{yz} \tag{39}$$

Analyses of variance (Table 6.4)

Tests of intercepts (Figure 6.1)

b. Multiple slopes model: $E(y_{ij}) = \mu_i + b_i z_{ij}$

$$\hat{b}_i = W_{i,\,yz}/W_{i,\,zz}, \qquad v(\hat{b}_i) = \sigma^2/W_{i,\,zz} \quad \text{(78) and (80)}$$

$$\hat{\mu}_i = \bar{y}_{i\cdot} - \hat{b}_i\bar{z}_{i\cdot} \tag{75}$$

$$R(\mu_i, b_i) = \Sigma_i n_i \bar{y}_{i\cdot}^2 + \Sigma_i \hat{b}_i W_{i,\,yz} \tag{96}$$

Analyses of variance (Tables 6.6 and 6.7)

Tests of intercepts (Figure 6.2)

6.5. EXERCISES

E6.1. Using the covariate w in Table 6.1 in place of z:
(a) Calculate Table 6.4. (b) Calculate Tables 6.6 and 6.7.

E6.2. Using w_{ij} of Table 6.1 in place of y_{ij} and z_{ij} as the covariate:
(a) Calculate Table 6.4. (b) Calculate Tables 6.6 and 6.7.

E6.3. Since $E(y_{ij}e_{ij}) = \sigma^2$ and $E(\bar{y}_{i\cdot}e_{ij}) = \sigma^2/n_i$, explain why

$$\theta = \Sigma_i\Sigma_j\left\{E\left[\Sigma_i\Sigma_j(y_{ij} - \bar{y}_{i\cdot})(z_{ij} - \bar{z}_{i\cdot})\right]e_{ij}(z_{ij} - \bar{z}_{i\cdot})\right\}$$

$$\neq \Sigma_i\Sigma_j(1 - 1/n_i)(z_{ij} - \bar{z}_{i\cdot})^2 .$$

E6.4. Corresponding to Table 6.4 in E6.1, calculate $\hat{\sigma}$, and all variances of and covariances between $\hat{\mu}_i$s and $\hat{b}$.

E6.5. Derive $E(\mathrm{SSE}_i'') = (n_i - 2)\sigma^2$ of (89).

E6.6. In Table 6.4, show that the hypotheses in lines 2 and 3 imply that of line 1—as is to be expected.

E6.7. (a) Fit the model $E(y_{ij}) = \mu_i + bz_{ij}$, where $\mu_i + b\mathring{z}_i = \lambda \;\; \forall\; i$ and for given $\mathring{z}_1, \mathring{z}_2, \ldots, \mathring{z}_a$.

(b) Derive the reduction in sum of squares due to fitting the model in (a).

(c) Confirm Q of (70).

E6.8. (a) Fit the model $E(y_{ij}) = \mu_i + b_i z_{ij}$, where $\mu_i + b_i\mathring{z}_i = \lambda \;\; \forall\; i$ and given $\mathring{z}_1, \mathring{z}_2, \ldots, \mathring{z}_a$.

(b) Derive the reduction in sum of squares due to fitting the model in (a).

(c) Confirm Q of (138).

CHAPTER 7

MATRIX ALGEBRA AND QUADRATIC FORMS (A Prelude to Chapter 8)

This chapter contains results that support the general matrix description of linear model methodology given in Chapter 8. It includes results in matrix algebra, solving linear equations, χ^2- and F-distributions, distributional properties of quadratic forms, and testing linear hypotheses. A basic one-semester course in matrix algebra is assumed of the reader although some of what will be found here would also be part of such a course. The topics in this chapter are not dealt with comprehensively: the object is to provide a reader who has the assumed prerequisite with a bridge between that and Chapter 8. The account of the mathematical and statistical preliminaries which follows is therefore both brief and confined to topics that are deemed particularly pertinent to a practitioner's knowledge of linear models. Readers wanting to assimilate Chapter 8 without being concerned about details therein that depend on referring back to Chapter 7 can well overlook this chapter, especially at a first reading.

7.1. MATRIX ALGEBRA

a. Notation

$\mathbf{A}_{p\times q} = \{a_{ij}\}$ for $i = 1,\ldots,p$ and $j = 1,\ldots,q$ is a matrix of order $p \times q$, with a_{ij} being the element in row i and column j. Equivalent notations that will be used are

$$\mathbf{A} = \{_m\, a_{ij}\}_{i=1,\, j=1}^{i=p,\, j=q} = \{_m\, a_{ij}\}_{i,j} = \{_m\, a_{ij}\},$$

where m indicates that the elements inside the braces are being arrayed as a matrix; and sufficient detail of subscripts follows the braces as is necessary, depending on context.

This notation is extended to row and column vectors and to diagonal matrices with the use of r, c and d, as follows. First, a column vector is

$$\mathbf{u} = \begin{bmatrix} u_1 \\ u_2 \\ \vdots \\ u_t \end{bmatrix} = \{_c\, u_i\}_{i=1}^{i=t} = \{_c\, u_i\},$$

the c being used to show that it is a column vector. Similarly

$$\mathbf{u}' = \{_r\, u_i\}_{i=t}^{i=1} = \{_r\, u_i\}$$

is a row vector, and a diagonal matrix is

$$\begin{bmatrix} a_1 & 0 & 0 & \cdots & 0 \\ 0 & a_2 & 0 & \cdots & 0 \\ \vdots & & & \ddots & \vdots \\ 0 & 0 & 0 & \cdots & a_k \end{bmatrix} = \{_d\, a_i\}_{i=1}^{i=k} = \{_d\, a_i\} = \mathbf{D}\{a_i\},$$

where each of the last two symbols are often used interchangeably. Extension to partitioned matrices is straightforward. For example, the direct sum is given by

$$\bigoplus_{i=1}^{3} \mathbf{A}_i = \begin{bmatrix} \mathbf{A}_1 & \mathbf{0} & \mathbf{0} \\ \mathbf{0} & \mathbf{A}_2 & \mathbf{0} \\ \mathbf{0} & \mathbf{0} & \mathbf{A}_3 \end{bmatrix} = \{_d\, \mathbf{A}_i\}_{i=1}^{i=3} = \mathbf{D}\{\mathbf{A}_i\} .$$

b. The rank and trace of a matrix

The rank of $\mathbf{A}$ is denoted by $r_{\mathbf{A}}$ or $r(\mathbf{A})$, usually the former.

The trace of a matrix $\mathbf{A}$ is defined only when $\mathbf{A}$ is square, of order n say, $\mathbf{A}_{n\times n}$, and then it is $\text{tr}(\mathbf{A}_{n\times n}) = \Sigma_{i=1}^{n} a_{ii}$. Further, provided a product $\mathbf{PQ}$ is square, $\text{tr}(\mathbf{PQ}) = \text{tr}(\mathbf{QP})$. A particularly useful application of this commutative property under the trace operator is that for $\mathbf{x}$ being a vector, $\mathbf{x}'\mathbf{Ax} = \text{tr}(\mathbf{x}'\mathbf{Ax}) = \text{tr}(\mathbf{Axx}')$, the first equality being true because $\mathbf{x}'\mathbf{Ax}$ is a scalar.

c. Eigenvalues and eigenvectors

For a square matrix $\mathbf{A}$ and $\mathbf{u} \neq \mathbf{0}$ the equation $\mathbf{Au} = \lambda\mathbf{u}$ defines λ as an eigenvalue of $\mathbf{A}$, and $\mathbf{u}$ as the corresponding eigenvector. For $\mathbf{A}$ being symmetric, every λ is real.

In particular, for a symmetric matrix (of real elements, i.e., not involving $\sqrt{-1}$), there exists an orthogonal matrix $\mathbf{P}$ such that

$$\mathbf{PAP}' = \begin{pmatrix} \mathbf{D}_r & \mathbf{0} \\ \mathbf{0} & \mathbf{0} \end{pmatrix}$$

where $\mathbf{D}_r$ is diagonal, having r nonzero real diagonal elements that are the eigenvalues of $\mathbf{A}$ [e.g., Searle (1982), Section 11.6]; r is the rank of $\mathbf{A}$.

d. Idempotent matrices

A square matrix $\mathbf{P}$ of order n having the property $\mathbf{P}^2 = \mathbf{P}$ is said to be idempotent. It has $r_{\mathbf{P}}$ eigenvalues that are unity and $n - r_{\mathbf{P}}$ that are zero, and $\text{tr}(\mathbf{P}) = r_{\mathbf{P}}$.

e. Summing vectors and J matrices

$\mathbf{1}'_n$ is defined as a row vector, of order n, with every element unity. It is called a *summing vector* because

$$\mathbf{1}'_n\mathbf{x} = \sum_{i=1}^{n} x_i = \mathbf{x}'\mathbf{1}, \quad \text{with} \quad \mathbf{1}'_n\mathbf{1}_n = n$$

being a special case. With the definitions

$$\mathbf{J}_n = \mathbf{1}_n\mathbf{1}'_n \quad \text{and} \quad \bar{\mathbf{J}}_n = \frac{1}{n}\mathbf{J}_n,$$

$\mathbf{J}_n$ and $\bar{\mathbf{J}}_n$ are both square matrices of order n, with $\mathbf{J}_n$ having every element unity, and $\bar{\mathbf{J}}_n$ having every element $1/n$. With $\mathbf{I}_n$ being the identity matrix of order n, the *centering matrix*

$$\mathbf{C}_n = \mathbf{I}_n - \bar{\mathbf{J}}_n$$

is so called because

$$\mathbf{C}_n\mathbf{x} = \{_c\, x_i - \bar{x}.\} \quad \text{for } \bar{x}. = \sum_{i=1}^{n} x_i/n \;.$$

is a vector of deviations of the x_i-values from their mean. $\mathbf{J}_n$, $\bar{\mathbf{J}}_n$ and $\mathbf{C}_n$ are symmetric; $\mathbf{J}_n$ and $\bar{\mathbf{J}}_n$ have rank 1, $\mathbf{C}_n$ has rank $n - 1$, and $\bar{\mathbf{J}}_n$ and $\mathbf{C}_n$ are both idempotent and their products are null: $\mathbf{C\bar{J}} = \mathbf{\bar{J}C} = \mathbf{0}$. Cox (1980) suggests that $\mathbf{J}$ be called the *unitform* matrix.

f. Positive definite and allied matrices

Consider $\mathbf{x'Ax}$ for $\mathbf{A}$ of order n and symmetric. Then $\mathbf{x'Ax} = 0$ for $\mathbf{x} = \mathbf{0}$. Other than this, for $\mathbf{x}$ being real

$\mathbf{x'Ax} > 0 \quad \forall\ \mathbf{x}$ defines $\mathbf{A}$ as positive definite (p.d.);

$\mathbf{x'Ax} \geq 0 \quad \forall\ \mathbf{x}$ with $\mathbf{x'Ax} = 0$ for some $\mathbf{x} \neq \mathbf{0}$ defines $\mathbf{A}$ as positive semi-definite (p.s.d.);

$\mathbf{x'Ax} \geq 0 \quad \forall\ \mathbf{x}$, i.e., including $\mathbf{A}$ as either p.d. or p.s.d., defines $\mathbf{A}$ as non-negative definite (n.n.d.): and then elements of $\mathbf{D}_r$ in Section 7.1c are positive, so that $\mathbf{D}_r = \mathbf{\Delta}_r^2$ for some diagonal matrix $\mathbf{\Delta}_r$.

g. Dispersion matrices

Denoting the variance of a random variable y_i by $v(y_i)$, and the covariance of y_i with y_j by $\text{cov}(y_i, y_j)$ we have the definitions

$$v(y_i) = E[y_i - E(y_i)]^2 \text{ and } \text{cov}(y_i, y_j) = E[y_i - E(y_i)][y_j - E(y_j)] \ .$$

Therefore on defining $\mathbf{V}$ as the matrix of variances of, and covariances between, elements of $\mathbf{y}' = [y_1\ y_2\ \cdots\ y_n]$ we have

$$\begin{aligned}\mathbf{V} &= \text{var}(\mathbf{y}) \\ &= \begin{bmatrix} v(y_1) & \text{cov}(y_1, y_2) & \text{cov}(y_1, y_3) & \cdots & \text{cov}(y_1, y_n) \\ \text{cov}(y_1, y_2) & v(y_2) & \text{cov}(y_2, y_3) & \cdots & \text{cov}(y_2, y_n) \\ \vdots & \vdots & \vdots & & \vdots \\ \text{cov}(y_1, y_n) & \text{cov}(y_2, y_n) & \text{cov}(y_3, y_n) & \cdots & v(y_n) \end{bmatrix} \\ &= E[\{\mathbf{y} - E(\mathbf{y})\}\{\mathbf{y} - E(\mathbf{y})\}'] \ .\end{aligned}$$

$\mathbf{V}$ is called the *dispersion* matrix or *variance–covariance* matrix of $\mathbf{y}$. The variance of a scalar is always represented as $v(\cdot)$, and the dispersion matrix of a vector by $\text{var}(\cdot)$.

The variance of a linear combination of the elements of $\mathbf{y}$, $\mathbf{t'y}$ say, is

$$v(\mathbf{t'y}) = E\{\mathbf{t'y} - E(\mathbf{t'y})\}\{\mathbf{t'y} - E(\mathbf{t'y})\}' = \mathbf{t'Vt} \tag{1}$$

and since $v(\mathbf{t'y}) \geq 0$, we see that $\mathbf{V}$ is n.n.d.

h. Generalized inverse matrices

Readers will be familiar with a nonsingular matrix **T** being a square matrix that has an inverse $\mathbf{T}^{-1}$ such that $\mathbf{TT}^{-1} = \mathbf{T}^{-1}\mathbf{T} = \mathbf{I}$. More generally, for any matrix **A**, be it rectangular, or square and singular, there are always matrices $\mathbf{A}^-$ satisfying

$$\mathbf{AA}^-\mathbf{A} = \mathbf{A}\ . \tag{2}$$

When **A** is nonsingular, (2) leads to $\mathbf{A}^- = \mathbf{A}^{-1}$, but otherwise there is an infinite number of matrices $\mathbf{A}^-$ that for each matrix **A** satisfy (2). Each such $\mathbf{A}^-$ is called a *generalized inverse* of **A**.

Example. For

$$\mathbf{A} = \begin{bmatrix} 1 & 2 & 3 & 2 \\ 3 & 7 & 11 & 4 \\ 4 & 9 & 14 & 6 \end{bmatrix}, \qquad \mathbf{A}^- = \begin{bmatrix} 7-t & -2-t & t \\ -3+2t & 1+2t & -2t \\ -t & -t & t \\ 0 & 0 & 0 \end{bmatrix}. \tag{3}$$

Calculation of $\mathbf{AA}^-\mathbf{A}$ yields **A** no matter what value is used for t, thus illustrating the existence of infinitely many matrices $\mathbf{A}^-$ satisfying (2).

Two useful matrices involving products of **A** and $\mathbf{A}^-$ are

$$\mathbf{A}^-\mathbf{A}, \quad \text{idempotent, of rank } r_{\mathbf{A}}, \tag{4}$$

and

$$\mathbf{A}^{\sim} = \mathbf{A}^-\mathbf{AA}^- \quad \text{for which} \quad \mathbf{AA}^{\sim}\mathbf{A} = \mathbf{A} \quad \text{and} \quad \mathbf{A}^{\sim}\mathbf{AA}^{\sim} = \mathbf{A}^{\sim}\ . \tag{5}$$

Any matrix $\mathbf{A}^*$ satisfying $\mathbf{AA}^*\mathbf{A} = \mathbf{A}$ and $\mathbf{A}^*\mathbf{AA}^* = \mathbf{A}$ is called a *reflexive generalized inverse* of **A**, and $\mathbf{A}^{\sim} = \mathbf{A}^-\mathbf{AA}^-$ of (5) provides a simple way of deriving a generalized inverse of **A** that is reflexive from one that is not.

An important special case of both $\mathbf{A}^-$ and $\mathbf{A}^{\sim}$ is the unique (for given **A**) matrix $\mathbf{A}^+$, which satisfies what are known as the four Penrose conditions:

$$\begin{array}{ll} \text{(i)} \ \ \mathbf{AA}^+\mathbf{A} = \mathbf{A} & \text{(iii)} \ \mathbf{AA}^+ \text{ symmetric} \\ \text{(ii)} \ \mathbf{A}^+\mathbf{AA}^+ = \mathbf{A}^+ & \text{(iv)} \ \mathbf{A}^+\mathbf{A} \text{ symmetric}\ . \end{array} \tag{6}$$

Named after its originators, Moore (1920) and Penrose (1955), the matrix $\mathbf{A}^+$ is called the Moore–Penrose inverse. Matrices $\mathbf{A}^-$ satisfying (2) are matrices that satisfy just Penrose condition (i), in (6), and reflexive generalized inverses $\mathbf{A}^{\sim}$ of (5) satisfy (i) and (ii). The satisfying of all four conditions in (6) produces the matrix $\mathbf{A}^+$ that is not only unique for given

A but which also plays a role for rectangular and for square singular matrices that is similar to that played by the regular inverse of nonsingular (square) matrices. A convenient derivation of $\mathbf{A}^+$ is

$$\mathbf{A}^+ = \mathbf{A}'(\mathbf{A}\mathbf{A}')^-\mathbf{A}(\mathbf{A}'\mathbf{A})^-\mathbf{A}', \tag{7}$$

where $\mathbf{A}'$ represents the transpose of $\mathbf{A}$.

A variety of names for matrices satisfying various combinations of the Penrose conditions of (6) are to be found in the literature, but we shall have occasion to use only the three already mentioned, the most frequently used being matrices that satisfy just (i), namely generalized inverses $\mathbf{A}^-$ of (2). Occasional use is made of a reflexive generalized inverse $\mathbf{A}^{\sim}$ satisfying (i) and (ii), as in (5); and of the Moore–Penrose inverse $\mathbf{A}^+$ in (7), which satisfies all of (6).

i. Generalized inverses of X′X

Matrices of the form $\mathbf{X}'\mathbf{X}$ play an important role in linear models. Clearly, $\mathbf{X}'\mathbf{X}$ is square and symmetric and for $\mathbf{X}$ having elements that are real numbers (i.e., do not involve $\sqrt{-1}$), $\mathbf{X}'\mathbf{X}$ is positive semi-definite (p.s.d.). Solving linear equations of the form $\mathbf{X}'\mathbf{X}\boldsymbol{\beta} = \mathbf{X}'\mathbf{y}$ is considered at length and in terms of using generalized inverses of $\mathbf{X}'\mathbf{X}$, which we denote as $(\mathbf{X}'\mathbf{X})^-$ and $\mathbf{G}$ interchangeably. Then $\mathbf{G}$ is defined by

$$\mathbf{X}'\mathbf{X}\mathbf{G}\mathbf{X}'\mathbf{X} = \mathbf{X}'\mathbf{X} . \tag{8}$$

We also define $\mathbf{H}$ as

$$\mathbf{H} = \mathbf{G}\mathbf{X}'\mathbf{X}, \quad \text{idempotent, of rank } r_{\mathbf{X}} . \tag{9}$$

Note that although $\mathbf{X}'\mathbf{X}$ is symmetric, $\mathbf{G}$ need not be symmetric. For example,

$$\mathbf{X}'\mathbf{X} = \begin{bmatrix} 7 & 3 & 2 & 2 \\ 3 & 3 & \cdot & \cdot \\ 2 & \cdot & 2 & \cdot \\ 2 & \cdot & \cdot & 2 \end{bmatrix} \quad \text{has} \quad \mathbf{G} = \begin{bmatrix} 9 & 0 & 2 & 3 \\ 5 & -13\frac{2}{3} & -16 & -17 \\ 1 & -10 & -11\frac{1}{2} & -13 \\ 0 & -9 & -11 & -11\frac{1}{2} \end{bmatrix} \tag{10}$$

as a generalized inverse, and $\mathbf{G}$ is certainly not symmetric. Despite this, transposing (8) shows that when $\mathbf{G}$ is a generalized inverse of $\mathbf{X}'\mathbf{X}$, then so also is $\mathbf{G}'$. As a consequence, as may be easily verified,

$$(\mathbf{X}'\mathbf{X})^{\sim} = \mathbf{G}\mathbf{X}'\mathbf{X}\mathbf{G}' \tag{11}$$

is a symmetric, reflexive generalized inverse of $\mathbf{X}'\mathbf{X}$ as defined in (5).

The following theorem is a cornerstone for developing many results in linear model theory:

Theorem 7.1. When $\mathbf{G}$ is a generalized inverse of $\mathbf{X}'\mathbf{X}$:

$$\mathbf{G}' \text{ is also a generalized inverse of } \mathbf{X}'\mathbf{X}, \tag{12}$$

$$\mathbf{X}\mathbf{G}\mathbf{X}'\mathbf{X} = \mathbf{X}, \tag{13}$$

$$\mathbf{X}\mathbf{G}\mathbf{X}' \text{ is invariant to } \mathbf{G}, \tag{14}$$

$$\mathbf{X}\mathbf{G}\mathbf{X}' \text{ is symmetric, whether } \mathbf{G} \text{ is or not}, \tag{15}$$

$$\mathbf{X}\mathbf{G}\mathbf{X}'\mathbf{1} = \mathbf{1} \quad \text{when } \mathbf{1} \text{ is a column of } \mathbf{X}, \tag{16}$$

$$\mathbf{X}\mathbf{G}\mathbf{X}' = \mathbf{X}\mathbf{X}^+, \quad \text{where } \mathbf{X}^+ \text{ is the Moore–Penrose inverse of } \mathbf{X}\,. \tag{17}$$

Proof. Condition (12) comes from transposing (8). Result (13) is true because for real matrices there is a theorem [e.g., Searle (1982), p. 63] indicating that if $\mathbf{P}\mathbf{X}'\mathbf{X} = \mathbf{Q}\mathbf{X}'\mathbf{X}$ then $\mathbf{P}\mathbf{X} = \mathbf{Q}\mathbf{X}$; applying this to the transpose of (8) and then transposing yields (13); and applying it to $\mathbf{X}\mathbf{G}\mathbf{X}'\mathbf{X} = \mathbf{X} = \mathbf{X}\mathbf{F}\mathbf{X}'\mathbf{X}$ for $\mathbf{F}$ being any other generalized inverse of $\mathbf{X}'\mathbf{X}$ yields (14). Using $(\mathbf{X}'\mathbf{X})^-$ of (11) in place of $\mathbf{G}$ in $\mathbf{X}\mathbf{G}\mathbf{X}'$ demonstrates the symmetry of (15) which, by (14), therefore holds for any $\mathbf{G}$. Finally, (16) follows from considering an individual column of $\mathbf{X}$ in (13), and (17) is established by using (7) for $\mathbf{X}^+$. **Q.E.D.**

Notice that (12) and (13) spawn three other results similar to (13): $\mathbf{X}\mathbf{G}'\mathbf{X}'\mathbf{X} = \mathbf{X}$, $\mathbf{X}'\mathbf{X}\mathbf{G}\mathbf{X}' = \mathbf{X}'$ and $\mathbf{X}'\mathbf{X}\mathbf{G}'\mathbf{X}' = \mathbf{X}'$. These and (12)–(17) are used frequently in Chapter 8. They have the effect of making $\mathbf{G}$ behave very like (but not exactly the same as) a regular inverse.

A particularly useful matrix is $\mathbf{M} = \mathbf{I} - \mathbf{X}\mathbf{G}\mathbf{X}'$. Theorem 7.1 provides the means for verifying that $\mathbf{M}$ has the following properties: $\mathbf{M}$ is symmetric, idempotent, invariant to $\mathbf{G}$, of rank $N - r_\mathbf{X}$ when $\mathbf{X}$ has N rows, and its products with $\mathbf{X}$ and $\mathbf{X}'$ are null. Thus, with $\mathbf{M}$ having three equivalent forms,

$$\mathbf{M} = \mathbf{I} - \mathbf{X}\mathbf{G}\mathbf{X}' = \mathbf{I} - \mathbf{X}(\mathbf{X}'\mathbf{X})^-\mathbf{X}' = \mathbf{I} - \mathbf{X}\mathbf{X}^+, \tag{18}$$

we have

$$\mathbf{M} = \mathbf{M}' = \mathbf{M}^2, \qquad r_\mathbf{M} = N - r_\mathbf{X}, \qquad \mathbf{M}\mathbf{X} = \mathbf{0} \quad \text{and} \quad \mathbf{X}'\mathbf{M} = \mathbf{0}\,.$$

j. Calculating a generalized inverse

When **A** can be partitioned as

$$\mathbf{A} = \begin{bmatrix} \mathbf{A}_{11} & \mathbf{A}_{12} \\ \mathbf{A}_{21} & \mathbf{A}_{22} \end{bmatrix} \quad \text{with } \mathbf{A}_{11} \text{ nonsingular of order } r_{\mathbf{A}}$$

then a convenient way to calculate a generalized inverse is as

$$\mathbf{A}^{-} = \begin{bmatrix} \mathbf{A}_{11}^{-1} & \mathbf{0} \\ \mathbf{0} & \mathbf{0} \end{bmatrix}, \tag{19}$$

where the null matrices, represented as **0**, have order such that for **A** of order $p \times q$, the order of $\mathbf{A}^{-}$ is $q \times p$, as necessitated by (2).

Example (continued).

$$A^{-} = \left[\begin{array}{cc|cc} 1 & 2 & 3 & 2 \\ 3 & 7 & 11 & 4 \\ \hline 4 & 9 & 14 & 6 \end{array}\right]^{-} = \begin{bmatrix} \begin{pmatrix} 1 & 2 \\ 3 & 7 \end{pmatrix}^{-1} & \begin{matrix} 0 \\ 0 \end{matrix} \\ \begin{matrix} 0 & 0 \\ 0 & 0 \end{matrix} & \begin{matrix} 0 \\ 0 \end{matrix} \end{bmatrix} = \begin{bmatrix} 7 & -2 & 0 \\ -3 & 1 & 0 \\ 0 & 0 & 0 \\ 0 & 0 & 0 \end{bmatrix},$$

as may be confirmed by using $t = 0$ in (3).

Other than (19), little will be said about the computation of $\mathbf{A}^{-}$ for a given **A**. This is because the contribution of generalized inverses to linear model work is more important in the simplification they bring to the derivation and understanding of results than it is in the arithmetic of calculating those results from data. In fact, in many important situations, analytic expressions for elements of required generalized inverses are readily available, so that specific computation is no problem. Otherwise, given the rank of **A**, a broadly applicable algorithm is available that is a generalization of (19). The algorithm, which depends on finding in **A** any nonsingular submatrix of order $r_{\mathbf{A}}$, is described in detail in Searle (1982, Section 8.4). The special form of the algorithm suitable for $\mathbf{X}'\mathbf{X}$ is as follows (*loc. cit.*, Section 8.6b).

(i) In $\mathbf{X}'\mathbf{X}$ find a nonsingular, principal, submatrix of order $r_{\mathbf{X}}$; call it **S**.

(ii) In $\mathbf{X}'\mathbf{X}$ replace each element s_{ij} of **S** by its corresponding element s^{ij} in $\mathbf{S}^{-1}$, the inverse of **S**.

(iii) Replace all other elements of $\mathbf{X}'\mathbf{X}$ by zero.

(iv) The result is a $\mathbf{G} = (\mathbf{X}'\mathbf{X})^{-}$, one that is a symmetric, reflexive generalized inverse [defined in (5)] of $\mathbf{X}'\mathbf{X}$ that satisfies (8) and $\mathbf{G} = \mathbf{G}\mathbf{X}'\mathbf{X}\mathbf{G}$.

7.2. SOLVING LINEAR EQUATIONS

a. Solutions

Linear equations of the form

$$\mathbf{X'X}\boldsymbol{\beta}^\circ = \mathbf{X'y} \tag{20}$$

are the basis of estimation using linear models. When $\mathbf{X'X}$ is nonsingular and $(\mathbf{X'X})^{-1}$ exists, the solution for $\boldsymbol{\beta}^\circ$ is straightforward: $\boldsymbol{\beta}^\circ = (\mathbf{X'X})^{-1}\mathbf{X'y}$. But when $\mathbf{X'X}$ is singular, $(\mathbf{X'X})^{-1}$ does not exist. Then $\mathbf{X'X}$ has rank less than its order, its rows are not linearly independent and so some of its rows are linear combinations of others; and similar relationships hold for the elements of $\mathbf{X'y}$. Thus some equations in (20) are just linear combinations of others. As a result, (20) is basically a set of equations with more unknowns than equations. A consequence of this is that there is not just one unique solution for $\boldsymbol{\beta}^\circ$ to (20), but there is an infinite number of solutions.

Example. Using $\mathbf{X'X}$ of (10), consider the four scalar equations implicit in

$$\begin{bmatrix} 7 & 3 & 2 & 2 \\ 3 & 3 & \cdot & \cdot \\ 2 & \cdot & 2 & \cdot \\ 2 & \cdot & \cdot & 2 \end{bmatrix} \begin{bmatrix} \mu^\circ \\ \alpha_1^\circ \\ \alpha_2^\circ \\ \alpha_3^\circ \end{bmatrix} = \begin{bmatrix} 553 \\ 219 \\ 156 \\ 178 \end{bmatrix}. \tag{21}$$

It is clear that the last three add to the first. As a result there is not just a single solution to (21) but many solutions. Four of them, for example, are

$$\begin{bmatrix} 0 \\ 73 \\ 78 \\ 89 \end{bmatrix}, \quad \begin{bmatrix} 82.25 \\ -9.25 \\ -4.25 \\ 6.75 \end{bmatrix}, \quad \begin{bmatrix} 79 \\ -6 \\ -1 \\ 10 \end{bmatrix} \quad \text{and} \quad \begin{bmatrix} 89 \\ -16 \\ -11 \\ 0 \end{bmatrix}. \tag{22}$$

Equations (21) come from (23) of Section 8.2b. They are an example of (20) and as such, at least for overparameterized models, are not atypical of least squares equations in linear model theory. In light of the multiple solutions (22), and not only for the example but also for (20) in general, a first question therefore is "How do we derive at least one solution?" And, of course, second and subsequent questions relate to such things as deriving other solutions, ascertaining relationships among them and utilizing them in real-life situations. We begin with deriving a solution.

Rao (1962) gives a theorem which, in being applied to (20), shows that $\mathbf{X}'\mathbf{X}\boldsymbol{\beta}^\circ = \mathbf{X}'\mathbf{y}$ has a solution

$$\boldsymbol{\beta}^\circ = \mathbf{G}\mathbf{X}'\mathbf{y}$$

if and only if $\mathbf{X}'\mathbf{X}\mathbf{G}\mathbf{X}'\mathbf{X} = \mathbf{X}'\mathbf{X}$. This means that $\boldsymbol{\beta}^\circ = \mathbf{G}\mathbf{X}'\mathbf{y}$ is a solution for any generalized inverse $\mathbf{G}$ of $\mathbf{X}'\mathbf{X}$. This is the importance of generalized inverses to our purpose.

Thus to obtain a solution we find any $\mathbf{G}$ and use it in $\boldsymbol{\beta}^\circ = \mathbf{G}\mathbf{X}'\mathbf{y}$. But since there are many values for $\mathbf{G}$ and hence many solutions, it is of passing interest to note certain results [e.g., Searle (1971), pp. 8–16; Searle (1982), pp. 236–248] concerning those many solutions. First, if all possible generalized inverses of $\mathbf{X}'\mathbf{X}$ are used for $\mathbf{G}$, all possible solutions of $\mathbf{X}'\mathbf{X}\boldsymbol{\beta} = \mathbf{X}'\mathbf{y}$ are generated by $\mathbf{G}\mathbf{X}'\mathbf{y}$. For example, each of the solutions in (22) corresponds to a different $\mathbf{G}$ (see E7.9). Second, if for any particular $\mathbf{G}$ all possible values of an arbitrary vector $\mathbf{z}$ are used in

$$\dot{\boldsymbol{\beta}} = \mathbf{G}\mathbf{X}'\mathbf{y} + (\mathbf{I} - \mathbf{G}\mathbf{X}'\mathbf{X})\mathbf{z} \tag{23}$$

then $\dot{\boldsymbol{\beta}}$ generates all possible solutions (see E7.10). $\mathbf{z} = \mathbf{0}$ reduces (23) to $\boldsymbol{\beta}^\circ = \mathbf{G}\mathbf{X}'\mathbf{y}$. Third, if $\tilde{\boldsymbol{\beta}}_i$ for $i = 1, \ldots, t$ are any t solutions then $\Sigma_{i=1}^{t}\lambda_i\tilde{\boldsymbol{\beta}}_i$ is also a solution, provided $\Sigma_{i=1}^{t}\lambda_i = 1$. Of all the solutions that can be so generated by these results, no more than $\mathbf{p} - r_{\mathbf{X}} + 1$ of them (for $\mathbf{X}$ of p columns and $\mathbf{y}$ non-null) can be linearly independent.

b. An invariance property

A result that is extremely useful is the following. For a given set of equations (20) there are many solutions, but no matter which one of them is used as $\boldsymbol{\beta}^\circ$, the linear combination of its elements represented by $\mathbf{k}'\boldsymbol{\beta}^\circ$ has the following invariance property:

$$\mathbf{k}'\boldsymbol{\beta}^\circ \text{ is invariant to } \boldsymbol{\beta}^\circ \begin{cases} \text{if and only if } \mathbf{k}' = \mathbf{k}'\mathbf{H} \text{ for } \mathbf{H} = \mathbf{G}\mathbf{X}'\mathbf{X} \\ \text{if and only if } \mathbf{k}' = \mathbf{t}'\mathbf{X} \text{ for some } \mathbf{t}'. \end{cases} \tag{24}$$

The two conditions in (24) are equivalent and mean for any given $\mathbf{X}$, and for any particular $\mathbf{k}'$ such that $\mathbf{k}' = \mathbf{k}'\mathbf{G}\mathbf{X}'\mathbf{X}$, that no matter what solution $\boldsymbol{\beta}^\circ$ is used with that particular $\mathbf{k}'$, the value of $\mathbf{k}'\boldsymbol{\beta}^\circ$ is the same for all solutions $\boldsymbol{\beta}^\circ$. This result is useful in linear model work because the features of a solution to $\mathbf{X}'\mathbf{X}\boldsymbol{\beta}^\circ = \mathbf{X}'\mathbf{y}$ that are usually most interesting are precisely the linear combinations $\mathbf{k}'\boldsymbol{\beta}^\circ$ that are invariant to $\boldsymbol{\beta}^\circ$. As a result, there need be no concern about there being many different solutions: we take any one of them as $\boldsymbol{\beta}^\circ$ and concentrate attention on values of $\mathbf{k}'\boldsymbol{\beta}^\circ$ for different values of $\mathbf{k}' = \mathbf{k}'\mathbf{H}$.

Example (continued). For the equations (21) it is easily verified that a value of $\mathbf{G}$ is

$$\mathbf{G} = \begin{bmatrix} \cdot & \cdot & \cdot & \cdot \\ \cdot & \frac{1}{3} & \cdot & \cdot \\ \cdot & \cdot & \frac{1}{2} & \cdot \\ \cdot & \cdot & \cdot & \frac{1}{2} \end{bmatrix} \quad \text{with} \quad \mathbf{H} = \mathbf{GX'X} = \begin{bmatrix} \cdot & \cdot & \cdot & \cdot \\ 1 & 1 & \cdot & \cdot \\ 1 & \cdot & 1 & \cdot \\ 1 & \cdot & \cdot & 1 \end{bmatrix} \tag{25}$$

and that this corresponds to the first solution shown in (22):

$$\boldsymbol{\beta}^\circ = [0 \quad 73 \quad 78 \quad 89]' \,. \tag{26}$$

Suppose

$$\mathbf{k}' = [7 \quad 2 \quad 1 \quad 4] \,. \tag{27}$$

Using $\mathbf{H}$ of (25) it is clear that the product $\mathbf{k}'\mathbf{H}$ gives $\mathbf{k}'$, i.e., $\mathbf{k}'\mathbf{H} = \mathbf{k}'$. Hence (24) is satisfied. Then for $\mathbf{k}'$ of (27) with $\boldsymbol{\beta}^\circ$ of (26),

$$\mathbf{k}'\boldsymbol{\beta}^\circ = 7(0) + 2(73) + 1(78) + 4(89) = 580 \,. \tag{28}$$

Result (24) means that with $\mathbf{k}'$ of (27) the linear combination $\mathbf{k}'\tilde{\boldsymbol{\beta}}$ will be 580 of (28) for any solution $\tilde{\boldsymbol{\beta}}$. For example, using the other solutions in (22), the values of $\mathbf{k}'\boldsymbol{\beta}^\circ$ for $\mathbf{k}'$ of (27) are

$$\begin{aligned} &7(82\tfrac{1}{4}) + 2(-9\tfrac{1}{4}) + 1(-4\tfrac{1}{4}) + 4(6\tfrac{3}{4}) = 575\tfrac{3}{4} - 18\tfrac{1}{2} - 4\tfrac{1}{4} + 27 = 580 \\ &7(79) + 2(-6) + 1(-1) + 4(10) = 553 + -12 + -1 + 40 = 580 \\ &7(89) + 2(-16) + 1(-11) + 4(0) = 623 - 32 - 11 = 580. \end{aligned}$$

For a $\mathbf{k}'$ different from (27) but one that still satisfies (24), the value of $\mathbf{k}'\tilde{\boldsymbol{\beta}}$ will (usually) be different from the 580 obtained when using (27), but whatever its value it will be the same for all solutions $\tilde{\boldsymbol{\beta}}$. Thus for $\mathbf{k}' = [10 \quad 7 \quad 1 \quad 2]$, the condition $\mathbf{k}' = \mathbf{k}'\mathbf{H}$ in (24) is satisfied and using (26) gives

$$\mathbf{k}'\boldsymbol{\beta}^\circ = 10(0) + 7(73) + 1(78) + 2(89) = 767 \,.$$

The reader should confirm that in using the other solutions in (22), or indeed any other solution of (21), the value of $\mathbf{k}'\tilde{\boldsymbol{\beta}}$ for this second $\mathbf{k}'$ is 767.

A feature of the condition on $\mathbf{k}'$ imposed by (24) is that when it is satisfied for $\mathbf{H} = \mathbf{GX'X}$ based on one generalized inverse of $\mathbf{X'X}$, then it is also satisfied for the $\mathbf{H}$ based on each other generalized inverse of $\mathbf{X'X}$. This is so because if $\dot{\mathbf{H}} = \dot{\mathbf{G}}\mathbf{X'X}$ for $\dot{\mathbf{G}}$ being a generalized inverse of $\mathbf{X'X}$

different from $\mathbf{G}$, then $\mathbf{k}' = \mathbf{k}'\mathbf{H}$ implies

$$\mathbf{k}' = \mathbf{k}'\mathbf{H} = \mathbf{k}'\mathbf{G}\mathbf{X}'\mathbf{X} = \mathbf{k}'\mathbf{G}\mathbf{X}'\mathbf{X}\dot{\mathbf{G}}\mathbf{X}'\mathbf{X} = (\mathbf{k}'\mathbf{G}\mathbf{X}'\mathbf{X})\dot{\mathbf{G}}\mathbf{X}'\mathbf{X} = \mathbf{k}'\dot{\mathbf{H}};$$

i.e., when $\mathbf{k}' = \mathbf{k}'\mathbf{H} = \mathbf{k}'\mathbf{G}\mathbf{X}'\mathbf{X}$ is true for one $\mathbf{G}$ it is also true for all $\mathbf{G}$s.

Another outcome of (24) is that

$$\mathbf{k}' = \mathbf{k}'\mathbf{H} \quad \text{if and only if} \quad \mathbf{k}' = \mathbf{w}'\mathbf{H} \;\text{ for some }\; \mathbf{w}' \;. \tag{29}$$

This means that one way of deriving vectors $\mathbf{k}'$ that satisfy $\mathbf{k}' = \mathbf{k}'\mathbf{H}$ is to form $\mathbf{k}' = \mathbf{w}'\mathbf{H}$ for arbitrary $\mathbf{w}'$ and then give any desired values to elements of $\mathbf{w}$.

Example (continued). Using $\mathbf{G}$ of and $\mathbf{X}'\mathbf{X}$ (10) gives $\mathbf{H} = \mathbf{G}\mathbf{X}'\mathbf{X}$ as

$$\mathbf{H} = \begin{bmatrix} 9 & 0 & 2 & 3 \\ 5 & -13\frac{2}{3} & -16 & -17 \\ 1 & -10 & -11\frac{1}{2} & -13 \\ 0 & -9 & -11 & -11\frac{1}{2} \end{bmatrix} \begin{bmatrix} 7 & 3 & 2 & 2 \\ 3 & 3 & \cdot & \cdot \\ 2 & \cdot & 2 & \cdot \\ 2 & \cdot & \cdot & 2 \end{bmatrix}$$

$$= \begin{bmatrix} 73 & 27 & 22 & 24 \\ -72 & -26 & -22 & -24 \\ -72 & -27 & -21 & -24 \\ -72 & -27 & -22 & -23 \end{bmatrix}, \tag{30}$$

for which it will be found that $\mathbf{k}'$ of (27) satisfies $\mathbf{k}' = \mathbf{k}'\mathbf{H}$:

$$\mathbf{k}'\mathbf{H} = [7 \quad 2 \quad 1 \quad 4]\mathbf{H} = [7 \quad 2 \quad 1 \quad 4] = \mathbf{k}' \;. \tag{31}$$

As illustration of (29), using $\mathbf{H}$ of (25),

$$\mathbf{k}' = \mathbf{w}'\mathbf{H} = [w_1 \quad w_2 \quad w_3 \quad w_4] \begin{bmatrix} \cdot & \cdot & \cdot & \cdot \\ 1 & 1 & \cdot & \cdot \\ 1 & \cdot & 1 & \cdot \\ 1 & \cdot & \cdot & 1 \end{bmatrix}$$

$$= [w_2 + w_3 + w_4 \quad w_2 \quad w_3 \quad w_4] \tag{32}$$

and, for example, $w_2 = 2$, $w_3 = 1$, and $w_4 = 4$, gives the $\mathbf{k}'$ used in (31). Although the same procedure applied to $\mathbf{H}$ of (30) might appear to yield a result different from (32), it does not in fact do so. Thus

$$[w_1 \quad w_2 \quad w_3 \quad w_4]\begin{bmatrix} 73 & 27 & 22 & 24 \\ -72 & -26 & -22 & -24 \\ -72 & -27 & -21 & -24 \\ -72 & -27 & -22 & -23 \end{bmatrix}$$

$$= [73t + w_2 + w_3 + w_4 \quad 27t + w_2 \quad 22t + w_3 \quad 24t + w_4]\ , \quad (33)$$

for $t = w_1 - w_2 - w_3 - w_4$. This has the same property as does (32), namely that its first element is the sum of its other three elements. Further, since w_1 occurs only in t, and w_2, w_3 and w_4 can be chosen at will, as can w_1, we can always choose w_1 to equal $w_2 + w_3 + w_4$, whereupon $t = 0$; and so (33) is identical to (32).

7.3. PARTITIONING X′X

a. Generalized inverses

We have occasion to partition $\mathbf{X}$ as

$$\mathbf{X} = [\mathbf{X}_1 \quad \mathbf{X}_2] \quad (34)$$

with

$$\mathbf{X}'\mathbf{X} = \begin{bmatrix} \mathbf{X}_1'\mathbf{X}_1 & \mathbf{X}_1'\mathbf{X}_2 \\ \mathbf{X}_2'\mathbf{X}_1 & \mathbf{X}_2'\mathbf{X}_2 \end{bmatrix}. \quad (35)$$

Then one form of generalized inverse of (35) is, as taken from Searle (1982, p. 263),

$$\begin{bmatrix} \mathbf{X}_1'\mathbf{X}_1 & \mathbf{X}_1'\mathbf{X}_2 \\ \mathbf{X}_2'\mathbf{X}_1 & \mathbf{X}_2'\mathbf{X}_2 \end{bmatrix}^- = \begin{bmatrix} (\mathbf{X}_1'\mathbf{X}_1)^- & \mathbf{0} \\ \mathbf{0} & \mathbf{0} \end{bmatrix} + \begin{bmatrix} -(\mathbf{X}_1'\mathbf{X}_1)^-\mathbf{X}_1'\mathbf{X}_2 \\ \mathbf{I} \end{bmatrix}$$

$$\times [\mathbf{X}_2'\mathbf{X}_2 - \mathbf{X}_2'\mathbf{X}_1(\mathbf{X}_1'\mathbf{X}_1)^-\mathbf{X}_1'\mathbf{X}_2]^-[-\mathbf{X}_2'\mathbf{X}_1(\mathbf{X}_1'\mathbf{X}_1)^- \quad \mathbf{I}]\ . \quad (36)$$

Define

$$\mathbf{M}_1 = \mathbf{I} - \mathbf{X}_1(\mathbf{X}_1'\mathbf{X}_1)^-\mathbf{X}_1', \quad (37)$$

with $\mathbf{M}_1$ being the same function of $\mathbf{X}_1$ as $\mathbf{M}$ of (18) is of $\mathbf{X}$, and hence $\mathbf{M}_1 = \mathbf{M}_1' = \mathbf{M}_1^2$ and $\mathbf{M}_1\mathbf{X}_1 = \mathbf{0}$. Then writing $\mathbf{G}$ for (36) gives

$$\mathbf{G} = \begin{bmatrix} (\mathbf{X}_1'\mathbf{X}_1)^- & \mathbf{0} \\ \mathbf{0} & \mathbf{0} \end{bmatrix} + \begin{bmatrix} -(\mathbf{X}_1'\mathbf{X}_1)^-\mathbf{X}_1'\mathbf{X}_2 \\ \mathbf{I} \end{bmatrix}(\mathbf{X}_2'\mathbf{M}_1\mathbf{X}_2)$$

$$\times [-\mathbf{X}_2'\mathbf{X}_1(\mathbf{X}_1\mathbf{X}_1)^-\mathbf{I}] \quad (38)$$

as a generalized inverse of $\mathbf{X'X}$. [The reader is warned that the symmetry of $\mathbf{X'X}$ contributes to $\mathbf{G}$ of (38) being a generalized inverse of $\mathbf{X'X}$. Replacing the matrices $\mathbf{X}_1'\mathbf{X}_1$, $\mathbf{X}_1'\mathbf{X}_2$, $\mathbf{X}_2'\mathbf{X}_1$ and $\mathbf{X}_2'\mathbf{X}_2$ in (36) by arbitrary matrices $\mathbf{A}$, $\mathbf{B}$, $\mathbf{C} \neq \mathbf{B}'$ and $\mathbf{D}$ of appropriate order does not, in general, lead to the resulting form of (36) being valid—see Searle (1982, Section 10.5).]

Another form of generalized inverse of (35), analogous to (38), utilizes

$$\mathbf{M}_2 = \mathbf{I} - \mathbf{X}_2(\mathbf{X}_2'\mathbf{X}_2)^{-}\mathbf{X}_2' \tag{39}$$

and is

$$\mathbf{F} = \begin{bmatrix} \mathbf{0} & \mathbf{0} \\ \mathbf{0} & (\mathbf{X}_2'\mathbf{X}_2)^{-} \end{bmatrix} + \begin{bmatrix} \mathbf{I} \\ (-\mathbf{X}_2'\mathbf{X}_2)^{-}\mathbf{X}_2'\mathbf{X}_1 \end{bmatrix} (\mathbf{X}_1'\mathbf{M}_2\mathbf{X}_1)^{-} \times \begin{bmatrix} \mathbf{I} & -\mathbf{X}_1'\mathbf{X}_2(\mathbf{X}_2'\mathbf{X}_2)^{-} \end{bmatrix} . \tag{40}$$

Verification that (38) and (40) are each generalized inverses of $\mathbf{X'X}$ of (35) demands using (13); and although we find that

$$\mathbf{XGX}' = \mathbf{X}_1(\mathbf{X}_1'\mathbf{X}_1)^{-}\mathbf{X}_1' + \mathbf{M}_1\mathbf{X}_2(\mathbf{X}_2'\mathbf{M}_1\mathbf{X}_2)^{-}\mathbf{X}_2'\mathbf{M}_1 \tag{41}$$

and

$$\mathbf{XFX}' = \mathbf{X}_2(\mathbf{X}_2'\mathbf{X}_2)^{-}\mathbf{X}_2' + \mathbf{M}_2\mathbf{X}_1(\mathbf{X}_1'\mathbf{M}_2\mathbf{X}_1)^{-}\mathbf{X}_1'\mathbf{M}_2, \tag{42}$$

which look different, we know from (14) that they are the same. That each is invariant to the generalized inverses it involves is nevertheless clear. In (41), the first term is invariant to the choice of $(\mathbf{X}_1'\mathbf{X}_1)^{-}$—by (14); and by the symmetry and idempotency of $\mathbf{M}_1$ the second term is $\mathbf{M}_1\mathbf{X}_2[(\mathbf{M}_1\mathbf{X}_2)'\mathbf{M}_1\mathbf{X}_2]^{-}(\mathbf{M}_1\mathbf{X}_2)'$ and so it too, by (14), has the invariance property. Nevertheless, a direct development of the equality of (41) to (42) without appealing to (14) seems difficult.

The preceding results of this section are all in terms of generalized inverses. When $\mathbf{X'X}$ of (35) is nonsingular all of those results still apply, with the generalized inverses being regular inverses.

b. Rank

The standard result for the rank of a product matrix is $r_{\mathbf{AB}} \leq r_{\mathbf{A}}$ and $r_{\mathbf{AB}} \leq r_{\mathbf{B}}$. Thus $r(\mathbf{AA}^{-}) \leq r_{\mathbf{A}}$; and since $\mathbf{A} = \mathbf{AA}^{-}\mathbf{A}$, $r_{\mathbf{A}} \leq r(\mathbf{AA}^{-})$. Therefore $r(\mathbf{AA}^{-}) = r_{\mathbf{A}}$. Also, because $\mathbf{AA}^{-}$ is idempotent its trace and rank are equal. In particular, $\text{tr}(\mathbf{AA}^{+}) = r_{\mathbf{A}}$. Therefore from (17)

$$\text{tr}\left[\mathbf{A}(\mathbf{A}'\mathbf{A})^{-}\mathbf{A}'\right] = \text{tr}(\mathbf{AA}^{+}) = r_{\mathbf{A}} . \tag{43}$$

Applying (43) to each term in (41), using the idempotency and symmetry of $\mathbf{M}_1$ in doing so, gives

$$r_{\mathbf{X}} = r_{\mathbf{X}_1} + r_{\mathbf{X}_2'\mathbf{M}_1\mathbf{X}_2},$$

which, on using $r_{\mathbf{AA'}} = r_{\mathbf{A}}$ for $\mathbf{A}$ being real, leads to

$$r_{\mathbf{X}_2'\mathbf{M}_1\mathbf{X}_2} = r_{\mathbf{M}_1\mathbf{X}_2} = r_{[\mathbf{X}_1\ \mathbf{X}_2]} - r_{\mathbf{X}_1} . \tag{44}$$

This result is, it will be seen, most useful in the context of degrees of freedom for sums of squares based on (41). A particular case of (44) is that when $\mathbf{X}$ has full column rank then so too does $\mathbf{M}_1\mathbf{X}_2$; and, of course, $\mathbf{M}_2\mathbf{X}_1$ also.

7.4. NON-CENTRAL χ^2 AND f

Probability densities are the very heart of statistical inference; yet detailed knowledge of their characteristics such as moments, marginal and conditional densities, generating functions, quantiles and so on usually demands familiarity with numerous mathematical details: calculus, transformations, Jacobians and much more. For many practicing statisticians, even those who once had this familiarity, much of this detail ultimately becomes just background and what is important is an understanding of principles and a feel for the importance of selected features of certain distributions. With this in mind, we here give but the briefest description of relevant characteristics of those distributions that are important to understanding linear models. Few of the attendant details [e.g., Searle (1971), Chapter 2] are shown. The reader desiring details can find them in any number of sources such as Arnold (1981), Graybill (1976), Rao (1973), Searle (1971) and Johnson and Kotz (1970), to mention but a few.

a. Normal distributions

Familiarity with the normal distribution is assumed. When a random variable X is normally distributed with mean μ and variance σ^2 we write $X \sim \mathcal{N}(\mu, \sigma^2)$, or X is $\mathcal{N}(\mu, \sigma^2)$.

Notation. Customary statistical notation uses capital letters for random variables and lowercase letters for realizations thereof. Extending this to several variables would lead to using $\mathbf{X}$ as a vector of those variables and $\mathbf{x}$ as a realization. Since this use of $\mathbf{X}$ would clash with its traditional mathematical use as a matrix, we retain the latter and make $\mathbf{x}$ do double duty as a vector of either random variables or a realization thereof, depending on context. Then $\mathbf{X}$ always represents a matrix.

Consider $x_i \sim \mathcal{N}(\mu, \sigma^2)$ for $i = 1, 2, \ldots, n$. We use x_i rather than y_i so as to prompt no questions as to which of the models in Chapters 2–6 is being used. None of them are; x_i is just a general variable. Likewise, μ is just a symbol for a general mean, unrelated to any μ_i or μ_{ij} symbols used earlier. Arraying the x_is in $\mathbf{x}' = [x_1 \quad x_2 \quad \cdots \quad x_n]$ we summarize their being normally distributed by writing $\mathbf{x} \sim \mathcal{N}(\mu\mathbf{1}, \sigma^2\mathbf{I})$. The mean of $\mathbf{x}$, using E as the expectation operator, is $E(\mathbf{x}') = [\mu \quad \mu \quad \cdots \quad \mu] = \mu\mathbf{1}'$, and the dispersion (variance-covariance) matrix of $\mathbf{x}$ is

$$\text{var}(\mathbf{x}) = E(\mathbf{x} - \mu\mathbf{1})(\mathbf{x} - \mu\mathbf{1})' = \sigma^2\mathbf{I} \; .$$

With $E(\mathbf{x}) = \mu\mathbf{1}$ and $\text{var}(\mathbf{x}) = \sigma^2\mathbf{I}$ and each x_i normally distributed we say that the x_is are independently and identically distributed (i.i.d.) $\mathcal{N}(\mu, \sigma^2)$.

More generally, the means of the individual x_is can be represented as elements of the vector $\boldsymbol{\mu}$; and when the variances of the x_is are not necessarily equal, and to allow for covariances between them, the dispersion matrix (which is symmetric) can be represented generally by $\mathbf{V} = \mathbf{V}'$. Hence

$$E(\mathbf{x}) = \boldsymbol{\mu} \quad \text{and} \quad E(\mathbf{x} - \boldsymbol{\mu})(\mathbf{x} - \boldsymbol{\mu})' = \mathbf{V} \; . \tag{45}$$

As in Section 7.1g, $\mathbf{V}$ is always non-negative definite and in most practical situations is positive definite, whereupon $\mathbf{V}$ is nonsingular and $\mathbf{V}^{-1}$ exists. We take this to be the case throughout, and summarize (45) and the normality of the xs by writing

$$\mathbf{x} \sim \mathcal{N}(\boldsymbol{\mu}, \mathbf{V}) \quad \text{with} \quad \mathbf{V} \text{ nonsingular} \; . \tag{46}$$

$\mathbf{x}$ is then described as having a multivariate normal (or multinormal) distribution.

Of the numerous properties of the multivariate normal distribution we state but one that is pertinent: linear combinations of multinormally distributed variables are themselves normally distributed [see, e.g., E8.21 and Searle (1982, p. 353) for an extension]. Thus, for $\mathbf{x}$ of (46) the linear combinations represented by $\mathbf{Tx}$ are normally distributed with mean $\mathbf{T}\boldsymbol{\mu}$ and dispersion matrix $\mathbf{TVT}'$:

$$\mathbf{Tx} \sim \mathcal{N}(\mathbf{T}\boldsymbol{\mu}, \mathbf{TVT}') \; . \tag{47}$$

Developing this result begins by appealing to one's knowledge of the univariate normal distribution for an understanding of the multivariate normal distribution of (46). Then a consequence is (47). But, on starting with (47) one can show that a consequence is (46). More particularly, if a vector of random variables is such that every linear combination of them is normally distributed then one can show that that vector has a multivariate normal distribution. This is sometimes taken as a formal definition of the multivariate normal distribution.

b. χ^2-distributions

The reader will know that when $x_i \sim$ i.i.d. $\mathcal{N}(0,1)$ then $\Sigma_{i=1}^n x_i^2$ has a χ^2-distribution with n degrees of freedom. We write this as

$$\mathbf{x} \sim \mathcal{N}(\mathbf{0}, \mathbf{I}_n) \quad \text{implies} \quad \sum_{i=1}^{n} x_i^2 \sim \chi_n^2 . \tag{48}$$

A familiar adaptation of this [a special case of (67) in Chapter 2] is

$$\mathbf{x} \sim \mathcal{N}(\mu\mathbf{1}, \mathbf{I}_n) \quad \text{implies} \quad \sum_{i=1}^{n} (x_i - \bar{x}.)^2 \sim \chi_{n-1}^2 \tag{49}$$

for $\bar{x}. = \Sigma_{i=1}^n x_i/n$; and an extension of this is

$$\mathbf{x} \sim \mathcal{N}(\mu\mathbf{1}, \sigma^2\mathbf{I}_n) \quad \text{implies} \quad \sum_{i=1}^{n} (x_i - \bar{x}.)^2/\sigma^2 \sim \chi_{n-1}^2 . \tag{50}$$

χ^2-distributions of this nature are known as central χ^2-distributions in contrast to the non-central χ^2-distribution, which is now described.

A feature of the basic χ_n^2-distribution of (48) is that it is the distribution of the sum of squares of i.i.d. normal variables that have zero mean (and unit variance). The distribution of that same sum of squares when those variables have nonzero means is different from χ_n^2 and is called the non-central χ^2, denoted by $\chi^{2\prime}(n, \lambda)$. In addition to its being a function of the degrees of freedom n, it also depends on what is called the non-centrality parameter λ, which is $\lambda = \frac{1}{2}\boldsymbol{\mu}'\boldsymbol{\mu}$ where $\boldsymbol{\mu}$ is the mean of the $\mathbf{x}$. Then, just like (48) we have the following definition.

Definition 7.1.

$$\mathbf{x} \sim \mathcal{N}(\boldsymbol{\mu}, \mathbf{I}_n) \quad \text{implies} \quad \sum_{i=1}^{n} x_i^2 \sim \chi^{2\prime}(n, \lambda) \ \text{with} \ \lambda = \tfrac{1}{2}\boldsymbol{\mu}'\boldsymbol{\mu} . \tag{51}$$

One relationship between the central and non-central χ^2s of (48) and (51), respectively, is very simple: when $\boldsymbol{\mu} = \mathbf{0}$ then $\lambda = 0$ and the non-central $\chi^{2\prime}(n, \lambda)$-distribution reduces to the central χ_n^2-distribution.

c. *F*-distributions

The ratio of two independently distributed χ^2-variables (central χ^2) each divided by their degrees of freedom, has, as is well known, an F-distribution, with degrees of freedom equal to those of the χ^2-variables.

We write this as a definition:

Definition 7.2.

$$\left.\begin{array}{l} u \sim \chi^2_n \\ v \sim \chi^2_d, \text{ independently of } u \end{array}\right\} \text{ implies } \frac{u/n}{v/d} \sim \mathscr{F}_{n,d}, \tag{52}$$

where $\mathscr{F}_{n,d}$ represents the F-distribution on n and d degrees of freedom. In this context u and v are often referred to as the numerator and denominator χ^2s, respectively, and, when they are sums of squares, as the numerator and denominator sums of squares. The label "numerator sum of squares" has been used repeatedly in earlier chapters.

The F-distribution in (52) is called the central F-distribution and, as with χ^2s, there is analogously a non-central F-distribution. Its distinction from (52) is simply that the numerator χ^2 is a non-central χ^2. The resulting non-central F-distribution then depends not only on the two degrees of freedom n and d but also on the non-centrality parameter of the numerator non-central χ^2.

Definition 7.3.

$$\left.\begin{array}{l} u \sim \chi^{2\prime}(n, \lambda) \\ v \sim \chi^2_d, \text{ independently of } u \end{array}\right\} \text{ implies } \frac{u/n}{v/d} \sim \mathscr{F}'(n, d, \lambda)\ . \tag{53}$$

Corollary. Just as the non-central $\chi^{2\prime}(n, \lambda)$ reduces to the central χ^2_n when $\lambda = 0$, so also does the non-central $\mathscr{F}'(n, d, \lambda)$ reduce to the central $\mathscr{F}_{n,d}$ for $\lambda = 0$. This is the crux of the procedure to be described for testing hypotheses using F-statistics.

Other non-central distributions can be found in the literature, such as the non-central t-distribution and the doubly non-central F-distribution; but we need not be concerned about them here. But we do mention the familiar t-distribution.

Definition 7.4.

$$\left.\begin{array}{l} x \sim \mathscr{N}(0,1) \\ u \sim \chi^2_n, \text{ independently of } x \end{array}\right\} \text{ implies } \frac{x}{\sqrt{u/n}} \sim t_n\ . \tag{54}$$

A well-known result connecting (52) and (54) is that the square of a variable having a t-distribution has an F-distribution:

$$w \sim t_n \text{ implies } w^2 \sim \mathscr{F}_{1,n}\ . \tag{55}$$

Examples of this have already been used at equations (87) and (22) of Chapters 2 and 4, respectively.

7.5. QUADRATIC FORMS

In applications of linear models we use F-statistics calculated from various sums of squares obtained from data. On the basis of normality assumptions we are therefore interested in whether or not certain sums of squares have χ^2-distributions, and whether or not they are distributed independently. These are, by Definition 7.2, the conditions under which sums of squares can be used in the form of F-statistics. But every sum of squares calculated from a vector of observations $\mathbf{x}$ is a quadratic function of those observations and so can be expressed as a quadratic form $\mathbf{x}'\mathbf{A}\mathbf{x}$ for $\mathbf{A}$ symmetric and non-negative definite. Furthermore, for $\mathbf{x}$ being a vector of normally distributed variables, two general theorems provide exactly what we want: conditions under which (i) $\mathbf{x}'\mathbf{A}\mathbf{x}$ has a χ^2-distribution and (ii) two quadratic forms $\mathbf{x}'\mathbf{A}\mathbf{x}$ and $\mathbf{x}'\mathbf{B}\mathbf{x}$ are independent. There are also two other theorems giving first and second moments of a quadratic form, and conditions under which $\mathbf{x}'\mathbf{A}\mathbf{x}$ and a set of linear functions $\mathbf{B}\mathbf{x}$ are independent. Before stating these theorems some comments about them are in order.

The theorems can be stated in varying forms, depending on the degree of generality required. For example, in this book we use $\mathbf{x} \sim \mathcal{N}(\boldsymbol{\mu}, \mathbf{V})$, where $\boldsymbol{\mu}$ is any vector and where $\mathbf{V}$ is assumed positive definite and hence nonsingular; but the theorems also can be stated much more broadly, with $\mathbf{V}$ non-negative definite and singular, and also in terms of $\mathbf{x}'\mathbf{A}\mathbf{x} + \mathbf{t}'\mathbf{x} + c$ rather than just $\mathbf{x}'\mathbf{A}\mathbf{x}$. Furthermore, complete statement of the theorems includes both necessary and sufficient conditions; e.g., a necessary and sufficient condition for $\mathbf{x}'\mathbf{A}\mathbf{x}$ to be distributed as a χ^2-variable. Although this generality is attractive mathematically, it is the sufficient condition that is most useful in practice. For example, given a quadratic form $\mathbf{x}'\mathbf{A}\mathbf{x}$ knowing $\mathbf{A}$, the useful thing is to know what (sufficient) condition $\mathbf{A}$ has to satisfy in order for $\mathbf{x}'\mathbf{A}\mathbf{x}$ to have a χ^2-density. The reverse procedure is seldom used: we scarcely ever know that $\mathbf{x}'\mathbf{A}\mathbf{x}$ has a χ^2-density and want to use this to impute some (necessary) condition on $\mathbf{A}$. The statements of the theorems that follow are therefore restricted to sufficient conditions.

There is one theorem for ascertaining whether or not a quadratic form has a χ^2-distribution, and one for ascertaining whether or not two quadratic forms are distributed independently. There also exists a lengthy and comprehensive theorem that provides opportunity for ascertaining both the χ^2- and the independence properties of a set of quadratic forms. This theorem, known as a generalization of Cochran's Theorem, is not given here because of its length, even though, after partitioning a total sum of squares, we do deal with sets of quadratic forms. Nevertheless, the two individual theorems are stated with limited proof—because several of the proofs are very lengthy. Readers interested in details of these proofs will find them in a variety of texts [e.g., Searle (1971), Section 2.5]. In particular, for one necessity condition that has so often in the literature been "proven" erroneously [including Searle (1971), pp. 59–60], three recently-published proofs are given in Searle (1984a). The history of establishing this condition is very interestingly recounted by Driscoll and Gundberg (1986), who also explain why the proof of necessity must always remain so difficult.

Thus our statements of these theorems are made with the limitations (i) nonsingular $\mathbf{V}$, (ii) sufficient conditions only, (iii) no generalization of Cochran's Theorem and (iv) limited proofs. Although these limitations disrobe the theorems of their wide mathematical generality, it is hoped that sufficient generality has nevertheless been retained to make the theorems (and using them) appealing to a broad readership.

The starting point is

$$\mathbf{x} \sim \mathcal{N}(\boldsymbol{\mu}, \mathbf{V}), \tag{56}$$

where $\boldsymbol{\mu} = E(\mathbf{x})$ is quite general, the vector of means of the elements of $\mathbf{x}$; and where $\mathbf{V}$ is positive definite.

a. Mean and variance

Theorem 7.2a. The expected value of a quadratic form $\mathbf{x}'\mathbf{A}\mathbf{x}$ in $\mathbf{x} \sim (\boldsymbol{\mu}, \mathbf{V})$ is

$$E(\mathbf{x}'\mathbf{A}\mathbf{x}) = \text{tr}(\mathbf{A}\mathbf{V}) + \boldsymbol{\mu}'\mathbf{A}\boldsymbol{\mu} .$$

Proof of this is straightforward, as follows.

Proof.

$$E(\mathbf{x}'\mathbf{A}\mathbf{x}) = E[\text{tr}(\mathbf{x}'\mathbf{A}\mathbf{x})] = E[\text{tr}(\mathbf{A}\mathbf{x}\mathbf{x}')] = \text{tr}[\mathbf{A}E(\mathbf{x}\mathbf{x}')] .$$

But from (45)

$$\mathbf{V} = E[(\mathbf{x} - \boldsymbol{\mu})(\mathbf{x} - \boldsymbol{\mu})'] = E(\mathbf{x}\mathbf{x}') - \boldsymbol{\mu}\boldsymbol{\mu}' .$$

Therefore

$$\begin{aligned} E(\mathbf{x}'\mathbf{A}\mathbf{x}) &= \text{tr}[\mathbf{A}(\mathbf{V} + \boldsymbol{\mu}\boldsymbol{\mu}')] = \text{tr}(\mathbf{A}\mathbf{V}) + \text{tr}(\mathbf{A}\boldsymbol{\mu}\boldsymbol{\mu}') \\ &= \text{tr}(\mathbf{A}\mathbf{V}) + \boldsymbol{\mu}'\mathbf{A}\boldsymbol{\mu} . \end{aligned} \tag{57}$$

Q.E.D.

Note that this result does *not* require normality. The next two results do.

Theorem 7.2b. For $\mathbf{x} \sim \mathcal{N}(\boldsymbol{\mu}, \mathbf{V})$

$$v(\mathbf{x}'\mathbf{A}\mathbf{x}) = 2\,\mathrm{tr}(\mathbf{A}\mathbf{V})^2 + 4\boldsymbol{\mu}'\mathbf{A}\mathbf{V}\mathbf{A}\boldsymbol{\mu} \tag{58}$$

and

$$\mathrm{cov}(\mathbf{x}'\mathbf{A}\mathbf{x}, \mathbf{x}) = 2\mathbf{V}\mathbf{A}\boldsymbol{\mu} \ . \tag{59}$$

A general derivation of (58) is that it is the second cumulant of $\mathbf{x}'\mathbf{A}\mathbf{x}$. [The rth cumulant is derived, for example, in Searle (1971, Section 2.5a).] Derivation of (59) is straightforward:

$$\begin{aligned}
\mathrm{cov}(\mathbf{x}, \mathbf{x}'\mathbf{A}\mathbf{x}) &= E\,(\mathbf{x} - \boldsymbol{\mu})[\mathbf{x}'\mathbf{A}\mathbf{x} - E(\mathbf{x}'\mathbf{A}\mathbf{x})] \\
&= E\,(\mathbf{x} - \boldsymbol{\mu})[\mathbf{x}'\mathbf{A}\mathbf{x} - \boldsymbol{\mu}'\mathbf{A}\boldsymbol{\mu} - \mathrm{tr}(\mathbf{A}\mathbf{V})] \\
&= E\,(\mathbf{x} - \boldsymbol{\mu})[(\mathbf{x} - \boldsymbol{\mu})'\mathbf{A}(\mathbf{x} - \boldsymbol{\mu}) + 2(\mathbf{x} - \boldsymbol{\mu})'\mathbf{A}\boldsymbol{\mu} - \mathrm{tr}(\mathbf{A}\mathbf{V})] \\
&= \mathbf{0} + 2\mathbf{V}\mathbf{A}\boldsymbol{\mu} - \mathbf{0} = 2\mathbf{V}\mathbf{A}\boldsymbol{\mu}
\end{aligned}$$

because the first and third moments of $(\mathbf{x} - \boldsymbol{\mu})$ are zero.

b. The χ^2-distribution of a quadratic form

Theorem 7.3. For $\mathbf{x} \sim \mathcal{N}(\boldsymbol{\mu}, \mathbf{V})$, if $\mathbf{A}\mathbf{V}$ is idempotent, then

$$\mathbf{x}'\mathbf{A}\mathbf{x} \sim \chi^{2\prime}\left(r_{\mathbf{A}}, \tfrac{1}{2}\boldsymbol{\mu}'\mathbf{A}\boldsymbol{\mu}\right) \ .$$

Thus the quadratic form $\mathbf{x}'\mathbf{A}\mathbf{x}$ in a normally distributed vector $\mathbf{x}$ has a χ^2-distribution when $\mathbf{A}\mathbf{V}$, the product of the matrix $\mathbf{A}$ of the quadratic form and the dispersion matrix $\mathbf{V}$, is idempotent; and the degrees of freedom of that $\chi^{2\prime}$-distribution are then the rank of $\mathbf{A}$, and the non-centrality parameter is $\frac{1}{2}\boldsymbol{\mu}'\mathbf{A}\boldsymbol{\mu}$. A proof is available in Searle (1971, Section 2.5b).

Corollary. If $\boldsymbol{\mu}'\mathbf{A}\boldsymbol{\mu} = 0$, and $\mathbf{A}\mathbf{V}$ is idempotent then $\mathbf{x}'\mathbf{A}\mathbf{x} \sim \chi^2_{r_{\mathbf{A}}}$, a central χ^2-distribution.

c. Independence of two quadratic forms

Theorem 7.4. Two quadratic forms $\mathbf{x}'\mathbf{A}\mathbf{x}$ and $\mathbf{x}'\mathbf{B}\mathbf{x}$ in the normally distributed vector $\mathbf{x} \sim \mathcal{N}(\boldsymbol{\mu}, \mathbf{V})$ are stochastically independent if $\mathbf{A}\mathbf{V}\mathbf{B} = \mathbf{0}$ (or, equivalently, $\mathbf{B}\mathbf{V}\mathbf{A} = \mathbf{0}$).

This theorem applies whether or not the quadratic forms have χ^2-distributions. In other words, whether both, either or neither of $\mathbf{x}'\mathbf{A}\mathbf{x}$ and $\mathbf{x}'\mathbf{B}\mathbf{x}$ satisfy Theorem 7.3, they are independent if $\mathbf{A}\mathbf{V}\mathbf{B} = \mathbf{0}$.

Proof. From Sections 7.1c and f, $\mathbf{A}$ can be written as $\mathbf{A} = \mathbf{L}\mathbf{L}'$ for $\mathbf{L}' = [\mathbf{D}_r \quad \mathbf{0}]\mathbf{P}$ of full row rank r. Similarly, $\mathbf{B} = \mathbf{K}\mathbf{K}'$ for some full row rank matrix $\mathbf{K}'$. Therefore $\mathbf{AVB} = \mathbf{0}$ implies $\mathbf{LL'VMM'} = \mathbf{0}$ and thus $\mathbf{L'LL'VMM'M} = \mathbf{0}$. But $\mathbf{L'L}$ and $\mathbf{M'M}$ are nonsingular [because $\mathbf{L}'$ and $\mathbf{M}'$ are full row rank—see Searle (1982, Section 7.7e)] and so $\mathbf{L'VM} = \mathbf{0}$. Therefore, because $\mathbf{x}$ is normally distributed, so are $\mathbf{L'x}$ and $\mathbf{x'M}$, and so $\text{cov}(\mathbf{L'x}, \mathbf{x'M}) = \mathbf{L'VM} = \mathbf{0}$ implies $\mathbf{L'x}$ and $\mathbf{x'M}$ being independent. Therefore $\mathbf{x'Ax} = (\mathbf{L'x})'\mathbf{L'x}$ and $\mathbf{x'Bx} = (\mathbf{M'x})'\mathbf{M'x}$ are independent. **Q.E.D.**

It is the necessity condition re-statement of this theorem that is so difficult to prove—see Driscoll and Gundberg (1986).

Theorem 7.5. If $\mathbf{x} \sim N(\boldsymbol{\mu}, \mathbf{V})$ the quadratic form $\mathbf{x'Ax}$ and the set of linear forms $\mathbf{Bx}$ are stochastically independent if $\mathbf{BVA} = \mathbf{0}$.

Again, this is a theorem that applies whether $\mathbf{x'Ax}$ has a χ^2-distribution or not. But note, the condition is just that $\mathbf{BVA} = \mathbf{0}$. There is no equivalent use of $\mathbf{AVB}$, as in Theorem 7.4, because by the very nature of $\mathbf{Bx}$, the product $\mathbf{VB}$ may not exist.

Proof. Using the arguments of the preceding proof, $\mathbf{BVA} = \mathbf{0}$ implies $\mathbf{BVLL'L} = \mathbf{0}$, i.e., $\mathbf{BVL} = \mathbf{0}$. Hence $\text{cov}(\mathbf{Bx}, \mathbf{x'L}) = \mathbf{BVL} = \mathbf{0}$. Thus $\mathbf{Bx}$ and $\mathbf{x'L}$ are independent, and so are $\mathbf{Bx}$ and $\mathbf{x'Ax}$. **Q.E.D.**

7.6. HYPOTHESIS TESTING

a. A general procedure

Theorems 7.3 and 7.4 determine when quadratic forms have χ^2-distributions and when they are independent; and Definition 7.3 indicates how and when to combine them into an F-statistic that has a non-central $\mathcal{F}$-distribution, the corollary to which indicates when it is a central $\mathcal{F}$. This is the framework on which hypotheses are tested using F-statistics.

Suppose we have two quadratic forms $\mathbf{y'Ay}$ and $\mathbf{y'By}$ for

$$\mathbf{y} \sim \mathcal{N}(\boldsymbol{\theta}, \mathbf{V}) \; . \tag{60}$$

Then, from Theorem 7.3,

$$\text{if } \mathbf{AV} \text{ is idempotent,} \quad \mathbf{y'Ay} \sim \chi^{2\prime}\left(r_{\mathbf{A}}, \tfrac{1}{2}\boldsymbol{\theta}'\mathbf{A}\boldsymbol{\theta}\right) \tag{61}$$

and from its corollary

$$\text{if } \mathbf{B}\boldsymbol{\theta} = \mathbf{0} \text{ and if } \mathbf{BV} \text{ is idempotent,} \quad \mathbf{y'By} \sim \chi^2_{r_{\mathbf{B}}}; \tag{62}$$

and from Theorem 7.4,

$$\text{if } \mathbf{AVB} = \mathbf{0}, \quad \mathbf{y'Ay} \text{ and } \mathbf{y'By} \text{ are independent} \tag{63}$$

so that Definition 7.3 then gives

$$F = \frac{\mathbf{y}'\mathbf{A}\mathbf{y}/r_{\mathbf{A}}}{\mathbf{y}'\mathbf{B}\mathbf{y}/r_{\mathbf{B}}} \sim \mathscr{F}_{r_{\mathbf{A}},\, r_{\mathbf{B}},\, \lambda} \quad \text{with} \quad \lambda = \tfrac{1}{2}\boldsymbol{\theta}'\mathbf{A}\boldsymbol{\theta} \ . \tag{64}$$

Now consider the hypothesis

$$\text{H: } \boldsymbol{\theta}'\mathbf{A}\boldsymbol{\theta} = 0 \ . \tag{65}$$

Since $\lambda' = \frac{1}{2}\boldsymbol{\theta}'\mathbf{A}\boldsymbol{\theta}$, the corollary to Definition 7.3 indicates that under the hypothesis (65), the F-statistic of (64) has a central $\mathscr{F}_{r_{\mathbf{A}},\, r_{\mathbf{B}}}$-distribution. Thus comparing the calculated F of (64) with tabulated values of the $\mathscr{F}_{r_{\mathbf{A}},\, r_{\mathbf{B}}}$-distribution provides a test of H of (65).

b. Alternative forms of a hypothesis

H of (65) is of limited use as it stands. It needs to be stated not as a quadratic form of the parameters in $\boldsymbol{\theta} = E(\mathbf{y})$ but as a set of linear statements about the parameters.

-i. Scrutinizing $\boldsymbol{\theta}'\mathbf{A}\boldsymbol{\theta}$. One way of developing a linear hypothesis from H: $\boldsymbol{\theta}'\mathbf{A}\boldsymbol{\theta} = 0$ is, for each particular case, to just simplify $\boldsymbol{\theta}'\mathbf{A}\boldsymbol{\theta}$ and from that simplification derive, by observation, a condition on the elements of $\boldsymbol{\theta}$ that makes $\boldsymbol{\theta}'\mathbf{A}\boldsymbol{\theta}$ zero. For example, if $\boldsymbol{\theta}'\mathbf{A}\boldsymbol{\theta}$ reduces to $\sum_i(\mu_i - \bar{\mu})^2$ then clearly $\mu_1 = \mu_2 = \cdots = \mu_a$ is a condition that makes $\sum_i(\mu_i - \bar{\mu})^2$ zero, and so H: μ_i *all equal* would be the hypothesis. This is the method of simplifying H of (65) used in the example that follows.

-ii. Using expected values. From (57) of Theorem 7.2 we have

$$E(\mathbf{y}'\mathbf{A}\mathbf{y}) = \text{tr}(\mathbf{A}\mathbf{V}) + \boldsymbol{\theta}'\mathbf{A}\boldsymbol{\theta} \ . \tag{66}$$

Extensive use is made of

$$\mathbf{V} = \sigma^2\mathbf{I} \quad \text{and} \quad \mathbf{A} \text{ being idempotent} \ . \tag{67}$$

Under these conditions $\mathbf{A}\mathbf{V}/\sigma^2 = \mathbf{A}$ is idempotent and so

$$\text{tr}(\mathbf{A}\mathbf{V}) = \sigma^2\text{tr}(\mathbf{A}) = \sigma^2 r_{\mathbf{A}} \ .$$

Hence (66) is

$$\boldsymbol{\theta}'\mathbf{A}\boldsymbol{\theta} = E(\mathbf{y}'\mathbf{A}\mathbf{y}) - r_{\mathbf{A}}\sigma^2 \ .$$

Thus H of (65) can be expressed as

$$\text{H: } E(\mathbf{y}'\mathbf{A}\mathbf{y}) - r_{\mathbf{A}}\sigma^2 = 0$$

or

$$\text{H: } \left[E(\mathbf{y}'\mathbf{A}\mathbf{y}) \textit{ ignoring term in } \sigma^2\right] = 0 \ . \tag{68}$$

This is not really any more useful than (65) except that we often derive $\mathbf{E}(\mathbf{y}'\mathbf{A}\mathbf{y})$ directly from the model at hand, with no use of matrix formulations. But only by observation does that produce a hypothesis in terms of linear statements about the parameters.

-iii. Too many statements. Another adaptation of (65) starts from observing that because $\mathbf{y}'\mathbf{A}\mathbf{y}$ is a sum of squares, $\mathbf{A}$ is symmetric and non-negative definite, and so $\mathbf{A} = \mathbf{L}\mathbf{L}'$ for some real matrix $\mathbf{L}'$ (i.e., no element involves $\sqrt{-1}$) of full-row rank $r_{\mathbf{A}}$. Hence $\boldsymbol{\theta}'\mathbf{A}\boldsymbol{\theta} = 0$ is $\boldsymbol{\theta}'\mathbf{L}\mathbf{L}'\boldsymbol{\theta} = 0$, which implies $\mathbf{L}'\boldsymbol{\theta} = \mathbf{0}$ [Searle (1982), p. 62]. Hence (65) is equivalent to

$$\text{H: } \mathbf{L}'\boldsymbol{\theta} = \mathbf{0}\ . \tag{69}$$

From the full-row rank property of $\mathbf{L}'$, this is $r_{\mathbf{A}}$ LIN statements about elements of $\boldsymbol{\theta}$—just the form we want, except that in many cases $\mathbf{L}'$ may not be easy to calculate.

An alternative to (69) comes from observing that $\mathbf{L}'\boldsymbol{\theta} = \mathbf{0}$ implies $\mathbf{L}\mathbf{L}'\boldsymbol{\theta} = \mathbf{0}$ and hence $\mathbf{A}\boldsymbol{\theta} = \mathbf{0}$. Thus

$$\text{H: } \mathbf{A}\boldsymbol{\theta} = \mathbf{0} \tag{70}$$

is also feasible. This is easy to calculate, but it consists of N statements about elements of $\boldsymbol{\theta}$, of which only $r_{\mathbf{A}}$ can be linearly independent. To use (70), therefore, we need to be able to extract from it just $r_{\mathbf{A}}$ linearly independent statements $\mathbf{a}'\boldsymbol{\theta} = 0$. Sometimes this may be easy; and sometimes not.

-iv. Using eigenvectors of A. The F-statistic of (64) used in the form $\mathbf{y}'\mathbf{A}\mathbf{y}/r_{\mathbf{A}}\hat{\sigma}^2$ relies on $(N - r_X)\hat{\sigma}^2 = \mathbf{y}'(\mathbf{I} - \mathbf{X}\mathbf{G}\mathbf{X}')\mathbf{y}/\sigma^2$ having a χ^2-distribution resulting from $\mathbf{I} - \mathbf{X}\mathbf{G}\mathbf{X}'$ being idempotent. It also relies on $\mathbf{y}'\mathbf{A}\mathbf{y}$ (for symmetric $\mathbf{A}$) and $\hat{\sigma}^2$ being independent, consonant with $\mathbf{A}(\mathbf{I} - \mathbf{X}\mathbf{G}\mathbf{X}') = \mathbf{0}$, and on $\mathbf{y}'\mathbf{A}\mathbf{y}/\sigma^2$ having a χ^2-distribution, so that $\mathbf{A}$ is idempotent. Hence $\mathbf{A}$ can be expressed [e.g., Searle (1982), p. 302] as $\mathbf{A} = \mathbf{T}\mathbf{T}'$ for $\mathbf{T}'\mathbf{T} = \mathbf{I}$, where the columns of $\mathbf{T}$ are the $r_{\mathbf{A}}$ orthogonal eigenvectors of $\mathbf{A}$ corresponding to its $r_{\mathbf{A}}$ nonzero eigenvalues. Hence (70) becomes H: $\mathbf{T}\mathbf{T}'\boldsymbol{\theta} = \mathbf{0}$. Pre-multiplying $\mathbf{T}\mathbf{T}'\boldsymbol{\theta} = \mathbf{0}$ by $\mathbf{T}'$ and using $\mathbf{T}'\mathbf{T} = \mathbf{I}$ gives

$$\text{H: } \mathbf{T}'\boldsymbol{\theta} = \mathbf{0}\ . \tag{71}$$

This contains precisely $r_{\mathbf{A}}$ LIN statements about the parameters in $\boldsymbol{\theta}$. It is therefore a useful alternative to (70). True, it involves calculating the eigenvectors of $\mathbf{A}$, but when one is unsuccessful at trying, by observation, to reduce $\boldsymbol{\theta}'\mathbf{A}\boldsymbol{\theta} = 0$ of (65), or $E(\mathbf{y}'\mathbf{A}\mathbf{y}) - r_{\mathbf{A}}\sigma^2 = 0$ of (68), or $\mathbf{A}\boldsymbol{\theta} = \mathbf{0}$ of (70) to $r_{\mathbf{A}}$ LIN statements about elements of $\boldsymbol{\theta}$, then (71) can be a viable alternative. [We return to (71) in Chapter 12, when discussing estimable functions in the SAS computing package output.]

Notation. The notations of Section 7.1a and e are here combined so that, for example, $\mathbf{D}\{\mathbf{J}_i\}$ is a diagonal matrix of sub-matrices $\mathbf{J}_i$ for $i = 1, \ldots, a$, where $\mathbf{J}_i$ represents $\mathbf{J}_{n_i}$ of order n_i; $\mathbf{1}_i$ and $\bar{\mathbf{J}}_i$ will be used similarly, but $\mathbf{J}_N$ and $\bar{\mathbf{J}}_N$ are of order N.

Example. In the 1-way classification of Chapter 2

$$Q = \mathscr{R}(\mu, \mu_i \mid \mu) = \Sigma_i n_i (\bar{y}_{i\cdot} - \bar{y}_{\cdot\cdot})^2 = \mathbf{y}'(\mathbf{D}\{\bar{\mathbf{J}}_i\} - \bar{\mathbf{J}}_N)\mathbf{y}$$

and $$\text{SSE} = \Sigma_i \Sigma_j (y_{ij} - \bar{y}_{i\cdot})^2 = \mathbf{y}'(\mathbf{I} - \mathbf{D}\{\bar{\mathbf{J}}_i\})\mathbf{y} \ .$$

Then $$Q = \mathbf{y}'\mathbf{A}\mathbf{y} \quad \text{has} \quad \mathbf{A} = \mathbf{D}\{\bar{\mathbf{J}}_i\} - \bar{\mathbf{J}}_N \quad \text{with} \quad r_{\mathbf{A}} = a - 1 \tag{72}$$

and $$\text{SSE} = \mathbf{y}'\mathbf{B}\mathbf{y} \quad \text{has} \quad \mathbf{B} = \mathbf{I} - \mathbf{D}\{\bar{\mathbf{J}}_i\} \quad \text{with} \quad r_{\mathbf{B}} = N - a \ . \tag{73}$$

And for $\mathbf{y} \sim [\mathbf{D}\{\mathbf{1}_i\}\boldsymbol{\mu}, \sigma^2\mathbf{I}]$ represented as $\mathbf{y} \sim (\boldsymbol{\theta}, \mathbf{V})$ of (60) we have

$$\boldsymbol{\theta} = \mathbf{D}\{\mathbf{1}_i\}\boldsymbol{\mu} \quad \text{and} \quad \mathbf{V} = \sigma^2\mathbf{I} \ . \tag{74}$$

Properties of the matrices in (72) and (73) useful in (61)–(65) are

$$\bar{\mathbf{J}}_i^2 = \bar{\mathbf{J}}_i, \quad \mathbf{D}\{\bar{\mathbf{J}}_i\}\bar{\mathbf{J}}_N = \bar{\mathbf{J}}_N = \bar{\mathbf{J}}_N\mathbf{D}\{\bar{\mathbf{J}}_i\} \quad \text{and} \quad \bar{\mathbf{J}}_N^2 = \bar{\mathbf{J}}_N, \tag{75}$$

$$\mathbf{B} = \mathbf{I}_N - \mathbf{D}\{\bar{\mathbf{J}}_i\} = \mathbf{D}\{\mathbf{I}_i - \bar{\mathbf{J}}_i\} = \mathbf{D}\{\mathbf{I}_i - \bar{\mathbf{J}}_i\}^2 = \mathbf{B}^2 \tag{76}$$

$$\mathbf{D}\{\bar{\mathbf{J}}_i\}\mathbf{D}\{\mathbf{1}_i\} = \mathbf{D}\{\bar{\mathbf{J}}_i\mathbf{1}_i\} = \mathbf{D}\{\mathbf{1}_i\} \ .$$

Now consider (61) for $\mathbf{A}$ of (72):

$$\mathbf{AV} = \mathbf{A}\sigma^2 \quad \text{and} \quad \mathbf{A}^2 = \mathbf{D}\{\bar{\mathbf{J}}_i\} - 2\bar{\mathbf{J}}_N + \bar{\mathbf{J}}_N = \mathbf{D}\{\bar{\mathbf{J}}_i\} - \bar{\mathbf{J}}_N = \mathbf{A},$$

thus $(\mathbf{A}/\sigma^2)\mathbf{V} = \mathbf{A}$ is idempotent; and $r_{\mathbf{A}/\sigma^2} = r_{\mathbf{A}} = a - 1$. Also $\mathbf{BV} = \mathbf{B}\sigma^2$ and, from (76), $\mathbf{B}^2 = \mathbf{B}$; thus $(\mathbf{B}/\sigma^2)\mathbf{V} = \mathbf{B}$ is idempotent and $r_{\mathbf{B}/\sigma^2} = r_{\mathbf{B}} = N - a$; and for (62),

$$\mathbf{B}\boldsymbol{\theta} = (\mathbf{I}_N - \mathbf{D}\{\bar{\mathbf{J}}_i\})\mathbf{D}\{\mathbf{1}_i\}\boldsymbol{\mu} = (\mathbf{D}\{\mathbf{1}_i\} - \mathbf{D}\{\mathbf{1}_i\})\boldsymbol{\mu} = \mathbf{0} \ .$$

Finally, for (63)

$$\mathbf{AVB} = \sigma^2\mathbf{AB} = \sigma^2(\mathbf{D}\{\bar{\mathbf{J}}_i\} - \bar{\mathbf{J}}_N)(\mathbf{I} - \mathbf{D}\{\bar{\mathbf{J}}_i\})$$

$$= \sigma^2(\mathbf{D}\{\bar{\mathbf{J}}_i\} - \bar{\mathbf{J}}_N - \mathbf{D}\{\bar{\mathbf{J}}_i\} + \bar{\mathbf{J}}_N) = \mathbf{0} \ .$$

Hence $\mathbf{A}/\sigma^2$ and $\mathbf{B}/\sigma^2$ for $\mathbf{A}$ and $\mathbf{B}$ of (72) and (73) satisfy all the conditions of $\mathbf{A}$ and $\mathbf{B}$ in (61)–(63). Therefore by (64) the F-statistic is

$$F = \frac{(\mathbf{y}'\mathbf{A}\mathbf{y}/\sigma^2)/(a-1)}{(\mathbf{y}'\mathbf{B}\mathbf{y}/\sigma^2)/(N-a)} = \frac{\Sigma_i(\bar{y}_{i\cdot} - \bar{y}_{\cdot\cdot})^2/(a-1)}{\Sigma_i\Sigma_j(y_{ij} - \bar{y}_{i\cdot})^2/(N-a)} \sim \mathscr{F}'_{a-1,\, N-a,\, \lambda} \tag{77}$$

for

$$\lambda = \boldsymbol{\theta}'\mathbf{A}\boldsymbol{\theta} = \boldsymbol{\mu}'\mathbf{D}\{\mathbf{1}'_i\}(\mathbf{D}\{\bar{\mathbf{J}}_i\} - \bar{\mathbf{J}}_N)\mathbf{D}\{\mathbf{1}_i\}\boldsymbol{\mu} = \boldsymbol{\mu}'\left(\mathbf{D}\{\mathbf{n}_i\} - \frac{1}{N}\mathbf{nn}'\right)\boldsymbol{\mu}$$

for $\mathbf{n}' = [n_1 \quad n_2 \quad \cdots \quad n_a]$. Therefore

$$\lambda = \boldsymbol{\theta}'\mathbf{A}\boldsymbol{\theta} = \Sigma_i n_i \mu_i^2 - \frac{1}{N}(\Sigma n_i \mu_i)^2 = \Sigma_i n_i(\mu_i - \Sigma_i n_i \mu_i/N)^2 \ . \tag{78}$$

Because, by observation, it is clear that $\mu_1 = \mu_2 = \cdots = \mu_a$ makes $\lambda = 0$, the hypothesis H: $\boldsymbol{\theta}'\mathbf{A}\boldsymbol{\theta} = 0$ of (65) is equivalent to

$$\text{H:}\ \mu_i \textit{ all equal} \ . \tag{79}$$

Thus (79) is the hypothesis tested by F of (77) [see (68)–(70)].

This is a lengthy development of a result obtained much more easily in Chapter 2. It is given purely as illustration of the procedure summarized in (60)–(64). An example of deriving (79) from using eigenvectors as in (71) is given in Section 12.3g-i.

The value of all these procedures is that they apply with broad generality. First, as in Section 8.6, they apply very generally to the sums of squares from partitioned models, as shown in Table 8.4. Once that table is established, we no longer need to consider particular cases of it individually. The interpretive features of any sum of squares is then the hypothesis of (65) and its equivalent forms in (68)–(71). Since it is sums of squares of this nature that occur in much computer output, this also provides the wherewithal to decide on the utility of those sums of squares. A second important use for these procedures is that they enable us to derive the hypothesis tested by any F-statistic found in computer-package output, which, by its nature and title therein might appear to be useful, but for which there is otherwise no explicit statement. And finally, and with emphasis, the conditions of (60)–(63) are essential when we approach hypothesis testing properly, that is, by setting up a hypothesis and testing it.

7.7. EXERCISES

E7.1. Calculate generalized inverses of

$$\mathbf{R} = \begin{bmatrix} 2 & 3 & 5 & 1 \\ 5 & 8 & 13 & 3 \\ 1 & 2 & 3 & 1 \end{bmatrix} \quad \text{and} \quad \mathbf{S} = \begin{bmatrix} 3 & 4 & 0 & 1 \\ 5 & 7 & 0 & 2 \\ 1 & 3 & 2 & 2 \\ 4 & 4 & -2 & 0 \\ 3 & 1 & -4 & -2 \end{bmatrix}.$$

E7.2. From the results obtained in E7.1, calculate reflexive generalized inverses of $\mathbf{R}$ and $\mathbf{S}$.

E7.3. Prove that $\mathbf{A}^-\mathbf{A}$ has rank $r_\mathbf{A}$, and $\mathbf{I} - \mathbf{A}^-\mathbf{A}$ has rank $q - r_\mathbf{A}$ for $\mathbf{A}$ having q columns.

E7.4. Show that:

(a) The definition of $\mathbf{A}^-$ is satisfied by (3) and by (19).

(b) $\mathbf{A}^-\mathbf{A}$ is idempotent.

(c) $\mathbf{A}^-\mathbf{A}\mathbf{A}^-$ and $\mathbf{G}\mathbf{X}'\mathbf{X}\mathbf{G}'$ are reflexive for $\mathbf{X}'\mathbf{X}\mathbf{G}\mathbf{X}'\mathbf{X} = \mathbf{X}'\mathbf{X}$.

E7.5. Show that (7) satisfies (6). Calculate $\mathbf{A}^+$ for $\mathbf{A}$ of E7.1.

E7.6. For

$$\mathbf{X}' = \begin{bmatrix} 1 & 1 & 1 & 1 & 1 & 1 & 1 & 1 \\ 1 & 1 & \cdot & \cdot & \cdot & \cdot & \cdot & \cdot \\ \cdot & \cdot & 1 & 1 & 1 & 1 & \cdot & \cdot \\ \cdot & \cdot & \cdot & \cdot & \cdot & \cdot & 1 & 1 \end{bmatrix}$$

calculate two generalized inverses of $\mathbf{X}'\mathbf{X}$, and with them verify (12)–(17).

E7.7. With results of E7.6, verify (17) using (7) to calculate $\mathbf{X}^+$.

E7.8. Show that

$$\mathbf{X'X} = \begin{bmatrix} 7 & 3 & 2 & 2 \\ 3 & 3 & \cdot & \cdot \\ 2 & \cdot & 2 & \cdot \\ 2 & \cdot & \cdot & 2 \end{bmatrix} \quad \text{has} \quad \mathbf{G} = \begin{bmatrix} 0 & 0 & 0 & 0 \\ 0 & \frac{1}{3} & 0 & 0 \\ 0 & 0 & \frac{1}{2} & 0 \\ 0 & 0 & 0 & \frac{1}{2} \end{bmatrix}$$

as a generalized inverse, and with it verify that

$$\mathbf{F} = \mathbf{G} + \begin{bmatrix} a_1 \\ a_2 \\ a_3 \\ a_4 \end{bmatrix} [1 \quad -1 \quad -1 \quad -1] + \begin{bmatrix} 1 \\ -1 \\ -1 \\ -1 \end{bmatrix} [b_1 \quad b_2 \quad b_3 \quad b_4]$$

is a generalized inverse of $\mathbf{X'X}$ for any values of the a_is and b_is for $i = 1, \ldots, 4$. In particular, show that

$$\mathbf{F}_1 = \begin{bmatrix} 0 & \frac{1}{9} & \frac{1}{6} & \frac{1}{6} \\ 0 & \frac{2}{9} & -\frac{1}{6} & -\frac{1}{6} \\ 0 & -\frac{1}{9} & \frac{1}{3} & -\frac{1}{6} \\ 0 & -\frac{1}{9} & -\frac{1}{6} & \frac{1}{3} \end{bmatrix} \quad \text{and} \quad \mathbf{F}_2 = \begin{bmatrix} \frac{1}{7} & 0 & 0 & 0 \\ -\frac{1}{7} & \frac{1}{3} & 0 & 0 \\ -\frac{1}{7} & 0 & \frac{1}{2} & 0 \\ -\frac{1}{7} & 0 & 0 & \frac{1}{2} \end{bmatrix}$$

are special cases of $\mathbf{F}$.

E7.9. Use $\mathbf{F}_1$ and $\mathbf{F}_2$ of E7.8 as $\mathbf{G}$ in $\boldsymbol{\beta}^\circ = \mathbf{GX'y}$ for solving (21).

E7.10. For $\mathbf{G}$ of (25) calculate $\dot{\boldsymbol{\beta}} = \mathbf{GX'y} + (\mathbf{I} - \mathbf{GX'X})\mathbf{z}$ for equations (21), and then find the values of $\mathbf{z}$ that yield the solutions in (22).

E7.11. With solutions (22) verify (24) using $[13 \quad -4 \quad 8 \quad 9]$ and $[5 \quad -3 \quad 2 \quad 6]$ as values for $\mathbf{k}'$.

E7.12. Solve $\mathbf{X'X}\boldsymbol{\beta} = [88 \quad 22 \quad 52 \quad 12]'$ using $\mathbf{X'X}$ of E7.6. With $\mathbf{G}$ obtained there, develop a $\mathbf{k}'$ suited to (24) and verify the invariance of $\mathbf{k}'\boldsymbol{\beta}^\circ$.

E7.13. Prove (29).

E7.14. (a) Derive (41) and (42).

(b) Using (41) and (42), show that $\mathbf{XGX'X}_1 = \mathbf{X}_1 = \mathbf{XFX'X}_1$.

(c) Verify (38) and (40).

E7.15. For

$$\mathbf{x} \sim \mathcal{N}\left(\begin{bmatrix} 3 \\ 12 \\ 24 \end{bmatrix}, \begin{bmatrix} 4 & 12 & 20 \\ 12 & 45 & 78 \\ 20 & 78 & 137 \end{bmatrix}\right)$$

describe the distribution of x_1, $x_2 - 3x_1$ and $x_1 - 2x_2 + x_3$.

E7.16. Use Theorem 7.3 to establish (48), (49) and (50); and use Theorem 7.4 to show in (50) that $\bar{x}^2$ and $\Sigma_i(x_i - \bar{x})^2$ are independent.

E7.17. Use Theorem 7.3 to establish that $n\bar{x}^2/\sigma^2$ has a χ^2-distribution.

E7.18. Using Theorem 7.2b, show for $\mathbf{x} \sim \mathcal{N}(\boldsymbol{\mu}, \mathbf{V})$ that

$$\operatorname{cov}(\mathbf{x}'\mathbf{A}\mathbf{x}, \mathbf{x}'\mathbf{B}\mathbf{x}) = 2\operatorname{tr}(\mathbf{AVBV}) + 4\boldsymbol{\mu}'\mathbf{AVB}\boldsymbol{\mu} .$$

E7.19. For $\mathbf{x} \sim \mathcal{N}(0, \sigma^2\mathbf{I}_n)$ show that $n\bar{x}$ and Σx_i^2 are uncorrelated but not independent.

E7.20. For $\mathbf{x} \sim \mathcal{N}(\mu_1\mathbf{1}, \sigma_1^2\mathbf{I}_n)$ and $\mathbf{y} \sim \mathcal{N}(\mu_2\mathbf{1}, \sigma_2^2\mathbf{I}_n)$, and with $\operatorname{cov}(x_i y_j) = \delta_{ij}\sigma_{12}$ for δ_{ij} being the Kronecker δ, show that

$$v[\Sigma(x_i - \bar{x})(y_i - \bar{y})] = (n - 1)(\sigma_1^2\sigma_2^2 + \sigma_{12}^2) .$$

E7.21. For $\mathbf{X} = [\mathbf{1}_N \quad \mathbf{W}]$ and $\bar{\mathbf{w}}'$, the row vector of column means of $\mathbf{W}$, give two distinct proofs of the equality

$$[1 \quad \bar{\mathbf{w}}'](\mathbf{X}'\mathbf{X})^{-1}\begin{bmatrix} 1 \\ \bar{\mathbf{w}} \end{bmatrix} = 1/N .$$

CHAPTER 8

A GENERAL LINEAR MODEL

Chapters 2–6 deal with six particular cases: the 1-way classification, the 2-way nested classification, the 2-way crossed classification with and without interaction, and the 1-way classification with a covariate, both the single slope and the multiple slopes models. Readers of those chapters will by this point have sensed that there are certain principles and procedures common to all of them. Examples are (i) minimizing a sum of squares to obtain least squares equations [e.g., Chapter 2, equation (23), which we denote by 2/(23); and 4/(64), 6/(3) and 6/(73)]; (ii) using $\text{SSE} = \Sigma_{ijk}(y_{ijk} - \hat{y}_{ijk})^2$ [e.g., 2/(30), 4/(17), 6/(17) and 6/(83)]; (iii) estimating the residual variance as $\text{SSE}/(N - f)$ for degrees of freedom f [e.g., 2/(34), 3/(8), 4/(17), 4/(82), 5/(10), 6/(25) and 6/(91)]; and (iv) partitioning of SST, the total sum of squares of observations [e.g., Tables 2.4, 3.3, 4.8, 4.11, 6.4 and 6.6]. Other examples could be cited.

There is, indeed, a variety of procedures throughout the preceding chapters that have a common thread from chapter to chapter; and thereby hangs a dilemma in learning about linear models. One learning path is that followed in Chapters 2–6: start with a simple case and from it build towards more and more complicated cases. This is the traditional route. It has the advantage of starting at a point that is easily understood and of progressing by relatively small steps to dealing with more complicated situations. Furthermore, both the starting point and the progression require minimal mathematics. But the disadvantage is that after say, five chapters of this method of learning, one begins to be overwhelmed by numerous details which, although different for the different cases, appear to have something in common—and this can haunt one in terms of questions of the form "Is there an underlying principle that unifies these numerous details?". The answer is resoundingly "Yes". But (and, as is often the case, it is a rather large "but"), learning this unified approach puts a demand on the reader that the start-easy-and-progress-slowly approach does not require, namely knowing a little mathematics. In this case one requirement is an appreciation for the abstraction that is implicit in handling one, single, formulation of a linear model which can encompass any particular model that one might subsequently be interested in; and another requirement is to know some matrix algebra.

Fortunately, in today's world, the realization is well established that basic matrix algebra is as important and useful a mathematical tool as has been acknowledged for basic calculus for generations. This is the prerequisite for understanding the unifying principles of linear model theory: a basic knowledge of matrix algebra. In exchange, one gains a set of unifying methods that are widely applicable to all linear models, and from which, for example, all of Chapters 2–6 are just special cases. More than that, though, the unifying methods also provide procedures for topics that have not been dealt with in those chapters—situations that would, using the methods of those chapters, involve a great deal of tedious and detailed algebra. Such topics include testing multi-part linear hypotheses, establishing χ^2 and independence properties of sums of squares, and (for evaluating the utility of computer-output F-statistics) the derivation of the hypothesis generated by a sum of squares, Q, of known form and with properties such that $Q/r\hat{\sigma}^2$ is an F-statistic. All of these and more exist in very succinct form in the matrix algebra development of linear model theory—and they have wide applicability. This chapter therefore presents a somewhat detailed outline of linear model theory using matrices; some of the requisite matrix algebra is given in Chapter 7, as a prelude to this chapter; and ensuing chapters utilize the matrix notation to supply certain derivations omitted from Chapters 2–6 and to present additional topics pertinent to those chapters. The material in this chapter provides a basic understanding of linear model theory; in no way does it embrace up-to-the-minute research findings or all the nuances of results that are of more interest mathematically than they are for understanding and carrying out data analysis using linear models.

Linear model theory is described in a variety of texts. Seven examples are: Searle (1971), which emphasizes unbalanced data and variance components; Rao (1973), with its broad-based mathematical generality; Graybill (1976), which emphasizes balanced data; Seber (1977), with its concentration on the full-rank model; Arnold (1981), which uses a coordinate free approach and emphasizes similarities between univariate and multivariate analyses; Guttman (1982), which is mainly an introduction; and Hocking (1985), which is very wide-ranging. All of these present theory quite extensively, and overall they cater well to the reader whose prime interests are the theory of linear models and the techniques of establishing that theory. In contrast, this chapter summarizes elements of the theory that are deemed important to understanding the consequences of using and relying on theory in analyzing data. Specific references are given to sources where development of theoretical results can be found, although in some cases that development is given here, in small-print paragraphs that can be overlooked by readers not wishing to be concerned with mathematical details.

8.1. THE MODEL

We consider N items of data arrayed as a vector $\mathbf{y}$ of order $N \times 1$. Using E to represent expectation over repeated sampling we take the basic (vector) equation of our model as

$$E(\mathbf{y}) = \mathbf{X}\boldsymbol{\beta} \tag{1}$$

where $\boldsymbol{\beta}$ is a $p \times 1$ vector of unknown parameters and $\mathbf{X}$ is a known matrix (of order $N \times p$). In many instances, each element of $\mathbf{X}$ is either zero or unity, in which case $\mathbf{X}$ is called a *design matrix* (especially when the data come from a designed and well-executed experiment) or an *incidence* matrix. The latter name arises because those elements of $\mathbf{X}$ that are unity correspond to the incidence of the parameters of $\boldsymbol{\beta}$ through the data $\mathbf{y}$. Elements of $\mathbf{X}$ can also be observed or measured values of variables other than those represented in $\mathbf{y}$—such as regressor variables in regression, and concomitant variables or covariates in analysis of covariance (see Chapter 11). In general, to cover all eventualities, the matrix $\mathbf{X}$ is nowadays called the *model matrix* (Kempthorne, 1980).

Of course $\mathbf{y}$ and $E(\mathbf{y})$ will not, in general, be equal. We define their difference as

$$\mathbf{e} = \mathbf{y} - E(\mathbf{y}) \tag{2}$$

and consider it as a vector of random errors. Then (2) gives $\mathbf{y} = E(\mathbf{y}) + \mathbf{e}$ and in combination with (1) this is

$$\mathbf{y} = \mathbf{X}\boldsymbol{\beta} + \mathbf{e}\ . \tag{3}$$

This is the *model equation* of the general linear model. It is linear in its parameters, the elements of $\boldsymbol{\beta}$.

TABLE 8.1. HEIGHT OF TOMATO PLANTS (IN INCHES)

		Treatment			
		1	2	3	
		74	76	87	
		68	80	91	
		77			
Total:	$y_{i\cdot}$	219	156	178	$553 = y_{\cdot\cdot}$
Number:	(n_i)	(3)	(2)	(2)	$7 = n_{\cdot} = N$
Mean:	$\bar{y}_{i\cdot}$	73	78	89	$79 = \bar{y}_{\cdot\cdot} = y_{\cdot\cdot}/N$

Example 1. Consider the example from Table 2.1, given in Table 8.1. As usual, y_{ij} is the jth observation on the ith treatment, with model

$$y_{ij} = \mu_i + e_{ij}$$

as in (3) of Section 2.2a. Then equations (3) for the data of Table 8.1 are

$$\begin{aligned}
y_{11} &= 74 = \mu_1 \qquad\qquad + e_{11} \\
y_{12} &= 68 = \mu_1 \qquad\qquad + e_{12} \\
y_{13} &= 77 = \mu_1 \qquad\qquad + e_{13} \\
y_{21} &= 76 = \qquad \mu_2 \qquad + e_{21} \\
y_{22} &= 80 = \qquad \mu_2 \qquad + e_{22} \\
y_{31} &= 87 = \qquad\qquad \mu_3 + e_{31} \\
y_{32} &= 91 = \qquad\qquad \mu_3 + e_{32}
\end{aligned} \tag{4}$$

i.e.,

$$\begin{bmatrix} 74 \\ 68 \\ 77 \\ 76 \\ 80 \\ 87 \\ 91 \end{bmatrix} = \begin{bmatrix} 1 & 0 & 0 \\ 1 & 0 & 0 \\ 1 & 0 & 0 \\ 0 & 1 & 0 \\ 0 & 1 & 0 \\ 0 & 0 & 1 \\ 0 & 0 & 1 \end{bmatrix} \begin{bmatrix} \mu_1 \\ \mu_2 \\ \mu_3 \end{bmatrix} + \mathbf{e}, \tag{5}$$

where $\mathbf{e}$ in (5) is the vector of elements e_{ij} in (4). Now define

$$\mathbf{y} = \begin{bmatrix} 74 \\ 68 \\ 77 \\ 76 \\ 80 \\ 87 \\ 91 \end{bmatrix}, \qquad \mathbf{X} = \begin{bmatrix} 1 & \cdot & \cdot \\ 1 & \cdot & \cdot \\ 1 & \cdot & \cdot \\ \cdot & 1 & \cdot \\ \cdot & 1 & \cdot \\ \cdot & \cdot & 1 \\ \cdot & \cdot & 1 \end{bmatrix} \quad \text{and} \quad \boldsymbol{\beta} = \begin{bmatrix} \mu_1 \\ \mu_2 \\ \mu_3 \end{bmatrix}, \tag{6}$$

where in $\mathbf{X}$ each dot represents an element of zero (a frequent notational convenience). Then (5) can be written as $\mathbf{y} = \mathbf{X}\boldsymbol{\beta} + \mathbf{e}$, and is an example of the model equation (3).

The model equation is only part of a model (see Section 2.2b). A model is its equation (3) together with statements about probability distributional properties assumed for the random error terms that are elements of $\mathbf{e}$ defined in (2). First, it is clear that they have zero expectation:

$$E(\mathbf{e}) = E[\mathbf{y} - E(\mathbf{y})] = E(\mathbf{y}) - E(\mathbf{y}) = \mathbf{0}, \tag{7}$$

where $E(\mathbf{e})$ is the vector of elements $E(e_{ij})$. Second, at this stage we do no more than define the second moments of the random variables e_{ij}. We assume they all have the same variance, σ^2, and that they are all uncorrelated with each other. In this way the variance–covariance matrix (see Section 7.1g) of the elements of $\mathbf{e}$ is assumed as being $\sigma^2\mathbf{I}$:

$$\text{var}(\mathbf{e}) = E[\mathbf{e} - E(\mathbf{e})][\mathbf{e} - E(\mathbf{e})]' = E(\mathbf{e}\mathbf{e}') = \sigma^2\mathbf{I} \,. \tag{8}$$

Note that no assumption about normality is yet made.

The model is now (3), (7) and (8) taken together:

$$\mathbf{y} = \mathbf{X}\boldsymbol{\beta} + \mathbf{e} \tag{9}$$

and $$E(\mathbf{y}) = \mathbf{X}\boldsymbol{\beta}, \qquad E(\mathbf{e}) = \mathbf{0} \quad \text{and} \quad \text{var}(\mathbf{e}) = \sigma^2\mathbf{I}_N \,. \tag{10}$$

We use the notation $$\mathbf{y} \sim \left(\mathbf{X}\boldsymbol{\beta}, \sigma^2\mathbf{I}_N\right) \tag{11}$$

as a summary statement that $\mathbf{y}$ is distributed with mean $\mathbf{X}\boldsymbol{\beta}$ and variance–covariance matrix $\sigma^2\mathbf{I}_N$. This last result is so because on using (1) and (8)

$$\text{var}(\mathbf{y}) = E[\mathbf{y} - E(\mathbf{y})][\mathbf{y} - E(\mathbf{y})]' = E(\mathbf{e}\mathbf{e}') = \sigma^2\mathbf{I} \,. \tag{12}$$

8.2. NORMAL EQUATIONS AND THEIR SOLUTIONS

a. The general case

The usual method of estimating $\boldsymbol{\beta}$ is by least squares, which requires minimizing the sum of squares of elements of $\mathbf{y} - E(\mathbf{y}) = \mathbf{y} - \mathbf{X}\boldsymbol{\beta}$, i.e, of

$$S = (\mathbf{y} - \mathbf{X}\boldsymbol{\beta})'(\mathbf{y} - \mathbf{X}\boldsymbol{\beta}) \ . \tag{13}$$

Minimizing (13) with respect to $\boldsymbol{\beta}$ leads to equations

$$\mathbf{X}'\mathbf{X}\boldsymbol{\beta}^\circ = \mathbf{X}'\mathbf{y}, \tag{14}$$

to be solved for $\boldsymbol{\beta}^\circ$. Then $\boldsymbol{\beta}^\circ$ is a value for $\boldsymbol{\beta}$ that minimizes (14), and the estimation procedure is based on $\boldsymbol{\beta}^\circ$.

Establishing (14) can be achieved either by differentiating (13) with respect to $\boldsymbol{\beta}$ [e.g., Searle (1982), Section 12.8] or, with benefit of hindsight, by verifying that (14) does minimize (13). This verification proceeds as follows.

Define (13) as $S(\boldsymbol{\beta})$. Then $S(\boldsymbol{\beta}^\circ) = \mathbf{y}'\mathbf{y} - \boldsymbol{\beta}^{\circ\prime}\mathbf{X}'\mathbf{X}\boldsymbol{\beta}^\circ$, and for $\dot{\boldsymbol{\beta}}$ different from $\boldsymbol{\beta}^\circ$ but such that $\mathbf{X}'\mathbf{y} = \mathbf{X}'\mathbf{X}\boldsymbol{\beta}^\circ = \mathbf{X}'\mathbf{X}\dot{\boldsymbol{\beta}}$,

$$\begin{aligned} S(\dot{\boldsymbol{\beta}}) - S(\boldsymbol{\beta}^\circ) &= \mathbf{y}'\mathbf{y} - 2\dot{\boldsymbol{\beta}}'\mathbf{X}'\mathbf{y} + \dot{\boldsymbol{\beta}}\mathbf{X}'\mathbf{X}\dot{\boldsymbol{\beta}} - (\mathbf{y}'\mathbf{y} - \boldsymbol{\beta}^{\circ\prime}\mathbf{X}'\mathbf{X}\boldsymbol{\beta}^\circ) \\ &= -2\dot{\boldsymbol{\beta}}'\mathbf{X}'\mathbf{X}\boldsymbol{\beta}^\circ + \dot{\boldsymbol{\beta}}'\mathbf{X}'\mathbf{X}\dot{\boldsymbol{\beta}} + \boldsymbol{\beta}^{\circ\prime}\mathbf{X}'\mathbf{X}\boldsymbol{\beta}^\circ \\ &= (\dot{\boldsymbol{\beta}}' - \boldsymbol{\beta}^{\circ\prime})\mathbf{X}'\mathbf{X}(\dot{\boldsymbol{\beta}} - \boldsymbol{\beta}^\circ) \ . \end{aligned}$$

This is never less than 0, because $\mathbf{X}'\mathbf{X}$ is non-negative definite (see Section 7.1f). Therefore $S(\dot{\boldsymbol{\beta}}) \geq S(\boldsymbol{\beta}^\circ)$, and so $\boldsymbol{\beta}^\circ$ satisfying (14) used in place of $\boldsymbol{\beta}$ in (13) minimizes (13).

Equations (14) are a direct outcome of the method of least squares [and, in view of (12), also of the method of generalized least squares using (8)—see Section 8.10], without involving normality assumptions. When such assumptions are made, then (14) is also the outcome of the method of maximum likelihood.

Equations (14) are known as the *normal equations*, even though "normal" in this context has nothing to do with the Normal distribution. When $\mathbf{X}$ has full column rank, $\mathbf{X}'\mathbf{X}$ has an inverse and the solution to (14) is then the regression result

$$\hat{\boldsymbol{\beta}} = (\mathbf{X}'\mathbf{X})^{-1}\mathbf{X}'\mathbf{y}, \quad \text{for } (\mathbf{X}'\mathbf{X})^{-1} \text{ existing} \ . \tag{15}$$

In all cases we take $\mathbf{X}$ as having order $N \times p$, for $p \leq N$, and so (15) holds only when $\mathbf{X}$ has full column rank. More generally, though, we need to provide for $\mathbf{X}$ being of less than full column rank, because with many forms of linear models (especially for the traditional, overparameterized models) $\mathbf{X}$ is of this nature. When this is so, the model is known as a *model of less than full rank*, or a *non-full-rank model*. In that case $\mathbf{X}'\mathbf{X}$ has no inverse and so, as discussed in Sections 7.1h and 7.1i, we use any matrix $\mathbf{G}$ that satisfies

$$\mathbf{X}'\mathbf{X}\mathbf{G}\mathbf{X}'\mathbf{X} = \mathbf{X}'\mathbf{X} \ . \tag{16}$$

This yields a solution (see Section 7.2a) to (13) as

$$\boldsymbol{\beta}^{\circ} = \mathbf{G}\mathbf{X}'\mathbf{y} \ . \tag{17}$$

The symbol $\hat{\boldsymbol{\beta}}$ is traditionally used as an estimator of $\boldsymbol{\beta}$, and is appropriate in (15) where $(\mathbf{X}'\mathbf{X})^{-1}$ exists, because then $\hat{\boldsymbol{\beta}}$ is the only solution of (14) and is the estimator of $\boldsymbol{\beta}$. It is also, of course, just a special case of (17), because when $(\mathbf{X}'\mathbf{X})^{-1}$ exists, $\mathbf{G} = (\mathbf{X}'\mathbf{X})^{-1}$ is the sole choice for $\mathbf{G}$.

In contrast to the uniqueness property reserved for the symbol $\hat{\boldsymbol{\beta}}$, the symbol $\boldsymbol{\beta}^{\circ}$ of (17) is used to emphasize that any $\boldsymbol{\beta}^{\circ}$ of that form is just one of the infinite number of solutions to the normal equations (14); and it is not an unbiased estimator of $\boldsymbol{\beta}$. Nevertheless, through both Theorem 7.1 concerning $\mathbf{G}$ in Section 7.1i and the invariance property of Section 7.2b we are able to categorize numerous features of linear model analysis that are impervious to there being many solution vectors $\boldsymbol{\beta}^{\circ}$. Some of these features are described in Section 8.3. First are some examples.

b. Examples

We begin with an example where $(\mathbf{X}'\mathbf{X})^{-1}$ exists and (15) is used as the solution to (14).

Example 1 (continued). The normal equations $\mathbf{X}'\mathbf{X}\boldsymbol{\beta}^{\circ} = \mathbf{X}'\mathbf{y}$ of (14) for Example 1 are, from (6),

$$\begin{bmatrix} 3 & \cdot & \cdot \\ \cdot & 2 & \cdot \\ \cdot & \cdot & 2 \end{bmatrix} \begin{bmatrix} \mu_1^{\circ} \\ \mu_2^{\circ} \\ \mu_3^{\circ} \end{bmatrix} = \begin{bmatrix} 74 + 68 + 77 \\ 76 + 80 \\ 87 + 91 \end{bmatrix} = \begin{bmatrix} 219 \\ 156 \\ 178 \end{bmatrix} . \tag{18}$$

Since the matrix $\mathbf{X}'\mathbf{X}$ on the left-hand side is nonsingular we use (15) to solve (14) as

$$\begin{bmatrix} \hat{\mu}_1 \\ \hat{\mu}_2 \\ \hat{\mu}_3 \end{bmatrix} = \begin{bmatrix} 3 & \cdot & \cdot \\ \cdot & 2 & \cdot \\ \cdot & \cdot & 2 \end{bmatrix}^{-1} \begin{bmatrix} 219 \\ 156 \\ 178 \end{bmatrix} = \begin{bmatrix} \frac{1}{3} & \cdot & \cdot \\ \cdot & \frac{1}{2} & \cdot \\ \cdot & \cdot & \frac{1}{2} \end{bmatrix} \begin{bmatrix} 219 \\ 156 \\ 178 \end{bmatrix} = \begin{bmatrix} 73 \\ 76 \\ 89 \end{bmatrix} . \tag{19}$$

This is, of course, the same set of estimates as at the end of Section 2.3. The values in (19) are, in fact, the cell means, $\bar{y}_{i\cdot}$ as is discussed in subsection *c*.

The next two examples are cases of $\mathbf{X}'\mathbf{X}$ being singular that have arisen in earlier chapters.

Example 2 (overparameterized model). Suppose we use the overparameterized model

$$E(y_{ij}) = \mu + \alpha_i \tag{20}$$

of (129) in Section 2.13. Then, similar, to (5), the model equations for the data are

$$\begin{bmatrix} y_{11} \\ y_{12} \\ y_{13} \\ y_{21} \\ y_{22} \\ y_{31} \\ y_{32} \end{bmatrix} = \begin{bmatrix} 74 \\ 68 \\ 77 \\ 76 \\ 80 \\ 87 \\ 91 \end{bmatrix} = \begin{bmatrix} \mu + \alpha_1 \\ \mu + \alpha_1 \\ \mu + \alpha_1 \\ \mu + \alpha_2 \\ \mu + \alpha_2 \\ \mu + \alpha_3 \\ \mu + \alpha_3 \end{bmatrix} + \mathbf{e} = \begin{bmatrix} 1 & 1 & \cdot & \cdot \\ 1 & 1 & \cdot & \cdot \\ 1 & 1 & \cdot & \cdot \\ 1 & \cdot & 1 & \cdot \\ 1 & \cdot & 1 & \cdot \\ 1 & \cdot & \cdot & 1 \\ 1 & \cdot & \cdot & 1 \end{bmatrix} \begin{bmatrix} \mu \\ \alpha_1 \\ \alpha_2 \\ \alpha_3 \end{bmatrix} + \mathbf{e}\,. \tag{21}$$

Hence

$$\mathbf{X} = \begin{bmatrix} 1 & 1 & \cdot & \cdot \\ 1 & 1 & \cdot & \cdot \\ 1 & 1 & \cdot & \cdot \\ 1 & \cdot & 1 & \cdot \\ 1 & \cdot & 1 & \cdot \\ 1 & \cdot & \cdot & 1 \\ 1 & \cdot & \cdot & 1 \end{bmatrix} \tag{22}$$

and so the normal equations $\mathbf{X}'\mathbf{X}\boldsymbol{\beta}^\circ = \mathbf{X}'\mathbf{y}$ of (14) are

$$\begin{bmatrix} 7 & 3 & 2 & 2 \\ 3 & 3 & \cdot & \cdot \\ 2 & \cdot & 2 & \cdot \\ 2 & \cdot & \cdot & 2 \end{bmatrix} \begin{bmatrix} \mu^\circ \\ \alpha_1^\circ \\ \alpha_2^\circ \\ \alpha_3^\circ \end{bmatrix} = \begin{bmatrix} 553 \\ 219 \\ 156 \\ 178 \end{bmatrix}. \tag{23}$$

It is clear in (22) that the first column of $\mathbf{X}$ equals the sum of its last three columns. Therefore $\mathbf{X}$ is not of full column rank and so $\mathbf{X}'\mathbf{X}$, the matrix on the left-hand side of (23), is singular. Hence in accord with (16) we find a generalized inverse, $\mathbf{G}$, of $\mathbf{X}'\mathbf{X}$, and use (17) as a solution. Using

(19) of Section 7.1j, we calculate a $\mathbf{G}$ as

$$\mathbf{G} = \begin{bmatrix} \begin{pmatrix} 7 & 3 & 2 \\ 3 & 3 & \cdot \\ 2 & \cdot & 2 \end{pmatrix}^{-1} & \begin{matrix} 0 \\ 0 \\ 0 \end{matrix} \\ \begin{matrix} 0 & 0 & 0 \end{matrix} & 0 \end{bmatrix} = \tfrac{1}{6}\begin{bmatrix} 3 & -3 & -3 & 0 \\ -3 & 5 & 3 & 0 \\ -3 & 3 & 6 & 0 \\ 0 & 0 & 0 & 0 \end{bmatrix} \tag{24}$$

and so the corresponding value of $\boldsymbol{\beta}^{\circ}$, using (24) and the right-hand side of (23) in (17) is

$$\boldsymbol{\beta}^{\circ} = \begin{bmatrix} \mu^{\circ} \\ \alpha_1^{\circ} \\ \alpha_2^{\circ} \\ \alpha_3^{\circ} \end{bmatrix} = \tfrac{1}{6}\begin{bmatrix} 3 & -3 & -3 & 0 \\ -3 & 5 & 3 & 0 \\ -3 & 3 & 6 & 0 \\ 0 & 0 & 0 & 0 \end{bmatrix}\begin{bmatrix} 553 \\ 219 \\ 156 \\ 178 \end{bmatrix} = \begin{bmatrix} 89 \\ -16 \\ -11 \\ 0 \end{bmatrix}. \tag{25}$$

Direct substitution shows that (25) is a solution of (23).

Since there are many matrices $\mathbf{G}$ that satisfy (16), we develop one different from that in (24) and find the solution (17) corresponding thereto. Thus we use

$$\begin{bmatrix} 7 & 2 & 2 \\ 2 & 2 & \cdot \\ 2 & \cdot & 2 \end{bmatrix}^{-1} = \frac{1}{6}\begin{bmatrix} 2 & -2 & -2 \\ -2 & 5 & 2 \\ -2 & 2 & 5 \end{bmatrix}$$

and take (based on the algorithm of Section 7.1j)

$$\dot{\mathbf{G}} = \frac{1}{6}\begin{bmatrix} 2 & 0 & -2 & -2 \\ 0 & 0 & 0 & 0 \\ -2 & 0 & 5 & 2 \\ -2 & 0 & 2 & 5 \end{bmatrix} \tag{26}$$

and then (17) is

$$\dot{\boldsymbol{\beta}} = \begin{bmatrix} \dot{\mu} \\ \dot{\alpha}_1 \\ \dot{\alpha}_2 \\ \dot{\alpha}_3 \end{bmatrix} = \frac{1}{6}\begin{bmatrix} 2 & 0 & -2 & -2 \\ 0 & 0 & 0 & 0 \\ -2 & 0 & 5 & 2 \\ -2 & 0 & 2 & 5 \end{bmatrix}\begin{bmatrix} 553 \\ 219 \\ 156 \\ 178 \end{bmatrix} = \begin{bmatrix} 73 \\ 0 \\ 5 \\ 16 \end{bmatrix}. \tag{27}$$

Despite $\dot{\boldsymbol{\beta}}$ being different from $\boldsymbol{\beta}^{\circ}$ of (25), various functions of $\boldsymbol{\beta}^{\circ}$ and $\dot{\boldsymbol{\beta}}$ are equal, as is discussed more generally for (53) in Section 8.3b, for (57) in Section 8.3c and for (130) in Section 8.7c.

TABLE 8.2. DATA FOR A 2-WAY CROSSED CLASSIFICATION

	y_{ijk}		$y_{i\cdot\cdot}(n_{i\cdot})\bar{y}_{i\cdot\cdot}$
	8	11	
	12	16	
		18	
	20(2)10	45(3)15	65(5)13
	13	47	
	15		
	28(2)14	47(1)47	75(3)25
$y_{\cdot j\cdot}(n_{\cdot j})\bar{y}_{\cdot j\cdot}$	48(4)12	92(4)23	$140(8)23\frac{1}{2}$

Example 3. Table 8.2 shows data for a 2-way crossed classification with two rows and two columns. As in Chapters 4 and 5, each triplet of numbers in Table 8.2 represents a total, the number of observations therein and the corresponding mean.

Suppose, now, that we fit the no-interaction model

$$E(y_{ijk}) = \mu_i + \tau_j \tag{28}$$

of (62) in Section 4.9a, without putting τ_b to zero as is done later in that section. Then for the data of Table 8.2 the model equations are

$$\begin{aligned}
y_{111} &= 8 = \mu_1 + \tau_1 + e_{111} \\
y_{112} &= 12 = \mu_1 + \tau_1 + e_{112} \\
y_{121} &= 11 = \mu_1 + \tau_2 + e_{121} \\
y_{122} &= 16 = \mu_1 + \tau_2 + e_{122} \\
y_{123} &= 18 = \mu_1 + \tau_2 + e_{123} \\
y_{211} &= 13 = \mu_2 + \tau_1 + e_{211} \\
y_{212} &= 15 = \mu_2 + \tau_1 + e_{212} \\
y_{221} &= 47 = \mu_2 + \tau_2 + e_{221},
\end{aligned} \tag{29}$$

i.e.,

$$\mathbf{y} = \begin{bmatrix} 1 & \cdot & 1 & \cdot \\ 1 & \cdot & 1 & \cdot \\ 1 & \cdot & \cdot & 1 \\ 1 & \cdot & \cdot & 1 \\ 1 & \cdot & \cdot & 1 \\ \cdot & 1 & 1 & \cdot \\ \cdot & 1 & 1 & \cdot \\ \cdot & 1 & \cdot & 1 \end{bmatrix} \begin{bmatrix} \mu_1 \\ \mu_2 \\ \tau_1 \\ \tau_2 \end{bmatrix} + \mathbf{e} \tag{30}$$

$$= \mathbf{X}\boldsymbol{\beta} + \mathbf{e}, \tag{31}$$

where $\mathbf{y}$ and $\mathbf{e}$ in (30) and (31) are the vectors, respectively, of the y_{ijk}s and e_{ijk}s in (29); and $\mathbf{X}$ and $\boldsymbol{\beta}$ of (30) are

$$\mathbf{X} = \begin{bmatrix} 1 & \cdot & 1 & \cdot \\ 1 & \cdot & 1 & \cdot \\ 1 & \cdot & \cdot & 1 \\ 1 & \cdot & \cdot & 1 \\ 1 & \cdot & \cdot & 1 \\ \cdot & 1 & 1 & \cdot \\ \cdot & 1 & 1 & \cdot \\ \cdot & 1 & \cdot & 1 \end{bmatrix} \quad \text{and} \quad \boldsymbol{\beta} = \begin{bmatrix} \mu_1 \\ \mu_2 \\ \tau_1 \\ \tau_2 \end{bmatrix}. \tag{32}$$

With these values the normal equations $\mathbf{X}'\mathbf{X}\boldsymbol{\beta}^\circ = \mathbf{X}'\mathbf{y}$ of (14) are

$$\begin{bmatrix} 5 & 0 & 2 & 3 \\ 0 & 3 & 2 & 1 \\ 2 & 2 & 4 & 0 \\ 3 & 1 & 0 & 4 \end{bmatrix} \begin{bmatrix} \mu_1^\circ \\ \mu_2^\circ \\ \tau_1^\circ \\ \tau_2^\circ \end{bmatrix} = \begin{bmatrix} 65 \\ 75 \\ 48 \\ 92 \end{bmatrix}, \tag{33}$$

where

$$\mathbf{X}'\mathbf{X} = \begin{bmatrix} 5 & 0 & 2 & 3 \\ 0 & 3 & 2 & 1 \\ 2 & 2 & 4 & 0 \\ 3 & 1 & 0 & 4 \end{bmatrix}. \tag{34}$$

In (32) we see that the sum of the first two columns of $\mathbf{X}$ is the same as the sum of the second two. Therefore $\mathbf{X}'\mathbf{X}$ of (34) is singular and, in accord with (16), we find a generalized inverse of (34) and use it in (17) to solve the normal equations. On this occasion we use $\mathbf{F}$ to represent a generalized inverse of (34), so as to avoid confusion with (24) and (26), and using (3) of Section 7.1h calculate an $\mathbf{F}$ as

$$\mathbf{F} = \begin{bmatrix} 5 & 0 & 2 & 3 \\ 0 & 3 & 2 & 1 \\ 2 & 2 & 4 & 0 \\ 3 & 1 & 0 & 4 \end{bmatrix}^{-} = \begin{bmatrix} \begin{pmatrix} 5 & 0 & 2 \\ 0 & 3 & 2 \\ 2 & 2 & 4 \end{pmatrix}^{-1} & \begin{matrix} 0 \\ 0 \\ 0 \end{matrix} \\ \begin{matrix} 0 & 0 & 0 \end{matrix} & 0 \end{bmatrix}$$

$$= \tfrac{1}{28}\begin{bmatrix} 8 & 4 & -6 & 0 \\ 4 & 16 & -10 & 0 \\ -6 & -10 & 15 & 0 \\ 0 & 0 & 0 & 0 \end{bmatrix}. \tag{35}$$

Hence using (35) and the right-hand side of (33) in (17) gives

$$\boldsymbol{\beta}^{\circ} = \begin{bmatrix} \mu_1^{\circ} \\ \mu_2^{\circ} \\ \tau_1^{\circ} \\ \tau_2^{\circ} \end{bmatrix} = \mathbf{FX'y} = \tfrac{1}{28}\begin{bmatrix} 8 & 4 & -6 & 0 \\ 4 & 16 & -10 & 0 \\ -6 & -10 & 15 & 0 \\ 0 & 0 & 0 & 0 \end{bmatrix}\begin{bmatrix} 65 \\ 75 \\ 48 \\ 92 \end{bmatrix} = \begin{bmatrix} 19 \\ 35 \\ -15 \\ 0 \end{bmatrix}. \tag{36}$$

We also develop an $\mathbf{F}$ different from (35) and find the corresponding solution of (33).

$$\dot{\mathbf{F}} = \begin{bmatrix} 5 & 0 & 2 & 3 \\ 0 & 3 & 2 & 1 \\ 2 & 2 & 4 & 0 \\ 3 & 1 & 0 & 4 \end{bmatrix}^{-} = \begin{bmatrix} 0 & \begin{matrix} 0 & 0 & 0 \end{matrix} \\ \begin{matrix} 0 \\ 0 \\ 0 \end{matrix} & \begin{pmatrix} 3 & 2 & 1 \\ 2 & 4 & 0 \\ 1 & 0 & 4 \end{pmatrix}^{-1} \end{bmatrix}$$

$$= \tfrac{1}{28}\begin{bmatrix} 0 & 0 & 0 & 0 \\ 0 & 16 & -8 & -4 \\ 0 & -8 & 11 & 2 \\ 0 & -4 & 2 & 8 \end{bmatrix} \tag{37}$$

for which, akin to (36),

$$\dot{\boldsymbol{\beta}} = \begin{bmatrix} \dot{\mu}_1 \\ \dot{\mu}_2 \\ \dot{\tau}_1 \\ \dot{\tau}_2 \end{bmatrix} = \dot{\mathbf{F}}\mathbf{X'y} = \tfrac{1}{28}\begin{bmatrix} 0 & 0 & 0 & 0 \\ 0 & 16 & -8 & -4 \\ 0 & -8 & 11 & 2 \\ 0 & -4 & 2 & 8 \end{bmatrix}\begin{bmatrix} 65 \\ 75 \\ 48 \\ 92 \end{bmatrix} = \begin{bmatrix} 0 \\ 16 \\ 4 \\ 19 \end{bmatrix}. \tag{38}$$

Once more, we could highlight functions of $\boldsymbol{\beta}^{\circ}$ and $\dot{\boldsymbol{\beta}}$ in (36) and (38) that are the same for both, but such comments are deferred.

The important feature is that when $\mathbf{X}'\mathbf{X}$ is singular there are (very) many solutions to the normal equations $\mathbf{X}'\mathbf{X}\boldsymbol{\beta}^{\circ} = \mathbf{X}'\mathbf{y}$.

c. The 1-way classification

In equations (4) for the data of Example 1, the observations y_{ij} are listed in what is known as lexicon order, which can be described as ordered according to j within i. This is denoted by $\mathbf{y}$. Similarly, for each i, with $j = 1, 2, \ldots, n_i$ we use the notation of Section 7.1a to define

$$\mathbf{y}_i' = \{_r\, y_{ij}\}_{j=1}^{j=n_i} \quad \text{and} \quad \boldsymbol{\mu} = \{_c\, \mu_i\} \ . \tag{39}$$

Then the vector of observations is

$$\mathbf{y} = \{_c\, \mathbf{y}_i\}_{i=1}^{i=a} = \left\{_c \{_c\, y_{ij}\}_{j=1}^{j=n_i}\right\}_{i=1}^{i=a} = \{_c\, y_{ij}\}_{i=1,\, j=1}^{i=a,\, j=n_i}, \tag{40}$$

where, by the last expression in (40) we adopt the convention that for each i, the subscript j takes values $j = 1, 2, \ldots n_i$, so that the elements y_{ij} are in lexicon order.

Corresponding to $\mathbf{y}$ of (4) we see in (5) that the 8×3 matrix is

$$\mathbf{X} = \begin{bmatrix} \mathbf{1}_3 & \mathbf{0} & \mathbf{0} \\ \mathbf{0} & \mathbf{1}_2 & \mathbf{0} \\ \mathbf{0} & \mathbf{0} & \mathbf{1}_2 \end{bmatrix} = \{_d\, \mathbf{1}_n\}_{i=1}^{i=3} \tag{41}$$

where, for example, $\mathbf{1}_3 = [1 \quad 1 \quad 1]'$ is a summing vector of order 3 (see Section 7.1e). For data consisting of a classes with n_i observations in the ith class, for $i = 1, \ldots, a$, the form of $\mathbf{X}$ in (41) generalizes very easily to

$$\mathbf{X} = \mathbf{D}\{\mathbf{1}_{n_i}\} \ . \tag{42}$$

With $\mathbf{y}$ of (40) and $\mathbf{X}$ of (42), keeping in mind that $i = 1, 2, \ldots, a$, the normal equations $\mathbf{X}'\mathbf{X}\boldsymbol{\beta}^{\circ} = \mathbf{X}'\mathbf{y}$ of (14) take the form

$$\{_d\, \mathbf{1}_{n_i}'\}\{_d\, \mathbf{1}_{n_i}\}\boldsymbol{\mu}^{\circ} = \{_d\, \mathbf{1}_{n_i}'\}\{_c\, \mathbf{y}_i\} \tag{43}$$

which reduces to

$$\{_d\, n_i\}\boldsymbol{\mu}^{\circ} = \{_c\, y_{i\cdot}\} \ . \tag{44}$$

This is

$$\begin{bmatrix} n_1 & \cdot & \cdots & \cdot \\ \cdot & n_2 & & \cdot \\ \vdots & & \ddots & \vdots \\ \cdot & \cdot & \cdots & n_a \end{bmatrix} \begin{bmatrix} \mu_1^\circ \\ \mu_2^\circ \\ \vdots \\ \mu_a^\circ \end{bmatrix} = \begin{bmatrix} y_{1\cdot} \\ y_{2\cdot} \\ \vdots \\ y_{a\cdot} \end{bmatrix}. \tag{45}$$

Since the diagonal matrix has an inverse, (15) gives $\mu_i^\circ = \hat{\mu}_i$ with

$$\begin{bmatrix} \hat{\mu}_1 \\ \hat{\mu}_2 \\ \vdots \\ \hat{\mu}_a \end{bmatrix} = \begin{bmatrix} \dfrac{1}{n_1} & \cdot & \cdots & \cdot \\ \cdot & \dfrac{1}{n_2} & & \cdot \\ \vdots & & \ddots & \vdots \\ \cdot & \cdots & & \dfrac{1}{n_a} \end{bmatrix} \begin{bmatrix} y_{1\cdot} \\ y_{2\cdot} \\ \vdots \\ y_{a\cdot} \end{bmatrix} \tag{46}$$

and so $$\hat{\mu}_{i\cdot} = \bar{y}_{i\cdot},$$

as in Chapter 2. A convenient vector form of (46), coming from (44), is

$$\hat{\boldsymbol{\mu}} = \mathbf{D}\{1/n_i\}\{_c\, y_{i\cdot}\} = \bar{\mathbf{y}} = \{_c\, \bar{y}_{i\cdot}\}\ . \tag{47}$$

This is the form exemplified in (19).

8.3. USING A SOLUTION TO THE NORMAL EQUATIONS

Although for nonsingular $\mathbf{X'X}$ there is only one solution $\hat{\boldsymbol{\beta}} = (\mathbf{X'X})^{-1}\mathbf{X'y}$ to the normal equations, we consider the more general solution $\boldsymbol{\beta}^\circ = \mathbf{GX'y}$ of (17) which provides for $\mathbf{X'X}$ being singular, with a generalized inverse $\mathbf{G}$.

a. Mean, and dispersion matrix, of $\boldsymbol{\beta}^\circ$

The expected value of $\boldsymbol{\beta}^\circ$ is

$$E(\boldsymbol{\beta}^\circ) = \mathbf{GX'}E(\mathbf{y}) = \mathbf{GX'X}\boldsymbol{\beta} = \mathbf{H}\boldsymbol{\beta} \tag{48}$$

on defining $$\mathbf{H} = \mathbf{GX'X} \tag{49}$$

as in (9) of Section 7.1i. Thus $\boldsymbol{\beta}^\circ$ is an unbiased estimator of $\mathbf{H}\boldsymbol{\beta}$, and $\mathbf{H}$ clearly depends on $\mathbf{G}$. Hence $\boldsymbol{\beta}^\circ$ is not an unbiased estimator of $\boldsymbol{\beta}$.

Example 2 (continued). From (23), (24) and (26) we have

$$\mathbf{H} = \mathbf{GX'X} = \begin{bmatrix} 1 & \cdot & \cdot & 1 \\ \cdot & 1 & \cdot & -1 \\ \cdot & \cdot & 1 & -1 \\ \cdot & \cdot & \cdot & \cdot \end{bmatrix} \text{ and } \dot{\mathbf{H}} = \dot{\mathbf{G}}\mathbf{X'X} = \begin{bmatrix} 1 & 1 & \cdot & \cdot \\ \cdot & \cdot & \cdot & \cdot \\ \cdot & -1 & 1 & \cdot \\ \cdot & -1 & \cdot & 1 \end{bmatrix} \tag{50}$$

and so for $\boldsymbol{\beta}^\circ$ of (25) and $\dot{\boldsymbol{\beta}}$ of (27), $E(\boldsymbol{\beta}^\circ) = \mathbf{H}\boldsymbol{\beta}$ and $E(\dot{\boldsymbol{\beta}}) = \dot{\mathbf{H}}\boldsymbol{\beta}$ gives

$$E(\boldsymbol{\beta}^\circ) = \begin{bmatrix} \mu + \alpha_3 \\ \alpha_1 - \alpha_3 \\ \alpha_2 - \alpha_3 \\ 0 \end{bmatrix} \quad \text{and} \quad E(\dot{\boldsymbol{\beta}}) = \begin{bmatrix} \mu + \alpha_1 \\ 0 \\ -\alpha_1 + \alpha_2 \\ -\alpha_1 + \alpha_3 \end{bmatrix}. \tag{51}$$

The dispersion matrix of $\boldsymbol{\beta}^\circ$ is

$$\text{var}(\boldsymbol{\beta}^\circ) = \text{var}(\mathbf{GX'y}) = \mathbf{GX'}(\sigma^2\mathbf{I})\mathbf{XG'} = \mathbf{GX'XG'}\sigma^2 . \tag{52}$$

This result is also dependent on whatever $\mathbf{G}$ is being used.

Note that both (48) and (52) reduce to the results in regression when $(\mathbf{X'X})^{-1}$ exists. Then, as already discussed, $\boldsymbol{\beta}^\circ = \hat{\boldsymbol{\beta}}$, and (48) and (52) become $E(\hat{\boldsymbol{\beta}}) = \boldsymbol{\beta}$ and $\text{var}(\hat{\boldsymbol{\beta}}) = (\mathbf{X'X})^{-1}\sigma^2$, respectively. Also, if $\mathbf{G}$ is a symmetric, reflexive generalized inverse of $\mathbf{X'X}$, as in (11) of Section 7.1i, then (52) reduces to $\text{var}(\boldsymbol{\beta}^\circ) = \mathbf{G}\sigma^2$, a result more similar to $\text{var}(\hat{\boldsymbol{\beta}}) = (\mathbf{X'X})^{-1}\sigma^2$ than is (52) itself.

b. Estimating $E(\mathbf{y})$

Corresponding to $\mathbf{y}$ we have what is often called the predicted $\mathbf{y}$ (denoted by $\hat{\mathbf{y}}$) or what is, more correctly, the estimated value of $E(\mathbf{y})$:

$$\text{Estimate of } E(\mathbf{y}) \equiv \hat{\mathbf{y}} = \mathbf{X}\boldsymbol{\beta}^\circ = \mathbf{XGX'y} . \tag{53}$$

By Theorem 7.1 of Section 7.1i, $\mathbf{XGX'}$ is invariant to $\mathbf{G}$. Hence (53) is also. This means that for every possible $\mathbf{G}$ used in (53), i.e., for every possible $\boldsymbol{\beta}^\circ$, the value of (53) is, for given $\mathbf{X}$ and $\mathbf{y}$, the same. This is reflected in the symbol $\hat{\mathbf{y}}$, where the "hat" indicates uniqueness, just as it does in $\hat{\boldsymbol{\beta}}$ of (15). We return to this later, in Section 8.7. In the meantime note that $\hat{\mathbf{y}}$ is a traditional symbol, meaning "predicted $\mathbf{y}$". But (53) shows what $\hat{\mathbf{y}}$ truly is

—it is the estimate of the expected value of the observation vector **y**. It is left as an exercise for the reader to show that

$$\text{var}(\hat{\mathbf{y}}) = \mathbf{XGX}'\sigma^2 . \tag{54}$$

We note in passing that **XGX′** is known as the "hat" matrix [e.g., Hoaglin and Welsch (1978)], because it transforms **y** to $\hat{\mathbf{y}} = \mathbf{XGX'y}$.

Example 2 (continued). Using $\boldsymbol{\beta}^\circ$ of (25) and $\dot{\boldsymbol{\beta}}$ of (27) with **X** of (22) gives

$$\mathbf{y} = \mathbf{X}\boldsymbol{\beta}^\circ = \begin{bmatrix} 1 & 1 & \cdot & \cdot \\ 1 & 1 & \cdot & \cdot \\ 1 & 1 & \cdot & \cdot \\ 1 & \cdot & 1 & \cdot \\ 1 & \cdot & 1 & \cdot \\ 1 & \cdot & \cdot & 1 \\ 1 & \cdot & \cdot & 1 \end{bmatrix} \begin{bmatrix} 89 \\ -16 \\ -11 \\ 0 \end{bmatrix} = \begin{bmatrix} 73 \\ 73 \\ 73 \\ 78 \\ 78 \\ 89 \\ 89 \end{bmatrix}$$

$$= \mathbf{X}\dot{\boldsymbol{\beta}} = \begin{bmatrix} 1 & 1 & \cdot & \cdot \\ 1 & 1 & \cdot & \cdot \\ 1 & 1 & \cdot & \cdot \\ 1 & \cdot & 1 & \cdot \\ 1 & \cdot & 1 & \cdot \\ 1 & \cdot & \cdot & 1 \\ 1 & \cdot & \cdot & 1 \end{bmatrix} \begin{bmatrix} 73 \\ 0 \\ 5 \\ 16 \end{bmatrix} . \tag{55}$$

The mean of the elements in $\hat{\mathbf{y}}$ is almost always $\bar{y}$. This is so because in most applications of $E(\mathbf{y}) = \mathbf{X}\boldsymbol{\beta}$ the matrix **X** is such that there is some linear combination of its columns that is a vector of unities, i.e., is $\mathbf{1}_N$. Examples are (6), (22) and (32). Therefore there exists a vector **t** such that $\mathbf{Xt} = \mathbf{1}_N$. Hence

$$\begin{aligned} \bar{\hat{y}} &= \mathbf{1}'\hat{\mathbf{y}}/N = \mathbf{1'XGX'y/N} = \mathbf{t'X'XGX'y}/N \\ &= \mathbf{t'X'y}/N, \quad \text{from (13) in Theorem 7.1} \\ &= \mathbf{1'y}/N, \quad \text{using} \quad \mathbf{Xt} = \mathbf{1} \\ &= \bar{y} . \end{aligned} \tag{56}$$

c. Residual sum of squares

The residual sum of squares is the sum of squares of the deviations of the elements of **y** from their corresponding values in $\hat{\mathbf{y}}$. We use the symbol SSE and thus have

$$\text{SSE} = (\mathbf{y} - \hat{\mathbf{y}})'(\mathbf{y} - \hat{\mathbf{y}}) \ .$$

Using (53) for $\hat{\mathbf{y}}$ together with Theorem 7.1 readily reduces SSE to

$$\text{SSE} = \mathbf{y}'(\mathbf{I} - \mathbf{XGX}')\mathbf{y} = \mathbf{y}'\mathbf{My} \quad \text{for } \mathbf{M} = \mathbf{I} - \mathbf{XGX}' \tag{57}$$

of (18) of Section 7.1i. Since $\mathbf{M}$ is invariant to $\mathbf{G}$, so also is SSE; i.e., for every $\mathbf{G}$ and hence every $\boldsymbol{\beta}^\circ$, SSE is the same (for given $\mathbf{X}$ and $\mathbf{y}$, of course).

SSE of (57) also has another useful form: using $\boldsymbol{\beta}^\circ$ of (17), SSE can be expressed as

$$\text{SSE} = \mathbf{y}'\mathbf{y} - \boldsymbol{\beta}^{\circ\prime}\mathbf{X}'\mathbf{y} \ . \tag{58}$$

On observing that $\mathbf{X}'\mathbf{y}$ in (58) is the right-hand side of the normal equations (14), we can describe (58) in words:

SSE = (sum of squares of all observations)

− (sum of products of elements of solution, $\boldsymbol{\beta}^\circ$, with elements of right-hand side of normal equations, $\mathbf{X}'\mathbf{y}$)

$$= \sum_{t=1}^{N} y_t^2 - \sum_{r=1}^{p} \beta_r^\circ (\mathbf{X}'\mathbf{y})_r, \tag{59}$$

where y_t is the tth observation, and β_r° and $(\mathbf{X}'\mathbf{y})_r$ are the rth elements of $\boldsymbol{\beta}^\circ$ and $(\mathbf{X}'\mathbf{y})$, respectively.

Example 2 (continued). From (21) and (55)

$$\mathbf{y} - \hat{\mathbf{y}} = \begin{bmatrix} 74 \\ 68 \\ 77 \\ 76 \\ 80 \\ 87 \\ 91 \end{bmatrix} - \begin{bmatrix} 73 \\ 73 \\ 73 \\ 78 \\ 78 \\ 89 \\ 89 \end{bmatrix} = \begin{bmatrix} 1 \\ -5 \\ 4 \\ -2 \\ 2 \\ -2 \\ 2 \end{bmatrix} .$$

The sum of squares of elements of this vector is

$$\text{SSE} = 1^2 + 5^2 + 4^2 + 4(2^2) = 58 \ .$$

And with

$$\text{SST} = \mathbf{y}'\mathbf{y} = 74^2 + 68^2 + 77^2 + 76^2 + 80^2 + 87^2 + 91^2 = 44055$$

using $\boldsymbol{\beta}^\circ$ of (25) in (58) gives

$$\text{SSE} = 44055 - [89(553) + (-16)219 + (-11)(156) + 0(178)] = 58;$$

and using $\dot{\boldsymbol{\beta}}$ of (27) in (58) gives the same result:

$$\text{SSE} = 44055 - [73(553) + 0(219) + 5(156) + 16(178)] = 58\ .$$

An abbreviated form of (59) is

$$\text{SSE} = \text{SST} - (\text{solution vector}) \cdot (\text{r.h.s. vector of normal equations}) \quad (60)$$

where the dot notation represents the inner product (sum of products of corresponding elements) of the two vectors.

Summary.

$$\text{SSE} = \mathbf{y}'\mathbf{y} - \boldsymbol{\beta}^{\circ\prime}\mathbf{X}'\mathbf{y} = \Sigma_t y_t^2 - \Sigma_r \beta_r^\circ(\mathbf{X}'\mathbf{y})_r$$

of (58) and (59) are our first examples of the broad applicability that comes from using the matrix algebra description of linear models. These formulae hold for *any* linear model that is written as $E(\mathbf{y}) = \mathbf{X}\boldsymbol{\beta}$. This includes all the examples in Chapters 2–6 as well as designed experiments, regression analysis and extensions of the with-covariate analyses of Chapter 6 to more complicated models than the 1-way classification, with or without more than one covariate. For every linear model formulated as $E(\mathbf{y}) = \mathbf{X}\boldsymbol{\beta}$, the residual sum of squares is $\text{SSE} = \mathbf{y}'\mathbf{y} - \boldsymbol{\beta}^{\circ\prime}\mathbf{X}'\mathbf{y}$. It is a result that is universally applicable.

Important, too, is the feature mentioned following (57): that because $\mathbf{XGX}'$ is invariant to $\mathbf{G}$ (Theorem 7.1),

$$\text{SSE} = \mathbf{y}'\mathbf{y} - \mathbf{y}'\mathbf{XGX}'\mathbf{y} = \mathbf{y}'\mathbf{y} - \boldsymbol{\beta}^{\circ\prime}\mathbf{X}'\mathbf{y}$$

is invariant to $\boldsymbol{\beta}^\circ$. This means, as illustrated in the example, that no matter what $\boldsymbol{\beta}^\circ$ is used in $\text{SSE} = \mathbf{y}'\mathbf{y} - \boldsymbol{\beta}^{\circ\prime}\mathbf{X}'\mathbf{y}$, the value (for given $\mathbf{y}$ and $\mathbf{X}$) is always the same.

d. Estimating residual error variance

The expected value of SSE is derived by applying Section 7.5 to (57). Since (57) is $\text{SSE} = \mathbf{y}'\mathbf{M}\mathbf{y}$ and $\mathbf{y} \sim (\mathbf{X}\boldsymbol{\beta}, \sigma^2\mathbf{I})$ from (11), Section 7.5 gives

$$\begin{aligned} E(\text{SSE}) &= \text{tr}(\mathbf{M}\sigma^2\mathbf{I}) + (\mathbf{X}\boldsymbol{\beta})'\mathbf{M}\mathbf{X}\boldsymbol{\beta} \\ &= \sigma^2\text{tr}(\mathbf{M}), \quad \text{because } \mathbf{MX} = \mathbf{0} \text{ from Section 7.1i}, \\ &= \sigma^2 r_{\mathbf{M}} \qquad (61) \end{aligned}$$

because, since $\mathbf{M}$ is idempotent [as in (18) of Section 7.1i], the trace of $\mathbf{M}$ equals its rank (Section 7.1d). Then, since $r_{\mathbf{M}} = N - r_{\mathbf{X}}$ from Section 7.1i, equation (61) becomes

$$E(\text{SSE}) = \sigma^2(N - r_{\mathbf{X}}) \ . \tag{62}$$

Hence an unbiased estimator of σ^2 is

$$\hat{\sigma}^2 = \frac{\text{SSE}}{N - r_{\mathbf{X}}} = \frac{\mathbf{y}'(\mathbf{I} - \mathbf{XGX}')\mathbf{y}}{N - r_{\mathbf{X}}} \ . \tag{63}$$

For the example this is $\hat{\sigma}^2 = 58/(7 - 3) = 14\frac{1}{2}$. This, like (60), is another result that is universally applicable to any linear model $E(\mathbf{y}) = \mathbf{X}\boldsymbol{\beta}$.

Examples. In Example 1, the rank of $\mathbf{X}$ in (6) is 3; more generally, for a 1-way classification of a classes, $r_{\mathbf{X}} = a$. Hence (63) is $\hat{\sigma}^2 = \text{SSE}/(N - a)$, just as in (34) of Section 2.6. Similarly, in Example 2, we get the same result because the model there, (21), is just an overparameterized reformulation of (3): $\mathbf{X}$ of (22) has rank 3, for a classes it has rank a and so (62) is $\hat{\sigma}^2 = \text{SSE}/(N - a)$. In Example 3, $\mathbf{X}$ of (30) has its first two columns corresponding to the rows of the 2-way classification layout, and its second two columns corresponding to the columns of the layout. And in $\mathbf{X}$ we see that the sum of its first two columns equals the sum of its second two. More generally, for a data layout of a rows and b columns, where the first a parameters are for rows and the last b are for columns, the sum of the first a columns of the resulting $\mathbf{X}$ always equals the sum of its last b columns. Hence $r_{\mathbf{X}} = a + b - 1$ and so (62) is $\hat{\sigma}^2 = \text{SSE}/(N - a - b + 1)$ just as in (104) of Section 4.10b, for the no-interaction two-way crossed classification.

8.4. PARTITIONING THE TOTAL SUM OF SQUARES

a. Reduction in sum of squares

The total sum of squares of the observations is (for y_t being a typical observation)

$$\text{SST} = \sum_{t=1}^{N} y_t^2 = \mathbf{y}'\mathbf{y} \ . \tag{64}$$

And the error sum of squares SSE is the residual sum of squares after fitting the model. Therefore SST − SSE is the reduction in sum of squares due to fitting the model, and with the model being $E(\mathbf{y}) = \mathbf{X}\boldsymbol{\beta}$ we call this reduction $R(\boldsymbol{\beta})$. Thus from (64), (57) and (58) we have

$$R(\boldsymbol{\beta}) = \text{SST} - \text{SSE} = \mathbf{y}'\mathbf{y} - \mathbf{y}'\mathbf{My} = \mathbf{y}'\mathbf{XGX}'\mathbf{y} = \boldsymbol{\beta}^{\circ\prime}\mathbf{X}'\mathbf{y} \ . \tag{65}$$

In this way the total sum of squares has been partitioned into two portions:

$$
\begin{aligned}
&R(\boldsymbol{\beta}) = \mathbf{y}'\mathbf{X}\mathbf{G}\mathbf{X}'\mathbf{y} \\
&\underline{\text{SSE} = \text{SST} - R(\boldsymbol{\beta}) = \mathbf{y}'\mathbf{y} - \mathbf{y}'\mathbf{X}\mathbf{G}\mathbf{X}'\mathbf{y}} \\
&\underline{\text{SST} = \mathbf{y}'\mathbf{y}.}
\end{aligned}
\tag{66}
$$

The result $R(\boldsymbol{\beta}) = \boldsymbol{\beta}^{\circ\prime}\mathbf{X}'\mathbf{y}$ of (65) is not only useful but like SSE $= \mathbf{y}'(\mathbf{I} - \mathbf{X}\mathbf{G}\mathbf{X}')\mathbf{y}$ and $\hat{\sigma}^2 = \text{SSE}/(N - r_{\mathbf{X}})$ it is universal in applying to all linear models of the form $E(\mathbf{y}) = \mathbf{X}\boldsymbol{\beta}$. We call $R(\boldsymbol{\beta}) = \boldsymbol{\beta}^{\circ\prime}\mathbf{X}'\mathbf{y}$ of (65) the *R-algorithm*, as do Searle *et al.* (1981). In this form, $R(\boldsymbol{\beta})$ is the inner product of the solution vector $\boldsymbol{\beta}^{\circ}$ and the vector $\mathbf{X}'\mathbf{y}$ of the right-hand sides of the normal equations, i.e., it is the sum of products of each element of the solution vector multiplied by the corresponding element of the vector of right-hand sides of the normal equations. This algorithm is used repeatedly in what follows. An example of it has already been noted, in (115) of Section 6.2e-i.

b. Sum of squares due to the mean

The simplest linear model is where every element of $\mathbf{y}$ has the same expected value, μ. Then

$$E(\mathbf{y}) = \mu\mathbf{1}_N \tag{67}$$

where $\mathbf{1}_N$ is a summing vector of N unities (see Section 7.1e). Hence, with $\mathbf{X}$ of the general case being $\mathbf{X} = \mathbf{1}_N$, we have $R(\mu)$ from (65) given by

$$R(\mu) = \mathbf{y}'\mathbf{1}(\mathbf{1}'\mathbf{1})^{-}\mathbf{1}'\mathbf{y} = N\bar{y}(N)^{-1}N\bar{y} = N\bar{y}^2 \tag{68}$$

where $\bar{y}$ is the mean of all elements of $\mathbf{y}$:

$$\bar{y} = \sum_{t=1}^{N} y_t/N = \mathbf{1}'\mathbf{y}/N\ . \tag{69}$$

Thus we see that the familiar "correction for the mean", or sum of squares due to the mean, is simply the reduction in sum of squares due to fitting $E(\mathbf{y}) = \mu\mathbf{1}$, namely $R(\mu)$.

It is widely-accepted practice to take $R(\mu)$ into account in the partitioning (66) by giving it a line to itself and subtracting it from $R(\boldsymbol{\beta})$. This gives the following partitioning of SST.

$$R(\mu) = N\bar{y}^2$$

$$\mathscr{R}(\boldsymbol{\beta} \mid \mu) = R(\boldsymbol{\beta}) - R(\mu) = \mathbf{y}'\mathbf{XGX}'\mathbf{y} - N\bar{y}^2 \tag{70}$$

$$\text{SSE} = \mathbf{y}'\mathbf{y} - \mathbf{y}'\mathbf{XGX}'\mathbf{y}$$

$$\text{SST} = \mathbf{y}'\mathbf{y}$$

c. Coefficient of determination

The square of the product–moment correlation between the elements of $\mathbf{y}$ and those of $\hat{\mathbf{y}}$ is called the coefficient of determination:

$$R^2 = \frac{\left[\sum_{t=1}^{N} (y_t - \bar{y})(\hat{y}_t - \bar{\hat{y}})\right]^2}{\sum_{t=1}^{N} (y_t - \bar{y})^2 \sum_{t=1}^{N} (\hat{y}_t - \bar{\hat{y}})^2} . \tag{71}$$

On substituting $\bar{\hat{y}} = \bar{y}$ from (56) this becomes

$$R^2 = \frac{\left(\Sigma_t y_t \hat{y}_t - N\bar{y}^2\right)^2}{\left(\Sigma_t y_t^2 - N\bar{y}^2\right)\left(\Sigma_t \hat{y}_t^2 - N\bar{y}^2\right)} = \frac{\left(\mathbf{y}'\hat{\mathbf{y}} - N\bar{y}^2\right)^2}{\left(\mathbf{y}'\mathbf{y} - N\bar{y}^2\right)\left(\hat{\mathbf{y}}'\hat{\mathbf{y}} - N\bar{y}^2\right)}$$

and on using $\hat{\mathbf{y}} = \mathbf{XGX}'\mathbf{y}$ with $\mathbf{XGX}'$ symmetric and idempotent this reduces to

$$R^2 = \frac{\mathbf{y}'\mathbf{XGX}'\mathbf{y} - N\bar{y}^2}{\mathbf{y}'\mathbf{y} - N\bar{y}^2} = \frac{R(\boldsymbol{\beta}) - R(\mu)}{\text{SST} - R(\mu)} = \frac{\text{SSR}_m}{\text{SST}_m} \tag{72}$$

where

$$\text{SSR}_m = R(\boldsymbol{\beta}) - R(\mu) = \mathscr{R}(\boldsymbol{\beta} \mid \mu) \quad \text{and} \quad \text{SST}_m = \text{SST} - R(\mu)$$

are the reduction, and total, sums of squares, respectively, each of them adjusted for the mean. Equation (72) is the same result as (52) in Section 2.8b, but (72) is true quite generally.

Example 2 (continued). Following (59) are the derivations of SST = 44055 and SSE = 58. Implicit therein is, from (25),

$$R(\boldsymbol{\beta}) = 89(553) + (-16)219 + (-11)156 + 0(178) = 43997$$

and, from Table 7.1, $\bar{y}_{..} = 79$ with $N = 7$ so that

$$N\bar{y}^2 = 7(79)^2 = 43687 .$$

Hence the partitioning in (70) is as follows:

$$
\begin{array}{rr}
R(\mu) & = 43687 \\
\mathscr{R}(\boldsymbol{\beta} \mid \mu) = R(\boldsymbol{\beta}) - R(\mu) = 43997 - 43687 = & 310 \\
\text{SSE} = 44055 - 43997 = & 58 \\
\hline
\text{SST} & = 44055 \\
\hline
\end{array}
$$

Apart from the inclusion of $R(\mu)$, this is the same as Table 2.3 of Section 2.8a; there the partitioning is of $\text{SST}_m = \text{SST} - R(\mu) = 368$ whereas here it is of SST = 44055.

8.5. PARTITIONING THE MODEL

a. More than one kind of parameter

Most sections of Chapters 2–5 deal with models that contain only one kind of parameter. Examples are $E(y_{ij}) = \mu_i$ of Chapter 2, which has only class means μ_i; also $E(y_{ij}) = \mu$, which has just μ; and $E(y_{ijk}) = \mu_{ij}$ of Chapters 4 and 5, the with-interaction model for the 2-way classification, which has just cell means, the μ_{ij}s. In contrast, we have also dealt with models that have two kinds of parameters: $E(y_{ij}) = \mu + \alpha_i$ of the over-parameterized model for the 1-way classification in Section 2.13, $E(y_{ijk}) = \mu_i + \tau_j$ of the no-interaction model in Section 4.9, and $E(y_{ij}) = \mu_i + bz_{ij}$ of the with-covariate model in Section 6.1. In writing these latter models as $E(\mathbf{y}) = \mathbf{X}\boldsymbol{\beta}$ it is therefore often useful to partition the model into two parts, i.e., to partition $\boldsymbol{\beta}$ into two subvectors,

$$\boldsymbol{\beta} = \begin{bmatrix} \boldsymbol{\beta}_1 \\ \boldsymbol{\beta}_2 \end{bmatrix},$$

and partition $\mathbf{X}$ correspondingly. Then we have the model as

$$E(\mathbf{y}) = \mathbf{X}\boldsymbol{\beta} = [\mathbf{X}_1 \quad \mathbf{X}_2]\begin{bmatrix} \boldsymbol{\beta}_1 \\ \boldsymbol{\beta}_2 \end{bmatrix} = \mathbf{X}_1\boldsymbol{\beta}_1 + \mathbf{X}_2\boldsymbol{\beta}_2 . \tag{73}$$

Example 1 (continued). For $\mathbf{X}$ of (22), the model equation for (21) can be written as (73) in the form

$$E(\mathbf{y}) = \mathbf{X}\boldsymbol{\beta} = \begin{bmatrix} 1 \\ 1 \\ 1 \\ 1 \\ 1 \\ 1 \\ 1 \end{bmatrix} \mu + \begin{bmatrix} 1 & \cdot & \cdot \\ 1 & \cdot & \cdot \\ 1 & \cdot & \cdot \\ \cdot & 1 & \cdot \\ \cdot & 1 & \cdot \\ \cdot & \cdot & 1 \\ \cdot & \cdot & 1 \end{bmatrix} \begin{bmatrix} \alpha_1 \\ \alpha_2 \\ \alpha_3 \end{bmatrix} .$$

The partitioned model (73) is useful not only for models that have just two kinds of parameters, as already referred to, but also when there are more than two kinds such as, for example, analysis with covariates when there is more than one covariate. It is also useful when, as shall sometimes be done, we combine two (or more) of the one-kind-of-parameter-only models into a single model; e.g., $E(y_{ij}) = \mu + \mu_i$ is an overparameterized combining of $E(y_{ij}) = \mu$ and $E(y_{ij}) = \mu_i$. For dealing with models that have two or more kinds of parameters we therefore consider (73).

b. Estimation

First, the normal equations for (73) are

$$\begin{bmatrix} \mathbf{X}_1' \\ \mathbf{X}_2' \end{bmatrix} [\mathbf{X}_1 \quad \mathbf{X}_2] \begin{bmatrix} \boldsymbol{\beta}_1^\circ \\ \boldsymbol{\beta}_2^\circ \end{bmatrix} = \begin{bmatrix} \mathbf{X}_1' \\ \mathbf{X}_2' \end{bmatrix} \mathbf{y}$$

i.e.,

$$\begin{bmatrix} \mathbf{X}_1'\mathbf{X}_1 & \mathbf{X}_1'\mathbf{X}_2 \\ \mathbf{X}_1'\mathbf{X}_1 & \mathbf{X}_2'\mathbf{X}_2 \end{bmatrix} \begin{bmatrix} \boldsymbol{\beta}_1^\circ \\ \boldsymbol{\beta}_2^\circ \end{bmatrix} = \begin{bmatrix} \mathbf{X}_1'\mathbf{y} \\ \mathbf{X}_2'\mathbf{y} \end{bmatrix} . \tag{74}$$

Then, from (38) of Section 7.3 a solution to (74) is

$$\begin{bmatrix} \boldsymbol{\beta}_1^\circ \\ \boldsymbol{\beta}_2^\circ \end{bmatrix} = \left\{ \begin{bmatrix} (\mathbf{X}_1'\mathbf{X}_1)^- & \mathbf{0} \\ \mathbf{0} & \mathbf{0} \end{bmatrix} + \begin{bmatrix} -(\mathbf{X}_1'\mathbf{X}_1)^- \mathbf{X}_1'\mathbf{X}_2 \\ \mathbf{I} \end{bmatrix} \times \mathbf{U}_{21}^- [-\mathbf{X}_2'\mathbf{X}_1(\mathbf{X}_1'\mathbf{X}_1)^- \quad \mathbf{I}] \right\} \begin{bmatrix} \mathbf{X}_1'\mathbf{y} \\ \mathbf{X}_2'\mathbf{y} \end{bmatrix} \tag{75}$$

for $\quad \mathbf{M}_1 = \mathbf{I} - \mathbf{X}_1(\mathbf{X}_1'\mathbf{X}_1)^- \mathbf{X}_1' = \mathbf{M}_1' = \mathbf{M}_1^2 \quad$ and $\quad \mathbf{U}_{21} = \mathbf{X}_2'\mathbf{M}_1\mathbf{X}_2$. (76)

Simplifying (75) with the use of (76) gives

$$\boldsymbol{\beta}_1^\circ = (\mathbf{X}_1'\mathbf{X}_1)^- \mathbf{X}_1'\mathbf{y} - (\mathbf{X}_1'\mathbf{X}_1)^- \mathbf{X}_1'\mathbf{X}_2\boldsymbol{\beta}_2^\circ \tag{77}$$

and

$$\boldsymbol{\beta}_2^\circ = (\mathbf{X}_2'\mathbf{M}_1\mathbf{X}_2)^- \mathbf{X}_2'\mathbf{M}_1\mathbf{y} . \tag{78}$$

c. Sums of squares

The reduction in sum of squares due to fitting the partitioned model is denoted $R(\beta_1, \beta_2)$. Its value, from the R-algorithm of (60), is

$$R(\beta_1, \beta_2) = \begin{bmatrix} \beta_1^\circ \\ \beta_2^\circ \end{bmatrix}' \begin{bmatrix} \mathbf{X}_1'\mathbf{y} \\ \mathbf{X}_2'\mathbf{y} \end{bmatrix} = \beta_1^{\circ\prime}\mathbf{X}_1'\mathbf{y} + \beta_2^{\circ\prime}\mathbf{X}_2'\mathbf{y}$$

and from (77) and (78) this is

$$\begin{aligned} R(\beta_1, \beta_2) &= \mathbf{y}'\mathbf{X}_1(\mathbf{X}_1'\mathbf{X}_1)^{-\prime}\mathbf{X}_1'\mathbf{y} + \beta_2^{\circ\prime}\mathbf{X}_2'\left[\mathbf{I} - \mathbf{X}_1(\mathbf{X}_1'\mathbf{X}_1)^{-\prime}\mathbf{X}_1'\right]\mathbf{y} \\ &= \mathbf{y}'\mathbf{X}_1(\mathbf{X}_1'\mathbf{X}_1)^{-}\mathbf{X}_1'\mathbf{y} + \mathbf{y}'\mathbf{M}_1\mathbf{X}_2(\mathbf{X}_2'\mathbf{M}_1\mathbf{X}_2)^{-}\mathbf{X}_2'\mathbf{M}_1\mathbf{y} \ . \quad (79) \end{aligned}$$

The first term in (79) is clearly, by analogy with the penultimate equality in (65),

$$R(\beta_1) = \mathbf{y}'\mathbf{X}_1(\mathbf{X}_1'\mathbf{X}_1)^{-}\mathbf{X}_1'\mathbf{y} \quad \text{due to fitting} \quad E(\mathbf{y}) = \mathbf{X}_1\beta_1 \ . \quad (80)$$

Since $\mathbf{X}_1\beta_1$ is part of $E(\mathbf{y}) = \mathbf{X}_1\beta_1 + \mathbf{X}_2\beta_2$ for which (79) has been derived, we call $E(\mathbf{y}) = \mathbf{X}_1\beta_1$ of (80) a *sub-model* of $E(\mathbf{y}) = \mathbf{X}_1\beta_1 + \mathbf{X}_2\beta_2$; and the latter is called the *full model*. Then the difference between the sums of squares in (79) and (80) is the difference between the reductions for the full model and sub-model and is

$$\mathscr{R}(\beta_1, \beta_2 \mid \beta_1) = R(\beta_2 \mid \beta_1) = R(\beta_1, \beta_2) - R(\beta_1), \quad (81)$$

$$= \mathbf{y}'\mathbf{M}_1\mathbf{X}_2(\mathbf{X}_2'\mathbf{M}_1\mathbf{X}_2)^{-}\mathbf{X}_2'\mathbf{M}_1\mathbf{y} \ . \quad (82)$$

Equation (82) is a general expression for $\mathscr{R}(\beta_1, \beta_2 \mid \beta_1)$, the reduction in sum of squares for fitting $E(\mathbf{y}) = \mathbf{X}_1\beta_1 + \mathbf{X}_2\beta_2$ after fitting $E(\mathbf{y}) = \mathbf{X}_1\beta_1$. It is the sum of squares due to fitting β_1 and β_2, minus that due to fitting just β_1. It is therefore the increase in sum of squares that can be attributed to fitting β_1 and β_2 over and above fitting just β_1; and so it is the increase in sum of squares attributable to fitting β_2, having already fitted β_1. In this sense $R(\beta_2|\beta_1)$ is conventionally called the sum of squares due to "β_2 adjusted for β_1". This is a conveniently brief description, but it may not always convey the important sense of $R(\beta_2 \mid \beta_1)$ being the sum of squares due to β_2 *after* (having already fitted) β_1. Thus $R(\beta_2 \mid \beta_1)$ being called the sum of squares due to "β_2 after β_1" may be more apt.

The utility of $R(\beta_2 \mid \beta_1)$ is not that all sums of squares of this form are necessarily of interest, but that many sums of squares available from computer output are readily describable in this manner—and it is a description that succinctly identifies the nature of a sum of squares so labeled.

It is not particularly informative to illustrate (82) with examples because doing so usually involves tedious matrix algebra only to end up with a known expression for whatever sum of squares is being illustrated. Nevertheless we do give two examples; the first is relatively brief, and the second demonstrates how extensive the algebra can get.

d. Examples

-i. The 1-way classification with covariate. The model for this is, as in Section 6.1,

$$E(y_{ij}) = \mu_i + bz_{ij}$$

for $i = 1, \ldots, a$ and $j = 1, \ldots, n_i$. In the form of (73) it is

$$E(\mathbf{y}) = \mathbf{D}\{\mathbf{1}_i\}\boldsymbol{\mu} + b\mathbf{z},$$

where, for convenience, $\mathbf{1}_i$ is $\mathbf{1}_{n_i}$ and where $\boldsymbol{\mu}' = [\mu_1 \quad \mu_2 \quad \cdots \quad \mu_a]$ and $\mathbf{z}$ is the vector of z_{ij}s, similar to $\mathbf{y}$. Then for $\mathbf{X}_1$ and $\mathbf{X}_2$ of (73) we have

$$\mathbf{X}_1 = \mathbf{D}\{\mathbf{1}_i\} \quad \text{and} \quad \mathbf{X}_2 = \mathbf{z}\ .$$

Therefore
$$\mathbf{M}_1 = \mathbf{I} - \mathbf{X}_1(\mathbf{X}_1'\mathbf{X}_1)^- \mathbf{X}_1' = \mathbf{I} - \mathbf{D}\{\bar{\mathbf{J}}_i\}$$

for $\bar{\mathbf{J}}_i$ as used in (72) of Section 7.6 (a square matrix of order n_i, with every element $1/n_i$), whereupon

$$\mathbf{M}_1\mathbf{X}_2 = (\mathbf{I} - \mathbf{D}\{\bar{\mathbf{J}}_i\})\mathbf{z} = \mathbf{z} - \{_c\, \bar{z}_{i\cdot}\mathbf{1}_i\}\ .$$

Hence, recalling $W_{yz} = \Sigma_i\Sigma_j(y_{ij} - \bar{y}_{i\cdot})(z_{ij} - \bar{z}_{i\cdot})$ of (30) in Section 6.1d, we find that

$$\mathbf{y}'\mathbf{M}_1\mathbf{X}_2 = \mathbf{y}'\mathbf{M}_1(\mathbf{M}_1\mathbf{X}_2) = \mathbf{y}'\mathbf{M}_1(\mathbf{M}_1\mathbf{z}) = W_{yz}$$

and

$$\mathbf{X}_2'\mathbf{M}_1\mathbf{X}_2 = \mathbf{X}_2'\mathbf{M}_1(\mathbf{M}_1\mathbf{X}_2) = \mathbf{z}'\mathbf{M}_1(\mathbf{M}_1\mathbf{z}) = W_{zz}$$

and so (82) is
$$\mathscr{R}(\mu_i, b\,|\,\mu_i) = W_{yz}^2/W_{zz}$$

as in (48) and (35) of Chapter 6.

-ii. The 1-way classification. The overparameterized model for the 1-way classification is $E(y_{ij}) = \mu + \alpha_i$ with $i = 1, \ldots, a$ and $j = 1, \ldots, n_i$ as in (20). We are interested in deriving $\mathscr{R}(\mu, \alpha_i \mid \mu)$. In the form of (73) the model is $E(\mathbf{y}) = \mu \mathbf{1}_N + \mathbf{D}\{\mathbf{1}_i\}\boldsymbol{\alpha}$, where $\boldsymbol{\alpha}' = [\alpha_1 \quad \cdots \quad \alpha_a]$; and its coefficient matrix is the same as in the preceding example. Then for (73), $\mathbf{X}_1 = \mathbf{1}_N$ and $\mathbf{X}_2 = \mathbf{D}\{\mathbf{1}_i\}$. Therefore $\mathbf{M}_1 = \mathbf{I} - \mathbf{X}_1(\mathbf{X}_1'\mathbf{X}_1)^-\mathbf{X}_1' = \mathbf{I} - \bar{\mathbf{J}}_N$ and

$$\mathbf{M}_1\mathbf{X}_2 = (\mathbf{I} - \bar{\mathbf{J}}_N)\mathbf{D}\{\mathbf{1}_i\} = \mathbf{X}_2 - \mathbf{1}_N[n_1 \quad \cdots \quad n_a]/N .$$

Hence

$$\mathbf{X}_2'\mathbf{M}_1\mathbf{X}_2 = \mathbf{X}_2'\mathbf{X}_2 - \mathbf{X}_2'\mathbf{1}[n_1 \quad \cdots \quad n_a]/N = \mathbf{D}\{n_i\} - \begin{bmatrix} n_1 \\ \vdots \\ n_a \end{bmatrix}[n_1 \quad \cdots \quad n_a]/N .$$

This matrix is singular: elements in row i sum to zero since $n_i - n_i\Sigma_i n_i/N = 0$. Define

$$\boldsymbol{\Delta} = \mathbf{D}\{n_i\}_{i=1}^{i=a-1} \quad \text{and} \quad \mathbf{t}' = [n_1 \quad \cdots \quad n_{a-1}] .$$

Then

$$(\boldsymbol{\Delta} - \mathbf{t}\mathbf{t}'/N)^{-1} = \boldsymbol{\Delta}^{-1} + \boldsymbol{\Delta}^{-1}\mathbf{t}\mathbf{t}'\boldsymbol{\Delta}^{-1}/(N - \mathbf{t}'\boldsymbol{\Delta}^{-1}\mathbf{t}) = \boldsymbol{\Delta}^{-1} + \mathbf{J}_{a-1}/n_a,$$

where $\mathbf{J}_{a-1}$ is a J-matrix of order $a - 1$. Thus with

$$\mathbf{y}'\mathbf{M}_1\mathbf{X}_2 = [n_1(\bar{y}_{1\cdot} - \bar{y}_{\cdot\cdot}) \quad \cdots \quad n_a(\bar{y}_{a\cdot} - \bar{y}_{\cdot\cdot})]$$

we have (82) as

$$R(\alpha \mid \mu) = \mathscr{R}(\mu, \alpha_i \mid \mu) = \mathbf{y}'\mathbf{M}_1\mathbf{X}_2 \begin{bmatrix} \boldsymbol{\Delta}^{-1} + \mathbf{J}_{a-1}/n_a & \mathbf{0}\mathbf{1}_{a-1} \\ \mathbf{0}\mathbf{1}'_{a-1} & 0 \end{bmatrix} \mathbf{M}_1\mathbf{X}_2'\mathbf{y}$$

$$= \sum_{i=1}^{a-1} n_i^2(\bar{y}_{i\cdot} - \bar{y}_{\cdot\cdot})^2/n_i + \left[\sum_{i=1}^{a-1} n_i(\bar{y}_{i\cdot} - \bar{y}_{\cdot\cdot})\right]^2 \Big/ n_a .$$

But

$$\sum_{i=1}^{a-1} n_i(\bar{y}_{i\cdot} - \bar{y}_{\cdot\cdot}) = N\bar{y}_{\cdot\cdot} - n_a\bar{y}_{a\cdot} - (N - n_a)\bar{y}_{\cdot\cdot} = -n_a(\bar{y}_{a\cdot} - \bar{y}_{\cdot\cdot}),$$

and so

$$R(\alpha \mid \mu) = \mathscr{R}(\mu, \alpha_i \mid \mu) = \sum_{i=1}^{a} n_i(\bar{y}_{i\cdot} - \bar{y}_{\cdot\cdot})^2,$$

as is well known.

Despite the lengthy algebra in these examples, (82) is a useful result. It is of interest in itself, it is available as the basis for a computing algorithm, and it is a general expression that can be applied in any particular cases of interest to derive specific expressions for sums of squares that may otherwise not be so readily available.

e. Expected sums of squares

Expected values of $R(\boldsymbol{\beta}_1)$ and of $R(\boldsymbol{\beta}_2 | \boldsymbol{\beta}_1) = \mathscr{R}(\boldsymbol{\beta}_1, \boldsymbol{\beta}_2 | \boldsymbol{\beta}_1)$ are obtained from

$$E(\mathbf{y}'\mathbf{A}\mathbf{y}) = \boldsymbol{\mu}'\mathbf{A}\boldsymbol{\mu} + \text{tr}(\mathbf{A}\mathbf{V})$$

of Section 7.5, with $\mathbf{V} = \sigma^2\mathbf{I}$ and $\boldsymbol{\mu} = \mathbf{X}_1\boldsymbol{\beta}_1 + \mathbf{X}_2\boldsymbol{\beta}_2$. For $R(\boldsymbol{\beta}_1)$ of (80) the $\text{tr}(\mathbf{A}\mathbf{V})$ part is

$$\text{tr}\left[\mathbf{X}_1(\mathbf{X}_1'\mathbf{X}_1)^{-}\mathbf{X}_1\sigma^2\mathbf{I}\right] = \sigma^2 r(\mathbf{X}_1'\mathbf{X}_1) = \sigma^2 r_{\mathbf{X}_1}$$

and so, on denoting $r_{\mathbf{X}_1}$ by r_1 (and later, $r_{[\mathbf{X}_1\ \mathbf{X}_2]}$ by r_{12}), we have

$$E_{\boldsymbol{\beta}_1, \boldsymbol{\beta}_2} R(\boldsymbol{\beta}_1) = (\boldsymbol{\beta}_1'\mathbf{X}_1' + \boldsymbol{\beta}_2'\mathbf{X}_2')\mathbf{X}_1(\mathbf{X}_1'\mathbf{X}_1)^{-}\mathbf{X}_1'(\mathbf{X}_1\boldsymbol{\beta}_1 + \mathbf{X}_2\boldsymbol{\beta}_2) + r_1\sigma^2 . \tag{83}$$

The notation $E_{\boldsymbol{\beta}_1, \boldsymbol{\beta}_2}$ in (83) emphasizes that expectation is under the model $E(\mathbf{y}) = \mathbf{X}_1\boldsymbol{\beta}_1 + \mathbf{X}_2\boldsymbol{\beta}_2$, in contrast to expectation under other models to be considered in Section 8.5g.

Similarly for $R(\boldsymbol{\beta}_2 | \boldsymbol{\beta}_1)$ of (82) the $\text{tr}(\mathbf{A}\mathbf{V})$ term of $E(\mathbf{y}'\mathbf{A}\mathbf{y})$ is

$$\text{tr}\left[\mathbf{M}_1\mathbf{X}_2(\mathbf{X}_2'\mathbf{M}_1\mathbf{X}_2)^{-}\mathbf{X}_2'\mathbf{M}_1\sigma^2\mathbf{I}\right] = (r_{12} - r_1)\sigma^2,$$

on using (44) of Section 7.3b; and the $\boldsymbol{\mu}'\mathbf{A}\boldsymbol{\mu}$ term is

$$\begin{aligned}(\mathbf{X}_1\boldsymbol{\beta}_1 + \mathbf{X}_2\boldsymbol{\beta}_2)'&\left[\mathbf{M}_1\mathbf{X}_2(\mathbf{X}_2'\mathbf{M}_1\mathbf{X}_2)^{-}\mathbf{X}_2'\mathbf{M}_1\right](\mathbf{X}_1\boldsymbol{\beta}_1 + \mathbf{X}_2\boldsymbol{\beta}_2)\\ &= \boldsymbol{\beta}_2'\mathbf{X}_2'\mathbf{M}_1\mathbf{X}_2\boldsymbol{\beta}_2\end{aligned}$$

because $\mathbf{M}_1\mathbf{X}_1 = \mathbf{0}$. Hence

$$E_{\boldsymbol{\beta}_1, \boldsymbol{\beta}_2} R(\boldsymbol{\beta}_2 | \boldsymbol{\beta}_1) = \boldsymbol{\beta}_2'\mathbf{X}_2'\mathbf{M}_1\mathbf{X}_2\boldsymbol{\beta}_2 + (r_{12} - r_1)\sigma^2 . \tag{84}$$

f. The "invert part of the inverse" algorithm

*-i. **For the full rank case.*** The full rank case is when $\mathbf{X} = [\mathbf{X}_1 \quad \mathbf{X}_2]$ has full column rank, $r_{\mathbf{X}_1} + r_{\mathbf{X}_2}$. Define

$$\begin{bmatrix} \mathbf{X}_1'\mathbf{X}_1 & \mathbf{X}_1'\mathbf{X}_2 \\ \mathbf{X}_2'\mathbf{X}_1 & \mathbf{X}_2'\mathbf{X}_2 \end{bmatrix}^{-1} = \mathbf{T} = \begin{bmatrix} \mathbf{T}_{11} & \mathbf{T}_{12} \\ \mathbf{T}_{21} & \mathbf{T}_{22} \end{bmatrix} \quad \text{with} \quad \mathbf{T}_{21} = (\mathbf{T}_{12})' . \tag{85}$$

Then $\mathbf{T}_{22}$ of (85) is $\mathbf{T}_{22} = \mathbf{U}_{21}^{-} = (\mathbf{X}_2'\mathbf{M}_1\mathbf{X}_2)^{-}$ of (76) is nonsingular, and the sole solution of the normal equations (74) is, from (77) and (78),

$$\hat{\boldsymbol{\beta}}_1 = (\mathbf{X}_1'\mathbf{X}_1)^{-1}\mathbf{X}_1'\mathbf{y} - (\mathbf{X}_1'\mathbf{X}_1)^{-1}\mathbf{X}_1'\mathbf{X}_2\hat{\boldsymbol{\beta}}_2 \tag{86}$$

and

$$\hat{\boldsymbol{\beta}}_2 = \mathbf{T}_{22}\left[\mathbf{X}_2'\mathbf{y} - \mathbf{X}_2'\mathbf{X}_1(\mathbf{X}_1'\mathbf{X}_1)^{-1}\mathbf{X}_1'\mathbf{y}\right] . \tag{87}$$

Hence

$$\mathbf{T}_{22}^{-1}\hat{\boldsymbol{\beta}}_2 = \mathbf{X}_2'\left[\mathbf{I} - \mathbf{X}_1(\mathbf{X}_1'\mathbf{X}_1)^{-1}\mathbf{X}_1'\right]\mathbf{y} = \mathbf{X}_2'\mathbf{M}_1\mathbf{y},$$

and so (82) is

$$\mathscr{R}(\boldsymbol{\beta}_2 \,|\, \boldsymbol{\beta}_1) = \hat{\boldsymbol{\beta}}_2'\mathbf{T}_{22}^{-1}\hat{\boldsymbol{\beta}}_2 . \tag{88}$$

This has been called [Searle (1971), p. 115] the "invert part of the inverse" algorithm due to the occurrence in (88) of $\mathbf{T}_{22}^{-1}$, the inverse of the $\mathbf{T}_{22}$ part of $(\mathbf{X}'\mathbf{X})^{-1}$ in (85).

Example 4. Consider fitting the full rank regression model $E(y_i) = \mu + \beta_1 x_{1i} + \beta_2 x_{2i}$ to the data of Table 8.3. The model equations are

$$\begin{bmatrix} 3 \\ 1 \\ 2 \\ 6 \end{bmatrix} = \begin{bmatrix} 1 & 2 & 3 \\ 1 & 4 & 5 \\ 1 & 3 & 4 \\ 1 & 3 & 8 \end{bmatrix} \begin{bmatrix} \mu \\ \beta_1 \\ \beta_2 \end{bmatrix} + \mathbf{e} .$$

These give normal equations

$$\begin{bmatrix} 4 & 12 & 20 \\ 12 & 38 & 62 \\ 20 & 62 & 114 \end{bmatrix} \begin{bmatrix} \hat{\mu} \\ \hat{\beta}_1 \\ \hat{\beta}_2 \end{bmatrix} = \begin{bmatrix} 12 \\ 34 \\ 70 \end{bmatrix} \quad \text{with solution}$$

$$\begin{bmatrix} \hat{\mu} \\ \hat{\beta}_1 \\ \hat{\beta}_2 \end{bmatrix} = \frac{1}{12} \begin{bmatrix} 61 & -16 & -2 \\ -16 & 7 & -1 \\ -2 & -1 & 1 \end{bmatrix} \begin{bmatrix} 12 \\ 34 \\ 70 \end{bmatrix} = \begin{bmatrix} 4 \\ -2 \\ 1 \end{bmatrix} .$$

TABLE 8.3. DATA FOR A REGRESSION EXAMPLE

y_i	x_{1i}	x_{2i}
3	2	3
1	4	5
2	3	4
6	3	8

Hence, using the R-algorithm of (65), the reduction in sum of squares is

$$R(\mu, \beta_1, \beta_2) = 4(12) - 2(34) + 1(70) = 50 \ .$$

For fitting the sub-model $E(y_i) = \mu + \beta_1 x_{1i}$ the normal equations are

$$\begin{bmatrix} 4 & 12 \\ 12 & 38 \end{bmatrix} \begin{bmatrix} \tilde{\mu} \\ \tilde{\beta}_1 \end{bmatrix} = \begin{bmatrix} 12 \\ 34 \end{bmatrix}$$

with solution $$\begin{bmatrix} \tilde{\mu} \\ \tilde{\beta}_1 \end{bmatrix} = \frac{1}{8} \begin{bmatrix} 38 & -12 \\ -12 & 4 \end{bmatrix} \begin{bmatrix} 12 \\ 34 \end{bmatrix} = \begin{bmatrix} 6 \\ -1 \end{bmatrix} .$$

Hence $$R(\mu, \beta_1) = 6(12) - 1(34) = 38 \ .$$

Therefore

$$R(\beta_2 \mid \mu, \beta_1) = R(\mu, \beta_1, \beta_2) - R(\mu, \beta_1) = 50 - 38 = 12 \ .$$

We confirm this from the "invert part of the inverse" algorithm of (88):

$$R(\beta_2 \mid \mu, \beta_1) = \hat{\beta}_2' \mathbf{T}^{-1} \tilde{\beta}_2 = 1\left(\tfrac{1}{12}\right)^{-1} 1 = 12 \ .$$

Likewise,

$$R(\beta_1 \mid \mu, \beta_2) = \hat{\beta}_1 \mathbf{T}_{11}^{-1} \hat{\beta}_1 = (-2)\left(\tfrac{7}{12}\right)^{-1}(-2) = 6\tfrac{6}{7},$$

as can be confirmed by the reader.

-ii. ***For the non full rank case—a caution.*** Essentially the same principle appears to exist for the non full rank case when substituting (78) into (82). As in (75), the matrix $(\mathbf{X}_2'\mathbf{M}_1\mathbf{X}_2)^-$ is a submatrix of $(\mathbf{X}'\mathbf{X})^-$. Hence, provided $(\mathbf{X}_2'\mathbf{M}_1\mathbf{X}_2)^-$ is taken as a symmetric reflexive generalized inverse of $\mathbf{X}_2'\mathbf{M}_1\mathbf{X}_2$ [see (11) of Section 7.1i], then $\mathbf{X}_2'\mathbf{M}_1\mathbf{X}_2$ is a generalized inverse of $(\mathbf{X}_2'\mathbf{M}_1\mathbf{X}_2)^-$ and the description "generalized inverse of part of the generalized inverse" applies to (82):

$$R(\boldsymbol{\beta}_2 \mid \boldsymbol{\beta}_1) = \boldsymbol{\beta}_2^{\circ\prime}\left[(\mathbf{X}_2'\mathbf{M}_1\mathbf{X}_2)^-\right]^- \boldsymbol{\beta}_2^{\circ} \ . \tag{89}$$

But this applies only if $(\mathbf{X}'\mathbf{X})^-$ is derived exactly as inside the braces of (75), and then (89) applies *only* for $R(\boldsymbol{\beta}_2 \mid \boldsymbol{\beta}_1)$. With that generalized inverse (89) does not apply to $R(\boldsymbol{\beta}_1 \mid \boldsymbol{\beta}_2)$. This severe limitation greatly restricts the usefulness of the algorithm for non full rank models. Yet this fact itself is important because (88) and its companion

$$R(\boldsymbol{\beta}_1 | \boldsymbol{\beta}_2) = \hat{\boldsymbol{\beta}}_1' \mathbf{T}_{11}^{-1} \hat{\boldsymbol{\beta}}_1$$

have on occasion been used in computing packages. That is successful for the full rank case but its counterpart, (89), is not always successful for the non full rank case.

Example 3 (continued). For data of 2 rows and 2 columns in Table 8.2,

$$R(\mu_i) = \Sigma_i n_{i\cdot} \bar{y}_{i\cdot\cdot}^2 = 5(13^2) + 3(25^2) = 2720$$

and
$$R(\tau_j) = \Sigma_j n_{\cdot j} \bar{y}_{\cdot j\cdot}^2 = 4(12^2) + 4(23^2) = 2692 \ .$$

And for the no-interaction model the normal equations of (33) with solution (36) give

$$R(\mu_i, \tau_j) = 19(65) + 35(75) + (-15)48 = 3140;$$

and, of course, using solution (38) gives the same value. Hence

$$R(\tau_j | \mu_i) = \mathscr{R}(\mu_i, \tau_j | \mu_i) = 3140 - 2720 = 420$$

and
$$R(\mu_i | \tau_j) = \mathscr{R}(\mu_i, \tau_j | \tau_j) = 3140 - 2692 = 448 \ .$$

Now **F** of (35), which yields solution (36), is precisely the form of $(\mathbf{X}'\mathbf{X})^-$ in (75) that is suitable for using (89) to calculate

$$R(\tau_j | \mu_i) = \mathscr{R}(\mu_i, \tau_j | \mu_i)$$
$$= \begin{bmatrix} -15 & 0 \end{bmatrix} \begin{bmatrix} 15/28 & 0 \\ 0 & 0 \end{bmatrix}^- \begin{bmatrix} -15 \\ 0 \end{bmatrix} = 15(28) = 420 \ .$$

But (35) and (36) do not yield $R(\mu_i | \tau_j) = \mathscr{R}(\mu_i, \tau_j | \tau_j)$ in the form

$$\begin{bmatrix} 19 & 35 \end{bmatrix} \left[\frac{1}{28} \begin{pmatrix} 8 & 4 \\ 4 & 16 \end{pmatrix} \right]^{-1} \begin{bmatrix} 19 \\ 35 \end{bmatrix} = 2564 \neq 448 \ .$$

Despite the last result, (82) and its mate are correct:

$$R(\boldsymbol{\beta}_2 | \boldsymbol{\beta}_1) = \mathbf{y}'\mathbf{X}_2\mathbf{M}_1(\mathbf{X}_2'\mathbf{M}_1\mathbf{X}_2)^- \mathbf{M}_1\mathbf{X}_2'\mathbf{y}$$

and
$$R(\boldsymbol{\beta}_1 | \boldsymbol{\beta}_2) = \mathbf{y}'\mathbf{X}_1\mathbf{M}_2(\mathbf{X}_1'\mathbf{M}_2\mathbf{X}_1)^- \mathbf{M}_2\mathbf{X}_1'\mathbf{y} \ .$$

To use these in the example we get from (30) that

$$\mathbf{X}_1 = \begin{bmatrix} 1 & \cdot \\ 1 & \cdot \\ 1 & \cdot \\ 1 & \cdot \\ 1 & \cdot \\ \cdot & 1 \\ \cdot & 1 \\ \cdot & 1 \end{bmatrix} \quad \text{and} \quad \mathbf{X}_2 = \begin{bmatrix} 1 & \cdot \\ 1 & \cdot \\ \cdot & 1 \\ \cdot & 1 \\ \cdot & 1 \\ 1 & \cdot \\ 1 & \cdot \\ \cdot & 1 \end{bmatrix}.$$

These give

$$\mathbf{M}_1 = \mathbf{I} - \begin{bmatrix} \bar{\mathbf{J}}_5 & \cdot \\ \cdot & \bar{\mathbf{J}}_3 \end{bmatrix} \quad \text{and} \quad \mathbf{M}_2 = \mathbf{I} - \frac{1}{4}\begin{bmatrix} \mathbf{J}_2 & \cdot & \mathbf{J}_2 & \cdot \\ \cdot & \mathbf{J}_3 & \cdot & \mathbf{J}_{3,1} \\ \mathbf{J}_2 & \cdot & \mathbf{J}_2 & \cdot \\ \cdot & \mathbf{J}_{1,3} & \cdot & \mathbf{J}_1 \end{bmatrix},$$

where $\mathbf{J}_{3,1}$ is a 3×1 matrix with every element unity. Then

$$\mathbf{X}_2'\mathbf{M}_1 = \begin{bmatrix} \mathbf{t}' \\ -\mathbf{t}' \end{bmatrix} \quad \text{for} \quad \mathbf{t}' = [9 \quad 9 \quad -6 \quad -6 \quad -6 \quad 5 \quad 5 \quad -10]/15$$

and

$$\mathbf{X}_1'\mathbf{M}_2 = \begin{bmatrix} \mathbf{u}' \\ -\mathbf{u}' \end{bmatrix} \quad \text{for} \quad \mathbf{u}' = [2 \quad 2 \quad 1 \quad 1 \quad 1 \quad -2 \quad -2 \quad -3]/4\,.$$

Hence, with $\mathbf{y}' = [8 \quad 12 \quad 11 \quad 16 \quad 18 \quad 13 \quad 15 \quad 47]$

$$\mathbf{X}_2'\mathbf{M}_1\mathbf{y} = [\mathbf{t}'\mathbf{y} \quad -\mathbf{t}'\mathbf{y}]' \quad \text{for} \quad \mathbf{t}'\mathbf{y} = -28,$$

$$\mathbf{X}_1'\mathbf{M}_2\mathbf{y} = [\mathbf{u}'\mathbf{y} \quad -\mathbf{u}'\mathbf{y}] \quad \text{for} \quad \mathbf{u}'\mathbf{y} = -28,$$

and

$$\mathbf{X}_2'\mathbf{M}_1\mathbf{X}_2 = \mathbf{X}_2'\mathbf{M}_1(\mathbf{X}_2'\mathbf{M}_1)' = \tfrac{1}{15}\begin{bmatrix} 28 & -28 \\ -28 & 28 \end{bmatrix}$$

and

$$\mathbf{X}_1'\mathbf{M}_2\mathbf{X}_1 = \mathbf{X}_1'\mathbf{M}_2(\mathbf{X}_1'\mathbf{M}_2)' = \tfrac{1}{4}\begin{bmatrix} 7 & -7 \\ -7 & 7 \end{bmatrix}.$$

Hence:

$$\mathscr{R}(\mu_i, \tau_j \mid \mu_i) = [-28 \quad 28]\left[\tfrac{1}{15}\begin{pmatrix} 28 & -28 \\ -28 & 28 \end{pmatrix}\right]^{-}\begin{bmatrix} -28 \\ 28 \end{bmatrix} = 420$$

$$\mathscr{R}(\mu_i, \tau_j \mid \tau_j) = [-28 \quad 28]\left[\tfrac{1}{4}\begin{pmatrix} 7 & -7 \\ -7 & 7 \end{pmatrix}\right]^{-}\begin{bmatrix} -28 \\ 28 \end{bmatrix} = \frac{(-28)^2}{7/4} = 448\,.$$

g. Extended partitioning

The derivation of $R(\boldsymbol{\beta}_2|\boldsymbol{\beta}_1)$ in (82) came from the partitioned model $E(\mathbf{y}) = \mathbf{X}_1\boldsymbol{\beta}_1 + \mathbf{X}_2\boldsymbol{\beta}_2$ having in it the same two $\boldsymbol{\beta}$s as are in $R(\boldsymbol{\beta}_2|\boldsymbol{\beta}_1)$. But this is not necessary. $R(\boldsymbol{\beta}_2|\boldsymbol{\beta}_1)$ of (82) can be taken as a stand-alone definition; it does not depend on $\boldsymbol{\beta}$ being partitioned into just $\boldsymbol{\beta}_1$ and $\boldsymbol{\beta}_2$. Even if $\boldsymbol{\beta}$ is partitioned as $\boldsymbol{\beta}' = [\boldsymbol{\beta}_1' \quad \boldsymbol{\beta}_2' \quad \boldsymbol{\beta}_3']$, the definition in (82) still holds. It is applicable for any vectors $\boldsymbol{\beta}_1$ and $\boldsymbol{\beta}_2$, provided only that they are disjoint, in the sense of having no element in common. No parameter can be in both $\boldsymbol{\beta}_1$ and $\boldsymbol{\beta}_2$ for (82) to hold. Thus

$$R(\boldsymbol{\beta}_2|\boldsymbol{\beta}_1) = R(\boldsymbol{\beta}_1, \boldsymbol{\beta}_2) - R(\boldsymbol{\beta}_1) = \mathbf{y}'\mathbf{M}_1\mathbf{X}_2(\mathbf{X}_2'\mathbf{M}_1\mathbf{X}_2)^-\mathbf{X}_2'\mathbf{M}_1\mathbf{y} \quad (90)$$

of (82), even when the model is partitioned into more parts than just $\mathbf{X}_1\boldsymbol{\beta}_1$ and $\mathbf{X}_2\boldsymbol{\beta}_2$, e.g., even when the model is

$$E(\mathbf{y}) = \mathbf{X}_1\boldsymbol{\beta}_1 + \mathbf{X}_2\boldsymbol{\beta}_2 + \mathbf{X}_2\boldsymbol{\beta}_3 \ . \quad (91)$$

Denoting expectation under this model by $E_{\boldsymbol{\beta}_1, \boldsymbol{\beta}_2, \boldsymbol{\beta}_3}$ we can, by applying Theorem 7.2a to (82) and using (44) of Section 7.3b, derive

$$\begin{aligned} &E_{\boldsymbol{\beta}_1, \boldsymbol{\beta}_2, \boldsymbol{\beta}_3} R(\boldsymbol{\beta}_2|\boldsymbol{\beta}_1) \\ &\quad = (\mathbf{X}_2\boldsymbol{\beta}_2 + \mathbf{X}_3\boldsymbol{\beta}_3)'\mathbf{M}_1\mathbf{X}_2(\mathbf{X}_2'\mathbf{M}_1\mathbf{X}_2)^-\mathbf{X}_2'\mathbf{M}_1(\mathbf{X}_2\boldsymbol{\beta}_2 + \mathbf{X}_3\boldsymbol{\beta}_3) \\ &\qquad + (r_{12} - r_1)\sigma^2 \ . \end{aligned} \quad (92)$$

h. Partitioning the total sum of squares

Partitioning a model is what leads to partitioning a total sum of squares, as has been done in a variety of ways in preceding chapters (e.g., Tables 2.4, 3.3, 4.8, 4.11, 6.4 and 6.6). Each such partitioning is one or other of the general forms shown in Table 8.4, or an extension thereof when the model has been partitioned into more than three parts. Notation in Table 8.4 is

$$r_1, r_{12} \text{ and } r_{123} \quad \text{as rank of} \quad \mathbf{X}_1, \quad [\mathbf{X}_1 \quad \mathbf{X}_2] \text{ and } [\mathbf{X}_1 \quad \mathbf{X}_2 \quad \mathbf{X}_3] \quad (93)$$

respectively; and, based on Section 2.7, we have

$$\begin{aligned} R(\boldsymbol{\beta}_1) &= \mathcal{R}(\boldsymbol{\beta}_1) &&= R(\boldsymbol{\beta}_1) \\ R(\boldsymbol{\beta}_2|\boldsymbol{\beta}_1) &= \mathcal{R}(\boldsymbol{\beta}_1, \boldsymbol{\beta}_2|\boldsymbol{\beta}_1) &&= R(\boldsymbol{\beta}_1, \boldsymbol{\beta}_2) - R(\boldsymbol{\beta}_1) \\ R(\boldsymbol{\beta}_3|\boldsymbol{\beta}_1, \boldsymbol{\beta}_2) &= \mathcal{R}(\boldsymbol{\beta}_1, \boldsymbol{\beta}_2, \boldsymbol{\beta}_3|\boldsymbol{\beta}_1, \boldsymbol{\beta}_2) &&= R(\boldsymbol{\beta}_1, \boldsymbol{\beta}_2, \boldsymbol{\beta}_3) - R(\boldsymbol{\beta}_1, \boldsymbol{\beta}_2) \ . \end{aligned}$$

TABLE 8.4. PARTITIONING SST FOR THREE DIFFERENT PARTITIONED MODELS

	Partitioning of SST		
Model for $E(\mathbf{y})$	d.f.	Sum of Squares	
$\mathbf{X}_1\boldsymbol{\beta}_1$	r_1	$R(\boldsymbol{\beta}_1)$	$= R(\boldsymbol{\beta}_1)$
	$N - r_1$	SSE_1	$= \text{SST} - R(\boldsymbol{\beta}_1)$
	N	SST	$= \mathbf{y}'\mathbf{y}$
$\mathbf{X}_1\boldsymbol{\beta}_1 + \mathbf{X}_2\boldsymbol{\beta}_2$	r_1	$R(\boldsymbol{\beta}_1)$	$= R(\boldsymbol{\beta}_1)$
	$r_{12} - r_1$	$R(\boldsymbol{\beta}_2 \mid \boldsymbol{\beta}_1)$	$= R(\boldsymbol{\beta}_1, \boldsymbol{\beta}_2) + R(\boldsymbol{\beta}_1)$
	$N - r_{12}$	SSE_{12}	$= \text{SST} - R(\boldsymbol{\beta}_1, \boldsymbol{\beta}_2)$
	N	SST	$= \mathbf{y}'\mathbf{y}$
$\mathbf{X}_1\boldsymbol{\beta}_1 + \mathbf{X}_2\boldsymbol{\beta}_2 + \mathbf{X}_3\boldsymbol{\beta}_3$	r_1	$R(\boldsymbol{\beta}_1)$	$= R(\boldsymbol{\beta}_1)$
	$r_{12} - r_1$	$R(\boldsymbol{\beta}_2 \mid \boldsymbol{\beta}_1)$	$= R(\boldsymbol{\beta}_1, \boldsymbol{\beta}_2) - R(\boldsymbol{\beta}_1)$
	$r_{123} - r_{12}$	$R(\boldsymbol{\beta}_3 \mid \boldsymbol{\beta}_1, \boldsymbol{\beta}_2)$	$= R(\boldsymbol{\beta}_1, \boldsymbol{\beta}_2, \boldsymbol{\beta}_3) - R(\boldsymbol{\beta}_1, \boldsymbol{\beta}_2)$
	$N - r_{123}$	SSE_{123}	$= \text{SST} - R(\boldsymbol{\beta}_1, \boldsymbol{\beta}_2, \boldsymbol{\beta}_3)$
	N	SST	$= \mathbf{y}'\mathbf{y}$

For the model $E(\mathbf{y}) = \mathbf{X}_1\boldsymbol{\beta}_1 + \mathbf{X}_2\boldsymbol{\beta}_2$ in Table 8.4, $R(\boldsymbol{\beta}_1)$ is for fitting $E(\mathbf{y}) = \mathbf{X}_1\boldsymbol{\beta}_1$ and $R(\boldsymbol{\beta}_2 \mid \boldsymbol{\beta}_1)$ is for fitting $E(\mathbf{y}) = \mathbf{X}_1\boldsymbol{\beta}_1 + \mathbf{X}_2\boldsymbol{\beta}_2$ after fitting $E(\mathbf{y}) = \mathbf{X}_1\boldsymbol{\beta}_1$. This amounts to fitting first $\boldsymbol{\beta}_1$ and then $\boldsymbol{\beta}_2$. We could just as well consider first $\boldsymbol{\beta}_2$ and then $\boldsymbol{\beta}_1$. But whatever analytic results apply to the former can be adapted to the latter simply by interchanging subscripts. Hence there is no loss of generality, as far as considering methodology is concerned, in having Table 8.4 as it is. The same applies to its third part, for fitting $\boldsymbol{\beta}_1$, then $\boldsymbol{\beta}_2$ and then $\boldsymbol{\beta}_3$.

i. Summary

The partitioned model (73) provides a useful way of comparing a model with its sub-models. Its normal equations are (74), with solution (77) and (78), and (79) is the reduction in sum of squares due to fitting the model. (82) is an expression for the reduction sum of squares due to some part of a partitioned model adjusted for some other part, an expression that is useful both as a calculation formula and for providing analytic results in cases not otherwise available (usually involving tedious algebra). The "invert part of the inverse" algorithm form of (82) in (88) is very useful for full rank models, but its non full rank form (89) is of limited use. Extended partitioning such as (91) does not affect (82) but it does alter its expected value—e.g., (92) compared to (84). Finally, the partitioning of a model leads to partitionings of the total sum of squares, as in Table 8.4.

8.6. F-STATISTICS FROM PARTITIONED MODELS

The utility of partitioning a total sum of squares is to use the sums of squares into which it is partitioned for testing hypotheses. With balanced data we know that many of those sums of squares test hypotheses of broad use and interest. And in earlier chapters we have seen that with some cases of unbalanced data there are useful hypotheses, but with other cases there are hypotheses that are of little use whatever. What we do now is to look at the general case, the sums of squares in Table 8.4, and investigate their properties vis-à-vis using them in F-statistics. We find that the F-statistic calculated from any of the three parts of Table 8.4 does have an $\mathscr{F}$-distribution; and we have a general expression for the corresponding associated hypothesis. The benefit of these results is that they apply to a broad array of widely-used F-statistics; for example, every F-statistic developed in Chapters 2–6 has an $\mathscr{F}$-distribution; and its associated hypothesis can be established.

The model we use is $E(\mathbf{y}) = \mathbf{X}\boldsymbol{\beta}$ and $\text{var}(\mathbf{y}) = \mathbf{V} = \sigma^2\mathbf{I}$ with normality

$$\mathbf{y} \sim \mathcal{N}(\mathbf{X}\boldsymbol{\beta}, \sigma^2\mathbf{I}) \ . \tag{94}$$

Using this in the general procedure of hypothesis testing described in equations (60)–(65) of Section 7.6a gives the following procedure.

For quadratic forms $\mathbf{y}'\mathbf{A}\mathbf{y}/\sigma^2$ and $\mathbf{y}'\mathbf{B}\mathbf{y}/\sigma^2$ for $\mathbf{y}$ of (94):

$$\text{if } \mathbf{A} \text{ is idempotent,} \quad \mathbf{y}'\mathbf{A}\mathbf{y}/\sigma^2 \sim \chi^{2\prime}\left(r_{\mathbf{A}}, \tfrac{1}{2}\boldsymbol{\beta}'\mathbf{X}'\mathbf{A}\mathbf{X}\boldsymbol{\beta}/\sigma^2\right); \tag{95}$$

$$\text{if } \mathbf{B} \text{ is idempotent} \tag{96}$$

$$\text{and} \quad \text{if } \mathbf{B}\mathbf{X} = \mathbf{0}, \quad \text{then} \quad \mathbf{y}'\mathbf{B}\mathbf{y}/\sigma^2 \sim \chi^2_{r_{\mathbf{B}}}; \tag{97}$$

$$\text{if } \mathbf{A}\mathbf{V}\mathbf{B} = \mathbf{0} \quad \text{then} \quad \mathbf{y}'\mathbf{A}\mathbf{y} \text{ and } \mathbf{y}'\mathbf{B}\mathbf{y} \text{ are independent;} \tag{98}$$

then, comparing

$$F = \frac{\mathbf{y}'\mathbf{A}\mathbf{y}/r_{\mathbf{A}}}{\mathbf{y}'\mathbf{B}\mathbf{y}/r_{\mathbf{B}}} \quad \text{against} \quad \mathscr{F}_{r_{\mathbf{A}},\, r_{\mathbf{B}}} \text{ tabulations,} \tag{99}$$

tests, from (70) of Section 7.6b with $\mathbf{X}\boldsymbol{\beta}$ for $\boldsymbol{\theta}$,

$$\text{H: } \mathbf{A}\mathbf{X}\boldsymbol{\beta} = \mathbf{0} \ . \tag{100}$$

The sums of squares of Table 8.4 are now shown to satisfy (95)–(98); and so they can be used in (99) to test (100).

a. Error sums of squares

Each SSE in Table 8.4 is of the form

$$\text{SSE} = \mathbf{y}'\mathbf{B}\mathbf{y} \quad \text{for} \quad \mathbf{B} = \mathbf{I} - \mathbf{X}\mathbf{G}\mathbf{X}' = \mathbf{M}, \tag{101}$$

where, in the three sections of that table, respectively, $\mathbf{X}$ has the values

$$\mathbf{X} = \mathbf{X}_1, \qquad \mathbf{X} = [\mathbf{X}_1 \quad \mathbf{X}_2], \quad \text{and} \quad \mathbf{X} = [\mathbf{X}_1 \quad \mathbf{X}_2 \quad \mathbf{X}_3] \; . \tag{102}$$

Since $\mathbf{M}$ is symmetric and idempotent, no matter which $\mathbf{X}$ is involved, $\text{SSE} = \mathbf{y}'\mathbf{M}\mathbf{y} = (\mathbf{M}\mathbf{y})'\mathbf{M}\mathbf{y}$ and so it *is* a sum of squares. Also, $\mathbf{B} = \mathbf{M}$ is idempotent, thus satisfying (96), and because, as in (18) of Section 7.1i, $\mathbf{BX} = \mathbf{MX} = \mathbf{0}$, (97) is also satisfied. Therefore, using $r_{\mathbf{B}} = r_{\mathbf{M}} = N - r_{\mathbf{X}}$ from (18) of Section 7.1, we have from (97)

$$\text{SSE}/\sigma^2 \sim \chi^2_{N-r_{\mathbf{X}}} \; . \tag{103}$$

Note the generality of this result. It applies to *any* error sum of squares calculated as $\text{SSE} = \mathbf{y}'\mathbf{y} - \boldsymbol{\beta}^{\circ\prime}\mathbf{X}'\mathbf{y}$ of Section 8.4a, using the normality assumption $\mathbf{y} \sim N(\mathbf{X}\boldsymbol{\beta}, \sigma^2\mathbf{I})$. With this result in hand we therefore know, for example, that for every SSE considered in earlier chapters SSE/σ^2 has a χ^2-distribution. Result (103) thus eliminates the need to consider each SSE individually. As long as $\mathbf{y}'\mathbf{y} - \boldsymbol{\beta}^{\circ\prime}\mathbf{X}'\mathbf{y}$ is its form, for $\boldsymbol{\beta}^{\circ\prime} = \mathbf{G}\mathbf{X}'\mathbf{y}$ and $\mathbf{y} \sim N(\mathbf{X}\boldsymbol{\beta}, \sigma^2\mathbf{I})$, the result (103) applies.

b. Reductions in sums of squares

-i. Two sums of squares. Table 8.4 contains only two different forms of reduction in sums of squares: $R(\boldsymbol{\beta}_1)$ due to fitting $\boldsymbol{\beta}_1$, and $R(\boldsymbol{\beta}_2 \,|\, \boldsymbol{\beta}_1)$ due to fitting $\boldsymbol{\beta}_2$ adjusted for $\boldsymbol{\beta}_1$. Although there is also $R(\boldsymbol{\beta}_3 \,|\, \boldsymbol{\beta}_1, \boldsymbol{\beta}_2)$, it is just a special case of $R(\boldsymbol{\beta}_2 \,|\, \boldsymbol{\beta}_1)$, with $\boldsymbol{\beta}_2$ having the value $\boldsymbol{\beta}_3$, and $\boldsymbol{\beta}_1$ having the value $[\boldsymbol{\beta}_1' \quad \boldsymbol{\beta}_2']'$. Similar substitutions also apply for $R(\boldsymbol{\beta}_4 \,|\, \boldsymbol{\beta}_1, \boldsymbol{\beta}_2, \boldsymbol{\beta}_3)$, for example. Thus only the two forms $R(\boldsymbol{\beta}_1)$ and $R(\boldsymbol{\beta}_2 \,|\, \boldsymbol{\beta}_1)$ need be considered.

From (65) and (90), using $\mathbf{X}_1^+$ of Section 7.1h,

$$R(\boldsymbol{\beta}_1) = \mathbf{y}'\mathbf{X}_1\mathbf{G}_1\mathbf{X}_1'\mathbf{y} = \mathbf{y}'\mathbf{X}_1\mathbf{X}_1^+\mathbf{y}$$

and
$$R(\boldsymbol{\beta}_2 \,|\, \boldsymbol{\beta}_1) = \mathbf{y}'\mathbf{M}_1\mathbf{X}_2(\mathbf{X}_2'\mathbf{M}_1\mathbf{X}_2)^-\mathbf{X}_2'\mathbf{M}_1\mathbf{y} \; .$$

First, we observe that each of these expressions is a sum of squares because each is $\mathbf{y}'\mathbf{A}\mathbf{y}$ for an $\mathbf{A}$ that is symmetric and idempotent, so that $\mathbf{y}'\mathbf{A}\mathbf{y} = (\mathbf{A}\mathbf{y})'\mathbf{A}\mathbf{y}$ and is thus a sum of squares.

-ii. ***χ^2-properties.*** Write

$$R(\boldsymbol{\beta}_1) = \mathbf{y}'\mathbf{A}_1\mathbf{y} \tag{104}$$

with $\mathbf{A}_1 = \mathbf{X}_1(\mathbf{X}_1'\mathbf{X}_1)^-\mathbf{X}_1' = \mathbf{I} - \mathbf{M}_1,$ which is idempotent . (105)

Therefore applying (95) to (104) using (105) gives

$$\begin{aligned} R(\boldsymbol{\beta}_1)/\sigma^2 &\sim \chi^{2\prime}\left(r_{\mathbf{A}}, \tfrac{1}{2}\boldsymbol{\beta}'\mathbf{X}'\mathbf{A}_1\mathbf{X}\boldsymbol{\beta}/\sigma^2\right) \\ &\sim \chi^{2\prime}\left(r_1, \tfrac{1}{2}\boldsymbol{\beta}'\mathbf{X}'\mathbf{A}_1\mathbf{X}\boldsymbol{\beta}/\sigma^2\right) \end{aligned} \tag{106}$$

where, from Section 7.1d, we get $r_{\mathbf{A}_1} = r_{\mathbf{X}_1}$, for which r_1 is used, as in (93). Note in (106) that $\mathbf{X}\boldsymbol{\beta}$ has not been specified; it will depend upon which part of Table 8.4 is used.

Also write

$$R(\boldsymbol{\beta}_2 | \boldsymbol{\beta}_1) = \mathbf{y}'\mathbf{A}_{12}\mathbf{y} \quad \text{with } \mathbf{A}_{12} = \mathbf{M}_1\mathbf{X}_2(\mathbf{X}_2'\mathbf{M}_1\mathbf{X}_2)^-\mathbf{X}_2'\mathbf{M}_1 \ . \tag{107}$$

Then $\mathbf{M}_1 = \mathbf{M}_1^2 = \mathbf{M}_1' = \mathbf{M}_1'\mathbf{M}_1$ means that $\mathbf{A}_{12} = \mathbf{L}(\mathbf{L}'\mathbf{L})^-\mathbf{L}'$ for $\mathbf{L} = \mathbf{M}_1\mathbf{X}_2$ and so $\mathbf{A}_{12}$ is idempotent. Therefore (95) also gives, similar to (106),

$$R(\boldsymbol{\beta}_2 | \boldsymbol{\beta}_1) \sim \chi^{2\prime}\left(r_{\mathbf{A}_{12}}, \tfrac{1}{2}\boldsymbol{\beta}'\mathbf{X}'\mathbf{A}_{12}\mathbf{X}\boldsymbol{\beta}/\sigma^2\right) \ . \tag{108}$$

Again, $\mathbf{X}\boldsymbol{\beta}$ is not yet specified; and applying (44) of Section 7.3b to $\mathbf{A}_{12}$ gives (108) as

$$R(\boldsymbol{\beta}_2 | \boldsymbol{\beta}_1) \sim \chi^{2\prime}\left(r_{12} - r_1, \tfrac{1}{2}\boldsymbol{\beta}'\mathbf{X}'\mathbf{A}_{12}\mathbf{X}\boldsymbol{\beta}/\sigma^2\right) \ . \tag{109}$$

-iii. ***Independence of SSE.*** We now have to be more specific about the forms of SSE in Table 8.4:

$$\text{SSE}_1 = \mathbf{y}'\mathbf{y} - R(\boldsymbol{\beta}_1) = \mathbf{y}'\mathbf{B}_1\mathbf{y} \quad \text{for } \mathbf{B}_1 = \mathbf{M}_1 \tag{110}$$

$$\begin{aligned} \text{SSE}_{12} &= \mathbf{y}'\mathbf{y} - \mathbf{R}(\boldsymbol{\beta}_1, \boldsymbol{\beta}_2) \\ &= \mathbf{y}'\mathbf{y} - R(\boldsymbol{\beta}_1) - R(\boldsymbol{\beta}_2 | \boldsymbol{\beta}_1) \\ &= \mathbf{y}'[\mathbf{I} - (\mathbf{I} - \mathbf{M}_1) - \mathbf{A}_{12}]\mathbf{y} \\ &= \mathbf{y}'(\mathbf{M}_1 - \mathbf{A}_{12})\mathbf{y} \\ &= \mathbf{y}'\mathbf{B}_{12}\mathbf{y} \end{aligned}$$

for

$$\mathbf{B}_{12} = \mathbf{M}_1(\mathbf{I} - \mathbf{T}_{12}) \quad \text{and} \quad \mathbf{T}_{12} = \mathbf{X}_2(\mathbf{X}_2'\mathbf{M}_1\mathbf{X}_2)^-\mathbf{X}_2'\mathbf{M}_1 \ . \tag{111}$$

Now consider $R(\boldsymbol{\beta}_1)$ of (104). Observing that the idempotency of $\mathbf{M}_1$ implies $(\mathbf{I} - \mathbf{M}_1)\mathbf{M}_1 = \mathbf{0}$ we have, for ascertaining the independence of $R(\boldsymbol{\beta}_1)$ from SSE terms, using $\mathbf{A}_1$, $\mathbf{B}_1$ and $\mathbf{B}_{12}$ of (105), (110) and (111), respectively,

$$\mathbf{A}_1\mathbf{B}_1 = (\mathbf{I} - \mathbf{M}_1)\mathbf{M}_1 = \mathbf{0}; \quad \mathbf{A}_1\mathbf{B}_{12} = (\mathbf{I} - \mathbf{M}_1)\mathbf{M}_1(\mathbf{I} - \mathbf{T}_{12}) = \mathbf{0} \ . \tag{112}$$

Therefore by (98)

$$R(\boldsymbol{\beta}_1) \text{ and } \text{SSE}_1 \text{ are independent} \tag{113}$$

and

$$R(\boldsymbol{\beta}_1) \text{ and } \text{SSE}_{12} \text{ are independent} \ ; \tag{114}$$

and also, by analogy with the latter,

$$R(\boldsymbol{\beta}_1) \text{ and } \text{SSE}_{123} \text{ are independent} \ . \tag{115}$$

Thus $R(\boldsymbol{\beta}_1)$ is independent of SSE in models that contain $\boldsymbol{\beta}_1$. Also,

$$R(\boldsymbol{\beta}_1, \boldsymbol{\beta}_2) \text{ and } \text{SSE}_{12} \text{ are independent} \tag{116}$$

and

$$R(\boldsymbol{\beta}_1, \boldsymbol{\beta}_2) \text{ and } \text{SSE}_{123} \text{ are independent,} \tag{117}$$

by analogy with (113) and (114), respectively.

Now consider

$$R(\boldsymbol{\beta}_2 | \boldsymbol{\beta}_1) = R(\boldsymbol{\beta}_1, \boldsymbol{\beta}_2) - R(\boldsymbol{\beta}_1) \ . \tag{118}$$

From (116) and (114), each term on the right-hand side of (118) is independent of SSE_{12}; and by (117) and (115) it is also independent of SSE_{123}. Thus $R(\boldsymbol{\beta}_2 | \boldsymbol{\beta}_1)$ is independent of SSE_{12} and of SSE_{123}, i.e., of the SSE for any model that contains $\boldsymbol{\beta}_1$ and $\boldsymbol{\beta}_2$.

Thus in summary, any $R(\boldsymbol{\beta}_1)$ and $R(\boldsymbol{\beta}_2 | \boldsymbol{\beta}_1)$ are independent of the SSE for any model that contains $\boldsymbol{\beta}_1$, and $\boldsymbol{\beta}_1$ and $\boldsymbol{\beta}_2$, respectively.

*-iv. **Independence of each other.*** From $R(\boldsymbol{\beta}_1) = \mathbf{y}'\mathbf{A}_1\mathbf{y}$ and from $R(\boldsymbol{\beta}_2 | \boldsymbol{\beta}_1) = \mathbf{y}'\mathbf{A}_{12}\mathbf{y}$ of (104) and (107), respectively, we see from (98) that these two sums of squares are independent if $\mathbf{A}_1\mathbf{A}_{12} = \mathbf{0}$. But from (105), (107) and (112), with $\mathbf{M}_1 = \mathbf{M}_1^2$,

$$\mathbf{A}_1\mathbf{A}_{12} = (\mathbf{I} - \mathbf{M}_1)\mathbf{M}_1\mathbf{T}_{12} = \mathbf{0}$$

and so $R(\boldsymbol{\beta}_1)$ and $R(\boldsymbol{\beta}_2 | \boldsymbol{\beta}_1)$ *are* independent. Finally we consider $R(\boldsymbol{\beta}_2 | \boldsymbol{\beta}_1)$

and $R(\beta_3 | \beta_1, \beta_2)$, and first note that

$$\begin{aligned} R(\beta_3 | \beta_1, \beta_2) &= R(\beta_1, \beta_2, \beta_3) - R(\beta_1, \beta_2) \\ &= \mathbf{y}'\mathbf{y} - R(\beta_1, \beta_2) - [\mathbf{y}'\mathbf{y} - R(\beta_1, \beta_2, \beta_3)] \\ &= \text{SSE}_{12} - \text{SSE}_{123} \ . \end{aligned} \tag{119}$$

Then by (116) and (117) the first term of $R(\beta_2 | \beta_1)$ in (118) is independent of both terms in (119); and by (113) and (114) so is the second term. Hence $R(\beta_2 | \beta_1)$ and $R(\beta_3 | \beta_1, \beta_2)$ are independent.

-v. ***Summary.*** For each part of Table 8.4 we have proven:

I: Each term is a sum of squares.
II: Each SSE has a central χ^2-distribution.
III: Each term other than SSE has a non-central χ^2-distribution.
IV: Each pair of terms is independent.

As a result, any F-statistic calculated in the usual way from any part of Table 8.4 does have a non-central $\mathscr{F}$-distribution; and we proceed to consider the hypotheses so tested.

c. A general hypothesis

The most general F-statistic available from Table 8.4 is

$$F = \frac{R(\beta_2 | \beta_1)/(r_{12} - r_1)}{[\mathbf{y}'\mathbf{y} - R(\beta_1, \beta_2, \beta_3)]/(N - r_{123})} \ . \tag{120}$$

It includes as special cases all of the five other F-statistics that are available. So consideration is given first to the hypothesis tested by (120). From $R(\beta_2 | \beta_1) = \mathbf{y}'\mathbf{A}_{12}\mathbf{y}$ of (107) it is, from (100),

$$\text{H: } \mathbf{A}_{12}\mathbf{X}\beta = \mathbf{0}, \tag{121}$$

i.e.,

$$\text{H: } \mathbf{A}_{12}(\mathbf{X}_1\beta_1 + \mathbf{X}_2\beta_2 + \mathbf{X}_3\beta_3) = \mathbf{0} \ . \tag{122}$$

But (106) gives $\mathbf{A}_{12}\mathbf{X}_1 = \mathbf{0}$, because $\mathbf{M}_1\mathbf{X}_1 = \mathbf{0}$; and $\mathbf{A}_{12}\mathbf{X}_2 = \mathbf{M}_1\mathbf{X}_2$ from using Theorem 7.1 on $\mathbf{M}_1\mathbf{X}_2$. Hence (122) is

$$\text{H: } \mathbf{M}_1\mathbf{X}_2\beta_2 + \mathbf{M}_1\mathbf{X}_2(\mathbf{X}_2'\mathbf{M}_1\mathbf{X}_2)^{-}\mathbf{X}_2'\mathbf{M}_1\mathbf{X}_3\beta_3 = \mathbf{0} \ . \tag{123}$$

The hypothesis (123) has the advantage of complete generality: $R(\boldsymbol{\beta}_2 | \boldsymbol{\beta}_1)$ is a reduction due to some part of a model, $\boldsymbol{\beta}_2$, adjusted for some other part of a model, $\boldsymbol{\beta}_1$; and the model includes not only those two parts but also a third part, $\boldsymbol{\beta}_3$. Nothing can be more general. That is the merit of (123). The demerit is that (123) represents N statements about the parameters and we need to have the hypothesis in terms of $r_{12} - r_1$ LIN functions of the hypothesis. This can often be easily achieved either by observation, as in the example of Section 7.6b, or by any other of the methods described in Section 7.6b.

d. Specific hypotheses

Derivation of the hypotheses associated with the sums of squares in Table 8.4 is now indicated. This is simply a matter of using (123) for special cases. For example, the hypothesis associated with $R(\boldsymbol{\beta}_1)$ in the model $E(\mathbf{y}) = \mathbf{X}_1\boldsymbol{\beta}_1$, i.e., the hypothesis tested by

$$F = \frac{R(\boldsymbol{\beta}_1)/r_1}{[\mathbf{y}'\mathbf{y} - R(\boldsymbol{\beta}_1)]/(N - r_1)}$$

comes from amending (123) by replacing (i) $\mathbf{X}_1$ by $\mathbf{0}$, (ii) $\mathbf{X}_2$ and $\boldsymbol{\beta}_2$ by $\mathbf{X}_1$ and $\boldsymbol{\beta}_1$, respectively, and (iii) $\mathbf{X}_3$ by $\mathbf{0}$. This reduces (123) to

$$\text{H: } (\mathbf{I} - \mathbf{0})\mathbf{X}_1\boldsymbol{\beta}_1 + \mathbf{0} = \mathbf{0},$$

i.e., to

$$\text{H: } \mathbf{X}_1\boldsymbol{\beta}_1 = \mathbf{0}\ . \tag{124}$$

This, as is well known, is indeed the hypothesis tested by $R(\boldsymbol{\beta}_1)$ in the model $E(\mathbf{y}) = \mathbf{X}_1\boldsymbol{\beta}_1$.

The hypothesis (124) is shown as line 1 of Table 8.5, which includes the special value of the vector $[\boldsymbol{\beta}_1' \quad \boldsymbol{\beta}_2' \quad \boldsymbol{\beta}_3']$ corresponding to replacements (i), (ii) and (iii) that led to (124). Similar information is shown in Table 8.5 for the other five F-statistics available from Table 8.4, including (123), shown as line 5.

A requirement of Tables 8.4 and 8.5, when considering the hypothesis tested by any general $R(\boldsymbol{\beta}_1)$ or $R(\boldsymbol{\beta}_2 | \boldsymbol{\beta}_1)$, is that the model within which these are being considered must include, respectively, the $\boldsymbol{\beta}_1$ of $R(\boldsymbol{\beta}_1)$ or the $\boldsymbol{\beta}_1$ and $\boldsymbol{\beta}_2$ of $R(\boldsymbol{\beta}_2 | \boldsymbol{\beta}_1)$. This means that for applying these tables to cell means models we sometimes have to rewrite those models in a form that specifically includes both $\boldsymbol{\beta}_1$ and $\boldsymbol{\beta}_2$.

TABLE 8.5. SUMS OF SQUARES AND ASSOCIATED HYPOTHESES IN PARTITIONED LINEAR MODELS
[The general result (123) is line 5, of which the other lines are special cases.]

	Model[1] for $E(\mathbf{y})$	Special Value of the General Vector $[\boldsymbol{\beta}_1' \; \boldsymbol{\beta}_2' \; \boldsymbol{\beta}_3']$	Sum of Squares[2]	Associated Hypothesis[3]
1.	$\mathbf{X}_1\boldsymbol{\beta}_1$	$[\;\mathbf{0} \quad \boldsymbol{\beta}_1' \quad \mathbf{0}\;]$	$R(\boldsymbol{\beta}_1)$	H: $\mathbf{X}_1\boldsymbol{\beta}_1 = \mathbf{0}$
2.	$\mathbf{X}_1\boldsymbol{\beta}_1 + \mathbf{X}_2\boldsymbol{\beta}_2$	$[\;\mathbf{0} \quad \boldsymbol{\beta}_1' \quad \boldsymbol{\beta}_2'\;]$	$R(\boldsymbol{\beta}_1)$	H: $\mathbf{X}_1\boldsymbol{\beta}_1 + \mathbf{X}_1\mathbf{X}_1^+\mathbf{X}_2\boldsymbol{\beta}_2 = \mathbf{0}$
3.		$[\;\boldsymbol{\beta}_1' \quad \boldsymbol{\beta}_2' \quad \mathbf{0}\;]$	$R(\boldsymbol{\beta}_2 \mid \boldsymbol{\beta}_1)$	H: $\mathbf{M}_1\mathbf{X}_2\boldsymbol{\beta}_2 = \mathbf{0}$
4.	$\mathbf{X}_1\boldsymbol{\beta}_1 + \mathbf{X}_2\boldsymbol{\beta}_2 + \mathbf{X}_3\boldsymbol{\beta}_3$	$[\;\mathbf{0} \quad \boldsymbol{\beta}_1' \quad (\boldsymbol{\beta}_2' \quad \boldsymbol{\beta}_3')]$	$R(\boldsymbol{\beta}_1)$	H: $\mathbf{X}_1\boldsymbol{\beta}_1 + \mathbf{X}_1\mathbf{X}_1^+(\mathbf{X}_2\boldsymbol{\beta}_2 + \mathbf{X}_3\boldsymbol{\beta}_3) = \mathbf{0}$
5.		$[\;\boldsymbol{\beta}_1' \quad \boldsymbol{\beta}_2' \quad \boldsymbol{\beta}_3'\;]$	$R(\boldsymbol{\beta}_2 \mid \boldsymbol{\beta}_1)$	H: $\mathbf{M}_1\mathbf{X}_2\boldsymbol{\beta}_2 + \mathbf{M}_1\mathbf{X}_2(\mathbf{M}_1\mathbf{X}_2)^+\mathbf{X}_3\boldsymbol{\beta}_3 = \mathbf{0}$
6.		$[(\boldsymbol{\beta}_1' \quad \boldsymbol{\beta}_2') \quad \boldsymbol{\beta}_3' \quad \mathbf{0}\;]$	$R(\boldsymbol{\beta}_3 \mid \boldsymbol{\beta}_1, \boldsymbol{\beta}_2)$	H: $\mathbf{M}_{12}\mathbf{X}_3\boldsymbol{\beta}_3 = \mathbf{0}$

[1] In each model, $\hat{\sigma}^2 = \mathbf{y}'(\mathbf{I} - \mathbf{X}\mathbf{X}^+)\mathbf{y}/(N - r_{\mathbf{X}})$, where $\mathbf{X}$ is $\mathbf{X}_1$, $[\mathbf{X}_1 \quad \mathbf{X}_2]$ and $[\mathbf{X}_1 \quad \mathbf{X}_2 \quad \mathbf{X}_3]$, respectively; and $\mathbf{X}\mathbf{X}^+ = \mathbf{X}(\mathbf{X}'\mathbf{X})^-\mathbf{X}'$ as in Theorem 7.1.

[2] The F-statistic in each case is the sum of squares divided by $s\hat{\sigma}^2$, for s being the degrees of freedom of the sum of squares, which for $R(\boldsymbol{\beta}_2 \mid \boldsymbol{\beta}_1)$ is $r_{[\mathbf{X}_1 \; \mathbf{X}_2]} - r_{\mathbf{X}_1}$.

[3] $\mathbf{M}_1 = \mathbf{I} - \mathbf{X}_1\mathbf{X}_1^+$ and $\mathbf{M}_{12} = \mathbf{I} - (\mathbf{X}_1 \quad \mathbf{X}_2)(\mathbf{X}_1 \quad \mathbf{X}_2)^+ = \mathbf{M}_1 - \mathbf{M}_1\mathbf{X}_2(\mathbf{M}_1\mathbf{X}_2)^+$.

e. The full rank case

In Table 8.5 we see in the model $E(\mathbf{y}) = \mathbf{X}_1\boldsymbol{\beta}_1 + \mathbf{X}_2\boldsymbol{\beta}_2$ that $R(\boldsymbol{\beta}_2 | \boldsymbol{\beta}_1)$ tests H: $\mathbf{M}_1\mathbf{X}_2\boldsymbol{\beta}_2 = \mathbf{0}$. But (44) of Chapter 7 shows that $\mathbf{M}_1\mathbf{X}_2$ has full column rank in the full rank model. Hence the hypothesis becomes H: $\boldsymbol{\beta}_2 = \mathbf{0}$. Thus, in the full rank model $E(\mathbf{y}) = \mathbf{X}_1\boldsymbol{\beta}_1 + \mathbf{X}_2\boldsymbol{\beta}_2$ we have $R(\boldsymbol{\beta}_2 | \boldsymbol{\beta}_1)$ testing H: $\boldsymbol{\beta}_2 = \mathbf{0}$. This has important consequences—see Sections 8.9c-iii and 9.4b.

f. Examples

-i. The 1-way classification. Consider using Table 8.4 to derive the hypothesis tested by $R(\mu)$ in the model $E(y_{ij}) = \mu_i$. The model can be written as $E(\mathbf{y}) = \mathbf{D}\{\mathbf{1}_{n_i}\}\boldsymbol{\mu}$ for $\boldsymbol{\mu}' = [\mu_1 \quad \mu_2 \quad \cdots \quad \mu_a]$. But that model does not include μ. This is achieved by writing the model as

$$E(y_{ij}) = \mu_i = \mu + \mu_i - \mu = \mu + \theta_i \quad \text{for } \theta_i = \mu_i - \mu \,. \tag{125}$$

Hence $E(\mathbf{y})$ is

$$E(\mathbf{y}) = \mathbf{1}_N\mu + \mathbf{D}\{\mathbf{1}_i\}\boldsymbol{\theta} \tag{126}$$

for $\boldsymbol{\theta} = \{_c\ \theta_i\}$. Then, from line 2 of Table 8.5 and from (126)

$$\boldsymbol{\beta}_1 = \mu, \qquad \mathbf{X}_1 = \mathbf{1}_N, \qquad \boldsymbol{\beta}_2 = \boldsymbol{\theta} \ \text{ and } \ \mathbf{X}_2 = \mathbf{D}\{\mathbf{1}_i\} \tag{127}$$

and the hypothesis is

$$\text{H: } \mathbf{X}_1\boldsymbol{\beta}_1 + \mathbf{X}_1\mathbf{X}_1^+\mathbf{X}_2\boldsymbol{\beta}_2 = \mathbf{0}$$

This, with $\mathbf{1}_N^+$ being $\mathbf{1}'/N$, is

$$\text{H: } \mathbf{1}_N(\mu + \mathbf{1}'_N\mathbf{X}_2\boldsymbol{\theta}/N) = \mathbf{0}$$

which, from the nature of $\mathbf{X}_2$ and hence of $\mathbf{1}'_N\mathbf{X}_2$, is

$$\text{H: } (\mu + \Sigma_i n_i\theta_i/N)\mathbf{1}_N = \mathbf{0} \,.$$

This is, as noted for the general case, N statements. But clearly they are all just one, $\mu + \Sigma_i n_i\theta_i/N = 0$, which on using θ_i from (125) gives the hypothesis as H: $\Sigma_i n_i\mu_i = 0$ just as it should be, in Table 2.7.

-ii. The 1-way classification with a covariate. The model equation is

$$y_{ij} = \mu + \theta_i + bz_{ij},$$

similar to (125) and to Chapter 6. The vector equivalent of this is

$$\mathbf{y} = \mathbf{1}_N\mu + \mathbf{D}\{\mathbf{1}_i\}\boldsymbol{\theta} + b\mathbf{z} .$$

To use Table 8.5 for finding the hypothesis associated with $R(\theta_i\,|\,\mu)$ we have

$$\boldsymbol{\beta}_1 = \mu,\quad \mathbf{X}_1 = \mathbf{1}_N,\quad \boldsymbol{\beta}_2 = \boldsymbol{\theta},\quad \mathbf{X}_2 = \mathbf{D}\{\mathbf{1}_i\},\quad \boldsymbol{\beta}_3 = b \quad\text{and}\quad \mathbf{X}_3 = \mathbf{z} .$$

Then $\qquad \mathbf{M}_1 = \mathbf{I} - \mathbf{X}_1(\mathbf{X}_1'\mathbf{X}_1)^-\mathbf{X}_1' = \mathbf{I} - \bar{\mathbf{J}}_N,$

$$\mathbf{M}_1\mathbf{X}_2 = \mathbf{X}_2 - \mathbf{1}_N\mathbf{n}'/N \quad\text{for}\quad \mathbf{n}' = [n_1 \quad n_2 \quad \cdots \quad n_a],$$

and $\qquad \mathbf{X}_2'\mathbf{M}_1\mathbf{X}_2 = \mathbf{D}\{n_i\} - \mathbf{n}\mathbf{n}'/N.$

The hypothesis of line 5 in Table 8.5 is

$$H\colon\ \mathbf{M}_1\mathbf{X}_2\boldsymbol{\beta}_2 + \mathbf{M}_1\mathbf{X}_2(\mathbf{X}_2'\mathbf{M}_1\mathbf{X}_2)^-\mathbf{X}_2'\mathbf{M}_1\mathbf{X}_3\boldsymbol{\beta}_3 = \mathbf{0} .$$

Pre-multiplying it by $\mathbf{X}_2'$ gives it as

$$H\colon\ \mathbf{X}_2'\mathbf{M}_1\mathbf{X}_2\boldsymbol{\beta}_2 + \mathbf{X}_2'\mathbf{M}_1\mathbf{X}_3\boldsymbol{\beta}_3 = \mathbf{0} .$$

i.e.,

$$H\colon\ (\mathbf{D}\{n_i\} - \mathbf{n}\mathbf{n}'/N)\boldsymbol{\theta} + (\mathbf{X}_2' - \mathbf{n}\mathbf{1}_N'/N)b\mathbf{z} = \mathbf{0},$$

which is

$$H\colon\ \{_c\, n_i\theta_i - n_i\Sigma n_i\theta_i/N\} + \{_c\, bz_{i\cdot} - n_i b\bar{z}_{\cdot\cdot}\} = 0 .$$

But this is

$$H\colon\ \theta_i + b\bar{z}_{i\cdot} - (\Sigma_i n_i\theta_i/N + b\bar{z}_{\cdot\cdot}) = 0$$

and since $\theta_i = \mu_i - \mu$, this is H: $\mu_i + b\bar{z}_{\cdot\cdot}$ *equal* $\forall\ i$, as in Chapter 6.

8.7. ESTIMABLE FUNCTIONS

a. Invariance to solutions of the normal equations

We saw in Section 8.2 how it is that when for normal equations $\mathbf{X}'\mathbf{X}\boldsymbol{\beta}^\circ = \mathbf{X}'\mathbf{y}$ the matrix $\mathbf{X}'\mathbf{X}$ has no inverse, there are then many solution

TABLE 8.6. FIVE SOLUTIONS TO NORMAL EQUATIONS (23) AND, FOR EACH SOLUTION, THE VALUE OF CERTAIN LINEAR FUNCTIONS OF ELEMENTS OF THE SOLUTION

Elements of Solution and Functions Thereof	Solution				
	1	2	3	4	5
Elements					
μ°	0	$82\frac{1}{4}$	79	89	5283
α_1°	73	$-9\frac{1}{4}$	-6	-16	-5210
α_2°	78	$-4\frac{1}{4}$	-1	-11	-5205
α_3°	89	$6\frac{3}{4}$	10	0	-5194
Functions					
Group I: $\alpha_1^\circ + \alpha_2^\circ$	151	$-13\frac{1}{2}$	-7	-27	-10415
$\frac{1}{3}(\alpha_1^\circ + \alpha_2^\circ + \alpha_3^\circ)$	80	$-2\frac{1}{4}$	1	-9	-5203
Group II: $\alpha_1^\circ - \alpha_2^\circ$	-5	-5	-5	-5	-5
$\mu^\circ + \alpha_1^\circ$	73	73	73	73	73
$\mu^\circ + \alpha_3^\circ$	89	89	89	89	89

vectors $\boldsymbol{\beta}^\circ$. Under these circumstances one cannot think of any particular $\boldsymbol{\beta}^\circ$ as being an estimator of $\boldsymbol{\beta}$. Hence, if those solution vectors are to be of any use insofar as estimation is concerned, then some property of invariance to those many solutions is sorely needed. This is now considered, in terms of the idea of estimable functions, a concept that was introduced in Section 8.3b in an elementary manner.

We begin with an example, using an overparameterized model in order to have $\mathbf{X'X}$ singular. Table 8.6 shows five solution vectors for the normal equations (23) for the overparameterized model $E(y_{ij}) = \mu + \alpha_i$ of the example in Table 8.1. Three of the solutions come from (19), (25) and (27) and the others have been derived otherwise. Table 8.6 also shows, for each solution, the value of certain linear functions of the elements of solution.

Two features of Table 8.6 are striking. First, values for each element of the solution vector are seen to vary greatly from solution to solution; e.g., in one solution $\alpha_1^\circ = 73$ and in another $\alpha_1^\circ = -5210$. Second, for the functions of elements of the solutions, those of Group I vary from solution to solution whereas those of Group II are the same for all solutions; each is invariant to whatever solution is used. For example, $\alpha_1^\circ - \alpha_2^\circ = -5$ for all solution vectors.

In the presence of numerous solutions to the normal equations, the distinction between the Group I and Group II functions apparent in Table 8.6 is the key to estimating functions of the parameters of a linear model. To begin with, note that attention is confined solely to linear functions of parameters, i.e., to functions $\mathbf{q}'\boldsymbol{\beta}$ for any vector $\mathbf{q}'$. Suppose in this example we want to estimate $\alpha_1 + \alpha_2$. It would seem natural to use $\alpha_1^\circ + \alpha_2^\circ$ as the estimator, except that in Table 8.6 this is seen to vary, depending on what solution vector is used. Clearly, $\alpha_1^\circ + \alpha_2^\circ$ as an estimator of $\alpha_1 + \alpha_2$ is therefore not satisfactory: a user of solution 1 in Table 8.6 would estimate $\alpha_1 + \alpha_2$ as 151 whereas a user of solution 5 would have -10415 as an estimate. In contrast, no matter which of the five solution vectors of Table 7.6 is used in estimating $\alpha_1 - \alpha_2$ by $\alpha_1^\circ - \alpha_2^\circ$, the estimate is -5. More than that, and as shall be shown subsequently, for each of the infinite number of solutions $\boldsymbol{\beta}^\circ$ to the normal equations (23), the value of $\alpha_1^\circ - \alpha_2^\circ$ is -5. Likewise the value of $\mu^\circ + \alpha_1^\circ$ is 73, and the value of $\mu^\circ + \alpha_3^\circ$ is 89. These are the examples labeled Group II in Table 8.6, and there are many others that could be included. Obviously, the linear functions $\mathbf{q}'\boldsymbol{\beta}^\circ$ that fall in Group II have an important characteristic: for each of them, the value of $\mathbf{q}'\boldsymbol{\beta}^\circ$ is invariant to the solution vector $\boldsymbol{\beta}^\circ$. Therefore each such function $\mathbf{q}'\boldsymbol{\beta}^\circ$ would seem to be a satisfactory estimator of the corresponding function of parameters, $\mathbf{q}'\boldsymbol{\beta}$. It is, in fact, the least squares estimator. It is written as

$$\text{BLUE}(\mathbf{q}'\boldsymbol{\beta}) = \mathbf{q}'\boldsymbol{\beta}^\circ$$

and has several useful properties, as shall be shown.

b. Definitions

Definition 8.1. [See Seber (1977, Exercise 5, p. 81).] Any function $\mathbf{q}'\boldsymbol{\beta}$ for which $\mathbf{q}'\boldsymbol{\beta}^\circ$ is invariant to $\boldsymbol{\beta}^\circ$ is said to be an *estimable function*; and is estimated by $\mathbf{q}'\boldsymbol{\beta}^\circ$.

This definition emphasizes our understanding of what an estimable function is. It is a function of the parameters for which the same function of the elements of a solution vector is invariant to whatever solution vector is used. But as a definition it does not explicitly indicate how to ascertain what form $\mathbf{q}'$ must take in order for $\mathbf{q}'\boldsymbol{\beta}$ to be estimable. This is done by the mathematically more formal (and more commonly seen) definition.

Definition 8.2. When $\mathbf{q}' = \mathbf{t}'\mathbf{X}$ for any $\mathbf{t}'$, the function $\mathbf{q}'\boldsymbol{\beta}$ is said to be an *estimable function*.

The two definitions are equivalent. Each implies the other.

When $\mathbf{q}' = \mathbf{t}'\mathbf{X}$ then $\mathbf{q}'\boldsymbol{\beta}^\circ = \mathbf{q}'\mathbf{G}\mathbf{X}'\mathbf{y} = \mathbf{t}'\mathbf{X}\mathbf{G}\mathbf{X}'\mathbf{y}$, which is invariant to $\mathbf{G}$ and hence to $\boldsymbol{\beta}^\circ = \mathbf{G}\mathbf{X}'\mathbf{y}$. Thus Definition 8.2 implies Definition 8.1. Showing the converse is more lengthy. First, although $\boldsymbol{\beta}^\circ = \mathbf{G}\mathbf{X}'\mathbf{y}$ is, for given $\mathbf{G}$, the solution being used, it is known from (23) of Section 7.2a that $\dot{\boldsymbol{\beta}} = \mathbf{G}\mathbf{X}'\mathbf{y} + (\mathbf{I} - \mathbf{G}\mathbf{X}'\mathbf{X})\mathbf{z}$ is a solution for any $\mathbf{z}$. Hence $\mathbf{q}'\dot{\boldsymbol{\beta}} = \mathbf{q}'\mathbf{GX}'\mathbf{y} + \mathbf{q}'(\mathbf{I} - \mathbf{GX}'\mathbf{X})\mathbf{z}$, and this will not depend on the arbitrary $\mathbf{z}$ if and only if $\mathbf{q}'(\mathbf{I} - \mathbf{GX}'\mathbf{X}) = \mathbf{0}$, i.e., $\mathbf{q}' = \mathbf{t}'\mathbf{X}$ for $\mathbf{t}' = \mathbf{q}'\mathbf{GX}'$. And then $\mathbf{q}'\dot{\boldsymbol{\beta}} = \mathbf{q}'\mathbf{GX}'\mathbf{y} = \mathbf{t}'\mathbf{XGX}'\mathbf{y}$ is the same for all $\mathbf{G}$, i.e., for all $\boldsymbol{\beta}^\circ$. Hence Definition 8.1 implies Definition 8.2.

Of the two definitions, 8.2 is the most useful technically and it leads to the invariance property that is the crux of 8.1. But, although 8.2 is nearly always the starting point for discussing estimable functions, it does not draw on our being interested in only those linear functions of elements of $\boldsymbol{\beta}^\circ$ that are invariant to $\boldsymbol{\beta}^\circ$. That is a direct outcome of Definition 8.1.

c. BLU estimation

When

$$\mathbf{q}' = \mathbf{t}'\mathbf{X} \tag{128}$$

then

$$\mathbf{q}'\boldsymbol{\beta} \quad \text{is an estimable function .} \tag{129}$$

The estimator to be used is

$$\mathbf{q}'\boldsymbol{\beta}^\circ, \quad \text{which is invariant to} \quad \boldsymbol{\beta}^\circ, \tag{130}$$

with

$$E(\mathbf{q}'\boldsymbol{\beta}^\circ) = \mathbf{q}'\boldsymbol{\beta}, \tag{131}$$

and

$$v(\mathbf{q}'\boldsymbol{\beta}^\circ) = \mathbf{q}'\mathbf{Gq}\sigma^2, \quad \text{invariant to} \quad \mathbf{G} . \tag{132}$$

Then

$$\text{BLUE}(\mathbf{q}'\boldsymbol{\beta}) = \mathbf{q}'\boldsymbol{\beta}^\circ \quad \text{is the BLUE of} \quad \mathbf{q}'\boldsymbol{\beta}, \tag{133}$$

meaning that it is the *b*est, *l*inear, *u*nbiased *e*stimator (BLUE) of $\mathbf{q}'\boldsymbol{\beta}$.

This estimator, the BLUE, has attractive properties: it is unbiased, because of (131); it is linear in the sense of being a linear combination, $\mathbf{q}'\mathbf{GX}'\mathbf{y}$, of the observations; and it is "best" in the sense of being, among all linear and unbiased estimators of $\mathbf{q}'\boldsymbol{\beta}$, that one which has the smallest variance, i.e., $\mathbf{q}'\mathbf{Gq}\sigma^2$ of (132) is less than the variance of any other linear function of $\mathbf{y}$ that has expected value equal to $\mathbf{q}'\boldsymbol{\beta}$.

Verification. For verifying (130)–(133), repeated use is made of Theorem 7.1.

$$\mathbf{q}'\boldsymbol{\beta}^\circ = \mathbf{t}'\mathbf{XGX}'\mathbf{y} \quad \text{is invariant to} \quad \mathbf{G} .$$

$$E(\mathbf{q}'\boldsymbol{\beta}^\circ) = \mathbf{t}'\mathbf{XGX}'E(\mathbf{y}) = \mathbf{t}'\mathbf{XGX}'\mathbf{X}\boldsymbol{\beta} = \mathbf{t}'\mathbf{X}\boldsymbol{\beta} = \mathbf{q}'\boldsymbol{\beta} .$$

$$v(\mathbf{q}'\boldsymbol{\beta}^\circ) = \mathbf{t}'\mathbf{XGX}'\sigma^2\mathbf{IXG}'\mathbf{X}'\mathbf{t} = \mathbf{t}'\mathbf{XG}'\mathbf{X}'\mathbf{t}\sigma^2 = \mathbf{t}'\mathbf{XGX}'\mathbf{t}\sigma^2 = \mathbf{q}'\mathbf{Gq}\sigma^2.$$

Suppose $\mathbf{k}'\mathbf{y}$ is an unbiased estimator of $\mathbf{q}'\boldsymbol{\beta}$ that is different from $\mathbf{q}'\boldsymbol{\beta}^\circ$. Its unbiasedness implies $\mathbf{k}'\mathbf{X} = \mathbf{q}'$ and so

$$\text{cov}(\mathbf{q}'\boldsymbol{\beta}^\circ, \mathbf{k}'\mathbf{y}) = \text{cov}(\mathbf{q}'\mathbf{GX}'\mathbf{y}, \mathbf{k}'\mathbf{y}) = \mathbf{q}'\mathbf{GX}'\mathbf{k}\sigma^2 = \mathbf{q}'\mathbf{Gq}\sigma^2 .$$

Hence

$$v(\mathbf{q}'\boldsymbol{\beta}^\circ - \mathbf{k}'\mathbf{y}) = \mathbf{q}'\mathbf{G}\mathbf{q}\sigma^2 + v(\mathbf{k}'\mathbf{y}) - 2\mathbf{q}'\mathbf{G}\mathbf{q}\sigma^2 = v(\mathbf{k}'\mathbf{y}) - \mathbf{q}'\mathbf{G}\mathbf{q}\sigma^2 .$$

But

$$v(\mathbf{q}'\boldsymbol{\beta}^\circ - \mathbf{k}'\mathbf{y}) \geq 0 \quad \text{and so} \quad v(\mathbf{q}'\boldsymbol{\beta}^\circ) = \mathbf{q}'\mathbf{G}\mathbf{q}\sigma^2 \leq v(\mathbf{k}'\mathbf{y}) .$$

d. Confidence intervals

The procedures of Section 2.12a apply directly to calculating confidence intervals for $\mathbf{q}'\boldsymbol{\beta}$. Thus, for example,

$$\mathbf{q}'\boldsymbol{\beta}^\circ \pm \hat{\sigma}\sqrt{\mathbf{q}'\mathbf{G}\mathbf{q}}\, t_{N-r_{\mathbf{X}},\frac{1}{2}\alpha} \tag{134}$$

is a symmetric, $100(1-\alpha)\%$ confidence interval on the estimable function $\mathbf{q}'\boldsymbol{\beta}$, similar to (82) of Section 2.12a, wherein the t-value is defined using $N - a$, which here is $N - r_{\mathbf{X}}$ in (134). Extension to basing the confidence interval on the normal distribution when $N - r_{\mathbf{X}}$ is large (exceeding 100, say) is as in (83) of Section 2.12a.

e. Other properties

It is clear from (128)–(133) that the concept of estimable function given by either of the definitions carries with it far more than is stated in the definitions. Not only is $\mathbf{q}'\boldsymbol{\beta}^\circ$ invariant to $\boldsymbol{\beta}^\circ$ for $\mathbf{q}'\boldsymbol{\beta}$ being estimable, but as an estimator it also has the useful properties of unbiasedness and minimum variance among all linear functions of $\mathbf{y}$ that are unbiased for $\mathbf{q}'\boldsymbol{\beta}$. We therefore have cause to make considerable use of estimable functions and their BLUEs.

Four other useful properties of estimable functions are as follows.

$$\mathbf{q}'\boldsymbol{\beta} \text{ is estimable if and only if } \mathbf{q}' = \mathbf{q}'\mathbf{H}, \tag{135}$$

with a special case being

$$\mathbf{q}'\boldsymbol{\beta} \text{ is estimable for } \mathbf{q}' = \mathbf{w}'\mathbf{H} \text{ for any } \mathbf{w}' . \tag{136}$$

Linear combinations of estimable functions are estimable . (137)

A set of linearly independent estimable functions contains at most $r_{\mathbf{X}}$ such functions . (138)

Each of these is a consequence of $\mathbf{q}' = \mathbf{t}'\mathbf{X}$ which, although it prescribes a suitable form of $\mathbf{q}'$ for $\mathbf{q}'\boldsymbol{\beta}$ to be estimable, is not as easy to use as (136). This is because in $\mathbf{q}' = \mathbf{t}'\mathbf{X}$ the vector $\mathbf{t}'$ has to have order N, the number of observations, whereas in $\mathbf{q}' = \mathbf{w}'\mathbf{H}$ of (136), $\mathbf{w}'$ has order p, the number of parameters; and usually p is much less than N. Statement (136) may not be the easiest condition, either. Alalouf and Styan (1973) give seven conditions based on $\mathbf{X}$ and ten based on $\mathbf{X}'\mathbf{X}$, each of which is necessary and sufficient for $\mathbf{q}'\boldsymbol{\beta}$ to be estimable.

f. Basic estimable functions

The prescription $\mathbf{q}' = \mathbf{t}'\mathbf{X}$ for $\mathbf{q}'\boldsymbol{\beta}$ to be estimable means that $\mathbf{q}'$ is a linear combination of rows of $\mathbf{X}$. The simplest such $\mathbf{q}'$ is therefore a row of $\mathbf{X}$, which we can call $\mathbf{x}'$. Then $\mathbf{x}'\boldsymbol{\beta}$ is estimable. Hence, since $\mathbf{X}\boldsymbol{\beta} = E(\mathbf{y})$ the estimable function $\mathbf{x}'\boldsymbol{\beta}$ is an element of $E(\mathbf{y})$. And such is the case for every element of $E(\mathbf{y})$. Thus we have the simple result:

The expected value of every observation is an estimable function . (139)

We call this the *basic estimable function*. For most models it provides, in combination with (137), an easy way of deriving estimable functions.

Example. The model equation for the overparameterized 1-way classification is $E(y_{ij}) = \mu + \alpha_i$. Therefore $\mu + \alpha_i$ is, by (139), an estimable function, and by (133) its BLUE is

$$\text{BLUE}(\mu + \alpha_i) = \mu^{\circ} + \alpha_i^{\circ} .$$

Furthermore, by (137), any linear combination of the $\mu + \alpha_i$ expressions are estimable. Thus

$$\alpha_1 - \alpha_2 = (\mu + \alpha_1) - (\mu + \alpha_2)$$

is estimable and so by (133)

$$\text{BLUE}(\alpha_1 - \alpha_2) = \alpha_1^{\circ} - \alpha_2^{\circ} .$$

Similarly,

$$\mu + \tfrac{1}{3}(\alpha_1 + \alpha_2 + \alpha_3) = \tfrac{1}{3}[(\mu + \alpha_1) + (\mu + \alpha_2) + (\mu + \alpha_3)]$$

is estimable with BLUE:

$$\text{BLUE}[\mu + \tfrac{1}{3}(\alpha_1 + \alpha_2 + \alpha_3)] = \mu^{\circ} + \tfrac{1}{3}(\alpha_1^{\circ} + \alpha_2^{\circ} + \alpha_3^{\circ}) .$$

With $\mu_i = \mu + \alpha_i$, this corresponds precisely to ρ of (98) in Section 2.12d; and in Section 2.12c the ρ of (94) corresponds to

$$\mu + \tfrac{1}{6}(3\alpha_1 + 2\alpha_2 + \alpha_3) = \tfrac{1}{6}[3(\mu + \alpha_1) + 2(\mu + \alpha_2) + (\mu + \alpha_3)],$$

which is estimable with BLUE:

$$\text{BLUE}\left[\mu + \tfrac{1}{6}(3\alpha_1 + 2\alpha_2 + \alpha_3)\right] = \mu^\circ + \tfrac{1}{6}\left(3\alpha_1^\circ + 2\alpha_2^\circ + \alpha_3^\circ\right).$$

g. Full rank models

It is to be emphasized that none of the problems arising from numerous $\boldsymbol{\beta}^\circ$s satisfying the normal equations occur when $\mathbf{X}'\mathbf{X}$ is nonsingular, i.e., has full rank. For then there is only one solution $\hat{\boldsymbol{\beta}} = (\mathbf{X}'\mathbf{X})^{-1}\mathbf{X}'\mathbf{y}$ to the normal equations and every element of $\hat{\boldsymbol{\beta}}$ is the BLUE of the corresponding element of $\boldsymbol{\beta}$. This is precisely the situation with the cell means models of Chapters 2 and 3 and of the with-interaction cell means model of Chapter 4. The first occurrence of the idea of an estimable function is in Section 5.2 when dealing with the no-interaction model $E(y_{ijk}) = \mu_i + \tau_j$ in the 2-way classification. This is dealt with further in Sections 9.1d and 9.2d. But with cell means models that implicitly include all interactions, the formal need for estimable functions does not arise. The cell mean of every filled cell is estimable.

h. Summary

Since estimable functions are the only linear combinations of parameters that can be estimated invariantly to the numerous solutions of the normal equations they are the only functions to which attention is paid insofar as estimation is concerned. The basic properties are:

$\mathbf{q}'\boldsymbol{\beta}$ is estimable when $\mathbf{q}' = \mathbf{t}'\mathbf{X}$ for some $\mathbf{t}'$;

$\text{BLUE}(\mathbf{q}'\boldsymbol{\beta}) = \mathbf{q}'\boldsymbol{\beta}^\circ$ with $v(\mathbf{q}'\boldsymbol{\beta}^\circ) = \mathbf{q}'\mathbf{G}\mathbf{q}\sigma^2$;

all elements of $E(\mathbf{y})$ are estimable, as are linear functions of them.

With cell means models (that implicitly include interactions) every cell mean of a filled cell is estimable.

8.8. THE GENERAL LINEAR HYPOTHESIS

Much space has been devoted to discussing hypotheses that correspond to F-statistics coming from the partitioning of a model, as shown quite generally in Sections 8.5 and 8.6. And many specific examples of this are given in Chapters 2–6. All of this is useful for the dilemma often encountered in using statistical computing packages for linear models calculations, namely evaluating the utility of computed F-statistics. When those statistics are based on a sum of squares of the form $R(\boldsymbol{\beta}_2 | \boldsymbol{\beta}_1)$ its associated hypothesis is given by (123), and that is the basis of assessing the usefulness of the F. But, as has been stated so often, the real use of hypothesis testing is not to look at F-statistics that are, in some sense standard, but to set up a hypothesis and develop a test statistic for it. In the context of linear models, attention is confined to hypotheses that concern linear functions of the parameters. This has been done in an elementary way in Section 2.12b-i for a one-part hypothesis H: $\Sigma_i k_i \mu_i = m$, that is concerned with just a single linear function of the parameters. Cookbook extension is given in Section 2.12b-ii to a two-part hypothesis consisting of two linear functions of parameters. And extensions to an array of linear functions is now described.

Example 2 (continued). The overparameterized model $E(y_{ij}) = \mu + \alpha_i$ of the 1-way classification for the example of Table 8.1 continues to be used so as to maintain the necessary generalization that copes with $\mathbf{X}'\mathbf{X}$ being singular. For that example, suppose we wished to test the hypothesis

$$\text{H:} \begin{cases} \alpha_1 - \alpha_2 = 2 \\ \alpha_1 - \alpha_3 = 5 \end{cases} . \tag{140}$$

This is equivalent to

$$\text{H:} \begin{bmatrix} 0 & 1 & -1 & 0 \\ 0 & 1 & 0 & -1 \end{bmatrix} \begin{bmatrix} \mu \\ \alpha_1 \\ \alpha_2 \\ \alpha_3 \end{bmatrix} = \begin{bmatrix} 2 \\ 5 \end{bmatrix} . \tag{141}$$

And it is in this form, written in general terms as

$$\text{H: } \mathbf{K}'\boldsymbol{\beta} = \mathbf{m},$$

that we deal with the general linear hypothesis.

a. A general form

The general linear hypothesis is taken to be of the form

$$\text{H: } \mathbf{K}'\boldsymbol{\beta} = \mathbf{m},$$

where $\mathbf{K}'$ is a known matrix and $\mathbf{m}$ is a known vector (often null). Any linear hypothesis can be written in this form, as exemplified in (140) and (141).

A hypothesis is said to be testable when we can calculate an F-statistic that is suitable for testing it. There are three limitations that $\mathbf{K}'$ must satisfy in order for this to be possible. All three are simultaneously mathematical and practical.

(i) $\mathbf{K}'\boldsymbol{\beta}$ must be estimable, i.e., $\mathbf{K} = \mathbf{T}'\mathbf{X}$ for some $\mathbf{T}'$. This means that only hypotheses which are in terms of estimable functions can be tested; but since they are the only functions for which $\mathbf{K}'\boldsymbol{\beta}^\circ$ is invariant to $\boldsymbol{\beta}^\circ$, this is eminently reasonable. [An associated condition given by Seely (1977) concerns disjoint classes of distributions.]

(ii) $\mathbf{K}'$ must have full row rank; i.e., $(\mathbf{K}')_{s\times p}$ must have $r_{\mathbf{K}'} = s$. This ensures that the hypothesis contains no redundant statements (see example 2 that follows).

(iii) $\mathbf{K}'$ must have no more than $r_{\mathbf{X}}$ rows. This is a consequence of (ii) and of (138).

Limitation (ii) on $\mathbf{K}'$ means that in stating the hypothesis as $\mathbf{K}'\boldsymbol{\beta} = \mathbf{m}$ it must not contain any redundant statements, i.e., statements that are linear combinations of others.

Example 2 (continued). The hypothesis

$$\text{H:} \begin{cases} \alpha_1 - \alpha_2 = 2 \\ \alpha_1 - \alpha_3 = 5, \\ \alpha_3 - \alpha_2 = -3 \end{cases}$$

equivalent to

$$\text{H:} \begin{bmatrix} 0 & 1 & -1 & 0 \\ 0 & 1 & 0 & -1 \\ 0 & 0 & -1 & 1 \end{bmatrix} \begin{bmatrix} \mu \\ \alpha_1 \\ \alpha_2 \\ \alpha_3 \end{bmatrix} = \begin{bmatrix} 2 \\ 5 \\ -3 \end{bmatrix},$$

has a redundant statement: its last statement is its first statement minus its second. Thus its $\mathbf{K}'$ has 3 rows and rank 2. The hypothesis is stated just as satisfactorily by omitting its last statement.

b. The *F*-statistic

We use the estimated variance

$$\hat{\sigma}^2 = \frac{\mathbf{y}'(\mathbf{I} - \mathbf{X}\mathbf{G}\mathbf{X}')\mathbf{y}}{N - r_{\mathbf{X}}} = \frac{\mathbf{y}'\mathbf{M}\mathbf{y}}{N - r_{\mathbf{x}}} \tag{142}$$

of (63), together with

$$r_{\mathbf{K}'} = \text{rank of } \mathbf{K}' = r_{\mathbf{K}} = s = \text{number of rows in } \mathbf{K}' \tag{143}$$

and

$$Q = (\mathbf{K}'\boldsymbol{\beta}^\circ - \mathbf{m})'(\mathbf{K}'\mathbf{G}\mathbf{K})^{-1}(\mathbf{K}'\boldsymbol{\beta}^\circ - \mathbf{m}), \tag{144}$$

and the normality assumptions $\mathbf{y} \sim \mathcal{N}(\mathbf{X}\boldsymbol{\beta}, \sigma^2\mathbf{I})$. Then the F-statistic for testing

$$\text{H: } \mathbf{K}'\boldsymbol{\beta} = \mathbf{m} \tag{145}$$

is

$$F(\text{H}) = \frac{Q}{r_{\mathbf{K}}\hat{\sigma}^2} = \frac{(\mathbf{K}'\boldsymbol{\beta} - \mathbf{m})'(\mathbf{K}'\mathbf{G}\mathbf{K})(\mathbf{K}'\boldsymbol{\beta} - \mathbf{m})}{r_{\mathbf{K}}\hat{\sigma}^2} \tag{146}$$

with

$$F(\text{H}) \sim \mathscr{F}'\left[r_{\mathbf{K}}, N - r_{\mathbf{X}}, (\mathbf{K}'\boldsymbol{\beta} - \mathbf{m})'(\mathbf{K}'\mathbf{G}\mathbf{K})^{-1}(\mathbf{K}\boldsymbol{\beta} - \mathbf{m})/2\sigma^2\right] . \tag{147}$$

Verification. We have

$$\mathbf{y} \sim \mathcal{N}(\mathbf{X}\boldsymbol{\beta}, \sigma^2\mathbf{I})$$

$$\boldsymbol{\beta}^\circ = \mathbf{G}\mathbf{X}'\mathbf{y} \sim \mathcal{N}(\mathbf{G}\mathbf{X}'\mathbf{X}\boldsymbol{\beta}, \mathbf{G}\mathbf{X}'\mathbf{X}\mathbf{G}'\sigma^2)$$

$$\mathbf{K}'\boldsymbol{\beta}^\circ - \mathbf{m} \sim \mathcal{N}(\mathbf{K}'\boldsymbol{\beta} - \mathbf{m}, \mathbf{K}'\mathbf{G}\mathbf{K}\sigma^2).$$

Because, as in (i), $\mathbf{K}'\boldsymbol{\beta}$ is estimable, we have $\mathbf{K}' = \mathbf{T}'\mathbf{X}$ for some $\mathbf{T}'$ and so

$$\text{var}(\mathbf{K}'\boldsymbol{\beta}^\circ) = \mathbf{K}'\mathbf{G}\mathbf{X}'\mathbf{X}\mathbf{G}'\mathbf{K}\sigma^2 = \mathbf{K}'\mathbf{G}\mathbf{X}'\mathbf{X}\mathbf{G}'\mathbf{X}'\mathbf{T}\sigma^2 = \mathbf{K}'\mathbf{G}\mathbf{X}'\mathbf{T}\sigma^2 = \mathbf{K}'\mathbf{G}\mathbf{K}\sigma^2 .$$

Therefore applying Theorem 7.3 to Q as a quadratic form in $\mathbf{K}'\boldsymbol{\beta}^\circ - \mathbf{m}$ gives

$$Q/\sigma^2 \sim \chi^{2\prime}\left[r_{\mathbf{K}}, (\mathbf{K}'\boldsymbol{\beta} - \mathbf{m})(\mathbf{K}'\mathbf{G}\mathbf{K})^{-1}(\mathbf{K}'\boldsymbol{\beta} - \mathbf{m})/2\sigma^2\right] .$$

Also, because $\mathbf{K}'$ has full row rank, $(\mathbf{K}'\mathbf{K})^{-1}$ exists, and with $\mathbf{K}' = \mathbf{T}'\mathbf{X}$

$$\mathbf{K}'\mathbf{G}\mathbf{X}'\mathbf{X}\mathbf{K}(\mathbf{K}'\mathbf{K})^{-1} = \mathbf{I} .$$

Hence Q of (144) can also be written as

$$Q = \left[\mathbf{y} - \mathbf{X}\mathbf{K}(\mathbf{K}'\mathbf{K})^{-1}\mathbf{m}\right]'\mathbf{X}\mathbf{G}'\mathbf{K}(\mathbf{K}'\mathbf{G}\mathbf{K})^{-1}\mathbf{K}'\mathbf{G}\mathbf{X}'\left[\mathbf{y} - \mathbf{X}\mathbf{K}(\mathbf{K}'\mathbf{K})^{-1}\mathbf{m}\right];$$

and since $\mathbf{MX} = \mathbf{0}$, SSE is

$$\text{SSE} = \left[\mathbf{y} - \mathbf{XK}(\mathbf{K'K})^{-1}\mathbf{m}\right]'\mathbf{M}\left[\mathbf{y} - \mathbf{XK}(\mathbf{K'K})^{-1}\mathbf{m}\right] .$$

Thus Q and SSE are both expressed as quadratic forms in the same vector, namely $\mathbf{y} - \mathbf{XK}(\mathbf{K'K})^{-1}\mathbf{m}$, for which the variance–covariance matrix is $\sigma^2\mathbf{I}$. Hence Q and SSE will, by Theorem 7.4, be independent if the product of the two matrices of those quadratic forms is null: which it is, because $\mathbf{MXG'K}(\mathbf{K'GK})^{-1}\mathbf{K'GX'} = \mathbf{0}$ since $\mathbf{MX} = \mathbf{0}$. Thus $F(\mathrm{H})$ of (146) is established, with its non-centrality parameter being that of the $\chi^{2\prime}$-distribution of Q/σ^2. Derivation can also be had from the likelihood ratio test [see Appendix (Section 8.13)].

A number of observations are warranted.

(i) In (147) the non-centrality parameter is zero if $\mathbf{K'\beta} = \mathbf{m}$, whereupon $F(\mathrm{H}) \sim \mathscr{F}_{r_{\mathbf{K}},\, N-r_{\mathbf{X}}}$. Thus is the test of H: $\mathbf{K'\beta} = \mathbf{m}$ provided.
(ii) Q of (144) is the numerator sum of squares of $F(\mathrm{H})$ of (146).
(iii) $(\mathbf{K'GK})^{-1}$ exists because $\mathbf{K'}$ has full row rank and $\mathbf{K'} = \mathbf{T'X}$.
(iv) $\mathbf{K'GK} = \mathrm{var}(\mathbf{K'\beta^\circ})/\sigma^2$.
(v) $\mathbf{K'\beta^\circ}$ and $\mathbf{K'GK}$ are invariant to $\boldsymbol{\beta}^\circ$ and $\mathbf{G}$ because $\mathbf{K\beta}$ is estimable with $\mathbf{K'} = \mathbf{T'X}$.

Verification of (iii). We use two standard results concerning rank: $r(\mathbf{A'A}) = r(\mathbf{A'})$ and $r(\mathbf{AB}) \geq r(\mathbf{A})$. Then, with $\mathbf{K'\beta}$ estimable implying $\mathbf{K'} = \mathbf{T'X}$ for some $\mathbf{T'}$, and on using a symmetric $\mathbf{G}$ [which is always possible—see (11) of Section 7.1i] so that $\mathbf{G} = \mathbf{L'L}$ for some $\mathbf{L}$, we have

$$r(\mathbf{K'GK}) = r(\mathbf{K'L'LK}) = r(\mathbf{K'L'}) \geq r(\mathbf{K'L'L}) = r(\mathbf{K'G}) = r(\mathbf{T'XG}) .$$

Therefore $r(\mathbf{K'GK}) \geq r(\mathbf{T'XG}) \geq r(\mathbf{T'XGX'X}) = r(\mathbf{T'X}) = r(\mathbf{K}) = s$. Hence $r(\mathbf{K'GK}) \geq s$; but the order of $\mathbf{K'GK}$ is s. Therefore $\mathbf{K'GK}$ is nonsingular.

Example 2 (continued). A value of $\mathbf{G}$ and its corresponding $\boldsymbol{\beta}^\circ$ are given in (24) and (25) as

$$\mathbf{G} = \tfrac{1}{6}\begin{bmatrix} 3 & -3 & -3 & 0 \\ -3 & 5 & 3 & 0 \\ -3 & 3 & 6 & 0 \\ 0 & 0 & 0 & 0 \end{bmatrix} \quad \text{and} \quad \boldsymbol{\beta}^\circ = \begin{bmatrix} 89 \\ -16 \\ -11 \\ 0 \end{bmatrix} . \tag{148}$$

To test the hypothesis (141) where

$$\mathbf{K'} = \begin{bmatrix} 0 & 1 & -1 & 0 \\ 0 & 1 & 0 & -1 \end{bmatrix} \quad \text{and} \quad \mathbf{m} = \begin{bmatrix} 2 \\ 5 \end{bmatrix}$$

we find

$$\mathbf{K}'\boldsymbol{\beta}^{\circ} - \mathbf{m} = \begin{bmatrix} -16 + 11 - 2 \\ -16 + 0 - 5 \end{bmatrix} = \begin{bmatrix} -7 \\ -21 \end{bmatrix}$$

and

$$(\mathbf{K}'\mathbf{G}\mathbf{K})^{-1} = \left[\tfrac{1}{6}\begin{pmatrix} 5 & 2 \\ 2 & 5 \end{pmatrix}\right]^{-1} = \tfrac{2}{7}\begin{bmatrix} 5 & -2 \\ -2 & 5 \end{bmatrix} .$$

Hence from (144)

$$Q = [-7 \quad -21]\tfrac{2}{7}\begin{bmatrix} 5 & -2 \\ -2 & 5 \end{bmatrix}\begin{bmatrix} -7 \\ -21 \end{bmatrix} = 532 .$$

Thus, using $\hat{\sigma}^2 = 14\frac{1}{2}$, as following (63), the F-statistic (146) is

$$F = \frac{532}{2(14\frac{1}{2})} = 18\tfrac{10}{29} .$$

The reader should confirm this result using a different $\mathbf{G}$ and $\boldsymbol{\beta}^{\circ}$. The value $Q = 532$ will also be obtained from (89) of Section 2.12b.

c. The hypothesis H: $\mathbf{K}'\boldsymbol{\beta} = \mathbf{0}$

Many hypotheses have $\mathbf{m}$ null, so that the hypothesis is then

$$\text{H:}\ \mathbf{K}'\boldsymbol{\beta} = \mathbf{0} .$$

On denoting Q for this case as Q_0, its value from (144) becomes

$$Q_0 = \boldsymbol{\beta}^{\circ\prime}\mathbf{K}(\mathbf{K}'\mathbf{G}\mathbf{K})^{-1}\mathbf{K}'\boldsymbol{\beta}^{\circ} = \mathbf{y}'\mathbf{X}\mathbf{G}'\mathbf{K}(\mathbf{K}'\mathbf{G}\mathbf{K})^{-1}\mathbf{K}'\mathbf{G}\mathbf{X}'\mathbf{y} . \quad (149)$$

Since, in practice, hypotheses having $\mathbf{m} = \mathbf{0}$ occur more frequently than those having $\mathbf{m} \neq \mathbf{0}$ the question might be raised, "Why start with (144) for $\mathbf{m} \neq \mathbf{0}$ and then develop (149) for $\mathbf{m} = \mathbf{0}$?" The answer is easy: (149) is a simplification of (144), whereas without benefit of hindsight it could be difficult to see (144) as the generalization of (149).

At least two special cases of Q_0 of (149) are important. They concern situations when $r_{\mathbf{K}} = r_{\mathbf{X}}$ and when $r_{\mathbf{K}} = r_{\mathbf{X}} - 1$. First, when $r_{\mathbf{K}} = r_{\mathbf{X}}$ and $\mathbf{m} = \mathbf{0}$, we have $Q = R(\boldsymbol{\beta})$, because the corresponding hypothesis is then equivalent to H: $\mathbf{X}\boldsymbol{\beta} = \mathbf{0}$. Second, when $r_{\mathbf{K}} = r_{\mathbf{X}} - 1$, $\mathbf{m} = \mathbf{0}$ and $\mathbf{K}'\mathbf{1} = \mathbf{0}$ we have $Q = R(\boldsymbol{\beta}) - R(\mu)$.

Direct verification[1] of $Q_0 = \text{SSR}$ when $r_{\mathbf{K}} = r_{\mathbf{X}}$ and $\mathbf{m} = \mathbf{0}$ is as follows. Start with (149). The symmetry of $\mathbf{X}'\mathbf{X}$ implies the existence of a matrix $\mathbf{R}$ of full column

[1] I am grateful to W. H. Swallow for this.

rank r_X such that $\mathbf{X}'\mathbf{X} = \mathbf{RR}'$ with $(\mathbf{R}'\mathbf{R})^{-1}$ existing. Also, the Moore–Penrose inverse [(6) in Section 7.1h] of $\mathbf{X}'\mathbf{X}$ is then $\mathbf{R}(\mathbf{R}'\mathbf{R})^{-2}\mathbf{R}'$; using it as $\mathbf{G}$ in Q_0 gives

$$
\begin{aligned}
Q_0 &= \mathbf{y}'\mathbf{XR}(\mathbf{R}'\mathbf{R})^{-2}\mathbf{R}'\mathbf{K}\left[\mathbf{K}'\mathbf{R}(\mathbf{R}'\mathbf{R})^{-2}\mathbf{R}'\mathbf{K}\right]^{-1}\mathbf{K}'\mathbf{R}(\mathbf{R}'\mathbf{R})^{-2}\mathbf{R}'\mathbf{X}'\mathbf{y} \\
&= \mathbf{y}'\mathbf{XR}(\mathbf{R}'\mathbf{R})^{-1}\mathbf{L}'(\mathbf{LL}')^{-1}\mathbf{L}(\mathbf{R}'\mathbf{R})^{-1}\mathbf{RX}'\mathbf{y} \quad \text{for} \quad \mathbf{L} = \mathbf{K}'\mathbf{R}(\mathbf{R}'\mathbf{R})^{-1} .
\end{aligned}
$$

Because $\mathbf{K}'$ has full row rank r_X and $\mathbf{R}$ has full column rank r_X, $\mathbf{L}$ is square; and since $(\mathbf{L}'\mathbf{L})^{-1}$ exists, $|\mathbf{L}| \neq \mathbf{0}$ and so $\mathbf{L}^{-1}$ exists. Hence

$$
Q_0 = \mathbf{y}'\mathbf{XR}(\mathbf{R}'\mathbf{R})^{-2}\mathbf{RX}'\mathbf{y} = \mathbf{y}'\mathbf{XGX}'\mathbf{y} = \text{SSR} .
$$

d. Estimation under the hypothesis

When the conclusion from using $F(H)$ is to not reject the hypothesis H: $\mathbf{K}'\boldsymbol{\beta} = \mathbf{m}$ one might want to estimate $\boldsymbol{\beta}$ under the condition of taking that hypothesis to be true. Denoting this estimator by $\boldsymbol{\beta}_H^\circ$, it is derived by minimizing $(\mathbf{y} - \mathbf{X}\boldsymbol{\beta}_H^\circ)'(\mathbf{y} - \mathbf{X}\boldsymbol{\beta}_H^\circ)$ subject to $\mathbf{K}'\boldsymbol{\beta}_H^\circ = \mathbf{m}$. This leads [see Appendix (Section 8.13)] to

$$
\boldsymbol{\beta}_H^\circ = \boldsymbol{\beta}^\circ - \mathbf{GK}(\mathbf{K}'\mathbf{GK})^{-1}(\mathbf{K}'\boldsymbol{\beta}^\circ - \mathbf{m}) \tag{150}
$$

with

$$
\text{var}(\boldsymbol{\beta}_H^\circ) = \mathbf{G}\left[\mathbf{X}'\mathbf{X} - \mathbf{K}(\mathbf{K}'\mathbf{GK})^{-1}\mathbf{K}'\right]\mathbf{G}'\sigma^2 . \tag{151}
$$

Furthermore, on using (150) for $\boldsymbol{\beta}_H^\circ$ it is readily shown that the residual sum of squares under the hypothesis is

$$
\begin{aligned}
\text{SSE}_H &= (\mathbf{y} - \mathbf{X}\boldsymbol{\beta}_H^\circ)'(\mathbf{y} - \mathbf{X}\boldsymbol{\beta}_H^\circ) = \text{SSE} + Q \\
&= \text{SSE} + (\mathbf{K}'\boldsymbol{\beta}^\circ - \mathbf{m})'(\mathbf{K}'\mathbf{GK})^{-1}(\mathbf{K}'\boldsymbol{\beta}^\circ - \mathbf{m}) \qquad (152)
\end{aligned}
$$

with expected value

$$
E(\text{SSE}_H) = (N - r_X + r_{K'})\sigma^2 . \tag{153}
$$

e. Calculating Q

There are now two ways to calculate the numerator sum of squares Q: from (144)

$$
Q = (\mathbf{K}'\boldsymbol{\beta}^\circ - \mathbf{m})'(\mathbf{K}'\mathbf{GK})^{-1}(\mathbf{K}'\boldsymbol{\beta}^\circ - \mathbf{m})
$$

and from (152)

$$Q = \text{SSE}_\text{H} - \text{SSE} .$$

In connection with the latter the terms *full model* and *reduced model* are often used. As in Section 8.5c, immediately preceding (81), the full model is the model as specified, in this case $\mathbf{y} = \mathbf{X}\boldsymbol{\beta} + \mathbf{e}$. The reduced model is the full model as amended ("reduced") by the hypothesis H; thus it is

$$\mathbf{y} = \mathbf{X}\boldsymbol{\beta}_\text{H} + \mathbf{e} \quad \text{and} \quad \mathbf{K}'\boldsymbol{\beta}_\text{H} = \mathbf{m}.$$

Attaching these names to the sums of squares in Q we have

$$\text{SSE} = \text{SSE(full model)} \quad \text{and} \quad \text{SSE}_\text{H} = \text{SSE(reduced model)}$$

and so

$$Q = \text{SSE(reduced model)} - \text{SSE(full model)} . \tag{154}$$

Only when $\mathbf{m} = \mathbf{0}$ can (154) be converted to a difference between reductions in sums of squares, i.e.,

$$Q = \text{SSR(full model)} - \text{SSR(reduced model)} \quad \text{only for} \quad \mathbf{m} = \mathbf{0} .$$

This is so because, from (154)

$$Q = \mathbf{y}'\mathbf{y} - \text{SSE(full model)} - [\mathbf{y}'\mathbf{y} - \text{SSE(reduced model)}] .$$

Then, from (36)

$$Q = \text{SSR(full model)} - [\mathbf{y}'\mathbf{y} - \text{SSE(reduced model)}] .$$

The second term in this last expression is R(reduced model) only for $\mathbf{m} = \mathbf{0}$; otherwise, for $\mathbf{m} \neq \mathbf{0}$, the total sum of squares for the reduced model is not $\mathbf{y}'\mathbf{y}$ and so that second term is therefore not R(reduced model).

Example 2 (continued). Consider the hypothesis H: $\alpha_1 - \alpha_2 = 4$. The reduced model is $E(y_{ij}) = \mu + \alpha_i$ *and* $\alpha_1 - \alpha_2 = 4$. The model equations are (21) adapted by using $\alpha_1 - \alpha_2 = 4$. Therefore they are

$$\left.\begin{array}{l} 74 = \mu + \alpha_1 + e_{11} \\ 68 = \mu + \alpha_1 + e_{12} \\ 77 = \mu + \alpha_1 + e_{13} \\ 76 = \mu + \alpha_1 - 4 + e_{21} \\ 80 = \mu + \alpha_1 - 4 + e_{22} \\ 87 = \mu + \alpha_3 + e_{31} \\ 91 = \mu + \alpha_3 = e_{32} \end{array}\right\} \text{ equivalent to } \left\{\begin{array}{l} 74 = \mu + \alpha_1 + e_{11} \\ 68 = \mu + \alpha_1 + e_{12} \\ 77 = \mu + \alpha_1 + e_{13} \\ 80 = \mu + \alpha_1 + e_{21} \\ 84 = \mu + \alpha_1 + e_{22} \\ 87 = \mu + \alpha_3 + e_{31} \\ 91 = \mu + \alpha_3 + e_{32} \end{array}\right.$$

TABLE 8.7. ANALYSIS OF VARIANCE FOR TESTING THE HYPOTHESIS $\mathbf{K}'\boldsymbol{\beta} = \mathbf{0}$

Source of Variation	d.f.	Sum of Squares
Full model	$r_{\mathbf{X}}$	$\text{SSR} = \boldsymbol{\beta}^{\circ\prime}\mathbf{X}'\mathbf{y}$
Mean	1	$\text{SSM} = N\bar{y}^2$
Full model (a.f.m.)	$r_{\mathbf{X}} - 1$	$\text{SSR}_m = \text{SSR} - \text{SSM}$
Hypothesis	$r_{\mathbf{K}}$	$Q = \boldsymbol{\beta}^{\circ\prime}\mathbf{K}(\mathbf{K}'\mathbf{G}\mathbf{K})^{-1}\mathbf{K}'\boldsymbol{\beta}^{\circ}$
Reduced model (a.f.m.)	$r_{\mathbf{X}} - r_{\mathbf{K}} - 1$	$\text{SSR}_m - Q$
Residual error	$N - r_{\mathbf{X}}$	$\text{SSE} = \mathbf{y}'\mathbf{M}\mathbf{y}$
Total	N	$\text{SST} = \mathbf{y}'\mathbf{y}$

and immediately we see that the vector of effective observations for the reduced model is different from $\mathbf{y}$. Hence the total sum of squares is different from $\mathbf{y}'\mathbf{y}$. Therefore $\mathbf{y}'\mathbf{y}$ − SSE(reduced model) does not equal R(reduced model).

f. Analysis of variance

The analysis of variance table for testing the hypothesis H: $\mathbf{K}'\boldsymbol{\beta} = \mathbf{0}$ is shown in Table 8.7 (where a.f.m. is "adjusted for the mean"). With $\hat{\sigma}^2 = \text{SSE}/(N - r)$ we have the following tests.

$$\text{SSR}/r\hat{\sigma}^2 \quad \text{tests} \quad \text{H: } \mathbf{X}\boldsymbol{\beta} = \mathbf{0}, \quad \text{i.e., tests the full model} .$$

$$\text{SSM}/\hat{\sigma}^2 \quad \text{tests} \quad \text{H: } E(\bar{y}) = 0 .$$

$$\text{SSR}_m/(r-1)\hat{\sigma}^2 \quad \text{tests the full model, adjusted for the mean} .$$

$$Q/s\hat{\sigma}^2 \quad \text{tests} \quad \text{H: } \mathbf{K}'\boldsymbol{\beta} = \mathbf{0} .$$

$$(\text{SSR}_m - Q)/(r - s - 1)\hat{\sigma}^2 \quad \text{tests the reduced model, a.f.m.,}$$

and $$(\text{SSR} - Q)/(r - s)\hat{\sigma}^2 \quad \text{tests the reduced model} .$$

g. Non-testable hypotheses

Formulating the general linear hypothesis as H: $\mathbf{K}'\boldsymbol{\beta} = \mathbf{m}$ demands that $\mathbf{K}'$ have full row rank and that $\mathbf{K}'\boldsymbol{\beta}$ be estimable. These are sufficient conditions for $(\mathbf{K}'\mathbf{G}\mathbf{K})^{-1}$ in Q to exist. The full rank condition is also a necessary condition, but the estimability requirement is not. This means there are matrices $\mathbf{K}'$ of full row rank that, although not of the form $\mathbf{T}'\mathbf{X}$, are such that $(\mathbf{K}'\mathbf{G}\mathbf{K})^{-1}$ exists. When this occurs, Q and $F(\text{H})$ for H: $\mathbf{K}'\boldsymbol{\beta} = \mathbf{m}$ can be calculated, even though $\mathbf{K}'$ does not have the form $\mathbf{T}'\mathbf{X}$

and so the hypothesis is not testable. Thus $F(\mathrm{H})$ is not testing H: $\mathbf{K}'\boldsymbol{\beta} = \mathbf{m}$. This prompts the question "If we calculate F for a hypothesis that we think is testable, but which is actually not testable, then what hypothesis *is* being tested?" The answer to this question is developed by first defining a non-testable hypothesis.

The definition of H: $\mathbf{K}'\boldsymbol{\beta} = \mathbf{m}$ being testable has effectively been given, namely that every element of $\mathbf{K}'\boldsymbol{\beta}$ be estimable. In contrast, H is defined as non-testable when not only every element of $\mathbf{K}'\boldsymbol{\beta}$ is non-estimable but also when every linear combination of elements of $\mathbf{K}'\boldsymbol{\beta}$ is non-estimable. In a situation where, for example, α_1 and α_2 are each non-estimable but $\alpha_1 - \alpha_2$ is estimable, this precludes describing

$$\mathrm{H}: \begin{cases} \alpha_1 = 7 \\ \alpha_2 = 4 \end{cases}$$

as non-testable because it can equally as well be written as

$$\mathrm{H}: \begin{cases} \alpha_1 = 7 \\ \alpha_1 - \alpha_2 = 3 \end{cases} .$$

This second form is in terms of a mixture of non-estimable and estimable functions. We call such a hypothesis *partially testable* and consider it more generally in subsection h.

A formal proof that a non-testable hypothesis cannot be tested is given by Searle *et al.* (1984). It amounts to showing that the non-estimability of every element of $\mathbf{K}'\boldsymbol{\beta}$ leads to SSE_H of (152) being equal to SSE. Hence $\mathrm{SSE}_\mathrm{H} - \mathrm{SSE}$, the numerator of the F-statistic, is identically zero and so there is no test of the hypothesis. Seely (1977) also discusses the need for a hypothesis to be testable.

Note in passing that a non-testable hypothesis, as defined, can involve no more than $p - r_\mathbf{X}$ statements about the p parameters in $\boldsymbol{\beta}$. This is so because for every linear combination of elements of $\mathbf{K}'\boldsymbol{\beta}$ to be non-estimable, every linear combination of rows of $\mathbf{K}'$ must not be a linear combination of rows of $\mathbf{X}$. Since rows of $\mathbf{K}'$ have order p and $\mathbf{X}$ has rank $r_\mathbf{X}$, this means $\mathbf{K}'$ for a non-testable hypothesis can have no more than $p - r_\mathbf{X}$ rows.

Despite being unable to test H: $\mathbf{K}'\boldsymbol{\beta} = \mathbf{m}$ when every element of $\mathbf{K}'\boldsymbol{\beta}$ is non-estimable, $\mathbf{K}'$ is sometimes such that $(\mathbf{K}'\mathbf{G}\mathbf{K})^{-1}$ exists and so Q of (144) can then be calculated. The pertinent question is then "If H: $\mathbf{K}'\boldsymbol{\beta} = \mathbf{m}$ is not the hypothesis being tested, then what is?" The answer is as follows. Provided the generalized inverse of $\mathbf{X}'\mathbf{X}$ that is being used for $\mathbf{G}$ is both symmetric and reflexive [see (5) of Section 7.1h], the hypothesis tested by $F = Q/\hat{\sigma}^2 r_{\mathbf{K}}$ is not H: $\mathbf{K}'\boldsymbol{\beta} = \mathbf{m}$ but is

$$\text{H: } \mathbf{K}'\mathbf{H}\boldsymbol{\beta} = \mathbf{m} . \tag{155}$$

The reason for this is that even without $\mathbf{K}' = \mathbf{T}'\mathbf{X}$, we have $\mathbf{K}'\mathbf{H}\boldsymbol{\beta}^\circ = \mathbf{K}'\mathbf{G}\mathbf{X}'\mathbf{X}\mathbf{G}\mathbf{X}'\mathbf{y} = \mathbf{K}'\mathbf{G}\mathbf{X}'\mathbf{y} = \mathbf{K}'\boldsymbol{\beta}^\circ$, and, on using $\mathbf{G} = \mathbf{G}' = \mathbf{G}\mathbf{X}'\mathbf{X}\mathbf{G}'$, it is easily shown that $\mathbf{K}'\mathbf{H}\mathbf{G}\mathbf{H}'\mathbf{K} = \mathbf{K}'\mathbf{G}\mathbf{K}$. Result (155) corrects an error implicit in Searle (1971, p. 195–196), which fails to mention that $\mathbf{G}$ must be symmetric and reflexive.

Of course the question will be asked, "When, for a non-testable hypothesis, will $\mathbf{K}'\mathbf{G}\mathbf{K}$ nevertheless be nonsingular?", thus permitting computation of Q and interpretation thereof by means of (155). There seems to be no universal answer to this: nonsingularity of $\mathbf{K}'\mathbf{G}\mathbf{K}$ depends completely on the particular forms of $\mathbf{K}'$ and $\mathbf{G}$ being used.

Example 2 (continued). Because α_1 is not estimable, H: $\alpha_1 = 7$ is not testable. Nevertheless, it can be written as $\mathbf{K}'\boldsymbol{\beta} = 7$ with $\mathbf{K}' = [0 \;\; 1 \;\; 0 \;\; 0]$. Using $\mathbf{G}$ of (148), which equals $\mathbf{G}' = \mathbf{G}\mathbf{X}'\mathbf{X}\mathbf{G}$, we then find $\mathbf{K}'\mathbf{G}\mathbf{K} = \frac{5}{6}$, which has an inverse and so, with $\mathbf{K}'\boldsymbol{\beta}^\circ = \mathbf{m} = -16 - 7 = -23$, $Q = 23(\frac{5}{6})^{-1}(23) = 634.8$. The hypothesis being tested is not H: $\alpha_1 = 7$, but is H: $\mathbf{K}'\mathbf{H}\boldsymbol{\beta} = 7$ with, using $\mathbf{H}$ from (50),

$$\mathbf{K}'\mathbf{H} = \mathbf{K}'\mathbf{G}\mathbf{X}'\mathbf{X} = [0 \;\; 1 \;\; 0 \;\; 0]\begin{bmatrix} 1 & \cdot & \cdot & 1 \\ \cdot & 1 & \cdot & -1 \\ \cdot & \cdot & 1 & -1 \\ \cdot & \cdot & \cdot & \cdot \end{bmatrix} = [0 \;\; 1 \;\; 0 \;\; -1] .$$

Thus the hypothesis being tested is

$$\text{H: } [0 \;\; 1 \;\; 0 \;\; -1]\boldsymbol{\beta} = 7 \quad \text{namely} \quad \text{H: } \alpha_1 - \alpha_3 = 7 .$$

Using Q of (144) on this hypothesis, based on (148), confirms $Q = 634.8$.

h. Partially testable hypotheses

As illustrated in the preceding subsection, hypotheses can be formulated with $\mathbf{K}'\boldsymbol{\beta}$ having both estimable and non-estimable elements. To exclude this mixture from the definition of a non-testable hypothesis we have defined that as a hypothesis for which no linear combination of elements of $\mathbf{K}'\boldsymbol{\beta}$ is estimable. Since linear combinations of estimable functions are estimable a similar definition, equivalent to that given earlier, can be made for testable hypotheses. H: $\mathbf{K}'\boldsymbol{\beta} = \mathbf{m}$ is defined as testable when all linear combinations of elements of $\mathbf{K}'\boldsymbol{\beta}$ are estimable. The remaining forms of $\mathbf{K}'\boldsymbol{\beta} = \mathbf{m}$ lead naturally to a definition of partially-testable hypotheses: H:

$\mathbf{K}'\boldsymbol{\beta} = \mathbf{m}$ is defined as partially testable when at least one linear combination of elements of $\mathbf{K}'\boldsymbol{\beta}$ is estimable and at least one linear combination of elements of $\mathbf{K}'\boldsymbol{\beta}$ is non-estimable. In this way we have a trichotomy that categorizes all possible forms of $\mathbf{K}'\boldsymbol{\beta}$ vis-à-vis estimable and non-estimable elements.

Note that the definition of a partially-testable hypothesis demands that at least one element of $\mathbf{K}'\boldsymbol{\beta}$ be non-estimable. This being so, a partially-testable hypothesis H: $\mathbf{K}'\boldsymbol{\beta} = \mathbf{m}$ can be partitioned into

$$\text{H: } \begin{cases} \mathbf{K}_1'\boldsymbol{\beta} = \mathbf{m}_1 \\ \mathbf{K}_2'\boldsymbol{\beta} = \mathbf{m}_2 \end{cases}, \quad \text{i.e.,} \quad \text{H: } \begin{bmatrix} \mathbf{K}_1' \\ \mathbf{K}_2' \end{bmatrix}\boldsymbol{\beta} = \begin{bmatrix} \mathbf{m}_1 \\ \mathbf{m}_2 \end{bmatrix},$$

where $\mathbf{K}_1'\boldsymbol{\beta}$ is testable and $\mathbf{K}_2'\boldsymbol{\beta}$ is not. Then application of (155) indicates that the hypothesis being tested is

$$\text{H: } \begin{bmatrix} \mathbf{K}_1' \\ \mathbf{K}_2' \end{bmatrix}\mathbf{H}\boldsymbol{\beta} = \begin{bmatrix} \mathbf{m}_1 \\ \mathbf{m}_2 \end{bmatrix}, \quad \text{i.e.,} \quad \text{H: } \begin{cases} \mathbf{K}_1'\mathbf{H}\boldsymbol{\beta} = \mathbf{m}_1 \\ \mathbf{K}_2'\mathbf{H}\boldsymbol{\beta} = \mathbf{m}_2 \end{cases} \quad \text{or} \quad \text{H: } \begin{cases} \mathbf{K}_1'\boldsymbol{\beta} \phantom{\mathbf{H}} = \mathbf{m}_1 \\ \mathbf{K}_2'\mathbf{H}\boldsymbol{\beta} = \mathbf{m}_2 \end{cases}. \tag{156}$$

The last form of the hypothesis arises because estimability of $\mathbf{K}_1'\boldsymbol{\beta}$ implies $\mathbf{K}_1' = \mathbf{T}'\mathbf{X}$ for some $\mathbf{T}'$ and so $\mathbf{K}_1'\mathbf{H} = \mathbf{T}'\mathbf{X}\mathbf{G}\mathbf{X}'\mathbf{X} = \mathbf{T}'\mathbf{X} = \mathbf{K}_1'$. Thus we see that the result of testing a hypothesis that is partially testable and partially non-testable is that we are testing the testable part and the amended form $\mathbf{K}_2'\mathbf{H}\boldsymbol{\beta} = \mathbf{m}_2$ of the non-testable part $\mathbf{K}_2'\boldsymbol{\beta} = \mathbf{m}_2$. This, too, corrects errors in Searle [(1971), pp. 194–195 and exercise 7(a), p. 225].

Example 2 (continued). By means of (156), the partially-testable hypothesis

$$\text{H: } \begin{cases} \alpha_1 - \alpha_2 = 3 \\ \phantom{\alpha_1 - {}} \alpha_1 = 7 \end{cases}, \quad \text{is converted to} \quad \text{H: } \begin{cases} \alpha_1 - \alpha_2 = 3 \\ \mu + \alpha_1 = 7 \end{cases},$$

which is testable.

i. Independent and orthogonal contrasts

-i. An example (Example 4). Suppose in a 1-way classification of three observations on each of four treatments that the four treatment means are

$$\bar{y}_{1\cdot} = 2, \qquad \bar{y}_{2\cdot} = 4, \qquad \bar{y}_{3\cdot} = 6 \quad \text{and} \quad \bar{y}_{4\cdot} = 8\ . \tag{157}$$

A traditional set of what are called contrasts among these means is

$$\begin{bmatrix} 1 & -1 & 0 & 0 \\ 0 & 0 & 1 & -1 \\ 1 & 1 & -1 & -1 \end{bmatrix} \begin{bmatrix} \bar{y}_{1\cdot} \\ \bar{y}_{2\cdot} \\ \bar{y}_{3\cdot} \\ \bar{y}_{4\cdot} \end{bmatrix} = \begin{bmatrix} 2-4 \\ 6-8 \\ 6-14 \end{bmatrix} = \begin{bmatrix} -2 \\ -2 \\ -8 \end{bmatrix}. \tag{158}$$

In the context of the model $E(y_{ij}) = \mu + \alpha_i$ this is simply the BLUE of the estimable function

$$\mathbf{K}'\boldsymbol{\beta} = \begin{bmatrix} 0 & 1 & -1 & 0 & 0 \\ 0 & 0 & 0 & 1 & -1 \\ 0 & 1 & 1 & -1 & -1 \end{bmatrix} \begin{bmatrix} \mu \\ \alpha_1 \\ \alpha_2 \\ \alpha_3 \\ \alpha_4 \end{bmatrix} = \begin{bmatrix} \alpha_1 - \alpha_2 \\ \alpha_3 - \alpha_4 \\ \alpha_1 + \alpha_2 - \alpha_3 - \alpha_4 \end{bmatrix}; \tag{159}$$

i.e., (158) is

$$\text{BLUE}(\mathbf{K}'\boldsymbol{\beta}) = \mathbf{K}'\boldsymbol{\beta}^\circ = [-2 \quad -2 \quad -8]' . \tag{160}$$

-ii. The general case. We consider properties of contrasts of this nature in the context of the general linear model $E(\mathbf{y}) = \mathbf{X}\boldsymbol{\beta}$.

The F-statistic $F(\text{H}) = Q_0/s\hat{\sigma}^2$ for Q_0 of (149) is testing H: $\mathbf{K}'\boldsymbol{\beta} = \mathbf{0}$. Hence with $\mathbf{K}' = \{_c\ \mathbf{k}'_i\}_{i=1}^{i=s}$ it is testing

$$\text{H: } \{_c\ \text{H}_i\text{: } \mathbf{k}'_i\boldsymbol{\beta} = \mathbf{0}\}_{i=1}^{i=s};$$

i.e., H is the hypothesis consisting of all the H_i simultaneously; and $s = r_{\mathbf{K}}$. Suppose we individually test each H_i using $F(\text{H}_i) = q_i/\hat{\sigma}^2$ with

$$q_i = \boldsymbol{\beta}^{\circ\prime}\mathbf{k}_i(\mathbf{k}'_i\mathbf{G}\mathbf{k}_i)^{-1}\mathbf{k}'_i\boldsymbol{\beta}^\circ = \frac{(\mathbf{k}'_i\boldsymbol{\beta}^\circ)^2}{\mathbf{k}'_i\mathbf{G}\mathbf{k}_i} . \tag{161}$$

Two comments can be made about Q_0 for testing H and the q_is for testing the individual H_is.

(A) The q_is are distributed pairwise independently if and only if $\mathbf{k}'_i\mathbf{G}\mathbf{k}_j = 0$ for $i \neq j = 1, \ldots, s$. This is established by applying Theorem 7.4 to q_i and q_j. When $\mathbf{k}'_i\mathbf{G}\mathbf{k}_j = 0$ we say that $\mathbf{k}'_i\boldsymbol{\beta}$ and $\mathbf{k}'_j\boldsymbol{\beta}$ are orthogonal. With balanced data (as in Example 4), $\mathbf{G}$ and the $\mathbf{k}'_i$s usually partition so that $\mathbf{k}'_i\mathbf{G}\mathbf{k}_j = 0$ reduces to the familiar definition of orthogonal vectors, e.g., $\mathbf{h}'_i\mathbf{h}_j = 0$, where $\mathbf{h}_i$ and $\mathbf{h}_j$ are sub-vectors of $\mathbf{k}_i$ and $\mathbf{k}_j$, respectively. Thus in

Example 4 with $\mathbf{k}_1' = [0 \quad 1 \quad -1 \quad 0 \quad 0]$ from (159)

$$\mathbf{G} = \begin{bmatrix} \mathbf{0} & \mathbf{0} \\ \mathbf{0} & \frac{1}{3}\mathbf{I}_4 \end{bmatrix}, \quad \text{and} \quad \mathbf{h}_1' = [1 \quad -1 \quad 0 \quad 0] \; . \tag{162}$$

(B) If the q_is are distributed independently, they "add up" to Q_0:

$$\text{if} \quad \mathbf{k}_i'\mathbf{G}\mathbf{k}_j = 0 \quad \forall \; i \neq j, \quad \text{then} \quad Q_0 = \sum_{i=1}^{s} q_i \; . \tag{163}$$

This is so because when $\mathbf{k}_i'\mathbf{G}\mathbf{k}_j = 0$, $\mathbf{K}'\mathbf{G}\mathbf{K}$ is diagonal with diagonal elements $\mathbf{k}_i'\mathbf{G}\mathbf{k}_i$ and so Q_0 of (149) simplifies as in (163).

Example 4 (continued). From (157), $\bar{y}_{..} = 20/4 = 5$ and $\mathscr{R}(\mu_i \mid \mu) = \Sigma_{i=1}^4 3(\bar{y}_{i.} - \bar{y}_{..})^2 = 3(3^2 + 1^2 + 1^2 + 3^2) = 60$. After using $\mathbf{K}'\boldsymbol{\beta}^\circ$ of (160) and, from (159) and (162)

$$\mathbf{K}'\mathbf{G}\mathbf{K} = \begin{bmatrix} \frac{2}{3} & \cdot & \cdot \\ \cdot & \frac{2}{3} & \cdot \\ \cdot & \cdot & \frac{4}{3} \end{bmatrix},$$

$$Q_0 = (\mathbf{K}'\boldsymbol{\beta}^\circ)'(\mathbf{K}'\mathbf{G}\mathbf{K})^{-1}\mathbf{K}'\boldsymbol{\beta}^\circ = (-2)^2(\tfrac{3}{2}) + (-2)^2(\tfrac{3}{2}) + (-8)^2(\tfrac{3}{4}) = 60 \; .$$

The individual q_is are $q_1 = (-2)^2/(\frac{2}{3}) = 6$, $q_2 = (-2)^2/(\frac{2}{3}) = 6$ and $q_3 = (-8)^2/(\frac{4}{3}) = 48$ so that, since $\mathbf{k}_i'\mathbf{G}\mathbf{k}_j = 0 \quad \forall \; i \neq j$, we have $q_1 + q_2 + q_3 = 6 + 6 + 48 = 60 = Q_0$, so illustrating (163).

-iii. A necessary condition. The condition $\mathbf{k}_i'\mathbf{G}\mathbf{k}_j = 0$ in (163) for having $Q_0 = \Sigma q_i$ is written there as a sufficient condition. This is because it is sufficient, no matter what the data are, i.e., for all data sets. It is also necessary in the sense that if $Q_0 = \Sigma q_i$ *for all data sets*, then $\mathbf{k}_i'\mathbf{G}\mathbf{K}_j = 0$ for all $i \neq j$.

The proof of this (thanks to C. E. McCulloch) is as follows. Define $\boldsymbol{\alpha} = \{\alpha_i\}$ for $i = 1, \ldots, s$. Then a necessary condition stemming from the equality

$$(\mathbf{K}'\boldsymbol{\beta}^\circ)'(\mathbf{K}'\mathbf{G}\mathbf{K})^{-1}\mathbf{K}'\boldsymbol{\beta}^\circ = \sum_{i=1}^{s} \frac{(\mathbf{k}_i'\boldsymbol{\beta}^\circ)^2}{\mathbf{k}_i'\mathbf{G}\mathbf{k}_i}$$

being true for all $\mathbf{y}$ will be developed by writing $\boldsymbol{\alpha} = \mathbf{K}'\boldsymbol{\beta}^\circ$. For then, for every $\mathbf{y}$ there is an $\boldsymbol{\alpha} = \mathbf{K}'\mathbf{G}\mathbf{X}'\mathbf{y}$; and with $(\mathbf{K}'\mathbf{K})^{-1}$ existing because $\mathbf{K}'$ has full row rank there is, for any $s \times 1$ vector $\boldsymbol{\alpha}_*$, a $\mathbf{y}_* = \mathbf{X}\mathbf{K}(\mathbf{K}'\mathbf{K})^{-1}\boldsymbol{\alpha}_*$ such that, since the (tacitly presumed) estimability of $\mathbf{K}'\boldsymbol{\beta}$ implies $\mathbf{K}' = \mathbf{T}'\mathbf{X}$ for some $\mathbf{T}'$,

$$\mathbf{K}'\mathbf{G}\mathbf{X}'\mathbf{y}_* = \mathbf{T}'\mathbf{X}\mathbf{G}\mathbf{X}'\mathbf{X}\mathbf{K}(\mathbf{K}'\mathbf{K})^{-1}\boldsymbol{\alpha}_* = \mathbf{T}'\mathbf{X}\mathbf{K}(\mathbf{K}'\mathbf{K})^{-1}\boldsymbol{\alpha}_* = \mathbf{K}'\mathbf{K}(\mathbf{K}'\mathbf{K})^{-1}\boldsymbol{\alpha}_* = \boldsymbol{\alpha}_* .$$

Therefore for every $\mathbf{y}$ there is a corresponding $\boldsymbol{\alpha}$ and for every $\boldsymbol{\alpha}$ there is a corresponding $\mathbf{y}$. Hence if the equality in terms of $\boldsymbol{\beta}^\circ$ is true for all $\mathbf{y}$ then it is true for all $\boldsymbol{\alpha}$, i.e.,

$$\boldsymbol{\alpha}'(\mathbf{K}'\mathbf{G}\mathbf{K})^{-1}\boldsymbol{\alpha} = \sum_{i=1}^{s} \frac{\alpha_i^2}{\mathbf{k}_i'\mathbf{G}\mathbf{k}_i} \quad \forall\ \boldsymbol{\alpha} .$$

Now write $\boldsymbol{\alpha} = [\alpha_1 \quad \alpha_2 \quad \boldsymbol{\alpha}_0']'$ and partition $(\mathbf{K}'\mathbf{G}\mathbf{K})^{-1} = \mathbf{A}$, say, conformably for $\boldsymbol{\alpha}'(\mathbf{K}'\mathbf{G}\mathbf{K})^{-1}\boldsymbol{\alpha}$, and so write the equality as

$$[\alpha_1 \quad \alpha_2 \quad \boldsymbol{\alpha}_0'] \begin{bmatrix} a_{11} & a_{12} & \mathbf{A}_{10} \\ a_{21} & a_{22} & \mathbf{A}_{20} \\ \mathbf{A}_{10}' & \mathbf{A}_{20}' & \mathbf{A}_{00} \end{bmatrix} \begin{bmatrix} \alpha_1 \\ \alpha_2 \\ \boldsymbol{\alpha}_0 \end{bmatrix} = \frac{\alpha_1^2}{\mathbf{k}_1'\mathbf{G}\mathbf{k}_1} + \frac{\alpha_2^2}{\mathbf{k}_2'\mathbf{G}\mathbf{k}_2} + \sum_{i=3}^{s} \frac{\alpha_i^2}{\mathbf{k}_i'\mathbf{G}\mathbf{k}_i} .$$

Since this is true for all $\boldsymbol{\alpha}$, the principle of equating coefficients can be used and it yields $a_{11} = 1/\mathbf{k}_1'\mathbf{G}\mathbf{k}_1$, $a_{22} = 1/\mathbf{k}_2'\mathbf{G}\mathbf{k}_2$ and $a_{12} = 0$. This argument holds for every i, j pair with $i \neq j$. Hence $\mathbf{K}'\mathbf{G}\mathbf{K} = \mathbf{D}\{\mathbf{k}_i'\mathbf{G}\mathbf{k}_i\}$. Thus $\mathbf{k}_i'\mathbf{G}\mathbf{k}_j = 0\ \forall\ i \neq j$ is necessary for having $Q_0 = \Sigma q_i$ *for all data sets*.

The necessity condition is a condition resulting from requiring that $Q_0 = \Sigma q_i$ be true for all data sets. But that does not preclude the possibility of there being some data sets for which $Q_0 = \Sigma q_i$ without having $\mathbf{k}_i'\mathbf{G}\mathbf{k}_j = 0\ \forall\ i \neq j$. In such cases, the necessity condition obviously does not hold, because it *is* necessary only if $Q_0 = \Sigma q_i$ for *all data sets*. This means that for some $\mathbf{k}_i$s for which $\mathbf{k}_i'\mathbf{G}\mathbf{k}_j = 0\ \forall\ i \neq j$ does not hold, there may be some data sets for which $Q_0 = \Sigma q_i$; i.e., we may have q_is that are not independent, but which nevertheless "add up".

Example 4 (continued). With $\boldsymbol{\beta}^\circ = [0 \quad 2 \quad 4 \quad 6 \quad 8]'$, for

$$\text{H:} \begin{cases} a_1 - a_4 = 0 \\ a_2 - a_3 = 0, \\ a_1 - 2a_2 + a_3 = 0 \end{cases} \text{i.e., H:} \begin{bmatrix} 0 & 1 & 0 & 0 & -1 \\ 0 & 0 & 1 & -1 & 0 \\ 0 & 1 & -2 & 1 & 0 \end{bmatrix} \begin{bmatrix} \mu \\ a_1 \\ a_2 \\ a_3 \\ a_4 \end{bmatrix} = \mathbf{0},$$

$$\mathbf{K}'\boldsymbol{\beta}^\circ = \begin{bmatrix} 2-8 \\ 4-6 \\ 2-8+6 \end{bmatrix} = \begin{bmatrix} -6 \\ -2 \\ 0 \end{bmatrix} \quad \text{and}$$

$$(\mathbf{K}'\mathbf{G}\mathbf{K})^{-1} = \left[\tfrac{1}{3}\begin{pmatrix} 2 & 0 & 1 \\ 0 & 2 & -3 \\ 1 & -3 & 6 \end{pmatrix}\right]^{-1} = \tfrac{3}{4}\begin{bmatrix} 3 & -3 & -2 \\ -3 & 11 & 6 \\ -2 & 6 & 4 \end{bmatrix} . \quad (164)$$

Hence

$$Q_0 = [-6 \quad -2 \quad 0]\tfrac{3}{4}\begin{bmatrix} 3 & -3 & -2 \\ -3 & 11 & 6 \\ -2 & 6 & 4 \end{bmatrix}\begin{bmatrix} -6 \\ -2 \\ 0 \end{bmatrix} = 60 \ .$$

Clearly, from $\mathbf{K'GK}$ given in (164), the q_is are not independent, and yet, using the diagonal elements of $\mathbf{K'GK}$,

$$q_1 = (-6)^2/(\tfrac{2}{3}) = 54, \qquad q_2 = (-2)^2/(\tfrac{2}{3}) = 6, \qquad q_3 = 0^2/2 = 0$$

and $$q_1 + q_2 + q_3 = 54 + 6 + 0 = 60 = Q_0 \ .$$

-iv. Contrasts are linearly independent of the mean. Hypotheses about contrasts always have $\mathbf{K'1} = \mathbf{0}$ and $\mathbf{m} = \mathbf{0}$. Therefore, when $r_{\mathbf{K}} = r_{\mathbf{X}} - 1$ as at the end of Subsection c, $Q = R(\boldsymbol{\beta}) - R(\mu)$ when $\mathbf{m} = \mathbf{0}$. This is evident in the example, wherein for any values used for a, b and c in

$$\mathbf{K'} = \begin{bmatrix} 0 & 1 & -1 & 0 & 0 \\ 0 & 0 & 0 & 1 & -1 \\ 0 & a & b & c & -(a+b+c) \end{bmatrix},$$

the value of $Q_0 = (\mathbf{K'}\boldsymbol{\beta}^\circ)(\mathbf{K'GK})^{-1}\mathbf{K'}\boldsymbol{\beta}^\circ$ is always 60. Thus, on defining $\lambda = a^2 + b^2 + c^2 + (a+b+c)^2$, it will be found that

$$Q_0 = [-2 \quad -2 \quad -2(3a+2b+c)]$$

$$\times \left[\tfrac{1}{3}\begin{pmatrix} 2 & 0 & a-b \\ 0 & 2 & a+b+2c \\ a-b & a+b+2c & \lambda \end{pmatrix}\right]^{-1}\begin{bmatrix} -2 \\ -2 \\ -2(3a+2b+c) \end{bmatrix}$$

reduces to

$$Q_0 = 3(80)(a+b)^2/4(a+b)^2 = 60 \ .$$

8.9. RESTRICTED MODELS

The most straightforward models are those of full rank, where each parameter can be estimated. But sometimes overparameterization seems unavoidable—such as in Section 4.9a. The difficulties arising from overparameterization can, as we have just seen, be avoided by the use of estimable functions; or by defining new parameters that are, in effect, estimable functions of the parameters in the overparameterized model. Yet one more alternative, which has long been used with balanced data, is to implicitly define new parameters through imposing restrictions on the parameters to be used. Thus in Example 2, which is the 1-way classification with 3 classes, we might implicitly define

$$\dot{\mu} = \tfrac{1}{3}[(\mu + \alpha_1) + (\mu + \alpha_2) + (\mu + \alpha_3)] = \mu + \tfrac{1}{3}(\alpha_1 + \alpha_2 + \alpha_3)$$

and $$\dot{\alpha}_i = \mu + \alpha_i - \dot{\mu} = \alpha_i - \tfrac{1}{3}(\alpha_1 + \alpha_2 + \alpha_3)$$

and as a result have the restriction

$$\sum_{i=1}^{3} \dot{\alpha}_i = 0\ . \tag{165}$$

The model is then

$$E(y_{ij}) = \dot{\mu} + \dot{\alpha}_i \quad \text{and} \quad \sum_{i=1}^{3} \dot{\alpha}_i = 0\ . \tag{166}$$

This is what we call a *restricted model*, with (165) being the restriction. It is an example of what are nowadays called the *Σ-restrictions*.

The Σ-restrictions have been widely used with balanced data for many years. In that context they are useful, easy to use, and successfully overcome the problem of overparameterization. But they are not so useful with unbalanced data—especially when some subclasses of a data structure are empty, as is shown in Section 9.1c. For developing those details some general results are described.

a. Using restrictions explicitly

The easiest way of using the restriction $\Sigma_i \dot{\alpha}_i = 0$ in the model (166) is to write the restriction as

$$\dot{\alpha}_3 = -\dot{\alpha}_1 - \dot{\alpha}_2 \tag{167}$$

and use this explicitly in the model equations so that they contain no terms $\dot{\alpha}_3$. Thus, in Example 2, the model equations of (21) would become

$$\mathbf{y} = \begin{bmatrix} 74 \\ 68 \\ 77 \\ 76 \\ 80 \\ 87 \\ 91 \end{bmatrix} = \begin{bmatrix} 1 & 1 & \cdot \\ 1 & 1 & \cdot \\ 1 & 1 & \cdot \\ 1 & \cdot & 1 \\ 1 & \cdot & 1 \\ 1 & -1 & -1 \\ 1 & -1 & -1 \end{bmatrix} \begin{bmatrix} \dot{\mu} \\ \dot{\alpha}_1 \\ \dot{\alpha}_2 \end{bmatrix} + \mathbf{e}, \tag{168}$$

This procedure is sometimes called "regression on dummy variables", the dummy variable being the 0s, 1s and -1s of the matrix $\mathbf{X}$ like that in (168). In this example the corresponding normal equations are

$$\begin{bmatrix} 7 & 1 & 0 \\ 1 & 5 & 2 \\ 0 & -2 & 4 \end{bmatrix} \begin{bmatrix} \hat{\mu} \\ \hat{\alpha}_1 \\ \hat{\alpha}_2 \end{bmatrix} = \begin{bmatrix} 553 \\ 41 \\ -22 \end{bmatrix} \quad \text{with solution} \quad \begin{bmatrix} \hat{\mu} \\ \hat{\alpha}_1 \\ \hat{\alpha}_2 \end{bmatrix} = \begin{bmatrix} 80 \\ -7 \\ -2 \end{bmatrix} . \qquad (169)$$

Then from (167), $\hat{\alpha}_3 = -(-7-2) = 9$.

Notice, of course, that the solution (169) is, along with $\hat{\alpha}_3 = 9$, just a solution to the normal equations (23) of the unrestricted model; and it yields BLUEs of estimable functions just as does any other solution.

In general, if, after making substitutions from the restrictions [e.g., (167)] into the model equations, we write those amended model equations [e.g., (168)] as $E(\mathbf{y}) = \mathbf{X}_R\boldsymbol{\beta}_R$, then $\mathbf{X}_R$ is of full column rank $r_{\mathbf{X}}$ and the resulting normal equations have but the single solution

$$\hat{\boldsymbol{\beta}}_R = (\mathbf{X}'_R\mathbf{X}_R)^{-1}\mathbf{X}'_R\mathbf{y} . \qquad (170)$$

From here on, application of general results is easy. All elements of $\boldsymbol{\beta}_R$ are estimable, with their BLUEs given by (170); and estimation of functions of elements of $\boldsymbol{\beta}_R$ and testing of hypotheses can proceed in a quite straightforward manner. And, in the example, from (167) we estimate $\dot{\alpha}_3$ and then have $\hat{\alpha}_1 + \hat{\alpha}_2 + \hat{\alpha}_3 = 0$.

The only difficulty with the preceding method of using restrictions is that using them in the manner of (167) and (168) is not always as easy as it appears there. In situations of two or more factors, with interactions and covariates, the procedure can be quite complicated, and even though it can be computerized the meanings to be associated with the outcome of that are not always clear. We therefore need a more generalized method.

b. General methodology

The general restricted model (with linear restrictions) is taken as

$$\mathbf{y} = \mathbf{X}\boldsymbol{\beta} + \mathbf{e} \quad \text{and} \quad \mathbf{P}'\boldsymbol{\beta} = \boldsymbol{\delta} \qquad (171)$$

where $\mathbf{P}'\boldsymbol{\beta} = \boldsymbol{\delta}$ are the restrictions; $\mathbf{P}'$ is assumed to be of full row rank and $\boldsymbol{\delta}$ is often null but does not have to be. Using a vector of Lagrange multipliers, $2\boldsymbol{\theta}$, minimization of $(\mathbf{y} - \mathbf{X}\boldsymbol{\beta})'(\mathbf{y} - \mathbf{X}\boldsymbol{\beta})$ subject to $\mathbf{P}'\boldsymbol{\beta} = \boldsymbol{\delta}$ leads to equations

$$\mathbf{X}'\mathbf{X}\boldsymbol{\beta}_r^{\circ} + \mathbf{P}\boldsymbol{\theta} = \mathbf{X}'\mathbf{y} \qquad (172)$$

and

$$\mathbf{P}'\boldsymbol{\beta}_r^{\circ} = \boldsymbol{\delta} . \qquad (173)$$

In solving (172) and (173) we distinguish two cases: (i) when $\mathbf{P}'\boldsymbol{\beta}$ of the restrictions is not estimable and (ii) when it is estimable.

c. Non-estimable restrictions

Restrictions $\mathbf{P}'\boldsymbol{\beta} = \boldsymbol{\delta}$ are defined as non-estimable when $\mathbf{P}'\boldsymbol{\beta}$ is non-estimable in the same sense that $\mathbf{K}'\boldsymbol{\beta}$ is non-estimable for a non-testable hypothesis (see Section 8.8g), namely every linear combination of elements of $\mathbf{P}'\boldsymbol{\beta}$ is non-estimable.

-i. A solution vector. A solution for $\boldsymbol{\beta}_r$ to (172) and (173) is obtained in this case simply by selecting a particular solution of the normal equations $\mathbf{X}'\mathbf{X}\boldsymbol{\beta}^\circ = \mathbf{X}'\mathbf{y}$ of the unrestricted model. To see this, start with the general form of solution from (23) of Section 7.2a, namely

$$\boldsymbol{\beta}^{\circ\circ} = \mathbf{G}\mathbf{X}'\mathbf{y} + (\mathbf{I} - \mathbf{H})\mathbf{z} = \boldsymbol{\beta}^\circ + (\mathbf{I} - \mathbf{H})\mathbf{z} \tag{174}$$

for $\mathbf{H} = \mathbf{G}\mathbf{X}'\mathbf{X}$ and for $\mathbf{z}$ being any arbitrary vector. Then consider the equations

$$\mathbf{P}'(\mathbf{I} - \mathbf{H})\mathbf{w} = \boldsymbol{\delta} - \mathbf{P}'\mathbf{G}\mathbf{X}'\mathbf{y}\ . \tag{175}$$

These represent $p - r_\mathbf{X}$ (at most) equations in $p - r_\mathbf{X}$ terms and so (Searle, 1984b) they have a solution.

From (9) of Section 7.1i, the rank of $\mathbf{I} - \mathbf{H}$ is $p - r_\mathbf{X}$ and so $(\mathbf{I} - \mathbf{H})\mathbf{w}$ has $p - r_\mathbf{X}$ linearly independent elements. Also $\mathbf{P}'$, of full row rank, has no more than $p - r_\mathbf{X}$ rows.

Using a solution for $\mathbf{w}$ obtained from (175) as the value for $\mathbf{z}$ in (174) gives a solution (along with $\boldsymbol{\theta} = \mathbf{0}$) to (172) and (173):

$$\boldsymbol{\beta}_r^\circ = \boldsymbol{\beta}^\circ + (\mathbf{I} - \mathbf{H})[\mathbf{P}'(\mathbf{I} - \mathbf{H})]^-(\boldsymbol{\delta} - \mathbf{P}'\boldsymbol{\beta}^\circ)\ . \tag{176}$$

Example 2 (continued). We use $\boldsymbol{\beta}^\circ$ and $\mathbf{G}$ of (148) and $\mathbf{H}$ of (50):

$$\boldsymbol{\beta}^\circ = \begin{bmatrix} 89 \\ -16 \\ -11 \\ 0 \end{bmatrix}, \quad \mathbf{G} = \tfrac{1}{6}\begin{bmatrix} 3 & -3 & -3 & 0 \\ -3 & 5 & 3 & 0 \\ -3 & 3 & 6 & 0 \\ 0 & 0 & 0 & 0 \end{bmatrix}$$

$$\mathbf{H} = \begin{bmatrix} 1 & \cdot & \cdot & 1 \\ \cdot & 1 & \cdot & -1 \\ \cdot & \cdot & 1 & -1 \\ \cdot & \cdot & \cdot & \cdot \end{bmatrix}.$$

Then the restriction $\dot{\alpha}_1 + \dot{\alpha}_2 + \dot{\alpha}_3 = 0$ has $\mathbf{P}' = [0 \quad 1 \quad 1 \quad 1]$ and $\boldsymbol{\delta} = 0$, so that for (176)

$$\mathbf{P}'(\mathbf{I}-\mathbf{H}) = [0 \quad 1 \quad 1 \quad 1]\begin{bmatrix} \cdot & \cdot & \cdot & -1 \\ \cdot & \cdot & \cdot & 1 \\ \cdot & \cdot & \cdot & 1 \\ \cdot & \cdot & \cdot & 1 \end{bmatrix} = [0 \quad 0 \quad 0 \quad 3]$$

and

$$\mathbf{P}'\boldsymbol{\beta}^\circ = [0 \quad 1 \quad 1 \quad 1]\boldsymbol{\beta}^\circ = -16 + 11 + 0 = -27 .$$

Hence (176) is

$$\boldsymbol{\beta}_r^\circ = \boldsymbol{\beta}^\circ + \begin{bmatrix} \cdot & \cdot & \cdot & -1 \\ \cdot & \cdot & \cdot & 1 \\ \cdot & \cdot & \cdot & 1 \\ \cdot & \cdot & \cdot & 1 \end{bmatrix}\begin{bmatrix} 0 \\ 0 \\ 0 \\ \frac{1}{3} \end{bmatrix}(27) = \boldsymbol{\beta}^\circ + \begin{bmatrix} -9 \\ 9 \\ 9 \\ 9 \end{bmatrix} = \begin{bmatrix} 80 \\ -7 \\ -2 \\ 9 \end{bmatrix} . \quad (177)$$

An important consequence of (176) is that $\boldsymbol{\beta}_r^\circ$ is just one of the solutions of the normal equations $\mathbf{X}'\mathbf{X}\boldsymbol{\beta}^\circ = \mathbf{X}'\mathbf{y}$ of the unrestricted model—as seen in the example, and because (176) is just (174) with $\mathbf{z}$ being given a value of $\mathbf{w}$ that satisfies (175). Hence $\boldsymbol{\beta}_r^\circ$ is just one form of $\boldsymbol{\beta}^\circ$. Therefore, because the methods already described for estimating estimable functions, for calculating $\hat{\sigma}^2$, and for testing hypotheses are invariant to whatever solution is used for $\boldsymbol{\beta}^\circ$, these methods are unaffected by using $\boldsymbol{\beta}_r^\circ$ as that solution; e.g., SSE_r in the restricted model is SSE of the unrestricted model, $\mathrm{SSE}_r = \mathrm{SSE}$.

That $\boldsymbol{\beta}_r^\circ$ is just a form of $\boldsymbol{\beta}^\circ$ has been demonstrated specifically by Searle *et al.* (1984), who write (176) as

$$\boldsymbol{\beta}_r^\circ = \mathbf{G}_r\mathbf{X}'\mathbf{y} + (\mathbf{I} - \mathbf{H}_r)\mathbf{z}$$

for $$\mathbf{z} = (\mathbf{I}-\mathbf{H})[\mathbf{P}'(\mathbf{I}-\mathbf{H})]^-\boldsymbol{\delta} \quad \text{and} \quad \mathbf{H}_r = \mathbf{I} - \mathbf{G}_r\mathbf{X}'\mathbf{X}$$

with $$\mathbf{G}_r = \mathbf{G} - (\mathbf{I}-\mathbf{H})[\mathbf{P}'(\mathbf{I}-\mathbf{H})]^-\mathbf{P}'\mathbf{G} .$$

Then $\mathbf{G}_r$ is a generalized inverse of $\mathbf{X}'\mathbf{X}$ that yields a solution of the normal equations that satisfy the restrictions, i.e., $\mathbf{P}'\boldsymbol{\beta}_r^\circ = \boldsymbol{\delta}$ (see Searle, 1984b). That $\mathbf{P}'\boldsymbol{\beta}_r^\circ$ does equal $\boldsymbol{\delta}$ is easily confirmed from (175) and (176); confirmation using $\mathbf{G}_r$ utilizes (175) and $\mathbf{H}_r\mathbf{H} = \mathbf{H}_r$.

-ii. Estimable functions. With restricted models, the form of function that is estimable is affected by the form of the restrictions. When restrictions are in terms of non-estimable functions, some functions that are non-estimable in the unrestricted model will become estimable in the restricted model. We begin with an example.

Example 2 (continued). With $\boldsymbol{\beta}^\circ$ as before, the estimable function $\mu + \frac{1}{3}(\alpha_1 + \alpha_2 + \alpha_3)$ has

$$\begin{aligned}\text{BLUE}\left[\mu + \tfrac{1}{3}(\alpha_1 + \alpha_2 + \alpha_3)\right] &= \mu^\circ + \tfrac{1}{3}(\alpha_1^\circ + \alpha_2^\circ + \alpha_3^\circ)\\ &= 0 + \tfrac{1}{3}(89 - 16 - 11) = 20\tfrac{2}{3}\ .\end{aligned}$$

Consider the restriction $\alpha_1 + \alpha_2 + \alpha_3 = 0$. It changes the estimable function $\mu + \frac{1}{3}(\alpha_1 + \alpha_2 + \alpha_3)$ with BLUE $20\frac{2}{3}$ to be the estimable function μ with BLUE $20\frac{2}{3}$. Thus we have

$$\text{in the } \Sigma\text{-restricted model, } \mu \text{ is estimable with BLUE}(\mu) = 20\tfrac{2}{3}. \qquad (178)$$

Notice how different (178) is from the unrestricted model, wherein μ is not estimable. In the Σ-restricted model μ is estimable. There is nothing inconsistent about these statements because they are statements about different models. Emphasis to this is given by using symbols $\dot{\mu}$ and $\dot{\alpha}_i$ in restricted models, as in (167)–(169). Indeed since there are restrictions other than the Σ-restrictions, notation such as $\dot{\mu}_\Sigma$ and $\dot{\alpha}_{i,\Sigma}$ clarifies the situation still more. Then (178) can be written succinctly as

$$\dot{\mu}_\Sigma \quad \text{is estimable with} \quad \hat{\dot{\mu}}_\Sigma = 20\tfrac{2}{3}\ . \qquad (179)$$

The generalization of the example is that, with $\mathbf{q}'\boldsymbol{\beta}$ being an estimable function in the unrestricted model, then in the restricted model $\mathbf{y} = \mathbf{XB} + \mathbf{e}$ and $\mathbf{P}'\boldsymbol{\beta} = \boldsymbol{\delta}$ with non-estimable $\mathbf{P}'\boldsymbol{\beta}$, an estimable function is

$$\mathbf{q}'\boldsymbol{\beta} + \boldsymbol{\lambda}'(\mathbf{P}'\boldsymbol{\beta} - \boldsymbol{\delta}) \text{ for } \boldsymbol{\lambda}' \text{ such that } \boldsymbol{\lambda}'\boldsymbol{\delta} = \mathbf{0};\ \text{any } \boldsymbol{\lambda}' \text{ if } \boldsymbol{\delta} = \mathbf{0}\ . \qquad (180)$$

$\boldsymbol{\lambda}'$ in (180) can be null; i.e., functions $\mathbf{q}'\boldsymbol{\beta}$ that are estimable in the unrestricted model are also estimable in the restricted model. When $\boldsymbol{\lambda}'$ in (180) is not null the resulting function, although estimable in the restricted model, will generally be a function that is not estimable in the unrestricted model. The μ in the example is just such a case.

-iii. Hypothesis testing. In having $\boldsymbol{\beta}_r^\circ$ as a $\boldsymbol{\beta}^\circ$ and hence $\text{SSE}_r = \text{SSE}$, we also have, for any testable hypothesis, $\text{SSE}_{r,\text{H}} = \text{SSE}_\text{H}$. Therefore the F-statistic is the same in the restricted model as in the unrestricted model. SSE and SEE_H can be calculated in the usual way and so, therefore, can

$$Q = \text{SSE}_\text{H} - \text{SSE} = (\mathbf{K}'\boldsymbol{\beta}^\circ - \mathbf{m})'(\mathbf{K}'\mathbf{GK})^{-1}(\mathbf{K}'\boldsymbol{\beta}^\circ - \mathbf{m})$$

from (144). To whatever extent $\mathbf{K}'\boldsymbol{\beta}$ is affected by $\mathbf{P}'\boldsymbol{\beta} = \boldsymbol{\delta}$, the hypothesis will be similarly affected—in the same way that $\mu + \frac{1}{3}(\alpha_1 + \alpha_2 + \alpha_3)$ became $\dot{\mu}_\Sigma$ in the example. In general, for any H: $\mathbf{K}'\boldsymbol{\beta} = \mathbf{m}$ that is testable in the unrestricted model, H: $(\mathbf{K}' + \mathbf{L}\mathbf{P}')\boldsymbol{\beta} = \mathbf{m} + \mathbf{L}\boldsymbol{\delta}$ is testable in the restricted model—for any $\mathbf{L}$ of order $r_\mathbf{K} \times p$. Conversely, any non-testable hypothesis in the unrestricted model will be testable in the restricted model only if it can be put in the form H: $(\mathbf{K}' + \mathbf{L}\mathbf{P}')\boldsymbol{\beta} = \mathbf{m} + \mathbf{L}\boldsymbol{\delta}$ for some $\mathbf{L}$.

Restrictions are often used as a method of converting a model that is not of full rank to one that is of full rank. Then all its elements, e.g., $\dot{\mu}$, $\dot{\alpha}_1$, $\dot{\alpha}_2$ and $\dot{\alpha}_3$ in the example, are estimable, as are all linear functions of them; and all hypotheses about such functions are testable. Furthermore, in terms of writing this full rank, restricted model as $E(\mathbf{y}) = \dot{\mathbf{X}}_1\dot{\boldsymbol{\beta}}_1 + \dot{\mathbf{X}}_2\dot{\boldsymbol{\beta}}_2$, we know from Section 8.6e that $R(\dot{\boldsymbol{\beta}}_2 \,|\, \dot{\boldsymbol{\beta}}_1)$ tests H: $\dot{\boldsymbol{\beta}}_2 = \mathbf{0}$. Hence in the example we have

$$R(\dot{\mu} \,|\, \dot{\boldsymbol{\alpha}})_\Sigma \quad \text{tests} \quad \text{H: } \dot{\mu} = 0,$$

where by $R(\dot{\mu} \,|\, \dot{\boldsymbol{\alpha}})_\Sigma$ we mean $R(\dot{\mu}, \dot{\boldsymbol{\alpha}})_\Sigma - R(\dot{\boldsymbol{\alpha}})_\Sigma$ for $\dot{\mu}$ and $\dot{\boldsymbol{\alpha}}$ being the terms in the Σ-restricted model. With $\dot{\mu} = \mu + \frac{1}{3}(\alpha_1 + \alpha_2 + \alpha_3)$ in the example, as shown just prior to (165), we can generalize and so have, for the 1-way classification, that

$$R(\dot{\mu} \,|\, \dot{\boldsymbol{\alpha}})_\Sigma \quad \text{tests} \quad \text{H: } \mu + \Sigma_i \alpha_i / a = 0,$$

which is also H: $\Sigma_i \mu_i / \text{a} = 0$. This contrasts with

$$R(\mu) = N\bar{y}_{..}^2 \quad \text{testing} \quad \text{H: } \mu + \Sigma_i n_i \alpha_i / N = 0 \ .$$

Moreover, since

$$R(\mu \,|\, \alpha) = R(\mu, \alpha) - R(\alpha) = \Sigma_i n_i \bar{y}_{i\cdot}^2 - \Sigma_i n_i \bar{y}_{i\cdot}^2 = 0,$$

we also have
$$R(\dot{\mu} \,|\, \dot{\alpha})_\Sigma \neq R(\mu \,|\, \alpha) \ .$$

Furthermore, if in (87) of Section 2.12b-i we use $k_i = 1/a$ and $m = 0$ we can conclude that

$$R(\dot{\mu} \,|\, \dot{\alpha})_\Sigma = (\Sigma_i \bar{y}_{i\cdot})^2 / \Sigma_i (1/n) \ .$$

d. Estimable restrictions

-i. A solution vector. When $\mathbf{P}'\boldsymbol{\beta}$ is estimable, $\mathbf{P}' = \mathbf{T}'\mathbf{X}$ for some $\mathbf{T}'$ and so $\mathbf{P}'(\mathbf{I} - \mathbf{H}) = \mathbf{T}'(\mathbf{X} - \mathbf{X}\mathbf{G}\mathbf{X}'\mathbf{X}) = \mathbf{0}$. Hence (175) does not exist. Therefore (172) and (173) have to be solved other than by using (175).

From (172) comes $\boldsymbol{\beta}_r^\circ = \mathbf{G}(\mathbf{X}'\mathbf{y} - \mathbf{P}\boldsymbol{\theta})$ and using (173) gives

$$\boldsymbol{\theta} = (\mathbf{P}'\mathbf{G}\mathbf{P})^{-1}(\mathbf{P}'\boldsymbol{\beta}^\circ - \boldsymbol{\delta})$$

and thus
$$\boldsymbol{\beta}_r^\circ = \boldsymbol{\beta}^\circ - \mathbf{G}\mathbf{P}(\mathbf{P}'\mathbf{G}\mathbf{P})^{-1}(\mathbf{P}'\boldsymbol{\beta}^\circ - \boldsymbol{\delta})\ . \tag{181}$$

This is exactly the same form of solution as given in (150) for $\boldsymbol{\beta}_{\mathrm{H}}^\circ$, the solution vector under the hypothesis H: $\mathbf{K}'\boldsymbol{\beta} = \mathbf{m}$ in the unrestricted model, where $\mathbf{K}'$ and $\mathbf{m}$, respectively, play the same roles there as $\mathbf{P}$ and $\boldsymbol{\delta}$ do here. In contrast to the case of non-estimable restrictions, notice with estimable restrictions that $\boldsymbol{\beta}_r^\circ$ of (181) is not a solution of the normal equations $\mathbf{X}'\mathbf{X}\boldsymbol{\beta}^\circ = \mathbf{X}'\mathbf{y}$ for the unrestricted model. Also, from (152), replacing $\mathbf{K}$ and $\mathbf{m}$ by $\mathbf{P}$ and $\boldsymbol{\delta}$, respectively, the error sum of squares in the restricted model is not SSE, but is

$$\mathrm{SSE}_r = \mathrm{SSE} + (\mathbf{P}'\boldsymbol{\beta}^\circ - \boldsymbol{\delta})'(\mathbf{P}'\mathbf{G}\mathbf{P})^{-1}(\mathbf{P}'\boldsymbol{\beta}^\circ - \boldsymbol{\delta}) \tag{182}$$

with expected value, similar to (153),

$$E(\mathrm{SSE}_r) = (N - r_{\mathbf{X}} + r_{\mathbf{P}})\sigma_r^2 \quad \text{yielding} \quad \hat{\sigma}_r^2 = \frac{\mathrm{SSE}_r}{N - r_{\mathbf{X}} + r_{\mathbf{P}}} \tag{183}$$

as an unbiased estimator of σ_r^2. Note too, that whereas in the case of non-estimable restrictions $\mathbf{P}'$ usually has rank $p - r_{\mathbf{X}}$, in the case of estimable restrictions the rank can be any value not exceeding $r_{\mathbf{X}}$.

-ii. ***Estimable functions.*** With estimable restrictions the form of estimable function in the restricted model is the same as in the case of non-estimable restrictions, given in (180). But now, because $\mathbf{P}'\boldsymbol{\beta}$ is estimable so is every $\mathbf{q}'\boldsymbol{\beta} + \boldsymbol{\lambda}'\mathbf{P}'\boldsymbol{\beta}$ for estimable $\mathbf{q}'\boldsymbol{\beta}$. Therefore estimable functions in a restricted model having estimable restrictions are the same as in the unrestricted model.

-iii. ***Hypothesis testing.*** The form of estimable functions is not affected by using estimable restrictions. But whereas hypothesis testing when using non-estimable restrictions is the same as in the unrestricted model, with estimable restrictions it is different. With restrictions $\mathbf{P}'\boldsymbol{\beta} = \boldsymbol{\delta}$ and $\mathbf{P}'\boldsymbol{\beta}$ estimable, the hypothesis H_r: $\mathbf{K}'\boldsymbol{\beta} = \mathbf{m}$ can be tested only if $\mathbf{K}'$ has the same limitations as previously ($\mathbf{K}'\boldsymbol{\beta}$ estimable and $\mathbf{K}'$ of full row rank) and also only if rows of $\mathbf{K}'$ are linearly independent of those of $\mathbf{P}'$. (One could also formulate a situation where some rows of $\mathbf{K}'$ are linear combinations of those of $\mathbf{P}'$—but it seems of little practical interest.) Under the hypothesis,

and with the restrictions, on defining

$$\mathbf{Q}' = \begin{bmatrix} \mathbf{P}' \\ \mathbf{K}' \end{bmatrix} \quad \text{and} \quad \ell = \begin{bmatrix} \boldsymbol{\delta} \\ \mathbf{m} \end{bmatrix}$$

the error sum of squares is, similar to (182),

$$\text{SSE}_{r,\text{H}} = \text{SSE} + (\mathbf{Q}'\boldsymbol{\beta}^\circ - \ell)'(\mathbf{Q}'\mathbf{G}\mathbf{Q})^{-1}(\mathbf{Q}'\boldsymbol{\beta}^\circ - \ell)\ .$$

Hence for testing H: $\mathbf{K}'\boldsymbol{\beta} = \mathbf{m}$ in the presence of estimable restrictions the F-statistic is

$$F(\text{H}_r) = \frac{\text{SSE}_{r,\text{H}} - \text{SSE}_r}{r_{\mathbf{K}}\hat{\sigma}_r^2},$$

for $\hat{\sigma}_r^2$ given by (183) and where

$$\text{SSE}_{r,\text{H}} - \text{SSE}_r = (\mathbf{Q}'\boldsymbol{\beta}^\circ - \ell)'(\mathbf{Q}'\mathbf{G}\mathbf{Q})^{-1}(\mathbf{Q}'\boldsymbol{\beta}^\circ - \ell)$$
$$-(\mathbf{P}'\boldsymbol{\beta}^\circ - \boldsymbol{\delta})'(\mathbf{P}'\mathbf{G}\mathbf{P})^{-1}(\mathbf{P}'\boldsymbol{\beta}^\circ - \boldsymbol{\delta})\ . \qquad (184)$$

The negative term in (184) [erroneously omitted from early printings of Searle (1971, see p. 210)] is SSE − SSE_r, available from (182). The need for it was pointed out by Timm and Carlson (1973). A more readily computable form of (184), obtained by using a partitioned inverse for $(\mathbf{Q}'\mathbf{G}\mathbf{Q})^{-1}$, is

$$\text{SSE}_{r,\text{H}} - \text{SSE}_r$$
$$= (\mathbf{K}'\boldsymbol{\beta}_r^\circ - \mathbf{m})'\left[\mathbf{K}'\mathbf{G}\mathbf{K} - \mathbf{K}'\mathbf{G}\mathbf{P}(\mathbf{P}'\mathbf{G}\mathbf{P})^{-1}\mathbf{P}'\mathbf{G}\mathbf{K}\right]^{-1}(\mathbf{K}'\boldsymbol{\beta}_r^\circ - \mathbf{m}) \quad (185)$$

for $\boldsymbol{\beta}_r^\circ$ of (181). Finally, estimation under the hypothesis, when there are estimable restrictions, gives

$$\boldsymbol{\beta}^\circ_{\text{H},r} = \boldsymbol{\beta}^\circ - \mathbf{G}\mathbf{Q}(\mathbf{Q}'\mathbf{G}\mathbf{Q})^{-1}(\mathbf{Q}'\boldsymbol{\beta}^\circ - \ell)\ . \qquad (186)$$

e. The full rank model

When $\mathbf{X}'\mathbf{X}$ is nonsingular, the normal equations have solution $\hat{\boldsymbol{\beta}} = (\mathbf{X}'\mathbf{X})^{-1}\mathbf{X}'\mathbf{y}$, and every element of $\hat{\boldsymbol{\beta}}$ is the BLUE of the corresponding element of $\boldsymbol{\beta}$. Under these circumstances there is no need to impose restrictions on the elements of $\boldsymbol{\beta}$ for the sake of reparameterization in order to derive a full rank model. That already exists. Nevertheless, if the situation is such that it is appropriate to have restrictions on parameters of the (already full rank) model, then the methods already described for estimable restrictions are available. One just uses $(\mathbf{X}'\mathbf{X})^{-1}$ in place of $\mathbf{G}$ in (181).

Mantell (1973) considers this more directly by partitioning the restrictions (which are now taken as $\mathbf{R}\boldsymbol{\beta} = \boldsymbol{\delta}$) and the model conformably as

$$\mathbf{R}_1\boldsymbol{\beta}_1 + \mathbf{R}_2\boldsymbol{\beta}_2 = \boldsymbol{\delta} \quad \text{and} \quad \mathbf{y} = \mathbf{X}_1\boldsymbol{\beta}_1 + \mathbf{X}_2\boldsymbol{\beta}_2 + \mathbf{e} \tag{187}$$

where $\mathbf{R}_1^{-1}$ exists. Then

$$\boldsymbol{\beta}_1 = \mathbf{R}_1^{-1}(\boldsymbol{\delta} - \mathbf{R}_2\boldsymbol{\beta}_2) \tag{188}$$

and

$$\mathbf{y} = \mathbf{X}_1\mathbf{R}_1^{-1}\boldsymbol{\delta} + (\mathbf{X}_2 - \mathbf{X}_1\mathbf{R}_1^{-1}\mathbf{R}_2)\boldsymbol{\beta}_2 + \mathbf{e}$$

can be written as

$$\mathbf{y} - \mathbf{X}_1\mathbf{R}_1^{-1}\boldsymbol{\delta} = \mathbf{S}\boldsymbol{\beta}_2 + \mathbf{e} \quad \text{for } \mathbf{S} = \mathbf{X}_2 - \mathbf{X}_1\mathbf{R}_1^{-1}\mathbf{R}_2, \tag{189}$$

so that the estimator of $\boldsymbol{\beta}$ is

$$\hat{\boldsymbol{\beta}} = \begin{bmatrix} \hat{\boldsymbol{\beta}}_1 = \mathbf{R}_1^{-1}(\boldsymbol{\delta} - \mathbf{R}_2\hat{\boldsymbol{\beta}}_2) \\ \hat{\boldsymbol{\beta}}_2 = (\mathbf{S}'\mathbf{S})^{-1}\mathbf{S}'(\mathbf{y} - \mathbf{X}_1\mathbf{R}_1^{-1}\boldsymbol{\delta}) \end{bmatrix}. \tag{190}$$

It is left to the reader (as E8.26) to show that (190) is identical to (181). In the non full rank case, Bittner (1974) shows that $\hat{\boldsymbol{\beta}}$ is unique provided $[\mathbf{R} \quad \mathbf{X}']'$ has full column rank. This results from the fact that

$$[\mathbf{R} \quad \mathbf{X}']' \text{ has full column rank } p \text{ if, and only if, } r_{\mathbf{S}} = p - r_{\mathbf{R}}\,. \tag{191}$$

Discussion of such rank results is in Marsaglia and Styan (1974).

8.10. APPLICATION TO A CELL MEANS MODEL

Application of the general results of this chapter to the normal equations of the 1-way classification cell means model is shown in Section 8.2c. Extension to the 2-way classification is straightforward. The model equation is that of Chapters 4 and 5:

$$E(y_{ijk}) = \mu_{ij}$$

for $i = 1, \dots, a$ and $j = 1, \dots, b$ and, for the s filled cells, for which $n_{ij} > 0$, $k = 1, \dots, n_{ij}$.

Definition. Consider all-cells-filled data as just a special case of some-cells-empty data and define everything as pertaining *only to filled cells*, i.e., those for which $n_{ij} > 0$.

In using vectors and matrices defined by the notation of Section 7.1a, their elements in all cases are assumed to be in lexicon order, that is, ordered by (k—if it occurs—within) j within i. For example, we write simply

$$\{_c\, \mu_{ij}\} \quad \text{for} \quad \boldsymbol{\mu} = \left\{_c \left\{_c\, \mu_{ij}\right\}_{j=1}^{j=b}\right\}_{i=1}^{i=a},$$

and use
$$\{_c\, y_{ijk}\} \quad \text{for} \quad \mathbf{y} = \left\{_c \left\{_c \left\{_c\, y_{ijk}\right\}_{k=1}^{n_{ij}}\right\}_{j=1}^{j=b}\right\}_{i=1}^{i=a},$$

which are easy extensions of (39) and (40).

Example. In the example of Table 8.8, each triplet represents a total, the corresponding number of observations (in parenthesis), and the resultant average.The model equations $y_{ijk} = \mu_{ij} + e_{ijk}$ for the data of Table 8.8 are

$$\mathbf{y} = \begin{bmatrix} 3 \\ 4 \\ 8 \\ 5 \\ 8 \\ 8 \\ 6 \\ 12 \\ 18 \\ 27 \end{bmatrix} = \begin{bmatrix} 1 & \cdot & \cdot & \cdot & \cdot \\ 1 & \cdot & \cdot & \cdot & \cdot \\ 1 & \cdot & \cdot & \cdot & \cdot \\ \cdot & 1 & \cdot & \cdot & \cdot \\ \cdot & 1 & \cdot & \cdot & \cdot \\ \cdot & 1 & \cdot & \cdot & \cdot \\ \cdot & \cdot & 1 & \cdot & \cdot \\ \cdot & \cdot & \cdot & 1 & \cdot \\ \cdot & \cdot & \cdot & 1 & \cdot \\ \cdot & \cdot & \cdot & \cdot & 1 \end{bmatrix} \boldsymbol{\mu} + \mathbf{e} \qquad (192)$$

TABLE 8.8. EXAMPLE: DATA IN 2 ROWS AND 3 COLUMNS, WITH ONE EMPTY CELL

3, 4, 8 15(3)5	5, 8, 8 21(3)7	6 6(1)6	42(7)6
12, 18 30(2)15	27 27(1)27	—	57(3)19
45(5)9	48(4)12	6(1)6	99(10)9.9

for $\boldsymbol{\mu}' = [\mu_{11} \quad \mu_{12} \quad \mu_{13} \quad \mu_{21} \quad \mu_{22}]$ and for $\mathbf{e}$ having elements e_{ijk} in the same lexicon order as are the elements of $\mathbf{y}$.

The 10×5 matrix in (192) has the same form as $\mathbf{X}$ of (41) from the example in (5). For the general model $\mathbf{y} = \mathbf{X}\boldsymbol{\mu} + \mathbf{e}$ the form of $\mathbf{X}$ is like (42):

$$\mathbf{X} = \{_d \, \mathbf{1}_{n_{ij}}\} \quad \text{with} \quad \mathbf{X}'\mathbf{X} = \{_d \, n_{ij}\} \,. \tag{193}$$

Using (193), the normal equations are $\mathbf{X}'\mathbf{X}\boldsymbol{\mu}^{\circ} = \mathbf{X}'\mathbf{y}$, namely

$$\{_d \, n_{ij}\}\boldsymbol{\mu}^{\circ} = \{_c \, y_{ij\cdot}\},$$

similar to (44), with solution

$$\boldsymbol{\mu}^{\circ} = \hat{\boldsymbol{\mu}} = \mathbf{D}\{1/n_{ij}\}\{_c \, y_{ij\cdot}\} = \{_c \, \bar{y}_{ij\cdot}\} = \bar{\mathbf{y}}, \tag{194}$$

on defining $\bar{\mathbf{y}}$ as the vector of cell means, $\{_c \, \bar{y}_{ij\cdot}\}$. This is similar to (47) and gives $\hat{\mu}_{ij} = \bar{y}_{ij\cdot}$, as in Chapters 4 and 5. Further, from (52) with $\mathbf{X}'\mathbf{X}$ being nonsingular,

$$\text{var}(\hat{\boldsymbol{\mu}}) = \{_d \, n_{ij}\}^{-1}\sigma^2 = \{_d \, \sigma^2/n_{ij}\}, \tag{195}$$

giving $v(\hat{\mu}_{ij}) = \sigma^2/n_{ij}$ and $\text{cov}(\hat{\mu}_{ij}, \hat{\mu}_{hk}) = 0$, as is to be expected. Also, from (63) we get (10) of Chapter 5:

$$\begin{aligned}\hat{\sigma}^2 &= \frac{\mathbf{y}'(\mathbf{I} - \mathbf{X}\mathbf{G}\mathbf{X}')\mathbf{y}}{N - r_X} = \frac{\mathbf{y}'\left[\mathbf{I} - \{_d \, \mathbf{1}_{n_{ij}}\}\{_d \, n_{ij}\}^{-1}\{_d \, \mathbf{1}'_{n_{ij}}\}\right]\mathbf{y}}{N - s} \\ &= \frac{\mathbf{y}'\mathbf{y} - \mathbf{y}'\{_d \, \bar{\mathbf{J}}_{n_{ij}}\}\mathbf{y}}{N - s} = \frac{\Sigma_i\Sigma_j\Sigma_k(y_{ijk} - \bar{y}_{ij\cdot})^2}{N - s} \,. \end{aligned} \tag{196}$$

Based on (194) and (196), estimating linear functions of μ_{ij}s and calculating confidence intervals (assuming normality) proceeds in the usual manner; e.g., Section 2.12. And from (146) we test H: $\mathbf{K}'\boldsymbol{\mu} = \mathbf{m}$ with

$$F(\text{H}) = \frac{(\mathbf{K}'\bar{\mathbf{y}} - \mathbf{m})'\left[\mathbf{K}'\{_d \, 1/n_{ij}\}\mathbf{K}\right]^{-1}(\mathbf{K}'\bar{\mathbf{y}} - \mathbf{m})}{r_{\mathbf{K}}\hat{\sigma}^2}, \tag{197}$$

where $\mathbf{K}'$ is any full row rank matrix of s columns, where $\hat{\sigma}^2$ is given by (196) and where $F(\text{H})$ is an F-statistic with $r_{\mathbf{K}}$ and $N - s$ degrees of freedom.

Example (continued). The cell means in Table 8.1 give the vector of means as $\hat{\boldsymbol{\mu}}' = \bar{\mathbf{y}}' = [5 \quad 7 \quad 6 \quad 15 \quad 27]$. To test the hypothesis

$$\text{H: } \mu_{11} - \mu_{12} - \mu_{21} + \mu_{22} = 0, \tag{198}$$

$\mathbf{K}' = [1 \quad -1 \quad 0 \quad -1 \quad 1]$ and $\mathbf{m} = 0$ so that

$$\mathbf{K}'\hat{\boldsymbol{\mu}} - \mathbf{m} = 5 - 7 - 15 + 27 - 0 = 10$$

and $\mathbf{K}'\{_d\, n_{ij}\}^{-1}\mathbf{K}$

$$= [1 \quad -1 \quad 0 \quad -1 \quad 1]\begin{bmatrix} 3 & \cdot & \cdot & \cdot & \cdot \\ \cdot & 3 & \cdot & \cdot & \cdot \\ \cdot & \cdot & 1 & \cdot & \cdot \\ \cdot & \cdot & \cdot & 2 & \cdot \\ \cdot & \cdot & \cdot & \cdot & 1 \end{bmatrix}^{-1}\begin{bmatrix} 1 \\ 1 \\ 0 \\ -1 \\ 1 \end{bmatrix} = 13/6 \ .$$

From (196) it will be found that $\hat{\sigma}^2 = 7.6$, so that the F-statistic for testing (198) is, from (197),

$$F = 10\left(\tfrac{13}{6}\right)^{-1} 10/1(7.6) = 1500/247 \ . \tag{199}$$

The F in (199) illustrates the special case of (197) for null m:

$$\text{H: } \mathbf{K}'\boldsymbol{\mu} = \mathbf{0} \quad \text{is tested by} \quad F = \bar{\mathbf{y}}'\mathbf{K}\left[\mathbf{K}'\{_d\, 1/n_{ij}\}\mathbf{K}\right]^{-1}\mathbf{K}'\bar{\mathbf{y}}/r_{\mathbf{K}}\hat{\sigma}^2 \ . \tag{200}$$

Other special cases are those when $\mathbf{K}'$ is either a row vector or has just two rows, as in (87) and (90) of Chapter 2—see E8.31.

8.11. FOUR METHODS OF ESTIMATION

The method of estimation that has been used up to this point is least squares: minimize $(\mathbf{y} - \mathbf{X}\boldsymbol{\beta})'(\mathbf{y} - \mathbf{X}\boldsymbol{\beta})$ with respect to $\boldsymbol{\beta}$ and solve the resulting equations for $\boldsymbol{\beta}$ as the basis of estimation. This means solving $\mathbf{X}'\mathbf{X}\boldsymbol{\beta}^{\circ} = \mathbf{X}'\mathbf{y}$. To contrast this with three other methods of estimation we call it *ordinary least squares estimation*, OLSE, and estimate $\mathbf{X}\boldsymbol{\beta}$ by

$$\text{OLSE}(\mathbf{X}\boldsymbol{\beta}) = \mathbf{X}(\mathbf{X}'\mathbf{X})^{-}\mathbf{X}'\mathbf{y} \ .$$

A more general least squares procedure is to use a weighting factor in the sum of squares that is minimized, and to minimize $(\mathbf{y} - \mathbf{X}\boldsymbol{\beta})'\mathbf{W}(\mathbf{y} - \mathbf{X}\boldsymbol{\beta})$, for $\mathbf{W}$ of some known value. This can be called *weighted least squares* or *generalized least squares* estimation, WLSE or GLSE, and leads to

$$\text{WLSE}(\mathbf{X}\boldsymbol{\beta}) = \mathbf{X}(\mathbf{X}'\mathbf{W}\mathbf{X})^{-}\mathbf{X}'\mathbf{W}\mathbf{y} \ .$$

Notice that neither OLSE nor WLSE involve the variance–covariance matrix of $\mathbf{y}$. This, which we denote by $\mathbf{V}$, is utilized in two other well-known and widely-used methods of estimation. One is *best linear unbiased estimation* (BLUE), which we have already used extensively in this book with $\mathbf{V} = \sigma^2\mathbf{I}$. In general it involves deriving an estimator $\boldsymbol{\lambda}'\mathbf{y}$ of a linear function of the elements of $\boldsymbol{\beta}$, $\mathbf{k}'\boldsymbol{\beta}$ say, such that $\boldsymbol{\lambda}'\mathbf{y}$ is unbiased for $\mathbf{k}'\boldsymbol{\beta}$, i.e., $E(\boldsymbol{\lambda}'\mathbf{y}) = \mathbf{k}'\boldsymbol{\beta}$, and with $\boldsymbol{\lambda}'\mathbf{y}$ having its variance $\boldsymbol{\lambda}'\mathbf{V}\boldsymbol{\lambda}$ be a minimum. This leads to equations

$$\text{BLUE}(\mathbf{X}\boldsymbol{\beta}) = \mathbf{X}(\mathbf{X}'\mathbf{V}^{-1}\mathbf{X})^{-}\mathbf{X}\mathbf{V}^{-1}\mathbf{y} \ .$$

Notice that although the method of BLUE utilizes $\mathbf{V}$, none of the preceding three methods make any demands on the form of the probability density function assumed for $\mathbf{y}$, only that it have mean vector $\mathbf{X}\boldsymbol{\beta}$ and, for BLUE, that $\mathbf{V}$, being a variance–covariance matrix, be non-negative definite. Maintaining those two conditions we also have *maximum likelihood estimation* (MLE). This demands specifying a probability density function for $\mathbf{y}$. Using normality, $\mathbf{y} \sim \mathcal{N}(\mathbf{X}\boldsymbol{\beta}, \mathbf{V})$, maximum likelihood estimation of $\boldsymbol{\beta}$, assuming $\mathbf{V}$ is known, then yields equations

$$\text{MLE}(\mathbf{X}\boldsymbol{\beta}) = \mathbf{X}(\mathbf{X}'\mathbf{V}^{-1}\mathbf{X})^{-}\mathbf{X}'\mathbf{V}^{-1}\mathbf{y} \ .$$

A great deal has been written about these four methods of estimation—but this is no place for a lengthy presentation. The object is simply to put into perspective the place of OLSE (extensively used in this book) vis-à-vis three other methods of estimation that sometimes appear similar to OLSE and that, in some situations, do yield the same estimators. In this regard, the following simple observations are in order.

(i) MLE (under normality) is the same as BLUE.

(ii) $\mathbf{V}^{-1}$ is often used for the weight matrix $\mathbf{W}$ in WLSE (or GLSE), so much so that WLSE using $\mathbf{V}^{-1}$, which is also GLSE using $\mathbf{V}^{-1}$, is often referred to as WLSE (or GLSE).

(iii) WLSE using $\mathbf{V}^{-1}$ is the same as MLE (under normality) and BLUE.

(iv) When $\mathbf{V}$ is taken as $\sigma^2\mathbf{I}$, all four methods are the same:

$$\begin{aligned}\text{OLSE} &= \text{WLSE} \quad \text{with } \mathbf{W} = \sigma^2\mathbf{I}\\ &= \text{BLUE} \quad \text{with } \mathbf{V} = \sigma^2\mathbf{I}\\ &= \text{MLE} \quad \text{under normality, with } \mathbf{V} = \sigma^2\mathbf{I}\ .\end{aligned}$$

(v) OLSE and BLUE are the same when $\mathbf{VX} = \mathbf{XQ}$ for some $\mathbf{Q}$.

This is a useful condition, from among many equivalent forms of it given by Zyskind (1967), for the equality of OLSE and BLUE, even when $\mathbf{V}$ is singular. Readers interested in details of this equality and of the numerous other ramifications of these four methods of estimation should consult current research literature, some recent references being, for example, Mathew and Bhimasankaram (1983) and Baksalary and Kala (1983). BLUE and MLE (under normality) are immensely important for one particular form of linear model which is not in the mainstream of this book, but which is considered briefly in Chapter 13. This is the mixed model. Essentially, in its cell means form, it is $E(\mathbf{y}) = \mathbf{X}\boldsymbol{\mu}$ with $\text{var}(\mathbf{y}) = \mathbf{V} \neq \sigma^2\mathbf{I}$ so that, on assuming that $\mathbf{V}$ is known, BLUE (and MLE under normality) gives

$$\mathbf{X}'\mathbf{V}^{-1}\mathbf{X}\boldsymbol{\mu}^\circ = \mathbf{X}'\mathbf{V}^{-1}\mathbf{y}\ . \tag{201}$$

We have cause to rely heavily on this in Chapter 13. Despite $\mathbf{V}$ not being the simple form $\sigma^2\mathbf{I}$, a function $\mathbf{q}'\boldsymbol{\mu}$ that is estimable under OLSE, namely when $\mathbf{q}' = \mathbf{t}'\mathbf{X}$, is also estimable under BLUE with the same condition, $\mathbf{q}' = \mathbf{t}'\mathbf{X}$.

8.12. SUMMARY

Model: $$\mathbf{y} = \mathbf{X}\boldsymbol{\beta} + \mathbf{e} \sim \left(\mathbf{X}\boldsymbol{\beta}, \sigma^2\mathbf{I}_N\right) \qquad (9), (11)$$

Normal equations: $$\mathbf{X}'\mathbf{X}\boldsymbol{\beta} = \mathbf{X}'\mathbf{y} \qquad (14)$$

Solution: $$\boldsymbol{\beta}^\circ = \mathbf{G}\mathbf{X}'\mathbf{y} \quad \text{for} \quad \mathbf{X}'\mathbf{X}\mathbf{G}\mathbf{X}'\mathbf{X} = \mathbf{X}'\mathbf{X} \qquad (16), (17)$$

$$E(\boldsymbol{\beta}^\circ) = \mathbf{H}\boldsymbol{\beta} \quad \text{for} \quad \mathbf{H} = \mathbf{G}\mathbf{X}'\mathbf{X}; \qquad (48), (49)$$

$$\text{var}(\boldsymbol{\beta}^\circ) = \mathbf{G}\mathbf{X}'\mathbf{X}\mathbf{G}'\sigma^2. \qquad (52)$$

Estimating $E(\mathbf{y})$:

$$\hat{\mathbf{y}} = \mathbf{X}\boldsymbol{\beta}^{\circ} = \mathbf{XGX'y} \quad \text{and} \quad \text{var}(\hat{\mathbf{y}}) = \mathbf{XGX'}\sigma^2. \qquad (53), (54)$$

Sum of squares: $R(\boldsymbol{\beta}) = \boldsymbol{\beta}^{\circ\prime}\mathbf{X'y} = \mathbf{y'XGX'y}$. (65)

$$\text{SSE} = \mathbf{y}'(\mathbf{I} - \mathbf{XGX'})\mathbf{y}\ . \qquad (57)$$

Coefficient of determination: $R^2 = \dfrac{R(\boldsymbol{\beta}) - N\bar{y}^2}{\Sigma_t(y_t - \bar{y})^2}$. (72)

Partitioned models: $\mathbf{y} = \mathbf{X}_1\boldsymbol{\beta}_1 + \mathbf{X}_2\boldsymbol{\beta}_2 + \mathbf{X}_3\boldsymbol{\beta}_3$. (91)

$$R(\boldsymbol{\beta}_2 | \boldsymbol{\beta}_1) = \mathbf{y}'\mathbf{M}_1\mathbf{X}_2(\mathbf{X}_2'\mathbf{M}_1\mathbf{X}_2)^{-}\mathbf{X}_2'\mathbf{M}_1\mathbf{y} = \mathbf{y}'\mathbf{M}_1\mathbf{X}_2(\mathbf{M}_1\mathbf{X}_2)^{+}\mathbf{y}\ . \qquad (90)$$

$$\mathbf{M}_1 = \mathbf{I} - \mathbf{X}_1(\mathbf{X}_1'\mathbf{X}_1)^{-}\mathbf{X}_1 = \mathbf{I} - \mathbf{X}_1\mathbf{X}_1^{+}\ . \qquad (76)$$

$$R(\boldsymbol{\beta}_2 | \boldsymbol{\beta}_1) \text{ tests H: } \mathbf{M}_1\mathbf{X}_2\boldsymbol{\beta}_2 + \mathbf{M}_1\mathbf{X}_2(\mathbf{M}_1\mathbf{X}_2)^{+}\mathbf{X}_3\boldsymbol{\beta}_3 = \mathbf{0}\ . \qquad (123)$$

Estimable function: $\mathbf{q}'\boldsymbol{\beta}$ for $\mathbf{q}' = \mathbf{t}'\mathbf{X}$ for some $\mathbf{t}'$. (128), (129)

$$\text{BLUE}(\mathbf{q}'\boldsymbol{\beta}) = \mathbf{q}'\boldsymbol{\beta}^{\circ} \text{ invariant to } \boldsymbol{\beta}^{\circ}\ . \qquad (130), (133)$$

$$E(\mathbf{q}'\boldsymbol{\beta}^{\circ}) = \mathbf{q}'\boldsymbol{\beta} \quad \text{and} \quad v(\mathbf{q}'\boldsymbol{\beta}^{\circ}) = \mathbf{q}'\mathbf{G}\mathbf{q}\sigma^2\ . \qquad (131), (132)$$

Basic estimable function: E(of any observation). (139)

Hypothesis testing: H: $\mathbf{K}'\boldsymbol{\beta} = \mathbf{m}$ (145)

$$\mathbf{K}' = \mathbf{T}'\mathbf{X} \text{ for some } \mathbf{T}', \text{ and } \mathbf{K}' \text{ of full row rank}, \quad r_{\mathbf{K}'} = s\ .$$

$$F = \frac{Q}{r_{\mathbf{K}}\hat{\sigma}^2} \sim \mathscr{F}_{r_{\mathbf{K}'}, N - r_{\mathbf{X}}} \quad \text{under H}\ . \qquad (146)\ (147)$$

$$Q = \text{SSE}_{\text{H}} - \text{SSE} = (\mathbf{K}'\boldsymbol{\beta}^{\circ} - \mathbf{m})'(\mathbf{K}'\mathbf{G}\mathbf{K})^{-1}(\mathbf{K}'\boldsymbol{\beta}^{\circ} - \mathbf{m})\ . \qquad (144)$$

$$\text{H: } \mathbf{K}'\boldsymbol{\beta} = \mathbf{0} \text{ has } Q = Q_0 = \mathbf{y}'\mathbf{XGK}(\mathbf{K}'\mathbf{GK})^{-}\mathbf{K}'\mathbf{GX}'\mathbf{y}\ . \qquad (149)$$

Estimation under hypothesis: H: $\mathbf{K}'\boldsymbol{\beta} = \mathbf{m}$.

$$\boldsymbol{\beta}^{\circ}_{\mathrm{H}} = \boldsymbol{\beta}^{\circ} - \mathbf{GK}(\mathbf{K}'\mathbf{GK})^{-1}(\mathbf{K}'\boldsymbol{\beta}^{\circ} - \mathbf{m}) \ . \tag{150}$$

$$\mathrm{var}(\boldsymbol{\beta}^{\circ}_{\mathrm{H}}) = \mathbf{G}\big[\mathbf{X}'\mathbf{X} - \mathbf{K}(\mathbf{K}'\mathbf{GK})^{-1}\mathbf{K}\big]\mathbf{G}'\sigma^2 \ . \tag{151}$$

Non-testable hypotheses: If $(\mathbf{K}'\mathbf{GK})^{-1}$ exists but every element (and every linear combination of elements) of $\mathbf{K}'\boldsymbol{\beta}$ is non-estimable, then F of (146) tests

$$\mathrm{H}: \ \mathbf{K}'\mathbf{H}\boldsymbol{\beta} = \mathbf{m} \ . \tag{155}$$

Independent and orthogonal contrasts: For

$$\mathrm{H}: \ \{_c\,\mathrm{H}_i: \mathbf{k}_i'\boldsymbol{\beta} = 0\} \ \mathrm{H}_i \text{ is tested by } \frac{q_i}{\hat{\sigma}^2} \text{ for } q_i = \frac{(\mathbf{k}_i'\boldsymbol{\beta}^{\circ})^2}{\mathbf{k}_i'\mathbf{G}\mathbf{k}_i} \tag{161}$$

$$q_i/\sigma^2 \sim \text{i.i.d. } \chi^2_1 \quad \text{if and only if} \quad \mathbf{k}_i'\mathbf{G}\mathbf{k}_j = 0$$

$$\sum_{i=1}^{r_{\mathbf{X}}} q_i = Q_0 \quad \text{if} \quad q_i/\sigma^2 \sim \text{i.i.d. } \chi^2_1.$$

Restricted model: with restriction $\mathbf{P}'\boldsymbol{\beta} = \boldsymbol{\delta}$

Non-estimable $\mathbf{P}'\boldsymbol{\beta}$: $\boldsymbol{\beta}^{\circ}_r = \boldsymbol{\beta}^{\circ} + (\mathbf{I} - \mathbf{H})[\mathbf{P}'(\mathbf{I} - \mathbf{H})]^{-}(\boldsymbol{\delta} - \mathbf{P}'\boldsymbol{\beta}^{\circ}),$ (176)

$$\mathrm{SSE}_r = \mathrm{SSE}$$

If $\mathbf{q}'\boldsymbol{\beta}$ is estimable in the unrestricted model then in the restricted model an estimable function is

$$\mathbf{q}'\boldsymbol{\beta} + \boldsymbol{\lambda}'(\mathbf{P}'\boldsymbol{\beta} - \boldsymbol{\delta}) \text{ for } \boldsymbol{\lambda}' \text{such that } \boldsymbol{\lambda}'\boldsymbol{\delta} = \mathbf{0}; \text{ any } \boldsymbol{\lambda}' \text{ if } \boldsymbol{\delta} = \mathbf{0} \ . \tag{180}$$

If H: $\mathbf{K}'\boldsymbol{\beta} = \mathbf{m}$ is testable in the unrestricted model then in the restricted model a testable hypothesis is

$$\mathrm{H}: \ (\mathbf{K}' + \mathbf{LP}')\boldsymbol{\beta} = \mathbf{m} + \mathbf{L}\boldsymbol{\delta} \quad \text{with} \quad \mathrm{SSE}_{r,\mathrm{H}} = \mathrm{SSE}_{\mathrm{H}} \ . \qquad \text{Sec. 8.9c-iii}$$

Estimable $\mathbf{P}'\boldsymbol{\beta}$: $\boldsymbol{\beta}^{\circ}_r = \boldsymbol{\beta}^{\circ} - \mathbf{GP}(\mathbf{P}'\mathbf{GP})^{-1}(\mathbf{P}'\boldsymbol{\beta}^{\circ} - \boldsymbol{\delta})$ (181)

$$\mathrm{SSE}_r = \mathrm{SSE} + (\mathbf{P}'\boldsymbol{\beta}^{\circ} - \boldsymbol{\delta})'(\mathbf{P}'\mathbf{GP})^{-1}(\mathbf{P}'\boldsymbol{\beta}^{\circ} - \boldsymbol{\delta}) \tag{182}$$

$$E(\mathrm{SSE}_r) = (N - r_{\mathbf{X}} + r_{\mathbf{P}})\sigma^2 \tag{183}$$

Estimable functions in the unrestricted model are estimable in the restricted model.

For testable H: $\mathbf{K}'\boldsymbol{\beta} = \mathbf{m}$, rows of $\mathbf{K}'$ must be linearly independent of rows of $\mathbf{P}'$:

$$\boldsymbol{\beta}^{\circ}_{r,\mathrm{H}} = \boldsymbol{\beta}^{\circ} - \mathbf{GQ}(\mathbf{Q}'\mathbf{GQ})^{-1}(\mathbf{Q}'\boldsymbol{\beta}^{\circ} - \boldsymbol{\ell}) \tag{186}$$

for $\mathbf{Q} = [\mathbf{P} \quad \mathbf{K}]$ and $\boldsymbol{\ell}' = [\boldsymbol{\delta}' \quad \mathbf{m}']$. And for $\boldsymbol{\beta}^{\circ}_r$ of (181) $\mathrm{SSE}_{r,\,\mathrm{H}} - \mathrm{SSE}_r =$

$$(\mathbf{K}'\boldsymbol{\beta}^{\circ}_r - \mathbf{m})'\left[\mathbf{K}'\mathbf{GK} - \mathbf{K}'\mathbf{GP}(\mathbf{P}'\mathbf{GP})^{-1}\mathbf{P}'\mathbf{GK}\right]^{-1}(\mathbf{K}'\boldsymbol{\beta}^{\circ}_r - \mathbf{m}) \ . \tag{185}$$

8.13. APPENDIX

a. Estimation under the hypothesis (Section 8.8d)

We seek a value of $\boldsymbol{\beta}$, call it $\boldsymbol{\beta}^{\circ}_{\mathrm{H}}$, that minimizes $(\mathbf{y} - \mathbf{X}\boldsymbol{\beta})'(\mathbf{y} - \mathbf{X}\boldsymbol{\beta})$ subject to $\mathbf{K}'\boldsymbol{\beta} = \mathbf{m}$. This is achieved with the use of $2\boldsymbol{\theta}$ as a vector of Lagrange multipliers, by minimizing $(\mathbf{y} - \mathbf{X}\boldsymbol{\beta})'(\mathbf{y} - \mathbf{X}\boldsymbol{\beta}) + 2\boldsymbol{\theta}'(\mathbf{K}'\boldsymbol{\beta} - \mathbf{m})$ with respect to elements of $\boldsymbol{\beta}$ and $\boldsymbol{\theta}$. This leads to equations

$$\mathbf{X}'\mathbf{X}\boldsymbol{\beta}^{\circ}_{\mathrm{H}} + \mathbf{K}\boldsymbol{\theta} = \mathbf{X}'\mathbf{y} \tag{A1}$$

and

$$\mathbf{K}'\boldsymbol{\beta}^{\circ}_{\mathrm{H}} = \mathbf{m} \ . \tag{A2}$$

From (A1),

$$\boldsymbol{\beta}^{\circ}_{\mathrm{H}} = \mathbf{G}(\mathbf{X}'\mathbf{y} - \mathbf{K}\boldsymbol{\theta})$$

which, on substituting into (A2), gives $\boldsymbol{\theta} = (\mathbf{K}'\mathbf{GK})^{-1}(\mathbf{K}'\mathbf{GX}'\mathbf{y} - \mathbf{m})$ and so, with $\boldsymbol{\beta}^{\circ} = \mathbf{GX}'\mathbf{y}$, gives (150):

$$\boldsymbol{\beta}^{\circ}_{\mathrm{H}} = \boldsymbol{\beta}^{\circ} - \mathbf{GK}(\mathbf{K}'\mathbf{GK})^{-1}(\mathbf{K}'\boldsymbol{\beta}^{\circ} - \mathbf{m}) \ .$$

b. The likelihood ratio test

Tests of linear hypotheses $\mathbf{K}'\boldsymbol{\beta} = \mathbf{m}$ have been developed from the starting point of the *F*-statistic. This, in turn, can be shown to stem from the likelihood ratio test.

For a sample of N observations $\mathbf{y}$, where $\mathbf{y} \sim \mathcal{N}(\mathbf{X}\boldsymbol{\beta}, \sigma^2\mathbf{I})$, the likelihood function is

$$L(\boldsymbol{\beta}, \sigma^2) = (2\pi\sigma^2)^{-\frac{1}{2}N}\exp\{-[(\mathbf{y} - \mathbf{X}\boldsymbol{\beta})'(\mathbf{y} - \mathbf{X}\boldsymbol{\beta})/2\sigma^2]\} \ .$$

The likelihood ratio test utilizes two values of $L(\boldsymbol{\beta}, \sigma^2)$:

(i) $\text{Max}(L_w)$, the maximum value of $L(\boldsymbol{\beta}, \sigma^2)$ over the complete range of parameters, $0 < \sigma^2 < \infty$, and $-\infty < \beta_i < \infty$ for all i.

(ii) $\text{Max}(L_{\text{H}})$, the maximum value of $L(\boldsymbol{\beta}, \sigma^2)$ over the range of parameters limited (restricted or defined) by the hypothesis H.

The likelihood ratio is then $L = \max(L_{\text{H}})/\max(L_w)$. Each maximum is found in the usual manner; differentiate $L(\boldsymbol{\beta}, \sigma^2)$ with respect to σ^2 and the elements of $\boldsymbol{\beta}$, equate the differentials to zero, solve the resulting equations for $\boldsymbol{\beta}$ and σ^2 and use these solutions in place of $\boldsymbol{\beta}$ and σ^2 in $L(\boldsymbol{\beta}, \sigma^2)$. In the case of $\max(L_{\text{H}})$ the maximization procedure is carried out within the limitations of the hypothesis.

First, $\partial L(\boldsymbol{\beta}, \sigma^2)/\partial\boldsymbol{\beta} = \mathbf{0}$ gives, as in (14) and (17), $\boldsymbol{\beta}^\circ = \mathbf{GX'y}$; and $\partial L(\boldsymbol{\beta}, \sigma^2)/\partial\sigma^2 = 0$ leads to $\hat{\sigma}^2 = (\mathbf{y} - \mathbf{X}\boldsymbol{\beta}^\circ)'(\mathbf{y} - \mathbf{X}\boldsymbol{\beta}^\circ)/N$. Thus

$$\max(L_w) = L(\boldsymbol{\beta}^\circ, \hat{\sigma}^2) = (2\pi\hat{\sigma}^2)^{-\frac{1}{2}N}\exp\{-[(\mathbf{y} - \mathbf{X}\boldsymbol{\beta}^\circ)'(\mathbf{y} - \mathbf{X}\boldsymbol{\beta}^\circ)/2\hat{\sigma}^2]\}$$

$$= \frac{e^{-\frac{1}{2}N}N^{\frac{1}{2}N}}{(2\pi)^{\frac{1}{2}N}[(\mathbf{y} - \mathbf{X}\boldsymbol{\beta}^\circ)'(\mathbf{y} - \mathbf{X}\boldsymbol{\beta}^\circ)]^{\frac{1}{2}N}} .$$

This is the denominator of L.

The numerator comes from amending $L(\boldsymbol{\beta}, \sigma^2)$ by the hypothesis, which is done by maximizing $L(\boldsymbol{\beta}, \sigma^2) + 2\boldsymbol{\theta}'(\mathbf{K}'\boldsymbol{\beta} - \mathbf{m})$, where $2\boldsymbol{\theta}'$ is a vector of Lagrange multipliers. This leads to equations (A1) and (A2) and to

$$\tilde{\sigma}^2 = (\mathbf{y} - \mathbf{X}\boldsymbol{\beta}^\circ_{\text{H}})'(\mathbf{y} - \mathbf{X}\boldsymbol{\beta}^\circ_{\text{H}})/N = (\text{SSE} + Q)/N,$$

from (152). Then

$$\max(L_{\text{H}}) = L(\boldsymbol{\beta}^\circ_{\text{H}}, \tilde{\sigma}^2) = (2\pi\tilde{\sigma}^2)^{-\frac{1}{2}N}\exp\{-[(\text{SSE} + Q)/2\tilde{\sigma}^2]\}$$

$$= \frac{e^{-\frac{1}{2}N}N^{\frac{1}{2}n}}{(2\pi)^{\frac{1}{2}N}(\text{SSE} + Q)^{\frac{1}{2}N}} .$$

With these values for the maxima, the likelihood ratio is

$$L = \frac{\max(L_{\text{H}})}{\max(L_w)} = \left[\frac{(\mathbf{y} - \mathbf{X}\boldsymbol{\beta}^\circ)'(\mathbf{y} - \mathbf{X}\boldsymbol{\beta}^\circ)}{\text{SSE} + Q}\right]^{\frac{1}{2}N} = \left[\frac{1}{1 + Q/\text{SSE}}\right]^{\frac{1}{2}N} .$$

Clearly L is a single-valued function of Q/SSE, montonic decreasing when Q/SSE increases. Therefore Q/SSE can be used as a test statistic in place of L. By the same reasoning so can $(Q/\text{SSE})[(N-r)/r] = (Q/r)/[\text{SSE}(N-r)]$, whose use as the F-statistic has already been discussed. Thus is the use of the F-statistic established as an outcome of the likelihood ratio test.

8.14. EXERCISES

E8.1. Suppose that data for a 1-way classification of three observations in each of two classes are as follows.

Response: y_{ij}		Covariate: z_{ij}	
I	II	I	II
34	32	2	11
36	40	8	3
47	51	5	7

For these data write down, in matrix and vector form, the model equation for

(a) the regression model $y_{ij} = \alpha + \beta z_{ij} + e_{ij}$,

(b) the analysis of covariance model $y_{ij} = \mu + \alpha_i + bz_{ij} + e_{ij}$,

(c) the intra-class regression model $y_{ij} = \mu + \alpha_i + b_i z_{ij} + e_{ij}$.

E8.2. Treat the data of E8.1 as being from a 2-way classification of three rows and two columns. Write down, in matrix and vector form, the model equation for the models

(a) $y_{ij} = \mu + \alpha_i + \beta_j + e_{ij}$,

(b) $y_{ij} = \mu + \alpha_i + \beta_j + bz_{ij} + e_{ij}$,

(c) $y_{ij} = \mu + \alpha_i + \beta_j + b_i z_{ij} + e_{ij}$.

E8.3. Derive normal equations and a solution of them for models (a), (b) and (c) of E8.1. Find expected values and dispersion matrices of the solution vectors you obtain.

E8.4. Repeat E8.3 using the models of E8.2.

E8.5. What is the hypothesis tested by $F(M)$ in each of the models given in E8.1 and E8.2?

E8.6. Using three different solutions to the normal equations for Example 3 at the end of Section 8.2b, calculate (a) $\hat{\mathbf{y}}$ and (b) $\mathbf{y}'\mathbf{y} - \boldsymbol{\beta}^{\circ\prime}\mathbf{X}'\mathbf{y}$.

E8.7. Demonstrate the "invert part of the inverse" rule for E8.1(a).

E8.8. From line 5 of Table 8.5, develop the other five lines.

E8.9. Prove (131), (132), and (135)–(138).

E8.10. What is the basic estimable function in each of the models of E8.1 and E8.2?

E8.11. Calculate the numerator sum of squares for testing H: $\mu_1 = \mu_2$ in Example 3 using two different solution vectors.

E8.12. Derive (151), (152) and (153).

E8.13. For Example 2, calculate Q for

$$\text{H:} \begin{cases} \alpha_1 + \alpha_2 = 2\alpha_3 \\ 3\alpha_1 + 4\alpha_3 = 7\alpha_2 \end{cases}.$$

E8.14. For Example 2, find two independent and orthogonal contrasts and test the hypothesis that they are zero.

E8.15. Verify $Q_0 = 240(a + b)^2/4(a + b)^2$ at the end of Section 8.8.

E8.16. For Example 2, use the restriction $3\alpha_1 + 2\alpha_2 + \alpha_3 = 0$ and solve the normal equations.

E8.17. Derive (184) and (185).

E8.18. Identify $\boldsymbol{\beta}^\circ$, $\mathbf{G}$ and SSE as functions of $\mathbf{A}$, $\mathbf{u}$ and λ in

$$\begin{bmatrix} \mathbf{X}'\mathbf{X} & \mathbf{X}'\mathbf{y} \\ \mathbf{y}'\mathbf{X} & \mathbf{y}'\mathbf{y} \end{bmatrix}^- = \begin{bmatrix} \mathbf{A} & \mathbf{u} \\ \mathbf{u}' & \lambda \end{bmatrix}.$$

E8.19. Through expressing $F = \mathscr{R}(\boldsymbol{\beta} \mid \mu)/[(r_{\mathbf{X}} - 1)\hat{\sigma}^2]$ as a function of R^2 of (72), show that $F > 1$ only if $R^2 > (r_{\mathbf{X}} - 1)/(N - 1)$.

E8.20. For $y_{ij} = \mu + i\alpha + e_{ij}$ for $i = 1, \ldots, a$ and $j = 1, \ldots, n$:

(a) Find least squares estimates $\hat{\mu}$ and $\hat{\alpha}$ of μ and α.

(b) Show that $\hat{\mu} = \bar{y} - \frac{1}{2}(a + 1)\hat{\alpha}$.

(c) Explain why (b) is no surprise.

(d) Show that Q for H: $\alpha = 0$ is $an(a^2 - 1)\hat{\alpha}^2/12$.

E8.21. Repeat E8.11, 8.13, 8.14 and 8.16 for the following data in a 1-way classification of 3 classes:

$$12, 16; \quad 13, 15, 17; \quad \text{and} \quad 12, 20, 31 \ .$$

E8.22. Fit the model that has $E(\mathbf{y}_1) = \mathbf{X}_1\boldsymbol{\beta}$ and then the model that has $E[\mathbf{y}_1' \ \ \mathbf{y}_2']' = [\mathbf{X}_1' \ \ \mathbf{X}_2']'\boldsymbol{\beta}$. From the first model find $\hat{\mathbf{e}}_1 = \mathbf{y}_1 - \hat{\mathbf{y}}_1$ and from the second find $[\tilde{\mathbf{e}}_1' \ \ \tilde{\mathbf{e}}_2'] = [\mathbf{y}_1' - \tilde{\mathbf{y}}_1' \ \ \mathbf{y}_2' - \tilde{\mathbf{y}}_2']$, where $[\tilde{\mathbf{y}}_1' \ \ \tilde{\mathbf{y}}_2']$ is the vector of predicted y-values in the second model. Show that $\text{cov}(\hat{\mathbf{e}}_1, \tilde{\mathbf{e}}_1) = \text{var}(\hat{\mathbf{e}}_1)$ but $\text{cov}(\hat{\mathbf{e}}_1, \tilde{\mathbf{e}}_2) = \mathbf{0}$.

E8.23. Prove that diagonal elements of $\mathbf{X}(\mathbf{X}'\mathbf{X})^{-}\mathbf{X}'$ do not exceed unity.

E8.24. Suppose for the regression model $E(\mathbf{y}) = \mathbf{X}\boldsymbol{\beta}$, with $\mathbf{X}$ of full column rank, that there are two independent data sets $\mathbf{y}_1, \mathbf{X}_1$ and $\mathbf{y}_2, \mathbf{X}_2$ of n_1 and n_2 observations, respectively, gathered at two distinctly different locations. Within locations the estimator of $\boldsymbol{\beta}$ is $\hat{\boldsymbol{\beta}}_i = \mathbf{A}_i^{-1}\mathbf{X}_i'\mathbf{y}_i$ for $i = 1, 2$ and $\mathbf{A}_i = \mathbf{X}_i'\mathbf{X}_i$.

(a) Write down a model that permits testing the hypothesis that $\boldsymbol{\beta}$ is the same in both locations. Derive, in the simplest terms possible, the F-statistic for testing that hypothesis.

(b) Under the hypothesis of (a), derive the estimator $\boldsymbol{\beta}$ in as simple form as possible.

(c) Confirm your result in (b) by an alternative derivation.

(d) Why can simplifications in (b) not be done when $\mathbf{X}_1$ and $\mathbf{X}_2$ are not of full rank?

E8.25. Define $\boldsymbol{\delta} = \{_c\, y_{ij} - \bar{y}_{i\cdot}\}_{i=1,\ j=1}^{i=a,\ j=n_i}$ and $\mathbf{w} = \{_c\, z_{ij} - \bar{z}_{i\cdot}\}_{i=1,\ j=1}^{i=a,\ j=n_i}$. Show that SSE of (38) in Chapter 6 is $\boldsymbol{\delta}'\mathbf{A}\boldsymbol{\delta}$, where $\mathbf{A}$ is idempotent, and consider applying Theorems 7.2 and 7.3 to $\boldsymbol{\delta}'\mathbf{A}\boldsymbol{\delta}$.

E8.26. Show that (181) and (190) are equivalent.

E8.27. Prove $\mathbf{S}$ in (190) has full column rank.

E8.28. Use the fact that $\mathbf{X}$ has $\mathbf{1}$ as one of its columns to show that elements of $\hat{\mathbf{e}} = \mathbf{y} - \hat{\mathbf{y}}$ sum to zero. *Note*: Zyskind and Johnson (1973) point out that residuals sum to zero only when $\mathbf{V1}$ is in the column space of $\mathbf{X}$, where $\mathbf{V}$ is the dispersion matrix of $\mathbf{y}$.

E8.29. For Example 2, using $\mathbf{G}$ and $\boldsymbol{\beta}^\circ$ of (24) and (25), verify that Q is the same for

$$\text{H:} \begin{cases} \alpha_1 - \alpha_2 = 3 \\ \alpha_1 \phantom{{}-\alpha_2} = 7 \end{cases} \quad \text{as it is for} \quad \text{H:} \begin{cases} \alpha_1 - \alpha_2 = 3 \\ \alpha_1 - \alpha_3 = 7 \end{cases} .$$

E8.30. Using (71) of Section 7.6b-iv, show that the numerator sum of squares for testing H: $\mathbf{LT'X\beta} = \mathbf{0}$, where $\mathbf{L}$ is any nonsingular matrix is $\mathbf{y'Ay}$ for $\mathbf{A} = \mathbf{TT'}$.

E8.31. Use F of (146) for the 1-way classification to derive (87) and (90) of Chapter 2.

E8.32. For the 2-way crossed classification with interaction, use (144) to show that with all-cells-filled data the numerator sum of squares for testing

$$\text{H: } \frac{1}{b}\Sigma_j\mu_{ij} \textit{ equal } \forall\ i \text{ is } Q = \text{SSA}_w = \Sigma_i w_i(\tilde{y}_{i\cdot\cdot} - \Sigma w_i\tilde{y}_{i\cdot\cdot}/\Sigma w_i)^2$$

of (41) and (42) of Chapter 4 where, in (29) of the same chapter,

$$\tilde{y}_{i\cdot\cdot} = \frac{1}{b}\Sigma_j\bar{y}_{ij\cdot} \quad \text{and} \quad \frac{1}{w_i} = \frac{1}{b^2}\Sigma_j\frac{1}{n_{ij}} .$$

E8.33. For the cell means model of the 2-way classification of two rows and two columns, with all cells filled, use (181) to show that in the no-interaction case

$$\text{BLUE}(\mu_{ij}) = \bar{y}_{ij\cdot} - \frac{(-1)^{i+j}(\bar{y}_{11\cdot} - \bar{y}_{12\cdot} - \bar{y}_{21\cdot} - \bar{y}_{22\cdot})}{n_{ij}\left(\dfrac{1}{n_{11}} + \dfrac{1}{n_{12}} + \dfrac{1}{n_{21}} + \dfrac{1}{n_{22}}\right)} .$$

E8.34. Use (146) to confirm (70) and (135) of Chapter 6.

CHAPTER 9

THE 2-WAY CROSSED CLASSIFICATION: OVERPARAMETERIZED MODELS

This chapter deals with the traditional overparameterized form of model for the 2-way crossed classification. It begins with showing relationships between the cell means model and the overparameterized model, and describes the latter in some detail. The effects of restrictions on the parameters of the model, particularly of what are often called the "usual restrictions", known nowadays as the Σ-restrictions, are also described; and so are the use of constraints on the solutions, and their connection to restrictions on the parameters.

9.1. THE WITH-INTERACTION MODEL

a. The model

A widely-used, with-interaction, overparameterized model for the 2-way crossed classification is

$$E(y_{ijk}) = \mu + \alpha_i + \beta_j + \phi_{ij} \tag{1}$$

where μ is a general mean, α_i is an effect due to the ith level of (what we call) the row factor, β_j is an effect due to the jth level of the column factor and ϕ_{ij} represents an effect for the interaction of row i with column j. In place of ϕ_{ij} many writers use $(\alpha\beta)_{ij}$—but ϕ_{ij} is maintained here.

The number of parameters (p, say) involved in data from (1) is 1 for μ, plus a for the α_is, plus b for the β_js, plus s for the ϕ_{ij}s corresponding to the $s \leq ab$ filled cells. Thus

$$p = \text{number of parameters} = 1 + a + b + s\ . \tag{2}$$

But there are only $s \leq ab$ observed all means $\bar{y}_{ij\cdot}$ from which to try to estimate those $s + a + b + 1$ parameters. Thus we have

$$\text{excess number of parameters} = s + a + b + 1 - s = a + b + 1\ . \tag{3}$$

It is because there is this excess that the model (1) is called overparameterized: there are more parameters in (1) than there are cell means to estimate them from. This is what is meant by the model being overparameterized—as discussed in Section 2.13.

In using (1), definition of a residual error term is exactly as in (10) and (11) of Section 4.2,

$$e_{ijk} = y_{ijk} - E(y_{ijk}), \tag{4}$$

so that

$$y_{ijk} = \mu + \alpha_i + \beta_j + \phi_{ij} + e_{ijk} \tag{5}$$

with

$$E(e_{ijk}) = 0\ . \tag{6}$$

On assembling the e_{ijk}s in lexicon order (by k within j, within i) in the vector $\mathbf{e}$, we take its dispersion matrix to be

$$\text{var}(\mathbf{e}) = \sigma^2\mathbf{I}, \tag{7}$$

just like (8) of Section 8.1. Normality, $\mathbf{e} \sim \mathcal{N}(\mathbf{0}, \sigma^2\mathbf{I})$, is assumed for hypothesis testing and calculating confidence intervals.

b. Relationships to cell means

It is clear that in the model (1) the population mean for cell i, j is being taken as $\mu + \alpha_i + \beta_j + \phi_{ij}$. This is μ_{ij} of Chapters 4 and 5. Thus the basic relationship between the parameters of the overparameterized model and μ_{ij} of the cell means model is

$$\mu_{ij} = \mu + \alpha_i + \beta_j + \phi_{ij}\ . \tag{8}$$

In Section 4.8b the interaction of rows i and i' with columns j and j' is defined as

$$\theta_{ij,i'j'} = \mu_{ij} - \mu_{ij'} - \mu_{i'j} + \mu_{i'j'}\ . \tag{9}$$

Since the somewhat glib description of ϕ_{ij} following (1) is that it is an interaction effect and yet (9) is what we mean by interaction, there is a need for relating ϕ_{ij} to (9). This is easily achieved by substituting (8) into (9), which gives

$$\theta_{ij,i'j'} = \phi_{ij} - \phi_{ij'} - \phi_{i'j} + \phi_{i'j'} \ . \tag{10}$$

Thus the interaction term $\theta_{ij,i'j'}$, defined in (9) in terms of μs, has in (10) exactly the same definition in terms of ϕs.

The use of ϕ_{ij} in (1) represents a very general form of interaction effect. In recent years a variety of forms have been considered for ϕ_{ij}, such as $\phi_{ij} = \delta\alpha_i\beta_j$ of Graybill (1976, p. 596) or his generalization $\mathbf{F}\boldsymbol{\delta}$, where $\mathbf{F}$ has elements that are "nonadditive functions of estimable functions" of μ, α_i and β_j (*loc. cit.*); or $\phi_{ij} = \lambda\tau_i\eta_j$ in $\mu + \alpha_i + \beta_j + \lambda\tau_i\eta_j$ of Marasinghe and Johnson (1981, 1982) and Emerson *et al.* (1984); and $\phi_{ij} = \Sigma_k\theta_k\tau_{ki}\eta_{kj}$ of Snee (1982). We leave it to the interested reader to pursue these specialized forms of interaction effects as desired, e.g., the agricultural example in Martin (1980).

c. Reparameterization

The relationship (8) between μ_{ij} and $\mu + \alpha_i + \beta_j + \phi_{ij}$ is unsatisfying in the sense that it provides no explicit definition of μ, α_i, β_j and ϕ_{ij} in terms of the μ_{ij}s. One commonly-used relationship comes from starting with $E(y_{ijk}) = \mu_{ij}$ and defining

$$\bar{\mu}_{i\cdot} = \Sigma_j\mu_{ij}/b \quad \text{and} \quad \bar{\mu}_{\cdot j} = \Sigma_i\mu_{ij}/a \tag{11}$$

and

$$\bar{\mu}_{\cdot\cdot} = \Sigma_i\bar{\mu}_{i\cdot}/a = \Sigma_j\bar{\mu}_{\cdot j}/b \ . \tag{12}$$

Then $E(y_{ijk}) = \mu_{ij}$ can be written as

$$E(y_{ijk}) = \bar{\mu}_{\cdot\cdot} + (\bar{\mu}_{i\cdot} - \bar{\mu}_{\cdot\cdot}) + (\bar{\mu}_{\cdot j} - \bar{\mu}_{\cdot\cdot}) + (\mu_{ij} - \bar{\mu}_{i\cdot} - \bar{\mu}_{\cdot j} + \bar{\mu}_{\cdot\cdot}) \ .$$

On defining

$$\dot{\mu} = \bar{\mu}_{\cdot\cdot}, \qquad \dot{\alpha}_i = \bar{\mu}_{i\cdot} - \bar{\mu}_{\cdot\cdot}, \qquad \dot{\beta}_j = \bar{\mu}_{\cdot j} - \bar{\mu}_{\cdot\cdot} \tag{13}$$

and

$$\dot{\phi}_{ij} = \mu_{ij} - \bar{\mu}_{i\cdot} - \bar{\mu}_{\cdot j} + \bar{\mu}_{\cdot\cdot} \tag{14}$$

we then have, analogous to (1),

$$E(y_{ijk}) = \dot{\mu} + \dot{\alpha}_i + \dot{\beta}_j + \dot{\phi}_{ij} \ . \tag{15}$$

But in contrast to the parameters in (1), those in (15) now satisfy certain linear relationships, as a consequence of (13) and (14):

$$\Sigma_i \dot{\alpha}_i = 0 \quad \text{and} \quad \Sigma_j \dot{\beta}_j = 0 \tag{16}$$

and

$$\Sigma_i \dot{\phi}_{ij} = 0 \quad \forall \; j \quad \text{and} \quad \Sigma_j \dot{\phi}_{ij} = 0 \quad \forall \; i \; . \tag{17}$$

As derived here, equations (16) and (17) are a direct outcome of the definitions (11)–(14) implicit in the parameters of (15). As such, (16) and (17) are called restrictions on the model (in particular, the Σ-restrictions of Section 8.9, due to their summation nature). The complete model is (15), (16) and (17); it is a restricted model, the Σ-restricted model.

The Σ-restrictions (16), as part of the definition of the main-effect parameters, make eminent sense. They effectively redefine the α_is and β_js to be deviations from their own means; e.g., defining

$$\dot{\alpha}_i = \alpha_i - \Sigma_i \alpha_i / a \quad \text{gives} \quad \Sigma_i \dot{\alpha}_i = 0$$

and $\dot{\alpha}_i$ is the deviation of α_i from the mean of the α_is. Furthermore, $\Sigma_i \dot{\alpha}_i = 0$ is consistent with the $a - 1$ degrees of freedom among the a rows. This is so because, in having $\Sigma_i \dot{\alpha}_i = 0$ as part of the definition, $a - 1$ of the $\dot{\alpha}_i$s can be any values at all and then the value of the a'th one is determined from them by $\Sigma_i \dot{\alpha}_i = 0$, i.e., as $-\Sigma_{i=1}^{a-1} \dot{\alpha}_i$ for the other $a - 1$ $\dot{\alpha}_i$-values. Similarly for the $\dot{\beta}_j$s, with $\Sigma_j \dot{\beta}_j = 0$ of (16).

For the $\dot{\phi}_{ij}$s, the first equation in (17), namely $\Sigma_i \dot{\phi}_{ij} = 0 \; \forall \; j$, allows free choice of $b(a - 1)\dot{\phi}_{ij}$s, but the further restriction $\Sigma_j \dot{\phi}_{ij} = 0 \; \forall \; i$ reduces this to $b(a - 1) - (a - 1) = (a - 1)(b - 1)$, the degrees of freedom for interaction when using all-cells-filled data.

The degrees of freedom (for connected data) is always $a - 1$ for rows; and data with a rows always have data in every row. (Were a row to be empty, it would not be part of the data.) The association of $\Sigma_i \dot{\alpha}_i = 0$ with $a - 1$ degrees of freedom for rows therefore always holds true, even for some-cells-empty data; similarly for columns. But this is not the case with degrees of freedom for interaction; with some-cells-empty data, (38) of Section 5.4c shows that the degrees of freedom for interaction is $s - a - b + 1$, which is less than $(a - 1)(b - 1)$ when $s < ab$. Thus the degrees of freedom are less than the number of different $\dot{\phi}_{ij}$s available from (17). This occurs because, with some-cells-empty data, (17) involves $\dot{\phi}_{ij}$s that are not present in the data. One way out of this predicament, if one wants to persist with some kind of Σ-restrictions, would be to adapt (17) to apply only to those $\dot{\phi}_{ij}$s occurring in the data. Denote this kind of summation by Σ^*. A difficulty associated with it is that, for example, the columns included in

$\Sigma_j^* \dot{\phi}_{ij} = 0 \ \forall \ i$ are then different for different values of i. (And, similarly for rows, in $\Sigma_i^* \dot{\phi}_{ij} = 0 \ \forall \ j$.)

Example. Suppose available data were as indicated in Grid 9.1. Then the Σ-restriction $\Sigma_j \phi_{1j} = 0$ is $\phi_{11} + \phi_{12} + \phi_{13} + \phi_{14} = 0$ which involves ϕ_{12} and ϕ_{14} that do not occur in the data. How, therefore, are we likely to be able to estimate ϕ_{11} and ϕ_{13} when they are defined in terms of parameters that do not occur in the data? The possible alternative, $\Sigma_j^* \phi_{ij} = 0$, of adapting the Σ-restrictions (17) to involve only the ϕ_{ij}s corresponding to filled cells then leads, for Grid 9.1, to part of (17) being $\phi_{11} + \phi_{13} = 0$ and $\phi_{22} + \phi_{23} = 0$, which, through involving columns 1 and 3, and 2 and 3, respectively, do not involve the same two sets of columns.

Grid 9.1

	1	2	3	4
1	√		√	
2		√	√	
3		√		√
4			√	√

Nevertheless, what we are calling Σ*-restrictions do lead to a definition of interaction terms that is consistent with whatever hypotheses about interactions can be tested from some-cells-empty data. For example, on writing down all eight of the Σ*-restrictions for Grid 9.1, one finds that they reduce in a manner that allows the $\dot{\phi}_{ij}$s to be represented as shown in Grid 9.1A, where $\dot{\phi}$ is arbitrary; thus, $\dot{\phi}_{22} = \dot{\phi}$ and $\dot{\phi}_{23} = -\dot{\phi}$. Since the data are connected (see Section 5.3), there is $s - a - b + 1 =$

Grid 9.1A

0		0	
	$\dot{\phi}$	$-\dot{\phi}$	
	$-\dot{\phi}$		$\dot{\phi}$
		$\dot{\phi}$	$-\dot{\phi}$

$8 - 4 - 4 + 1 = 1$ degree of freedom for interaction. Scrutiny of Grid 9.1 suggests that, in accord with (42) of Section 5.4e, the hypothesis tested by the sum of squares corresponding to this one degree of freedom is

$$\text{H:}\ \mu_{22} - \mu_{23} - \mu_{32} + \mu_{33} - (\mu_{33} - \mu_{34} - \mu_{43} + \mu_{44}) = 0, \tag{18}$$

i.e.,

$$\text{H:}\ \mu_{22} - \mu_{23} - \mu_{32} + \mu_{34} + \mu_{43} - \mu_{44} = 0$$

which from (10) is

$$\text{H:}\ \phi_{22} - \phi_{23} - \phi_{32} + \phi_{34} + \phi_{43} - \phi_{44} = 0\ .$$

And from Grid 9.1A it is

$$\text{H:}\ \dot\phi + \dot\phi + \dot\phi + \dot\phi + \dot\phi + \dot\phi = 0, \quad \text{i.e.,} \quad \text{H:}\ \dot\phi = 0\ .$$

Thus the Σ^*-restrictions that led to Grid 9.1A also lead to the hypothesis for the interaction sum of squares of Table 4.9 being that $\dot\phi$ is zero.

Readers who find hypotheses stated with the simplicity of H: $\dot\phi = 0$ to be appealing are free to use the Σ^*-restrictions that lead to a parameterization that gives such simplicity. But for many, hypothesis statements such as (18) provide a clearer description of what is being tested. Those who are attracted to the reparameterization should consider it in detail for examples other than Grid 9.1; e.g., that grid but with data in cell 3, 1 (see E9.1).

d. Estimable functions

Having $\mu_{ij} = \mu + \alpha_i + \beta_j + \phi_{ij}$ of (8), and knowing from Chapters 4 and 5 that μ_{ij} is estimable for $n_{ij} \neq 0$, it is clear that in the overparameterized model (1),

$$\mu + \alpha_i + \beta_j + \phi_{ij} \quad \text{for} \quad n_{ij} \neq 0 \quad \text{is estimable}\ . \tag{19}$$

This is also evident from (139) of Section 8.7f, in that the basic estimable function, $E(y_{ijk})$, is equal in this case to $\mu + \alpha_i + \beta_j + \phi_{ij}$ of (1). Therefore (19) follows.

An important consequence of (19) is that differences such as $\alpha_i - \alpha_h$ and $\beta_j - \beta_k$ (for $i \neq h$ and $j \neq k$) are not estimable. This unhappy state of affairs arises because, for example,

$$\mu + \alpha_i + \beta_j + \phi_{ij} - (\mu + \alpha_h + \beta_j + \phi_{hj}) = \alpha_i - \alpha_h + \phi_{ij} - \phi_{hj} \tag{20}$$

and although the right-hand side of (20) contains $\alpha_i - \alpha_h$ it also has $\phi_{ij} - \phi_{hj}$; and there is no way of getting rid of such ϕ_{ij}s from any linear function of terms $(\mu + \alpha_i + \beta_j + \phi_{ij})$. This is further compounded by the $n_{ij} \neq 0$ proviso of (19), that μ_{ij} is estimable only when cell i, j contains data, and that (20) itself is estimable only when n_{ij} and n_{hj} are both nonzero. Another consequence is that only for every row (or column) that has data in every cell is ρ_i (or γ_j) estimable—see Sections 5.2a and 5.4a.

A case of particular interest is when there are data in all cells in rows i and h:

$$\alpha_i - \alpha_h + \Sigma_j \phi_{ij}/b - \Sigma_j \phi_{hj}/b \quad \text{is estimable;} \tag{21}$$

and, more particularly, when there are data in all cells in all rows, i.e., all-cells-filled data, then (21) is true for every pair $i \neq h$.

e. Estimation

Knowing from Chapters 4 and 5 that the BLUE of μ_{ij} is

$$\hat{\mu}_{ij} = \bar{y}_{ij\cdot}, \quad \text{with} \quad v(\hat{\mu}_{ij}) = \sigma^2/n_{ij}, \quad \text{for } n_{ij} \neq 0, \tag{22}$$

we have at once that

$$\text{BLUE}(\mu + \alpha_i + \beta_j + \phi_{ij}) = \bar{y}_{ij\cdot} \quad \text{for } n_{ij} \neq 0, \tag{23}$$

with the same variance as in (22). Estimation of, and confidence intervals on, linear combinations of $(\mu + \alpha_i + \beta_j + \phi_{ij})$-terms therefore follow the same procedures as described in Section 4.4.

f. Hypothesis testing

The result given in (197) of Chapter 8 for testing linear hypotheses about μ_{ij}s is directly usable for hypotheses about the terms $\mu + \alpha_i + \beta_j + \phi_{ij}$; simply use $\mu + \alpha_i + \beta_j + \phi_{ij}$ for μ_{ij} for filled cells in the formulation $\mathbf{K}'\boldsymbol{\mu} = \mathbf{m}$ of a hypothesis. Thus any hypothesis stated in terms of μ_{ij}s (for filled cells) can be re-stated in terms of $(\mu + \alpha_i + \beta_j + \phi_{ij})$-terms and its F-statistic can be calculated from (197) of Chapter 8.

A most important aspect of hypothesis testing that arises from the nature of what functions are estimable is that hypotheses about differences between row effects (and column effects) are not testable. This means that hypotheses about linear combinations of differences such as $\alpha_i - \alpha_h$ (and $\beta_j - \beta_k$) are not testable; i.e., hypotheses concerning contrasts of αs (and of βs) are not testable. In particular, the hypotheses H: α_i *all equal* and H: β_j *all equal* cannot be tested. The reason for this is that the only hypotheses that can be tested are hypotheses H: $\mathbf{K}'\boldsymbol{\beta} = \mathbf{m}$, where the elements of $\mathbf{K}'\boldsymbol{\beta}$ are estimable—and in the preceding subsection we saw that $\alpha_i - \alpha_h$ and $\beta_j - \beta_k$ are not estimable. And, of course, it is rarely of scientific impor-

TABLE 9.1. SUMS OF SQUARES AND ASSOCIATED HYPOTHESES OF TABLE 4.9 EXPRESSED IN TERMS OF THE OVERPARAMETERIZED WITH-INTERACTION MODEL $E(y_{ijk}) = \mu + \alpha_i + \beta_j + \phi_{ij}$

Sum of Squares		Associated Hypothesis
$R(\mu)$	$= R(\mu)$	H: $N\mu + \Sigma_i n_{i\cdot}\alpha_i + \Sigma_j n_{\cdot j}\beta_j + \Sigma_i\Sigma_j n_{ij}\phi_{ij} = 0$
$\mathscr{R}(\mu_i \mid \mu)$	$= R(\boldsymbol{\alpha} \mid \mu)$	H: $\alpha_i + \Sigma_j n_{ij}(\beta_j + \phi_{ij})/n_{i\cdot}$ *equal* $\forall\ i$
$\mathscr{R}(\mu_i, \tau_j \mid \mu_i)$	$= R(\boldsymbol{\beta} \mid \mu, \boldsymbol{\alpha})$	H: $\Sigma_i n_{ij}(\beta_j + \phi_{ij}) = \Sigma_i\Sigma_s \dfrac{n_{ij}n_{is}}{n_{i\cdot}}(\beta_s + \phi_{is})\ \ \forall\ j$
$\mathscr{R}(\mu_{ij} \mid \mu_i, \tau_j)$	$= R(\boldsymbol{\phi} \mid \mu, \boldsymbol{\alpha}, \boldsymbol{\beta})$	H: Table 4.9, with ϕ_{ij} in place of μ_{ij}
$\mathscr{R}(\tau_j \mid \mu)$	$= R(\boldsymbol{\beta} \mid \mu)$	H: $\beta_j + \Sigma_i n_{ij}(\alpha_i + \phi_{ij})/n_{\cdot j}$ *equal* $\forall\ j$
$\mathscr{R}(\mu_i, \tau_j \mid \tau_j)$	$= R(\boldsymbol{\alpha} \mid \mu, \boldsymbol{\beta})$	H: $\Sigma_j n_{ij}(\alpha_i + \phi_{ij}) = \Sigma_j\Sigma_t \dfrac{n_{ij}n_{tj}}{n_{\cdot j}}(\alpha_t + \phi_{tj})\ \ \forall\ i$

tance to estimate $\alpha_i - \alpha_h$ or $\beta_j - \beta_k$ in the presence of interactions, although there may be occasions (Section 10.3) when it can be informative to do so.

g. Analysis of variance

The sums of squares to be calculated for the analysis of variance for the overparameterized model (5) are exactly the same as those for the cell means model of Table 4.9. The differences are in the notation to be used for those sums of squares and their associated hypotheses. These are shown in Table 9.1.

Derivation of Table 9.1 stems directly from replacing μ_{ij} in Table 4.9 by $\mu + \alpha_i + \beta_j + \phi_{ij}$. Thus $R(\mu)$ tests H: $\Sigma_i\Sigma_j n_{ij}\mu_{ij} = 0$ in the cell means model, but in the overparameterized model this hypothesis is

$$\text{H: } n_{\cdot\cdot}\mu + \Sigma_i n_{i\cdot}\alpha_i + \Sigma_i n_{\cdot j}\beta_j + \Sigma_i\Sigma_j n_{ij}\phi_{ij} = 0\ .$$

Similarly $\mathscr{R}(\mu_i \mid \mu)$, which is now

$$\mathscr{R}(\mu_i \mid \mu) = \mathscr{R}(\mu, \alpha_i \mid \mu) = R(\boldsymbol{\alpha} \mid \mu)$$

tests H: $\Sigma_j n_{ij}u_{ij}/n_{i\cdot}$ *all equal*, which is

$$\text{H: } \Sigma_j n_{ij}(\mu + \alpha_i + \beta_j + \phi_{ij})/n_{i\cdot}\ \textit{equal}\ \ \forall\ i\ .$$

Since $\Sigma_j n_{ij}\mu/n_{i\cdot} = \mu$ occurs in every term in this hypothesis it can be dropped, and the hypothesis is

$$\text{H: } \alpha_i + \Sigma_j n_{ij}(\beta_j + \phi_{ij})/n_{i\cdot} \textit{ equal } \forall\ i\ .$$

In this fashion

$$\mathscr{R}(\mu_i, \tau_j | \mu_i) = \mathscr{R}(\mu, \alpha_i, \beta_j | \mu, \alpha_i) = R(\boldsymbol{\beta} | \mu, \boldsymbol{\alpha})$$

tests, from Table 4.9,

$$\text{H: } \Sigma_i n_{ij}\mu_{ij} = \Sigma_i n_{ij}(\Sigma_k n_{ik}\mu_{ik})/n_{i\cdot} \quad \forall\ j,$$

which becomes

$$\text{H: } \Sigma_i n_{ij}(\mu + \alpha_i + \beta_j + \phi_{ij}) = \Sigma_i n_{ij}[\Sigma_k n_{ik}(\mu + \alpha_i + \beta_k + \phi_{ik})]/n_{i\cdot} \quad \forall\ j$$

and this reduces to

$$\text{H: } \Sigma_i n_{ij}(\beta_j + \phi_{ij}) = \Sigma_i \Sigma_k \frac{n_{ij}n_{ik}}{n_{i\cdot}}(\beta_k + \phi_{ik})\ .$$

Finally, since the hypothesis associated with

$$\mathscr{R}(\mu_{ij} | \mu_i, \tau_j) = \mathscr{R}(\mu, \alpha_i, \beta_j, \phi_{ij} | \mu, \alpha_i, \beta_j) = R(\boldsymbol{\phi} | \mu, \boldsymbol{\alpha}, \boldsymbol{\beta})$$

is stated in terms of $\theta_{ij,i'j'}$s, which in (10) are the same functions of ϕs as they are of the μ_{ij}s, the same is true of the hypothesis: whatever function it is of the μ_{ij}s in Table 4.9 it is the same function of the ϕ_{ij}s in the overparameterized model of Table 9.1.

Notice, too, that each sum of squares in Table 9.1 has the same form as one or other of the sums of squares of Table 8.5 in Section 8.6d. This equivalence is evident from the $R(\cdot\,|\,\cdot)$ notation used in Table 9.1 where, for example, with the model having μ, α_i, β_j and ϕ_{ij} in it, $R(\boldsymbol{\beta} | \mu, \boldsymbol{\alpha})$ is just a special case of line 5 of Table 8.5. Hence the sums of squares of Table 9.1 do satisfy the χ^2 and independence properties discussed at length in Section 8.6, and so are suitable for testing the hypotheses shown. Those same properties therefore also apply to Table 4.9.

None of the hypotheses in Table 9.1 are of the form H: α_i *equal* $\forall\ i$ or H: β_j *equal* $\forall\ j$. This is, of course, because neither $\alpha_i - \alpha_h$ nor $\beta_j - \beta_k$ are estimable. Something closer to this than what is in Table 9.1 comes from

SSA_w and SSB_w, the weighted squares of means, available for all-cells-filled data. In the cell means model, as in (43) of Section 4.6,

$$SSA_w \quad \text{tests} \quad H\colon\ \Sigma_j \mu_{ij}/b \ \textit{equal} \quad \forall\ i\ .$$

With $\mu_{ij} = \mu + \alpha_i + \beta_j + \phi_{ij}$ this becomes, in the overparameterized model,

$$SSA_w \quad \text{tests} \quad H\colon\ \alpha_i + \Sigma_j \phi_{ij}/b \ \textit{equal} \quad \forall\ i$$

and

$$SSB_w \quad \text{tests} \quad H\colon\ \beta_j + \Sigma_i \phi_{ij}/a \ \textit{equal} \quad \forall\ j\ .$$

Thus SSA_w can be described as testing equality of row effects in the presence of averaged interaction effects—averaged over each row. And a similar description pertains to SSB_w.

h. Model equations

All the preceding results concerning estimable functions, estimation and hypothesis testing have been developed for the overparameterized model directly from using $\mu + \alpha_i + \beta_j + \phi_{ij}$ in place of μ_{ij} of the cell means model. Those results can, of course, also be derived from applying the general methods of Chapter 8 directly to the overparameterized model of (8) itself. Extensive details of this are given in Searle (1971, Chapter 7). They are not all repeated here; a briefer account is given of the model equations, and of the normal equations and a solution of them that coincides with results in Chapters 4 and 5.

Example (continued). The model equations for the data of Table 8.8 for the overparameterized model (5) are

$$\mathbf{y} = \begin{bmatrix} 3 \\ 4 \\ 8 \\ 5 \\ 8 \\ 8 \\ 6 \\ 12 \\ 18 \\ 27 \end{bmatrix} = \left[\begin{array}{c|cc|ccc|ccccc} 1 & 1 & \cdot & 1 & \cdot & \cdot & 1 & \cdot & \cdot & \cdot & \cdot \\ 1 & 1 & \cdot & 1 & \cdot & \cdot & 1 & \cdot & \cdot & \cdot & \cdot \\ 1 & 1 & \cdot & 1 & \cdot & \cdot & 1 & \cdot & \cdot & \cdot & \cdot \\ 1 & 1 & \cdot & \cdot & 1 & \cdot & \cdot & 1 & \cdot & \cdot & \cdot \\ 1 & 1 & \cdot & \cdot & 1 & \cdot & \cdot & 1 & \cdot & \cdot & \cdot \\ 1 & 1 & \cdot & \cdot & 1 & \cdot & \cdot & 1 & \cdot & \cdot & \cdot \\ 1 & 1 & \cdot & \cdot & \cdot & 1 & \cdot & \cdot & 1 & \cdot & \cdot \\ 1 & \cdot & 1 & 1 & \cdot & \cdot & \cdot & \cdot & \cdot & 1 & \cdot \\ 1 & \cdot & 1 & 1 & \cdot & \cdot & \cdot & \cdot & \cdot & 1 & \cdot \\ 1 & \cdot & 1 & \cdot & 1 & \cdot & \cdot & \cdot & \cdot & \cdot & 1 \end{array}\right] \begin{bmatrix} \mu \\ \hline \alpha_1 \\ \alpha_2 \\ \hline \beta_1 \\ \beta_2 \\ \beta_3 \\ \hline \phi_{11} \\ \phi_{12} \\ \phi_{13} \\ \phi_{21} \\ \phi_{22} \end{bmatrix} + \mathbf{e}, \tag{24}$$

$$\mathbf{X}_1 \quad \mathbf{X}_2 \qquad \mathbf{X}_3 \qquad\qquad \mathbf{X}_4$$

where $\mathbf{e}$ is the vector of elements e_{ijk} corresponding to y_{ijk} and arrayed in lexicon order. In terms of the general model $\mathbf{y} = \mathbf{X\beta} + \mathbf{e}$, the 10×11 matrix in (24) is $\mathbf{X}$, partitioned into a row of four submatrices as indicated below (24), the partitioning being in accord with the four different kinds of parameters μ, α_i, β_j and ϕ_{ij}. To specify $\mathbf{X}_1, \ldots, \mathbf{X}_4$ we set up some notation.

Notation. For the general case of $i = 1, \ldots, a$, and $j = 1, \ldots, b$, define, using the notation of Section 7.1a,

$$\boldsymbol{\alpha} = \{_c\, \alpha_i\}, \quad \boldsymbol{\beta}_b = \{_c\, \beta_j\} \quad \text{and} \quad \boldsymbol{\phi} = \{_c\, \phi_{ij}\} \quad \text{for} \quad n_{ij} \neq 0\ . \tag{25}$$

Note that the vector of β_js is $\boldsymbol{\beta}_b$ (of order b), to be distinguished from the general $\boldsymbol{\beta}$ of $E(\mathbf{y}) = \mathbf{X\beta}$ of Chapter 8.

Also use

$$\mathbf{D}^*\{\mathbf{1}_{n_{ij}}\} = \{_{d^*}\, \mathbf{1}_{n_{ij}}\} \quad \text{as equivalent adaptations of} \quad \{_d\, \mathbf{1}_{n_{ij}}\}, \tag{26}$$

where $\mathbf{D}^*$ and $\mathbf{d}^*$ play the same role as $\mathbf{D}$ and $\mathbf{d}$, respectively, but adapted by the convention of taking $\mathbf{1}_0$ to mean a column vector of no rows, i.e., a "matrix" of no rows and one column. For example, with $n_1 = 2$, $n_2 = 3$ and $n_3 = 0$

$$\mathbf{D}^*\{\mathbf{1}_{n_t}\}_{t=1}^{t=3} = \begin{bmatrix} \mathbf{1}_2 & \cdot & \cdot \\ \cdot & \mathbf{1}_3 & \cdot \\ \cdot & \cdot & \mathbf{1}_0 \end{bmatrix} = \begin{bmatrix} \mathbf{1}_2 & \cdot & \cdot \\ \cdot & \mathbf{1}_3 & \cdot \end{bmatrix}. \tag{27}$$

Another example is that for n-values of 2, 0, 3, 0, 0, 6

$$\begin{bmatrix} \mathbf{1}_2 & \cdot & \cdot & \cdot & \cdot & \cdot \\ \cdot & \mathbf{1}_0 & \cdot & \cdot & \cdot & \cdot \\ \cdot & \cdot & \mathbf{1}_3 & \cdot & \cdot & \cdot \\ \cdot & \cdot & \cdot & \mathbf{1}_0 & \cdot & \cdot \\ \cdot & \cdot & \cdot & \cdot & \mathbf{1}_0 & \cdot \\ \cdot & \cdot & \cdot & \cdot & \cdot & \mathbf{1}_6 \end{bmatrix} = \begin{bmatrix} \mathbf{1}_2 & \cdot & \cdot & \cdot & \cdot & \cdot \\ \cdot & \cdot & \mathbf{1}_3 & \cdot & \cdot & \cdot \\ \cdot & \cdot & \cdot & \cdot & \cdot & \mathbf{1}_6 \end{bmatrix}. \tag{28}$$

With the preceding notation we then write

$$E(\mathbf{y}) = \mathbf{X\beta} = [\mathbf{X}_1 \quad \mathbf{X}_2 \quad \mathbf{X}_3 \quad \mathbf{X}_4] \begin{bmatrix} \mu \\ \boldsymbol{\alpha} \\ \boldsymbol{\beta}_b \\ \boldsymbol{\phi} \end{bmatrix} + \mathbf{e} \tag{29}$$

$$= \mathbf{X}_1\mu + \mathbf{X}_2\boldsymbol{\alpha} + \mathbf{X}_3\boldsymbol{\beta}_b + \mathbf{X}_4\boldsymbol{\phi} + \mathbf{e}, \tag{30}$$

where, by generalizing from the example in (24), we find that

$$\mathbf{X}_1 = \mathbf{1}_N \quad \text{and} \quad \mathbf{X}_2 = \mathbf{D}\{\mathbf{1}_{n_{i\cdot}}\}_{i=1}^{i=a} \tag{31}$$

$$\mathbf{X}_3 = \{_c\, \mathbf{X}_{3i}\}_{i=1}^{i=a} \quad \text{for } \mathbf{X}_{3i} = \mathbf{D}^*\{\mathbf{1}_{n_{ij}}\}_{j=1}^{j=b} \tag{32}$$

and

$$\mathbf{X}_4 = \mathbf{D}\{\mathbf{1}_{n_{ij}}\}_{i=1,\, j=1}^{i=a,\, j=b} . \tag{33}$$

Example (continued). Writing the model equations (24) as

$$\mathbf{y} = \begin{bmatrix} 3\\4\\8\\5\\8\\8\\6\\12\\18\\27 \end{bmatrix} = \begin{bmatrix} 1\\1\\1\\1\\1\\1\\1\\1\\1\\1 \end{bmatrix}\mu + \begin{bmatrix} 1 & \cdot\\ 1 & \cdot\\ 1 & \cdot\\ 1 & \cdot\\ 1 & \cdot\\ 1 & \cdot\\ 1 & \cdot\\ \cdot & 1\\ \cdot & 1\\ \cdot & 1 \end{bmatrix}\begin{bmatrix} \alpha_1\\ \alpha_2 \end{bmatrix} + \begin{bmatrix} 1 & \cdot & \cdot\\ 1 & \cdot & \cdot\\ 1 & \cdot & \cdot\\ \cdot & 1 & \cdot\\ \cdot & 1 & \cdot\\ \cdot & 1 & \cdot\\ \cdot & \cdot & 1\\ 1 & \cdot & \cdot\\ 1 & \cdot & \cdot\\ \cdot & 1 & \cdot \end{bmatrix}\begin{bmatrix} \beta_1\\ \beta_2\\ \beta_3 \end{bmatrix} + \begin{bmatrix} 1 & \cdot & \cdot & \cdot & \cdot\\ 1 & \cdot & \cdot & \cdot & \cdot\\ 1 & \cdot & \cdot & \cdot & \cdot\\ \cdot & 1 & \cdot & \cdot & \cdot\\ \cdot & 1 & \cdot & \cdot & \cdot\\ \cdot & 1 & \cdot & \cdot & \cdot\\ \cdot & \cdot & 1 & \cdot & \cdot\\ \cdot & \cdot & \cdot & 1 & \cdot\\ \cdot & \cdot & \cdot & 1 & \cdot\\ \cdot & \cdot & \cdot & \cdot & 1 \end{bmatrix}\begin{bmatrix} \phi_{11}\\ \phi_{12}\\ \phi_{13}\\ \phi_{21}\\ \phi_{22} \end{bmatrix} + \mathbf{e} \tag{34}$$

illustrates the form of the general $\mathbf{X}$-matrices of (30) given in (31)–(33).

Those general forms show that, similar to (2), p = number of columns in $\mathbf{X}$ is $p = 1 + a + b + s$ and that the following relationships exist among the columns of these submatrices of $\mathbf{X}$.

$$\text{Columns of } \mathbf{X}_2 \text{ sum to } \mathbf{X}_1 = \mathbf{1}_N . \tag{35}$$

$$\text{Columns of } \mathbf{X}_3 \text{ sum to } \mathbf{X}_1 . \tag{36}$$

$$\sum_{j=1}^{b} (\text{columns of } \mathbf{X}_4 \text{ corresponding to } \phi_{ij}) = i\text{th column of } \mathbf{X}_2, \quad \forall\ i \tag{37}$$

$$\sum_{i=1}^{a} (\text{columns of } \mathbf{X}_4 \text{ corresponding to } \phi_{ij}) = j\text{th column of } \mathbf{X}_3, \quad \forall\ j \tag{38}$$

Given the $1 + 1 + a = 2 + a$ equations in (35)–(37), the b equations of (38) then contain only $b - 1$ linearly independent equations because the sum of those b equations is, by (38) and (36), equal to $\mathbf{1}_N$. Hence the number of linearly independent equations in (35)–(38) is $2 + a + b - 1 = 1 + a + b$. Thus, in general, the rank of $\mathbf{X}$ is

$$r_{\mathbf{X}} = 1 + a + b + s - (1 + a + b) = s\ . \tag{39}$$

i. The normal equations

The normal equations $\mathbf{X'X\beta}^{\circ} = \mathbf{X'y}$ of Section 8.2a are, for the model equations (24) and (34),

$$\left[\begin{array}{c:cc:ccc:ccccc} 10 & 7 & 3 & 5 & 4 & 1 & 3 & 3 & 1 & 2 & 1 \\ \hdashline 7 & 7 & \cdot & 3 & 3 & 1 & 3 & 3 & 1 & \cdot & \cdot \\ 3 & \cdot & 3 & 2 & 1 & \cdot & \cdot & \cdot & \cdot & 2 & 1 \\ \hdashline 5 & 3 & 2 & 5 & \cdot & \cdot & 3 & \cdot & \cdot & 2 & \cdot \\ 4 & 3 & 1 & \cdot & 4 & \cdot & \cdot & 3 & \cdot & \cdot & 1 \\ 1 & 1 & \cdot & \cdot & \cdot & 1 & \cdot & \cdot & 1 & \cdot & \cdot \\ \hdashline 3 & 3 & \cdot & 3 & \cdot & \cdot & 3 & \cdot & \cdot & \cdot & \cdot \\ 3 & 3 & \cdot & \cdot & 3 & \cdot & \cdot & 3 & \cdot & \cdot & \cdot \\ 1 & 1 & \cdot & \cdot & \cdot & 1 & \cdot & \cdot & 1 & \cdot & \cdot \\ 2 & \cdot & 2 & 2 & \cdot & \cdot & \cdot & \cdot & \cdot & 2 & \cdot \\ 1 & \cdot & 1 & \cdot & 1 & \cdot & \cdot & \cdot & \cdot & \cdot & 1 \end{array}\right] \left[\begin{array}{c} \mu^{\circ} \\ \hdashline \alpha_1^{\circ} \\ \alpha_2^{\circ} \\ \hdashline \beta_1^{\circ} \\ \beta_2^{\circ} \\ \beta_3^{\circ} \\ \hdashline \phi_{11}^{\circ} \\ \phi_{12}^{\circ} \\ \phi_{13}^{\circ} \\ \phi_{21}^{\circ} \\ \phi_{22}^{\circ} \end{array}\right] = \left[\begin{array}{r} 99 \\ \hdashline 42 \\ 57 \\ \hdashline 45 \\ 48 \\ 6 \\ \hdashline 15 \\ 21 \\ 6 \\ 30 \\ 27 \end{array}\right]. \tag{40}$$

The partitioning in (40) is based on that of (24). In general,

$$\mathbf{X'X} = \begin{bmatrix} \mathbf{X}_1'\mathbf{X}_1 & \mathbf{X}_1'\mathbf{X}_2 & \mathbf{X}_1'\mathbf{X}_3 & \mathbf{X}_1'\mathbf{X}_4 \\ & \mathbf{X}_2'\mathbf{X}_2 & \mathbf{X}_2'\mathbf{X}_3 & \mathbf{X}_2'\mathbf{X}_4 \\ & & \mathbf{X}_3'\mathbf{X}_3 & \mathbf{X}_3'\mathbf{X}_4 \\ & \text{Symmetric} & & \mathbf{X}_4'\mathbf{X}_4 \end{bmatrix}, \tag{41}$$

the normal equations are, using (30)–(33),

$$\begin{bmatrix} n_{..} & \{_r\, n_{i\cdot}\}_i & \{_r\, n_{\cdot j}\}_j & \{_r\, n_{ij}\}_{i,j} \\ & \{_d\, n_{i\cdot}\}_i & \{_m\, n_{ij}\}_{i,j} & \{_d\, \{_r\, n_{ij}\}_j,\ n_{ij} > 0\}_i \\ & & \{_d\, n_{\cdot j}\}_j & \{_r\, D^*\{n_{ij}\}_j\}_i \\ & \text{Symmetric} & & \{_d\, n_{ij}\}_{i,j} \end{bmatrix} \begin{bmatrix} \mu^{\circ} \\ \boldsymbol{\alpha}^{\circ} \\ \boldsymbol{\beta}_b^{\circ} \\ \boldsymbol{\phi}^{\circ} \end{bmatrix}$$

$$= \begin{bmatrix} y_{..} \\ \{_c\, y_{i..}\}_i \\ \{_c\, y_{\cdot j \cdot}\}_j \\ \{_c\, y_{ij\cdot}\}_{i,j} \end{bmatrix}, \tag{42}$$

where, on all occasions $i = 1, 2, \ldots, a$ and $j = 1, 2, \ldots, b$. The partitioning in (42) is the same as that in (41), and the submatrices in (42) are illustrated in (40).

j. Solving the normal equations

The rank of $\mathbf{X}$ is $r_{\mathbf{X}} = s$, from (39). We therefore solve (42) by using the algorithm of Section 7.1j to obtain a generalized inverse of the matrix $\mathbf{X}'\mathbf{X}$, namely the matrix on the left-hand side of (42). This algorithm demands finding in $\mathbf{X}'\mathbf{X}$ a nonsingular, principal submatrix of order $r_{\mathbf{X}} = s$. An obvious candidate is, by inspection, $\mathbf{D}\{n_{ij}\}_{i,j}$. Therefore, by that algorithm,

$$\mathbf{G} = \begin{bmatrix} 0 & \mathbf{0} & \mathbf{0} & \mathbf{0} \\ \mathbf{0} & \mathbf{0} & \mathbf{0} & \mathbf{0} \\ \mathbf{0} & \mathbf{0} & \mathbf{0} & \mathbf{0} \\ \mathbf{0} & \mathbf{0} & \mathbf{0} & \mathbf{D}\{1/n_{ij}\}_{i,j} \end{bmatrix}$$

is a generalized inverse. Pre-multiplying the right-hand side of (42) by $\mathbf{G}$ gives a solution vector as

$$\begin{bmatrix} \mu^\circ \\ \boldsymbol{\alpha}^\circ \\ \boldsymbol{\beta}_b^\circ \\ \boldsymbol{\phi}^\circ \end{bmatrix} = \begin{bmatrix} 0 \\ \mathbf{0} \\ \mathbf{0} \\ \{_c\, \bar{y}_{ij\cdot}\}_{i=1,\, j=1}^{i=a,\, j=b} \text{ for } n_{ij} > 0 \end{bmatrix}. \tag{43}$$

Hence

$$\phi_{ij}^\circ = \bar{y}_{ij\cdot} \quad \text{for} \quad n_{ij} > 0 \; .$$

Estimating estimable functions is then quite straightforward: using the standard result from (133) of Section 8.7c, namely that the BLUE of an estimable function $\mathbf{q}'\boldsymbol{\beta}$ is $\text{BLUE}(\mathbf{q}'\boldsymbol{\beta}) = \mathbf{q}'\boldsymbol{\beta}^\circ$, gives

$$\begin{aligned} \text{BLUE}(\mu + \alpha_i + \beta_j + \phi_{ij}) &= \mu^\circ + \alpha_i^\circ + \beta_j^\circ + \phi_{ij}^\circ \\ &= 0 + 0 + 0 + \bar{y}_{ij\cdot}, \quad \text{from (43)} \\ &= \bar{y}_{ij\cdot}, \end{aligned}$$

i.e., $\hat{\mu}_{ij} = \bar{y}_{ij\cdot}$, just as derived in (23) of the cell means model, where the derivation is much more direct than here.

k. A zero sum of squares

At equation (17) of Section 4.3 we have

$$\text{SSE}\, \Sigma_i \Sigma_j \Sigma_k (y_{ijk} - \bar{y}_{ij\cdot})^2 = \text{SST} - \Sigma_i \Sigma_j n_{ij} \bar{y}_{ij\cdot}^2 \; .$$

Therefore

$$R(\mu_{ij}) = R(\mu, \boldsymbol{\alpha}, \boldsymbol{\beta}, \boldsymbol{\phi}) = \text{SST} - \text{SSE} = \Sigma_i\Sigma_j n_{ij}\bar{y}_{ij\cdot}^2 \ .$$

Consider the possibility of calculating

$$R(\boldsymbol{\alpha} \mid \mu, \boldsymbol{\beta}, \boldsymbol{\phi}) = R(\mu, \boldsymbol{\alpha}, \boldsymbol{\beta}, \boldsymbol{\phi}) - R(\mu, \boldsymbol{\beta}, \boldsymbol{\phi})$$

$$= \Sigma_i\Sigma_j n_{ij}\bar{y}_{ij\cdot}^2 - R(\mu, \boldsymbol{\beta}, \boldsymbol{\phi}) \ .$$

The $R(\mu, \boldsymbol{\beta}, \boldsymbol{\phi})$ term comes from fitting

$$E(y_{ijk}) = \mu + \beta_i + \phi_{ij} \ .$$

This model is the same as $E(y_{ijk}) = \mu + \alpha_i + \beta_j + \phi_{ij}$ only without α_i. Its normal equations will therefore be exactly the same as (42) after deleting the a rows and columns corresponding to $\boldsymbol{\alpha}^{\circ}$ therein; i.e., after deleting the rows and columns through the principal submatrix $\{_d\ n_{i\cdot}\}_i$. As a result, the normal equations for $E(y_{ijk}) = \mu + \beta_j + \phi_{ij}$ have order $1 + b + s$ and rank s and hence the same solutions $\phi_{ij} = \bar{y}_{ij\cdot}$ as in (43). Therefore, on using the R-algorithm of Section 8.4a, the reduction in sum of squares due to fitting $E(y_{ijk}) = \mu + \beta_j + \phi_{ij}$ is

$$R(\mu, \boldsymbol{\beta}, \boldsymbol{\phi}) = \Sigma_i\Sigma_j n_{ij}\bar{y}_{ij\cdot}^2 .$$

and so

$$R(\alpha \mid \mu, \boldsymbol{\beta}, \boldsymbol{\phi}) \equiv 0 \ .$$

This result, which we later refer to when considering the use of Σ-restrictions with the model (1), is a consequence of what Nelder (1976, 1977) calls marginality.

9.2. THE WITHOUT-INTERACTION MODEL

a. The model

The model equation for the no-interaction version of the overparameterized model is simply that of the with-interaction version but with interaction parameters omitted. Thus it is (1) and (5) without ϕ_{ij}s:

$$E(y_{ijk}) = \mu + \alpha_i + \beta_j \quad \text{and} \quad y_{ijk} = \mu + \alpha_i + \beta_j + e_{ijk} \qquad (44)$$

with μ, α_i and β_j having the same meaning as previously and, as usual, $e_{ijk} = y_{ijk} - E(y_{ijk})$, $E(e_{ijk}) = 0$, and $\text{var}(\mathbf{e}) = \sigma^2\mathbf{I}_{n\cdot\cdot}$.

b. Relationships to cell means

Similar to (8) we now have

$$\mu_{ij} = \mu + \alpha_i + \beta_j \tag{45}$$

with, of course, interaction effects being zero,

$$\theta_{ij,i'j'} = \mu_{ij} - \mu_{ij'} - \mu_{i'j} + \mu_{i'j'} = 0, \tag{46}$$

after substitution from (45).

The no-interaction model used in Chapters 4 and 5 is

$$E(y_{ijk}) = \mu_i + \tau_j \ .$$

Relating this to (45) is easy:

$$\mu_i = \mu + \alpha_i \quad \text{and} \quad \tau_j = \beta_j \ . \tag{47}$$

An equally as easy alternative is $\mu_i = \alpha_i$ and $\tau_j = \mu + \beta_j$. We stay with (47).

c. Reparameterization

The reparameterization that redefines the α_is and β_js to be deviations from their own means is as before:

$$\dot{\alpha}_i = \alpha_i - \sum_{i=1}^{a} \alpha_i/a \quad \text{with} \quad \Sigma_i\dot{\alpha}_i = 0$$

and

$$\dot{\beta}_j = \beta_j - \sum_{j=1}^{b} \beta_j/b \quad \text{with} \quad \Sigma_j\dot{\beta}_j = 0 \ . \tag{48}$$

With interaction parameters being omitted from the model, we have none of the difficulties of summation restrictions that include ϕ_{ij}s that do not occur in the model equations for the data, e.g., ϕ_{ij}s for empty cells. This difficulty arose with the with-interaction model and led to the Σ^*-restrictions of Section 9.1c, but these are not needed with the no-interaction model.

d. Estimable functions

The basic estimable function [from (139) of Section 8.7f] is $\mu + \alpha_i + \beta_j$, the expected value of y_{ijk}. And, when data are connected (Section 5.3),

$$\alpha_i - \alpha_h = \mu + \alpha_i + \beta_j - (\mu + \alpha_h + \beta_j) \tag{49}$$

is also estimable. The existence of a column j in which there is data in both rows i and h, ensures that both $\mu + \alpha_i + \beta_j$ and $\mu + \alpha_h + \beta_j$ of (49) are estimable; if no such j exists, connectedness nevertheless ensures that $\alpha_i - \alpha_h$ is estimable; e.g., there will be a row t and columns k and m (or a succession of such rows and columns) such that each term on the right-hand side of

$$\alpha_i - \alpha_h = (\mu + \alpha_i + \beta_k) - (\mu + \alpha_t + \beta_k) + (\mu + \alpha_t + \beta_m) - (\mu + \alpha_h + \beta_m)$$

is estimable, and hence $\alpha_i - \alpha_h$ is estimable. Thus every difference $\alpha_i - \alpha_h$ and $\beta_j - \beta_k$ is estimable; i.e., differences between row effects and between column effects are estimable—in the no-interaction model.

The preceding result is a result that does not pertain for the with-interaction model. It is also a result that demands no use of Σ-restrictions or of any other restrictions. It applies directly to the parameters of the over-parameterized no-interaction model.

e. Estimation

Through adopting the identities $\mu_i = \mu + \alpha_i$ and $\tau_j = \beta_j$ of (47), we get

$$(\mu + \alpha_i)^\circ = \hat{\mu}_i \quad \text{and} \quad \beta_j^\circ = \hat{\tau}_j \tag{50}$$

for $\hat{\tau}_j$ and $\hat{\mu}_i$ of (65) and (68) in Section 4.9b. Then

$$\begin{aligned} \text{BLUE}(\mu + \alpha_i + \beta_j) &= (\mu + \alpha_i)^\circ + \beta_j^\circ \quad \forall\ i \text{ and } j, \\ \text{BLUE}(\alpha_i - \alpha_h) &= (\mu + \alpha_i)^\circ - (\mu + \alpha_h)^\circ \\ \text{BLUE}(\beta_j - \beta_k) &= \beta_j^\circ - \beta_k^\circ\ . \end{aligned} \tag{51}$$

Neither here nor in Chapters 4 and 5 have we provided sampling variances for these estimators. We do this in subsection k which follows.

f. Hypothesis testing

An important feature of the no-interaction model that distinguishes it from the with-interaction model is that, as in the two preceding subsections, differences between αs and between βs are estimable. Consequently, hypotheses about these differences are testable, and in particular the useful hypotheses

$$\text{H: } \alpha_i\text{s } \textit{all equal} \quad \text{and} \quad \text{H: } \beta_j\text{s } \textit{all equal} \tag{52}$$

TABLE 9.2. SUMS OF SQUARES AND ASSOCIATED HYPOTHESES FOR THE OVERPARAMETERIZED NO-INTERACTION MODEL (ADAPTED FROM TABLE 9.1).

Sum of Squares		Associated Hypothesis
$R(\mu)$	$= R(\mu)$	H: $n_{..}\mu + \Sigma_i n_{i.}\alpha_i + \Sigma_j n_{.j}\beta_j = 0$
$\mathscr{R}(\mu_i \mid \mu)$	$= R(\boldsymbol{\alpha} \mid \mu)$	H: $\alpha_i + \Sigma_j n_{ij}\beta_j/n_{i.}$ *equal* $\forall\ i$
$\mathscr{R}(\mu_i, \tau_j \mid \mu_i)$	$= R(\boldsymbol{\beta} \mid \mu, \boldsymbol{\alpha})$	H: β_j *equal* $\forall\ j$
$\mathscr{R}(\tau_j \mid \mu)$	$= R(\boldsymbol{\beta} \mid \mu)$	H: $\beta_j + \Sigma_i n_{ij}\alpha_i/n_{.j}$ *equal* $\forall\ i$
$\mathscr{R}(\mu_i, \tau_j \mid \tau_j)$	$= R(\boldsymbol{\alpha} \mid \mu, \boldsymbol{\beta})$	H: α_i *equal* $\forall\ i$

can be tested. Furthermore, the F-statistics for each of these hypotheses turn out to be based on sums of squares that are part of the analysis of variance table—see Table 9.2.

g. Analysis of variance

The sums of squares and associated hypotheses for the no-interaction version of the overparameterized model are the same as those of Table 9.1 for the with-interaction version, but with two omissions. One is the deletion of the interaction sum of squares, $\mathscr{R}(\mu_{ij} \mid \mu_i, \tau_j) = R(\boldsymbol{\phi} \mid \mu, \boldsymbol{\alpha}, \boldsymbol{\beta})$, and the other is deletion of ϕ_{ij}s from all the associated hypotheses. This produces Table 9.2.

Deletion of ϕ_{ij}s from the hypothesis associated with $R(\boldsymbol{\beta} \mid \mu, \boldsymbol{\alpha})$ in Table 9.1 leads to the straightforward hypothesis H: β_j *all equal* of Table 9.2 in a manner that merits explanation. In the hypothesis of Table 9.1, deletion of ϕ_{ij}s leaves

$$\text{H: } \Sigma_i n_{ij}\beta_j - \Sigma_i\Sigma_s \frac{n_{ij}n_{is}}{n_{i.}}\beta_s = 0 \quad \forall\ j,$$

which is

$$\text{H: } \left(n_{.j} - \Sigma_i \frac{n_{ij}^2}{n_{i.}}\right)\beta_j - \Sigma_{j' \neq j}\left(\Sigma_i \frac{n_{ij}n_{ij'}}{n_{i.}}\right)\beta_{j'} = 0 \quad \forall\ j\ . \tag{53}$$

The hypothesis statement in (53) uses all b of the β_js. On referring to c_{jj} and $c_{jj'}$ for $j' \neq j$ in (66) of Section 4.9b, we see in (53) that the coefficient of β_j is c_{jj} and that of $\beta_{j'}$ for $j' \neq j$ is $c_{jj'}$. Define

$$\mathbf{C}_b = \{_m\ c_{jj'}\}_{j=1,\, j'=1}^{j=b,\, j'=b}\ . \tag{54}$$

Then (53) is

$$\text{H: } \mathbf{C}_b\boldsymbol{\beta}_b = \mathbf{0}\ . \tag{55}$$

But the elements of a row of $\mathbf{C}_b$ sum to zero:

$$n_{\cdot j} - \Sigma_i \frac{n_{ij}^2}{n_{i\cdot}} - \Sigma_{j' \neq j}\Sigma_i \frac{n_{ij}n_{ij'}}{n_{i\cdot}} = n_{\cdot j} - \Sigma_i \frac{n_{ij}(n_{ij} + n_{i\cdot} - n_{ij})}{n_{i\cdot}}$$

$$= n_{\cdot j} - n_{\cdot j} = 0\ . \tag{56}$$

Therefore $\mathbf{C}_b\mathbf{1}_b = \mathbf{0}$. Hence $\boldsymbol{\beta}_b = \lambda\mathbf{1}_b$ for any λ satisfies the hypothesis in (53); but $\boldsymbol{\beta}_b = \lambda\mathbf{1}_b$ is equality of all β_js, and so (53) is equivalent to H: β_j *all equal*, as shown in the Table 9.2. And H: β_j *equal* is, from (47), equivalent to H: τ_j *all equal* shown in (106) of Section 4.10b.

h. Model equations

The model equations for the no-interaction overparameterized model are simply those in (30) for the with-interaction case, only with $\boldsymbol{\phi}$ omitted:

$$E(\mathbf{y}) = \mathbf{X}_1\mu + \mathbf{X}_2\boldsymbol{\alpha} + \mathbf{X}_3\boldsymbol{\beta}_b, \tag{57}$$

with its terms defined exactly as in (25), (31) and (32). The number of columns in $\mathbf{X} = [\mathbf{X}_1 \quad \mathbf{X}_2 \quad \mathbf{X}_3]$ is

$$p = 1 + a + b, \tag{58}$$

and with (35) and (36) being the only two relationships among the columns of $\mathbf{X}$, the rank of $\mathbf{X}$ is

$$r_{\mathbf{X}} = 1 + a + b - 2 = a + b - 1\ . \tag{59}$$

i. The normal equations

With $\boldsymbol{\phi}$ omitted from the model equations it is also omitted from the normal equations (42), as are the last s of those equations. This yields

$$\begin{bmatrix} n_{\cdot\cdot} & \{_r\, n_{i\cdot}\}_i & \{_r\, n_{\cdot j}\}_j \\ & \{_d\, n_{i\cdot}\}_i & \{_m\, n_{ij}\}_{i,j} \\ \text{Symmetric} & & \{_d\, n_{\cdot j}\}_j \end{bmatrix} \begin{bmatrix} \mu^{\circ} \\ \boldsymbol{\alpha}^{\circ} \\ \boldsymbol{\beta}_b^{\circ} \end{bmatrix} = \begin{bmatrix} y_{\cdots} \\ \{_c\, y_{i\cdot\cdot}\}_i \\ \{_c\, y_{\cdot j\cdot}\}_j \end{bmatrix} . \tag{60}$$

j. Solving the normal equations

We show explicitly that a solution of (60) is $\mu^{\circ} = 0$, $\alpha_i^{\circ} = \hat{\mu}_i$ and $\beta_j^{\circ} = \hat{\tau}_j$ for $\hat{\tau}_j$ and $\hat{\mu}_i$ of (65) and (68) in Section 4.9b.

Notation. New notation is introduced for the submatrices and subvectors of (60), in order to simplify deriving a solution:

$$
\begin{aligned}
&\mathbf{n}'_a = [n_{1\cdot} \quad \cdots \quad n_{a\cdot}] &\qquad& \mathbf{n}'_b = [n_{\cdot 1} \quad \cdots \quad n_{\cdot b}] \\
&\mathbf{D}_a = \{{}_d\, n_{i\cdot}\} && \mathbf{D}_b = \{{}_d\, n_{\cdot j}\} \\
&\mathbf{N} = \{{}_m\, n_{ij}\}_{i=1,\, j=1}^{i=a,\, j=b} \\
&\mathbf{y}'_a = [y_{1\cdot\cdot} \quad \cdots \quad y_{a\cdot\cdot}] && \mathbf{y}'_b = [y_{\cdot 1\cdot} \quad \cdots \quad y_{\cdot b\cdot}] \\
&\bar{\mathbf{y}}'_a = [\bar{y}_{1\cdot\cdot} \quad \cdots \quad \bar{y}_{a\cdot\cdot}] && \bar{\mathbf{y}}'_b = [\bar{y}_{\cdot 1\cdot} \quad \cdots \quad \bar{y}_{\cdot b\cdot}] \\
&\quad = (\mathbf{D}_a^{-1}\mathbf{y}_a)' && \quad = (\mathbf{D}_b^{-}\mathbf{y}_b)\ .
\end{aligned}
\tag{61}
$$

With the preceding notation, equations (60) can then be written as

$$
\begin{bmatrix} n_{\cdot\cdot} & \mathbf{n}'_a & \mathbf{n}'_b \\ \mathbf{n}_a & \mathbf{D}_a & \mathbf{N} \\ \mathbf{n}_b & \mathbf{N}' & \mathbf{D}_b \end{bmatrix}
\begin{bmatrix} \mu^{\circ} \\ \boldsymbol{\alpha}^{\circ} \\ \boldsymbol{\beta}_b^{\circ} \end{bmatrix}
=
\begin{bmatrix} y_{\cdots} \\ \mathbf{y}_a \\ \mathbf{y}_b \end{bmatrix} .
\tag{62}
$$

Definition of the terms in (62) makes their description quite straightforward: $\mathbf{n}'_a$ and $\mathbf{n}'_b$ are vectors of the numbers of observations in the rows and columns, respectively, of the 2-way layout, and $\mathbf{D}_a$ and $\mathbf{D}_b$ are diagonal matrices of those same numbers. $\mathbf{N}$ is the $a \times b$ matrix of the n_{ij}s, and $\mathbf{y}_a$ and $\mathbf{y}_b$ are vectors of the row and column totals, respectively, of the observations.

Equations (62) have order $1 + a + b$ and their rank is $r_{\mathbf{X}} = a + b - 1$ from (59). The a rows through $\mathbf{D}_a$ sum to the first row, as do the b rows through $\mathbf{D}_b$. Therefore a solution can be obtained from the algorithm of Section 7.1j, by finding in the matrix on the left-hand side of (62) a nonsingular, principal submatrix of order $a + b - 1$. A convenient way of doing this, and one which leads to the results of Section 4.9b, is to use the square matrix of order $a + b - 1$ that consists of that matrix after dropping its first and last rows and columns. This is achieved by defining

$\mathbf{N}_2$ as $\mathbf{N}$ with its bth column deleted;

$\mathbf{D}_2$ as $\mathbf{D}_b$ with its bth row and column deleted; (63)

$\mathbf{n}_2$, $\mathbf{y}_2$ and $\boldsymbol{\beta}_2$ as $\mathbf{n}_b$, $\mathbf{y}_b$ and $\boldsymbol{\beta}_b$ with their bth elements deleted.

Then equations (62) are

$$\begin{bmatrix} n_{..} & \mathbf{n}'_a & \mathbf{n}'_2 & n_{.b} \\ \mathbf{n}_a & \mathbf{D}_a & \mathbf{N}_2 & \mathbf{c}_b \\ \mathbf{n}_2 & \mathbf{N}'_2 & \mathbf{D}_2 & \mathbf{0} \\ n_{.b} & \mathbf{c}'_b & \mathbf{0} & n_{.b} \end{bmatrix} \begin{bmatrix} \mu^o \\ \boldsymbol{\alpha}^o \\ \boldsymbol{\beta}_2^o \\ \beta_b^o \end{bmatrix} = \begin{bmatrix} y_{...} \\ \mathbf{y}_a \\ \mathbf{y}_2 \\ y_{.b.} \end{bmatrix}. \tag{64}$$

where $\mathbf{c}_b$ is the bth column of $\mathbf{N}$.

Equations (64) are now solved using

$$\mathbf{G} = \begin{bmatrix} 0 & \mathbf{0} & \mathbf{0} & 0 \\ \mathbf{0} & \multicolumn{2}{c}{\begin{pmatrix} \mathbf{D}_a & \mathbf{N}_2 \\ \mathbf{N}'_2 & \mathbf{D}_2 \end{pmatrix}^{-1}} & 0 \\ \mathbf{0} & & & 0 \\ 0 & \mathbf{0} & \mathbf{0} & 0 \end{bmatrix} \tag{65}$$

as a generalized inverse of the matrix on the left-hand side of (64). The inverted matrix in $\mathbf{G}$ is, from (36) of Section 7.3,

$$\begin{bmatrix} \mathbf{D}_a & \mathbf{N}_2 \\ \mathbf{N}'_2 & \mathbf{D}_2 \end{bmatrix}^{-1} = \begin{bmatrix} \mathbf{D}_a^{-1} & \mathbf{0} \\ \mathbf{0} & \mathbf{0} \end{bmatrix}$$

$$+ \begin{bmatrix} -\mathbf{D}_a^{-1}\mathbf{N}_2 \\ \mathbf{I} \end{bmatrix} \left(\mathbf{D}_2 - \mathbf{N}'_2\mathbf{D}_a^{-1}\mathbf{N}_2\right)^{-1} \left[-\mathbf{N}'_2\mathbf{D}_a^{-1} \quad \mathbf{I}\right]. \tag{66}$$

To simplify (66) define

$$\mathbf{C} = \mathbf{D}_b - \mathbf{N}'\mathbf{D}_a^{-1}\mathbf{N} = \mathbf{D}\{n_{.j}\} - \{n_{ij}\}'\mathbf{D}\{1/n_{i.}\}\{n_{ij}\}$$

$$= \{_m\, c_{jj'}\}_{j=1,\, j'=1}^{j=b,\, j'=b} \quad \text{with} \quad \begin{cases} c_{jj} = n_{.j} - \Sigma_i n_{ij}^2/n_{i.} \\ c_{jj'} = -\Sigma_i n_{ij}n_{ij'}/n_{i.} \quad \text{for } j \neq j' \end{cases} \tag{67}$$

$$= \mathbf{C}_b \text{ of } (54),$$

with c_{jj} and $c_{jj'}$ in both (54) and (67) being the same as in (66) of Section 4.9b. Now from (67), comparable to (65) and for use in (66), define

$$\mathbf{C}_2 = \mathbf{C}_b \quad \text{with } b\text{th row and column deleted}$$
$$= \mathbf{D}_2 - \mathbf{N}'_2\mathbf{D}_a^{-1}\mathbf{N}_2\,. \tag{68}$$

Then substituting (68) into (66) gives

$$\begin{bmatrix} \mathbf{D}_a & \mathbf{N}_2 \\ \mathbf{N}_2' & \mathbf{D}_2 \end{bmatrix}^{-1} = \begin{bmatrix} \mathbf{D}_a^{-1} + \mathbf{M}'\mathbf{C}_2^{-1}\mathbf{M} & -\mathbf{M}'\mathbf{C}_2^{-1} \\ -\mathbf{C}_2^{-1}\mathbf{M} & \mathbf{C}_2^{-1} \end{bmatrix}, \tag{69}$$

where

$$\mathbf{M} = \mathbf{D}_a^{-1}\mathbf{N}_2 = \{_m\, n_{ij}/n_{i\cdot}\}_{i=1,\, j=1}^{i=a,\, j=b-1} . \tag{70}$$

The solution of (64) corresponding to (65) is obtained by pre-multiplying the right-hand side of (64) by **G** of (65). This gives

$$\mu^\circ = 0 \quad \text{and} \quad \beta_b^\circ = 0,$$

and although (69) could be used in (65) for deriving $\boldsymbol{\alpha}^\circ$ and $\boldsymbol{\beta}_2^\circ$, the use of (66) is more instructive. That gives

$$\begin{bmatrix} \boldsymbol{\alpha}^\circ \\ \boldsymbol{\beta}_2^\circ \end{bmatrix} = \begin{bmatrix} \mathbf{D}_a^{-1} & \mathbf{0} \\ \mathbf{0} & \mathbf{0} \end{bmatrix} \begin{bmatrix} \mathbf{y}_a \\ \mathbf{y}_2 \end{bmatrix} + \begin{bmatrix} -\mathbf{D}_a^{-1}\mathbf{N}_2 \\ \mathbf{I} \end{bmatrix} \mathbf{C}_2^{-1}(\mathbf{y}_2 - \mathbf{N}_2'\mathbf{D}_a^{-1}\mathbf{y}_a),$$

which is

$$\boldsymbol{\alpha}^\circ = \bar{\mathbf{y}}_a - \mathbf{D}_a^{-1}\mathbf{N}_2\boldsymbol{\beta}_2^\circ \tag{71}$$

and

$$\boldsymbol{\beta}_2^\circ = \mathbf{C}_2^{-1}(\mathbf{y}_2 - \mathbf{N}_2'\mathbf{D}_a^{-1}\mathbf{y}_a) = \mathbf{C}_2^{-1}(\mathbf{y}_2 - \mathbf{N}_2'\bar{\mathbf{y}}_a) . \tag{72}$$

But for (72), using (61) and (63) gives

$$\mathbf{y}_2 - \mathbf{N}_2'\mathbf{D}_a^{-1}\mathbf{y}_a = \left\{_c \left(y_{\cdot j \cdot} - \sum_{i=1}^{a} n_{ij}\bar{y}_{i\cdot\cdot} \right) \right\}_{j=1}^{j=b-1}, \tag{73}$$

$$= \{_c\, r_j\}_{j=1}^{j=b-1} = \mathbf{r}_2, \quad \text{say} \tag{74}$$

for r_j of (67) in Section 4.9b. Therefore (72) is

$$\boldsymbol{\beta}_2^\circ = \mathbf{C}_2^{-1}\mathbf{r}_2 . \tag{75}$$

This is, of course, $\mathbf{C}_2\boldsymbol{\beta}_2^\circ = \mathbf{r}_2$, which is identical to (65) of Section 4.9b.

Thus, as in (50),

$$\boldsymbol{\beta}_2^{\circ} = \{\hat{\tau}_j\}_{j=1}^{j=b-1} \quad \text{for } \hat{\tau}_j \text{ of (65) in Section 4.9b} . \tag{76}$$

And then (71) is

$$\boldsymbol{\alpha}^{\circ} = \bar{\mathbf{y}}_a - \mathbf{D}_a^{-1}\mathbf{N}_2\boldsymbol{\beta}_2^{\circ} = \left\{_c \ \bar{y}_{i\cdot\cdot} - \sum_{j=1}^{b-1} n_{ij}\beta_j^{\circ}/n_{i\cdot}\right\}_{i=1}^{i=a} . \tag{77}$$

But $\beta_j^{\circ} = \hat{\tau}_j$ of (76) means that (77) is the same as (68) of Section 4.9b. Thus, also as in (50),

$$\boldsymbol{\alpha}^{\circ} = \left\{_c \ \hat{\mu}_i\right\}_{i=1}^{i=a} \quad \text{for } \hat{\mu}_i \text{ of (68) in Section 4.9b} . \tag{78}$$

Thus the complete solution vector so obtained is

$$\begin{bmatrix} \mu_{\circ} \\ \boldsymbol{\alpha}^{\circ} \\ \boldsymbol{\beta}_2^{\circ} \\ \beta_b^{\circ} \end{bmatrix} = \begin{bmatrix} 0 \\ \left\{_c \left(\bar{y}_{i\cdot\cdot} - \sum_{j=1}^{b-1} n_{ij}\beta_j^{\circ}/n_{i\cdot}\right)\right\}_{i=1}^{i=a} \\ \mathbf{C}_2^{-1}\mathbf{r}_2 \\ 0 \end{bmatrix} = \begin{bmatrix} 0 \\ \{_c \ \hat{\mu}_i\}_{i=1}^{i=a} \\ \{_c \ \hat{\tau}_j\}_{j=1}^{j=b-1} \\ 0 \end{bmatrix} \tag{79}$$

for $\hat{\mu}_i$ and $\hat{\tau}_j$ of (68) and (65), respectively, in Section 4.9b.

k. Sampling variances

The sampling variance of the BLUE of an estimable function $\mathbf{q}'\boldsymbol{\beta}$ in the general linear model $E(\mathbf{y}) = \mathbf{X}\boldsymbol{\beta}$ is, from (132) of Section 8.7c, $v(\mathbf{q}'\boldsymbol{\beta}^{\circ}) = \mathbf{q}'\mathbf{G}\mathbf{q}\sigma^2$, where $\mathbf{G}$ is a generalized inverse of $\mathbf{X}'\mathbf{X}$. From using (65) and (69) we have

$$\mathbf{G} = \begin{bmatrix} 0 & 0 & 0 & 0 \\ \mathbf{0} & \mathbf{D}_a^{-1} + \mathbf{M}\mathbf{C}_2^{-1}\mathbf{M}' & -\mathbf{M}\mathbf{C}_2^{-1} & \mathbf{0} \\ \mathbf{0} & -\mathbf{C}_2^{-1}\mathbf{M}' & \mathbf{C}_2^{-1} & \mathbf{0} \\ 0 & \mathbf{0} & \mathbf{0} & 0 \end{bmatrix}, \tag{80}$$

from which $\mathbf{q}'\mathbf{G}\mathbf{q}$ can be obtained for any appropriate $\mathbf{q}$.

l. Estimating cell means and contrasts

The BLUE of the cell mean $\mu + \alpha_i + \beta_j$ is

$$\text{BLUE}(\mu + \alpha_i + \beta_j) = \mu^o + \alpha_i^o + \beta_j^o \tag{81}$$

$$= \bar{y}_{i..} - \sum_{j=1}^{b-1} n_{ij}\beta_j^o/n_{i.} + \beta_j^o, \tag{82}$$

from (79). The sampling variance of (81) comes from $\mathbf{q}'\mathbf{G}\mathbf{q}\sigma^2$ being the sampling variance of the BLUE of the estimable function $\mathbf{q}'\boldsymbol{\beta}$. To utilize this, note that $\mathbf{G}$ has order $1 + a + (b - 1) + 1 = 1 + a + b$, and define

$$\mathbf{G} = \{_m\, g_{t,t'}\}_{t,t'=1}^{t,t'=1+a+b} \tag{83}$$

with $g_{tt'} = 0$ whenever t and/or t' are 1 or $1 + a + b$. Then the variance of (81) is

$$v[\text{BLUE}(\mu + \alpha_i + \beta_j)] = v(\mu^o + \alpha_i^o + \beta_j^o)$$

$$= (g_{1+i,1+i} + g_{1+a+j,1+a+j} + 2g_{1+i,1+a+j})\sigma^2 \,. \tag{84}$$

Similarly, from (51), (79) and (83), $\text{BLUE}(\alpha_i - \alpha_h) = \alpha_i^o - \alpha_h^o$ with

$$v(\alpha_i^o - \alpha_h^o) = (g_{1+i,1+i} + g_{1+h,1+h} - 2g_{1+i,1+h})\sigma^2 \tag{85}$$

and $\text{BLUE}(\beta_j - \beta_k) = \beta_j^o - \beta_k^o$ with

$$v(\beta_j^o - \beta_k^o) = (g_{1+a+j,1+a+j} + g_{1+a+k,1+a+k} - 2g_{1+a+j,1+a+k})\sigma^2 \,. \tag{86}$$

The preceding variance results are particularly important when one comes to using certain statistical computing-package output. Some output gives the values $\hat{v}(\alpha_i^o) = g_{1+i,1+i}\hat{\sigma}^2$ [and $\hat{v}(\beta_j^o) = g_{1+a+j,1+a+j}\hat{\sigma}^2$]. But these alone are insufficient for then deriving $\hat{v}(\alpha_i^o - \alpha_h^o)$ because this, as is seen in (85), requires not only $g_{1+i,1+i}$ and $g_{1+h,1+h}$, but also $g_{1+i,1+h}$, and the latter is not part of computer output that lists just $\hat{v}(\alpha_i^o)$. Only if $\mathbf{G}$ is provided (or can be obtained) so as to yield $g_{1+i,1+h}$, can $\hat{v}(\alpha_i^o - \alpha_h^o)$ be derived.

It bears repetition to again emphasize that (84) holds for all cells in the data grid, even for those containing no data. Thus in the no-interaction model, all population cell means are estimable with estimator given by (81), whereas in the interaction model, cell means of only the filled cells are estimable, the BLUE there being $\bar{y}_{ij\cdot}$, as in (23). Similarly, in the no-interaction model, (85) and (86) apply for all pairs α_i, α_h and β_j, β_k; and this is not the case for the with-interaction model—see (21).

Example (continued). The normal equations (62) are (40) with the last five rows and columns omitted:

$$\left[\begin{array}{c|cc|ccc} 10 & 7 & 3 & 5 & 4 & 1 \\ \hline 7 & 7 & \cdot & 3 & 3 & 1 \\ 3 & \cdot & 3 & 2 & 1 & \cdot \\ \hline 5 & 3 & 2 & 5 & \cdot & \cdot \\ 4 & 3 & 1 & \cdot & 4 & \cdot \\ 1 & 1 & \cdot & \cdot & \cdot & 1 \end{array}\right] \left[\begin{array}{c} \mu^\circ \\ \hline \alpha_1^\circ \\ \alpha_2^\circ \\ \hline \beta_1^\circ \\ \beta_2^\circ \\ \beta_3^\circ \end{array}\right] = \left[\begin{array}{c} 99 \\ \hline 42 \\ 57 \\ \hline 45 \\ 48 \\ 6 \end{array}\right]. \tag{87}$$

The elements $c_{jj'}$ of (67) are as follows:

$$c_{11} = 5 - \left(\frac{3^2}{7} + \frac{2^2}{3}\right) = \frac{50}{21} \qquad c_{12} = -\frac{3(3)}{7} - \frac{2(1)}{3} = -\frac{41}{21} \qquad c_{13} = -\frac{3(1)}{7} = -\frac{9}{21}$$

$$c_{22} = 4 - \left(\frac{3^2}{7} + \frac{1^2}{3}\right) = \frac{50}{21} \qquad c_{23} = \frac{-3(1)}{7} = -\frac{9}{21}$$

$$c_{33} = 1 - \frac{1^2}{7} = \frac{18}{21}.$$

Hence $\mathbf{C}_b$ of (54) and (67), and $\mathbf{C}_2$ of (68) are

$$\mathbf{C}_b = \frac{1}{21}\begin{bmatrix} 50 & -41 & -9 \\ -41 & 50 & -9 \\ -9 & -9 & 18 \end{bmatrix} \quad \text{and} \quad \mathbf{C}_2 = \tfrac{1}{21}\begin{bmatrix} 50 & -41 \\ -41 & 50 \end{bmatrix}. \tag{88}$$

In $\mathbf{C}_b$ it is readily evident that the elements of each row sum to zero—as in (56). Also, applying terms from the right-hand side of (87) to (73) and (74) gives

$$\mathbf{r} = \begin{bmatrix} 45 - \{3(6) + 2(19)\} \\ 48 - \{3(6) + 1(19)\} \\ 6 - 1(6) \end{bmatrix} = \begin{bmatrix} -11 \\ 11 \\ 0 \end{bmatrix} \quad \text{and} \quad \mathbf{r}_2 = \begin{bmatrix} -11 \\ 11 \end{bmatrix}. \tag{89}$$

Hence (75) is

$$\boldsymbol{\beta}_2^\circ = \begin{bmatrix} \beta_1^\circ \\ \beta_2^\circ \end{bmatrix} = \left[\tfrac{1}{21}\begin{pmatrix} 50 & -41 \\ -41 & 50 \end{pmatrix}\right]^{-1}\begin{bmatrix} -11 \\ 11 \end{bmatrix}$$

$$= \tfrac{1}{39}\begin{bmatrix} 50 & 41 \\ 41 & 50 \end{bmatrix}\begin{bmatrix} -11 \\ 11 \end{bmatrix} = \begin{bmatrix} -\tfrac{33}{13} \\ \tfrac{33}{13} \end{bmatrix} \tag{90}$$

and then (77) is

$$\begin{bmatrix} \alpha_1^\circ \\ \alpha_2^\circ \end{bmatrix} = \begin{bmatrix} 6 - \{3(-\tfrac{33}{13}) + 3(\tfrac{33}{13})\}/7 \\ 19 - \{2(-\tfrac{33}{13}) + 1(\tfrac{33}{13})\}/3 \end{bmatrix} = \begin{bmatrix} 6 \\ 19\tfrac{11}{13} \end{bmatrix}. \tag{91}$$

Therefore the complete solution vector (79) is

$$[\mu^\circ \quad \alpha_1^\circ \quad \alpha_2^\circ \quad \beta_1^\circ \quad \beta_2^\circ \quad \beta_3^\circ] = [0 \quad 6 \quad 19\tfrac{11}{13} \quad -2\tfrac{7}{13} \quad 2\tfrac{7}{13} \quad 0] \tag{92}$$

Then, with **M** of (70) being

$$\mathbf{M} = \begin{bmatrix} \tfrac{1}{7} & \cdot \\ \cdot & \tfrac{1}{3} \end{bmatrix}\begin{bmatrix} 3 & 3 \\ 2 & 1 \end{bmatrix} = \begin{bmatrix} \tfrac{3}{7} & \tfrac{3}{7} \\ \tfrac{2}{3} & \tfrac{1}{3} \end{bmatrix} = \tfrac{1}{21}\begin{bmatrix} 9 & 9 \\ 14 & 7 \end{bmatrix}$$

and hence

$$\mathbf{MC}_2^{-1} = \tfrac{1}{21}\left(\tfrac{1}{39}\right)\begin{bmatrix} 9 & 9 \\ -14 & 7 \end{bmatrix}\begin{bmatrix} 50 & 41 \\ 41 & 50 \end{bmatrix} = \tfrac{1}{39}\begin{bmatrix} 39 & 39 \\ 47 & 44 \end{bmatrix},$$

we have

$$\mathbf{MC}_2^{-1}\mathbf{M}' = \tfrac{1}{39}\left(\tfrac{1}{21}\right)\begin{bmatrix} 39 & 39 \\ 47 & 44 \end{bmatrix}\begin{bmatrix} 9 & 14 \\ 9 & 7 \end{bmatrix} = \begin{bmatrix} \tfrac{6}{7} & 1 \\ 1 & \tfrac{46}{39} \end{bmatrix}.$$

Then, using $\frac{1}{7} + \frac{6}{7} = 1$ and $\frac{1}{3} + \frac{46}{39} = \frac{59}{39}$, **G** of (80) is

$$\mathbf{G} = \begin{bmatrix} 0 & 0 & 0 & 0 & 0 & 0 \\ 0 & 1 & 1 & -1 & -1 & 0 \\ 0 & 1 & \multicolumn{3}{c}{\multirow{3}{*}{$\tfrac{1}{39}\begin{bmatrix} 59 & -47 & -44 \\ -47 & 50 & 41 \\ -44 & 41 & 50 \end{bmatrix}$}} & 0 \\ 0 & -1 & & & & 0 \\ 0 & -1 & & & & 0 \\ 0 & 0 & 0 & 0 & 0 & 0 \end{bmatrix}. \tag{93}$$

Examples of estimating estimable functions using (92), and of variances of such estimators using (93), are as follows.

$$\text{BLUE}(\mu + \alpha_1 + \beta_1) = 0 + 6 - 2\tfrac{7}{13} = 3\tfrac{6}{13},$$

$$\text{has variance } \left(1 - 2 + \tfrac{50}{39}\right)\sigma^2 = \tfrac{11}{39}\sigma^2;$$

$$\text{BLUE}(\beta_1 - \beta_2) \quad = -2\tfrac{7}{13} - 2\tfrac{7}{13} = -5\tfrac{1}{13},$$

$$\text{has variance } (50 + 50 - 82)\sigma^2/39 = \tfrac{6}{13}\sigma^2 .$$

9.3. SUMS OF SQUARES FOR THE OVERPARAMETERIZED MODEL

Tables 9.1 and 9.2 show analysis-of-variance style sums of squares and their associated hypotheses for the with-interaction and the no-interaction models, respectively. The only difference in the sums of squares is that Table 9.2 for the no-interaction model has no term $R(\boldsymbol{\phi} \mid \mu, \boldsymbol{\alpha}, \boldsymbol{\beta})$ for interaction.

Table 9.1 shows the equivalence of all the sums of squares to those of the cell means model in Table 4.9, for which in Table 4.8, specific formulae for the sums of squares are indicated. Other expressions for the sums of squares are also worth noting.

a. $R(\mu)$, $R(\boldsymbol{\alpha} \mid \mu)$ and $R(\boldsymbol{\beta} \mid \mu)$

For the sake of completeness we begin with the three easy terms, the first being the standard

$$R(\mu) = n_{..}\bar{y}^2_{...} \ . \tag{94}$$

Next, similar to $\mathscr{R}(\mu_i \mid \mu)$ of Section 2.8 for the 1-way classification, we have

$$R(\boldsymbol{\alpha} \mid \mu) = R(\mu, \boldsymbol{\alpha}) - R(\mu) = \Sigma_i n_{i.}\bar{y}^2_{i..} - n_{..}\bar{y}^2_{...} = \Sigma_i n_{i.}(\bar{y}_{i..} - \bar{y}_{...})^2 \tag{95}$$

and, analogously, $$R(\boldsymbol{\beta} \mid \mu) = \Sigma_j n_{.j}(\bar{y}_{.j.} - \bar{y}_{...})^2 \ . \tag{96}$$

b. $R(\boldsymbol{\beta} \mid \mu, \boldsymbol{\alpha})$ and $R(\boldsymbol{\alpha} \mid \mu, \boldsymbol{\beta})$

By definition

$$R(\boldsymbol{\beta} \mid \mu, \boldsymbol{\alpha}) = R(\mu, \boldsymbol{\alpha}, \boldsymbol{\beta}) - R(\mu, \boldsymbol{\alpha}) = R(\mu, \boldsymbol{\alpha}, \boldsymbol{\beta}) - \Sigma_i n_{i.}\bar{y}^2_{i..} \ . \tag{97}$$

$R(\mu, \boldsymbol{\alpha}, \boldsymbol{\beta})$ is the sum of squares due to fitting the no-interaction model.

Therefore for that model it is, from Section 8.4a, the sum of products of solutions of the normal equations each multiplied by the corresponding right-hand side of those equations. These solutions and right-hand sides are available in (79) and (64), respectively. Therefore

$$\begin{aligned} R(\mu, \boldsymbol{\alpha}, \boldsymbol{\beta}) &= \Sigma_i \alpha_i^\circ y_{i\cdot\cdot} + \Sigma_j \beta_j^\circ y_{\cdot j\cdot} \qquad (98) \\ &= \boldsymbol{\alpha}^{\circ\prime}\mathbf{y}_a + \boldsymbol{\beta}_2^{\circ\prime}\,\mathbf{y}_2 \\ &= \left(\bar{\mathbf{y}}_a - \mathbf{D}_a^{-1}\mathbf{N}_2\boldsymbol{\beta}_2^\circ\right)'\mathbf{y}_a + \boldsymbol{\beta}_2^{\circ\prime}\,\mathbf{y}_2, \quad \text{from (77)} \\ &= \bar{\mathbf{y}}_a'\mathbf{y}_a + \boldsymbol{\beta}_2^{\circ\prime}\left(\mathbf{y}_2 - \mathbf{N}_2'\mathbf{D}_a^{-1}\mathbf{y}_a\right) \\ &= \Sigma_i n_{i\cdot}\bar{y}_{i\cdot\cdot}^2 + \boldsymbol{\beta}_2^{\circ\prime}\,\mathbf{r}_2 \qquad (99) \end{aligned}$$

from (73) and (74). Hence in (97)

$$R(\boldsymbol{\beta}\,|\,\mu, \boldsymbol{\alpha}) = \boldsymbol{\beta}_2^{\circ\prime}\,\mathbf{r}_2 = \mathbf{r}_2'\mathbf{C}_2^{-1}\mathbf{r}_2\;. \qquad (100)$$

This is a useful result.

There is also an analogous result for $R(\boldsymbol{\alpha}\,|\,\mu, \boldsymbol{\beta})$. In solving the normal equations (62) for the no-interaction model we chose to use a generalized inverse that had null rows (and columns) corresponding to μ° and β_b°. This led to $\mathbf{C}_2\boldsymbol{\beta}_2^\circ = \mathbf{r}_2$ for $\mathbf{C}_2$ and $\mathbf{r}_2$ of (68) and (74), respectively, and thence to $R(\boldsymbol{\beta}\,|\,\mu, \boldsymbol{\alpha}) = \mathbf{r}_2'\mathbf{C}_2^{-1}\mathbf{r}_2$ of (100). But those normal equations could just as well have been solved with a generalized inverse that had null rows and columns corresponding to μ° and α_a°. This would lead to $\mathbf{T}_1\boldsymbol{\alpha}_1^{\circ\circ} = \mathbf{u}_1$ for

$$\mathbf{T}_1 = \{_m\, t_{ii'}\}_{i,i'=1}^{a-1}, \qquad t_{ii} = n_{i\cdot} - \Sigma_j n_{ij}^2/n_{\cdot j}$$

$$t_{ii}' = -\,\Sigma_j n_{ij}n_{i'j}/n_{\cdot j},\; i \neq i', \qquad (101)$$

analogous to (67), and

$$\mathbf{u}_1 = \{_c\, u_i\}_{i=1}^{a-1} \quad \text{for } u_i = y_{i\cdot\cdot} - \sum_{j=1}^{b} n_{ij}\bar{y}_{\cdot j\cdot}, \qquad (102)$$

analogous to (74). Then

$$\boldsymbol{\alpha}_1^{\circ\circ} = \mathbf{T}_1^{-1}\mathbf{u}_1 \qquad (103)$$

and

$$R(\boldsymbol{\alpha}\,|\,\mu, \boldsymbol{\beta}) = \mathbf{u}_1'\mathbf{T}_1^{-1}\mathbf{u}_1\;. \qquad (104)$$

Of course, exactly the same thing would be achieved by interchanging the definition of rows and columns and then using the $\mathbf{C}_2\boldsymbol{\beta}_2 = \mathbf{r}_2$ form. But (104) provides a clear distinction between $R(\boldsymbol{\alpha}\,|\,\mu, \boldsymbol{\beta})$ and $R(\boldsymbol{\beta}\,|\,\mu, \boldsymbol{\alpha})$.

It is not essential to use (104). Once $R(\boldsymbol{\beta}\,|\,\mu, \boldsymbol{\alpha}) = \mathbf{r}_2'\mathbf{C}_2^{-1}\mathbf{r}_2$ is obtained then, because

$$R(\mu, \boldsymbol{\alpha}, \boldsymbol{\beta}) = R(\boldsymbol{\beta}\,|\,\mu, \boldsymbol{\alpha}) + R(\mu, \boldsymbol{\alpha}) = R(\boldsymbol{\alpha}\,|\,\mu, \boldsymbol{\beta}) + R(\mu, \boldsymbol{\beta}),$$

$R(\boldsymbol{\alpha}\,|\,\mu, \boldsymbol{\beta})$ is obtainable, without the need for (104), as

$$\begin{aligned} R(\boldsymbol{\alpha}\,|\,\mu, \boldsymbol{\beta}) &= R(\boldsymbol{\beta}\,|\,\mu, \boldsymbol{\alpha}) + R(\mu, \boldsymbol{\alpha}) - R(\mu, \boldsymbol{\beta}) \\ &= R(\boldsymbol{\beta}\,|\,\mu, \boldsymbol{\alpha}) + R(\boldsymbol{\alpha}\,|\,\mu) - R(\boldsymbol{\beta}\,|\,\mu)\,. \end{aligned} \tag{105}$$

c. $R(\boldsymbol{\phi}\,|\,\mu, \boldsymbol{\alpha}, \boldsymbol{\beta})$

This sum of squares is

$$R(\mu, \boldsymbol{\alpha}, \boldsymbol{\beta}, \boldsymbol{\phi}) - R(\mu, \boldsymbol{\alpha}, \boldsymbol{\beta}) = \Sigma_i\Sigma_j n_{ij}\bar{y}_{ij\cdot}^2 - R(\mu, \boldsymbol{\alpha}, \boldsymbol{\beta}), \tag{106}$$

where $R(\mu, \boldsymbol{\alpha}, \boldsymbol{\beta})$ is given by (99). Thus

$$\begin{aligned} R(\boldsymbol{\phi}\,|\,\mu, \boldsymbol{\alpha}, \boldsymbol{\beta}) &= \Sigma_i\Sigma_j n_{ij}\bar{y}_{ij\cdot}^2 - \Sigma_i n_{i\cdot}\bar{y}_{i\cdot\cdot}^2 - \boldsymbol{\beta}_2^{\circ\prime}\mathbf{r}_2 \\ &= \Sigma_i\Sigma_j n_{ij}\bar{y}_{ij\cdot}^2 - \Sigma_j n_{\cdot j}\bar{y}_{\cdot j\cdot}^2 - \boldsymbol{\alpha}_1^{\circ\circ\prime}\mathbf{u}_1, \end{aligned} \tag{107}$$

the latter by analogy with (107) and using (103).

There is yet one further method of calculation—and, in certain cases, the easiest one. We know from (42) of Section 5.4e that $R(\boldsymbol{\phi}|\mu, \boldsymbol{\alpha}, \boldsymbol{\beta})$ is the sum of squares for testing a hypothesis about interaction terms in the with-interaction model. That hypothesis is of the form H: $\mathbf{K}'\boldsymbol{\mu} = \mathbf{0}$ and so the numerator sum of squares for testing it is, from (149) of Section 8.8c,

$$Q = \bar{\mathbf{y}}'\mathbf{K}\left[\mathbf{K}'\{_d\, 1/n_{ij}\}\mathbf{K}\right]^{-1}\mathbf{K}'\bar{\mathbf{y}}\,. \tag{108}$$

Hence, if the pattern of filled cells is such that the interaction hypothesis of (42) in Section 5.4e can be easily specified as H: $\mathbf{K}'\boldsymbol{\mu} = \mathbf{0}$, then Q of (108) is $R(\boldsymbol{\phi}\,|\,\mu, \boldsymbol{\alpha}, \boldsymbol{\beta})$. For example, consider the case of 2 rows and 3 columns with all cells filled. The hypothesis tested by $R(\boldsymbol{\phi}\,|\,\mu, \boldsymbol{\alpha}, \boldsymbol{\beta})$ can be specified as

$$\text{H:}\ \begin{cases} \mu_{11} - \mu_{12} - \mu_{21} + \mu_{22} = 0 \\ \mu_{11} - \mu_{13} - \mu_{21} + \mu_{23} = 0 \end{cases}.$$

Hence

$$\mathbf{K}'\bar{\mathbf{y}} = \begin{bmatrix} \bar{y}_{11\cdot} - \bar{y}_{12\cdot} - \bar{y}_{21\cdot} + \bar{y}_{22\cdot} \\ \bar{y}_{11\cdot} - \bar{y}_{13\cdot} - \bar{y}_{21\cdot} + \bar{y}_{23\cdot} \end{bmatrix} \tag{109}$$

and $\mathbf{K}'\left\{_d\, 1/n_{ij}\right\}\mathbf{K}$

$$= \begin{bmatrix} \dfrac{1}{n_{11}} + \dfrac{1}{n_{12}} + \dfrac{1}{n_{21}} + \dfrac{1}{n_{22}} & \dfrac{1}{n_{11}} + \dfrac{1}{n_{21}} \\ \dfrac{1}{n_{11}} + \dfrac{1}{n_{21}} & \dfrac{1}{n_{11}} + \dfrac{1}{n_{13}} + \dfrac{1}{n_{21}} + \dfrac{1}{n_{23}} \end{bmatrix}. \tag{110}$$

Using (109) and (110) in (108) gives $Q = R(\boldsymbol{\phi} \mid \mu, \boldsymbol{\alpha}, \boldsymbol{\beta})$.

A variation on the preceding example would be if the (2, 3) cell was empty. Then $Q = R(\boldsymbol{\phi} \mid \mu, \boldsymbol{\alpha}, \boldsymbol{\beta})$

$$= (\bar{y}_{11\cdot} - \bar{y}_{12\cdot} - \bar{y}_{21\cdot} + \bar{y}_{22\cdot})^2 \Big/ \left(\frac{1}{n_{11}} + \frac{1}{n_{12}} + \frac{1}{n_{21}} + \frac{1}{n_{22}} \right). \tag{111}$$

Whenever calculating $R(\boldsymbol{\phi} \mid \mu, \boldsymbol{\alpha}, \boldsymbol{\beta})$ by means of Q is feasible, it has the advantage of our then being able to calculate $R(\mu, \boldsymbol{\alpha}, \boldsymbol{\beta})$ as

$$R(\mu, \boldsymbol{\alpha}, \boldsymbol{\beta}) = R(\mu, \boldsymbol{\alpha}, \boldsymbol{\beta}, \boldsymbol{\phi}) - R(\boldsymbol{\phi} \mid \mu, \boldsymbol{\alpha}, \boldsymbol{\beta}) = \Sigma_i \Sigma_j n_{ij} \bar{y}_{ij\cdot}^2 - Q \tag{112}$$

without needing $\mathbf{C}_2$ and $\mathbf{r}_2$; and then

$$R(\boldsymbol{\beta} \mid \mu, \boldsymbol{\alpha}) = R(\mu, \boldsymbol{\alpha}, \boldsymbol{\beta}) - \Sigma_i n_{i\cdot} \bar{y}_{i\cdot\cdot}^2$$

and

$$R(\boldsymbol{\alpha} \mid \mu, \boldsymbol{\beta}) = R(\mu, \boldsymbol{\alpha}, \boldsymbol{\beta}) - \Sigma_j n_{\cdot j} \bar{y}_{\cdot j\cdot}^2 \,. \tag{113}$$

Of course, for any but small data sets one would invariably use a computing package to do the arithmetic for these sums of squares, and all of these different expressions would be somewhat immaterial.

d. Example (continued)

Using Table 8.8 gives

$$(94) \text{ as } R(\mu) = 10(9.9)^2 = 980.1, \tag{114}$$

$$(95) \text{ as } R(\boldsymbol{\alpha} \mid \mu) = 7(6 - 9.9)^2 + 3(19 - 9.9)^2 = 354.9, \tag{115}$$

and (96) as

$$R(\boldsymbol{\beta}|\mu) = 5(9-9.9)^2 + 4(12-9.9)^2 + (6-9.9)^2 = 36.9 \ . \qquad (116)$$

Then $\boldsymbol{\beta}_2^{o\prime} = [-\frac{33}{13} \quad \frac{33}{13}]$ from (90) and $\mathbf{r}_2' = [-11 \quad 11]$ from (89) give

$$(97) \quad \text{as} \quad R(\boldsymbol{\beta}|\mu, \boldsymbol{\alpha}) = [-\tfrac{33}{13} \quad \tfrac{33}{13}]\begin{bmatrix} -11 \\ 11 \end{bmatrix} = 55\tfrac{11}{13} \ . \qquad (117)$$

Also, with (115) and (116)

$$(105) \quad \text{is} \quad R(\boldsymbol{\alpha}|\mu, \boldsymbol{\beta}) = 55\tfrac{11}{13} + 354.9 - 36.9 = 373\tfrac{11}{13} \ . \qquad (118)$$

This can be confirmed using (104) which in this case is, from (101), (102) and Table 8.8

$$\frac{u_1^2}{t_{11}} = \frac{\{42 - [3(9) + 3(12) + 1(6)]\}^2}{7 - (3^2/5 + 3^2/4 + 1^2/1)} = \frac{27^2}{\frac{39}{20}} = 373\tfrac{11}{13} \ .$$

Then from (114), (115) and (117)

$$R(\mu, \boldsymbol{\alpha}, \boldsymbol{\beta}) = R(\mu) + R(\boldsymbol{\alpha}|\mu) + R(\boldsymbol{\beta}|\mu, \boldsymbol{\alpha})$$
$$= 980.1 + 354.9 + 55\tfrac{11}{13} = 1390\tfrac{11}{13} \ . \qquad (119)$$

This can be further confirmed from the R-algorithm of (98), using the solution vector (92) and the vector of right-hand sides of the normal equations (87), as

$$R(\mu, \boldsymbol{\alpha}, \boldsymbol{\beta}) = 6(42) + 19\tfrac{11}{13}(57) - 2\tfrac{7}{13}(45) + 2\tfrac{7}{13}(48) = 1390\tfrac{11}{13} \ . \qquad (120)$$

Finally, with

$$R(\mu, \boldsymbol{\alpha}, \boldsymbol{\beta}, \boldsymbol{\phi}) = \Sigma n_{ij}\bar{y}_{ij}^2 = 3(5^2) + 3(7^2) + 6^2 + 2(15^2) + 27^2 = 1437$$

we have

$$(106) \quad \text{as} \quad R(\boldsymbol{\phi}|\mu, \boldsymbol{\alpha}, \boldsymbol{\beta}) = 1437 - 1390\tfrac{11}{13} = 46\tfrac{2}{13} \ . \qquad (121)$$

From observing the pattern of filled cells in Table 8.8 this can be confirmed as Q of (108), in this particular case it being (111), with the data of Table

8.8 giving

$$(111)\quad \text{as}\quad Q=\frac{(5-7-15+27)^2}{\frac{1}{3}+\frac{1}{3}+\frac{1}{2}+1}=\frac{100}{\frac{13}{6}}=46\tfrac{2}{13}\ .$$

e. Analyses of variance

Analysis of variance tables for the general with-interaction and no-interaction models are shown in Tables 4.9 and 4.11, for cell means models. The same sums of squares are used for overparameterized models, only with different labels—as shown in Table 9.1. And Tables 9.1 and 9.2 show the

TABLE 9.1E. ANALYSIS OF VARIANCE OF DATA IN TABLE 8.8, FOR THE OVERPARAMETERIZED WITH-INTERACTION MODEL

Label	d.f.	Sum of Squares	Associated Hypothesis
$R(\mu)$	1	$980\frac{1}{10}$	H: $\mu+7\alpha_1+3\alpha_2+5\beta_1+4\beta_2+\beta_3$ $+3\phi_{11}+3\phi_{12}+\phi_{13}+2\phi_{21}+\phi_{22}=0$
$R(\boldsymbol{\alpha}\mid\mu)$	1	$354\frac{9}{10}$	H: $\alpha_1+[3(\beta_1+\phi_{11})+3(\beta_2+\phi_{12})+\beta_3+\phi_{13}]/7$ $=\alpha_2+[2(\beta_1+\phi_{21})+\beta_2+\phi_{22}]/3$
$R(\boldsymbol{\beta}\mid\mu,\boldsymbol{\alpha})$	2	$55\frac{11}{13}$	H: $\begin{cases}3(\beta_1+\phi_{11})+2(\beta_1+\phi_{21})\\ \quad=[9(\beta_1+\phi_{11})+9(\beta_2+\phi_{12})+3(\beta_3+\phi_{13})]/7\\ \quad\quad+[4(\beta_1+\phi_{21})+2(\beta_2+\phi_{22})]/3\\ 3(\beta_2+\phi_{12})+(\beta_2+\phi_{22})\\ \quad=[9(\beta_1+\phi_{11})+9(\beta_2+\phi_{12})+3(\beta_3+\phi_{13})]/7\\ \quad\quad+[2(\beta_1+\phi_{21})+\beta_2+\phi_{22}]/3\end{cases}$
$R(\boldsymbol{\phi}\mid\mu,\boldsymbol{\alpha},\boldsymbol{\beta})$	1	$46\frac{2}{13}$	H: $\phi_{11}-\phi_{12}-\phi_{21}+\phi_{22}=0$
SSE	5	38	
SST	10	1475	
$R(\boldsymbol{\beta}\mid\mu)$	2	$36\frac{9}{10}$	H: $\begin{cases}\beta_1+[3(\alpha_1+\phi_{11})+2(\alpha_2+\phi_{21})]/5\\ \quad=\beta_2+[3(\alpha_1+\phi_{12})+\alpha_2+\phi_{22}]/4\\ \quad=\beta_3+\alpha_1+\phi_{13}\end{cases}$
$R(\boldsymbol{\alpha}\mid\mu,\boldsymbol{\beta})$	1	$373\frac{11}{13}$	H: $\begin{cases}3(\alpha_1+\phi_{11})+3(\alpha_1+\phi_{12})+\alpha_1+\phi_{13}\\ \quad=[9(\alpha_1+\phi_{11})+6(\alpha_2+\phi_{21})]/5\\ \quad\quad+[9(\alpha_1+\phi_{12})+3(\alpha_2+\phi_{22})]/4\\ \quad\quad+\alpha_1+\phi_{13}\end{cases}$

TABLE 9.2E. ANALYSIS OF VARIANCE OF DATA IN TABLE 8.8, FOR THE OVERPARAMETERIZED NO-INTERACTION MODEL

Label	d.f.	Sum of Squares	Associated Hypothesis
$R(\mu)$	1	$980\frac{1}{10}$	H: $\mu + 7\alpha_1 + 3\alpha_2 + 5\beta_1 + 4\beta_2 + \beta_3 = 0$
$R(\boldsymbol{\alpha} \mid \mu)$	1	$354\frac{9}{10}$	H: $\alpha_1 + (3\beta_1 + 3\beta_2 + \beta_3)/7 = \alpha_2 + (2\beta_1 + \beta_2)/3$
$R(\boldsymbol{\beta} \mid \mu, \boldsymbol{\alpha})$	2	$55\frac{11}{13}$	H: $\beta_1 = \beta_2 = \beta_3$
SSE	6	$84\frac{2}{13}$	
SST	10	1475	
$R(\boldsymbol{\beta} \mid \mu)$	2	$36\frac{9}{10}$	H: $\beta_1 + (3\alpha_1 + 2\alpha_2)/5 = \beta_2 + (3\alpha_1 + \alpha_2)/4 = \beta_3 + \alpha_1$
$R(\boldsymbol{\alpha} \mid \mu, \boldsymbol{\beta})$	1	$373\frac{11}{13}$	H: $\alpha_1 = \alpha_2$

associated hypotheses for the overparameterized model. For the example, Tables 9.1E and 9.2E show the sums of squares and the associated hypotheses derived from Tables 9.1 and 9.2. In Table 9.1 the hypothesis associated with $R(\boldsymbol{\beta} \mid \mu, \boldsymbol{\alpha})$ is a statement involving b equations; Table 9.1E has $b = 3$, but shows only two equations for that hypothesis. The third equation is not shown because, similar to the discussion in Section 4.9e-iii, all three equations add to an identity. The same kind of reason is behind the associated hypothesis for $R(\boldsymbol{\alpha} \mid \mu, \boldsymbol{\beta})$ having only one equation in Table 9.1E.

What has already been discussed in Section 4.9h is very evident in Table 9.1E for the with-interaction model: none of the associated hypotheses are of general interest, except that for $R(\boldsymbol{\phi} \mid \mu, \boldsymbol{\alpha}, \boldsymbol{\beta})$, the interaction sum of squares. And, as described at length in Sections 5.6f and g, the interest in this depends upon its specific form which is in turn dependent upon the pattern of empty cells. In contrast, for the no-interaction model in Table 9.2E, the hypotheses associated with $R(\boldsymbol{\beta} \mid \mu, \boldsymbol{\alpha})$ and $R(\boldsymbol{\alpha} \mid \mu, \boldsymbol{\beta})$, namely H: $\beta_1 = \beta_2 = \beta_3$ and H: $\alpha_1 = \alpha_2$, *are* of general interest, and are, of course, examples of the general hypotheses H: β_j *all equal* and H: α_i *all equal*, tested by these sums of squares (provided the data are connected).

9.4. MODELS WITH Σ-RESTRICTIONS

Although it is totally unnecessary to do so, some people still like to use what have been called in Sections 8.9, 9.1 and 9.2 the Σ-restrictions. In all cases, these restrictions convert overparameterized models that are not of full rank to models that are of full rank, and so every parameter of the Σ-restricted model is estimable and the normal equations have but a single solution. And with balanced data these parameters are easily understood and the arithmetic of solving the normal equations is in most cases very straightforward. But with unbalanced data one has to be a little wary because in some situations interpretation is not exactly what one would apparently expect. We illustrate such situations, in terms of examples.

a. The no-interaction model

The Σ-restricted, no-interaction model is

$$E(y_{ijk}) = \dot{\mu} + \dot{\alpha}_i + \dot{\beta}_j \tag{122}$$

and

$$\Sigma_i\dot{\alpha}_i = 0 \quad \text{and} \quad \Sigma_j\dot{\beta}_j = 0\ . \tag{123}$$

For data consisting of 2 rows and 3 columns as do those of Table 8.8, the Σ-restrictions of (123) can be utilized by writing

$$\dot{\alpha}_2 = -\dot{\alpha}_1 \quad \text{and} \quad \dot{\beta}_3 = -\dot{\beta}_1 - \dot{\beta}_2\ . \tag{124}$$

Then the model equations of (24), after deleting the ϕ_{ij}-terms and making use of (124), are

$$\begin{bmatrix} 3 \\ 4 \\ 8 \\ 5 \\ 8 \\ 8 \\ 6 \\ 12 \\ 18 \\ 27 \end{bmatrix} = \begin{bmatrix} 1 & 1 & 1 & \cdot \\ 1 & 1 & 1 & \cdot \\ 1 & 1 & 1 & \cdot \\ 1 & 1 & \cdot & 1 \\ 1 & 1 & \cdot & 1 \\ 1 & 1 & \cdot & 1 \\ 1 & 1 & -1 & -1 \\ 1 & -1 & 1 & \cdot \\ 1 & -1 & 1 & \cdot \\ 1 & -1 & \cdot & 1 \end{bmatrix} \begin{bmatrix} \dot{\mu} \\ \dot{\alpha}_1 \\ \dot{\beta}_1 \\ \dot{\beta}_2 \end{bmatrix} + \mathbf{e}\ . \tag{125}$$

Hence the normal equations are

$$\begin{bmatrix} 10 & 4 & 4 & 3 \\ 4 & 10 & 0 & 1 \\ 4 & 0 & 6 & 1 \\ 3 & 1 & 1 & 5 \end{bmatrix} \begin{bmatrix} \dot{\mu} \\ \dot{\alpha}_1 \\ \dot{\beta}_1 \\ \dot{\beta}_2 \end{bmatrix} = \begin{bmatrix} 99 \\ -15 \\ 39 \\ 42 \end{bmatrix} . \tag{126}$$

-i. ***Solutions.*** Solutions to equations (126) are

$$\begin{bmatrix} \hat{\dot{\mu}} \\ \hat{\dot{\alpha}}_1 \\ \hat{\dot{\beta}}_1 \\ \hat{\dot{\beta}}_2 \end{bmatrix} = \begin{bmatrix} 10 & 4 & 4 & 3 \\ 4 & 10 & 0 & 1 \\ 4 & 0 & 6 & 1 \\ 3 & 1 & 1 & 5 \end{bmatrix}^{-1} \begin{bmatrix} 99 \\ -15 \\ 39 \\ 42 \end{bmatrix}$$

$$= \tfrac{1}{702} \begin{bmatrix} 142 & -51 & -85 & -58 \\ -51 & 90 & 33 & 6 \\ -85 & 33 & 172 & 10 \\ -58 & 6 & 10 & 172 \end{bmatrix} \begin{bmatrix} 99 \\ -15 \\ 39 \\ 42 \end{bmatrix} = \tfrac{3}{13} \begin{bmatrix} 56 \\ -30 \\ -11 \\ 11 \end{bmatrix} . \quad (127)$$

These are, of course, different from solution (92) of the normal equations (87) of the unrestricted model. But they can be converted to a solution of (87) simply by using the Σ-restrictions (124) to obtain

$$\hat{\dot{\alpha}}_2 = -\hat{\dot{\alpha}}_1 = 3(30)/13 \quad \text{and} \quad \hat{\dot{\beta}}_3 = -\left(\hat{\dot{\beta}}_1 + \hat{\dot{\beta}}_2\right) = 3(-11 + 11)/13 = 0 \; .$$

Then

$$[\mu^{\circ\circ} \;\; \alpha_1^{\circ\circ} \;\; \alpha_2^{\circ\circ} \;\; \beta_1^{\circ\circ} \;\; \beta_2^{\circ\circ} \;\; \beta_3^{\circ\circ}] = \tfrac{3}{13}[56 \;\; -30 \;\; 30 \;\; -11 \;\; 11 \;\; 0] \quad (128)$$

is a solution of (87); and, for example, estimates of estimable functions are as before, e.g.,

$$\text{BLUE}(\alpha_1 - \alpha_2) = \alpha_1^{\circ\circ} - \alpha_2^{\circ\circ} = 3(-30 - 30)/13 = -13\tfrac{11}{13},$$

exactly as at the end of Section 9.2.

-ii. ***Analysis of variance sums of squares.*** If the sums of squares of Table 9.2E are calculated for the Σ-restricted model of (122) and (123) and sub-models thereof, the same results will be obtained as in Table 9.2E. To see this we make repeated use of the R-algorithm of Section 8.4, written in the following form: for the model $E(\mathbf{y}) = \mathbf{X}\boldsymbol{\beta}$, with normal equations $\mathbf{X}'\mathbf{X}\boldsymbol{\beta}^\circ = \mathbf{X}'\mathbf{y}$ the reduction in the sum of squares is

$$R(\boldsymbol{\beta}) = (\mathbf{X}'\mathbf{y})'(\mathbf{X}'\mathbf{X})^- \mathbf{X}'\mathbf{y} \; . \quad (129)$$

We give three examples. First, for the sub-model $E(\mathbf{y}) = \dot{\mu}\mathbf{1}$ of the model (122): from (126) its normal equation is $10\hat{\dot{\mu}} = 99$ and so by (129)

$$R(\dot{\mu}) = 99(\tfrac{1}{10})99 = 980.1 = R(\mu)$$

of Table 9.2E. Similarly for the sub-model $E(\mathbf{y}) = \dot{\mu}\mathbf{1} + \dot{\alpha}_1\mathbf{x}$, where $\mathbf{x}$ is the second column of the 10×4 matrix in (125),

$$R(\dot{\mu}, \dot{\boldsymbol{\alpha}}) = [99 \quad -15]\begin{bmatrix} 10 & 4 \\ 4 & 10 \end{bmatrix}^{-1}\begin{bmatrix} 99 \\ -15 \end{bmatrix} = 1335 \tag{130}$$

so that

$$R(\dot{\boldsymbol{\alpha}} \mid \dot{\mu}) = 1335 - 980.1 = 354.9 = R(\alpha \mid \mu)$$

of Table 9.2E. Also, using the inverse matrix in (127), $R(\dot{\mu}, \dot{\boldsymbol{\alpha}}, \dot{\boldsymbol{\beta}}) =$

$$[99 \quad -15 \quad 39 \quad 42]\tfrac{1}{702}\begin{bmatrix} 142 & -51 & -85 & -58 \\ -51 & 90 & 33 & 6 \\ -85 & 33 & 172 & 10 \\ -58 & 6 & 10 & 172 \end{bmatrix}\begin{bmatrix} 99 \\ -15 \\ 39 \\ 42 \end{bmatrix}$$

$$= 1390\tfrac{11}{13} = R(\mu, \boldsymbol{\alpha}, \boldsymbol{\beta}), \tag{131}$$

of (119). We also see from (130) and (131) that

$$\begin{aligned} R(\dot{\boldsymbol{\beta}} \mid \dot{\mu}, \dot{\boldsymbol{\alpha}}) &= R(\dot{\mu}, \dot{\boldsymbol{\alpha}}, \dot{\boldsymbol{\beta}}) - R(\dot{\mu}, \dot{\boldsymbol{\alpha}}) = 1390\tfrac{11}{13} - 1335 = 55\tfrac{11}{13} \\ &= R(\boldsymbol{\beta} \mid \mu, \boldsymbol{\alpha}), \end{aligned} \tag{132}$$

of (117).

-iii. Another sum of squares. $R(\mu)$ is, as always, $R(\mu) = n_{..}\bar{y}^2_{...}$; in this case, $10(9.9)^2 = 980.1$. But from the full rank, Σ-restricted model there is a different sum of squares that can be associated with the mean, and which is calculated by some computing packages as the sum of squares due to the mean. It is

$$R(\dot{\mu} \mid \dot{\boldsymbol{\alpha}}, \dot{\boldsymbol{\beta}}) = R(\dot{\mu}, \dot{\boldsymbol{\alpha}}, \dot{\boldsymbol{\beta}}) - R(\dot{\boldsymbol{\alpha}}, \dot{\boldsymbol{\beta}}) \, .$$

For the example, $R(\dot{\mu}, \dot{\boldsymbol{\alpha}}, \dot{\boldsymbol{\beta}}) = 1390\frac{11}{13}$ from (131); and from the normal equations (126) the R-algorithm (124) gives

$$R(\dot{\boldsymbol{\alpha}}, \dot{\boldsymbol{\beta}}) = [-15 \quad 39 \quad 42]\begin{bmatrix} 10 & 0 & 1 \\ 0 & 6 & 1 \\ 1 & 1 & 5 \end{bmatrix}^{-1}\begin{bmatrix} -15 \\ 39 \\ 42 \end{bmatrix} = 565\tfrac{16}{71} \, .$$

Hence

$$R(\dot{\mu} \mid \dot{\alpha}, \dot{\beta}) = 1390\tfrac{11}{13} - 565\tfrac{16}{71} = 825\tfrac{573}{923} \neq R(\mu)$$

since $R(\mu) = 980.1$, as immediately precedes (130). The different interpretations of the two sums of squares is best seen by considering their associated hypotheses.

-iv. ***Associated hypotheses.*** The associated hypothesis for $R(\mu)$ is H: $E(\bar{y}_{...}) = 0$ and, in terms of the unrestricted model, this is equivalent to

$$\text{H: } n_{..}\mu + \Sigma_i n_{i.}\alpha_i + \Sigma_j n_{.j}\beta_j = 0\ . \tag{133}$$

But for $R(\dot{\mu} \mid \dot{\alpha}, \dot{\beta})$ the associated hypothesis is H: $\dot{\mu} = 0$, which is equivalent to

$$\text{H: } ab\mu + b\Sigma_i\alpha_i + a\Sigma_j\beta_j = 0\ . \tag{134}$$

Confirmation of the corresponding sums of squares comes from writing each hypothesis as H: $\mathbf{k}'\boldsymbol{\beta} = 0$, and for each one calculating $q = (\mathbf{k}'\boldsymbol{\beta}^\circ)^2/\mathbf{k}'\mathbf{G}\mathbf{k}$ using $\boldsymbol{\beta}^\circ$ of (92) and $\mathbf{G}$ of (93). Thus for (133)

$$\mathbf{k}'\boldsymbol{\beta}^\circ = [10\quad 7\quad 3\quad 5\quad 4\quad 1][0\quad 6\quad 19\tfrac{11}{13}\quad -2\tfrac{7}{13}\quad 2\tfrac{7}{13}\quad 0]' = 99$$

and

$$\mathbf{k}'\mathbf{G}\mathbf{k} = [7\quad 3\quad 5\quad 4]\begin{bmatrix} 1 & 1 & -1 & -1 \\ 1 & & & \\ -1 & & & \\ -1 & & & \end{bmatrix}\!\!\!\!\begin{matrix}\\ \frac{1}{39}\begin{pmatrix} 59 & -47 & -44 \\ -47 & 50 & 41 \\ -44 & 41 & 50 \end{pmatrix}\end{matrix}\begin{bmatrix} 7 \\ 3 \\ 5 \\ 4 \end{bmatrix} = 10\ .$$

Hence $q = 99^2/10 = 980.1 = R(\mu)$. And for (134)

$$\mathbf{k}'\boldsymbol{\beta}^\circ = [6\quad 3\quad 3\quad 2\quad 2\quad 2][0\quad 6\quad 19\tfrac{11}{13}\quad -2\tfrac{7}{13}\quad 2\tfrac{7}{13}\quad 0]' = 77\tfrac{7}{13}$$

and $\mathbf{k}'\mathbf{G}\mathbf{k} = 7\tfrac{11}{39}$. Hence $q = (77\tfrac{7}{13})^2/7\tfrac{11}{39} = 825\tfrac{573}{923} = R(\dot{\mu} \mid \dot{\alpha}, \dot{\beta})$.

-v. ***The invert-part-of-the-inverse algorithm.*** Sums of squares can also be calculated from the invert-part-of-the-inverse algorithm of Section 8.5f, which we express in the following form: if the normal equations for the full rank model $E(\mathbf{y}) = \mathbf{X}_1\boldsymbol{\beta}_1 + \mathbf{X}_2\boldsymbol{\beta}_2$ have solution

$$\begin{bmatrix} \hat{\boldsymbol{\beta}}_1 \\ \hat{\boldsymbol{\beta}}_2 \end{bmatrix} = \begin{bmatrix} \mathbf{T}_{11} & \mathbf{T}_{12} \\ (\mathbf{T}_{12})' & \mathbf{T}_{22} \end{bmatrix}\begin{bmatrix} \mathbf{X}_1'\mathbf{y} \\ \mathbf{X}_2'\mathbf{y} \end{bmatrix} \tag{135}$$

then
$$R(\boldsymbol{\beta}_1 \mid \boldsymbol{\beta}_2) = \hat{\boldsymbol{\beta}}_1 \mathbf{T}_{11}^{-1} \hat{\boldsymbol{\beta}}_1 \; . \tag{136}$$

For example, applying (136) to (127) gives

$$R(\dot{\mu} \mid \dot{\boldsymbol{\alpha}}, \dot{\boldsymbol{\beta}}) = \left[3\left(\tfrac{56}{13}\right)\right]^2 / \tfrac{142}{702} = 825\tfrac{573}{923}$$

as just obtained. Likewise, the method also gives (132):

$$R(\dot{\boldsymbol{\beta}} \mid \dot{\mu}, \dot{\boldsymbol{\alpha}}) = \left(\tfrac{3}{13}\right)[-11 \quad 11]\left[\tfrac{1}{702}\begin{pmatrix} 172 & 10 \\ 10 & 172 \end{pmatrix}\right]^{-1} \tfrac{3}{13}\begin{bmatrix} -11 \\ 11 \end{bmatrix} = 55\tfrac{11}{13} \; .$$

b. The with-interaction model

The with-interaction, Σ-restricted model is

$$E(y_{ijk}) = \dot{\mu} + \dot{\alpha}_i + \dot{\beta}_j + \dot{\phi}_{ij}$$

and

$$\Sigma_i \dot{\alpha}_i = 0, \qquad \Sigma_j \dot{\beta}_j = 0, \qquad \Sigma_i \dot{\phi}_{ij} = 0 \;\; \forall \; j \;\; \text{and} \;\; \Sigma_j \dot{\phi}_{ij} = 0 \;\; \forall \; i, \tag{137}$$

and in (15)–(17), following which is a discussion of the effects of empty cells on the restrictions on the interaction effects.

The preceding subsection deals with the no-interaction model, wherein equalities such as $R(\dot{\boldsymbol{\alpha}} \mid \dot{\mu}) = R(\boldsymbol{\alpha} \mid \mu)$ and $R(\dot{\boldsymbol{\beta}} \mid \dot{\boldsymbol{\alpha}}) = R(\boldsymbol{\beta} \mid \mu, \boldsymbol{\alpha})$ are illustrated. Those equalities also apply for the with-interaction model because, of course, $R(\dot{\boldsymbol{\alpha}} \mid \dot{\mu})$ and $R(\dot{\boldsymbol{\beta}} \mid \dot{\mu}, \dot{\boldsymbol{\alpha}})$ involve no interaction terms. But, since the Σ-restricted model is a full-rank model, the sums of squares that are of real interest are those of each factor adjusted for all others—see Section 8.6e. These are shown in Table 9.3, with their associated hypotheses in both the Σ-restricted model and in the unrestricted model with all-cells-filled data.

The associated hypotheses in the Σ-restricted model, in terms of the parameters of that model, are disarmingly straightforward-looking. Nevertheless, they merit further examination, for which all-cells-filled data and some-cells-empty data need to be distinguished.

-i. All-cells-filled data. The Σ-restrictions, in combination with all-cells-filled data, lead to $\dot{\mu}$ and $\dot{\phi}_{ij}$ being definable in terms of parameters of the unrestricted model as

$$\dot{\mu} = \mu + \Sigma_i \alpha_i / a + \Sigma_j \beta_j / b + \Sigma_i \Sigma_j \phi_{ij} / ab$$

and
$$\dot{\phi}_{ij} = \phi_{ij} - \phi_{aj} - \phi_{ib} + \phi_{ab} \; . \tag{138}$$

Also, as derived in Searle, *et al.* (1981), two of the sums of squares of the Σ-restricted model are the same as SSA_w and SSB_w of the weighted squares of means analysis of Section 4.6d:

$$R(\dot{\boldsymbol{\alpha}} \mid \dot{\mu}, \dot{\boldsymbol{\beta}}, \dot{\boldsymbol{\phi}}) = \text{SSA}_w \quad \text{and} \quad R(\dot{\boldsymbol{\beta}} \mid \dot{\mu}, \dot{\boldsymbol{\alpha}}, \dot{\boldsymbol{\phi}}) = \text{SSB}_w \; . \tag{139}$$

TABLE 9.3. SUMS OF SQUARES FOR THE Σ-RESTRICTED MODEL AND ASSOCIATED HYPOTHESES

Sum of Squares	Associated Hypothesis: Σ-Restricted Model	Associated Hypothesis: Unrestricted Model, All Cells Filled
$R(\dot{\mu}\mid\dot{\alpha},\dot{\beta},\dot{\phi})_\Sigma$	H: $\dot{\mu}=0$	H: $\mu+\Sigma_i\alpha_i/a+\Sigma_j\beta_j/b+\Sigma\phi_{ij}/ab=0$
$R(\dot{\alpha}\mid\dot{\mu},\dot{\beta},\dot{\phi})_\Sigma$	H: $\dot{\alpha}_i=0\ \ \forall\ i$	H: $\alpha_i+\Sigma_j\phi_{ij}/b$ *equal* $\forall\ i$
$R(\dot{\beta}\mid\dot{\mu},\dot{\alpha},\dot{\phi})_\Sigma$	H: $\dot{\beta}_j=0\ \ \forall\ j$	H: $\beta_j+\Sigma_i\phi_{ij}/a$ *equal* $\forall\ j$
$R(\dot{\phi}\mid\dot{\mu},\dot{\alpha},\dot{\beta})_\Sigma$	H: $\dot{\phi}_{ij}=0\ \ \forall\ i,j$	H: $\phi_{ij}-\phi_{i'j}-\phi_{ij'}+\phi_{i'j'}=0\ \ \forall\ i,j$

Hence, for all-cells-filled data, the sums of squares $R(\dot{\alpha}\mid\dot{\mu},\dot{\beta},\dot{\phi})$ and $R(\dot{\beta}\mid\dot{\mu},\dot{\alpha},\dot{\phi})$ have associated hypotheses H: $\alpha_i+\bar{\phi}_{i\cdot}$ *equal* $\forall\ i$, and H: $\beta_j+\bar{\phi}_{\cdot j}$ *equal* $\forall\ j$, respectively, hypotheses that have already been discussed at the end of Section 9.1g. In this way (138) and (139) lead to the associated hypotheses for the sums of squares of the Σ-restricted models, as shown on the right of Table 9.3.

Example. A simple example of all-cells-filled data is shown in Table 9.4. As usual, each triplet represents $y_{ij\cdot}$, (n_{ij}), $\bar{y}_{ij\cdot}$.

From Section 4.6d we know that the associated hypothesis for SSA_w is H: $\Sigma_j\mu_{ij}/b$ *equal* $\forall\ i$. In the case of Table 9.4 this is

$$\text{H: } \mu_{11}+\mu_{12}+\mu_{13}-(\mu_{21}+\mu_{22}+\mu_{23})=0 \ .$$

Hence, using (87) of Section 2.12b and $\hat{\mu}_{ij}=\bar{y}_{ij\cdot}$ for the $y_{ij\cdot}$-values of Table 9.4, the numerator sum of squares for the F-statistic for this hypothesis, which is SSA_w, is

$$\text{SSA}_w=\frac{[8+6+2-(8+6+12)]^2}{\frac{1}{2}+\frac{1}{1}+\frac{1}{1}+\frac{1}{1}+\frac{1}{2}+\frac{1}{1}}=\frac{100}{5}=20 \ . \tag{140}$$

TABLE 9.4. EXAMPLE

7, 9	6	2
16(2)8	6(1)6	2(1)2
8	4, 8	12
8(1)8	12(2)6	12(1)12

Similarly, that for testing H: $\Sigma_i \mu_{ij}/a$ *equal* $\forall\, j$, namely

$$\text{H:}\ \left\{\begin{bmatrix} \mu_{11} + \mu_{21} - (\mu_{12} + \mu_{22}) \\ \mu_{11} + \mu_{21} - (\mu_{13} + \mu_{23}) \end{bmatrix} = \begin{bmatrix} 0 \\ 0 \end{bmatrix},\right.$$

is

$$\text{SSB}_w = \begin{bmatrix} 8+8-(6+\ 6) \\ 8+8-(2+12) \end{bmatrix}' \begin{bmatrix} \frac{1}{2}+\frac{1}{1}+\frac{1}{1}+\frac{1}{2} & \frac{1}{2}+\frac{1}{1} \\ \frac{1}{2}+\frac{1}{1} & \frac{1}{2}+\frac{1}{1}+\frac{1}{1}+\frac{1}{1} \end{bmatrix}^{-1}$$

$$\times \begin{bmatrix} 8+8-(6+\ 6) \\ 8+8-(2+12) \end{bmatrix}$$

$$= [4 \quad 2] \begin{bmatrix} 3 & 1\frac{1}{2} \\ 1\frac{1}{2} & 3\frac{1}{2} \end{bmatrix}^{-1} \begin{bmatrix} 4 \\ 2 \end{bmatrix} = \frac{16(3\frac{1}{2}) + 4(3) - 16(1\frac{1}{2})}{10\frac{1}{2} - 2\frac{1}{4}}$$

$$= 5\tfrac{1}{3}\ . \tag{141}$$

We now confirm these results using $R(\dot{\boldsymbol{\alpha}} \mid \dot{\mu}, \dot{\boldsymbol{\beta}}, \dot{\boldsymbol{\phi}})$ and $R(\dot{\boldsymbol{\beta}} \mid \dot{\mu}, \dot{\boldsymbol{\alpha}}, \dot{\boldsymbol{\phi}})$.

To begin, set out the normal equations for the data of Table 9.4, for an unrestricted, overparameterized model, $y_{ijk} = \mu + \alpha_i + \beta_j + \phi_{ij} + e_{ijk}$:

$$\begin{bmatrix} 7 \\ 9 \\ 6 \\ 2 \\ 8 \\ 4 \\ 8 \\ 12 \end{bmatrix} = \begin{bmatrix} 1 & 1 & \cdot & 1 & \cdot & \cdot & 1 & \cdot & \cdot & \cdot & \cdot & \cdot \\ 1 & 1 & \cdot & 1 & \cdot & \cdot & 1 & \cdot & \cdot & \cdot & \cdot & \cdot \\ 1 & 1 & \cdot & \cdot & 1 & \cdot & \cdot & 1 & \cdot & \cdot & \cdot & \cdot \\ 1 & 1 & \cdot & \cdot & \cdot & 1 & \cdot & \cdot & 1 & \cdot & \cdot & \cdot \\ 1 & \cdot & 1 & 1 & \cdot & \cdot & \cdot & \cdot & \cdot & 1 & \cdot & \cdot \\ 1 & \cdot & 1 & \cdot & 1 & \cdot & \cdot & \cdot & \cdot & \cdot & 1 & \cdot \\ 1 & \cdot & 1 & \cdot & 1 & \cdot & \cdot & \cdot & \cdot & \cdot & 1 & \cdot \\ 1 & \cdot & 1 & \cdot & \cdot & 1 & \cdot & \cdot & \cdot & \cdot & \cdot & 1 \end{bmatrix} \begin{bmatrix} \mu \\ \alpha_1 \\ \alpha_2 \\ \beta_1 \\ \beta_2 \\ \beta_3 \\ \phi_{11} \\ \phi_{12} \\ \phi_{13} \\ \phi_{21} \\ \phi_{22} \\ \phi_{23} \end{bmatrix} + \mathbf{e}\ . \tag{142}$$

The Σ-restrictions for this data set, indeed for any set of all-cells-filled data in two rows and three columns, are as follows:

$$\dot{\alpha}_1 + \dot{\alpha}_2 = 0 \quad \text{implying} \quad \dot{\alpha}_2 = -\dot{\alpha}_1,$$

$$\dot{\beta}_1 + \dot{\beta}_2 + \dot{\beta}_3 = 0 \quad \text{implying} \quad \dot{\beta}_3 = -\dot{\beta}_1 - \dot{\beta}_2,$$

$$\text{and} \quad \left.\begin{aligned} \dot{\phi}_{11} + \dot{\phi}_{12} + \dot{\phi}_{13} &= 0 \\ \dot{\phi}_{21} + \dot{\phi}_{22} + \dot{\phi}_{23} &= 0 \\ \dot{\phi}_{11} + \dot{\phi}_{21} &= 0 \\ \dot{\phi}_{12} + \dot{\phi}_{22} &= 0 \\ \dot{\phi}_{13} + \dot{\phi}_{23} &= 0 \end{aligned}\right\} \quad \text{implying} \quad \begin{aligned} \dot{\phi}_{11} &= \dot{\phi}_{11} \\ \dot{\phi}_{12} &= \dot{\phi}_{12} \\ \dot{\phi}_{13} &= -\dot{\phi}_{11} - \dot{\phi}_{12} \\ \dot{\phi}_{21} &= -\dot{\phi}_{11} \\ \dot{\phi}_{22} &= -\dot{\phi}_{12} \\ \dot{\phi}_{23} &= \dot{\phi}_{11} + \dot{\phi}_{12} \,. \end{aligned} \tag{143}$$

In (143), the right-hand statements include the obvious $\dot{\phi}_{11} = \dot{\phi}_{11}$ and $\dot{\phi}_{12} = \dot{\phi}_{12}$. This emphasizes that the Σ-restrictions, shown as the left-hand set of statements in (143), can be restated so that all the $\dot{\phi}$s are in terms of just $\dot{\phi}_{11}$ and $\dot{\phi}_{12}$. For the general case of a rows and b columns and all cells filled, there will be $a + b$ restrictions on the $\dot{\phi}$s, which can be restated so that all $\dot{\phi}$s are expressible in terms of just $(a - 1)(b - 1)$ of them.

The way in which the restrictions change the unrestricted model to the restricted model is seen by applying the restated restrictions of (143) to the model equations of the unrestricted model, shown in (142). The result is that the model equations for the restricted model are

$$\begin{bmatrix} 7 \\ 9 \\ 6 \\ 2 \\ 8 \\ 4 \\ 8 \\ 12 \end{bmatrix} = \left[\begin{array}{c|c|cc|cc} 1 & 1 & 1 & \cdot & 1 & \cdot \\ 1 & 1 & 1 & \cdot & 1 & \cdot \\ 1 & 1 & \cdot & 1 & \cdot & 1 \\ 1 & 1 & -1 & -1 & -1 & -1 \\ 1 & -1 & 1 & \cdot & -1 & \cdot \\ 1 & -1 & \cdot & 1 & \cdot & -1 \\ 1 & -1 & \cdot & 1 & \cdot & -1 \\ 1 & -1 & -1 & -1 & 1 & 1 \end{array}\right] \begin{bmatrix} \dot{\mu} \\ \dot{\alpha}_1 \\ \dot{\beta}_1 \\ \dot{\beta}_2 \\ \dot{\phi}_{11} \\ \dot{\phi}_{12} \end{bmatrix} + \mathbf{e} \,. \tag{144}$$

The normal equations resulting from (144) are

$$\left[\begin{array}{c|c|cc|cc} 8 & 0 & 1 & 1 & 1 & -1 \\ \hline 0 & 8 & 1 & -1 & 1 & 1 \\ \hline 1 & 1 & 5 & 2 & 1 & 0 \\ 1 & -1 & 2 & 5 & 0 & -1 \\ \hline 1 & 1 & 1 & 0 & 5 & 2 \\ -1 & 1 & 0 & -1 & 2 & 5 \end{array}\right] \left[\begin{array}{c} \hat{\dot{\mu}} \\ \hline \hat{\dot{\alpha}}_1 \\ \hline \hat{\dot{\beta}}_1 \\ \hat{\dot{\beta}}_2 \\ \hline \hat{\dot{\phi}}_{11} \\ \hat{\dot{\phi}}_{12} \end{array}\right] = \left[\begin{array}{c} 56 \\ \hline -8 \\ \hline 10 \\ 4 \\ \hline 18 \\ 4 \end{array}\right]$$

with solution

$$\begin{bmatrix} \hat{\mu} \\ \hat{\dot{\alpha}}_1 \\ \hat{\dot{\beta}}_1 \\ \hat{\dot{\beta}}_2 \\ \hat{\dot{\phi}}_{11} \\ \hat{\dot{\phi}}_{12} \end{bmatrix} = \tfrac{1}{72}\begin{bmatrix} 10 & 0 & -1 & -1 & -3 & 3 \\ 0 & 10 & -3 & 3 & -1 & -1 \\ -1 & -3 & 19 & -8 & -3 & 0 \\ -1 & 3 & -8 & 19 & 0 & 3 \\ -3 & -1 & -3 & 0 & 19 & -8 \\ 3 & -1 & 0 & 3 & -8 & 19 \end{bmatrix}\begin{bmatrix} 56 \\ -8 \\ 10 \\ 4 \\ 18 \\ 4 \end{bmatrix} = \begin{bmatrix} 7 \\ -\frac{10}{6} \\ 1 \\ -1 \\ \frac{10}{6} \\ \frac{10}{6} \end{bmatrix}. \tag{145}$$

From this solution a variety of results similar to those of the preceding subsection (see E9.6) can be illustrated. We show only the two equalities of (139), using the invert-part-of-the-inverse algorithm for each. From (145) we first have

$$R(\dot{\boldsymbol{\alpha}} \mid \dot{\mu}, \dot{\boldsymbol{\beta}}, \dot{\boldsymbol{\phi}}) = \left(-\tfrac{10}{6}\right)\left(\tfrac{10}{72}\right)^{-1}\left(-\tfrac{10}{6}\right) = 20 = \text{SSA}_w$$

of (140); and second is (141):

$$R(\dot{\boldsymbol{\beta}} \mid \dot{\mu}, \dot{\boldsymbol{\alpha}}, \dot{\boldsymbol{\phi}}) = [1 \quad -1]\left[\frac{1}{72}\begin{pmatrix} 19 & -8 \\ -8 & 19 \end{pmatrix}\right]^{-1}\begin{bmatrix} 1 \\ -1 \end{bmatrix} = 5\tfrac{1}{3} = \text{SSB}_w\,.$$

These relationships, for all-cells-filled data, between weighted squares of means and sums of squares for Σ-restricted models (with all interactions) are extended to models having more than two factors in Section 10.2.

-ii. Some-cells-empty data. When some cells of a data grid lack data, the weighted squares of means cannot be calculated, i.e., SSA_w and SSB_w do not exist. But $R(\dot{\boldsymbol{\alpha}} \mid \dot{\mu}, \dot{\boldsymbol{\beta}}, \dot{\boldsymbol{\phi}})$ and $R(\dot{\boldsymbol{\beta}} \mid \dot{\mu}, \dot{\boldsymbol{\alpha}}, \dot{\boldsymbol{\phi}})$ of the Σ-restricted model can still be calculated, and the associated hypotheses are still those of the middle column of Table 9.3, namely H: $\dot{\alpha}_i = 0 \ \forall \ i$ and H: $\dot{\beta}_j = 0 \ \forall \ j$, respectively. As has already been suggested, these hypotheses appear disarmingly simple in terms of the $\dot{\alpha}_i$ and $\dot{\beta}_j$, but that simplicity in the presence of empty cells can quickly disappear upon conversion to parameters of the unrestricted model. To make that conversion we have the following algorithm.

Let s be the number of filled cells in the data. Let $\boldsymbol{\beta}$ be the vector of parameters in the unrestricted model, and let $\boldsymbol{\beta}_s$ (of order s) be the vector of parameters in the restricted, full-rank model. If $\mathbf{X}$ and $\mathbf{X}_s$ are correspondingly defined by $E(\mathbf{y}) = \mathbf{X}\boldsymbol{\beta}$ and $E(\mathbf{y}) = \mathbf{X}_s\boldsymbol{\beta}_s$, then define $\mathbf{W}$ and $\mathbf{W}_s$ as matrices of the s distinctly different rows of $\mathbf{X}$ and $\mathbf{X}_s$, respectively. Then $\boldsymbol{\beta}$ and $\boldsymbol{\beta}_s$ are related by equating cell means: $\mathbf{W}_s\boldsymbol{\beta}_s = \mathbf{W}\boldsymbol{\beta}$. When all possible interactions exist, $\mathbf{W}_s^{-1}$ exists, and so

$$\boldsymbol{\beta}_s = \mathbf{W}_s^{-1}\mathbf{W}\boldsymbol{\beta} \; . \tag{146}$$

Example. For the example of Table 8.8 the grid of filled cells is that shown in Grid 9.2:

Grid 9.2

√	√	√
√	√	

Then (146) is

$$\begin{bmatrix} \dot{\mu} \\ \dot{\alpha}_1 \\ \dot{\beta}_1 \\ \dot{\beta}_2 \\ \dot{\phi}_{11} \end{bmatrix} = \begin{bmatrix} 1 & 1 & 1 & 0 & 1 \\ 1 & 1 & 0 & 1 & -1 \\ 1 & 1 & -1 & -1 & 0 \\ 1 & -1 & 1 & 0 & -1 \\ 1 & -1 & 0 & 1 & 1 \end{bmatrix}^{-1} \begin{bmatrix} 1 & 1 & \cdot & 1 & \cdot & \cdot & 1 & \cdot & \cdot & \cdot & \cdot \\ 1 & 1 & \cdot & \cdot & 1 & \cdot & \cdot & 1 & \cdot & \cdot & \cdot \\ 1 & 1 & \cdot & \cdot & \cdot & 1 & \cdot & \cdot & 1 & \cdot & \cdot \\ 1 & \cdot & 1 & 1 & \cdot & \cdot & \cdot & \cdot & \cdot & 1 & \cdot \\ 1 & \cdot & 1 & \cdot & 1 & \cdot & \cdot & \cdot & \cdot & \cdot & 1 \end{bmatrix} \begin{bmatrix} \mu \\ \alpha_1 \\ \alpha_2 \\ \beta_1 \\ \beta_2 \\ \beta_3 \\ \phi_{11} \\ \phi_{12} \\ \phi_{13} \\ \phi_{21} \\ \phi_{22} \end{bmatrix} \tag{147}$$

$$= \tfrac{1}{12} \begin{bmatrix} 1 & 1 & 4 & 3 & 3 \\ 3 & 3 & 0 & -3 & -3 \\ 5 & -1 & -4 & 3 & -3 \\ -1 & 5 & -4 & -3 & 3 \\ 3 & -3 & 0 & -3 & 3 \end{bmatrix} \begin{bmatrix} 1 & 1 & \cdot & 1 & \cdot & \cdot & 1 & \cdot & \cdot & \cdot & \cdot \\ 1 & 1 & \cdot & \cdot & 1 & \cdot & \cdot & 1 & \cdot & \cdot & \cdot \\ 1 & 1 & \cdot & \cdot & \cdot & 1 & \cdot & \cdot & 1 & \cdot & \cdot \\ 1 & \cdot & 1 & 1 & \cdot & \cdot & \cdot & \cdot & \cdot & 1 & \cdot \\ 1 & \cdot & 1 & \cdot & 1 & \cdot & \cdot & \cdot & \cdot & \cdot & 1 \end{bmatrix} \begin{bmatrix} \mu \\ \alpha_1 \\ \alpha_2 \\ \beta_1 \\ \beta_2 \\ \beta_3 \\ \phi_{11} \\ \phi_{12} \\ \phi_{13} \\ \phi_{21} \\ \phi_{22} \end{bmatrix} .$$

Hence

$$
\begin{aligned}
\dot{\mu} &= \mu + \tfrac{1}{2}(\alpha_1 + \alpha_2) + \tfrac{1}{3}(\beta_1 + \beta_2 + \beta_3) + \tfrac{1}{12}(\phi_{11} + \phi_{12} + 4\phi_{13} + 3\phi_{21} + 3\phi_{22}) \\
\dot{\alpha}_1 &= \tfrac{1}{2}(\alpha_1 - \alpha_2) + \tfrac{1}{4}(\phi_{11} + \phi_{12} - \phi_{21} - \phi_{22}) \\
\dot{\beta}_1 &= \tfrac{1}{3}(2\beta_1 - \beta_2 - \beta_3) + \tfrac{1}{12}(5\phi_{11} - \phi_{12} - 4\phi_{13} + 3\phi_{21} - 3\phi_{22}) \\
\dot{\beta}_2 &= \tfrac{1}{3}(-\beta_1 + 2\beta_2 - \beta_3) + \tfrac{1}{12}(-\phi_{11} + 5\phi_{12} - 4\phi_{13} - 3\phi_{21} + 3\phi_{22}) \\
\dot{\phi}_{11} &= \tfrac{1}{4}(\phi_{11} - \phi_{12} - \phi_{21} + \phi_{22}) .
\end{aligned}
\tag{148}
$$

What is most noticeable about equations (148) is the occurrence of ϕ_{ij}s in a manner that is not altogether expected—and is especially unexpected in view of the fact that the numbers of observations do not enter into this relationship at all, only the pattern of empty cells does.

One way of trying to see if (148) is generally useful would be to write each equation in terms of the cell means $\mu_{ij} = \mu + \alpha_i + \beta_j + \phi_{ij}$. For example, this leads to

$$\dot{\mu} = \tfrac{1}{12}(\mu_{11} + \mu_{12} + 4\mu_{13} + 3\mu_{21} + 3\mu_{22})$$

and

$$\dot{\alpha}_1 = \tfrac{1}{12}(\mu_{11} + \mu_{12} - \mu_{21} - \mu_{22}) .$$

These expressions can be displayed graphically by replacing, for each Σ-restricted parameter, the check marks in Grid 9.2 with the coefficients of the μ_{ij}-terms, as follows.

$12\dot{\mu}$		
1	1	4
3	3	—

$12\dot{\alpha}_1$		
1	1	0
−1	−1	—

Certainly the pattern for $\dot{\alpha}_1$ seems reasonable; it is the difference between the sum of the cell means for row 1 and row 2, summed over columns 1 and 2, the two columns in which there are data in both rows. But what of the pattern for $\dot{\mu}$, or of those for $\dot{\beta}_1$ and $\dot{\beta}_2$?

$12\dot{\beta}_1$		
5	−1	−4
3	−3	—

$12\dot{\beta}_2$		
−1	5	−4
−3	3	—

And again it is to be emphasized that these patterns arise solely from the manner in which empty cells occur—they in no way depend on the actual numbers of observations. That each $\dot{\beta}_j$ is a weighted sum of contrasts among the μ_{ij}-terms is patently clear; but one gains little insight as to why the weights might be generally useful, or as to what they would be in larger data grids with different and sparser patterns of empty cells. Yet, with Σ-restricted models with some-cells-empty data, the hypotheses associated with $R(\dot{\alpha} \mid \dot{\mu}, \dot{\beta}, \dot{\phi})$ and $R(\dot{\beta} \mid \dot{\mu}, \dot{\alpha}, \dot{\phi})$ are hypotheses that parametric functions of the nature just illustrated are zero. They seem to be of little interest.

Insofar as calculating sums of squares in the Σ-restricted model is concerned, we can either use the invert-part-of-the-inverse rule or calculate numerator sums of squares for testing hypotheses about the parametric functions in (148). For the former we derive and solve the normal equations, beginning with the model equations

$$\begin{bmatrix} 3 \\ 4 \\ 8 \\ 5 \\ 8 \\ 8 \\ 6 \\ 12 \\ 18 \\ 27 \end{bmatrix} = \begin{bmatrix} 1 & 1 & 1 & \cdot & 1 \\ 1 & 1 & 1 & \cdot & 1 \\ 1 & 1 & 1 & \cdot & 1 \\ 1 & 1 & \cdot & 1 & -1 \\ 1 & 1 & \cdot & 1 & -1 \\ 1 & 1 & \cdot & 1 & -1 \\ 1 & 1 & -1 & -1 & 0 \\ 1 & -1 & 1 & \cdot & -1 \\ 1 & -1 & 1 & \cdot & -1 \\ 1 & -1 & \cdot & 1 & 1 \end{bmatrix} \begin{bmatrix} \dot{\mu} \\ \dot{\alpha}_1 \\ \dot{\beta}_1 \\ \dot{\beta}_2 \\ \dot{\phi} \end{bmatrix} + \mathbf{e} \, .$$

From these the normal equations are

$$\begin{bmatrix} 10 & 4 & 4 & 3 & -1 \\ 4 & 10 & 0 & 1 & 1 \\ 4 & 0 & 6 & 1 & 1 \\ 3 & 1 & 1 & 5 & -2 \\ -1 & 1 & 1 & -2 & 9 \end{bmatrix} \begin{bmatrix} \hat{\dot{\mu}} \\ \hat{\dot{\alpha}}_1 \\ \hat{\dot{\beta}}_1 \\ \hat{\dot{\beta}}_2 \\ \hat{\dot{\phi}} \end{bmatrix} = \begin{bmatrix} 99 \\ -15 \\ 39 \\ 42 \\ -9 \end{bmatrix}$$

with solution

$$\begin{bmatrix} \hat{\dot{\mu}} \\ \hat{\dot{\alpha}}_1 \\ \hat{\dot{\beta}}_1 \\ \hat{\dot{\beta}}_2 \\ \hat{\dot{\phi}} \end{bmatrix} = \frac{1}{864}\begin{bmatrix} 181 & -69 & -115 & -61 & 27 \\ -69 & 117 & 51 & -3 & -27 \\ -115 & 51 & 229 & -5 & -45 \\ -61 & -3 & -5 & 229 & 45 \\ 27 & -27 & -45 & 45 & 117 \end{bmatrix}\begin{bmatrix} 99 \\ -15 \\ 39 \\ 42 \\ -9 \end{bmatrix}$$

$$= \tfrac{1}{2}[27 \quad -15 \quad -7 \quad 7 \quad 5]' . \tag{149}$$

Hence, by the invert-part-of-the-inverse algorithm

$$R(\dot{\boldsymbol{\alpha}} \mid \dot{\mu}, \dot{\boldsymbol{\beta}}, \dot{\boldsymbol{\phi}}) = \left(\tfrac{15}{2}\right)\left(\tfrac{117}{864}\right)^{-1}\left(\tfrac{15}{2}\right) = 415\tfrac{5}{13} .$$

Second, because the Σ-restricted model is a full rank model, $R(\dot{\boldsymbol{\alpha}} \mid \dot{\mu}, \dot{\boldsymbol{\beta}}, \dot{\boldsymbol{\phi}})$ tests H: $\dot{\alpha}_1 = 0$, and from (148) this is

$$\text{H: } \tfrac{1}{2}(\alpha_1 - \alpha_2) + \tfrac{1}{4}(\phi_{11} + \phi_{12} - \phi_{21} - \phi_{22}) = 0 . \tag{150}$$

For testing this hypothesis the numerator sum of squares, Q, comes from the solution vector (43) that has everything zero except $\phi^{\circ}_{ij} = \bar{y}_{ij\cdot}$ for $n_{ij} > 0$. From Table 8.8 the relevant data are shown in Table 9.5. Hence for (150)

$$Q = \frac{\left[\frac{1}{4}(5 + 7 - 15 - 27)\right]^2}{\left(\frac{1}{4}\right)^2\left(\frac{1}{3} + \frac{1}{3} + \frac{1}{2} + \frac{1}{1}\right)} = \frac{900}{\frac{13}{6}} = 415\tfrac{5}{13} = R(\dot{\alpha} \mid \dot{\mu}, \dot{\beta}, \dot{\phi}) .$$

But, of course, this cannot be calculated as SSA_w because not all cells are filled.

Finally, let it be noticed that although we might write

$$R(\dot{\boldsymbol{\alpha}} \mid \dot{\mu}, \dot{\boldsymbol{\beta}}, \dot{\boldsymbol{\phi}}) = R(\dot{\mu}, \dot{\boldsymbol{\alpha}}, \dot{\boldsymbol{\beta}}, \dot{\boldsymbol{\phi}}) - R(\dot{\mu}, \dot{\boldsymbol{\beta}}, \dot{\boldsymbol{\phi}})$$
$$= R(\mu, \boldsymbol{\alpha}, \boldsymbol{\beta}, \boldsymbol{\phi}) - R(\dot{\mu}, \dot{\boldsymbol{\beta}}, \dot{\boldsymbol{\phi}})$$

TABLE 9.5. $\bar{y}_{ij\cdot}(n_{ij})y_{ij}$

15(3) 5	21(3) 7	6(1)6
30(2)15	27(1)27	—

this is not the same as

$$R(\boldsymbol{\alpha} \mid \mu, \boldsymbol{\beta}, \boldsymbol{\phi}) = R(\mu, \boldsymbol{\alpha}, \boldsymbol{\beta}, \boldsymbol{\phi}) - R(\mu, \boldsymbol{\beta}, \boldsymbol{\phi}) = 0$$

of Section 9.1k. The reason is as follows: $R(\mu, \boldsymbol{\alpha}, \boldsymbol{\beta}, \boldsymbol{\phi})$ and $R(\dot{\mu}, \dot{\boldsymbol{\alpha}}, \dot{\boldsymbol{\beta}}, \dot{\boldsymbol{\phi}})$ are equal because the Σ-restricted $(\dot{\mu}, \dot{\boldsymbol{\alpha}}, \dot{\boldsymbol{\beta}}, \dot{\boldsymbol{\phi}})$-model is a reparameterization of the $(\mu, \boldsymbol{\alpha}, \boldsymbol{\beta}, \boldsymbol{\phi})$-model. But $R(\mu, \boldsymbol{\beta}, \boldsymbol{\phi})$ and $R(\dot{\mu}, \dot{\boldsymbol{\beta}}, \dot{\boldsymbol{\phi}})$ are not equal because the $\dot{\mu}$, $\dot{\boldsymbol{\beta}}$, and $\dot{\boldsymbol{\phi}}$ are from a Σ-restricted form not of the $(\mu, \boldsymbol{\beta}, \boldsymbol{\phi})$-model but of the $(\mu, \boldsymbol{\alpha}, \boldsymbol{\beta}, \boldsymbol{\phi})$-model. The $(\dot{\mu}, \dot{\boldsymbol{\beta}}, \dot{\boldsymbol{\phi}})$-model implicit in the $R(\dot{\mu}, \dot{\boldsymbol{\beta}}, \dot{\boldsymbol{\phi}})$ being used here therefore includes the restrictions $\Sigma_j \dot{\phi}_{ij} = 0 \;\forall\; i$ which it would not include if $R(\dot{\mu}, \dot{\boldsymbol{\beta}}, \dot{\boldsymbol{\phi}})$ were being calculated from a Σ-restricted form of the $(\mu, \boldsymbol{\beta}, \boldsymbol{\phi})$-model. This is what causes the $R(\dot{\mu}, \dot{\boldsymbol{\beta}}, \dot{\boldsymbol{\phi}})$ that *is* being used to be different from $R(\mu, \beta, \phi) = \Sigma_i \Sigma_j n_{ij} \bar{y}_{ij\cdot}^2$. Hence $R(\dot{\boldsymbol{\alpha}} \mid \dot{\boldsymbol{\mu}}, \dot{\boldsymbol{\beta}}, \dot{\boldsymbol{\phi}})$ is different from zero.

This distinction can be made notationally in the following manner. Use $\Sigma_{\beta, \phi}$ to denote Σ-restrictions

$$\Sigma_j \dot{\beta}_j = 0 \quad \text{and} \quad \Sigma_i \dot{\phi}_{ij} = 0 \quad \forall\; j,$$

and denote those restrictions, together with

$$\Sigma_j \dot{\alpha}_j = 0 \quad \text{and} \quad \Sigma_i \dot{\phi}_{ij} = 0 \quad \forall\; j,$$

by $\Sigma_{\alpha, \beta, \phi}$. Then what is not being calculated is

$$\begin{aligned} R(\dot{\alpha} \mid \dot{\mu}, \dot{\beta}, \dot{\phi})_{\Sigma_{\beta,\phi}} &= R(\dot{\mu}, \dot{\alpha}, \dot{\beta}, \dot{\phi})_{\Sigma_{\beta,\phi}} - R(\dot{\mu}, \dot{\beta}, \dot{\phi})_{\Sigma_{\beta,\phi}} \\ &= R(\mu, \alpha, \beta, \phi) - R(\mu, \beta, \phi) \\ &= R(\alpha \mid \mu, \beta, \phi) \\ &= 0, \end{aligned}$$

whereas what is being calculated is

$$\begin{aligned} R(\dot{\alpha} \mid \dot{\mu}, \dot{\beta}, \dot{\phi})_{\Sigma_{\alpha,\beta,\phi}} &= R(\dot{\mu}, \dot{\alpha}, \dot{\beta}, \dot{\phi})_{\Sigma_{\alpha,\beta,\phi}} - R(\dot{\mu}, \dot{\beta}, \dot{\phi})_{\Sigma_{\alpha,\beta,\phi}} \\ &= R(\mu, \alpha, \beta, \phi) - R(\dot{\mu}, \dot{\beta}, \dot{\phi})_{\Sigma_{\alpha,\beta,\phi}} \\ &\neq R(\mu, \alpha, \beta, \phi) - R(\mu, \beta, \phi) \\ &\neq 0\;. \end{aligned}$$

c. Other restrictions

There are, of course, alternatives to the Σ-restrictions; one that is occasionally used is what shall be called the W-restrictions, or weighted restrictions—using as weights the numbers of observations. For the no-interaction model these restrictions are

$$\Sigma_i n_{i.}\dot{\alpha}_i = 0 \quad \text{and} \quad \Sigma_j n_{.j}\dot{\beta}_j = 0, \tag{151}$$

and for the with-interaction model the restrictions are (151) and

$$\Sigma_i n_{ij}\dot{\phi}_{ij} = 0 \quad \forall\ j \quad \text{and} \quad \Sigma_j n_{ij}\phi_{ij} = 0 \quad \forall\ i\ . \tag{152}$$

9.5. CONSTRAINTS ON SOLUTIONS

a. Solving the normal equations

We have seen that for $\mathbf{X}'\mathbf{X}$ not of full rank the normal equations $\mathbf{X}'\mathbf{X}\boldsymbol{\beta}^\circ = \mathbf{X}'\mathbf{y}$ have many solutions $\boldsymbol{\beta}^\circ$. Whatever generalized inverse of $\mathbf{X}'\mathbf{X}$ is used for $\mathbf{G}$, the solution $\boldsymbol{\beta}^\circ = \mathbf{G}\mathbf{X}'\mathbf{y}$ provides, for an estimable function $\mathbf{q}'\boldsymbol{\beta}$, $\text{BLUE}(\mathbf{q}'\boldsymbol{\beta}) = \mathbf{q}'\boldsymbol{\beta}^\circ$ and $v(\mathbf{q}'\boldsymbol{\beta}^\circ) = \mathbf{q}'\mathbf{G}\mathbf{q}\sigma^2$, both invariant to $\mathbf{G}$ as is $\hat{\sigma}^2$ also, for $\hat{\sigma}^2 = \text{SSE}/(N - r_{\mathbf{X}})$ for $\text{SSE} = \mathbf{y}'(\mathbf{I} - \mathbf{X}\mathbf{G}\mathbf{X}')\mathbf{y} = \mathbf{y}'\mathbf{y} - \boldsymbol{\beta}^{\circ\prime}\mathbf{X}'\mathbf{y}$.

In developing these results we usually derive $\mathbf{G}$ first and then $\boldsymbol{\beta}^\circ$. But since $\boldsymbol{\beta}^\circ$ can be *any* solution, there is nothing to prevent our finding a solution any way we wish—maybe by a method easier than first finding a $\mathbf{G}$. For estimating estimable functions and σ^2 all that is needed is a solution, with or without its corresponding $\mathbf{G}$. Only when wanting to estimate the sampling variance $\mathbf{q}'\mathbf{G}\mathbf{q}\sigma^2$ of a BLUE, or to test a hypothesis using an F-statistic requiring $(\mathbf{K}'\mathbf{G}\mathbf{K})^{-1}$—only then do we need $\mathbf{G}$. But so long as it can be reconstructed from the manner in which $\boldsymbol{\beta}^\circ$ is derived, then $\boldsymbol{\beta}^\circ$ can be derived first.

One way of deriving a solution of the normal equations is to impose linear constraints on the solutions, sufficient in number and of appropriate nature, so that when combined with the normal equations those equations have but a single solution. That solution is then a $\boldsymbol{\beta}^\circ$, and there will be a corresponding $\mathbf{G}$ such that $\boldsymbol{\beta}^\circ = \mathbf{G}\mathbf{X}'\mathbf{y}$.

These constraints must be $p - r_{\mathbf{X}}$ LIN linear functions of the solution elements, with the same functions of the parameters being non-estimable; and they must be such that they do, when combined with the normal equations, yield a single solution to those equations. The easiest form of such constraints is that of setting $p - r_{\mathbf{X}}$ elements of the solution vector equal to zero. This eliminates columns from $\mathbf{X}'\mathbf{X}$, and also corresponding rows (because in the minimization procedure of least squares there will be no minimizing with respect to the elements put equal to zero). These constraints, when appropriately chosen, therefore lead to $\mathbf{X}'\mathbf{X}$ being reduced to a full rank matrix, of the same rank as $\mathbf{X}'\mathbf{X}$. This is what was done for the 2-way classification without interaction, following equation (60) in Section 9.2i. That led to $\mu^o = 0$, $\beta_b^o = 0$, (71) and (75) being the solution and to (80) as the corresponding $\mathbf{G}$. A simpler example is the 1-way classification.

Example. Consider the normal equations from (23) of Section 8.2b.

$$\begin{bmatrix} 7 & 3 & 2 & 2 \\ 3 & 3 & \cdot & \cdot \\ 2 & \cdot & 2 & \cdot \\ 2 & \cdot & \cdot & 2 \end{bmatrix} \begin{bmatrix} \mu^o \\ \alpha_1^o \\ \alpha_2^o \\ \alpha_3^o \end{bmatrix} = \begin{bmatrix} 553 \\ 219 \\ 156 \\ 178 \end{bmatrix}. \tag{153}$$

Using $\mu^o = 0$ as a constraint gives the solution

$$\begin{bmatrix} \mu^o \\ \alpha_1^o \\ \alpha_2^o \\ \alpha_3^o \end{bmatrix} = \boldsymbol{\beta}_1^o = \begin{bmatrix} 0 \\ 73 \\ 78 \\ 89 \end{bmatrix}$$

with

$$\mathbf{G}_1 = \begin{bmatrix} 0 & 0 & 0 & 0 \\ 0 & & & \\ 0 & & \begin{pmatrix} 3 & \cdot & \cdot \\ \cdot & 2 & \cdot \\ \cdot & \cdot & 2 \end{pmatrix}^{-1} & \\ 0 & & & \end{bmatrix} = \begin{bmatrix} 0 & \cdot & \cdot & \cdot \\ \cdot & \frac{1}{3} & \cdot & \cdot \\ \cdot & \cdot & \frac{1}{2} & \cdot \\ \cdot & \cdot & \cdot & \frac{1}{2} \end{bmatrix}.$$

A different constraint, $\alpha_3^o = 0$, gives the solution

$$\boldsymbol{\beta}_4^o = \begin{bmatrix} 89 \\ -16 \\ -11 \\ 0 \end{bmatrix}$$

with

$$\mathbf{G}_4 = \begin{bmatrix} \begin{pmatrix} 7 & 3 & 2 \\ 3 & 3 & 0 \\ 2 & 0 & 2 \end{pmatrix}^{-1} & \begin{matrix} 0 \\ 0 \\ 0 \end{matrix} \\ \begin{matrix} 0 & 0 & 0 \end{matrix} & 0 \end{bmatrix} = \frac{1}{2}\begin{bmatrix} 1 & -1 & -1 & 0 \\ -1 & 1\frac{2}{3} & 1 & 0 \\ -1 & 1 & 2 & 0 \\ 0 & 0 & 0 & 0 \end{bmatrix}.$$

These are solutions 1 and 4, respectively, in Table 8.6.

b. A general algorithm

The preceding examples are special cases of the following general algorithm [see Searle (1971), pp. 213–215].

1. For $\mathbf{X'X}$ of order p, find its rank $r_{\mathbf{X}}$.
2. Delete $p - r_{\mathbf{X}}$ rows and corresponding columns of $\mathbf{X'X}$ so as to leave a symmetric principal submatrix of full-rank $r_{\mathbf{X}}$; denote it by $(\mathbf{X'X})_m$.
3. Corresponding to rows deleted from $\mathbf{X'X}$, delete elements from $\mathbf{X'y}$, denoting the remaining vector of order $r_{\mathbf{X}}$ by $(\mathbf{X'y})_m$.
4. Calculate $\mathbf{b}_m^\circ = [(\mathbf{X'X})_m]^{-1}(\mathbf{X'y})_m$.
5. Construct $\mathbf{b}^\circ$ by putting equal to zero all those elements corresponding to rows deleted from $\mathbf{X'X}$ in deriving $(\mathbf{X'X})_m$; other elements are those of $\mathbf{b}_m^\circ$, in sequence.
6. In $\mathbf{X'X}$ replace each element of $(\mathbf{X'X})_m$ by the corresponding element of its inverse; and put all other elements equal to zero. The resulting matrix is a $\mathbf{G}$ corresponding to the $\mathbf{b}^\circ$ derived in step 5.

This algorithm always gives a symmetric and reflexive $\mathbf{G}$ (Section 7.1h).

c. Other procedures

The preceding algorithm is infallible provided it is carried out correctly, paying particular attention to step 2. Nevertheless, for some cases there may be other $\mathbf{G}$s that are more readily evident from the manner in which $\boldsymbol{\beta}^\circ$ is obtained from the normal equations. For example, after putting $\alpha_3^\circ = 0$ in (153) it is quite clear that, operationally, a $\mathbf{G}$ corresponding to the manner of then solving the equations is

$$\mathbf{G}_4^* = \begin{bmatrix} \cdot & \cdot & \cdot & \frac{1}{2} \\ \cdot & \frac{1}{3} & \cdot & -\frac{1}{2} \\ \cdot & \cdot & \frac{1}{2} & -\frac{1}{2} \\ \cdot & \cdot & \cdot & 0 \end{bmatrix}.$$

But this is not symmetric and reflexive. However, using (11) of Chapter 7, a $\mathbf{G}$ with those properties is obtained. In fact it is $\mathbf{G}_4$, for $\mathbf{G}_4 = \mathbf{G}_4^*\mathbf{X'XG}_4^*$. Both $\mathbf{G}_4^*$ and $\mathbf{G}_4$ used for $\mathbf{G}$ in $\boldsymbol{\beta}^\circ = \mathbf{GX'y}$ yield solution 4 in Table 8.6.

The algorithm uses the non-full rank properties of $\mathbf{X'X}$ in an easy way: put $p - r_{\mathbf{X}}$ elements of the solution vector equal to zero and derive a full-rank submatrix of order $r_{\mathbf{X}}$. Moreover, this is easily computerized. But other constraints can also be used. For example, using $3\alpha_1^\circ + 2\alpha_2^\circ + 2\alpha_3^\circ = 0$

yields

$$\boldsymbol{\beta}_3^\circ = \begin{bmatrix} 79 \\ -6 \\ -1 \\ 10 \end{bmatrix} \quad \text{with} \quad \mathbf{G}_3^* = \begin{bmatrix} \frac{1}{7} & \cdot & \cdot & \cdot \\ -\frac{1}{7} & \frac{1}{3} & \cdot & \cdot \\ -\frac{1}{7} & \cdot & \frac{1}{2} & \cdot \\ -\frac{1}{7} & \cdot & \cdot & \frac{1}{2} \end{bmatrix}.$$

This is solution 3 of Table 8.6. A corresponding symmetric reflexive **G** is

$$\mathbf{G}_3 = \mathbf{G}_3^*\mathbf{X}'\mathbf{X}\mathbf{G}_3^{*\prime} = \frac{1}{7}\begin{bmatrix} 1 & \cdot & \cdot & \cdot \\ \cdot & 1\frac{1}{3} & -1 & -1 \\ \cdot & -1 & 2\frac{1}{2} & -1 \\ \cdot & -1 & -1 & 2\frac{1}{2} \end{bmatrix}.$$

Of course, any solution can be used to estimate σ^2 and any estimable function $\mathbf{q}'\mathbf{b}$ as $\mathbf{q}'\boldsymbol{\beta}^\circ$; and by using a corresponding **G** the sampling variance $\mathbf{q}'\mathbf{G}\mathbf{q}\sigma^2$ of $\mathbf{q}'\boldsymbol{\beta}^\circ$ becomes available, as does $(\mathbf{K}'\mathbf{G}\mathbf{K})^{-1}$ for testing a hypothesis.

d. Constraints and restrictions

Although $\boldsymbol{\beta}_3^\circ$ is exactly the solution one obtains when using the restriction $3\alpha_1 + 2\alpha_2 + 2\alpha_3 = 0$ on the model, using the constraint $3\alpha_1^\circ + 2\alpha_2^\circ + 2\alpha_3^\circ = 0$ on the solution does not necessarily imply that the restriction on the model is being used. By the same token, in deriving $\boldsymbol{\beta}_1^\circ$ there is no tacit assumption that $\mu = 0$ (or in deriving $\boldsymbol{\beta}_4^\circ$ that $\alpha_3 = 0$). We are simply using the solution statements $\mu^\circ = 0$ and $\alpha_3^\circ = 0$ as a crutch, in each case, for calculating a solution. It is precisely to distinguish between the two procedures that we use the phrases "constraints on the solution" and "restrictions on the model". Constraints on the solution do *not* imply corresponding restrictions on the model. But the opposite is done, in the sense that restrictions on the model do lead to a solution whose elements satisfy the corresponding constraints. But in that case the constraints are a consequence of the restrictions; they are not chosen simply for deriving a solution.

One can, in fact, in a restricted model use constraints that are different from the restrictions (assumed to be non-estimable, see Section 8.9c). This would occur if one wanted to use constraints that lead to an easily-obtained solution, but then wanted the solution corresponding to the restrictions on the model. The mechanics for doing this are precisely the same as in deriving $\boldsymbol{\beta}_r^\circ$ from $\boldsymbol{\beta}^\circ$ in (176) of Chapter 8:

$$\boldsymbol{\beta}_r^\circ = \boldsymbol{\beta}_c^\circ + (\mathbf{I} - \mathbf{H}_c)[\mathbf{P}'(\mathbf{I} - \mathbf{H}_c)]^-(\boldsymbol{\delta} - \mathbf{P}'\boldsymbol{\beta}_c^\circ), \tag{154}$$

where $\boldsymbol{\beta}_c^\circ$ is the solution using the constraints, with $\mathbf{H}_c = \mathbf{G}_c\mathbf{X}'\mathbf{X}$ where $\mathbf{G}_c$ is the $\mathbf{G}$ corresponding to $\boldsymbol{\beta}_c^\circ$, and $\mathbf{P}'\boldsymbol{\beta} = \boldsymbol{\delta}$ are the restrictions.

Example. The without-interaction, overparameterized model for the 2-way crossed classification is dealt with in Section 9.2. The solution elements derived there in (75) and (77) are denoted by α_i° and β_j°, with $\mu^\circ = 0$ and $\beta_b^\circ = 0$. With $\mathbf{X}'\mathbf{X}$ from (64) and $\mathbf{G}$ corresponding to α_i° and β_j° from (80) of Section 9.2k, the product $\mathbf{G}\mathbf{X}'\mathbf{X}$ which we shall call $\mathbf{H}_c$ is

$$\mathbf{H}_c = \mathbf{G}\mathbf{X}'\mathbf{X} = \begin{bmatrix} 0 & \mathbf{0} & \mathbf{0} & 0 \\ \mathbf{1}_a & \mathbf{I}_a & \mathbf{0} & \mathbf{1}_a \\ \mathbf{0} & \mathbf{0} & \mathbf{I}_2 & -\mathbf{1}_2 \\ 0 & \mathbf{0} & \mathbf{0} & 0 \end{bmatrix},$$

where $\mathbf{I}_2$ and $\mathbf{1}_2$ are of order $b - 1$, in keeping with (63) of Section 9.2j. Therefore

$$\mathbf{I} - \mathbf{H}_c = \begin{bmatrix} 1 & \mathbf{0} & \mathbf{0} & 0 \\ -\mathbf{1}_a & \mathbf{0} & \mathbf{0} & -\mathbf{1}_a \\ \mathbf{0} & \mathbf{0} & \mathbf{0} & \mathbf{1}_2 \\ 0 & \mathbf{0} & \mathbf{0} & 1 \end{bmatrix}.$$

To obtain the solution corresponding to the Σ-restrictions, which have

$$\mathbf{P}' = \begin{bmatrix} 0 & \mathbf{1}'_a & \mathbf{0} \\ 0 & \mathbf{0} & \mathbf{1}'_b \end{bmatrix} \quad \text{and} \quad \boldsymbol{\delta} = \mathbf{0}$$

we have for (154)

$$\mathbf{P}'(\mathbf{I} - \mathbf{H}) = \begin{bmatrix} -a & \mathbf{0} & \mathbf{0} & -a \\ 0 & \mathbf{0} & \mathbf{0} & b \end{bmatrix} \quad \text{and} \quad \mathbf{P}'\boldsymbol{\beta}_c = \begin{bmatrix} \Sigma_i \alpha_i^\circ \\ \Sigma_j \beta_j^\circ \end{bmatrix}$$

with

$$[\mathbf{P}'(\mathbf{I} - \mathbf{H})]^- = \begin{bmatrix} \dfrac{-1}{a} & \dfrac{-1}{b} \\ \mathbf{0} & \mathbf{0} \\ \mathbf{0} & \mathbf{0} \\ 0 & \dfrac{1}{b} \end{bmatrix}.$$

Hence

$$\boldsymbol{\beta}_r^{\circ} = \boldsymbol{\beta}_c^{\circ} + \begin{bmatrix} 1 & \mathbf{0} & \mathbf{0} & 0 \\ -\mathbf{1}_a & \mathbf{0} & \mathbf{0} & -\mathbf{1}_a \\ \mathbf{0} & \mathbf{0} & \mathbf{0} & \mathbf{1}_2 \\ 0 & \mathbf{0} & \mathbf{0} & 1 \end{bmatrix} \begin{bmatrix} \dfrac{-1}{a} & \dfrac{-1}{b} \\ \mathbf{0} & \mathbf{0} \\ \mathbf{0} & \mathbf{0} \\ 0 & \dfrac{1}{b} \end{bmatrix} \begin{bmatrix} -\Sigma_i \alpha_i^{\circ} \\ -\Sigma_j \beta_j^{\circ} \end{bmatrix} \tag{155}$$

$$= \begin{bmatrix} 0 \\ \boldsymbol{\alpha}^{\circ} \\ \boldsymbol{\beta}_2^{\circ} \\ 0 \end{bmatrix} + \begin{bmatrix} \bar{\alpha}_{\cdot}^{\circ} + \bar{\beta}_{\cdot}^{\circ} \\ -\bar{\alpha}_{\cdot}^{\circ}\mathbf{1}_a \\ -\bar{\beta}_{\cdot}^{\circ}\mathbf{1}_2 \\ -\bar{\beta}_{\cdot}^{\circ} \end{bmatrix} \quad \text{for } \bar{\alpha}_{\cdot}^{\circ} = \frac{\Sigma_i \alpha_i^{\circ}}{a} \text{ and } \bar{\beta}_{\cdot}^{\circ} = \frac{\Sigma_j \beta_j^{\circ}}{b} .$$

This gives, using subscript Σ to denote the Σ-restrictions,

$$\dot{\mu}_{\Sigma}^{\circ} = \bar{\alpha}_{\cdot}^{\circ} + \bar{\beta}_{\cdot}^{\circ}, \qquad \dot{\alpha}_{i,\Sigma}^{\circ} = \alpha_i^{\circ} - \bar{\alpha}_{\cdot}^{\circ} \quad \text{and} \quad \dot{\beta}_{j,\Sigma}^{\circ} = \beta_j^{\circ} - \bar{\beta}_{\cdot}^{\circ} . \tag{156}$$

For the W-restrictions

$$\mathbf{P}' = \begin{bmatrix} 0 & \mathbf{n}'_a & 0 \\ 0 & 0 & \mathbf{n}'_b \end{bmatrix},$$

and on defining $\tilde{\alpha}_{\cdot}^{\circ} = \Sigma_i n_{i\cdot}\alpha_i^{\circ}/n_{\cdot\cdot}$ and $\tilde{\beta}_{\cdot}^{\circ} = \Sigma_j n_{\cdot j}\beta_j^{\circ}/n_{\cdot\cdot}$, this leads to

$$\hat{\mu}_w = \tilde{\alpha}_{\cdot}^{\circ} + \tilde{\beta}_{\cdot}^{\circ}, \qquad \hat{\alpha}_{i,w} = \alpha_i^{\circ} - \tilde{\alpha}_{\cdot}^{\circ} \quad \text{and} \quad \hat{\beta}_j = \beta_j^{\circ} - \tilde{\beta}_{\cdot}^{\circ} . \tag{157}$$

9.6. EXERCISES

E9.1. Suppose Grid 9.1 also has data in
(a) cell 3, 1;
(b) cells 3, 1 and 2, 4.
In each case use the adapted Σ-restrictions of Section 9.1c to describe the hypothesis tested by the interaction sum of squares.

E9.2. Starting from (61) and (74), show that the covariance of $\mathbf{y}_a$ with $\mathbf{r}_2$ is null; and then derive the dispersion matrix of $[\boldsymbol{\alpha}^{\circ\prime} \quad \boldsymbol{\beta}_2^{\circ\prime}]'$.

E9.3. Solve the normal equations (87) using (103) and confirm the results at the end of Section 9.3.

E9.4. In Section 9.5:

(a) Calculate $R(\mu, \dot{\boldsymbol{\beta}})$ in the same manner as (130) is derived.

(b) Thence calculate $R(\dot{\boldsymbol{\alpha}} \mid \dot{\mu}, \dot{\boldsymbol{\beta}})$ and confirm your result using the invert-part-of-the-inverse algorithm.

(c) Also use $q = (\mathbf{k}'\boldsymbol{\beta}^{\circ})^2/\mathbf{k}'\mathbf{G}\mathbf{k}$ for additional confirmation.

E9.5. Calculate $R(\dot{\mu} \mid \dot{\boldsymbol{\alpha}}, \dot{\boldsymbol{\beta}}, \dot{\boldsymbol{\phi}})$, $R(\dot{\boldsymbol{\beta}} \mid \dot{\mu}, \dot{\boldsymbol{\alpha}}, \dot{\boldsymbol{\phi}})$ and $R(\dot{\boldsymbol{\phi}} \mid \dot{\mu}, \dot{\boldsymbol{\alpha}}, \dot{\boldsymbol{\beta}})$ for the example of Table 8.8.

(a) Use the invert-part-of-the-inverse algorithm.

(b) Use Q, the numerator sum of squares of an F-statistic based on the overparameterized unrestricted model.

E9.6. For the example of Table 9.4:

(a) Calculate $R(\dot{\mu}, \dot{\boldsymbol{\alpha}}, \dot{\boldsymbol{\beta}}, \dot{\boldsymbol{\phi}})$ and confirm it from Chapter 4.

(b) Repeat (a) for $R(\dot{\boldsymbol{\alpha}} \mid \dot{\mu}, \dot{\boldsymbol{\beta}})$ and $R(\dot{\boldsymbol{\beta}} \mid \dot{\mu}, \dot{\boldsymbol{\alpha}})$.

(c) Calculate $R(\dot{\mu} \mid \dot{\boldsymbol{\alpha}}, \dot{\boldsymbol{\beta}}, \dot{\boldsymbol{\phi}})$ and confirm its value in the form of the numerator of an F-statistic.

E9.7. For the following data grid, calculate (146).

√	√	√
√	√	
	√	√

Special cases

E9.8. **Balanced data: all cells filled with $n_{ij} = n$**

(a) For the with-interaction model explain why one solution vector is

$$\mu^{\circ} = 0, \qquad \alpha_i^{\circ} = 0, \qquad \beta_j^{\circ} = 0 \quad \text{and} \quad \phi_{ij}^{\circ} = \bar{y}_{ij},$$

whereas that for the Σ-restricted model is

$$\dot{\mu}^{\circ} = \bar{y}_{\cdots} \qquad \dot{\alpha}_i^{\circ} = \bar{y}_{i\cdot\cdot} - \bar{y}_{\cdots}$$

$$\dot{\phi}_{ij}^{\circ} = \bar{y}_{ij\cdot} - \bar{y}_{i\cdot\cdot} - \bar{y}_{\cdot j\cdot} + \bar{y}_{\cdots} \qquad \dot{\beta}_j^{\circ} = \bar{y}_{\cdot j\cdot} - \bar{y}_{\cdots}\ .$$

(b) Derive solution vectors for the no-interaction model.

(c) Show that $R(\alpha \mid \mu, \beta) = R(\alpha \mid \mu)$.

(d) What hypothesis is tested by $R(\alpha \mid \mu)$ in (a) and in (b)?

E9.9. **$n_{ij} = 0$ or 1**

(a) Explain why the with-interaction model is not appropriate.

(b) Calculate Table 9.2E for the following data set in four rows and three columns.

5	5	11
—	—	4
2	—	6
8	3	19

E9.10. **Proportional subclass numbers.** Data are described as having proportional subclass numbers when

$$n_{ij}/n_{1j} = p'_i \quad \forall\ j, \quad \text{for } i = 2, \ldots, a\ .$$

(a) Show that the preceding equality implies $p'_i = n_{i\cdot}/n_{1\cdot}$.

(b) Show that the preceding equality implies:

$$n_{ij}/n_{i1} = q'_j \quad \forall\ i, \quad \text{for } j = 2, \ldots, b\ .$$

(c) With $p_1 = n_{1\cdot}/n_{\cdot\cdot}$, convert p'_i to $p_i = p_1 p'_i = n_{i\cdot}/n_{\cdot\cdot}$ and show that

$$R(\beta \mid \mu, \alpha) = \Sigma_j n_{\cdot\cdot} q_j (\bar{y}_{\cdot j\cdot} - \bar{y}_{\cdot\cdot\cdot})^2 = R(\beta \mid \mu)$$

with associated hypothesis

$$\text{H: } \beta_j + \Sigma_i p_i \phi_{ij} \textit{ equal } \quad \forall\ j\ .$$

(d) What is the associated hypothesis for $R(\mu)$ and for $R(\phi \mid \mu, \alpha, \beta)$?

(e) Derive the expression for $R(\dot{\beta} \mid \dot{\mu}, \dot{\alpha}, \dot{\phi})$ and its associated hypothesis.

(f) Calculate Table 9.1E for the following data.

3	11, 17	5
2, 6	3, 8, 9, 10	4, 6
4, 4, 5	4, 5, 7, 8, 11, 15	4, 8, 9

E9.11. **The 2 × b case**. For $a = 2$ and all cells filled, derive the following results.

(a) $\alpha_1^o = \dfrac{\Sigma_j h_j(\bar{y}_{1j\cdot} - \bar{y}_{2j\cdot})}{\Sigma_j h_j}$ for $\dfrac{1}{h_j} = \dfrac{1}{n_{1j}} + \dfrac{1}{n_{2j}}$

and $\beta_j^o = \bar{y}_{\cdot j\cdot} - n_{1j}\alpha_1^o / n_{\cdot j}$.

(b) $R(\boldsymbol{\alpha} \mid \mu) = h_r(\bar{y}_{1\cdot\cdot} - \bar{y}_{2\cdot\cdot})^2$ for $\dfrac{1}{h_r} = \dfrac{1}{n_{1\cdot}} + \dfrac{1}{n_{2\cdot}}$

$R(\boldsymbol{\alpha} \mid \mu, \boldsymbol{\beta}) = \alpha_1^{o2}\Sigma_j h_j$

$R(\boldsymbol{\phi} \mid \mu, \boldsymbol{\alpha}, \boldsymbol{\beta}) = \Sigma_j h_j(\bar{y}_{1j\cdot} - \bar{y}_{2j\cdot})^2 - R(\boldsymbol{\alpha} \mid \mu, \boldsymbol{\beta})$.

(c) Suggest a simplification of $R(\dot{\boldsymbol{\beta}} \mid \dot{\mu}, \dot{\boldsymbol{\alpha}}, \dot{\boldsymbol{\phi}})$.

(d) For the data of Table 4.1, calculate Table 9.2, using (b).

(e) Calculate Table 9.2 for the following data using (b):

2, 6	42	2, 4, 6, 8
4, 9, 14	17	5, 11

E9.12. **The 2 × 2 case**. The results in E9.11 simplify when $b = 2$ also.

(a) From $R(\boldsymbol{\alpha} \mid \mu)$ of E9.11, suggest a form for $R(\boldsymbol{\beta} \mid \mu)$ and confirm it.

(b) Show that $R(\boldsymbol{\phi} \mid \mu, \boldsymbol{\alpha}, \boldsymbol{\beta}) = \hat{\theta}_{11,22}^2 / \Sigma\Sigma(1/n_{ij})$.

(c) Calculate Table 9.2 for the following data set:

1, 1, 7	3, 3, 9
7	15, 17

E9.13. Define

$$\mathbf{G} = \begin{bmatrix} 0 & \mathbf{0} \\ \mathbf{0} & \mathbf{D}\{1/n_i\} \end{bmatrix} \quad \text{and} \quad \mathbf{G}_r = \begin{bmatrix} 0 & \{_r\, 1/an_i\} \\ \mathbf{0} & \mathbf{D}\{1/n_i\} - \mathbf{1}\{_r\, 1/an_i\} \end{bmatrix}.$$

For the 1-way classification show that, for the Σ-restrictions, (154) yields $\mathbf{G}_r$; and derive the corresponding solution vector.

E9.14. (a) Derive $\mathbf{GX'X}$ of (155).

(b) Use (154) to derive (157).

E9.15. For the overparameterized, with-interaction, 2-way crossed classification model with all-cells-filled data, use the Σ-restrictions to do the following.

(a) Confirm that

$$[\mathbf{P}'(\mathbf{I}-\mathbf{H})]^-$$

$$= \begin{bmatrix} -1/a & -1/b & -\mathbf{1}'_a/ab & \mathbf{0} & 0 \\ \mathbf{I}_a/a & \mathbf{0} & (-a\mathbf{I}_a+\mathbf{J}_a)/ab & \mathbf{0} & \mathbf{0} \\ \mathbf{0} & -\mathbf{1}_{b-1}/b & -\mathbf{J}_{(b-1)\times a}/ab & -\mathbf{I}_{b-1}/a & \mathbf{0} \\ 0 & -1/b & (b-1)\mathbf{1}'_a/ab & -\mathbf{1}'_{b-1}/a & 0 \\ 0 & \mathbf{0} & \mathbf{0} & \mathbf{0} & 0 \end{bmatrix}.$$

(b) Show that the solution vector using part (a) is

$$\hat{\dot{\mu}}_{\Sigma} = \tilde{y}_{..} \qquad \hat{\dot{\phi}}_{ij,\Sigma} = \bar{y}_{ij\cdot} - \tilde{y}_{i\cdot} - \tilde{y}_{\cdot j} + \tilde{y}_{..}$$

$$\hat{\dot{\alpha}}_{i,\Sigma} = \tilde{y}_{i\cdot} - \tilde{y}_{..} \qquad \hat{\dot{\beta}}_{j,\Sigma} = \tilde{y}_{\cdot j} - \tilde{y}_{..}$$

for $\tilde{y}_{i\cdot} = \Sigma_j \bar{y}_{ij\cdot}/b$, $\tilde{y}_{\cdot j} = \Sigma_i \bar{y}_{ij\cdot}/a$ and $\tilde{y}_{..} = \Sigma_i\Sigma_j \bar{y}_{ij\cdot}/ab$.

(c) For the solution in (b) use the R-algorithm to verify $R(\mu, \alpha, \beta, \phi)$.

(d) Why, using the solution in (b), does

$$R(\mu, \alpha, \beta, \phi) - R(\dot{\mu}, \dot{\beta}, \dot{\phi}) \neq \text{SSA}_w?$$

E9.16. For the following data in a 2×2 layout,

6, 14	10, 12, 13
11, 15, 18, 20	4, 6, 7, 9

(a) Calculate SSA_w and SSB_w.

(b) Write down and solve the normal equations for the Σ-restricted model. Note that for $\mathbf{B}^{-1} = \mathbf{A}^2 - k^2\mathbf{I}$

$$\begin{bmatrix} \mathbf{A} & k\mathbf{I} \\ k\mathbf{I} & \mathbf{A} \end{bmatrix}^{-1} = \begin{bmatrix} \mathbf{A}^{-1} & \mathbf{0} \\ \mathbf{0} & \mathbf{0} \end{bmatrix} + \begin{bmatrix} k^2\mathbf{B}\mathbf{A}^{-1} & -k\mathbf{B} \\ -k\mathbf{B} & \mathbf{A}\mathbf{B} \end{bmatrix}.$$

(c) Using the "invert part of the inverse" rule calculate

$$R(\dot{\alpha} \mid \dot{\mu}, \dot{\beta}, \dot{\gamma})_{\Sigma}, \quad R(\dot{\beta} \mid \dot{\mu}, \dot{\alpha}, \dot{\gamma})_{\Sigma} \text{ and } R(\dot{\gamma} \mid \dot{\mu}, \dot{\alpha}, \dot{\beta})_{\Sigma}.$$

(d) Calculate the numerator sum of squares for the hypothesis of no-interaction.

E9.17. For the data of E9.16, what are the least squares means for the rows

(a) for the with-interaction model?

(b) for the no-interaction model (to 2 decimal places)?

(c) for (a) and (b) when the data of cell 2, 1 are assumed not to be there?

E9.18. (Unweighted means analysis) With all cells filled,

$$\mathrm{SSA}_u = b\sum_{i=1}^{a} (\bar{x}_{i\cdot} - \bar{x}_{\cdot\cdot})^2$$

with $x_{ij} = \bar{y}_{ij\cdot}$, and $\bar{x}_{i\cdot} = \Sigma_{j=1}^{b} x_{ij}/b$ is sometimes considered an alternative to SSA_w.

(a) With $\mathbf{x} = \{x_{ij}\}$ and $\mathbf{y} = \{y_{ijk}\}$ having their elements in lexicon order, specify $\mathbf{T}$ such that $\mathbf{x} = \mathbf{T}\mathbf{y}$.

(b) Using $\mathbf{T}$ and assuming normality, show that SSA_u is not distributed as a χ^2, but is independent of SSE.

(c) Using the tr(**AV**) part of Theorem 7.2a, show that the term in σ^2 in the expected value of SSA_u is $(a-1)n_h\sigma^2$, where $n_h = (\Sigma_i\Sigma_j 1/n_{ij})/ab$.

(d) Using $E(y_{ijk}) = \mu + \alpha_i + \beta_j + \gamma_{ij}$ directly in SSA_u, show that the expected value of SSA_u also contains

$$b\sum_{i=1}^{a} (\alpha_i + \bar{\gamma}_{i\cdot} - \bar{\alpha}_{\cdot} - \bar{\gamma}_{\cdot\cdot})^2 .$$

(e) From (c) and (d) suggest an approximate F-test for the same hypothesis as is tested by SSA_w.

E9.19. Solve (40) using the W-restrictions and calculate $R(\dot{\boldsymbol{\alpha}} \mid \dot{\mu})$, $R(\dot{\boldsymbol{\beta}} \mid \dot{\mu}, \dot{\boldsymbol{\alpha}})$, $R(\dot{\boldsymbol{\beta}} \mid \dot{\mu})$, $R(\dot{\boldsymbol{\alpha}} \mid \dot{\mu}, \dot{\boldsymbol{\beta}})$ and $R(\dot{\boldsymbol{\phi}} \mid \dot{\mu}, \dot{\boldsymbol{\alpha}}, \dot{\boldsymbol{\beta}})$ for that W-restricted model.

E9.20. The 2-way nested model has equation $E(y_{ijk}) = \mu_i + \beta_{ij} + e_{ijk}$ with $i = 1, \ldots, a$, $j = 1, \ldots, b_i$ and $k = 1, \ldots, n_{ij}$. Suppose $\beta_{ij} \sim$ i.i.d. $(0, \sigma_\beta^2)$ and $e_{ijk} \sim$ i.i.d. $(0, \sigma_e^2)$ and that every β_{ij} is independent of every e_{ijk}. Find the GLSE of μ_i.

CHAPTER 10

EXTENDED CELL MEANS MODELS

10.1. MULTI-FACTOR DATA: BASIC RESULTS

Difficulties of analyzing unbalanced data using linear models have been illustrated and described in the preceding chapters largely in terms of models for the 2-way classification. Complexities associated with those models are simply aggravated when we come to deal with models for more than two factors. Nevertheless a number of general procedures are available, as are now described.

Although BLU estimation using cell means models is scarcely one whit more difficult for multi-factor data than it is for the 2-way classification, its description is brief, easy and given here for the sake of completeness. In the 2-way context, "cell" has been used to mean the intersection of a row and a column. Nevertheless, in Section 1.2a the broader concept was used, namely a subclassification of the data defined by one level of every factor, no matter how many factors there are. This is the meaning we now use; e.g., for three factors, A, B and C having 4, 3 and 2 levels, respectively, there will be $4 \times 3 \times 2 = 24$ cells. Sometimes the term "sub-most cell" is used in place of "cell", to emphatically distinguish those 24 cells from, for example, the 12 cells of the implicit A-by-B dichotomy.

a. The model

No matter how many classifications (factors) are being used, suppose s sub-most cells contain data. Define $\boldsymbol{\mu}$ as the $s \times 1$ vector of population means for those s filled, sub-most cells. In the tth such cell suppose there are n_t observations, to be denoted by $\mathbf{y}_t = [y_{t,1} \quad \cdots \quad y_{t,n_t}]'$ with mean

$\bar{y}_{t\cdot} = \Sigma_{j=1}^{n_t} y_{t,j}/n_t$. Define the vector of all observations as

$$\mathbf{y} = [\mathbf{y}_1' \quad \mathbf{y}_2' \quad \cdots \quad \mathbf{y}_t' \quad \cdots \quad \mathbf{y}_s']', \tag{1}$$

with the vectors of totals and means being, respectively,

$$\mathbf{y}. = \{_c\, \mathbf{y}_{t\cdot}\}_{t=1}^{t=s} \quad \text{and} \quad \bar{\mathbf{y}} = \{_c\, \bar{\mathbf{y}}_{t\cdot}\}_{t=1}^{t=s} . \tag{2}$$

Then the cell means model is

$$E(\mathbf{y}) = \mathbf{X}\boldsymbol{\mu}, \quad \text{or} \quad \mathbf{y} = \mathbf{X}\boldsymbol{\mu} + \mathbf{e} \quad \text{with } \mathbf{X} = \mathbf{D}\{\mathbf{1}_{n_t}\} . \tag{3}$$

Thus $\mathbf{X}$ is a diagonal matrix of s summing vectors, of order n_t for $t = 1, 2, \ldots, s$.

b. Estimation

The normal equations emanating from (3) are $\mathbf{X}'\mathbf{X}\hat{\boldsymbol{\mu}} = \mathbf{X}'\mathbf{y}$. Due to the nature of $\mathbf{X}$ in (3), these equations are

$$\mathbf{D}\{n_t\}\hat{\boldsymbol{\mu}} = \mathbf{y}. \quad \text{with solution} \quad \hat{\boldsymbol{\mu}} = \mathbf{D}\{1/n_t\}\mathbf{y}. = \bar{\mathbf{y}},$$

where $\mathbf{X}'\mathbf{X} = \mathbf{D}\{n_t\}$ is the diagonal matrix of diagonal elements n_t for $t = 1, \ldots, s$ for the filled cells. Hence the BLUE of μ_t is

$$\hat{\mu}_t = \bar{y}_{t\cdot} \quad \text{for each filled cell,} \qquad t = 1, \ldots, s . \tag{4}$$

The similarity of this to (22) of Chapter 2, to (4) of Chapter 3, to (12) of Chapter 4 and to (47) and (194) of Chapter 8 is readily apparent.

c. Residual variance

The estimated residual variance is the pooled within-cell variance:

$$\hat{\sigma}^2 = \sum_{t=1}^{s} \sum_{j=1}^{n_t} (y_{tj} - \bar{y}_{t\cdot})^2/(N - s)$$

for $N = \Sigma_{t=1}^{s} n_t = n.$ = total number of observations.

d. Estimable functions

The μ_t corresponding to every filled sub-most cell is estimable, with $\text{BLUE}(\mu_t) = \bar{y}_{t\cdot}$ of (4). Also, *every* linear combination of such μ_ts is estimable with BLUE:

$$\text{BLUE}\left(\sum_{t=1}^{s}\lambda_t\mu_t\right) = \sum_{t=1}^{s}\lambda_t\bar{y}_{t\cdot} \quad\text{and}\quad v\left[\text{BLUE}\left(\sum_{t=1}^{s}\lambda_t\mu_t\right)\right] = \sigma^2\sum_{t=1}^{s}\lambda_t^2/n_t\,. \tag{5}$$

This, too, is similar to results in earlier chapters: to (27) in Chapter 2, to (7) in Chapter 3 and to (16) in Chapter 4.

e. Hypothesis testing

Any linear hypothesis concerning cell means of filled cells is testable:

$$\text{H: } \mathbf{K}'\boldsymbol{\mu} = \mathbf{m}$$

is tested by

$$F = Q/\hat{\sigma}^2 r_{\mathbf{K}} = (\mathbf{K}'\bar{\mathbf{y}} - \mathbf{m})'[\mathbf{K}'\mathbf{D}\{1/n_t\}\mathbf{K}]^{-1}(\mathbf{K}'\bar{\mathbf{y}} - \mathbf{m})/\hat{\sigma}^2 r_{\mathbf{K}} \tag{6}$$

for any full row rank matrix $\mathbf{K}'$, and where $\mathbf{D}\{1/n_t\}$ is the diagonal matrix of diagonal elements $1/n_1, \ldots, 1/n_s$. The F-statistic in (6) is, of course, essentially the same as (197) of Section 8.10, each being just a special case of the general form (146) in Section 8.8b.

f. Analogy: the 1-way classification

The preceding analysis is quite straightforward. It is effectively just the analysis of a 1-way classification using the filled sub-most cells as levels of a 1-way classification. Indeed, this is one way in which computing packages designed for analysis of variance calculations can be persuaded to do such calculations for a cell means model: treat the sub-most cells as levels of a 1-way classification.

g. Some-cells-empty data

The occurrence of empty sub-most cells in multi-factor data demands the same approach to developing an analysis as is discussed at length in Chapter 5 for the 2-way classification—only oft-times more difficult; more difficult, not in concept but in deciding on useful applications of the concepts. The BLUE can be obtained for any linear combination of the population means of filled sub-most cells, and hypotheses about such combinations can be tested. Equations (4), (5) and (6) are the tools for doing this. The difficulty, in the presence of empty cells, is to decide which linear combinations of cell means (of filled cells) are of interest. The same kind of difficulty arises with deciding which subset analyses (Section 5.6) are of interest. Cooperation between investigator and statistician is essential to resolving these difficulties in a useful fashion—as discussed in Chapter 5.

10.2. MULTI-FACTOR DATA: ALL-CELLS-FILLED DATA

Analysis of all-cells-filled data is extended in this section from the 2-way classification of Chapter 4 to cases where there are more than two factors. We confine attention to situations having the following characteristics:

(i) two or more crossed factors, with no nested factors,
(ii) all cells filled, and
(iii) cell means models that tacitly have all possible interactions.

Cell means models for situations where some but not all interactions are to be included are dealt with in Section 10.4.

In the 2-factor case of Chapter 4 a hypothesis of interest at equations (40) and (41) of that chapter is, for

$$\bar{\mu}_{i\cdot} = \frac{1}{b}\sum_{j=1}^{b}\mu_{ij}, \tag{7}$$

$$\text{H: } \bar{\mu}_{i\cdot} \textit{ equal } \forall\ i\ . \tag{8}$$

This is the hypothesis that the average of the cell means over each row, $\bar{\mu}_{i\cdot}$, is the same for every row. It is a hypothesis dealing with equality of rows. And from (42) of Chapter 4 the numerator sum of squares for testing (8) is

$$\text{SSA}_w = \sum_{i=1}^{a} w_i\left(\tilde{y}_{i\cdot} - \sum_{i=1}^{a} w_i\tilde{y}_{i\cdot}\Big/\sum_{i=1}^{a} w_i\right)^2 \tag{9}$$

for

$$\tilde{y}_{i\cdot} = \frac{1}{b}\sum_{j=1}^{b}\bar{y}_{ij\cdot} \quad \text{and} \quad \frac{1}{w_i} = \frac{1}{b^2}\sum_{j=1}^{b}\frac{1}{n_{ij}}\ . \tag{10}$$

Although SSA_w is usually thought of as a sum of squares from the weighted squares of means analysis, it can be derived directly from applying the general results for testing H: $\mathbf{K}'\boldsymbol{\beta} = \mathbf{m}$ given in Section 8.8. To do this, rewrite the hypothesis (8) in the form

$$\text{H: } \mathbf{K}'\boldsymbol{\mu} = \mathbf{0} \tag{11}$$

with

$$\mathbf{K}' = \begin{bmatrix} \mathbf{1}'_b & -\mathbf{1}'_b & \cdot & \cdots & \cdot \\ \mathbf{1}'_b & \cdot & -\mathbf{1}'_b & & \\ \vdots & \vdots & & \ddots & \\ \mathbf{1}'_b & \cdot & & & -\mathbf{1}_b \end{bmatrix} = [\mathbf{1}_{a-1} \quad -\mathbf{I}_{a-1}] \otimes \mathbf{1}'_b, \tag{12}$$

where $\otimes$ is the direct product operator. Now define

$$\hat{\boldsymbol{\mu}} = \{\bar{y}_{ij\cdot}\} \quad \text{and} \quad \mathbf{G} = \mathbf{D}\{1/n_{ij}\}, \tag{13}$$

so that $\hat{\boldsymbol{\mu}}$ is the vector of observed cell means and $\mathbf{G}$ is the diagonal matrix of terms $1/n_{ij}$, each with its elements arrayed in lexicon order. Then (see E8.32), SSA_w of (9) can be derived from (6) using

$$Q = (\mathbf{K}'\hat{\boldsymbol{\mu}})'(\mathbf{K}'\mathbf{G}\mathbf{K})^{-1}\mathbf{K}'\hat{\boldsymbol{\mu}} \tag{14}$$

with $\mathbf{K}'$, $\hat{\boldsymbol{\mu}}$ and $\mathbf{G}$ of (12) and (13).

We show how hypotheses like (8) can be extended for multi-factor models to hypotheses about not only main effects but also about interactions, with main effects and interactions each being defined, as is (8), in terms of averages of cell means. For these hypotheses the sums of squares for testing those hypotheses can be derived from (14) using extensions of (12) and (13). We illustrate these extensions in terms of a 3-factor case.

a. The 3-way crossed classification

Consider three crossed factors, labeled A, B and C, having a, b and c levels, respectively. Let μ_{ijk} be the population mean of the cell defined by levels, i, j and k of factors A, B and C, respectively, having BLUE

$$\hat{\mu}_{ijk} = \bar{y}_{ijk\cdot} \ . \tag{15}$$

Then, just as in (8), the hypothesis of equality of levels of the A-factor will be, for $\bar{\mu}_{i\cdot\cdot} = \Sigma_j\Sigma_k\mu_{ijk}/bc$,

$$\text{H: } \bar{\mu}_{i\cdot\cdot} \ \textit{equal} \ \ \forall \ i \tag{16}$$

tested by SSA_w of (9) using

$$\tilde{y}_{i\cdot\cdot} = \frac{1}{bc}\sum_{j=1}^{b}\sum_{k=1}^{c}\bar{y}_{ijk\cdot} \quad \text{and} \quad \frac{1}{w_i} = \frac{1}{b^2c^2}\sum_{j=1}^{b}\sum_{j=1}^{c}\frac{1}{n_{ijk}}$$

in place of $\tilde{y}_{i\cdot}$ and w_i of (10).

The extension of (16) to a hypothesis about A-by-B interactions being zero is the hypothesis

$$\text{H: } \bar{\mu}_{ij\cdot} - \bar{\mu}_{i'j\cdot} - \bar{\mu}_{ij'\cdot} + \bar{\mu}_{i'j'\cdot} = 0 \quad \forall \; i \neq i' \text{ and } j \neq j' \;. \tag{17}$$

Rewriting this as H: $\mathbf{K}'\boldsymbol{\mu} = \mathbf{0}$ for an appropriate $\mathbf{K}'$ will yield a specific analytic expression for Q, the numerator sum of squares for testing (17). The derivation of such an expression has, no doubt, been made somewhere; the derivation would certainly be somewhat tedious. But, in the presence of today's computers and computing packages which will, for a wide variety of data sets, calculate numerical values of Q, there seems to be no great advantage in undertaking the tedium of deriving Q analytically. Instead, we indicate the basis on which the calculated Q is found to be available in computing-package output.

b. Computing, for Σ-restricted, overparameterized models

Chapter 9 deals with overparameterized models for the 2-way classification. Section 9.4b-i describes the Σ-restricted versions of such models, when interactions are included and the data have all cells filled; and at equation (139) we find for such a situation the identity

$$R(\dot{\boldsymbol{\alpha}} \mid \dot{\mu}, \dot{\boldsymbol{\beta}}, \dot{\boldsymbol{\phi}}) = \text{SSA}_w \;. \tag{18}$$

The symbol on the left is the reduction in sum of squares for $\dot{\alpha}$-effects in the full-rank, Σ-restricted, with-interaction model for the 2-way classification with all cells filled. Therefore, as in Table 9.3, the sum of squares (18) has

$$\text{H: } \dot{\alpha}_i = 0 \quad \forall \; i$$

as its associated hypothesis.

Now consider the hypothesis (17). In terms of a traditional overparameterized model for $E(y_{ijkl}) = \mu_{ijk}$ we have

$$\mu_{ijk} = \mu + \alpha_i + \beta_j + (\alpha\beta)_{ij} + \phi_k + (\alpha\phi)_{ik} + (\beta\phi)_{jk} + (\alpha\beta\phi)_{ijk} \tag{19}$$

where, for example, $(\alpha\beta)_{ij}$ is the interaction effect for the ith level of the A-factor and the jth level of the B-factor. On substituting (19) into (17) and using

$$\omega_{ij} = (\alpha\beta)_{ij} + \overline{(\alpha\beta\phi)}_{ij\cdot}, \tag{20}$$

the hypothesis becomes

$$\text{H: } \omega_{ij} - \omega_{ij'} - \omega_{i'j} + \omega_{i'j'} = 0 \quad \forall \; i \neq i' \text{ and } j \neq j' \;. \tag{21}$$

Introducing the Σ-restrictions we rewrite (20) as

$$\dot{\omega}_{ij} = (\alpha\dot{\beta})_{ij} + \sum_{k=1}^{c} (\alpha\dot{\beta}\phi)_{ijk}/c, \tag{22}$$

where a dot over a single letter of a parenthesized symbol represents a dot over the whole symbol. But one of the Σ-restrictions in the 3-way classification is

$$\sum_{k=1}^{c} (\alpha\dot{\beta}\phi)_{ijk} = 0 \quad \forall\ i \text{ and } j\ . \tag{23}$$

Hence
$$\dot{\omega}_{ij} = (\alpha\dot{\beta})_{ij}\ . \tag{24}$$

And others of the Σ-restrictions are

$$\sum_{i=1}^{a} \dot{\omega}_{ij} = 0 \quad \forall\ j \quad \text{and} \quad \sum_{j=1}^{b} \dot{\omega}_{ij} = 0 \quad \forall\ i\ . \tag{25}$$

These mean that the hypothesis (21) is equivalent to (e.g., E10.1)

$$\text{H: } \dot{\omega}_{ij} \textit{ all zero} \tag{26}$$

i.e., from (24)
$$\text{H: } (\dot{\alpha}\beta)_{ij} \textit{ all zero}\ . \tag{27}$$

But the Σ-restricted overparameterized model is of full rank, so that for (27), from Section 8.6e,

$$\begin{aligned} &\text{H: } (\alpha\dot{\beta})_{ij} \textit{ all zero} \text{ is tested by} \\ &\qquad R\left[(\dot{\alpha}\beta)| \dot{\mu}, \dot{\alpha}, \dot{\beta}, \dot{\phi}, (\dot{\alpha}\phi), (\dot{\beta}\phi), (\alpha\dot{\beta}\phi)\right]\ . \end{aligned} \tag{28}$$

Hence, if analogous to SSA_w we denote by $SSAB_w$ the sum of squares for testing (21), we then have that

$$\text{H: } \dot{\mu}_{ij\cdot} - \bar{\mu}_{ij'\cdot} - \bar{\mu}_{i'j\cdot} + \bar{\mu}_{i'j'\cdot} \textit{ all zero} \quad \text{is tested by } SSAB_w; \tag{29}$$

and from (28) we have

$$\begin{aligned} &\text{H: } (\alpha\dot{\beta})_{ij} \textit{ all zero} \quad \text{is tested by} \\ &\qquad R(\alpha\dot{\beta}, \text{adjusted for all other terms in the Σ-restricted model})\ . \end{aligned} \tag{30}$$

For simplicity, denote the $R(\cdot)$-term in (30) by $R(\boldsymbol{\alpha}\dot{\boldsymbol{\beta}} \mid \text{all other effects})$. Then (30) is

$$\text{H: } (\alpha\dot{\beta})_{ij} \textit{ all zero} \quad \text{is tested by } R(\boldsymbol{\alpha}\dot{\boldsymbol{\beta}} \mid \text{all other effects}) \ . \tag{31}$$

Since hypotheses (29) and (30) are the same, so are the sums of squares:

$$\text{SSAB}_w = R\left[(\boldsymbol{\alpha}\dot{\boldsymbol{\beta}}) \mid \text{all other effects}\right] \ . \tag{32}$$

The importance of this result lies in the fact that the right-hand side of (32) is known to be available in the output from certain computing packages (e.g., BMDP2V and SAS GLM Type III). Therefore, although we know that Q of (14) will yield a specific, analytic expression for the sum of squares SSAB_w for testing H: $\bar{\mu}_{ij\cdot} - \bar{\mu}_{ij'} - \bar{\mu}_{i'j\cdot} + \bar{\mu}_{i'j'\cdot}$ *all zero*, we do not need such an expression in order to obtain computed values of SSAB_w, because by (32) they can be obtained from computer output when identified in the form $R(\alpha\dot{\beta} \mid \text{all other effects})$.

The result (32) applies quite generally, not just to 2-factor interactions of which it is an example, but also to all higher-order interactions. Thus the hypothesis for the 3-factor interactions is

$$\text{H: } \mu_{ijk} - \mu_{ij'k} - \mu_{i'jk} + \mu_{i'j'k} - (\mu_{ijk} - \mu_{ij'k'} - \mu_{i'jk'} + \mu_{i'j'k'}) \textit{ all zero} \tag{33}$$

and it is tested by

$$\text{SSABC}_w = R\left[(\boldsymbol{\alpha}\dot{\boldsymbol{\beta}}\boldsymbol{\phi}) \mid \text{all other effects}\right] \ . \tag{34}$$

And, of course, the main effects result we started with is just a special case:

$$\text{H: } \bar{\mu}_{i\cdot\cdot} - \bar{\mu}_{i'\cdot\cdot} \textit{ all zero}, \quad \text{i.e.,} \quad \text{H: } \bar{\mu}_{i\cdot\cdot} \textit{ all equal}$$

is tested by
$$\text{SSA}_w = R(\dot{\boldsymbol{\alpha}} \mid \text{all other effects}) \ . \tag{35}$$

c. Example

Table 10.1 shows data for a 3-way classification having 4 levels of A, 3 levels of B and 2 levels of C. Each group of numbers in the table is

$$\frac{y_{ijk1}, y_{ijk2}, \ldots, y_{ijkn_{ijk}}}{y_{ijk\cdot}(n_{ijk})\bar{y}_{ijk\cdot}} \ .$$

We illustrate the calculation of Q of (14) for one main-effects hypothesis like (18) and for one interaction hypothesis like (32)—and leave one of each

TABLE 10.1. DATA IN A 3-WAY CLASSIFICATION

$i =$ Level of Factor A	k = Level of Factor C: 1			2		
	j = Level of Factor B			j = Level of Factor B		
	1	2	3	1	2	3
1	10	3, 5	1, 2, 3	4, 6	2, 3, 7	4, 5, 9
	10(1)10	8(2)4	6(3)2	10(2)5	12(3)4	18(3)6
2	5, 9	5	6, 10	8	6, 8	2
	14(2)7	5(1)5	16(2)8	8(1)8	14(2)7	2(1)2
3	2, 3, 3, 4	3, 4, 8	5	3, 4, 8	4	6
	12(4)3	15(3)5	5(1)5	15(3)5	4(1)4	6(1)6
4	4	1, 1, 3, 7	8	5, 7, 9	19	20, 24
	4(1)4	12(4)3	8(1)8	21(3)7	19(1)19	44(2)22

for the reader (E10.2 and E10.3). For all calculations we use two vectors and a diagonal matrix that have their elements in lexicon order:

$$\boldsymbol{\mu} = \{\mu_{ijk}\}, \qquad \hat{\boldsymbol{\mu}} = \{\bar{y}_{ijk\cdot}\} \quad \text{and} \quad \mathbf{G} = \mathbf{D}\left\{\frac{1}{n_{ijk}}\right\} = \frac{1}{12}\mathbf{D}\{r_{ijk}\},$$

where the means $\bar{y}_{ijk\cdot}$ from Table 10.1 and certain totals of them are shown in Table 10.2; values of n_{ijk} from Table 10.1 are shown in Table 10.3A; and $r_{ijk} = 12/n_{ijk}$ and totals thereof are given in Table 10.3B.

The hypothesis pertaining to the C-factor is, akin to (16),

$$\text{H: } \bar{\mu}_{\cdot\cdot k} \textit{ all equal} .$$

With there being only two levels of C this can be written as

$$\text{H: } \Sigma_i\Sigma_j\mu_{ij1} - \Sigma_i\Sigma_j\mu_{ij2} = 0 . \tag{36}$$

TABLE 10.2. MEANS, $\bar{y}_{ijk\cdot}$, FROM TABLE 10.1 AND TOTALS OF THOSE MEANS

	$k=1$				$k=2$				
	j				j				
	1	2	3		1	2	3		
i	$\bar{y}_{ijk\cdot}$			$\Sigma_j \bar{y}_{ijk\cdot}$	$\bar{y}_{ijk\cdot}$			$\Sigma_j \bar{y}_{ijk\cdot}$	$\Sigma_j \Sigma_k \bar{y}_{ijk\cdot}$
1	10	4	2	16	5	4	6	15	31
2	7	5	8	20	8	7	2	17	37
3	3	5	5	13	5	4	6	15	28
4	4	3	8	15	7	19	22	48	63
$\Sigma_i \bar{y}_{ijk\cdot}$	24	17	23	64	25	34	36	95	159
				↑ $\Sigma_i \Sigma_j \bar{y}_{ijk\cdot}$				↑ $\Sigma_i \Sigma_j \bar{y}_{ijk\cdot}$	↑ $\Sigma_i \Sigma_j \Sigma_k \bar{y}_{ijk\cdot}$

TABLE 10.3A. VALUES OF n_{ijk} FROM TABLE 10.1

	$k=1$			$k=2$		
	j			j		
	1	2	3	1	2	3
i	n_{ijk}			n_{ijk}		
1	1	2	3	2	3	3
2	2	1	2	1	2	1
3	4	3	1	3	1	1
4	1	4	1	3	1	2

TABLE 10.3B. VALUES OF $r_{ijk} = 12/n_{ijk}$ FROM TABLE 10.3A, AND OF TOTALS THEREOF

	$k=1$				$k=2$				
	j				j				
	1	2	3		1	2	3		
i	$r_{ijk} = 12/n_{ijk}$			$r_{i\cdot k}$	$r_{ijk} = 12/n_{ijk}$			$r_{i\cdot k}$	$r_{i\cdot\cdot}$
1	12	6	4	22	6	4	4	14	36
2	6	12	6	24	12	6	12	30	54
3	3	4	12	19	4	12	12	28	47
4	12	3	12	27	4	12	6	22	49
$r_{\cdot jk}$	33	25	34	92	26	34	34	94	186

Whereas Q for this could be derived by formally writing (36) as H: $\mathbf{K}'\boldsymbol{\mu} = \mathbf{0}$ for an appropriate $\mathbf{K}'$ and using Q of (14), we use the fact that $\mathbf{K}'\hat{\boldsymbol{\mu}}$ in (14) is the BLUE of $\mathbf{K}'\boldsymbol{\mu}$ of the hypothesis and $\mathbf{K}'\mathbf{G}\mathbf{K}'$ is (on ignoring σ^2) the variance of the BLUE. Applying these principles to (36) we get the sum of squares for that hypothesis as

$$\text{SSC}_w = \frac{(\Sigma_i\Sigma_j\hat{\mu}_{ij1} - \Sigma_i\Sigma_j\hat{\mu}_{ij2})^2}{v(\Sigma_i\Sigma_j\hat{\mu}_{ij1} - \Sigma_i\Sigma_j\hat{\mu}_{ij2})} = \frac{(\Sigma_i\Sigma_j\bar{y}_{ij1\cdot} - \Sigma_i\Sigma_j\bar{y}_{ij2\cdot})^2}{\Sigma_i\Sigma_j\left(\dfrac{1}{n_{ij1}} + \dfrac{1}{n_{ij2}}\right)} \ . \tag{37}$$

Hence on taking the necessary values from Tables 10.2 and 10.3B, where the latter shows

$$r_{ijk} = \frac{12}{n_{ijk}}$$

so that the denominator of (37) is $(r_{\cdot\cdot 1} + r_{\cdot\cdot 2})/12$, we get

$$\text{SSC}_w = \frac{(64 - 95)^2}{(92 + 94)/12} = 62 \ . \tag{38}$$

Factors B and C have three and two levels, respectively, and therefore there are two degrees of freedom for the BC-interactions. Hence the hypothesis for BC-interactions can, akin to (29), be represented as

$$\text{H:} \begin{cases} \bar{\mu}_{\cdot 11} - \bar{\mu}_{\cdot 12} - \mu_{\cdot 21} - \bar{\mu}_{\cdot 22} = 0 \\ \bar{\mu}_{\cdot 11} - \bar{\mu}_{\cdot 12} - \bar{\mu}_{\cdot 31} - \bar{\mu}_{\cdot 32} = 0 \end{cases} .$$

Therefore, for (14), using the same principles as used in (37),

$$\mathbf{K}'\hat{\boldsymbol{\mu}} = \frac{1}{a}\begin{bmatrix} \Sigma_i(\bar{y}_{i11\cdot} - \bar{y}_{i12\cdot} - \bar{y}_{i21\cdot} + \bar{y}_{i22\cdot}) \\ \Sigma_i(\bar{y}_{i11\cdot} - \bar{y}_{i12\cdot} - \bar{y}_{i31\cdot} + \bar{y}_{i32\cdot}) \end{bmatrix}$$

$$= \tfrac{1}{4}\begin{bmatrix} 24 - 25 - 17 + 34 \\ 24 - 25 - 23 + 36 \end{bmatrix}, \quad \text{from Table 10.2}$$

$$= [4 \quad 3]' \ .$$

Similarly $\mathbf{K}'\mathbf{G}\mathbf{K} = \text{var}(\mathbf{K}'\hat{\boldsymbol{\mu}})$

$$= \frac{1}{a^2}\begin{bmatrix} \Sigma_i\left(\dfrac{1}{n_{i11}} + \dfrac{1}{n_{i12}} + \dfrac{1}{n_{i21}} + \dfrac{1}{n_{i22}}\right) & \Sigma_i\left(\dfrac{1}{n_{i11}} + \dfrac{1}{n_{i12}}\right) \\ \Sigma_i\left(\dfrac{1}{n_{i11}} + \dfrac{1}{n_{i12}}\right) & \Sigma_i\left(\dfrac{1}{n_{i11}} + \dfrac{1}{n_{i12}} + \dfrac{1}{n_{i31}} + \dfrac{1}{n_{i32}}\right) \end{bmatrix}$$

$$= \tfrac{1}{16}\left(\tfrac{1}{12}\right)\begin{bmatrix} r_{\cdot 11} + r_{\cdot 12} + r_{\cdot 21} + r_{\cdot 22} & r_{\cdot 11} + r_{\cdot 12} \\ r_{\cdot 11} + r_{\cdot 12} & r_{\cdot 11} + r_{\cdot 12} + r_{\cdot 31} + r_{\cdot 32} \end{bmatrix}$$

$$= \tfrac{1}{16}\left(\tfrac{1}{12}\right)\begin{bmatrix} 33 + 26 + 25 + 34 & 33 + 26 \\ 33 + 26 & 33 + 26 + 34 + 34 \end{bmatrix} \quad \text{from Table 10.3B},$$

$$= \tfrac{1}{192}\begin{bmatrix} 118 & 59 \\ 59 & 127 \end{bmatrix}.$$

Therefore

$$Q = [4 \quad 3]\left\{\tfrac{1}{192}\begin{bmatrix} 118 & 59 \\ 59 & 127 \end{bmatrix}\right\}^{-1}\begin{bmatrix} 4 \\ 3 \end{bmatrix} = \frac{192(1678)}{11505} = 28.0031 \qquad (39)$$

i.e., $\text{SSBC}_w = 28.0031$.

Note, in passing, that $\text{SSC}_w = 62$ of (38) and $\text{SSBC}_w = 28.0031$ of (39) will be found as part of the computer output obtained from analyzing the data of Table 10.1 on either the routine P2V of the BMD statistical package or on GLM (as type III sums of squares) of the SAS package.

d. More than three factors

The extension of the preceding methods for the 3-way classification to 4, 5, 6 and higher-order classifications proceeds quite straightforwardly. Indeed, results such as (32), (34) and (35) apply not only as derived, for the 3-way classification, but also quite generally where, for example, in $R(\alpha\dot{\beta}\,|\,\text{all other effects})$ in the Σ-restricted, overparameterized model that includes all interactions, "all other effects" applies to everything except the $(\alpha\dot{\beta})_{ij}$ interaction effects. And that sum of squares will test

$$\text{H: } \bar{\mu}_{ij*} - \bar{\mu}_{ij'*} - \bar{\mu}_{i'j*} + \bar{\mu}_{i'j'*} = 0 \quad \forall\ i \neq i' \text{ and } j \neq j',$$

where the subscript $*$ represents $k - 2$ dots for a model having k crossed factors; i.e., $\bar{\mu}_{ij*}$ is the average of the cell means averaged over all levels of all factors save A and B.

And, of course, all this is applicable only to data in which all sub-most cells contain data; i.e., to all-cells-filled data.

10.3. MAIN-EFFECTS-ONLY MODELS

Any data that are to be analyzed by linear model techniques can be handled by the methods of Section 10.1. The sub-most cell population mean of any cell containing data is, by (4), estimated by the mean of those data, linear combinations of those subclass means can be estimated from (5), and hypotheses about such combinations can be tested using (6). All-cells-filled data can be further analyzed by the methods of Section 10.2—testing hypotheses about main effects and interactions defined in terms of averages of sub-most cell means; e.g., (8), (17) and (33). But for some-cells-empty data no such analysis is available; only the methods of (4), (5) and (6) can be used, in combination with subset analysis as described in Section 5.6. Of course, analysis-of-variance style tables that are extensions from the 2-way classification of Table 4.9 to $3, 4, \ldots$ and higher-order classifications are conceptually quite straightforward; and as long as the factors do not have so many levels as to exceed the capacity of available computing packages, the required calculations are effectively no problem either. Nevertheless, with some-cells-empty data, interpretation of the sums of squares is even more difficult than it is in Table 4.9, because the complexities evident in the hypotheses of Table 4.9 become even more complicated when there are more than two factors, along with some or all of their interactions.

Although the desire to use a with-interaction model may be strong, an alarming feature of doing so is that interactions in a multi-factor situation can be totally overwhelming, simply by virtue of their sheer number. The example in Searle (1971, Section 8.1) illustrates this. It is taken from the Bureau of Labor Statistics Survey of Consumer Expenditures, in which family investment patterns are classified by a total of 56 levels of nine different factors: the numbers of levels of the factors are 12, 11, 4, 3, 4, 6, 6, 6 and 4. This 9-way classification has 5,474,304 possible cells, with 1354 possible 2-factor interaction effects and 18,538 possible 3-factor interaction effects. Yet there were only 8577 observations in the study! Clearly, any desire to "study interactions" in some general sense would have to be tempered by the very paucity of the data relative to the number of possible interactions. Patterns of filled cells and the researcher's knowledge of the data would have to be used with great perspicacity to hopefully elucidate some subclass analyses that might provide even a little information about interactions.

For a k-way crossed classification where the rth factor has a_r levels, the number of possible t-factor interaction effects in a traditional, overparameterized model is the sum of all possible products of t values from the k numbers $a_1, a_2, \ldots, a_r, \ldots, a_k$. Call this number n'_t. Then n'_t can be calculated from the sums of powers $s_p = \Sigma_r a_r^p$ of the numbers of levels of the factors. For example,

$$n_2' = \Sigma a_{r_1} a_{r_2} = \tfrac{1}{2}\left[\left(\sum_{r=1}^{k} a_r\right)^2 - \sum_{r=1}^{k} a_r^2\right] = \tfrac{1}{2}\left(s_1^2 - s_2\right),$$

where, with $r_1 < r_2$, the first summation means the sum of all possible products of two numbers from the numbers $a_1, a_2, \ldots, a_k$. Similarly

$$n_3' = \Sigma a_{r_1} a_{r_2} a_{r_3} = \tfrac{1}{6}\left[\left(\sum_{r=1}^{k} a_r\right)^3 - 3\Sigma a_r \Sigma a_r^2 + 2\Sigma a_r^3\right] = \tfrac{1}{6}\left(s_1^3 - 3s_1 s_2 + 2s_3\right) .$$

These relationships between n_t' and $s_1, s_2, \ldots, s_t$ are the same as those between sums of powers and symmetric sums in the context of Newton's formulae used in solving polynomial equations. These are given in determinantal form by Turnbull (1947, Section 32) as

$$n_1' = s_1, \quad n_2' = \frac{1}{2!}\begin{vmatrix} s_1 & 1 \\ s_2 & s_1 \end{vmatrix}, \quad n_3' = \frac{1}{3!}\begin{vmatrix} s_1 & 1 & 0 \\ s_2 & s_1 & 2 \\ s_3 & s_2 & s_1 \end{vmatrix} \quad \text{and}$$

$$n_4' = \frac{1}{4!}\begin{vmatrix} s_1 & 1 & 0 & 0 \\ s_2 & s_1 & 2 & 0 \\ s_3 & s_2 & s_1 & 3 \\ s_4 & s_3 & s_2 & s_1 \end{vmatrix} .$$

The general form for n_t' in terms of $s_1, s_2, \ldots, s_t$ is clear.

A complication additional to that of the number of interaction effects is that the possible sequences in which factors can be fitted becomes much more than just the two sequences of a two-factor situation. For fitting k main-effect factors without any interactions there are $k!$ sequences, and even more if some of the interaction facts are included. Nevertheless, from amongst the many different sums of squares that are available from an overparameterized model there is one in particular that is of some general use and interest for the analysis of some-cells-empty data using main-effects-only models. It is an extension of $R(\boldsymbol{\alpha} \mid \mu, \boldsymbol{\beta})$ and $R(\boldsymbol{\beta} \mid \mu, \boldsymbol{\alpha})$ in the no-interaction case of the preceding chapter, equivalent to $\mathcal{R}(\mu_i, \tau_j \mid \tau_j)$ and $\mathcal{R}(\mu_i, \tau_j \mid \mu_i)$ of Chapters 4 and 5.

The extension is that for any k-way crossed classification ($k \geq 2$), we consider models that have no interactions. Such a model contains only main-effect factors and so can be called a *main-effects-only* model. It has the useful characteristic (stemming from Section 8.6e) that for a factor denoted by $\boldsymbol{\alpha}$, with a levels,

$$F = \frac{R(\text{any factor } \boldsymbol{\alpha} \mid \text{all other factors})}{(a-1)\hat{\sigma}^2}, \quad \text{for } \hat{\sigma}^2 = \frac{\mathbf{y}'\mathbf{y} - R(\text{all factors})}{(N - r_{\mathbf{X}})}$$

tests, in a main-effects-only model,

H: *effects of all levels of* $\boldsymbol{\alpha}$ *are equal* .

More formally, if the equation for such a model is written as

$$E(y_{i_1 i_2 \cdots i_t \cdots i_k}) = \mu + \alpha_{1i_1} + \alpha_{2i_2} + \cdots + \alpha_{ti_t} + \cdots + \alpha_{ki_k}, \tag{40}$$

where α_{ti_t} is the effect for the (i_t)th level of the tth main-effects factor that has a_t levels, i.e., $i_t = 1, 2, \ldots, a_t$ and $t = 1, 2, \ldots, k$, then for $\boldsymbol{\alpha}_t = [\alpha_{t1} \ \cdots \ \alpha_{ta_t}]'$,

$$F_t = \frac{R(\boldsymbol{\alpha}_t \mid \boldsymbol{\alpha}_1, \boldsymbol{\alpha}_2, \ldots, \boldsymbol{\alpha}_{t-1}, \boldsymbol{\alpha}_{t+1}, \ldots, \boldsymbol{\alpha}_k)}{(a_t - 1)\hat{\sigma}^2} \tag{41}$$

tests H: α_{ti_t} *equal for* $i_t = 1, \ldots, a_t$, (42)

where

$$\hat{\sigma}^2 = \frac{\mathbf{y}'\mathbf{y} - R(\boldsymbol{\alpha}_1, \boldsymbol{\alpha}_2, \ldots, \boldsymbol{\alpha}_k)}{N - \sum_{t=1}^{k} a_t + k - 1} . \tag{43}$$

Thus in main-effects-only models, the hypothesis of equality of effects of the levels of a main effect factor can be tested by (41) for each factor.

The tests provided by (41) are, of course, not independent of one another, from one factor to another; but then neither are the F-statistics in most analyses of variance tables (not even for balanced data). This is so because, although numerators of Fs may be independent, different Fs often have the same $\hat{\sigma}^2$ in their denominators and are therefore correlated. With balanced data it is usually the denominators of Fs that are correlated (see Hurlburt and Spiegel, 1976), whereas with unbalanced data, numerators are often correlated also. Nevertheless, in situations where no-interaction models can be deemed reasonable, (41) does provide a mechanism for making tests that have meaning and have a readily understood interpretation. True it is, that we are seldom in a situation in real life where we can firmly assume that a no-interaction model is an adequate mimicry of the real world; i.e., how can we truly know that there are no interactions? On the other hand, in today's computerized world, resources are often more readily made available for collecting and storing of data than they are for statistical analysis. Thus data are frequently collected without sufficient forethought for their analysis. As a result, statisticians are often asked for advice on analyzing data for which analysis of variance techniques have not been planned but, with the data and computing packages at hand, there is insistence that analysis of variance is what is wanted. In just such a situation (41) can be useful. When connected data involve many factors and have numerous empty cells (as they can when they come, for example, from surveys of human characteristics and behavior), then (41) is at least one analysis technique that is not totally unreasonable. It may not be the best statistical method to apply to any particular case, but within the framework of linear model analysis it does provide something that is easily understood and that has a clear interpretation: namely, for each factor, a method for testing the hypothesis that all levels of the factor occurring in the data are equal. (41) provides that test.

Estimation in main-effects-only models is, of course, based on the solution of the normal equations corresponding to the model equation (40). In the 2-way classification we have seen [in (65)–(68) of Section 4.9b, (23)–(26) of Section 5.4a, (50) of Section 9.2e (79) of Section 9.2j] that elements of that solution are not simple functions of cell means. This lack of simplicity is even more evident for k-way classifications. Nevertheless, in terms of the model equation (40), solution elements μ° and $\alpha^\circ_{ti_t}$ for $i_t = 1, \ldots, a_t$ and $t = 1, \ldots, k$ are always obtainable for connected data.

Also, for connected data [as an extension of (103) of Section 4.10a, (27) of Section 5.4a, (51) of Section 9.2e and (82) of Section 9.2j], the cell mean of every sub-most cell, whether it contains data or not, is estimable with

$$\text{BLUE}\left(\mu + \sum_{t=1}^{k} \alpha_{ti_t}\right) = \mu^\circ + \sum_{t=1}^{k} \alpha^\circ_{ti_t} \,. \tag{44}$$

Again, linear combinations of $\mu + \Sigma_{t=1}^{k}\alpha_{ti_t}$ are estimable, with BLUEs that are the same linear combinations of the right-hand sides of (44); and, for $\boldsymbol{\beta}$ being the vector of μ and all the αs ($\Sigma_t a_t$ of them), the hypothesis H: $\mathbf{K}'\boldsymbol{\beta} = \mathbf{m}$ is testable in the usual way using (6). In particular, differences between αs of each factor are estimable with

$$\text{BLUE}(\alpha_{ti_t} - \alpha_{ti_s}) = \alpha^\circ_{ti_t} - \alpha^\circ_{ti_s}, \tag{45}$$

as are linear combinations of these differences, namely contrasts among the αs of each factor.

10.4. MODELS WITH NOT ALL INTERACTIONS[1]

Cell means models, by their very nature, implicitly include all interactions in the population means of the sub-most cells; e.g., (19). In contrast, when a model is to have no interactions we use the main-effects-only model. But in some situations we may want a middle ground, where some but not all interactions are wanted in a model. This we now consider, and in doing so develop a method for the cell means model that can make it equivalent to the main-effects-only model. We begin with the 2-way classification.

a. The 2-way classification

With the 2-way classification cell means model, $E(y_{ijk}) = \mu_{ij}$, the cell mean μ_{ij} is estimable only for filled cells. But with the no-interaction model, the cell mean $\mu_{ij} = \mu_i + \tau_j$ is estimable for every cell, whether empty or not (Section 5.4a). The cell means model is therefore unsuitable in the no-interaction case because it utilizes (and estimates means for) only the filled cells. Consequently a more general model is needed. It must involve *all* the cell means and, for the no-interaction case, it must take account of the absence of interactions.

At first thought, involving all the cell means might seem easy. $\boldsymbol{\mu}$ would be defined as the vector of all cell means that could exist in the data, even if some cells are empty; e.g., for the 2-way crossed classification, $\boldsymbol{\mu}$ would contain ab elements. And then, in writing the model equation as

$$E(\mathbf{y}) = \mathbf{W}\boldsymbol{\mu}, \tag{46}$$

$\mathbf{W}$ would be a direct sum of $\mathbf{1}$-vectors, as is $\mathbf{X}$ of (3) in Section 10.1a, *except* that corresponding to each cell mean for the empty cells there would be a null column in $\mathbf{W}$.

Example. Suppose in a case of two rows and two columns that there are no data in cell 2,2, indicated in Grid 10.1 by having μ_{22} in parentheses:

Grid 10.1

μ_{11}	μ_{12}
μ_{21}	(μ_{22})

[1] I am greatly indebted to Dr. F. Michael Speed for motivating many of the ideas in this section—see Searle and Speed (1982).

Then the model equation (46) would be

$$E(\mathbf{y}) = \begin{bmatrix} \mathbf{1}_{n_{11}} & \cdot & \cdot & \mathbf{0} \\ \cdot & \mathbf{1}_{n_{12}} & \cdot & \mathbf{0} \\ \cdot & \cdot & \mathbf{1}_{n_{21}} & \mathbf{0} \end{bmatrix} \begin{bmatrix} \mu_{11} \\ \mu_{12} \\ \mu_{21} \\ \mu_{22} \end{bmatrix} = \mathbf{W}\boldsymbol{\mu}, \tag{47}$$

where the null column in $\mathbf{W}$ corresponds to μ_{22}, the mean of the empty cell.

b. Unrestricted models

The estimation of $\boldsymbol{\mu}$ from (46) by least squares leads, quite formally, to normal equations

$$\mathbf{W}'\mathbf{W}\hat{\boldsymbol{\mu}} = \mathbf{W}'\mathbf{y} \ . \tag{48}$$

Because $\mathbf{W}$ has null columns, as illustrated in (47), $\mathbf{W}'\mathbf{W}$ has null rows and $\mathbf{W}'\mathbf{y}$ has zero elements, corresponding to empty cells. A solution of (48) is therefore

$$\hat{\boldsymbol{\mu}} = (\mathbf{W}'\mathbf{W})^{-}\mathbf{W}'\mathbf{y} \ . \tag{49}$$

Since $\mathbf{W}'\mathbf{W}$ is diagonal with diagonal elements n_{ij} (including $n_{ij} = 0$ for each empty cell), a straightforward $(\mathbf{W}'\mathbf{W})^{-}$ is that it be diagonal with diagonal elements $1/n_{ij}$ for $n_{ij} \neq 0$ and 0 for $n_{ij} = 0$. Then (49) gives

$$\hat{\mu}_{ij} = \bar{y}_{ij\cdot} \ \text{ for each filled cell } \quad \text{and} \quad \hat{\mu}_{ij} = 0 \ \text{ for each empty cell.} \tag{50}$$

There is a deficiency of the estimation algorithm represented by (50); despite the empty cells having no data, their means are being estimated (as zero). This contradicts intuition, namely that without having data in those cells their means cannot be estimated and so are non-estimable. And one aspect of this which is particularly unedifying is that those zero estimates occur no matter what scale of measurement is being used: whether weighing elephants or mice the estimate of an empty cell would be zero. There is also an inconsistency in considering (48) as the normal equations for the model (46). They are, but only *pro forma*. Least squares is a procedure for estimating the parameters of a model that occur in a set of data. But (46) has null columns corresponding to μs for empty cells, as illustrated in (47), so that those μs do not appear explicitly in (46) at all. Therefore the least squares procedure of differentiating $(\mathbf{y} - \mathbf{W}\boldsymbol{\mu})'(\mathbf{y} - \mathbf{W}\boldsymbol{\mu})$ with respect to μs can be applied only to the μs of the filled cells. To execute this, re-sequence elements of $\boldsymbol{\mu}$ so that those for filled cells, to be denoted as $\boldsymbol{\mu}_f$, come first, followed by those for empty cells, μ_e. Then

$$\mu = \begin{bmatrix} \mu_f \\ \mu_e \end{bmatrix} . \tag{51}$$

Define $\mathbf{X}$ as in (193) of Section 8.10: a diagonal matrix of $\mathbf{1}$-vectors corresponding to filled cells. Then $\mathbf{W} = [\mathbf{X} \quad \mathbf{0}]$ and (46) is

$$E(\mathbf{y}) = \mathbf{X}\mu_f + \mathbf{0}\mu_e . \tag{52}$$

Least squares on (52) clearly involves differentiating $(\mathbf{y} - \mathbf{X}\mu_f)'(\mathbf{y} - \mathbf{X}\mu_f)$ with respect to elements of μ_f and so giving

$$\hat{\mu}_f = (\mathbf{X}'\mathbf{X})^{-1}\mathbf{X}'\mathbf{y} = \bar{\mathbf{y}} \tag{53}$$

as the estimator of μ_f, where $\bar{\mathbf{y}}$ is the vector of cell means $\bar{y}_{ij\cdot}$ for $n_{ij} \neq 0$, written in lexicon order.

The estimators in (53) are, of course, precisely the same for filled cells as those in (50). But the situation is not the same for empty cells. In (53), nothing is said about their means—and rightfully so, because those cells have no data and their means are not estimable in the model (46).

c. Restricted models

The preceding sentence may seem to imply that nothing can ever be said about the estimation of means of empty cells. This is clearly not so because, for example, we know with the no-interaction model for the 2-way classification that the mean of every empty cell is estimable (provided the data are connected). That result was achieved (Section 5.4a) by re-writing the cell means model $E(y_{ijk}) = \mu_{ij}$ as $E(y_{ijk}) = \mu_i + \tau_j$; as an alternative to that, we now retain the model $E(y_{ijk}) = \mu_{ij}$ and incorporate an absence of interactions through the use of restrictions, in the following manner.

-i. *Examples*

Example 1. Consider a 2-way classification with 2 rows and 2 columns, with all cells filled. The condition that there be no interaction is

$$\mu_{11} - \mu_{12} - \mu_{21} + \mu_{22} = 0 . \tag{54}$$

To adapt the cell means model $E(y_{ijk}) = \mu_{ij}$ to the no-interaction situation, we therefore use the restricted model

$$\begin{cases} E(y_{ijk}) = \mu_{ij} \quad \text{for } i = 1,2 \text{ and } j = 1,2 & \text{(55a)} \\ \text{and} & \\ \mu_{11} - \mu_{12} - \mu_{21} - \mu_{22} = 0. & \text{(55b)} \end{cases}$$

With all cells filled, the left-hand side of (54) is estimable, so that in using (55) as the model we are using what in Section 8.9d is called an estimable restriction. Hence, on writing $E(y_{ijk}) = \mu_{ij}$ as $E(\mathbf{y}) = \mathbf{X}\boldsymbol{\mu}$ and the restriction as $\mathbf{P}'\boldsymbol{\mu} = \mathbf{0}$, equation (181) in Section 8.9d provides

$$\hat{\boldsymbol{\mu}} = \bar{\mathbf{y}} - \mathbf{D}^{-1}\mathbf{P}(\mathbf{P}'\mathbf{D}^{-1}\mathbf{P})^{-1}\mathbf{P}'\bar{\mathbf{y}} \tag{56}$$

as the estimator in the restricted model—where

$$\mathbf{D}^{-1} = \mathbf{D}\{1/n_{ij}\} = (\mathbf{X}'\mathbf{X})^{-1}$$

and $\bar{\mathbf{y}}$ is the vector of observed cell means. The simplification of (56) is shown in E8.33.

The estimator in (56) can be used for any cell means model with all-cells-filled data when any of the interactions are to be taken as nonexistent through the use of restrictions, of which (54) is the simplest example.

Example 2. To illustrate the adaptation of (56) for some-cells-empty data a slightly bigger example is needed, and we consider the 2×3 case, with cell 2,3 empty, represented as in Grid 10.2:

Grid 10.2

μ_{11}	μ_{12}	μ_{13}
μ_{21}	μ_{22}	(μ_{23})

The no-interaction model for this case would *appear* to be

$$E(y_{ijk}) = \mu_{ij} \quad \text{for} \quad i = 1,2 \text{ and } j = 1,2,3 \tag{57a}$$

and

$$\left.\begin{aligned} \mu_{11} - \mu_{12} - \mu_{21} + \mu_{22} &= 0 \\ \mu_{11} - \mu_{13} - \mu_{21} + \mu_{23} &= 0. \end{aligned}\right. \tag{57b}$$

An important distinction between models (55) and (57) must be noted. It is this: every μ_{ij} occurring in the restriction (55b) also occurs in the model equations for the data, (55a). But in (57b) there is a μ_{ij}, namely μ_{23}, that does not occur in the model equations for the data, (57a). It is the μ_{ij} for the empty cell. Since that cell contains no data, its μ cannot be part of the model for the data. And yet we still want the μ_{ij}s to satisfy (57b). That being so, since there are data in all cells save the 2,3 cell, and since for that empty cell we have no information about its μ_{ij} other than that which is implicit in the restriction (57b), we can use (57b) to define μ_{23} (for the empty cell) in terms of the μ_{ij}s for the filled cells. This we now do; and although for this small example the arithmetic is very simple, we maintain a certain degree of matrix generality in order to lead up to the general case and a resulting theorem.

Define

$$\boldsymbol{\mu}' = [\mu_{11} \quad \mu_{12} \quad \mu_{13} \quad \mu_{21} \quad \mu_{22} \quad \mu_{23}] = [\boldsymbol{\mu}'_f \quad \boldsymbol{\mu}'_e]$$

as in (51), with

$$\boldsymbol{\mu}'_f = [\mu_{11} \quad \mu_{12} \quad \mu_{13} \quad \mu_{21} \quad \mu_{22}] \quad \text{and} \quad \boldsymbol{\mu}'_e = \mu_{23} \; .$$

Then the restriction (57b) is

$$\begin{bmatrix} 1 & -1 & 0 & -1 & 1 & 0 \\ 1 & 0 & -1 & -1 & 0 & 1 \end{bmatrix} \boldsymbol{\mu} = \mathbf{0}, \tag{58}$$

equivalent to

$$\begin{bmatrix} 1 & -1 & 0 & -1 & 1 \\ 1 & 0 & -1 & -1 & 0 \end{bmatrix} \boldsymbol{\mu}_f + \begin{bmatrix} 0 \\ 1 \end{bmatrix} \boldsymbol{\mu}_e = 0 \; . \tag{59}$$

Treating this as an equation to be solved for $\boldsymbol{\mu}_e$ gives

$$\begin{aligned} \boldsymbol{\mu}_e &= -\begin{bmatrix} 0 \\ 1 \end{bmatrix}^- \begin{bmatrix} 1 & -1 & 0 & -1 & 1 \\ 1 & 0 & -1 & -1 & 0 \end{bmatrix} \boldsymbol{\mu}_f \\ &= -[0 \quad 1] \begin{bmatrix} 1 & -1 & 0 & -1 & 1 \\ 1 & 0 & -1 & -1 & 0 \end{bmatrix} \boldsymbol{\mu}_f \; . \end{aligned} \tag{60}$$

Viewed as a restriction that is to be part of the model (57), substituting (60) into (58), i.e. (59), gives that restriction as

$$\left\{ \mathbf{I} - \begin{bmatrix} 0 \\ 1 \end{bmatrix} [0 \quad 1] \right\} \begin{bmatrix} 1 & -1 & 0 & -1 & 1 \\ 1 & 0 & -1 & -1 & 0 \end{bmatrix} \boldsymbol{\mu}_f = \mathbf{0}, \tag{61}$$

i.e., as

$$\mu_{11} - \mu_{12} - \mu_{21} + \mu_{22} = 0 \; . \tag{62}$$

And then (59) also defines μ_{23}, for the empty cell as

$$\mu_{23} = -\mu_{11} + \mu_{13} + \mu_{21}$$

Hence the restricted model (57) becomes

$$\begin{cases} E(y_{ijk}) = \mu_{ij} \quad \text{for} \quad i = 1,2 \text{ and} \\ j = 1,2,3 \text{ but not } i = 2 \text{ and } j = 3 \end{cases} \tag{63a}$$

and

$$\mu_{11} - \mu_{12} - \mu_{21} - \mu_{22} = 0. \tag{63b}$$

This is now a model in $\boldsymbol{\mu}_f$, the cell means corresponding to filled cells. And (56) provides its BLUE, $\hat{\boldsymbol{\mu}}_f$, which turns out to be the same as the results in E8.33, together with $\hat{\mu}_{13} = \bar{y}_{13\cdot}$ (see E10.4).

-ii. ***The general case.*** Let linear restrictions on $\boldsymbol{\mu}$ be represented as

$$\mathbf{R}\boldsymbol{\mu} = \mathbf{0}, \tag{64}$$

where $\boldsymbol{\mu}$, as in (51), involves all the cell means, both for filled cells and empty cells. Indeed, partition $\mathbf{R}$ as $\mathbf{R} = [\mathbf{R}_f \quad \mathbf{R}_e]$ conformably with (51), so that (64) is

$$\mathbf{R}_f\boldsymbol{\mu}_f + \mathbf{R}_e\boldsymbol{\mu}_e = \mathbf{0}, \tag{65}$$

which is a generalization of (59). The restricted cell means model is then (52) and (65) combined:

$$E(\mathbf{y}) = \mathbf{X}\boldsymbol{\mu}_f \quad \text{and} \quad \mathbf{R}_f\boldsymbol{\mu}_f + \mathbf{R}_e\boldsymbol{\mu}_e = \mathbf{0}\ . \tag{66}$$

For applying least squares there is now the logical difficulty that $E(\mathbf{y})$ involves only $\boldsymbol{\mu}_f$, whereas the restrictions (65) also involve $\boldsymbol{\mu}_e$. But do they? In terms of the model for the data, (65) can be viewed as simply defining some other parametric functions. And, like (60), these have the form, from (65) [using (23) of Section 7.2a],

$$\boldsymbol{\mu}_e = -\mathbf{R}_e^-\mathbf{R}_f\boldsymbol{\mu}_f, \tag{67}$$

where $\mathbf{R}_e^-$ is a generalized inverse of $\mathbf{R}_e$. Then, using (67) in (65), the restrictions on $\boldsymbol{\mu}_f$ are, like (61),

$$(\mathbf{I} - \mathbf{R}_e\mathbf{R}_e^-)\mathbf{R}_f\boldsymbol{\mu}_f = \mathbf{0}, \tag{68}$$

i.e.,

$$\mathbf{P}\boldsymbol{\mu}_f = \mathbf{0}, \quad \text{for } \mathbf{P} = (\mathbf{I} - \mathbf{R}_e\mathbf{R}_e^-)\mathbf{R}_f\ . \tag{69}$$

The model on which we carry out least squares is now (52) and (69):

$$E(\mathbf{y}) = \mathbf{X}\boldsymbol{\mu}_f \quad \text{and} \quad \mathbf{P}\boldsymbol{\mu}_f = \mathbf{0}\ . \tag{70}$$

The resulting normal equations are then

$$\begin{bmatrix} \mathbf{X'X} & \mathbf{P'} \\ \mathbf{P} & \mathbf{0} \end{bmatrix} \begin{bmatrix} \hat{\boldsymbol{\mu}}_f \\ \boldsymbol{\lambda} \end{bmatrix} = \begin{bmatrix} \mathbf{X'y} \\ \mathbf{0} \end{bmatrix}, \tag{71}$$

where $\boldsymbol{\lambda}$ is a vector of Lagrange multipliers for taking account of the restrictions $\mathbf{P}\boldsymbol{\mu}_f = \mathbf{0}$. Writing

$$\mathbf{X'X} = \mathbf{D}\{n_t\} = \mathbf{D} \quad \text{and} \quad \mathbf{X'y} = \mathbf{y}. = \mathbf{D}\bar{\mathbf{y}}$$

as in the derivation of (4) we then have (71) as

$$\begin{bmatrix} \mathbf{D} & \mathbf{P'} \\ \mathbf{P} & \mathbf{0} \end{bmatrix} \begin{bmatrix} \hat{\boldsymbol{\mu}}_f \\ \boldsymbol{\lambda} \end{bmatrix} = \begin{bmatrix} \mathbf{D}\bar{\mathbf{y}} \\ \mathbf{0} \end{bmatrix}. \tag{72}$$

Adapting the result for a partitioned matrix given in (38) of Section 7.3 to be

$$\begin{bmatrix} \mathbf{D} & \mathbf{P'} \\ \mathbf{P} & \mathbf{0} \end{bmatrix}^{-1} = \begin{bmatrix} \mathbf{D}^{-1} & \mathbf{0} \\ \mathbf{0} & \mathbf{0} \end{bmatrix} + \begin{bmatrix} -\mathbf{D}^{-1}\mathbf{P'} \\ \mathbf{I} \end{bmatrix} (-\mathbf{P}\mathbf{D}^{-1}\mathbf{P'})^{-1} [-\mathbf{P}\mathbf{D}^{-1} \quad \mathbf{I}] \tag{73}$$

gives the solution to (72) as

$$\hat{\boldsymbol{\mu}}_f = \bar{\mathbf{y}} - \mathbf{D}^{-1}\mathbf{P'}(\mathbf{P}\mathbf{D}^{-1}\mathbf{P'})^{-1}\mathbf{P}\bar{\mathbf{y}} \quad \text{and} \quad \boldsymbol{\lambda} = (\mathbf{P}\mathbf{D}^{-1}\mathbf{P'})^{-1}\mathbf{P}\bar{\mathbf{y}}\,. \tag{74}$$

It is easily verified that $\mathbf{P}\hat{\boldsymbol{\mu}}_f = \mathbf{0}$, corresponding to $\mathbf{P}\boldsymbol{\mu}_f = \mathbf{0}$ of (69), and that $E(\hat{\boldsymbol{\mu}}_f) = \boldsymbol{\mu}_f$. And then, corresponding to (67),

$$\hat{\boldsymbol{\mu}}_e = -\mathbf{R}_e^{-}\mathbf{R}_f\hat{\boldsymbol{\mu}}_f\,. \tag{75}$$

In the special case of $\mathbf{R}_e$ having full row rank, $\mathbf{R}_e\mathbf{R}_e^{-} = \mathbf{I}$ and so $\mathbf{P} = \mathbf{0}$ in (69), and (74) reduces to $\hat{\boldsymbol{\mu}}_f = \bar{\mathbf{y}}$. That also occurs in the unrestricted model wherein $\mathbf{R}$ does not exist and so $\hat{\boldsymbol{\mu}}_f = \bar{\mathbf{y}}$ as in (53). Hence

$$\hat{\boldsymbol{\mu}}_f = \bar{\mathbf{y}} \begin{cases} \text{when there are no restrictions,} \\ \text{or when } \mathbf{R}_e \text{ has full-row rank.} \end{cases} \tag{76}$$

When all cells are filled, there is no $\boldsymbol{\mu}_e$ nor $\mathbf{R}_e$, and so $\boldsymbol{\mu}_f \equiv \boldsymbol{\mu}$ and $\mathbf{P} = \mathbf{R}_f = \mathbf{R}$ and the normal equations are (71) with $\mathbf{P}$ replaced by $\mathbf{R}$, and so their solution for $\hat{\mu}_f$ is (74) with that same replacement, namely

$$\hat{\mu}_f = \hat{\mu} = \bar{y} - D^{-1}R'(RD^{-1}R')^{-1}R\bar{y} \ . \tag{77}$$

d. Estimability

We consider the estimability of

$$Q\mu = Q_f\mu_f + Q_e\mu_e \tag{78}$$

for matrices partitioned as $Q = [Q_f \quad Q_e]$.

-i. All cells filled. When all cells are filled, $\mu_f = \mu$ and the BLUE of $Q\mu$ is BLUE($Q\mu$) = $Q\hat{\mu}$. In the unrestricted model, $\hat{\mu} = \hat{\mu}_f = \bar{y}$, as in (76), and

$$\text{BLUE}(Q\mu) = Q\bar{y} \ . \tag{79}$$

In the restricted model, R_e does not exist, so that $R = R_f$ and $\hat{\mu}$ is given by (77) and

$$\text{BLUE}(Q\mu) = Q\hat{\mu} \quad \text{for } \hat{\mu} \text{ being } \hat{\mu}_f \text{ of (77).} \tag{80}$$

-ii. Some cells empty. When some cells are empty, and when there are no restrictions, only μ_f is estimable, with $\hat{\mu}_f = \bar{y}$, as in (76), and

$$\text{BLUE}(Q_f\mu_f) = Q_f\bar{y} \ . \tag{81}$$

And in the restricted model, the following theorem applies.

Theorem. When $\mu' = [\mu'_f \quad \mu'_e]$ is the vector of cell means in a linear model, with μ_f corresponding to filled cells and μ_e to empty cells, and with $R\mu \equiv R_f\mu_f + R_e\mu_e = 0$ being restrictions on μ, then with

$$Q_1 = Q_f - Q_eR_e^-R_f, \tag{82}$$

$$LQ\mu \text{ is estimable with } \quad \text{BLUE}(LQ\mu) = LQ_1\hat{\mu}_f, \tag{83}$$

$$\text{if and only if} \quad LQ_e = MR_e \quad \text{for some } L \text{ and } M. \tag{84}$$

Proof. A more general form of (67) is

$$\mu_e = -R_e^-R_f\mu_f + (I - R_e^-R_e)\theta \tag{85}$$

for arbitrary $\boldsymbol{\theta}$ [see Searle (1982), Section 9.4b]. From (85) and (78)

$$\mathbf{LQ\mu} = \mathbf{L}(\mathbf{Q}_f - \mathbf{Q}_e\mathbf{R}_e^-\mathbf{R}_f)\boldsymbol{\mu}_f + \mathbf{L}(\mathbf{Q}_e - \mathbf{Q}_e\mathbf{R}_e^-\mathbf{R}_e)\boldsymbol{\theta}; \tag{86}$$

and this contains no arbitrary vector $\boldsymbol{\theta}$ if and only if $\mathbf{LQ}_e = \mathbf{LQ}_e\mathbf{R}_e^-\mathbf{R}_e$, i.e., if and only if $\mathbf{LQ}_e = \mathbf{MR}_e$ for some $\mathbf{M}$. Then (86) yields (83) and (84). **Q.E.D.**

Corollaries.

(i) When $\mathbf{Q}_e = \mathbf{0}$, then $\text{BLUE}(\mathbf{Q\mu}) = \mathbf{Q}_f\hat{\boldsymbol{\mu}}_f$. ($\mathbf{Q}_e = \mathbf{0}$ implies $\mathbf{Q}_e = \mathbf{MR}_e$ for $\mathbf{M} = \mathbf{0}$, and $\mathbf{Q}_1 = \mathbf{Q}_f$.)

(ii) When $\mathbf{R}_e$ has full column rank, $\text{BLUE}(\mathbf{Q\mu}) = \mathbf{Q}_1\hat{\boldsymbol{\mu}}_f$ for any $\mathbf{Q}$ and, from (75), $\hat{\boldsymbol{\mu}}_e = -\mathbf{R}_e^-\mathbf{R}_f\hat{\boldsymbol{\mu}}_f$. [$\mathbf{Q}_e = \mathbf{MR}$ with $\mathbf{M} = \mathbf{Q}_e(\mathbf{R}'_e\mathbf{R}_e)^{-1}\mathbf{R}'_e$ for any $\mathbf{Q}_e$.]

(iii) When $\mathbf{R}_e$ is nonsingular, which is a special case of (ii), then $\mathbf{P} = \mathbf{0}$ from (69) and so $\hat{\boldsymbol{\mu}}_f = \bar{\mathbf{y}}$ from (74); and $\text{BLUE}(\mathbf{Q\mu}) = \mathbf{Q}_1\bar{\mathbf{y}}$ and $\hat{\boldsymbol{\mu}}_e = -\mathbf{R}_e^{-1}\mathbf{R}_f\bar{\mathbf{y}}$.

Actually, (83), (84) are quite general, with (79)–(81) being special cases thereof. In (79) and (80), all cells are filled, so that $\mathbf{R}_e$ and $\mathbf{Q}_e$ do not exist, and so (69) gives $\mathbf{P} = \mathbf{R}_f = \mathbf{R}$ and (82) gives $\mathbf{Q}_1 = \mathbf{Q}_f = \mathbf{Q}$. Thus (83) becomes $\text{BLUE}(\mathbf{Q\mu}) = \mathbf{Q}\hat{\boldsymbol{\mu}}$ of (80). Then for (79), $\mathbf{R} = \mathbf{0}$ so that $\hat{\boldsymbol{\mu}} = \bar{\mathbf{y}}$ and so (80) becomes $\text{BLUE}(\mathbf{Q\mu}) = \mathbf{Q}\bar{\mathbf{y}}$ of (79). When some cells are empty and there are no restrictions, $\mathbf{R}_f = \mathbf{0}$ so that (69) gives $\mathbf{P} = \mathbf{0}$ and (82) is $\mathbf{Q}_1 = \mathbf{Q}_f$, and so $\hat{\boldsymbol{\mu}}_f = \bar{\mathbf{y}}$ of (53). Hence (83) becomes $\text{BLUE}(\mathbf{Q}_f\boldsymbol{\mu}_f) = \mathbf{Q}_f\bar{\mathbf{y}}$ of (81). In essence, therefore, one only needs (83), along with (85); but since the special cases (79)–(81) readily occur in practice, it is useful to have them spelled out explicitly. All four results, (79)–(81), and (83), (84), are shown in Table 10.4.

e. Hypothesis testing

The general result (83) is useful for establishing a test of the hypothesis H: $\mathbf{K}'\boldsymbol{\mu} = \mathbf{m}$. Whenever $\mathbf{K}'\boldsymbol{\mu} = \mathbf{K}'_1\boldsymbol{\mu}_f$ is estimable (for $\mathbf{K}'_1 = \mathbf{K}'_f - \mathbf{K}'_e\mathbf{R}_e^-\mathbf{R}_f$), the hypothesis can, under the usual normality assumptions, be tested using

$$F = (\mathbf{K}'_1\hat{\boldsymbol{\mu}}_f - \mathbf{m})'(\mathbf{K}'_1\mathbf{V}^{-1}\mathbf{K}_1)^{-1}(\mathbf{K}'_1\hat{\boldsymbol{\mu}}_f - \mathbf{m})/s\hat{\sigma}^2, \tag{87}$$

where

$$\mathbf{V} = \text{var}(\hat{\boldsymbol{\mu}}_f)/\sigma^2 = \mathbf{D}^{-1} - \mathbf{D}^{-1}\mathbf{P}(\mathbf{P}\mathbf{D}^{-1}\mathbf{P}')^{-1}\mathbf{P}\mathbf{D}^{-1} \tag{88}$$

$$s = \text{full row rank of } \mathbf{K}'$$

$$\hat{\sigma}^2 = \text{residual mean square} = \mathbf{y}'[\mathbf{I} - \mathbf{X}\mathbf{D}^{-1}\mathbf{X}']\mathbf{y}/(N - r_{\mathbf{X}} + r_{\mathbf{R}_f})$$

TABLE 10.4. BEST LINEAR UNBIASED ESTIMATORS OF ESTIMABLE FUNCTIONS IN CELL MEANS MODELS[1]

All Cells Filled		Some Cells Empty	
Unrestricted Models			
$E(\mathbf{y}) = \mathbf{X}\boldsymbol{\mu}$		$E(\mathbf{y}) = \mathbf{X}_f\boldsymbol{\mu}_f$	
$\hat{\boldsymbol{\mu}} = \bar{\mathbf{y}}$	(4)	$\hat{\boldsymbol{\mu}}_f = \bar{\mathbf{y}}$	(53)
$\text{BLUE}(\mathbf{Q}\boldsymbol{\mu}) = \mathbf{Q}\bar{\mathbf{y}}$	(79)	$\text{BLUE}(\mathbf{Q}_f\boldsymbol{\mu}_f) = \mathbf{Q}_f\bar{\mathbf{y}}$	(81)
Restricted Models			
$\begin{cases} E(\mathbf{y}) = \mathbf{X}\boldsymbol{\mu} \\ \text{and} \\ \mathbf{R}\boldsymbol{\mu} = \mathbf{0} \end{cases}$		$\begin{cases} E(\mathbf{y}) = \mathbf{X}_f\boldsymbol{\mu}_f \\ \text{and} \\ \mathbf{R}_f\boldsymbol{\mu}_f + \mathbf{R}_e\boldsymbol{\mu}_e = \mathbf{0} \end{cases}$	
$\mathbf{D} = \mathbf{D}\{n_t\}$		$\mathbf{D} = \mathbf{D}\{n_t\}$	
		$\mathbf{P} = (\mathbf{I} - \mathbf{R}_e\mathbf{R}_e^-)\mathbf{R}_f$	(69)
		$\mathbf{Q}_1 = \mathbf{Q}_f - \mathbf{Q}_e\mathbf{R}_e^-\mathbf{R}_f$	(82)
$\hat{\boldsymbol{\mu}} = \bar{\mathbf{y}} - \mathbf{D}^{-1}\mathbf{R}'(\mathbf{R}\mathbf{D}^{-1}\mathbf{R}')^{-1}\mathbf{R}\bar{\mathbf{y}}$	(77)	$\hat{\boldsymbol{\mu}}_f = \bar{\mathbf{y}} - \mathbf{D}^{-1}\mathbf{P}'(\mathbf{P}\mathbf{D}^{-1}\mathbf{P}')^{-1}\mathbf{P}\bar{\mathbf{y}}$	(74)
$\text{BLUE}(\mathbf{Q}\boldsymbol{\mu}) = \mathbf{Q}\hat{\boldsymbol{\mu}}$	(80)	$\text{BLUE}(\mathbf{L}\mathbf{Q}\boldsymbol{\mu}) = \mathbf{L}\mathbf{Q}_1\hat{\boldsymbol{\mu}}_f$	(83)
		if $\mathbf{L}\mathbf{Q}_e = \mathbf{M}\mathbf{R}_e$ for some $\mathbf{L}$ and $\mathbf{M}$	(84)

[1] Numbers in parentheses are equation numbers.

where $r_{\mathbf{X}}$ and $r_{\mathbf{R}_f}$ are the ranks of $\mathbf{X}$ and $\mathbf{R}_f$, respectively, with $r_{\mathbf{X}}$ being the number of filled cells.

f. Examples of estimable functions

We give four examples. The first two are the 2 × 2 case of E4.6, first with all cells filled and then with an empty cell. The second two are illustrations of a 3-way classification.

-i. ***The 2 × 2 case with all cells filled.*** We use Grid 10.3:

Grid 10.3

μ_{11}	μ_{12}
μ_{21}	μ_{22}

All cells are filled and so (4) and (79), and also (77) and (80), apply: $\mathbf{Q}\boldsymbol{\mu}$ is estimable, whether there are restrictions on $\boldsymbol{\mu}$ or not. With no restrictions, (4) gives $\hat{\boldsymbol{\mu}} = \bar{\mathbf{y}}$, i.e., $\hat{\mu}_{ij} = \bar{y}_{ij\cdot}$, and every $\mathbf{Q}\boldsymbol{\mu}$ is estimable with, from (79),

BLUE($\mathbf{Q\mu}$) = $\mathbf{Q\bar{y}}$. With the restriction of no interaction,

$$\mu_{11} - \mu_{12} - \mu_{21} + \mu_{22} = 0, \tag{89}$$

the normal equations that yielded (77) are

$$\begin{bmatrix} n_{11} & \cdot & \cdot & \cdot & 1 \\ \cdot & n_{12} & \cdot & \cdot & -1 \\ \cdot & \cdot & n_{21} & \cdot & -1 \\ \cdot & \cdot & \cdot & n_{22} & 1 \\ 1 & -1 & -1 & 1 & 0 \end{bmatrix} \begin{bmatrix} \hat{\mu}_{11} \\ \hat{\mu}_{12} \\ \hat{\mu}_{21} \\ \hat{\mu}_{22} \\ \lambda \end{bmatrix} = \begin{bmatrix} y_{11\cdot} \\ y_{12\cdot} \\ y_{21\cdot} \\ y_{22\cdot} \\ 0 \end{bmatrix} \tag{90}$$

with solution that can be expressed as $\hat{\mu}_{ij} = \bar{y}_{ij\cdot} - (-1)^{i+j}\lambda/n_{ij}$ for

$$\lambda = (\bar{y}_{11\cdot} - \bar{y}_{12\cdot} - \bar{y}_{21\cdot} + \bar{y}_{22\cdot})/(1/n_{11} + 1/n_{12} + 1/n_{21} + 1/n_{22}) \ . \tag{91}$$

The BLUE($\mathbf{Q\mu}$) = $\mathbf{Q\hat{\mu}}$ of (80) uses $\hat{\mu}$ derived from (91).

-ii. The 2 × 2 case with an empty cell. With Grid 10.1 the occurrence of an empty cell means that in Table 10.4 results (53) and (81), and also (74) and (83), apply. In the unrestricted model, (53) gives $\hat{\boldsymbol{\mu}}_f = \bar{\mathbf{y}}$ for $\boldsymbol{\mu}'_f = (\mu_{11} \quad \mu_{12} \quad \mu_{21})$; and BLUE($\mathbf{Q}_f\boldsymbol{\mu}$) = $\mathbf{Q}_f\bar{\mathbf{y}}$ of (81). In the restricted model, with restriction $\mathbf{R}\boldsymbol{\mu} = \mu_{11} - \mu_{12} - \mu_{21} + \mu_{22} = 0$, $\mathbf{R}_e = 1$, a scalar, and so is nonsingular. Therefore Corollary (iii) applies, so that $\mathbf{P} = \mathbf{0}$; and then $\hat{\boldsymbol{\mu}}_f = \bar{\mathbf{y}}$, just as in the unrestricted model. But now μ_{22} is estimable, as $\hat{\mu}_{22} = -\hat{\mu}_{11} + \hat{\mu}_{12} + \hat{\mu}_{21}$ from $\hat{\boldsymbol{\mu}}_e = -\mathbf{R}_e^{-1}\mathbf{R}_f\bar{\mathbf{y}}$, and so from (83) the function $\mathbf{Q}\boldsymbol{\mu} = \mathbf{Q}_f\bar{\mathbf{y}} + \mathbf{q}_e(-\bar{y}_{11\cdot} + \bar{y}_{12\cdot} + \bar{y}_{21\cdot}) = \mathbf{Q}_1\hat{\boldsymbol{\mu}}_f = \mathbf{Q}_1\bar{\mathbf{y}}$ for $\mathbf{Q}_1 = \mathbf{Q}_f - \mathbf{q}_e(1 \quad -1 \quad -1)$. Thus, $\mu_{11} + \mu_{12} - \mu_{21} - \mu_{22} = [1 \quad -1 \quad 1]\,\boldsymbol{\mu}_f - \boldsymbol{\mu}_e$ is estimable as $\hat{\mu}_{11} + \hat{\mu}_{12} - \hat{\mu}_{21} - (-\hat{\mu}_{11} + \hat{\mu}_{12} + \hat{\mu}_{21}) = 2(\bar{y}_{11\cdot} - \bar{y}_{21\cdot})$.

-iii. A 3-factor case, with no 3-factor interactions. Consider a 3-way crossed classification with two levels of factor A, denoted $A1$ and $A2$, two levels of factor B, called $B1$ and $B2$, and three levels of factor C, labeled $C1$, $C2$ and $C3$. Two different cases of empty cells are considered, referred to as e_1 and e_2. The first is e_1: cells 111 and 121 are assumed empty, with (μ_{111}) and (μ_{121}) appearing in Grid 10.4. The second is e_2: cells 111 and 211 are assumed empty. To show both sets of empty cells in the same grid we use $[\mu_{111}]$ and $[\mu_{211}]$ for e_2 and hence have $[(\mu_{111})]$ in Grid 10.4:

Grid 10.4

	B1				B2		
	C1	C2	C3		C1	C2	C3
A1	$[(\mu_{111})]$	μ_{112}	μ_{113}	A1	(μ_{121})	μ_{122}	μ_{123}
A2	$[\mu_{211}]$	μ_{212}	μ_{213}	A2	μ_{221}	μ_{222}	μ_{223}

The vector of cell means is

$$\mu' = [\underset{\substack{\uparrow \\ (e_1) \\ [e_2]}}{\mu_{111}} \quad \mu_{112} \quad \mu_{113} \quad \underset{\substack{\uparrow \\ (e_1)}}{\mu_{121}} \quad \mu_{122} \quad \mu_{123} \quad \underset{\substack{\uparrow \\ [e_2]}}{\mu_{211}} \quad \mu_{212} \quad \mu_{213} \quad \mu_{221} \quad \mu_{222} \quad \mu_{223}] \,.$$

The arrows and symbols (e_1) and $[e_2]$ indicate occurrence of empty cells in the two different cases.

We now assume there are no 3-factor interactions, an assumption which can be stated as restrictions $\mathbf{R}\boldsymbol{\mu} = \mathbf{0}$ for

$$\mathbf{R} = \begin{bmatrix} 1 & -1 & 0 & -1 & 1 & 0 & -1 & 1 & 0 & 1 & -1 & 0 \\ 0 & 1 & -1 & 0 & -1 & 1 & 0 & -1 & 1 & 0 & 1 & -1 \end{bmatrix} . \tag{92}$$
$$\begin{matrix} \uparrow & & & \uparrow & & & \uparrow \\ (e_1) & & & (e_1) & & & \\ [e_2] & & & & & & [e_2] \end{matrix}$$

Therefore, for the two cases of empty cells, the values of $\mathbf{R}_e$ are

$$\mathbf{R}_{e_1} = \begin{bmatrix} 1 & -1 \\ 0 & 0 \end{bmatrix} \quad \text{and} \quad \mathbf{R}_{e_2} = \begin{bmatrix} 1 & -1 \\ 0 & 0 \end{bmatrix} . \tag{93}$$

Neither $\mathbf{R}_{e_1}$ nor $\mathbf{R}_{e_2}$ (which happen to be equal) have full column rank. Corollary (ii) of the theorem therefore does not apply.

Now consider linear combinations of μ_{ijk}s corresponding to traditional contrasts among levels of main effects and 2-factor interactions. For example, for factor A consider

$$\delta_A = \mu_{1\cdot\cdot} - \mu_{2\cdot\cdot}$$

TABLE 10.5. Q-MATRICES FOR MAIN EFFECT AND 2-FACTOR CONTRASTS, FOR GRID 10.4

	Q-matrices											
	Cell											
	μ_{111}	μ_{112}	μ_{113}	μ_{121}	μ_{122}	μ_{123}	μ_{211}	μ_{212}	μ_{213}	μ_{221}	μ_{222}	μ_{223}
Contrast	↑ e_1 e_2			↑ e_1			↑ e_2					
δ_A	[1	1	1	1	1	1	−1	−1	−1	−1	−1	−1]
δ_B	[1	1	1	−1	−1	−1	1	1	1	−1	−1	−1]
δ_C	[1	−1	0	1	−1	0	1	−1	0	1	−1	0
	1	0	−1	1	0	−1	1	0	−1	1	0	−1]
δ_{AB}	[1	1	1	−1	−1	−1	−1	−1	−1	1	1	1]
δ_{AC}	[1	−1	0	1	−1	0	−1	1	0	−1	1	0
	1	0	−1	1	0	−1	−1	0	1	−1	0	1]
δ_{BC}	[1	−1	0	−1	1	0	1	−1	0	−1	1	0
	1	0	−1	−1	0	1	1	0	−1	−1	0	1]

which is proportional to the main effect contrast $\bar{\mu}_{1\cdot\cdot} - \bar{\mu}_{2\cdot\cdot}$. We then have

$$\delta_A = [1 \quad 1 \quad 1 \quad 1 \quad 1 \quad 1 \quad -1 \quad -1 \quad -1 \quad -1 \quad -1 \quad -1]\boldsymbol{\mu} .$$

The matrix (vector in this case) multiplying $\boldsymbol{\mu}$ is, in terms of the theorem, a **Q**-matrix. This and the **Q**-matrices for the other such contrasts, δ_B, δ_C, δ_{AB}, δ_{AC} and δ_{BC}, are shown in Table 10.5.

To ascertain if a δ is estimable in each of the two cases, e_1 and e_2, of missing cells, its $\mathbf{Q}_e$ must be identified as a submatrix of its **Q**. This is done by scrutinizing the **Q**-matrices in Table 10.5, and the results are shown in Table 10.6. For example, in Table 10.5 in the **Q**-matrix (vector) for δ_A, the entries under e_1 are 1,1. Hence for δ_A the resultant $\mathbf{Q}_e$ is $\mathbf{Q}_{e_1} = [1 \quad 1]$, as shown in Table 10.6.

It so happens that there are only four different $\mathbf{Q}_e$-matrices in Table 10.6, where they are labeled I through IV. They are shown in Table 10.7, where the relationship of each to

$$\mathbf{R}_e = \begin{bmatrix} 1 & -1 \\ 0 & 0 \end{bmatrix} = \mathbf{R}_{e_1} = \mathbf{R}_{e_2}$$

of (93) is shown, in terms of (83) and (84) of the theorem. The resulting estimability of the corresponding $\boldsymbol{\delta}$ (or $\mathbf{L}\boldsymbol{\delta}$) is also shown in Table 10.7, the

TABLE 10.6. $\mathbf{Q}_e$-MATRICES FOR CONTRASTS IN TABLE 10.5 AND ESTIMABILITY OF THOSE CONTRASTS

	$\mathbf{Q}_e$-matrices		Type of $\mathbf{Q}_e$ (for Table 10.7)		Estimability of $\boldsymbol{\delta}$ (see Table 10.7)	
Contrast	$\mathbf{Q}_{e_1}$	$\mathbf{Q}_{e_2}$	$\mathbf{Q}_{e_1}$	$\mathbf{Q}_{e_2}$	Case e_1	Case e_2
δ_A	$[1 \quad 1]$	$[1 \quad -1]$	I	II	Not estimable	Estimable
δ_B	$[1 \quad -1]$	$[1 \quad 1]$	II	I	Estimable	Not estimable
$\boldsymbol{\delta}_C$	$\begin{bmatrix} 1 & 1 \\ 1 & 1 \end{bmatrix}$	$\begin{bmatrix} 1 & 1 \\ 1 & 1 \end{bmatrix}$	III	III	$[1 \quad -1]\boldsymbol{\delta}$ is estimable	$\mathbf{L}\boldsymbol{\delta}$ is estimable
δ_{AB}	$[1 \quad -1]$	$[1 \quad -1]$	II	II	Estimable	Estimable
$\boldsymbol{\delta}_{AC}$	$\begin{bmatrix} 1 & 1 \\ 1 & 1 \end{bmatrix}$	$\begin{bmatrix} 1 & -1 \\ 1 & -1 \end{bmatrix}$	III	IV	$[1 \quad -1]\boldsymbol{\delta}$ is estimable	Estimable
$\boldsymbol{\delta}_{BC}$	$\begin{bmatrix} 1 & -1 \\ 1 & -1 \end{bmatrix}$	$\begin{bmatrix} 1 & 1 \\ 1 & 1 \end{bmatrix}$	IV	III	Estimable	$\mathbf{L}\boldsymbol{\delta}$ is estimable

TABLE 10.7. THE FOUR VALUES OF $\mathbf{Q}_e$ IN TABLE 10.6 AND THE RESULTING ESTIMABILITY CONCLUSION

The four values of $\mathbf{Q}_e$		Relationship of $\mathbf{Q}_e$ with	Use	
Type	$\mathbf{Q}_e$	$\mathbf{R}_e = \begin{bmatrix} 1 & -1 \\ 0 & 0 \end{bmatrix}$	of Theorem	Estimability
I	$[1 \quad 1]$	$\mathbf{LQ}_e \neq \mathbf{MR}_e$ for all $\mathbf{L}$ and $\mathbf{M}$	No part applies	Neither $\boldsymbol{\delta}$ nor $\mathbf{L}\boldsymbol{\delta}$ is estimable
II	$[1 \quad -1]$	$\mathbf{Q}_e = [1 \quad 1]\mathbf{R}_e$	(83)	δ is estimable
III	$\begin{bmatrix} 1 & 1 \\ 1 & 1 \end{bmatrix}$	$[1 \quad -1]\mathbf{Q}_e = [0 \quad 1]\mathbf{R}_e$	(84)	$[1 \quad -1]\boldsymbol{\delta}$ is estimable
III	$\begin{bmatrix} 1 & -1 \\ 1 & -1 \end{bmatrix}$	$\mathbf{Q}_e = \begin{bmatrix} 1 & 1 \\ 1 & 1 \end{bmatrix}\mathbf{R}_e$	(83)	$\boldsymbol{\delta}$ is estimable

results of which are then applied to the δs in Table 10.6, for each of the two cases e_1 and e_2 of empty cells. It is clear from the last two columns of Table 10.6 that even when $\mathbf{R}_e$ is the same for both cases, estimability of contrasts can differ from one case of empty cells to another.

-iv. A 3-factor case with two sets of interactions absent. Consider the preceding example with the further assumption of no AC-interactions included with the assumption of no 3-factor interactions represented by $\mathbf{R}$ of (92). Then from $\mathbf{R}$ of (92) and $\mathbf{Q}$ for $\boldsymbol{\delta}_{AC}$ in Table 10.5, the $\mathbf{R}$-matrix for this combined case of no 3-factor interactions and no AC-interactions is

$$\mathbf{R} = \begin{bmatrix} 1 & -1 & 0 & -1 & 1 & 0 & -1 & 1 & 0 & 1 & -1 & 0 \\ 0 & 1 & -1 & 0 & -1 & 1 & 0 & -1 & 1 & 0 & 1 & -1 \\ 1 & -1 & 0 & 1 & -1 & 0 & -1 & 1 & 0 & -1 & 1 & 0 \\ 1 & 0 & -1 & 1 & 0 & -1 & -1 & 0 & 1 & -1 & 0 & 1 \end{bmatrix},$$
$$\begin{matrix} \uparrow & & & \uparrow & & & \uparrow \\ (e_1) & & & (e_1) & & & \\ [e_2] & & & & & & [e_2] \end{matrix}$$

so that

$$\mathbf{R}_{e_1} = \begin{bmatrix} 1 & -1 \\ 0 & 0 \\ 1 & 1 \\ 1 & 1 \end{bmatrix} \quad \text{and} \quad \mathbf{R}_{e_2} = \begin{bmatrix} 1 & -1 \\ 0 & 0 \\ 1 & -1 \\ 1 & -1 \end{bmatrix}.$$

In this case $\mathbf{R}_{e_1}$ has full column rank and Corollary (ii) of the theorem applies. Thus $\text{BLUE}(\mathbf{Q}\boldsymbol{\mu}) = \mathbf{Q}\hat{\boldsymbol{\mu}}_f$ holds for *any* $\mathbf{Q} = [\mathbf{Q}_{f_1} \quad \mathbf{Q}_{e_1}]$. Hence all $\boldsymbol{\delta}$s in Table 10.5 (save $\boldsymbol{\delta}_{AC}$) are estimable in the empty cell case e_1. This differs from the penultimate column of Table 10.6, thus illustrating that different restrictions on $\boldsymbol{\mu}$, even with the same pattern of empty cells (case e_1), lead to different estimability conclusions.

$\mathbf{R}_{e_2}$ does not have full column rank, and conclusions about estimability are the same as those in the last column of Table 10.7.

-v. Conclusions. Estimability of $\boldsymbol{\delta} = \mathbf{Q}\boldsymbol{\mu}$ in cell means models obviously depends on both $\mathbf{Q}$ and $\boldsymbol{\mu}$. But it also depends on both the pattern of empty cells and the restrictions (if any) on the elements of $\boldsymbol{\mu}$. Different patterns of empty cells with the same restrictions on $\boldsymbol{\mu}$ can lead to different conclusions regarding estimability, as in Section 10.4f-iii; and different restrictions on $\boldsymbol{\mu}$ with the same pattern of empty cells can yield different estimability conclusions, as illustrated by Sections 10.4f-iii and 10.4f-iv for the empty cell case e_1.

10.5. EXERCISES

E10.1. Suppose that values of ω_{ij} satisfying the Σ-restrictions

$$\sum_{j=1}^{b} \omega_{ij} = 0 \quad \forall\ i \quad \text{and} \quad \sum_{i=1}^{a} \omega_{ij} = 0 \quad \forall\ j$$

for $i = 1, 2, 3$ and $j = 1, 2, 3, 4$ are as follows:

$$\begin{array}{cccc} x & y & z & -x-y-z \\ u & v & w & -u-v-w \\ -x-u & -y-v & -z-w & x+y+z+u+v+w \end{array}$$

Show that H: $\omega_{ij} - \omega_{i'j} - \omega_{ij'} + \omega_{i'j'}$ *all zero* is equivalent to H: ω_{ij} *all zero.*

E10.2. For the example of Table 10.1, calculate
(a) SSB_w and (b) SSAC_w.

E10.3. For the example of Table 10.1, calculate
(a) $R(B \mid \mu)$ and (b) $R(AC \mid \mu, A, C)$
and confirm that these values do differ from those obtained in E10.2.

E10.4. For a 2-way classification of two rows and three columns with one empty cell, the 2,3 cell, use the restricted cell means model to derive the estimated cell means in the no-interaction model. Confirm your results using an overparameterized model.

E10.5. In the 2-way nested classification cell means model (Chapter 3) use (14) to show that
(a) H: $\Sigma_j n_{ij}\mu_{ij}/n_i$, *all equal* is tested by $\Sigma_i n_{i\cdot}(\bar{y}_{i\cdot\cdot} - \bar{y}_{\cdot\cdot\cdot})^2$.
(b) H: $\Sigma_j \mu_{ij}/b_i$ *all equal* is tested by
$\Sigma_i w_i(\tilde{y}_{i\cdot\cdot} - \Sigma w_i \tilde{y}_{i\cdot\cdot}/\Sigma w_i)^2$ for $w_i = b_i^2/\Sigma_j 1/n_{ij}$.

E10.6. Consider a 4-way crossed classification having 2, 2, 3 and 3 levels of the four factors. Suppose seven cells are empty, at least two in each of three of the four AB-subclasses. Use the theorem of Section 10.4 to consider the effect of these empty cells on the estimability of first-, second- and third-order interactions.

E10.7. For $\mathbf{M} = \mathbf{I} - \mathbf{X}(\mathbf{X}'\mathbf{X})^{-}\mathbf{X}$, prove that $\mathbf{LX} = \mathbf{0}$ if and only if $\mathbf{L} = \mathbf{TM}$ for some $\mathbf{T}$.

CHAPTER 11

MODELS WITH COVARIABLES: THE GENERAL CASE AND SOME APPLICATIONS

Chapter 6 describes several analyses of the 1-way classification with a covariable, showing in Tables 6.4, 6.6 and 6.7 a variety of sums of squares and their associated hypotheses. But no verification is given (see Section 6.1g) that any sum of squares Q, say, in those tables has properties such that $Q/r\hat{\sigma}^2$ is an F-statistic that tests the hypothesis shown in the tables as associated with Q. A portion of this chapter is therefore given to confirming, from results in Chapter 8, that the ratios implied as being F-statistics in Tables 6.4, 6.6 and 6.7 do, indeed, have that property. This is done in Section 11.3.

First, though, comes a brief presentation of a traditional approach to analysis with covariables, followed in Section 11.2 by a general (matrix notation) description based on the general linear model methodology of Chapter 8. This leads to confirming the associated hypotheses of Chapter 6; and the chapter ends with applying that linear model description to a number of special cases.

11.1. A TRADITIONAL DESCRIPTION

A frequently-seen description of the analysis of data in the presence of covariables (traditionally called analysis of covariance) is as a direct extension of analysis of variance. Consider the 1-way classification, for example. With balanced data of n observations on each of a classes, its analysis of variance is, from Table 6.3, based on the sums of squares shown in Table

TABLE 11.1. SUMS OF SQUARES FOR A 1-WAY CLASSIFICATION WITH BALANCED DATA

Source	d.f.	Sum of Squares
Between classes	$a - 1$	$B_{yy} = n \sum_{i=1}^{a} (\bar{y}_{i\cdot} - \bar{y}_{\cdot\cdot})^2$
Within classes	$a(n - 1)$	$W_{yy} = \sum_{i=1}^{a} \sum_{j=1}^{n} (y_{ij} - \bar{y}_{i\cdot})^2$
Total (adjusted for mean)	$an - 1$	$T_{yy} = \sum_{i=1}^{a} \sum_{j=1}^{n} (y_{ij} - \bar{y}_{\cdot\cdot})^2$

TABLE 11.2. SUMS OF SQUARES AND PRODUCTS FOR ANALYSIS OF COVARIANCE BASED ON TABLE 11.1

		Sums of Squares and Products		
Source	d.f.	y^2	z^2	yz
Between classes	$a - 1$	B_{yy}	B_{zz}	B_{yz}
Within classes	$a(n - 1)$	W_{yy}	W_{zz}	T_{yz}
Total	$an - 1$	T_{yy}	T_{zz}	T_{yz}

11.1. The extension of Table 11.1 to having a covariate, z_{ij}, is then based on the sums of squares shown in Table 11.2, where every B-, W- and T-term is as defined in (30) of Chapter 6, simplified here to the form indicated in Table 11.1 for balanced data.

The traditional presentation of analysis of covariance then involves making adjustments to W_{yy} and T_{yy} in the form

$$W'_{yy} = W_{yy} - W_{yz}^2/W_{zz} \quad \text{and} \quad T'_{yy} = T_{yy} - T_{yz}^2/T_{zz}, \tag{1}$$

together with

$$B'_{yy} = T'_{yy} - W'_{yy} \, . \tag{2}$$

Observe in (1) that T'_{yy} is just the residual sum of squares after fitting the simple regression of y on z; W'_{yy} is calculated similarly, and B'_{yy} is just the difference between T'_{yy} and W'_{yy} in the same way that $B_{yy} = T_{yy} - W_{yy}$. Then (1) and (2) are assembled in the analysis of covariance table shown as Table 11.3.

The F-statistic in Table 11.3 tests equality of class means adjusted for the covariate; and the average within-class regression coefficient

TABLE 11.3. ANALYSIS OF COVARIANCE FOR THE 1-WAY CLASSIFICATION WITH BALANCED DATA

Source	d.f.	Sum of Squares	Mean Square	F-statistic
Between classes	$a-1$	$B'_{yy} = T'_{yy} - W'_{yy}$	$\text{MSA}' = \dfrac{B'_{yy}}{a-1}$	$F = \dfrac{\text{MSA}'}{\text{MSE}'}$
Within classes	$a(n-1)-1$	$W'_{yy} = W_{yy} - \dfrac{W_{yz}^2}{W_{zz}}$	$\text{MSE}' = \dfrac{W'_{yy}}{an-a-1}$	
Total	$an-2$	$T'_{yy} = T_{yy} - \dfrac{T_{yz}^2}{T_{zz}}$		

is estimated as

$$\hat{b} = W_{yz}/W_{zz} \,. \tag{3}$$

The preceding manner of presenting analysis of covariance is similar to that used in many methods and design texts such as Federer (1955, p. 483), Snedecor and Cochran (1980, Section 18.2) and Winer (1971, Section 10.2). It also has extensions, to randomized complete blocks experiments (e.g., Steel and Torrie, 1980, p. 408, and Snedecor and Cochran, 1980, p. 378) and to latin square experiments (e.g., Federer, 1955, p. 491). Despite these extensions, the methodology embodied in Tables 11.1–11.3 does prompt some questions.

To begin with, (I), how do we recognize that the F of Table 11.3 does indeed have an F-distribution? And (II), how do we know that it provides a test of the hypothesis that class means adjusted for the covariate are equal? More importantly, given a definition for such adjusted class means, (III), how does the methodology lead us to B'_{yy} and W'_{yy} being the appropriate numerator and denominator sums of squares for that test? Moreover, (IV), how does it accommodate having a covariate that has a different coefficient for each class? And (V), how does it do this with unbalanced data, including empty cells, and perhaps even with a second covariate? For example, with (3) as a starting point, (VI), what form does $\hat{b}$ have for unbalanced data? Furthermore, (VII), how does this methodology extend to more complicated situations, such as a 2-way crossed classification? And (VIII), how does it extend to having more than one covariate? Answers to these questions do not emanate easily from the procedures of Tables 11.1–11.3. Maybe not all of those questions are equally useful, but there are situations in which some of them are certainly very useful. It therefore seems worthwhile to have a general procedure for analysis of covariance which can generate answers whenever they are needed. Such a procedure is available from treating analysis of covariance as just a particular application of the general linear model methodology of Chapter 8. This is now described. And from that application we will see how answers are provided to the preceding eight questions: the first six are considered in Section 11.2e, question VII is considered in Sections 11.6 and 11.7, and VIII in Sections 11.5 and 11.7.

11.2. A LINEAR MODEL DESCRIPTION

The formulation of the general linear model used in Chapter 8 is $E(\mathbf{y}) = \mathbf{X}\boldsymbol{\beta}$, where $\boldsymbol{\beta}$ is a vector of parameters representing either cell means or parameters of an overparameterized model, or the latter amended by restrictions designed to eliminate the overparameterization (or even of a combination of these three). Examples for the 1-way classification are $E(y_{ij}) = \mu_i$ in (1) of Section 2.2a, $E(y_{ij}) = \mu + \alpha_i$ of (129) in Section 2.13, and $E(y_{ij}) = \dot{\mu} + \dot{\alpha}_i$ with $\Sigma_i \dot{\alpha}_i = 0$ in (132) of Section 2.13.

Now consider the with-covariable models $E(y_{ij}) = \mu_i + bz_{ij}$, or $E(y_{ij}) = \mu_i + b_i z_{ij}$ of Sections 6.1 and 6.2; or, were there to be two covariables, z_{ij} and w_{ij}, $E(y_{ij}) = \mu_i + bz_{ij} + b^+ w_{ij}$. Based on the $E(\mathbf{y}) = \mathbf{X}\boldsymbol{\beta}$ representation of the non-covariate part of any of these models, the with-covariate model can always be represented by

$$E(\mathbf{y}) = \mathbf{X}\boldsymbol{\beta} + \mathbf{Z}\mathbf{b},$$

where $\mathbf{b}$ is the vector of slope parameters corresponding to the manner in which the covariables are to be represented in the model. Thus for the three preceding examples $\mathbf{b}$ is, respectively, $\mathbf{b} = b$, $\mathbf{b} = [b_1 \quad b_2 \quad \cdots \quad b_a]'$ and $\mathbf{b} = [b \quad b^+]'$; and were the model to be $E(y_{ij}) = \mu_i + b_i z_{ij} + b^+ w_{ij}$, then $\mathbf{b}$ would be $\mathbf{b} = [b_1 \quad \cdots \quad b_a \quad b^+]'$. With $\mathbf{b}$ determined thus, from the way in which the covariables are incorporated in the model, $\mathbf{Z}$ follows accordingly.

Examples. For the data of Table 6.1 the matrix form of the model equation for four different models is as follows.

A. For $y_{ij} = \mu_i + e_{ij}$,

$$\mathbf{y} = \mathbf{X}\boldsymbol{\beta} + \mathbf{e} \quad \text{is} \quad \begin{bmatrix} 74 \\ 68 \\ 77 \\ 76 \\ 80 \\ 87 \\ 91 \end{bmatrix} = \begin{bmatrix} 1 & \cdot & \cdot \\ 1 & \cdot & \cdot \\ 1 & \cdot & \cdot \\ \cdot & 1 & \cdot \\ \cdot & 1 & \cdot \\ \cdot & \cdot & 1 \\ \cdot & \cdot & 1 \end{bmatrix} \begin{bmatrix} \mu_1 \\ \mu_2 \\ \mu_3 \end{bmatrix} + \mathbf{e}\,. \tag{4}$$

B. For $y_{ij} = \mu_i + bz_{ij} + e_{ij}$,

$$\mathbf{y} = \mathbf{X}\boldsymbol{\beta} + \mathbf{Z}\mathbf{b} + \mathbf{e} \quad \text{is} \quad \begin{bmatrix} 74 \\ 68 \\ 77 \\ 76 \\ 80 \\ 87 \\ 91 \end{bmatrix} = \begin{bmatrix} 1 & \cdot & \cdot \\ 1 & \cdot & \cdot \\ 1 & \cdot & \cdot \\ \cdot & 1 & \cdot \\ \cdot & 1 & \cdot \\ \cdot & \cdot & 1 \\ \cdot & \cdot & 1 \end{bmatrix} \begin{bmatrix} \mu_1 \\ \mu_2 \\ \mu_3 \end{bmatrix} + \begin{bmatrix} 3 \\ 4 \\ 5 \\ 2 \\ 4 \\ 3 \\ 7 \end{bmatrix} b + \mathbf{e} \,. \tag{5}$$

C. For $y_{ij} = \mu_i + bz_{ij} + b^+ w_{ij} + e_{ij}$, equation $\mathbf{y} = \mathbf{X}\boldsymbol{\beta} + \mathbf{Z}\mathbf{b} + \mathbf{e}$ is

$$\begin{bmatrix} 74 \\ 68 \\ 77 \\ 76 \\ 80 \\ 87 \\ 91 \end{bmatrix} = \begin{bmatrix} 1 & \cdot & \cdot \\ 1 & \cdot & \cdot \\ 1 & \cdot & \cdot \\ \cdot & 1 & \cdot \\ \cdot & 1 & \cdot \\ \cdot & \cdot & 1 \\ \cdot & \cdot & 1 \end{bmatrix} \begin{bmatrix} \mu_1 \\ \mu_2 \\ \mu_3 \end{bmatrix} + \begin{bmatrix} 3 & 22 \\ 4 & 20 \\ 5 & 27 \\ 2 & 21 \\ 4 & 27 \\ 3 & 20 \\ 7 & 24 \end{bmatrix} \begin{bmatrix} b \\ b^+ \end{bmatrix} + \mathbf{e} \,. \tag{6}$$

D. For $y_{ij} = \mu_i + b_i z_{ij} + b^+ w_{ij} + e_{ij}$, equation $\mathbf{y} = \mathbf{X}\boldsymbol{\beta} + \mathbf{Z}\mathbf{b} + \mathbf{e}$ is

$$\begin{bmatrix} 74 \\ 68 \\ 77 \\ 76 \\ 80 \\ 87 \\ 91 \end{bmatrix} = \begin{bmatrix} 1 & \cdot & \cdot \\ 1 & \cdot & \cdot \\ 1 & \cdot & \cdot \\ \cdot & 1 & \cdot \\ \cdot & 1 & \cdot \\ \cdot & \cdot & 1 \\ \cdot & \cdot & 1 \end{bmatrix} \begin{bmatrix} \mu_1 \\ \mu_2 \\ \mu_3 \end{bmatrix} + \begin{bmatrix} 3 & \cdot & \cdot & 22 \\ 4 & \cdot & \cdot & 20 \\ 5 & \cdot & \cdot & 27 \\ \cdot & 2 & \cdot & 21 \\ \cdot & 4 & \cdot & 27 \\ \cdot & \cdot & 3 & 20 \\ \cdot & \cdot & 7 & 24 \end{bmatrix} \begin{bmatrix} b_1 \\ b_2 \\ b_3 \\ b^+ \end{bmatrix} + \mathbf{e} \,. \tag{7}$$

a. A general model

The general model for handling covariables is taken as

$$E(\mathbf{y}) = \mathbf{X}\boldsymbol{\beta} + \mathbf{Z}\mathbf{b} \,. \tag{8}$$

$\mathbf{X}\boldsymbol{\beta}$ represents the occurrence of main effects and interactions, be they cell means or parameters of an overparameterization (whether restricted or not). We call this the factors part of the model. In $\mathbf{Zb}$, $\mathbf{b}$ is the $c \times 1$ vector of coefficients ("slopes") corresponding to having c covariables; and, using

$$\mathbf{z}_t = [z_{1t} \cdots z_{Nt}]' \quad \text{leads to} \quad \mathbf{Z} = \{_r\, \mathbf{z}_t\}_{t=1}^{t=c} \tag{9}$$

being the matrix of N observations on each of those covariables. In this manner $\mathbf{Zb}$ is the covariables part of the model. We see in the examples (5), (6) and (7) that when the covariable part of a model changes, e.g., from $E(y_{ij}) = \mu_i + bz_{ij}$ to $E(y_{ij}) = \mu_i + b_i z_{ij}$, it is, of course, the $\mathbf{Zb}$ of the model that changes and, as has been said, it is the nature of $\mathbf{b}$ that determines the form of $\mathbf{Z}$.

When the factors part of the model includes a general mean μ, in the form $\mu\mathbf{1}$, we partition $\mathbf{X}\boldsymbol{\beta}$ as $\mathbf{X}\boldsymbol{\beta} = \mu\mathbf{1} + \mathbf{X}_1\boldsymbol{\beta}_1$ and have the model as

$$E(\mathbf{y}) = \mu\mathbf{1} + \mathbf{X}_1\boldsymbol{\beta}_1 + \mathbf{Zb}\ . \tag{10}$$

Without the $\mathbf{Zb}$ term, (10) is $E(\mathbf{y}) = \mu\mathbf{1} + \mathbf{X}_1\boldsymbol{\beta}_1$, which is a factors model; and without the $\mathbf{X}_1\boldsymbol{\beta}_1$ term, (10) is a multiple regression model $E(\mathbf{y}) = \mu\mathbf{1} + \mathbf{Zb}$ of $\mathbf{y}$ regressed on the covariables. Thus the general model for analysis of covariance, written as either (8) or (10), is simply a combination of factors (and interactions) and regression. Viewed in this light there is nothing difficult to understand about analysis of covariance. And the general linear model methodology of Chapter 8 provides the analysis procedures.

b. Estimation

-i. Normal equations. By writing the with-covariates model (8) as

$$E(\mathbf{y}) = [\mathbf{X}\quad \mathbf{Z}]\begin{bmatrix}\boldsymbol{\beta}\\ \mathbf{b}\end{bmatrix},$$

one can immediately write the normal equations as

$$\begin{bmatrix}\mathbf{X}'\mathbf{X} & \mathbf{X}'\mathbf{Z}\\ \mathbf{Z}'\mathbf{X} & \mathbf{Z}'\mathbf{Z}\end{bmatrix}\begin{bmatrix}\tilde{\boldsymbol{\beta}}\\ \hat{\mathbf{b}}\end{bmatrix} = \begin{bmatrix}\mathbf{X}'\mathbf{y}\\ \mathbf{Z}'\mathbf{y}\end{bmatrix}. \tag{11}$$

The notation $\tilde{\boldsymbol{\beta}}$ and $\hat{\mathbf{b}}$ has two distinctions: first, $\tilde{\boldsymbol{\beta}}$ is different from

$$\boldsymbol{\beta}^\circ = (\mathbf{X}'\mathbf{X})^-\mathbf{X}'\mathbf{y} \tag{12}$$

of the no-covariate model; and second, as shall be shown, $\hat{\mathbf{b}}$ is unique (for given $\mathbf{X}$, $\mathbf{y}$ and $\mathbf{Z}$), i.e., there is only one solution $\hat{\mathbf{b}}$ to (11).

-ii. Assumptions on the covariates. It is assumed that the covariates are such that columns of $\mathbf{Z}$ are linearly independent, and that they are also linearly independent of columns of $\mathbf{X}$. These assumptions are seldom violated, because each column of $\mathbf{Z}$ consists of observations on a covariable. Seldom will any of those columns be a linear combination of others, or of

the columns of zeros and unities that constitute **X** being, as it is, the incidence matrix corresponding to main effects and interactions.

The preceding linear independence properties of columns of **Z** mean for

$$\mathbf{M} = \mathbf{I} - \mathbf{X}(\mathbf{X}'\mathbf{X})^{-}\mathbf{X}' = \mathbf{M}' = \mathbf{M}^2 \quad \text{with} \quad \mathbf{MX} = \mathbf{0} \tag{13}$$

and

$$\mathbf{R} = \mathbf{MZ} \tag{14}$$

that, using (43) and (44) of Section 7.3,

$$\mathbf{R}'\mathbf{R} \text{ is nonsingular} . \tag{15}$$

This, as shall be seen, plays an important role in estimating **b**.

-iii. Estimators. The normal equations (11) are precisely the same form as those in (74) of Section 8.5b. Applying solutions (77) and (78) of that section to (11) therefore gives a solution to (11) as

$$\tilde{\boldsymbol{\beta}} = (\mathbf{X}'\mathbf{X})^{-}\mathbf{X}'(\mathbf{y} - \mathbf{Z}\hat{\mathbf{b}}) \tag{16}$$

and

$$\hat{\mathbf{b}} = (\mathbf{R}'\mathbf{R})^{-1}\mathbf{R}'\mathbf{y} . \tag{17}$$

Several aspects of (16) and (17) are worth noting.

First, $\tilde{\boldsymbol{\beta}}$ of (16) differs from $\boldsymbol{\beta}^{\circ}$ of (12), although $\boldsymbol{\beta}^{\circ}$ is a part of $\tilde{\boldsymbol{\beta}}$:

$$\tilde{\boldsymbol{\beta}} = (\mathbf{X}'\mathbf{X})^{-}\mathbf{X}'\mathbf{y} - (\mathbf{X}'\mathbf{X})^{-}\mathbf{X}'\mathbf{Z}\hat{\mathbf{b}} = \boldsymbol{\beta}^{\circ} - (\mathbf{X}'\mathbf{X})^{-}\mathbf{X}'\mathbf{Z}\hat{\mathbf{b}} . \tag{18}$$

Further, on writing $\boldsymbol{\beta}^{\circ}$ as $\boldsymbol{\beta}^{\circ}[\mathbf{y}]$, i.e.,

$$\boldsymbol{\beta}^{\circ} = (\mathbf{X}'\mathbf{X})^{-}\mathbf{X}'\mathbf{y} = \boldsymbol{\beta}^{\circ}[\mathbf{y}], \tag{19}$$

to emphasize the functional dependence on **y**, we can also usefully employ the same notation for $\tilde{\boldsymbol{\beta}}$:

$$\tilde{\boldsymbol{\beta}} = (\mathbf{X}'\mathbf{X})^{-1}\mathbf{X}'(\mathbf{y} - \mathbf{Z}\hat{\mathbf{b}}) = \boldsymbol{\beta}^{\circ}[\mathbf{y} - \mathbf{Z}\hat{\mathbf{b}}] = \boldsymbol{\beta}^{\circ}[\mathbf{y}] - \boldsymbol{\beta}^{\circ}[\mathbf{Z}\hat{\mathbf{b}}] . \tag{20}$$

Either form, (18) or (20), can be useful when, for given **y** and **X**, a change is made in **b**, e.g., from (5) to (6) or (7). Then only the second terms in (18) and (20) will change. Moreover, the final form in (20) is very convenient for deriving $\tilde{\boldsymbol{\beta}}$. With $\boldsymbol{\beta}^{\circ}[\mathbf{y}]$ being known from (19), $\boldsymbol{\beta}^{\circ}[\mathbf{Z}\hat{\mathbf{b}}]$ is obtained similarly and both are used in (20). For example, in $E(y_{ij}) = \mu_i + bz_{ij}$ we know that $\boldsymbol{\beta}^{\circ}[\mathbf{y}] = \boldsymbol{\mu}^{\circ}[y]$ yields $\tilde{\mu}_i = \bar{y}_{i\cdot}$. Therefore (20) yields $\tilde{\mu}_i = \bar{y}_{i\cdot} - \hat{b}\bar{z}_{i\cdot}$.

Second, because of the existence of $(\mathbf{R}'\mathbf{R})^{-1}$ from (15), the solution $\hat{\mathbf{b}}$ in (17) always exists and is the only solution for $\hat{\mathbf{b}}$. Therefore

$$\text{BLUE}(\mathbf{b}) = \hat{\mathbf{b}} = (\mathbf{R}'\mathbf{R})^{-1}\mathbf{R}'\mathbf{y} \ . \tag{21}$$

Thus $\hat{\mathbf{b}}$ is not one of many possible solutions, as is $\tilde{\boldsymbol{\beta}}$ through its dependence on $(\mathbf{X}'\mathbf{X})^{-}$: (21) is the only solution for $\hat{\mathbf{b}}$.

Third, in the no-covariable model $E(\mathbf{y}) = \mathbf{X}\boldsymbol{\beta}$ we know from (53) of Section 8.3b that the vector of residuals is $\mathbf{y} - \hat{\mathbf{y}} = \mathbf{M}\mathbf{y}$ for $\mathbf{M}$ of (13). Hence for $\mathbf{R}$ of (14) occurring in (21),

$$\begin{aligned}\mathbf{R} = \mathbf{M}\mathbf{Z} &= \mathbf{M}\{_r\, \mathbf{z}_t\}_{t=1}^{t=c} = \{_r\, \mathbf{M}\mathbf{z}_t\}_{t=1}^{t=c} \\ &= \{_r\, \mathbf{z}_t - \hat{\mathbf{z}}_t\}_{t=1}^{t=c} = \{_r\, \mathbf{z}_t - \mathbf{X}\boldsymbol{\beta}^{\circ}[\mathbf{z}_t]\}_{t=1}^{t=c} \ .\end{aligned} \tag{22}$$

In the first expression in (22) we see that the tth column of $\mathbf{R}$ is the vector of residuals after fitting the no-covariate part of the model to $\mathbf{z}_t$; i.e., after fitting $E(\mathbf{z}_t) = \mathbf{X}\boldsymbol{\beta}$. The second expression in (22) uses the notation of (19) and is the columns-of-residuals idea, evident in the first expression of (22), that is a convenient computational device for obtaining $\mathbf{R}$; but it is only that, a computational device. Nevertheless, it is useful, and quite general, being applicable to all (linear) covariance models no matter how many covariates there are nor whether the data are balanced or unbalanced. It always provides $\mathbf{R}$ and hence $\hat{\mathbf{b}} = (\mathbf{R}'\mathbf{R})^{-1}\mathbf{R}'\mathbf{y}$.

A final observation about $\hat{\mathbf{b}}$ is that it can always be computed using residual sums of squares and products of the zs, namely $\mathbf{R}'\mathbf{R}$, and residual sums of products of the ys and zs, $\mathbf{R}'\mathbf{y}$. This is so because

$$\hat{\mathbf{b}} = (\mathbf{R}'\mathbf{R})^{-1}\mathbf{R}'\mathbf{y} = (\mathbf{R}'\mathbf{R})^{-1}\mathbf{Z}'\mathbf{M}\mathbf{M}\mathbf{y} = (\mathbf{R}'\mathbf{R})^{-1}\mathbf{R}'(\mathbf{y} - \hat{\mathbf{y}}) \ . \tag{23}$$

This is evident in particular cases already considered, e.g., equations (7), (31), (55) and (78) of Chapter 6, and in (3) of this chapter.

-iv. ***Dispersion matrices of estimators.*** From (18) it is clear that

$$\text{var}(\hat{\mathbf{b}}) = (\mathbf{R}'\mathbf{R})^{-1}\sigma^2 \quad \text{and} \quad \text{cov}(\mathbf{y}, \hat{\mathbf{b}}') = \mathbf{R}(\mathbf{R}'\mathbf{R})^{-1}\sigma^2 \ . \tag{24}$$

Hence from (16), on using (25) and $\mathbf{X}'\mathbf{R} = \mathbf{0}$ (because $\mathbf{R} = \mathbf{M}\mathbf{Z}$ and $\mathbf{X}'\mathbf{M} = \mathbf{0}$), we get

$$\text{var}(\tilde{\boldsymbol{\beta}}) = (\mathbf{X}'\mathbf{X})^{-1}\mathbf{X}'\left[\mathbf{I} - \mathbf{Z}(\mathbf{R}'\mathbf{R})^{-1}\mathbf{Z}'\right]\mathbf{X}(\mathbf{X}'\mathbf{X})^{-1}\sigma^2 \ . \tag{25}$$

Similarly,

$$\operatorname{cov}(\tilde{\boldsymbol{\beta}}, \hat{\mathbf{b}}') = (\mathbf{X}'\mathbf{X})^{-1}\mathbf{X}'\left[\mathbf{I} - \mathbf{Z}(\mathbf{R}'\mathbf{R})^{-1}\mathbf{R}'\right]\mathbf{R}(\mathbf{R}'\mathbf{R})^{-1}\sigma^2$$

$$= -(\mathbf{X}'\mathbf{X})^{-1}\mathbf{X}'\mathbf{Z}(\mathbf{R}'\mathbf{R})^{-1}\sigma^2 . \tag{26}$$

c. Analysis of variance

Define

$$\mathscr{Z} = \text{matrix of deviations } z_{ij} - \bar{z}_{\cdot j}$$

$$= \left\{{}_m\, z_{ij} - \bar{z}_{\cdot j}\right\}_{i=1}^{i=N}\,{}_{j=1}^{j=c}$$

$$= \mathbf{CZ},$$

where $\mathbf{C} = \mathbf{I} - \bar{\mathbf{J}}_N$ is the centering matrix of order N, with $\bar{\mathbf{J}}_N$ being square, of order N, with every element $1/N$. Then

$\mathscr{Z}'\mathscr{Z}$ = matrix of sums of squares and products of the zs corrected for their means, $\bar{z}_{\cdot j} = \Sigma_{i=1}^{N} z_{ij}/N$ for $j = 1, \ldots, c$.

The model (10) with $\mathbf{X}_1\boldsymbol{\beta}_1$ omitted is $E(\mathbf{y}) = \mathbf{Zb}$. This is a regression model of y on the zs, and so $R(\mathbf{b}/\mu) = \mathbf{y}'\mathscr{Z}(\mathscr{Z}'\mathscr{Z})^{-1}\mathscr{Z}'\mathbf{y}$. Shown in Table 11.4 are this and the sums of squares that arise from fitting the with-covariables model, first after fitting the factors-only model $E(\mathbf{y}) = \mathbf{X}\boldsymbol{\beta}$, and second after fitting the regression model $E(\mathbf{y}) = \mu\mathbf{1} + \mathbf{Zb}$. In both cases, of course, it is readily shown from the R-algorithm that

$$R(\boldsymbol{\beta}, \mathbf{b}) = \tilde{\boldsymbol{\beta}}'\mathbf{X}'\mathbf{y} + \hat{\mathbf{b}}'\mathbf{Z}'\mathbf{y}$$

reduces to
$$R(\boldsymbol{\beta}, \mathbf{b}) = \mathbf{y}'\mathbf{X}(\mathbf{X}'\mathbf{X})^{-}\mathbf{X}'\mathbf{y} + \mathbf{y}'\mathbf{R}(\mathbf{R}'\mathbf{R})^{-1}\mathbf{R}'\mathbf{y} \tag{27}$$

and hence
$$R(\mathbf{b} \mid \boldsymbol{\beta}) = \mathbf{y}'\mathbf{R}(\mathbf{R}'\mathbf{R})^{-1}\mathbf{R}'\mathbf{y} .$$

d. Hypothesis testing

-i. Some general hypotheses. General formulation of the F-statistic for testing any (testable) linear hypothesis in a linear model is given in (146) of Section 8.8. Applying that to the with-covariates model $E(\mathbf{y}) = \mathbf{X}\boldsymbol{\beta} + \mathbf{Zb}$ leads (see exercise E11.3) to the following results. In each case $Q/r_{\mathbf{K}}\hat{\sigma}^2$ is the F-statistic.

(A) For H: $\mathbf{K}'\mathbf{b} = \mathbf{m}$,

$$Q = (\mathbf{K}'\hat{\mathbf{b}} - \mathbf{m})'\left[\mathbf{K}'(\mathbf{R}'\mathbf{R})^{-1}\mathbf{K}\right]^{-1}(\mathbf{K}'\hat{\mathbf{b}} - \mathbf{m}) .$$

TABLE 11.4. ANALYSES OF VARIANCE FOR MODELS WITH COVARIATES

Source of Variation	d.f.[1]	Sum of Squares
		(a) Fitting factors and then covariates
Factors	$r_{\mathbf{X}}$	$R(\boldsymbol{\beta}) = R(\mu, \boldsymbol{\beta}_1) = \mathbf{y}'\mathbf{X}(\mathbf{X}'\mathbf{X})^{-}\mathbf{X}'\mathbf{y}$
Mean	1	$R(\mu) = N\bar{y}^2$
Factors (adjusted for mean)	$r_{\mathbf{X}} - 1$	$R(\boldsymbol{\beta}_1 \mid \mu) = \mathbf{y}'\mathbf{X}(\mathbf{X}'\mathbf{X})^{-}\mathbf{X}'\mathbf{y} - N\bar{y}^2$
Covariates (adjusted for factors)	c	$R(\mathbf{b} \mid \boldsymbol{\beta}) = \mathbf{y}'\mathbf{R}(\mathbf{R}'\mathbf{R})^{-1}\mathbf{R}'\mathbf{y}$
Residual	$N - r_{\mathbf{X}} - c$	$\text{SSE} = \mathbf{y}'\mathbf{y} - R(\boldsymbol{\beta}, \mathbf{b})$ $= \mathbf{y}'\bar{\mathbf{y}} - \mathbf{y}'\mathbf{X}(\mathbf{X}'\mathbf{X})^{-}\mathbf{X}'\mathbf{y} - \mathbf{y}'\mathbf{R}(\mathbf{R}'\mathbf{R})^{-1}\mathbf{R}'\mathbf{y}$
Total	N	$\text{SST} = \mathbf{y}'\mathbf{y}$
		(b) Fitting covariates and then factors
Mean	1	$R(\mu) = N\bar{y}^2$
Covariates (adjusted for mean)	c	$R(\mathbf{b} \mid \mu) = \mathbf{y}'\mathscr{X}(\mathscr{X}'\mathscr{X})^{-1}\mathscr{X}'\mathbf{y}$
Factors (adjusted for mean and covariates)	$r_{\mathbf{X}} - 1$	$R(\boldsymbol{\beta}_1 \mid \mu, \mathbf{b}) = \mathbf{y}'\mathbf{X}(\mathbf{X}'\mathbf{X})^{-}\mathbf{X}'\mathbf{y} - N\bar{y}^2$ $+\mathbf{y}'\mathbf{R}(\mathbf{R}'\mathbf{R})^{-1}\mathbf{R}'\mathbf{y} - R(\boldsymbol{\beta} \mid \mu)$
Residual	$N - r_{\mathbf{X}} - c$	SSE = as above
Total	N	$\text{SST} = \mathbf{y}'\mathbf{y}$

[1] c = number of covariables = rank of $\mathbf{Z}_{N\times c}$, and $r_{\mathbf{X}}$ = rank of $\mathbf{X}$.

(B) For H: $\mathbf{K}'\boldsymbol{\beta} = \mathbf{m}$,

$$Q = (\mathbf{K}'\tilde{\boldsymbol{\beta}} - \mathbf{m})'\big[\mathbf{K}'(\mathbf{X}'\mathbf{X})^{-}\mathbf{K} + \mathbf{K}'(\mathbf{X}'\mathbf{X})^{-}\mathbf{X}'\mathbf{Z}(\mathbf{R}'\mathbf{R})^{-1}\mathbf{Z}'\mathbf{X}(\mathbf{X}'\mathbf{X})^{-}\mathbf{K}\big]^{-1}(\mathbf{K}'\tilde{\boldsymbol{\beta}} - \mathbf{m}) .$$

(C) For H: $\mathbf{K}'[\boldsymbol{\beta} + (\mathbf{X}'\mathbf{X})^{-}\mathbf{X}'\mathbf{Z}\mathbf{b}] = \mathbf{0}$,

$$Q = \boldsymbol{\beta}^{\circ\prime}\mathbf{K}[\mathbf{K}'(\mathbf{X}'\mathbf{X})^{-}\mathbf{K}]^{-1}\mathbf{K}'\boldsymbol{\beta}^{\circ} .$$

Hypothesis (A) provides opportunity for testing that elements of $\mathbf{b}$ equal pre-assigned values denoted by $\mathbf{m}$; $\mathbf{K}'$ can be any full row rank matrix of rank no greater than c, the number of covariables. Particular cases of (A) that are sometimes of interest are the hypotheses that some or all elements of $\mathbf{b}$ are equal; or that they are zero.

Hypotheses (B) and (C) require that $\mathbf{K}'$ have full row rank not exceeding $r_{\mathbf{X}}$, and that $\mathbf{K}'\boldsymbol{\beta}$ be estimable, i.e., $\mathbf{K}' = \mathbf{T}'\mathbf{X}$ for some $\mathbf{T}'$. In (C), the value of Q is the same as when testing H: $\mathbf{K}'\boldsymbol{\beta} = \mathbf{0}$ in the no-covariate model $E(\mathbf{y}) = \mathbf{X}\boldsymbol{\beta}$. Thus whenever a sum of squares tests H: $\mathbf{K}'\boldsymbol{\beta} = \mathbf{0}$ in the no-covariate model it tests H: $\mathbf{K}'[\boldsymbol{\beta} + (\mathbf{X}'\mathbf{X})^-\mathbf{X}'\mathbf{Z}\mathbf{b}] = \mathbf{0}$ in the with-covariate model.

-ii. Associated hypotheses in the analysis of variance. Applying (123) of Section 8.6c to the sums of squares in Table 11.4 yields the associated hypotheses shown in Table 11.5.

The hypothesis associated with $R(\mu)$ is derived as follows. First, $R(\mu) = \bar{\mathbf{y}}\bar{\mathbf{J}}\mathbf{y}$ so that using (70) of Section 7.6b-iii gives the associated hypothesis as H: $\bar{\mathbf{J}}\mathbf{X}\boldsymbol{\beta} = \mathbf{0}$, equivalent to H: $\mathbf{1}'E(\mathbf{y}) = 0$, or H: $E(\bar{y}) = 0$, i.e.,

$$R(\mu) \quad \text{tests} \quad \text{H: } E(\bar{y}) = 0 \ .$$

To simplify this we define $\boldsymbol{\beta}_1$ of (10) to have elements β_r for $r = 1, \ldots,$ and $\mathbf{Z}\mathbf{b}$ to be $\Sigma_{t=1}^c b_t\mathbf{z}_t$. Then, with $\bar{z}_t$ being the mean of the N elements of $\mathbf{z}_t$, and with n_r defined as the number of times β_r occurs in the data, we find that

$$E(\bar{y}) = \mu + \Sigma_r n_r \beta_r / N + \Sigma_t b_t \bar{z}_t \ .$$

The associated hypothesis for $R(\mu)$ follows accordingly, as in Table 11.5.

Two sums of squares in Table 11.5 are often of particular interest, by nature of their associated hypotheses. The first one is $R(\mathbf{b} \mid \boldsymbol{\beta})$ of part (a) of the table, which tests H: $\mathbf{b} = \mathbf{0}$; and so, from hypothesis (A) of Section 11.2d-i, we get

$$R(\mathbf{b} \mid \boldsymbol{\beta}) = \hat{\mathbf{b}}'\mathbf{R}'\mathbf{R}\hat{\mathbf{b}} \ .$$

The second one is $R(\boldsymbol{\beta}_1 \mid \mu, \mathbf{b})$ of part (b) of the table, which tests the hypothesis H: *all elements of* $\mathbf{X}_1\boldsymbol{\beta}$ *equal*. For example, in the 1-way classification $E(y_{ij}) = \mu + \alpha_i$ it tests H: α_i *all equal*.

e. Examples: 1-way classification, one covariable

-i. The single slope model. The model equation that is used in Section 6.1, $E(y_{ij}) = \mu_i + bz_{ij}$, can be written as

$$E(\mathbf{y}) = \{_d\, \mathbf{1}_{n_i}\}\boldsymbol{\mu} + b\mathbf{z}, \tag{28}$$

TABLE 11.5. ASSOCIATED HYPOTHESES FOR SUMS OF SQUARES OF TABLE 11.4

Source of Variation	d.f.	Sum of Squares	Associated Hypothesis
(a) Fitting factors and then covariates			
Factors	$r_{\mathbf{X}}$	$R(\boldsymbol{\beta})$	H: $\mathbf{X}[\boldsymbol{\beta} + (\mathbf{X}'\mathbf{X})^{-}\mathbf{X}'\mathbf{Z}\mathbf{b}] = \mathbf{0}$
Mean	1	$R(\mu)$	H: $\mu + \Sigma_r n_r \beta_r / N + \Sigma_t b_t \bar{z}_t = 0$
Factors (adjusted for mean)[1]	$r_{\mathbf{X}} - 1$	$R(\boldsymbol{\beta}_1 \mid \mu)$	H: $\mathbf{C}\mathbf{X}_1[\boldsymbol{\beta}_1 + (\mathbf{X}_1'\mathbf{C}\mathbf{X}_1)^{-}\mathbf{X}_1'\mathbf{C}\mathbf{Z}\mathbf{b}] = \mathbf{0}$
Covariates (adjusted for factors)	c	$R(\mathbf{b} \mid \boldsymbol{\beta})$	H: $\mathbf{b} = \mathbf{0}$
Residual	$N - r_{\mathbf{X}} - c$	SSE	
Total	N	SST	
(b) Fitting covariates and then factors			
Mean	1	$R(\mu)$	H: $\mu + \Sigma_r n_r \beta_r / N + \Sigma_t b_t \bar{z}_t = 0$
Covariates (adjusted for mean)	c	$R(\mathbf{b} \mid \mu)$	H: $(\mathscr{X}'\mathscr{X})^{-1}\mathscr{X}'\mathbf{X}_1\boldsymbol{\beta}_1 + \mathbf{b} = \mathbf{0}$
Factors (adjusted for mean and covariates)	$r_{\mathbf{X}} - 1$	$R(\boldsymbol{\beta}_1 \mid \mu, \mathbf{b})$	H: *all elements of* $\mathbf{X}_1\boldsymbol{\beta}_1$ *equal*
Residual	$N - r_{\mathbf{X}} - c$	SSE	
Total	N	SST	

[1] $\mathbf{C} = \mathbf{I}_N - \bar{\mathbf{J}}_N$.

where $\mathbf{z}$ is the vector of z_{ij}-values in lexicon order. In this case

$$\mathbf{M} = \mathbf{I} - \left\{ {}_d\, \bar{\mathbf{J}}_{n_i} \right\} . \tag{29}$$

Hence

$$\mathbf{R} = \mathbf{M}\mathbf{Z} = \mathbf{z} - \left\{ {}_c\, \bar{z}_{i\cdot} \mathbf{1}_{n_i} \right\}; \tag{30}$$

therefore $\mathbf{R}'\mathbf{R} = \sum_{i=1}^{a} \sum_{j=1}^{n_i} (z_{ij} - \bar{z}_{i\cdot})^2 = W_{zz}$ and

$$\mathbf{R}'\mathbf{y} = \sum_{i=1}^{a} \sum_{j=1}^{n_i} (z_{ij} - \bar{z}_{i\cdot}) y_{ij} = \sum_{i=1}^{a} \sum_{j=1}^{n_i} (z_{ij} - \bar{z}_{i\cdot})(y_{ij} - \bar{y}_{i\cdot}) = W_{yz},$$

where here, be it noted, W_{zz} and W_{yz} are defined for unbalanced data, as in Chapter 6, with the forms used in Section 11.1 being simply the special case of balanced data. Thus, using (17),

$$\hat{b} = (\mathbf{R}'\mathbf{R})^{-1}\mathbf{R}'\mathbf{y} = W_{yz}/W_{zz}, \tag{31}$$

similar to (3). And applying (18) to (28) gives $\tilde{\boldsymbol{\mu}} = \boldsymbol{\mu}^\circ - (\mathbf{X}'\mathbf{X})^-\mathbf{X}'\mathbf{Z}\hat{\mathbf{b}}$. In Sections 2.3 and 8.2c $\hat{\mu}_i = \bar{y}_{i\cdot}$, and so $\tilde{\boldsymbol{\mu}} = \{_c\ \bar{y}_{i\cdot}\}_{i=1}^{a} - \{_d\ 1/n_i\}\{_d\ \mathbf{1}'_{n_i}\}\mathbf{z}\hat{b}$ so that

$$\tilde{\mu}_i = \bar{y}_{i\cdot} - (1/n_i)z_{i\cdot}\hat{b} = \bar{y}_{i\cdot} - \hat{b}\bar{z}_{i\cdot}, \tag{32}$$

as in (6) of Section 6.1b-i.

This development is for unbalanced data and so includes the special case of balanced data; hence $\hat{b}$ of (3), in the traditional analysis-of-variance description of analysis of covariance, is a special case of $\hat{b}$ of (1). Thus question VI at the end of Section 11.1 has been answered: $\hat{b}$ for unbalanced data has, in (31), the same form as it does for balanced data. Moreover, both estimators (31) and (32) are the outcome of the general estimators (17) and (18) that can be used for any with-covariables model.

It is left to the reader (as E11.4) to show that Tables 11.4 and 11.5 yield Table 6.4 for the model (28). In doing this it will be found that in Table 11.5 the associated hypothesis for $R(\boldsymbol{\beta}_1 | \mu, b)$ reduces to H: μ_i *all equal.* This answers question II. And $R(\boldsymbol{\beta}_1 | \mu, b)$ is

$$\begin{aligned} R(\boldsymbol{\beta}_1 | \mu, b) &= \mathbf{y}'\mathbf{X}(\mathbf{X}'\mathbf{X})^-\mathbf{X}'\mathbf{y} - N\bar{y}_{\cdot\cdot}^2 + \mathbf{y}'\mathbf{R}(\mathbf{R}'\mathbf{R})^{-1}\mathbf{R}'\mathbf{y} - \mathbf{y}'\mathscr{Z}(\mathscr{Z}'\mathscr{Z})^{-1}\mathscr{Z}\mathbf{y} \\ &= \Sigma_i n_i \bar{y}_{i\cdot}^2 - N\bar{y}_{\cdot\cdot}^2 + W_{yz}^2/W_{zz} - T_{yz}^2/T_{zz}\ . \end{aligned}$$

Adding and subtracting $\Sigma_i\Sigma_j y_{ij}^2$ to this expression and re-arranging gives

$$\begin{aligned} R(\boldsymbol{\beta}_1 | \mu, b) &= \Sigma_i\Sigma_j y_{ij}^2 - N\bar{y}_{\cdot\cdot}^2 - T_{yz}^2/T_{zz} \\ &\qquad - \left[\left(\Sigma_i\Sigma_j y_{ij}^2 - \Sigma_i n_i \bar{y}_{i\cdot}^2\right) - W_{yz}^2/W_{zz}\right] \\ &= T_{yy} - T_{yz}^2/T_{zz} - \left(W_{yy} - W_{yz}^2/W_{yz}\right), \end{aligned}$$

which, using (1) and (2), corresponds precisely to $B'_{yy} = T'_{yy} - W'_{yy}$ of Table 11.3. Furthermore, SSE in Table 11.4 is

$$\begin{aligned} \text{SSE} &= \mathbf{y}'\mathbf{y} - \mathbf{y}'\mathbf{X}(\mathbf{X}'\mathbf{X})^-\mathbf{X}'\mathbf{y} - \mathbf{y}'\mathbf{R}(\mathbf{R}'\mathbf{R})^{-1}\mathbf{R}'\mathbf{y} \\ &= \Sigma_i\Sigma_j y_{ij}^2 - \Sigma_i n_i \bar{y}_{i\cdot}^2 - W_{yz}^2/W_{zz} = W_{yy} - W_{yz}^2/W_{zz} = W'_{yy} \end{aligned}$$

of Table 11.3. Therefore

$$F = \frac{\text{MSA}'}{\text{MSE}'} = \frac{B'_{yy}/(a-1)}{W'_{yy}/(an-a-1)} \quad \text{of Table 11.3}$$

$$= \frac{R(\beta_1 \mid \mu, b)/(r_{\mathbf{X}}-1)}{\text{SSE}/(N-r_{\mathbf{X}}-1)} \quad \text{of Table 11.4 .}$$

Thus Table 11.4 provides explanation of why the F-statistic of Table 11.3 (usually described in that context as testing equality of adjusted means) is testing H: μ_i *all equal*, so answering questions I and III. But the Table 11.4 formulation does more than Table 11.3 does.

Table 11.3 is for balanced data but Table 11.4 is more general. It is for unbalanced data. And, of course, this is part of a general methodology that can be applied to more complicated models—e.g., models with more than one covariable and/or covariate having different slopes for different levels of any factor in the model. To an example of the latter we now turn.

-ii. *The multiple slopes model.* The model in Section 6.2 is

$$E(y_{ij}) = \mu_i + b_i z_{ij},$$

which can be written as

$$E(\mathbf{y}) = \{_d\, \mathbf{1}_{n_i}\}\boldsymbol{\mu} + \{_r\, \mathbf{z}_i\}_{i=1}^{i=a}\mathbf{b} \tag{33}$$

where

$$\mathbf{z}_i = \begin{bmatrix}\mathbf{0}\mathbf{1}'_{f_i} & z_{i1} & z_{i2} & \cdots & z_{in_i} & \mathbf{0}\mathbf{1}'_{l_i}\end{bmatrix}', \quad \text{for } i = 1, \ldots, a$$

and

$$f_i = n_1 + n_2 + \cdots + n_{i-1} \quad \text{and} \quad l_i = n_{i+1} + n_{i+2} + \cdots + n_a .$$

An equivalent formulation is

$$E(\mathbf{y}) = \{_d\, \mathbf{1}_{n_i}\}\boldsymbol{\mu} + \{_d\, \mathbf{z}_i^*\}\mathbf{b}, \tag{34}$$

where $\mathbf{z}_i^* = [z_{i1} \quad z_{i2} \quad \cdots \quad z_{in_i}]'$.

Whether (33) or (34) is used, **M** is the same as (29) and

$$\mathbf{R} = \mathbf{MZ} = \left\{_d \left(\mathbf{z}_i^* - \bar{z}_{i\cdot}\mathbf{1}_{n_i}\right)\right\} .$$

Hence $$\mathbf{R}'\mathbf{R} = \left\{_d \Sigma_j\left(z_{ij} - \bar{z}_{i\cdot}\right)^2\right\} = \left\{_d W_{i,zz}\right\}$$

and $$\mathbf{R}'\mathbf{y} = \left\{_c \Sigma_i\left(z_{ij} - \bar{z}_{i\cdot}\right)y_{ij}\right\} = \left\{_c W_{i,yz}\right\} .$$

Therefore $$\hat{\mathbf{b}} = (\mathbf{R}'\mathbf{R})^{-1}\mathbf{R}'\mathbf{y} = \left\{_c W_{i,yz}/W_{,zz}\right\}$$

so giving $\hat{b}_i = W_{i,yz}/W_{i,zz}$ as in (78) of Section 6.2b; and answering questions IV and V in the process. And $\tilde{\boldsymbol{\mu}} = \hat{\boldsymbol{\mu}} - (\mathbf{X}'\mathbf{X})^{-1}\mathbf{X}'\mathbf{Z}\mathbf{b} = \{_c\, \bar{y}_{i\cdot}\} - \{_d\, 1/n_i\}\{_d\, \mathbf{1}'_{n_i}\}\{_d\, \mathbf{z}_i^*\}\hat{\mathbf{b}}$ gives, as in (75) of Section 6.2b,

$$\tilde{\mu}_i = \bar{y}_{i\cdot} - \hat{b}_i\bar{z}_{i\cdot} . \tag{35}$$

It is left to the reader (E11.4) to show, akin to the single slope model, that Tables 11.4 and 11.5 for (33) yield parts of Table 6.6.

-iii. ***Other hypotheses about slopes.*** The two traditional hypotheses about slopes in the multiple slopes model are either that they are all zero, or that they are all equal and nonzero. These are based on $\mathscr{R}(\mu_i, b_i \,|\, \mu_i)$ and on $\mathscr{R}(\mu_i, b_i \,|\, \mu_i, b)$, respectively, as in Table 6.6. Additional to these, though, from the general procedure for testing H: $\mathbf{K}'\mathbf{b} = \mathbf{m}$ given in Section 11.2d-i, it is clear that other less general tests about slopes can also be accommodated. For example, suppose for the data of Table 6.1 that one hypothesized H: $b_2 = b_3$, i.e., that the model equations, instead of being (7), are

$$\begin{bmatrix} 74 \\ 68 \\ 77 \\ 76 \\ 80 \\ 87 \\ 91 \end{bmatrix} = \begin{bmatrix} 1 & 1 & \cdot & \cdot \\ 1 & 1 & \cdot & \cdot \\ 1 & 1 & \cdot & \cdot \\ 1 & \cdot & 1 & \cdot \\ 1 & \cdot & 1 & \cdot \\ 1 & \cdot & \cdot & 1 \\ 1 & \cdot & \cdot & 1 \end{bmatrix} \begin{bmatrix} \mu \\ \alpha_1 \\ \alpha_2 \\ \alpha_3 \end{bmatrix} + \begin{bmatrix} 3 & \cdot \\ 4 & \cdot \\ 5 & \cdot \\ \cdot & 2 \\ \cdot & 4 \\ \cdot & 3 \\ \cdot & 7 \end{bmatrix} \begin{bmatrix} b_1 \\ b_2 \end{bmatrix} + \mathbf{e} .$$

In calculating the error sum of squares after fitting this model and calling it SSE_H, the F-statistic for testing H: $b_2 = b_3$ is then $(\text{SSE}_\text{H} - \text{SSE})/\hat{\sigma}^2$.

An alternative is to write H: $b_2 = b$ as H: $\mathbf{k}'\mathbf{b} = m$ for $m = 0$ and $\mathbf{k}' = [0 \quad 1 \quad -1]$, and then use (A) of Section 11.2d-i, which gives

$$F = \frac{(\mathbf{k}'\hat{\mathbf{b}})^2}{\mathbf{k}'(\mathbf{R}'\mathbf{R})^{-1}\mathbf{k}\hat{\sigma}^2} = \frac{\left(\hat{b}_2 - \hat{b}_3\right)}{\left(1/W_{2,zz} + 1/W_{3,zz}\right)\hat{\sigma}^2} .$$

Extensions to other hypotheses about elements of **b** follow similarly.

11.3. CONFIRMING ASSOCIATED HYPOTHESES OF CHAPTER 6

Section 8.6 contains results in terms of the model

$$E(\mathbf{y}) = \mathbf{X}_1\boldsymbol{\beta}_1 + \mathbf{X}_2\boldsymbol{\beta}_2 + \mathbf{X}_3\boldsymbol{\beta}_3 \tag{36}$$

that are pertinent to the conditions stated in Section 6.1g for $Q/r\hat{\sigma}^2$ to be an F-statistic. Summarized as follows, these results (except the first) also depend on $\mathbf{y}$ being normally distributed with dispersion matrix $\sigma^2\mathbf{I}$.

(i) Any $R(\boldsymbol{\beta}_1)$ or $R(\boldsymbol{\beta}_2 | \boldsymbol{\beta}_1)$ is a sum of squares (Section 8.6b-i).

(ii) Any $R(\boldsymbol{\beta}_1)$ or $R(\boldsymbol{\beta}_2 | \boldsymbol{\beta}_1)$, when divided by σ^2, has, under a specifiable hypothesis and with specifiable degrees of freedom, a central χ^2-distribution (Section 8.6b-ii).

(iii) Each of $R(\boldsymbol{\beta}_1)$, $R(\boldsymbol{\beta}_2 | \boldsymbol{\beta}_1)$ and $R(\boldsymbol{\beta}_3 | \boldsymbol{\beta}_1, \boldsymbol{\beta}_2)$ is independent of $\text{SSE} = \mathbf{y}'\mathbf{y} - \mathbf{y}'\mathbf{X}(\mathbf{X}'\mathbf{X})^-\mathbf{X}'\mathbf{y}$ (Section 8.6b-iii); and they are independent of each other (Section 8.6b-iv).

(iv) With $\hat{\sigma}^2 = \text{SSE}/(N - r_{\mathbf{X}})$ for SSE defined in (iii), the density of $\text{SSE}/\hat{\sigma}^2$ is a central $\chi^2_{n-r_{\mathbf{X}}}$, (Section 8.6a).

Results (i)–(iv) correspond precisely to requirements (i)–(iv) of Section 6.1g. The final result needed is:

(v) The associated hypothesis for $R(\boldsymbol{\beta}_2 | \boldsymbol{\beta}_1)$ in the model (36) is

$$\text{H: } \mathbf{M}_1\mathbf{X}_2\boldsymbol{\beta}_2 + \mathbf{M}_1\mathbf{X}_2(\mathbf{X}_2'\mathbf{M}_1\mathbf{X}_2)^-\mathbf{X}_2'\mathbf{M}_1\mathbf{X}_3\boldsymbol{\beta}_3 = \mathbf{0}; \tag{37}$$

$$\mathbf{M}_1 = \mathbf{I} - \mathbf{X}_1(\mathbf{X}_1'\mathbf{X}_1)^-\mathbf{X}_1' \,. \tag{38}$$

Hypothesis (37), which is (123) of Section 8.6c, is also the specifiable hypothesis referred to in (ii).

With the five preceding results in mind, all that has to be done to establish the legitimacy of the F-statistics in Tables 6.4, 6.6 and 6.7 is to show that every sum of squares in the partitioning of $\Sigma_i\Sigma_j y_{ij}^2$ given in those tables either has the form $R(\boldsymbol{\beta}_1)$ or $R(\boldsymbol{\beta}_2 | \boldsymbol{\beta}_1)$, or is an SSE. This is easy, as is shown in Tables 11.6, 11.7 and 11.8. Then (37) or its special cases shown in Table 8.5 can be used to establish the hypotheses shown in Tables 6.5, 6.6 and 6.7. In some cases this is a little tedious; details of three such derivations are shown in Sections 11.3a-ii and 11.3b-iii that follow. Others are left to the reader in exercises E11.1 and E11.2.

a. The single slope model: Table 6.4

-i. Sums of squares. The results just summarized in (i)–(v) from Section 8.6 involve $R(\boldsymbol{\beta}_1)$ and

$$R(\boldsymbol{\beta}_2 | \boldsymbol{\beta}_1) = R(\boldsymbol{\beta}_1, \boldsymbol{\beta}_2) - R(\boldsymbol{\beta}_1), \tag{39}$$

the latter rather than

$$\mathscr{R}(\boldsymbol{\beta}_2 | \boldsymbol{\beta}_1) = R(\boldsymbol{\beta}_2) - R(\boldsymbol{\beta}_1) . \tag{40}$$

But terms in Table 6.4 are in the form of (40); e.g.,

$$\mathscr{R}(\mu_i | \mu) = R(\mu_i) - R(\mu) . \tag{41}$$

In order to apply (i)–(v) to (41), it is therefore necessary to reformulate $E(y_{ij}) = \mu_i + bz_{ij}$ in a form that permits (41) to be rewritten in the manner of (39). This can be done by using the overparameterized model

$$E(y_{ij}) = \mu + \alpha_i + bz_{ij} .$$

Then (41) is

$$\mathscr{R}(\mu_i | \mu) = \mathscr{R}(\mu, \alpha_i | \mu) = R(\mu, \alpha_i) - R(\mu) = R(\alpha_i | \mu), \tag{42}$$

to which results (i)–(v) are directly applicable.

The reformulation of an $\mathscr{R}(\cdot\,|\,\cdot)$-term as an $R(\cdot\,|\,\cdot)$, as illustrated in (42), is all that has to be done for each of the $\mathscr{R}(\cdot\,|\,\cdot)$-terms in Tables 6.4, 6.6 and 6.7. Such reformulations for Table 6.4 are shown in Table 11.6 using the overparameterized form of (28), namely

$$E(\mathbf{y}) = \mu \mathbf{1}_N + \{_d\, \mathbf{1}_{n_i}\} \boldsymbol{\alpha} + b\mathbf{z} . \tag{43}$$

-ii. Associated hypotheses. The associated hypothesis for line 3 of Tables 6.4 and 11.6 is derived in Section 8.6f-ii. That for $R(\boldsymbol{\alpha} | \mu, b)$ of line 9 is now obtained, as illustration of using the general hypothesis associated with $R(\boldsymbol{\beta}_2 | \boldsymbol{\beta}_1)$ given in (37). To do this we first define $\boldsymbol{\beta}_1$ and $\boldsymbol{\beta}_2$ from

$$R(\boldsymbol{\beta}_2 | \boldsymbol{\beta}_1) \equiv R(\boldsymbol{\alpha} | \mu, b) : \tag{44}$$

$$\boldsymbol{\beta} \equiv [\mu \quad b]' \quad \text{and} \quad \boldsymbol{\beta}_2 \equiv \boldsymbol{\alpha} . \tag{45}$$

Then, since $\boldsymbol{\beta}_1$ and $\boldsymbol{\beta}_2$ of (45) constitute all the parameters in (44), $\boldsymbol{\beta}_3 = \mathbf{0}$.

TABLE 11.6. CORRESPONDENCE BETWEEN THE SUMS OF SQUARES OF TABLE 6.4 AND THOSE OF THE EQUIVALENT, PARTITIONED MODEL

$$E(\mathbf{y}) = \mu\mathbf{1}_N + \{_d\,\mathbf{1}_{n_i}\}\boldsymbol{\alpha} + b\mathbf{z}$$

Line	Source of Variation	Sums of Squares	
		Table 6.4	Partitioned Model
	(a) Fitting classes and then covariate		
1	Classes	$R(\mu_i)$	$R(\mu, \boldsymbol{\alpha})$
2	Mean	$R(\mu)$	$R(\mu)$
3	Classes (adjusted for mean)	$\mathscr{R}(\mu_i \mid \mu)$	$R(\boldsymbol{\alpha} \mid \mu)$
4	Covariate (adjusted for classes)	$\mathscr{R}(\mu_i, b \mid \mu_i)$	$R(b \mid \mu, \boldsymbol{\alpha})$
5	Residual	$W_{yy} - \mathscr{R}(\mu_i, b \mid \mu_i)$	$\mathbf{y}'\mathbf{y} - R(\mu, \boldsymbol{\alpha}, b)$
6	Total	$\Sigma_i\Sigma_j y_{ij}^2$	$\mathbf{y}'\mathbf{y}$
	(b) Fitting covariate and then classes		
7	Mean	$R(\mu)$	$R(\mu)$
8	Covariate (adjusted for mean)	$\mathscr{R}(\mu, b \mid \mu)$	$R(b \mid \mu)$
9	Classes (adjusted for covariate)	$\mathscr{R}(\mu_i \mid \mu, b)$	$R(\boldsymbol{\alpha} \mid \mu, b)$
10	Residual	$W_{yy} - \mathscr{R}(\mu_i b \mid \mu_i)$	$\mathbf{y}'\mathbf{y} - R(\mu, \boldsymbol{\alpha}, b)$
11	Total	$\Sigma_i\Sigma_j y_{ij}^2$	$\mathbf{y}'\mathbf{y}$

Therefore

$$\mathbf{X}_1 = [\mathbf{1}_N \quad \mathbf{z}], \qquad \mathbf{X}_2 = \left\{_d\,\mathbf{1}_{n_i}\right\} \quad \text{and} \quad \mathbf{X}_3 = \mathbf{0}\ . \tag{46}$$

Hence for (44) the hypothesis of (37) is

$$\text{H: } \mathbf{M}_1\mathbf{X}_2\boldsymbol{\alpha} = \mathbf{0}\ . \tag{47}$$

This consists of N statements about the elements of $\boldsymbol{\alpha}$; clearly, they need to be reduced to $a - 1$ linearly independent statements. To make this simplification we use the following lemma and properties of $\mathbf{M}_1 = \mathbf{I} - \mathbf{X}_1(\mathbf{X}_1)^-\mathbf{X}_1'$ and $\mathbf{X}_2$, for $\mathbf{X}_1$ and $\mathbf{X}_2$ of (46).

Lemma. For any $\mathbf{A}$ such that $\mathbf{A1} = \mathbf{0}$, a solution for $\mathbf{x}$ to equations $\mathbf{Ax} = \mathbf{0}$ is $\mathbf{x} = \lambda\mathbf{1}$ for any λ; i.e., a solution is that the elements of $\mathbf{x}$ are all equal.

Proof.

$$\mathbf{x} = \lambda\mathbf{1} \quad \text{gives} \quad \mathbf{Ax} = \mathbf{A}\lambda\mathbf{1} = \lambda\mathbf{A1} = \lambda\mathbf{0} = \mathbf{0}. \qquad \textbf{Q.E.D.}$$

To simplify (47) observe, similar to (13), that $\mathbf{M}_1\mathbf{X}_1 = \mathbf{0}$. Therefore $\mathbf{X}_1$ of (46) gives $\mathbf{M}_1\mathbf{1}_N = \mathbf{0}$. But with $\mathbf{X}_2$ of (46) we have $\mathbf{X}_2\mathbf{1}_a = \mathbf{1}_N$. Therefore $\mathbf{M}_1\mathbf{X}_2\mathbf{1}_a = \mathbf{M}_1\mathbf{1}_N = \mathbf{0}$ and so the lemma applies to the hypothesis equations $\mathbf{M}_1\mathbf{X}_2\boldsymbol{\alpha} = \mathbf{0}$ that are in (47). Therefore that hypothesis is equivalent to H: α_i *all equal* or, for $\mu_i = \mu + \alpha_i$. H: μ_i *all equal*, as in Table 6.4.

Example. Suppose $a = 2$, with two observations in one class with $z_{11} = 1$ and $z_{12} = 2$, and one observation in the second class with $z_{21} = 3$. Then $\mathbf{M}_1$ for $\mathbf{X}_1$ of (46) is

$$\mathbf{M}_1 = \mathbf{I} - \begin{bmatrix} 1 & 1 \\ 1 & 2 \\ 1 & 3 \end{bmatrix} \begin{bmatrix} 3 & 6 \\ 6 & 14 \end{bmatrix}^{-1} \begin{bmatrix} 1 & 1 & 1 \\ 1 & 2 & 3 \end{bmatrix} = \frac{1}{6} \begin{bmatrix} 1 & -2 & 1 \\ -2 & 4 & -2 \\ 1 & -2 & 1 \end{bmatrix}$$

and (47) is

$$\text{H:} \ \frac{1}{6} \begin{bmatrix} 1 & -2 & 1 \\ -2 & 4 & -2 \\ 1 & -2 & 1 \end{bmatrix} \begin{bmatrix} 1 & 0 \\ 1 & 0 \\ 0 & 1 \end{bmatrix} \begin{bmatrix} \alpha_1 \\ \alpha_2 \end{bmatrix} = \mathbf{0},$$

i.e.,

$$\text{H:} \ \begin{bmatrix} -1 & 1 \\ 2 & -2 \\ -1 & 1 \end{bmatrix} \begin{bmatrix} \alpha_1 \\ \alpha_2 \end{bmatrix} = \mathbf{0},$$

which is equivalent to H: $\alpha_1 = \alpha_2$, i.e., H: $\mu_1 = \mu_2$.

b. The multiple slopes model: Tables 6.6 and 6.7

-i. *Models.* The intra-class regression model of Tables 6.6 and 6.7 is

$$E(y_{ij}) = \mu_i + b_i z_{ij} \,. \tag{48}$$

Using the same overparameterization of μ_i as in Table 11.6, namely $\mu_i = \mu + \alpha_i$, together with $\mathbf{z}_i^*$ used in (34), the matrix–vector form of (48) is

$$E(\mathbf{y}) = \mu\mathbf{1}_N + \{_d\,\mathbf{1}_{n_i}\}\boldsymbol{\alpha} + \{_d\,\mathbf{z}_1^*\}\mathbf{b} \,. \tag{49}$$

This is shown in Table 11.7.

TABLE 11.7. CORRESPONDENCE BETWEEN THE SUMS OF SQUARES OF TABLE 6.6 AND THOSE OF THE EQUIVALENT, PARTITIONED MODELS

$$E(\mathbf{y}) = \mu\mathbf{1}_N + \{_d\ \mathbf{1}_{n_i}\}\boldsymbol{\alpha} + \{_d\ \mathbf{z}_i^*\}b$$

$$E(\mathbf{y}) = \mu\mathbf{1}_N + \{_d\ \mathbf{1}_{n_i}\}\boldsymbol{\alpha} + [\mathbf{z} \quad \{_d\ \mathbf{z}_i^*\}]\begin{bmatrix}\mathbf{b}\\ \mathbf{c}\end{bmatrix}$$

Line	Source of Variation	Sum of Squares: Table 6.6	Sum of Squares: Partitioned Model
	(a) Fitting classes and then covariate		
1	Classes	$R(\mu_i)$	$R(\mu, \boldsymbol{\alpha})$
2	Mean	$R(\mu)$	$R(\mu)$
3	Classes, adjusted for mean	$\mathscr{R}(\mu_i \mid \mu)$	$R(\boldsymbol{\alpha} \mid \mu)$
4	Covariate, adjusted for classes	$\mathscr{R}(\mu_i, b_i \mid \mu_i)$	$R(b, \mathbf{c} \mid \mu, \boldsymbol{\alpha})$
5	Single slope	$\mathscr{R}(\mu_i, b \mid \mu_i)$	$R(b \mid \mu, \boldsymbol{\alpha})$
6	Deviations	$\mathscr{R}(\mu_i, b_i \mid \mu_i, b)$	$R(\mathbf{c} \mid \mu, \boldsymbol{\alpha}, b)$
7	Residual	$W_{yy} - \Sigma_i \hat{b}_i W_{i,yz}$	$\mathbf{y}'\mathbf{y} - R(\mu, \alpha, b, \mathbf{c})$
8	Total	$\Sigma_i \Sigma_j y_{ij}^2$	$\mathbf{y}'\mathbf{y}$
	(b) Fitting covariate and then classes		
9	Mean and covariate	$R(\mu, b_i)$	$R(\mu, \mathbf{b})$
10	Classes, adjusted for covariate	$\mathscr{R}(\mu_i, b_i \mid \mu, b_i)$	$R(\boldsymbol{\alpha} \mid \mu, \mathbf{b})$
11	Residual	$W_{yy} - \mathscr{R}(\mu_i, b_i \mid \mu_i)$	$\mathbf{y}'\mathbf{y} - R(\mu, \boldsymbol{\alpha}, \mathbf{b})$
12	Total	$\Sigma_i \Sigma_i y_{ij}^2$	$\mathbf{y}'\mathbf{y}$

To compare (49) with the single-slope model (4) of Tables 6.4 and 11.6, wherein the covariable occurs in the form $b\mathbf{z}$, we overparameterize each slope parameter b_i of (49) in the same way as $\mu + \alpha_i$ in (49) is an overparameterization of μ_i of (48). Thus in (48) we write

$$b_i = b + c_i \tag{50}$$

for $i = 1, \ldots, a$. Then with (49) being

$$E(\mathbf{y}) = \mu\mathbf{1}_N + \{_d\ \mathbf{1}_{n_i}\}\boldsymbol{\alpha} + [\mathbf{z} \quad \{_d\ \mathbf{z}_i^*\}]\begin{bmatrix}b\\ c\end{bmatrix}, \tag{51}$$

$$\mathbf{c} = [c_1 \quad c_2 \quad \cdots \quad c_a]' . \tag{52}$$

TABLE 11.8. CORRESPONDENCE BETWEEN THE SUMS OF SQUARES OF TABLE 6.7 AND THOSE OF THE EQUIVALENT PARTITIONED MODELS OF TABLE 11.7

		Sum of Squares	
Line	Source of Variation	Table 6.7	Partitioned Model
	(a) Partitioning to account for $R(\mu)$		
1	Mean	$R(\mu)$	$R(\mu)$
2	Covariate	$\mathscr{R}(\mu, b_i \mid \mu)$	$R(\mathbf{b} \mid \mu)$
3	Single slope	$\mathscr{R}(\mu, b \mid \mu)$	$R(b \mid \mu)$
4	Deviations	$\mathscr{R}(\mu, b_i \mid \mu, b)$	$R(\mathbf{c} \mid \mu, b)$
5	Line 9 of Table 11.7	$R(\mu, b_i)$	$R(\mu, \mathbf{b})$
	(b) Partitioning to account for $R(\mathbf{b})$		
6	Slopes	$R(b_i)$	$R(\mathbf{b})$
7	Intercept, adjusted for slopes	$\mathscr{R}(\mu, b_i \mid b_i)$	$R(\mu \mid \mathbf{b})$
8	Line 9 of Table 11.7	$R(\mu, b_i)$	$R(\mu, \mathbf{b})$

*-ii. **Sums of squares.*** Comparing the multi-slope model (51) with the single slope model (43) can then be achieved using

$$\begin{aligned}\mathscr{R}(\mu_i, \mathbf{b} \mid \mu_i, b) &= \mathscr{R}(\mu, \boldsymbol{\alpha}, \mathbf{b} \mid \mu, \boldsymbol{\alpha}, b)\\ &= R(\mu, \boldsymbol{\alpha}, \mathbf{b}) - R(\mu, \boldsymbol{\alpha}, b)\\ &= R(\mu, \boldsymbol{\alpha}, b, \mathbf{c}) - R(\mu, \boldsymbol{\alpha}, b)\\ &= R(\mathbf{c} \mid \mu, \boldsymbol{\alpha}, b)\end{aligned}$$

as in line 6 of Table 11.7. Lines 4 and 5 of that table and lines 3 and 4 of Table 11.8 are established in similar manner, based on (51).

*-iii. **Associated hypotheses.*** To derive associated hypotheses based on (51) write it as

$$E(\mathbf{y}) = \mu \mathbf{1}_N + \{_d\, \mathbf{1}_{n_i}\}\boldsymbol{\alpha} + b\mathbf{z} + \{_d\, \mathbf{z}_i^*\}\mathbf{c}\ . \tag{53}$$

Then, for example, to use (37) for deriving the hypothesis associated with

$R(\mathbf{c} \mid \mu, \boldsymbol{\alpha}, \mathbf{b})$ of line 6 of Table 11.7 define

$$R(\boldsymbol{\beta}_2 \mid \boldsymbol{\beta}_1) \equiv R(\mathbf{c} \mid \mu, \boldsymbol{\alpha}, b)$$

so that with

$$\boldsymbol{\beta}_1 = [\mu \quad \boldsymbol{\alpha}' \quad b]', \qquad \boldsymbol{\beta}_2 = \mathbf{c} \quad \text{and} \quad \boldsymbol{\beta}_3 = \mathbf{0},$$

$$\mathbf{X}_1 = \left[\mathbf{1}_N \quad \{_d\, \mathbf{1}_{n_i}\} \quad \mathbf{z}\right], \qquad \mathbf{X}_2 = \{_d\, \mathbf{z}_i^*\} \quad \text{and} \quad \mathbf{X}_3 = \mathbf{0}\ .$$

Then (37) is

$$\text{H: } \mathbf{M}_1\mathbf{X}_2\mathbf{c} = \mathbf{0}\ . \tag{54}$$

Again $\mathbf{M}_1\mathbf{X}_1 = \mathbf{0}$, and with $\mathbf{z}$ being a column of $\mathbf{X}_1$ this implies $\mathbf{M}_1\mathbf{z} = \mathbf{0}$. But $\mathbf{X}_2\mathbf{1}_a = \mathbf{z}$ and so $\mathbf{M}_1\mathbf{X}_1\mathbf{1}_a = \mathbf{M}_1\mathbf{z} = \mathbf{0}$ and therefore the lemma applies to the equations in (54). Hence that hypothesis, namely that associated with $R(\mathbf{c} \mid \mu, \boldsymbol{\alpha}, b)$, is H: c_i *all equal*, i.e., H: b_i *all equal*, as shown in Table 6.6.

To derive the hypothesis for line 5 of Tables 6.6 and 11.7 define

$$R(\boldsymbol{\beta}_2 \mid \boldsymbol{\beta}_1) \equiv R(b \mid \mu, \boldsymbol{\alpha})$$

so that with $\boldsymbol{\beta}_1 = [\mu \quad \boldsymbol{\alpha}']'$, $\quad \boldsymbol{\beta}_2 = b$ and $\boldsymbol{\beta}_3 = c$,

$$\mathbf{X}_1 = \left[\mathbf{1}_N \quad \{_d\, \mathbf{1}_{n_i}\}\right], \qquad \mathbf{X}_2 = \mathbf{z} \quad \text{and} \quad \mathbf{X}_3 = \{_d\, \mathbf{z}_i^*\}\ . \tag{55}$$

With this notation the hypothesis from (37) is

$$\text{H: } \mathbf{M}_1\mathbf{X}_2 b + \mathbf{M}_1\mathbf{X}_2(\mathbf{X}_2'\mathbf{M}_1\mathbf{X}_2)^-\mathbf{X}_2'\mathbf{M}_1\mathbf{X}_3\mathbf{c} = \mathbf{0}, \tag{56}$$

for which

$$\mathbf{M}_1\mathbf{X}_2 = \left[\mathbf{I} - \mathbf{X}_1(\mathbf{X}_1'\mathbf{X}_1)^-\mathbf{X}_1'\right]\mathbf{z}\ . \tag{57}$$

Because $\mathbf{X}_1$ in (55) is precisely the $\mathbf{X}$-matrix in the model equation $E(\mathbf{y}) = \mathbf{X}[\mu \quad \boldsymbol{\alpha}']'$ of the 1-way classification without covariables, we know that for that model

$$\mathbf{M}_1\mathbf{y} = \mathbf{y} - \hat{\mathbf{y}} = \{_c\, y_{ij} - \bar{y}_{i\cdot}\}_{i=1}^{i=a}\ {}_{j=1}^{j=n_i}\ .$$

Therefore, using $\mathbf{z}_i^*$ defined following (34), we have (57) as

$$\mathbf{M}_1\mathbf{X}_2 = \{_c\, \mathbf{z}_i^* - \bar{z}_{i\cdot}\mathbf{1}_{n_i}\}_{i=1}^{i=a}\ . \tag{58}$$

Hence (56) becomes

$$\text{H: } \mathbf{M}_1\mathbf{X}_2\left[b + \frac{\Sigma_i\Sigma_j(z_{ij} - \bar{z}_{i\cdot})c_i z_{ij}}{\Sigma_i\Sigma_j(z_{ij} - \bar{z}_{i\cdot})^2}\right] = \mathbf{0}\ . \tag{59}$$

Since (58) shows that $\mathbf{M}_1\mathbf{X}_2$ is a vector, (59) is equivalent to the contents of its square brackets being zero; and using W_{zz} and $W_{i,zz}$ of (30) and (77) in Sections 6.1d and 6.2b (with $W_{zz} = \Sigma_i W_{i,zz}$) reduces (59) to

$$\text{H: } b + \frac{\Sigma_i c_i W_{i,zz}}{W_{zz}} = 0$$

or

$$\text{H: } \Sigma_i (b + c_i) W_{i,zz} = 0$$

since $W_{zz} \neq 0$; and this in turn is equivalent to

$$\text{H: } \Sigma_i b_i W_{i,zz} = 0 \ . \tag{60}$$

Thus is derived the hypothesis associated with $R(b \mid \mu, \boldsymbol{\alpha})$, as shown in line 5 of Table 6.6. The other associated hypotheses shown in Tables 6.6 and 6.7 can be derived in similar manner (see exercises E11.7 and E11.8). Certain sums of squares in those tables are shown without associated hypotheses. The reader might like trying to derive them (see exercise E11.3).

c. Usefulness of the hypotheses

It is to be emphasized that not all hypotheses derived from (37) in the manner illustrated will necessarily be useful. For example, that in (60) is unlikely to be of any general use whatever. What we are doing here is not setting up an interesting hypothesis and testing it; that is the useful thing to do. What we are doing is looking at a sum of squares that can be (and in some cases is) produced by a computing package which, by the label that that package attaches to it, might sound as if it were a useful sum of squares, i.e., useful in the sense of testing an interesting hypothesis. What the application of (37) does is to derive the exact form of that hypothesis. Then one can reliably judge how useful it is, rather than simply inferring that it is useful from some nice-sounding computer-output label, for that label may well be misleading. For instance, suppose the label attached to $R(b \mid \mu, \boldsymbol{\alpha})$ was "pooled slope," or "single slope" or some such. One could be very easily tempted to infer that the associated hypothesis was H: $b = 0$ (which indeed is often an interesting hypothesis). But (60) clearly shows that the correct associated hypothesis is something quite different—and something that is largely of dubious value. When this is, in fact, the situation, it is useful that we have a mechanism for elucidating it; that mechanism is (37).

11.4. THE 1-WAY CLASSIFICATION: RESTRICTED OVERPARAMETERIZED MODELS

All the preceding description of using covariables has utilized each one just as it is observed; for example, in the 1-way classification of Chapter 6 and Section 11.2e, the covariate is represented in the model as bz_{ij}. But some writers use it in the form $b(z_{ij} - \bar{z}_{..})$, i.e., expressing each covariate as a deviation from its observed overall mean, so that the corresponding model is

$$E(y_{ij}) = \mu_i + b(z_{ij} - \bar{z}_{..}) \; . \tag{61}$$

This involves the conceptual inconsistency of the population term $E(y_{ij})$ on the left-hand side of (61) being defined in terms of the sample mean $\bar{z}_{..}$ on the right-hand side of (61). Nevertheless, this has no effect on the estimation of either μ_i as $\tilde{\mu}_i = \bar{y}_{i\cdot}$ or of b as $\hat{b} = W_{yz}/W_{zz}$. But, in contrast, when $b(z_{ij} - \bar{z}_{..})$ is used in combination with an overparameterized modeling of μ_i as $\dot{\mu} + \dot{\alpha}_i$, along with the restriction $\Sigma\dot{\alpha}_i = 0$ or $\Sigma n_i\dot{\alpha}_i = 0$, it can lead to sums of squares "due to the mean" that differ both from $N\bar{y}^2$ and from sums of squares calculated using bz_{ij}.

The occurrence of non-standard sums of squares that can arise from having restrictions on terms of an overparameterized model has been

TABLE 11.9. SUMS OF SQUARES DUE TO THE MEAN (AND ASSOCIATED HYPOTHESES) IN TWO RESTRICTED OVERPARAMETERIZED FORMS OF THE 1-WAY CLASSIFICATION MODEL HAVING ONE COVARIATE (SINGLE SLOPE), WITH UNBALANCED DATA

Model for $E(y_{ij})$	Sum of Squares[1]	Associated Hypothesis in Models with $\mu_i = \dot{\mu} + \dot{\alpha}_i$
$\dot{\mu} + \dot{\alpha}_i$	$R(\dot{\mu} \mid \dot{\alpha})_\Sigma = \tilde{y}^2/h$	$H\colon \Sigma\mu_i/a = 0$
	$R(\dot{\mu} \mid \dot{\alpha})_W = N\bar{y}^2$	$H\colon \Sigma n_i\mu_i/N = 0$
$\dot{\mu} + \dot{\alpha}_i + b_z z_{ij}$	$R(\dot{\mu} \mid \dot{\alpha}, b_z)_\Sigma = \dfrac{(\tilde{y} - \hat{b}\tilde{z})^2}{h + \tilde{z}^2/S}$	$H\colon \Sigma\mu_i/a = 0$
	$R(\dot{\mu} \mid \dot{\alpha}, b_z)_W = \dfrac{(\bar{y} - \hat{b}\bar{z})^2}{1/N + \bar{z}^2/S}$	$H\colon \Sigma n_i\mu_i/N = 0$
$\dot{\mu} + \dot{\alpha}_i + b_\delta(z_{ij} - \bar{z})$	$R(\dot{\mu} \mid \dot{\alpha}, b_\delta)_\Sigma = \dfrac{\{\tilde{y} - \hat{b}(\tilde{z} - \bar{z})\}^2}{h + (\tilde{z} - \bar{z}^2/S)}$	$H\colon \Sigma\mu_i/a + b\bar{z} = 0$
	$R(\dot{\mu} \mid \dot{\alpha}, b_\delta)_W = N\bar{y}^2$	$H\colon \Sigma n_i\mu_i/N + b\bar{z} = 0$

[1] $N = \Sigma n_i$ and $h = \Sigma(1/n_i)/a^2$.
$\bar{y} = \bar{y}_{..} = \Sigma_i\Sigma_j y_{ij}/N$ and $\bar{z} = \bar{z}_{..} = \Sigma_i\Sigma_j z_{ij}/N$.
$\tilde{y} = \Sigma_i\bar{y}_{i\cdot}/a$ and $\tilde{z} = \Sigma_i\bar{z}_{i\cdot}/a$.
$b = W_{yz}/W_{zz} = \Sigma_i\Sigma_j(y_{ij} - \bar{y}_{i\cdot})(z_{ij} - \bar{z}_{i\cdot})/S$ and $S = W_{zz} = \Sigma_i\Sigma_j(z_{ij} - \bar{z}_{i\cdot})^2$.
Subscript on $R(\cdot)$ is for restrictions: Σ for $\Sigma_i\dot{\alpha}_i = 0$ and W for $\Sigma_i n_i\dot{\alpha}_i = 0$.

illustrated earlier for the no-covariable models of the 2-way classification in Sections 9.4a-iii and 9.4b-i. The latter shows the important connection of sums of squares from the Σ-restricted with-interaction 2-way model to those of the weighted squares of means analysis that test hypotheses of quite broad utility. It is therefore of interest to see the effect of having restrictions in the with-covariables model both when the covariate is treated as observed, z_{ij}, and when it is treated as the deviation from its observed mean, $z_{ij} - \bar{z}_{..}$.

To succinctly distinguish the two cases, subscripts z and δ are used on b to have the models as

$$E(y_{ij}) = \dot{\mu} + \dot{\alpha}_i + b_z z_{ij} \quad \text{and} \quad E(y_{ij}) = \dot{\mu} + \dot{\alpha}_i + b_\delta(z_{ij} - \bar{z}_{..}) \; . \qquad (62)$$

And for distinguishing between the two restricted models we use subscripts

$$\Sigma \text{ for } \Sigma_i\dot{\alpha}_i = 0 \quad \text{and} \quad W \text{ for } \Sigma_i n_i\dot{\alpha}_i = 0 \; . \qquad (63)$$

TABLE 11.10. SUMS OF SQUARES DUE TO THE MEAN (AND ASSOCIATED HYPOTHESES) IN TWO RESTRICTED OVERPARAMETERIZED FORMS OF THE 1-WAY CLASSIFICATION MODEL HAVING ONE COVARIATE (MULTIPLE SLOPES), WITH UNBALANCED DATA

Model for $E(y_{ij})$	Sum of Squares[1]	Associated Hypothesis in Models with $\mu_i = \dot{\mu} + \dot{\alpha}_i$
$\dot{\mu} + \dot{\alpha}_i + b_{i,z}z_{ij}$	$R(\mu \mid \alpha, b_z)_\Sigma = \dfrac{[\Sigma(\bar{y}_{i\cdot} - \hat{b}_i\bar{z}_{i\cdot})]^2}{\Sigma(t_i/n_i s_i)}$	H: $\Sigma\mu_i/a = 0$
	$R(\mu \mid \alpha, b_z)_W = \dfrac{(N\bar{y} - \Sigma\hat{b}_i z_{i\cdot})^2}{\Sigma(n_i t_i/s_i)}$	H: $\Sigma n_i\mu_i/N = 0$
$\dot{\mu} + \dot{\alpha}_i + b_{i,\delta}(z_{ij} - \bar{z})$	$R(\dot{\mu} \mid \dot{\alpha}, b_\delta)_\Sigma = \dfrac{\{\Sigma[\bar{y}_{i\cdot} - \hat{b}_i(\bar{z}_{i\cdot} - \bar{z})]\}^2}{\Sigma[1/n_i + (\bar{z}_{i\cdot} - \bar{z})^2/s_i]}$	H: $\dfrac{\Sigma\mu_i}{a} + \bar{z}\dfrac{\Sigma b_i}{a} = 0$
	$R(\mu \mid \alpha, b_\delta)_W = \dfrac{[N\bar{y} - \Sigma n_i\hat{b}_i(\bar{z}_{i\cdot} - \bar{z})]^2}{N + \Sigma[n_i^2(\bar{z}_{i\cdot} - \bar{z})^2/s_i]}$	H: $\dfrac{\Sigma n_i\mu_i}{N} + \bar{z}\dfrac{\Sigma n_i b_i}{N} = 0$

[1] $t_i = \Sigma_j z_{ij}^2$ and $s_i = W_{i,zz} = \Sigma_j(z_{ij} - \bar{z}_{i\cdot})^2$.
$\hat{b}_i = W_{i,yz}/W_{i,zz} = \Sigma_j(y_{ij} - \bar{y}_{i\cdot})(z_{ij} - \bar{z}_{i\cdot})/s_i$.
Subscript on $R(\cdot)$ is for restrictions: Σ for $\Sigma_i\dot{\alpha}_i = 0$ and W for $\Sigma n_i\dot{\alpha}_i = 0$.

For example, $R(\dot{\mu} \mid \dot{\alpha}, b_z)_\Sigma$ is the reduction in sum of squares due to $\dot{\mu}$ adjusted for $\dot{\alpha}$ in the first model shown in (62) when using the $\Sigma\dot{\alpha}_i = 0$ restriction. With this notation, Table 11.9 (adapted from Searle and Hudson, 1982) shows sums of squares due to the mean for the two models in (62) using the two different restrictions of (63). Also shown, for the sake of completeness, are comparable sums of squares for the no-covariate model. Table 11.10 shows the extension of Table 11.9 to the multiple slopes model.

The most noticeable feature of Tables 11.9 and 11.10 is that only two of the sums of squares therein are $N\bar{y}^2$. For balanced data (wherein $h = 1/N$ and $\tilde{y} = \bar{y}$ and $\tilde{z} = \bar{z}$) the Σ and W restrictions are the same, as are the corresponding sums of squares and hypotheses; in Table 11.9 those sums of squares are $N\bar{y}^2$ except for the model using b_z, for which it is the value shown in the fourth line of Table 11.9. In Table 11.10 none of the sums of squares reduces to $N\bar{y}^2$.

11.5. THE 1-WAY CLASSIFICATION: TWO COVARIATES

There are many potential extensions of the examples in Section 11.2e which follow quite naturally from the details already given. With data available, computing estimates of μ_i and of slopes can be accomplished quite straightforwardly from the general procedures given in Sections 11.2a–11.2d. No matter how many covariables there are, with whatever mix of single slopes and intra-class slopes that is appropriate, all linear combinations of μ_is and of slopes are estimable. To illustrate this, some basic results are now given for three different ways of incorporating two covariates in a 1-way classification. Enough is given to enable the reader to develop details as desired; and it provides some answers to question VIII posed at the end of Section 11.1.

a. Basic tools and notation

The estimation features of the general development in Section 11.2 that are used here are as follows. For the model

$$E(\mathbf{y}) = \mathbf{X\beta} + \mathbf{Zb} \tag{64}$$

of (8) we have from

$$(19): \quad \boldsymbol{\beta}^\circ = (\mathbf{X'X})^- \mathbf{X'y} = \boldsymbol{\beta}^\circ[\mathbf{y}], \tag{65}$$

$$(22): \quad \mathbf{R} = \{_r\, \mathbf{z}_t - \hat{\mathbf{z}}_t\}_{t=1}^{t=c} = \{_r\, \mathbf{z}_t - \mathbf{X}\boldsymbol{\beta}^\circ[\mathbf{z}_t]\}_{t=1}^{t=c}, \tag{66}$$

$$(23): \quad \hat{\mathbf{b}} = (\mathbf{R'R})^{-1}\mathbf{R'y} = (\mathbf{R'R})^{-1}\mathbf{R'}(\mathbf{y} - \hat{\mathbf{y}}), \tag{67}$$

$$(20): \quad \tilde{\boldsymbol{\beta}} = \boldsymbol{\beta}^\circ[\mathbf{y}] - \boldsymbol{\beta}^\circ[\mathbf{Z}\hat{\mathbf{b}}]\,. \tag{68}$$

Armed with the form of these expressions in any particular case, one has all the basic results needed for deriving for that case the dispersion properties of the estimators given in (24)–(26), the analyses of variance of Tables 11.4 and 11.5, and the hypothesis testing procedures of Section 11.2d. We therefore concentrate in this and the next section on deriving **R′R**, **R′y** and (67) and (68) for three special cases.

Notation. In all the cases to be dealt with, no more than two covariates are considered. They are denoted by z_{ij} and w_{ij} with corresponding slopes being represented by b and b^+ (or $\mathbf{b}$ and $\mathbf{b}^+$ for vectors of slopes in multiple slope models).

Up to this point W_{yy} has been the symbol for the within-class sum of squares (W as mnemonic for "within"). This is now changed to E_{yy} (E for "error"), so that

$$E_{yy} = W_{yy} = \sum_{i=1}^{a} \sum_{j=1}^{n_i} (y_{ij} - \bar{y}_{i\cdot})^2 \ . \tag{69}$$

E_{pq} is defined similarly for each of p and q being any of the symbols y, z or w; likewise for B_{pq} and T_{pq} of Chapter 6, and also

$$E_{i,\,yy} = W_{i,\,yy} = \sum_{j=1}^{n_i} (y_{ij} - \bar{y}_{i\cdot})^2 \ . \tag{70}$$

Using these symbols we then see, for example, from Sections 11.2e-i and -ii that with a single covariate in the single slope model

$$\mathbf{R'R} = E_{zz} \quad \text{and} \quad \mathbf{R'y} = E_{yz} \tag{71}$$

as shown following (30); and in the multiple slopes model

$$\mathbf{R'R} = \left\{ {}_d\, E_{i,\,zz} \right\}_{i=1}^{i=a} \quad \text{and} \quad \mathbf{R'y} = \left\{ {}_c\, E_{i,\,yz} \right\}_{i=1}^{i=a} \tag{72}$$

as shown following (34) – with E in place of W as in (70).

b. Single slope for each covariate

The first model considered has just a single slope for each covariate:

$$E(y_{ij}) = \mu_i + bz_{ij} + b^+ w_{ij} \ . \tag{73}$$

Since (71) applies to the single slope model with just one covariate, the corresponding values for (73) are

$$\mathbf{R'R} = \begin{bmatrix} E_{zz} & E_{zw} \\ E_{wz} & E_{ww} \end{bmatrix} \quad \text{and} \quad \mathbf{R'y} = \begin{bmatrix} E_{yz} \\ E_{yw} \end{bmatrix}. \tag{74}$$

Hence

$$\begin{bmatrix} \hat{b} \\ \hat{b}^+ \end{bmatrix} = \begin{bmatrix} E_{zz} & E_{wz} \\ E_{wz} & E_{ww} \end{bmatrix}^{-1} \begin{bmatrix} E_{yz} \\ E_{yw} \end{bmatrix}. \tag{75}$$

There are many equivalent expressions for the solutions in (75). For example, using

$$\hat{b}_{y:z} = E_{yz}/E_{zz} \quad \text{and} \quad \hat{\rho}^2_{yz} = E^2_{yz}/E_{yy}E_{zz} \tag{76}$$

(75) yields

$$\hat{b} = \frac{E_{ww}E_{yz} - E_{wy}E_{wz}}{E_{ww}E_{zz} - E^2_{wz}} = \frac{\hat{b}_{y:z} - \hat{b}_{y:w}\hat{b}_{w:z}}{1 - \hat{b}_{z:w}\hat{b}_{w:z}} = \frac{(\hat{\rho}_{yz} - \hat{\rho}_{yw}\hat{\rho}_{zw})}{(1 - \hat{\rho}^2_{zw})}\sqrt{\frac{E_{yy}}{E_{zz}}}$$

and (77)

$$\hat{b}^+ = \frac{E_{zz}E_{wy} - E_{wz}E_{yz}}{E_{ww}E_{zz} - E^2_{wz}} = \frac{\hat{b}_{y:w} - \hat{b}_{y:z}\hat{b}_{z:w}}{1 - \hat{b}_{w:z}\hat{b}_{w:z}} = \frac{(\hat{\rho}_{yw} - \hat{\rho}_{yz}\hat{\rho}_{wz})}{(1 - \hat{\rho}^2_{zw})}\sqrt{\frac{E_{yy}}{E_{ww}}}.$$

Then since (65) for (73) is

$$\mu^\circ = \{_c\, \mu^\circ_i\} = \{_c\, \bar{y}_{i\cdot}\} = \mu^\circ[\mathbf{y}], \tag{78}$$

application of this to (68) gives

$$\tilde{\mu} = \{_c\, \tilde{\mu}_i\} = \mu^\circ[\mathbf{y}] - \mu^\circ\left[(\mathbf{z} \quad \mathbf{w})\begin{pmatrix} \hat{b} \\ \hat{b}^+ \end{pmatrix}\right], \tag{79}$$

where **z** and **w** are vectors of the z_{ij} and w_{ij} values, in lexicon order, akin to **y**. Thus (79) yields, for $i = 1, \ldots, a$,

$$\tilde{\mu}_i = \bar{y}_{i\cdot} - \hat{b}\bar{z}_{i\cdot} - \hat{b}^+\bar{w}_{i\cdot}\ .$$

The reduction in sum of squares for fitting the model is

$$R(\mu, b, b^+) = \Sigma_i n_i y_i^2 + \hat{b}E_{yz} + \hat{b}^+E_{yw}\ . \tag{80}$$

c. Intra-class slopes and a single slope

The second model to be considered has intra-class slopes for one covariable and a single slope for the other. The model is

$$E(y_{ij}) = \mu_i + b_i z_{ij} + b^+ w_{ij} \,. \tag{81}$$

Since **R′R** and **R′y** are given for the single slope model by (71) and for the multiple slopes model by (72), the corresponding values for (81) are

$$\mathbf{R'R} = \begin{bmatrix} \{_d\, E_{i,zz}\} & \{_c\, E_{i,zw}\} \\ \{_r\, E_{i,zw}\} & E_{ww} \end{bmatrix} \quad \text{and} \quad \mathbf{R'y} = \begin{bmatrix} \{_c\, E_{i,yz}\} \\ E_{yw} \end{bmatrix}. \tag{82}$$

To invert **R′R** invoke a result for the inverse of a partitioned matrix [e.g., Searle (1982), p. 260, equation (12)] that is similar to (38) of Section 8.3: for any nonsingular matrix **A**, vector **c** and scalar λ such that $\lambda \neq \mathbf{c}'\mathbf{A}^{-1}\mathbf{c}$, and with $\theta = 1/(\lambda - \mathbf{c}'\mathbf{A}^{-1}\mathbf{c})$,

$$\begin{bmatrix} \mathbf{A} & \mathbf{c} \\ \mathbf{c}' & \lambda \end{bmatrix}^{-1} = \begin{bmatrix} \mathbf{A}^{-1} + \theta\mathbf{A}^{-1}\mathbf{c}\mathbf{c}'\mathbf{A}^{-1} & -\theta\mathbf{A}^{-1}\mathbf{c} \\ -\theta\mathbf{c}'\mathbf{A}^{-1} & \theta \end{bmatrix}. \tag{83}$$

Applying this to **R′R** in (82) gives

$$\begin{bmatrix} \hat{\mathbf{b}} \\ \hat{b}^+ \end{bmatrix} = \begin{bmatrix} \{_d\, 1/E_{i,zz}\} + \theta\{_c\, \hat{b}_{i,w:z}\}\{_r\, \hat{b}_{i,w:z}\} & -\theta\{_c\, \hat{b}_{i,w:z}\} \\ -\theta\{_r\, \hat{b}_{i,w:z}\} & \theta \end{bmatrix} \times \begin{bmatrix} \{_c\, E_{i,yz}\} \\ E_{yw} \end{bmatrix} \tag{84}$$

with

$$\hat{b}_{i,w:z} = E_{i,wz}/E_{i,zz} \quad \text{and} \quad \hat{\rho}^2_{i,wz} = E^2_{i,wz}/E_{i,ww}E_{i,zz} \tag{85}$$

similar to (76), but for each $i = 1, \ldots, a$; and with

$$\theta = \frac{1}{E_{ww} - \Sigma_i\left(E^2_{i,wz}/E_{i,zz}\right)} \tag{86}$$

$$= \frac{1}{\Sigma_i E_{i,ww}\left(1 - \hat{b}_{i,w:z}\hat{b}_{i,z:w}\right)} = \frac{1}{\Sigma_i E_{i,ww}\left(1 - \hat{\rho}^2_{i,wz}\right)}\,. \tag{87}$$

Hence from (84)

$$\hat{b}^+ = \theta\left(E_{yw} - \Sigma_i \hat{b}_{i,\,w:\,z} E_{i,\,yz}\right) = \frac{E_{yw} - \Sigma_i\left(E_{i,\,yz}E_{i,\,wz}/E_{i,\,zz}\right)}{E_{ww} - \Sigma_i\left(E^2_{i,\,wz}/E_{i,\,zz}\right)} \tag{88}$$

$$= \frac{\Sigma_i E_{i,\,ww}\left(\hat{b}_{i,\,y:\,w} - \hat{b}_{i,\,y:\,z}\hat{b}_{i,\,z:\,w}\right)}{\Sigma_i E_{i,\,ww}\left(1 - \hat{b}_{i,\,w:\,z}\hat{b}_{i,\,z:\,w}\right)} = \frac{\Sigma_i\left(\rho_{i,\,yw} - \hat{\rho}_{i,\,yz}\hat{\rho}_{i,\,zw}\right)\sqrt{E_{i,\,ww}E_{i,\,yy}}}{\Sigma_i\left(1 - \rho^2_{i,\,wz}\right)E_{i,\,ww}}$$

and then from (84)

$$\hat{b}_i = E_{i,\,yz}/E_{i,\,zz} + \theta\hat{b}_{i,\,w:\,z}\left(\Sigma_i \hat{b}_{i,\,w:\,z}E_{i,\,yz} - E_{yw}\right)$$

$$= \hat{b}_{i,\,y:\,z} - \hat{b}^+\hat{b}_{i,\,w:\,z}, \quad \text{for} \quad i = 1, \ldots, a\ . \tag{89}$$

And, similar to (79)

$$\tilde{\mu} = \{_c\, \tilde{\mu}_i\} = \mu^{\circ}[\mathbf{y}] - \mu^{\circ}\left[\left(\{_r\, \mathbf{z}_i\} \quad \mathbf{w}\right)\begin{pmatrix}\hat{b}\\ \hat{b}^+\end{pmatrix}\right] \tag{90}$$

for $\mathbf{z}_i$ of, and defined following, (33). Therefore, based on (78), from (90)

$$\tilde{\mu}_i = \bar{y}_{i\cdot} - \hat{b}_i\bar{z}_{i\cdot} - \hat{b}^+\bar{w}_{i\cdot}\ . \tag{91}$$

Lastly, the reduction in sum of squares due to fitting (73) is found to be

$$R(\mu, \mathbf{b}, b^+) = \Sigma_i n_i \bar{y}^2_{i\cdot} + \Sigma_i \hat{b}_{i,\,y:\,z}E_{i,\,yz} + \hat{b}^+E_{yw}\ . \tag{92}$$

d. Intra-class slopes for each covariate

The third and final model considered for the 1-way classification is where both covariables have intra-class slopes:

$$E(y_{ij}) = \mu_i + b_i z_{ij} + b_i^+ w_{ij}\ . \tag{93}$$

Since going from (75) to (82) extends directly to going from (82) to $\mathbf{R}'\mathbf{R}$ and $\mathbf{R}'\mathbf{y}$ for (93), we have

$$\mathbf{R}'\mathbf{R} = \begin{bmatrix}\{_d\, E_{i,\,zz}\} & \{_d\, E_{i,\,zw}\}\\ \{_d\, E_{i,\,zw}\} & \{d\, E_{i,\,ww}\}\end{bmatrix} \quad \text{and} \quad \mathbf{R}'\mathbf{y} = \begin{bmatrix}\{_c\, E_{i,\,yz}\}\\ \{_c\, E_{i,\,yw}\}\end{bmatrix}. \tag{94}$$

The form of the inverse of a partitioned matrix (*loc. cit.*) that we now use is

$$\begin{bmatrix} \mathbf{D}_1 & \mathbf{D}_2 \\ \mathbf{D}_2 & \mathbf{D}_3 \end{bmatrix}^{-1} = \begin{bmatrix} \mathbf{D}_1^{-1} + \mathbf{D}_1^{-1}\mathbf{D}_2\mathbf{U}^{-1}\mathbf{D}_2\mathbf{D}_1^{-1} & -\mathbf{D}_1^{-1}\mathbf{D}_2\mathbf{U}^{-1} \\ -\mathbf{U}^{-1}\mathbf{D}_2\mathbf{D}_1^{-1} & \mathbf{U}^{-1} \end{bmatrix}, \quad (95)$$

where $\mathbf{D}_1$, $\mathbf{D}_2$ and $\mathbf{D}_3$ are nonsingular diagonal matrices all of the same order, as is $\mathbf{U}^{-1} = (\mathbf{D}_3 - \mathbf{D}_2\mathbf{D}_1^{-1}\mathbf{D}_2)^{-1}$ also. Applying (95) to $\mathbf{R}'\mathbf{R}$ of (94) gives

$$\mathbf{U}^{-1} = \left\{_d\, E_{i,ww} - E_{i,zw}^2/E_{i,zz}\right\}^{-1}$$

$$= \left\{_d\, 1/\left[E_{i,ww}\left(1 - \hat{b}_{i,z:w}\hat{b}_{i,w:z}\right)\right]\right\} = \left\{_d\, 1/\left[E_{i,ww}\left(1 - \hat{\rho}_{i,zw}^2\right)\right]\right\}. \quad (96)$$

Therefore, using (96) in (95) and relating (95) to (94), used in

$$\begin{bmatrix} \hat{\mathbf{b}} \\ \hat{\mathbf{b}}^+ \end{bmatrix} = (\mathbf{R}'\mathbf{R})^- \mathbf{R}'\mathbf{y},$$

gives

$$\hat{b}_i^+ = \frac{E_{i,yw} - E_{i,yz}E_{i,zw}/E_{i,zz}}{E_{i,ww} - E_{i,zw}^2/E_{i,zz}}$$

$$= \frac{\hat{b}_{i,y:w} - \hat{b}_{i,y:z}\hat{b}_{i,z:w}}{1 - \hat{b}_{i,w:z}\hat{b}_{i,z:w}} = \frac{\left(\hat{\rho}_{i,yw} - \hat{\rho}_{i,yz}\hat{\rho}_{i,zw}\right)}{1 - \hat{\rho}_{i,wz}^2}\sqrt{\frac{E_{i,yy}}{E_{i,ww}}}\,. \quad (97)$$

And by symmetry b_i is b_i^+ with z and w interchanged, so that

$$\hat{b}_i = \frac{E_{i,yz} - E_{i,yw}E_{i,wz}/E_{i,ww}}{E_{i,zz} - E_{i,wz}^2/E_{i,ww}}$$

$$= \frac{\hat{b}_{i,y:z} - \hat{b}_{i,y:w}\hat{b}_{i,w:z}}{1 - \hat{b}_{i,z:w}\hat{b}_{i,w:z}} = \frac{\hat{\rho}_{i,yz} - \hat{\rho}_{i,yw}\hat{\rho}_{i,wz}}{1 - \hat{\rho}_{i,wz}^2}\sqrt{\frac{E_{i,yy}}{E_{i,zz}}}\,. \quad (98)$$

It is left to the reader (E11.12) to show that direct application of (96) to (95) and (94) yields (98). And, similar to (79) and (90),

$$\tilde{\boldsymbol{\mu}} = \{_c\, \tilde{\mu}_i\} = \boldsymbol{\mu}^\circ[\mathbf{y}] - \boldsymbol{\mu}^\circ\left[\left(\{_r\, \mathbf{z}_i\} \quad \{_r\, \mathbf{w}_i\}\right)\begin{pmatrix} \hat{\mathbf{b}} \\ \hat{\mathbf{b}}^+ \end{pmatrix}\right]$$

for $\mathbf{w}_i$ defined in the same way as is $\mathbf{z}_i$—see (33). Therefore, just like (91)

$$\tilde{\mu}_i = \bar{y}_{i\cdot} - \hat{b}_i \bar{z}_{i\cdot} - \hat{b}_i^+ \bar{w}_{i\cdot} \tag{99}$$

and

$$R(\boldsymbol{\mu}, \mathbf{b}, \mathbf{b}^+) = \Sigma_i \left(n_{i\cdot} y_{i\cdot}^2 + \hat{b}_i E_{i,\,yz} + \hat{b}_i E_{i,\,yw} \right) . \tag{100}$$

11.6. THE 2-WAY CLASSIFICATION: SINGLE SLOPE MODELS FOR ONE COVARIATE

In this section and the next we give $\mathbf{R}'\mathbf{R}$ and $\mathbf{R}'\mathbf{y}$ for four particular cases of the 2-way classification, and in so doing provide a substantive answer to question VII at the end of Section 11.1.

a. The with-interaction model

The no-covariate case of the with-interaction model discussed in Chapters 4 and 5 has model equation

$$E(y_{ijk}) = \mu_{ij} \tag{101}$$

for $i = 1, \ldots, a,\ j = 1, \ldots, b$ and $k = 1, \ldots, n_{ij} \geq 0$. From equation (4) of Chapter 5 the BLUE of μ_{ij} is $\bar{y}_{ij\cdot}$. Hence

$$\mathbf{y} - \hat{\mathbf{y}} = \left\{ {}_c\, y_{ijk} - \bar{y}_{ij\cdot} \right\}_{i=1\ \ j=1\ \ k=1}^{i=a\ \ j=b\ \ k=n_{ij}} . \tag{102}$$

The sum of squares of elements of (102), to be denoted E_{yy}, is

$$E_{yy} = \Sigma_i \Sigma_j \Sigma_k (y_{ijk} - y_{ij\cdot})^2, \tag{103}$$

as seen in Table 4.8 as SSE.

The traditional with-covariate form of (101) is

$$E(y_{ijk}) = \mu_{ij} + \lambda z_{ijk}, \tag{104}$$

where the "slope" parameter is now denoted λ to avoid confusion with using b as the number of β_js in the model. The estimator of λ is

$$\hat{\lambda} = (\mathbf{R}'\mathbf{R})^{-1}\mathbf{R}'\mathbf{y}, \tag{105}$$

where $\mathbf{R} = \mathbf{z} - \hat{\mathbf{z}}$ in the manner of (102). Therefore, similar to (103),

$$\mathbf{R}'\mathbf{R} = E_{zz} \tag{106}$$

and

$$\mathbf{R}'\mathbf{y} = E_{yz} = \Sigma_i \Sigma_j \Sigma_k (y_{ijk} - \bar{y}_{ij\cdot})(z_{ijk} - \bar{z}_{ij\cdot}) . \tag{107}$$

Hence $\hat{\lambda} =$

$$(\mathbf{R'R})^{-1}\mathbf{R'Y} = \frac{E_{yz}}{E_{zz}} = \frac{E_{yz}}{E_{zz}} = \frac{\Sigma_i\Sigma_j\Sigma_k(y_{ijk} - \bar{y}_{ij\cdot})(z_{ijk} - \bar{z}_{ij\cdot})}{\Sigma_i\Sigma_j\Sigma_k(z_{ijk} - \bar{z}_{ij\cdot})^2} \ . \tag{108}$$

Thus $\hat{\mu} = \mu^{\circ}[\mathbf{y}] - \mu^{\circ}[\mathbf{z}\hat{\lambda}]$ yields

$$\hat{\mu}_{ij} = \bar{y}_{ij\cdot} - \bar{z}_{ij\cdot} \ . \tag{109}$$

A useful hypothesis in the no-covariate model (101) when all cells are filled is

$$\text{H: } \bar{\mu}_{i\cdot} \textit{ all equal} \ . \tag{110}$$

It is tested by SSA_w, the sum of squares due to the α-factor in the weighted squares of means analysis (see Section 4.6d). We here denote SSA_w by $\dot{A}_{yy}$, and from (42) of Chapter 4 have its value as

$$\dot{A}_{yy} = \Sigma_i w_i\left\{\Sigma_j\bar{y}_{ij\cdot}/b - \left[\Sigma_i w_i(\Sigma_j\bar{y}_{ij\cdot}/b)\right]/\Sigma w_i\right\}^2 \tag{111}$$

for w_i (which is not a covariate) defined by

$$1/w_i = (\Sigma_j 1/n_{ij})/b^2 \ .$$

The symbol $\dot{A}_{yy}$ is introduced so that analogous values $\dot{A}_{zz}$ and $\dot{A}_{yz}$ can be identified. We are then able to show (see exercise E11.13) that with the covariate model of (104) the hypothesis (110) is tested by the sum of squares

$$Q = \dot{A}_{yy} + \frac{E_{yz}^2}{E_{zz}} - \frac{(E_{yz} + \dot{A}_{yz})^2}{E_{zz} + \dot{A}_{zz}} \ . \tag{112}$$

A special case of (111) and (112) is balanced data, when $\dot{A}_{yy}$ of (111) has then simplified to the familiar form

$$\dot{A}_{yy} = bn\Sigma_i(\bar{y}_{i\cdot\cdot} - \bar{y}_{\cdot\cdot\cdot})^2 \tag{113}$$

with comparable expressions for $\dot{A}_{yz}$ and $\dot{A}_{zz}$.

b. The no-interaction model

Estimation in the no-interaction model

$$E(y_{ijk}) = \mu_i + \tau_j \tag{114}$$

of Section 4.9 has, as seen in equations (65) and (68) there, no simple expressions for $\hat{\mu}_i$ and/or $\hat{\tau}_j$ when data are unbalanced. Therefore, although the general principles (64)–(68) of the analysis with a covariate apply, and can easily be used computationally, they do not lead to any appealing formulae; at least not for unbalanced data. For example, from (82) of Chapter 4 the error sum of squares for the no-interaction no-covariate model is

$$E_{yy} = \Sigma_i \Sigma_j \Sigma_k y_{ijk}^2 - \Sigma_i n_{i\cdot} \bar{y}_{i\cdot\cdot}^2 - \sum_{j=1}^{b-1} \hat{\tau}_j r_j$$

for $\hat{\tau}_j$ and r_j given in (65)–(67) of Chapter 4. An equivalent form, obtainable from (122) and (123) of Chapter 9, is

$$E_{yy} = \Sigma_i \Sigma_j \Sigma_k y_{ijk}^2 - \Sigma_i n_{i\cdot} \bar{y}_{i\cdot\cdot}^2 - \mathbf{r}_2' \mathbf{C}_2^{-1} \mathbf{r}_2, \tag{115}$$

for $\mathbf{C}_2$ and $\mathbf{r}_2$ given by (91) and (97) of that chapter. To emphasize that (115) is a function of observations on the y-variable, redefine $\mathbf{r}_2$ as

$$\mathbf{r}_{2,y} = \{_c\, y_{\cdot j \cdot} - \Sigma_i n_{ij} \bar{y}_{i\cdot\cdot}\}_{j=1}^{j=b-1}$$

and define

$$T_{yy} = \Sigma_i \Sigma_j \Sigma_k (y_{ijk} - \bar{y}_{\cdots})^2 \quad \text{and} \quad A_{yy} = \Sigma_i n_{i\cdot} (\bar{y}_{i\cdot\cdot} - y_{\cdots})^2 \,. \tag{116}$$

Then (115) becomes $E_{yy} = T_{yy} - A_{yy} - \mathbf{r}_{2,y}' \mathbf{C}_2^{-1} \mathbf{r}_{2,y}$. On making comparable definitions

$$E_{yz} = T_{yz} - A_{yz} - \mathbf{r}_{2,y}' \mathbf{C}_2^{-1} \mathbf{r}_{2,z}, \quad E_{zz} = T_{zz} - A_{zz} - \mathbf{r}_{2,z}' \mathbf{C}_2^{-1} \mathbf{r}_{2,z} \tag{117}$$

the estimator of the slope parameter in the covariable model

$$E(y_{ijk}) = \mu_i + \tau_j + \lambda z_{ijk} \tag{118}$$

is, using (117),

$$\hat{\lambda} = E_{yz}/E_{zz} \,. \tag{119}$$

Estimators of μ_i and τ_j for (118) depend, as usual, on (68). Denoting those estimators by $\tilde{\mu}_i$ and $\tilde{\tau}_j$, to distinguish them from $\hat{\mu}_i$ and $\hat{\tau}_j$ for (114), and using (98) and (100) of Chapter 9 in (68), gives

$$\tilde{\boldsymbol{\tau}}_2 = \{_c\,\tilde{\tau}_j\}_{j=1}^{j=b-1} = \mathbf{C}_2^{-1}(\mathbf{r}_{2,y} - \hat{\lambda}\mathbf{r}_{2,z}) \quad \text{with } \tilde{\tau}_b = 0; \tag{120}$$

and

$$\tilde{\mu}_i = \bar{y}_{i..} - \hat{\lambda}\bar{z}_{i..} - \sum_{j=1}^{b-1} n_{ij}\tilde{\tau}_j/n_{i.}, \quad \text{for } i = 1, \ldots, a\ . \tag{121}$$

Pursuing further details of this nature is not very revealing.

With balanced data though, of one observation per cell, results become quite tractable. We then have $\hat{\lambda} = E_{yz}/E_{zz}$ with

$$E_{yz} = \Sigma_i\Sigma_j(y_{ij} - \bar{y}_{i.} - \bar{y}_{.j} + \bar{y}_{..})(z_{ij} - \bar{z}_{i.} - \bar{z}_{.j} + \bar{z}_{..}) \tag{122}$$

and analogous (sums of squares) expressions E_{yy} and E_{zz}. Then (120) and (121) reduce to

$$\tilde{\mu}_i = \bar{y}_{i.} + \bar{y}_{.b} - \bar{y}_{..} - \hat{\lambda}(\bar{z}_{i.} + \bar{z}_{.b} - \bar{z}_{..}) \quad \text{for } i = 1, \ldots, a \tag{123}$$

$$\tilde{\tau}_j = \bar{y}_{.j} - \bar{y}_{.b} - \hat{\lambda}(\bar{z}_{.j} - \bar{z}_{.b}) \quad \text{for } j = 1, \ldots, b-1, \tag{124}$$

and $\tilde{\tau}_b = 0$.

It is also easily established that

$$\hat{\sigma}^2 = (E_{yy} - E_{yz}^2/E_{zz})/[a-1)(b-1) - 1], \tag{125}$$

and with this the numerator sum of squares for testing H: μ_i *all equal* is

$$Q = A_{yy} + \frac{E_{yz}^2}{E_{zz}} - \frac{(E_{yz} + A_{yz})^2}{E_{zz} + A_{zz}}\ . \tag{126}$$

This expression has the same form as (112) for the with-interaction model, with (for balanced data with all n_{ij} unity) $\dot{A}_{yy}$ of (111) and A_{yy} of (116) then being the same in both (112) and (126). But E_{yz} and E_{zz} in (112) are not the same as in (126). In (112) they have the form of (103), and in (126) they are based on (122).

11.7. THE 2-WAY CLASSIFICATION: MULTIPLE SLOPE MODELS

The traditional treatment of a covariate is that of the preceding section, of a single slope parameter regardless of rows or columns: i.e., the covariate is traditionally treated as λz_{ijk} as in (104) and (118). And this leads to the well-known form of estimator $\lambda = E_{yz}/E_{zz}$ for the appropriate definition of E_{yy} from the no-covariate model. But from the general result $\hat{\mathbf{b}} = (\mathbf{R}'\mathbf{R})^{-1}\mathbf{R}'\mathbf{y}$ of (99) we are freely able to consider other ways of handling covariates, with slope parameters related to rows and/or columns in a variety of ways. Thus, for example, an intra-row regression model would have $\lambda_i z_{ijk}$, and an intra-column regression model would have $\lambda_j z_{ijk}$. And for two covariates, one might want to use intra-row regression for one and intra-column regression for the other, so that the model would have $\lambda_i z_{ijk} + \lambda_j^+ w_{ijk}$. These and several other possibilities are considered in Searle (1979), from which the following selected results are summarized. They provide answers to questions VII and VIII asked at the end of Section 11.1.

Since $\mathbf{R}'\mathbf{R}$ and $\mathbf{R}'\mathbf{y}$ are the crux of analysis of covariance calculations, just these values are given, for the four cases considered, in both the with- and without-interaction models. In all cases, the ranges of the subscripts are those that have been used throughout: $i = 1, \ldots, a$, $j = 1, \ldots, b$ and $k = 1, \ldots, n_{ij}$.

a. Interaction models, with unbalanced data

Results for these models apply for both balanced and unbalanced data. The error sum of squares in the without-covariate model is (from Table 4.8) as in (103), which we write as

$$E_{yy} = \Sigma_i\Sigma_j\Sigma_k(y_{ijk} - \bar{y}_{ij\cdot})^2 = \Sigma_i\Sigma_j E_{ij,\,yy} \quad \text{for } E_{ij,\,yy} = \Sigma_k(y_{ijk} - \bar{y}_{ij\cdot})^2 .$$

$E_{ij,\,zz}$ is defined similarly, and then $E_{ij,\,yz}$ is the analogous sum of products. Sums of these are also used: e.g.,

$$E_{\cdot j,\,yy} = \Sigma_i E_{ij,\,yy} = \Sigma_i\Sigma_k(y_{ijk} - \bar{y}_{ij\cdot})^2 .$$

With these definitions, $\mathbf{R}'\mathbf{R}$ and $\mathbf{R}'\mathbf{y}$ for four different models are as follows.

-i. *Intra-row slopes for one covariate.* The model equation is

$$E(y_{ijk}) = \mu_{ij} + \lambda_i z_{ijk} .$$

$$\mathbf{R}'\mathbf{R} = \{_d\, E_{i\cdot,\,zz}\} \quad \text{and} \quad \mathbf{R}'\mathbf{y} = \{_c\, E_{i\cdot,\,yz}\}$$

so that $$\hat{\lambda}_i = E_{i\cdot,\,yz}/E_{i\cdot,\,zz} \quad \text{and} \quad \tilde{\mu}_{ij} = \bar{y}_{ij\cdot} - \hat{\lambda}_i \bar{z}_{ij} .$$

-ii. Intra-row slopes for one covariate, intra-column for another.

$$\text{Model:}\quad E(y_{ijk}) = \mu_{ij} + \lambda_i z_{ijk} + \lambda_j^+ w_{ijk}\ .$$

$$\mathbf{R'R} = \begin{bmatrix} \{_d\, E_{i\cdot,zz}\} & \{_m\, E_{ij,zw}\} \\ \{_m\, E_{ji,wz}\} & \{_d\, E_{\cdot j,ww}\} \end{bmatrix} \quad \text{and} \quad \mathbf{R'y} = \begin{bmatrix} \{_c\, E_{i\cdot,yz}\} \\ \{_c\, E_{\cdot j,yw}\} \end{bmatrix}.$$

-iii. Intra-row slopes for each of two covariates.

$$\text{Model:}\quad E(y_{ijk}) = \mu_{ij} + \lambda_i z_{ijk} + \lambda_i^+ w_{ijk}\ .$$

$$\mathbf{R'R} = \begin{bmatrix} \{_d\, E_{i\cdot,zz}\} & \{_d\, E_{i\cdot,zw}\} \\ \{_d\, E_{i\cdot,zw}\} & \{_d\, E_{i\cdot,ww}\} \end{bmatrix} \quad \text{and} \quad \mathbf{R'y} = \begin{bmatrix} \{_c\, E_{i\cdot,yz}\} \\ \{_c\, E_{i\cdot,yw}\} \end{bmatrix}.$$

Expressions for the estimated slopes $\hat{\lambda}_i$ and $\hat{\lambda}_i^+$ can be obtained from $(\mathbf{R'R})^{-1}\mathbf{R'y}$ in this case in exactly the same form as (97) and (98) were obtained from (94).

-iv. Intra-row plus intra-column slopes for one covariate.

$$\text{Model:}\quad E(y_{ijk}) = \mu_{ij} + (\lambda_i + \lambda_j^+) z_{ijk}\ .$$

$\mathbf{Z}$ in this model does not have full column rank (Searle, 1979). This difficulty is overcome by defining $\lambda_1 = 0$ and deleting the first column of what would otherwise be $\mathbf{Z}$. Then

$$\mathbf{R'R} = \begin{bmatrix} \{_d\, E_{i\cdot,zz}\}_{i=2}^{i=a} & \{_m\, E_{ij,zz}\}_{i=2,\ j=1}^{i=a,\ j=b} \\ \{_m\, E_{ji,zz}\}_{j=1,\ i=2}^{j=b,\ i=a} & \{_d\, E_{\cdot j,zz}\}_{j=1}^{j=b} \end{bmatrix}$$

and

$$\mathbf{R'y} = \begin{bmatrix} \{_c\, E_{i\cdot,yz}\}_{i=2}^{i=a} \\ \{_c\, E_{\cdot j,yz}\}_{j=1}^{j=b} \end{bmatrix}.$$

Note that although $\mathbf{R'R}$ and $\mathbf{R'y}$ have elements *defined* for i starting at 2 rather than 1, summations over i do start at 1; e.g., $E_{\cdot j,zz} = \Sigma_{i=1}^{a} E_{ij,zz}$.

b. No-interaction models with balanced data

Solutions of the normal equations for unbalanced data in the no-interaction model are given in (98) and (100) of Chapter 9. Their relative intractability makes adaption to with-covariate models difficult; e.g., equations (120) and (121). We therefore confine attention to balanced data for this model. The following matrix and vector forms are involved.

$$\mathbf{S}_{1,zw} = \left\{_d \Sigma_j (z_{ij} - \bar{z}_{i\cdot})(w_{ij} - \bar{w}_{i\cdot})\right\}_{i=1}^{i=a}$$

$$- (1/a)\left\{_m \Sigma_j (z_{ij} - \bar{z}_{i\cdot})(w_{hj} - \bar{w}_{h\cdot})\right\}_{i=1,\ \ h=1}^{i=a,\ \ h=a},$$

$$\mathbf{S}_{2,zw} = \left\{_d \Sigma_i (z_{ij} - \bar{z}_{\cdot i})(w_{ij} - \bar{w}_{\cdot j})\right\}_{j=1}^{j=b}$$

$$- (1/b)\left\{_m \Sigma_i (z_{ij} - \bar{z}_{\cdot j})(w_{ik} - \bar{w}_{\cdot k})\right\}_{j=1,\ \ k=1}^{j=b,\ \ k=b}$$

and their counterparts $\mathbf{S}_{1,zz}$, $\mathbf{S}_{1,ww}$, $\mathbf{S}_{2,zz}$ and $\mathbf{S}_{2,ww}$. Also the matrix

$$\mathbf{P}_{12,zw} = \left\{_m (z_{ij} - \bar{z}_{i\cdot})(w_{ij} - \bar{w}_{\cdot j})\right\}_{i=1,\ \ j=1}^{i=a,\ \ j=b}$$

and its counterparts $\mathbf{P}_{12,zz}$ and $\mathbf{P}_{12,ww}$, and the vectors

$$\mathbf{u}_{1,yz} = \left\{_c \Sigma_j (z_{ij} - \bar{z}_{i\cdot})(y_{ij} - \bar{y}_{i\cdot})\right\}_{i=1}^{i=a},$$

$$\mathbf{u}_{2,zz} = \left\{_c \Sigma_i (z_{ij} - \bar{z}_{\cdot j})(y_{ij} - \bar{y}_{\cdot j})\right\}_{j=1}^{j=b}$$

and their counterparts $\mathbf{u}_{1,yw}$ and $\mathbf{u}_{2,yw}$. Using these, $\mathbf{R}'\mathbf{R}$ and $\mathbf{R}'\mathbf{y}$ for the same four models as considered in the with-interaction case are as follows.

-i. *Intra-row slopes for one covariate.*

$$\text{Model:}\quad E(y_{ij}) = \mu_i + \tau_j + \lambda_i z_{ij}\ .$$

$$\mathbf{R}'\mathbf{R} = \mathbf{S}_{1,zz} \quad \text{and} \quad \mathbf{R}'\mathbf{y} = \mathbf{P}_{1,yz}\ .$$

-ii. *Intra-row slopes for one covariate, intra-column for another.*

$$\text{Model:}\quad E(y_{ij}) = \mu_i + \tau_j + \lambda_i z_{ij} + \lambda_j^+ w_{ij}\ .$$

$$\mathbf{R}'\mathbf{R} = \begin{bmatrix} \mathbf{S}_{1,zz} & \mathbf{P}_{12,zw} \\ (\mathbf{P}_{12,zw})' & \mathbf{S}_{2,ww} \end{bmatrix} \quad \text{and} \quad \mathbf{R}'\mathbf{y} = \begin{bmatrix} \mathbf{u}_{1,yz} \\ \mathbf{u}_{2,yw} \end{bmatrix}.$$

-iii. Intra-row slopes for each of two covariates.

$$\text{Model:}\quad E(y_{ij}) = \mu_i + \tau_j + \lambda_i z_{ij} + \lambda_i^+ w_{ij} .$$

$$\mathbf{R'R} = \begin{bmatrix} \mathbf{S}_{1,zz} & \mathbf{S}_{1,zw} \\ \mathbf{S}_{1,zw} & \mathbf{S}_{1,ww} \end{bmatrix} \quad \text{and} \quad \mathbf{R'y} = \begin{bmatrix} \mathbf{u}_{1,yz} \\ \mathbf{u}_{1,yw} \end{bmatrix} .$$

-iv. Intra-row plus intra-column slopes for one covariate.

$$\text{Model:}\quad E(y_{ij}) = \mu + \tau_j + (\lambda_i + \lambda_j^+) z_{ij} .$$

As in the preceding case of a model with $\lambda_i + \lambda_j^+$ as a slope parameter, $\mathbf{Z}$ does not have full column rank, but this is achieved by defining $\lambda_1 = 0$ and so amending $\mathbf{Z}$ through deleting its first column. Then with $\mathbf{S}_1^*$ being $\mathbf{S}_1$ with its first row and column deleted, and with $\mathbf{P}_{12}^*$ (and $\mathbf{u}_1^*$) being $\mathbf{P}_{12}$ (and $\mathbf{u}_1$) with their first row (element) deleted,

$$\mathbf{R'R} = \begin{bmatrix} \mathbf{S}_{1,zz}^* & \mathbf{P}_{12,zz}^* \\ \mathbf{P}^{*\prime}_{12,zz} & \mathbf{S}_{2,zz} \end{bmatrix} \quad \text{and} \quad \mathbf{R'y} = \begin{bmatrix} \mathbf{u}_{1,zz}^* \\ \mathbf{u}_{2,yz} \end{bmatrix} .$$

11.8. EXERCISES

E11.1. (a) Solve the normal equations of (11) to obtain the estimators given in (17) and (18).

(b) Using the data of Table 6.1, derive estimators of the parameters in the models (5), (6) and (7).

E11.2. Derive (24)–(27).

E11.3. Derive (A), (B) and (C) of Section 11.2d-i.

E11.4. Show that Tables 11.4 and 11.5 yield (a) Table 6.4 for (28), and (b) parts of Table 6.6 for (33).

E11.5. (a) Using (37), derive the associate hypotheses of Table 11.5.

(b) Calculate Tables 11.4 and 11.5 for the models (5), (6) and (7).

E11.6. Use (37) and Table 11.6 to derive the hypotheses shown in lines 1–4 and 7 and 8 of Table 6.4.

E11.7. Use (37) and Tables 11.7 and 11.8 to derive the hypotheses shown in lines 2–4 and 10 of Table 6.6 and in lines 1–3 of Table 6.7.

E11.8. Using (2) and Tables 11.2 and 11.3, try to derive the hypotheses associated with line 9 of Table 6.6 and with lines 4, 6 and 7 of Table 6.7.

E11.9. For (33), using the linear model formulation that follows it, show that

(a) $\hat{b}_i$ and $\bar{y}_i$ are uncorrelated;

(b) $v(\mathbf{q}'\tilde{\boldsymbol{\mu}}) = \Sigma_i q_i^2(1/n_i + \bar{z}_{i\cdot}^2/W_{i,zz})\sigma^2$; and

(c) $v[\text{BLUE}(\mathbf{q}'\boldsymbol{\mu} + \mathbf{p}'\mathbf{b})] = \Sigma_i[q_i^2/n_i + (q_i\bar{z}_{i\cdot} - p_i)^2/W_{i,zz}]\sigma^2$.

E11.10. (a) If the hypothesis in Section 11.2d-iii is not rejected, show that the estimator of b_2 is

$$b_2 = \frac{W_{2,yz} + W_{3,yz}}{W_{2,zz} + W_{3,zz}} \ .$$

(b) Why is this to be expected?

(c) What is the estimator of μ_i, $i = 1, 2, 3, \ldots$?

E11.11. Starting with the statement of a linear hypothesis, and using the standard formula for calculating the numerator sum of squares of the F-statistic for testing it, confirm the sums of squares shown in Tables 11.9 and 11.10.

E11.12. (a) Use (96) in (95) and (94) to derive (98).

(b) Describe the conditions under which (88) and (89) reduce to (77) and show that reduction.

(c) Suggest an adaptation of (89) that reduces it to (98), and show that reduction.

(d) Derive (80), (92) and (100).

E11.13. Derive (112).

E11.14. Confirm (120)–(126).

E11.15. For a 2-way classification of 4 rows and 3 columns, write down $\mathbf{Z}$ explicitly for the case of balanced data ($n = 1$) and confirm the results of Section 11.7b.

E11.16. (*The nested, mixed model*). The with-covariate mixed model of the 2-way nested classification has model equation

$$y_{ijk} = \mu_i + \beta_{ij} + bz_{ijk} + e_{ijk},$$

for $i = 1, \ldots, a$, $j = 1, \ldots, b_i$ and $k = 1, \ldots, n_{ij}$. For the mixed model, the μ_is are fixed effects and the β_{ij}s are taken as being i.i.d. random effects having zero mean, variance σ_β^2 and zero correlation with the error terms.

(a) For the no-covariate model show that μ_i has BLUE

$$\hat{\mu}_i = \tilde{y}_{i\cdot\cdot} = \sum_{j=1}^{b_i} w_{ij}\bar{y}_{ij\cdot} \bigg/ \sum_{j=1}^{b_i} w_{ij}$$

$$\text{for} \quad w_{ij} = 1\big/\left(\sigma_\beta^2 + \sigma_e^2/n_{ij\cdot}\right)$$

(b) For the with-covariate model, as above, derive the BLU estimators

$$\hat{\mu}_i = \tilde{y}_{i\cdot\cdot} - \hat{b}\tilde{z}_{i\cdot\cdot}$$

and

$$\hat{b} = \frac{E_{yz} + \sigma_e^2 \sum_{i=1}^{a} \sum_{j=1}^{b_i} w_{ij}(\bar{y}_{ij\cdot} - \tilde{y}_{i\cdot\cdot})(\bar{z}_{ij\cdot} - \tilde{z}_{i\cdot\cdot})}{E_{zz} + \sigma_e^2 \sum_{i=1}^{a} \sum_{j=1}^{b_i} w_{ij}(\bar{z}_{ij\cdot} - \tilde{z}_{i\cdot\cdot})^2}.$$

CHAPTER 12

COMMENTS ON COMPUTING PACKAGES

There is nowadays a wide array of computing package routines (with continual updates) available to statisticians, to say nothing of new routines being developed for both current and projected hardware. It is therefore invidious to isolate for comment any small group of routines for fear of unintentionally overlooking others that may be better in whatever manner one wishes to judge quality: speed, cost, numerical accuracy, documentation, ease of use, breadth of calculations, labeling of output, and so on. Furthermore, comments about packages made today are outdated by tomorrow. Hence, in offering any comments at all, one obviously risks both errors of omission and of instant outdatedness. Nevertheless, it is to be hoped that some brief attempt at indicating what calculations are available in five internationally used main-frame packages is likely to be of more use than nuisance. In this spirit we concentrate attention on sums of squares, since, by the variety of them that are available from unbalanced data, they are often a source of confusion to package users. Although most packages also contain regression routines, from which in some cases certain analysis of variance calculations for unbalanced data that cannot be readily obtained from an analysis of variance routine can, sometimes with considerable ingenuity, be obtained from a regression routine, such possibilities are not discussed here. Attention is confined to analysis of variance routines only. To developers chagrined by these errors of omission go my apologies. For users of the packages I trust the brief summary information will be helpful, albeit "warts and all".

As indicated in Section 1.4, a problem for many users of analysis of variance routines when processing unbalanced data is that labeling of sums

TABLE 12.1. STATISTICAL COMPUTING PACKAGES ON WHICH INFORMATION ON ANALYSIS OF VARIANCE ROUTINES IS PROVIDED IN TABLE 12.2

Acronym	Source
BMDP	BMDP Statistical Software P. O. Box 24A26 Los Angeles, California 90024
GENSTAT	Rothamsted Experiment Station Harpenden Hertfordshire, England
SAS GLM (82.3)	SAS Institute Inc. P. O. Box 8000 Cary, North Carolina 27511
SAS HARVEY	A user-supplied routine in SAS, from W. H. Harvey Dairy Science Department Ohio State University Columbus, Ohio 43210
SPSS-X (2.1)	SPSS Inc. Suite 3300 444 N. Michigan Avenue Chicago, Illinois 60611

of squares is seldom unequivocally clear in package output. For example, in analyzing data from a 2-way classification with-interaction model, we need to know the precise description of a sum of squares that is labeled simply as A; is it $R(\alpha|\mu)$, or $R(\alpha|\mu, \beta)$, or SSA_w, or $R(\dot{\alpha}|\mu, \dot{\beta}, \dot{\phi})$? Once it is identified, its associated hypothesis is available and we are in a position to know whether that hypothesis is useful to the study at hand. If it is useful, the F-statistic can be used to test the hypothesis. If not, the sum of squares can be ignored. It is with this *modus operandi* in mind that we highlight in Table 12.2 certain features of the sums of squares output from the five routines shown in Table 12.1.

12.1. SUMS OF SQUARES OUTPUT

The 2-way crossed classification with interaction is the simplest model with which to illustrate the vagaries of sums of squares for unbalanced data. It is therefore used in Table 12.2 for describing the sums of squares output of the different routines.

TABLE 12.2. SUMS OF SQUARES CALCULATED BY FIVE STATISTICAL COMPUTING ROUTINES FOR A 2-WAY CROSSED CLASSIFICATION WITH-INTERACTION MODEL: $E(y_{ijk}) = \mu + \alpha_i + \beta_j + \phi_{ij}$

BMDP2V uses Σ-restrictions; for with-interaction models, all cells must be filled.

GENSTAT ANOVA is designed for balanced data and handles unbalanced data using "missing value" techniques.

SAS GLM—There are four sets of sums of squares, named Types I–IV, for factors read as A, B and $A * B$.

Label	Type I[1] (Sequential)	Type II (Adjusted)	Type III (Σ-restricted)	Type IV (Hypotheses)
A	$R(\alpha \mid \mu)$	$R(\alpha \mid \mu, \beta)$	$R(\dot{\alpha} \mid \dot{\mu}, \dot{\beta}, \dot{\phi})_\Sigma$[3]	
B	$R(\beta \mid \mu, \alpha)$	$R(\beta \mid \mu, \alpha)$	$R(\dot{\beta} \mid \dot{\mu}, \dot{\alpha}, \dot{\beta})_\Sigma$	[4]
$A * B$[2]	$R(\phi \mid \mu, \alpha, \beta)$	$R(\phi \mid \mu, \alpha, \beta)$	$R(\phi \mid \mu, \alpha, \beta)$	$R(\phi \mid \mu, \alpha, \beta)$

SAS HARVEY uses Σ-restrictions and requires that data have at least one row and one column with all cells filled—and these must be the last row and last column (Section 12.5).

SPSS ANOVA uses Σ-restrictions and requires that data have at least one row and one column with all cells filled—and these must be the first row and first column.

[1] For balanced data Types I–IV are all the same.
[2] Without interaction, there is no $A * B$ line and Types II–IV are the same.
[3] For all-cells-filled, Types III and IV are the same, yielding SSA_w and SSB_w (Section 12.2c).
[4] For some-cells-empty data, hypotheses are selected by the routine (Section 12.2d).

The SAS GLM routine has four sets of sums of squares and so occupies the largest share of Table 12.2. Its sums of squares are described in terms of the overparameterized model

$$E(y_{ijk}) = \mu + \alpha_i + \beta_j + \phi_{ij} \tag{1}$$

to coincide with the nature of the output from that routine. Sums of squares symbolized in terms both of (1) and of cell means parameters are summarized in Table 12.3. For the same two model formulations, with and without interactions where relevant, the same table provides references to Chapters 4 and 9, where the corresponding associated hypotheses are discussed and displayed. Through describing computing-package sums of squares output in terms of the sums of squares in Table 12.2, we can use

TABLE 12.3. SUMS OF SQUARES IN A 2-WAY CROSSED CLASSIFICATION MODEL

Sums of Squares	Associated Hypotheses		
(a) From partitioning the total sum of squares (Calculation formulae in Table 4.8, Section 4.9f)			
	Location of Associated Hypotheses in Chapters 4 and 9		
	Interactions	Model	
		Overparameterized	Cell Means
$R(\alpha \mid \mu) = \mathcal{R}(\mu_i \mid \mu)$ $R(\beta \mid \mu, \alpha) = \mathcal{R}(\mu_i, \tau_j \mid \mu_i)$ $R(\phi \mid \mu, \alpha, \beta) = \mathcal{R}(\mu_{ij} \mid \mu_i, \tau_j)$	With interaction:	Table 9.1 (Section 9.1g)	Table 4.9 (Section 4.9g)
$R(\beta \mid \mu) = \mathcal{R}(\tau_j \mid \mu)$ $R(\alpha \mid \mu, \beta) = \mathcal{R}(\mu_i, \tau_j \mid \tau_j)$	No interaction:	Table 9.2 (Section 9.2g)	Table 4.11 (Section 4.10c)
(b) Using Σ-restrictions			
All-cells-filled data		Table 9.3 (Section 9.4b-i)	Eqs. (43) and (44) (Section 4.6d)
SSA_w SSB_w		H: $\alpha_i + \bar{\phi}_{i\cdot}$ *equal* $\forall\ i$ H: $\beta_j + \bar{\phi}_{\cdot j}$ *equal* $\forall\ j$	H: $\bar{\mu}_{i\cdot}$ *equal* $\forall\ i$ H: $\bar{\mu}_{\cdot j}$ *equal* $\forall\ j$
Some-cells-empty data		Section 9.4b-ii	Section 9.4b-ii
$R(\dot{\alpha} \mid \dot{\mu}, \dot{\beta}, \dot{\phi})_\Sigma$ $R(\dot{\beta} \mid \dot{\mu}, \dot{\alpha}, \dot{\phi})_\Sigma$		H: $\dot{\alpha}_i$ *equal* 0 $\forall\ i$ H: $\dot{\beta}_j$ *equal* 0 $\forall\ j$	The only general result is Eq. (146). Examples follow Eq. (148)

Table 12.3 to locate the associated hypothesis that corresponds to an individual sum of squares output item. And if a particular usage of a package does not coincide with the sums of squares in Table 12.3, that table should certainly provide the basis from which the specific associated hypothesis for any analysis of variance (fixed effects model) situation can be derived. Analysis with covariates is dealt with in Section 12.6.

Some comments about Table 12.2 are in order.

a. Σ-restrictions

Routines BMDP2V, SAS HARVEY and SPSS-X ANOVA are described as using the Σ-restrictions; so also are the Type III sums of squares of SAS GLM. In each case this does not mean that the relevant computing algorithm necessarily uses the Σ-restriction. But whatever algorithm is used has the effect of producing sums of squares that are those of a Σ-restricted model.

b. Weighted squares of means

In all cases where, in the sense just described, Σ-restrictions are used, the output sums of squares for all-cells-filled data are weighted squares of means such as SSA_w and SSB_w—see Sections 4.6 and 9.4b. For some-cells-empty data from with-interaction models, BMDP2V will not operate; SAS GLM operates, and it and SAS HARVEY and SPSS ANOVA (when they operate—see Section c which follows) produce sums of squares such as $R(\dot{\alpha} \mid \dot{\mu}, \dot{\beta}, \dot{\phi})_{\Sigma}$—see Section 9.4b-ii.

c. Patterns of filled cells

In order for SAS HARVEY and SPSS ANOVA to operate on some-cells-empty data from a with-interaction model, there must be at least one row of the data and at least one column of the data that each has all cells filled. Furthermore, the data must be sequenced for input in such a way that the required all-cells-filled row and column are, for SAS HARVEY, the last row and column, and, for SPSS ANOVA, the first row and column. For multi-factor models in general, each main effects factor must have at least one level in which all cells contain data, and these must be the last levels for SAS HARVEY and the first levels for SPSS ANOVA.

d. Balanced data

All routines operate on balanced data. This is what GENSTAT ANOVA (free-standing GLIM) is designed for, and it operates on unbalanced data by using estimated missing value techniques.

e. Four sets of output from SAS GLM

These are four sets of sums of squares in the output from SAS GLM. In some circumstances some sets are the same as each other and in other cases they are all different. These four sets are now described more fully than in Table 12.2.

12.2. SUMS OF SQUARES FROM SAS GLM

The objective of providing all of the different kinds of sums of squares for unbalanced data that originate from the underlying philosophy of analysis of variance (of partitioning the total sum of squares) is achieved by the SAS GLM routine through its output of four different sets of sums of squares available on a single run of each data set. The four sets are labeled Types I, II, III and IV, and as indicated in Table 12.2 they can be thought of, respectively, as Sequential, Each-after-all-others, Σ-restrictions and Hypotheses. Equalities among the types exist under the following circumstances:

$$I = II = III = IV \quad \text{with balanced data}$$

$$II = III = IV \quad \text{with no-interaction models}$$

$$III = IV \quad \text{with all-cells-filled data} .$$

Brief discussion of the four types follows.

a. Type I

The sequence in which the factors of the model are identified to the routine with the data input determines the sequence of sums of squares that constitute Type I. The output sequence of sums of squares corresponds precisely to the input sequence of the factors, which itself must have main effect factors preceding interactions and also, factors that have others nested within them must precede those nested factors. Thus when factors are rows, columns and interactions and are identified to the routine as A, B and AB, respectively, the Type I sums of squares are $R(\boldsymbol{\alpha} \mid \mu)$, $R(\boldsymbol{\beta} \mid \mu, \boldsymbol{\alpha})$ and $R(\boldsymbol{\alpha}^*\boldsymbol{\beta} \mid \mu, \boldsymbol{\alpha}, \boldsymbol{\beta})$, in that sequence, as in Table 12.2. This correspondence extends very naturally to more than two factors. For example, for a no-interaction 3-way classification model, the Type I sums of squares when sequencing the factor identities as

$$A, B, C \text{ are } R(\boldsymbol{\alpha} \mid \mu), \quad R(\boldsymbol{\beta} \mid \mu, \mathbf{a}) \quad \text{and} \quad R(\boldsymbol{\gamma} \mid \mu, \boldsymbol{\alpha}, \boldsymbol{\beta});$$

whereas the sums of squares for the sequence

$$C, B, A \text{ are } R(\boldsymbol{\gamma} \mid \mu), \quad R(\boldsymbol{\beta} \mid \mu, \boldsymbol{\gamma}) \quad \text{and} \quad R(\boldsymbol{\alpha} \mid \mu, \boldsymbol{\beta}, \boldsymbol{\gamma}) .$$

b. Type II

The sums of squares labeled Type II in the output are similar in kind to those of Type I, but only certain of them. They are of the kind described as R (each factor | all other factors) in Section 10.3, and illustrated in Sections 4.10b, 5.4f, 9.2f and 9.2g. Thus for the 2-way classification the Type II sums of squares are $R(\boldsymbol{\alpha} \mid \mu, \boldsymbol{\beta})$, $R(\boldsymbol{\beta} \mid \mu, \boldsymbol{\alpha})$ and $R(\boldsymbol{\phi} \mid \mu, \boldsymbol{\alpha}, \boldsymbol{\beta})$, as in Table 12.2. In this list we see that although Type II sums of squares can be described in a general way as R (each factor | all other factors), this description does in fact need to be made a little more carefully. The form

$$R(\text{a factor} \mid \text{all appropriate other factors}) \qquad (2)$$

is more accurate, where "appropriate" means that the sum of squares is, for some factor, adjusted for all other factors except interactions that involve that factor and except for factors nested within that factor. For example, a Type II sum of squares due to A is a sum of squares adjusted for all other factors and interactions, except for interactions involving A, and except for factors nested within A. Thus in Table 12.2 the sum of squares for factor A is $R(\boldsymbol{\alpha} \mid \mu, \boldsymbol{\beta})$ and not $R(\boldsymbol{\alpha} \mid \mu, \boldsymbol{\beta}, \boldsymbol{\alpha}^*\boldsymbol{\beta})$, if for no other reason than the latter is identically zero—see Section 9.1k. Likewise, for a three-factor case with the only interactions being A-by-B, the sums of squares would be $R(\boldsymbol{\alpha} \mid \mu, \boldsymbol{\beta}, \boldsymbol{\gamma})$, $R(\boldsymbol{\beta} \mid \mu, \boldsymbol{\alpha}, \boldsymbol{\gamma})$, $R(\boldsymbol{\gamma} \mid \mu, \boldsymbol{\alpha}, \boldsymbol{\beta}, \boldsymbol{\alpha}^*\boldsymbol{\beta})$ and $R(\boldsymbol{\alpha}^*\boldsymbol{\beta} \mid \mu, \boldsymbol{\alpha}, \boldsymbol{\beta}, \boldsymbol{\gamma})$. Similarly, for a model of two crossed factors A and B and a factor C nested within A, say $C: A$, the Type II sum of squares for A would be $R(\boldsymbol{\alpha} \mid \mu, \boldsymbol{\beta})$ and not $R(\boldsymbol{\alpha} \mid \mu, \boldsymbol{\beta}, \boldsymbol{\gamma}: \boldsymbol{\alpha})$. Keeping in mind these implications of "appropriate", one is then justified in labeling Type II sums of squares as being "each after all others".

c. Type III

These are the sums of squares for Σ-restricted models; e.g., $R(\dot{\boldsymbol{\alpha}} \mid \dot{\mu}, \dot{\boldsymbol{\beta}}, \dot{\boldsymbol{\phi}})$ discussed at the end of Section 9.4b-ii. Description is given in that section as to why

$$R(\dot{\boldsymbol{\alpha}} \mid \dot{\mu}, \dot{\boldsymbol{\beta}}, \dot{\boldsymbol{\phi}}) \neq R(\boldsymbol{\alpha} \mid \mu, \boldsymbol{\beta}, \boldsymbol{\gamma}) = 0 \ .$$

Also, it is to be remembered that for all-cells-filled data, these Type III sums of squares are identical to those from the weighted squares of means analysis, e.g., SSA_w and SSB_w. Extensions to more than two factors are dealt with in Section 10.2. And a faulty algorithm for calculating terms like $R(\dot{\alpha} \mid \dot{\mu}, \dot{\beta}, \dot{\phi})_\Sigma$ is discussed in Section 12.5.

d. Type IV

Sums of squares labeled Type I, II and III are those derived from fitting a model and different sub-models thereof. But Type IV sums of squares are for testing hypotheses that are determined by the SAS GLM routine itself. The hypotheses selected depend upon the pattern of filled cells; and then the F-statistics are in accord with the good statistical practice of setting up a hypothesis and testing it. This does mimic reality, but only as an algorithmic automaton looking at which cells contain data, and not as a knowledgeable scientist thinking about data. An example is that the Type IV sum of squares for rows is the numerator sum of squares for a hypothesis that consists of contrasts between cell means for filled cells that are in the same columns—and starting with those that are in the last row. Such a hypothesis therefore depends on which cells have data in them and on which level of the row factor has been defined as the last row. In general, hypotheses determined in this nature are not necessarily of any interest. Moreover, the Type IV sum of squares for rows, for example, can differ from one sequencing of rows to another, for the same set of data. Hence what is called a Type IV sum of squares for rows is not unique, for a given set of some-cells-empty data, and so the corresponding hypothesis is not unique, either.

Example. Consider the data pattern of Grid 12.1, where the occurrence of a μ_{ij} represents presence of data:

Grid 12.1

μ_{11}		μ_{13}	μ_{14}
μ_{21}	μ_{22}		
	μ_{32}	μ_{33}	μ_{34}

The Type IV hypothesis for rows set up by SAS GLM would be

$$H_1: \begin{cases} \mu_{33} + \mu_{34} - (\mu_{13} + \mu_{14}) = 0 \\ \mu_{32} - \mu_{22} = 0 \end{cases},$$

and the corresponding Type IV sum of squares would be the value of Q calculated from (149) of Section 8.8c. Three important features of a sum of squares calculated from a hypothesis determined in this way apply to each Type IV sum of squares.

(i) It is not necessarily part of any traditional analysis-of-variance partitioning of the total sum of squares.
(ii) It does not necessarily involve all the data.
(iii) Altering the sequence of rows can lead to a different Type IV sum of squares for rows.

We demonstrate (iii) for Grid 12.1. Suppose its second and third rows are interchanged to yield Grid 12.2, wherein the μ_{ij}s retain their subscripts from Grid 12.1.

Grid 12.2

μ_{11}		μ_{13}	μ_{14}
	μ_{32}	μ_{33}	μ_{34}
μ_{21}	μ_{22}		

For Grid 12.2 the Type IV sum of squares for rows would be the numerator sum of squares for testing the hypothesis

$$H_2: \begin{cases} \mu_{22} - \mu_{32} = 0 \\ \mu_{21} - \mu_{11} = 0 \end{cases} .$$

Since H_2 differs from H_1, the two Type IV sums of squares differ. This characteristic of Type IV sums of squares for rows depending on the sequence of rows establishes their non-uniqueness, and this in turn emphasizes that the hypotheses they are testing are by no means necessarily of any general interest. Of course, the same is true of Type IV sums of squares for columns; and for other factors and their interactions in k-way classifications ($k > 2$) when there are empty cells. More examples are given in Hudson and Searle (1982).

12.3. ESTIMABLE FUNCTIONS IN SAS GLM OUTPUT

Output from SAS GLM includes a special form of estimable function for each sum of squares in each of the Type I, II, III and IV listings. The utility of these estimable functions is that each one provides a basis for the user to be able to display the associated hypothesis for the corresponding sum of squares.

a. Examples

Table 12.4 shows estimable functions for the Type I sum of squares for A in two different situations: (a) a 1-way classification of 3 classes and (b) a 2-way classification without interaction, of 2 rows and 3 columns with the n_{ij}-values of Table 12.5, taken from Table 8.8.

b. Output is parameter labels and coefficients

Parts (a) and (b) of Table 12.4 show output from SAS GLM and interpretation thereof. In each case, as is true generally, the output consists of two columns. One is a column titled "Effect": it contains parameter labels for the appropriate overparameterized model, using "Intercept" for μ, and labels such as $A1$, $A2$, and $B1$, $B2$ for α_1, α_2, β_1 and β_2, respectively. The second column of output is labeled "Coefficient". It contains symbols that, with + or − signs, are either $L1, L2, L3, \ldots$ (some of which may be zero and so do not occur, such as $L1$ in Table 12.4) or they may be linear combinations of such symbols with coefficients that in some instances are integers [as in part (a) of Table 12.4] and in other instances are decimal numbers [as in part (b) of Table 12.4].

TABLE 12.4. EXAMPLES OF ESTIMABLE FUNCTIONS FROM SAS GLM OUTPUT FOR TYPE I SUM OF SQUARES

Output		Interpretation
Effect	**Coefficient**	
(a) 1-way classification of 3 classes (see Grid 12.1)		
Intercept		
$A1$	$L2$	$f = l_2\alpha_1 + l_3\alpha_2 - (l_2 + l_3)\alpha_3$
$A2$	$L3$	
$A3$	$-L2-L3$	
(b) 2-way classification of 2 rows and 3 classes, without interaction (see Table 12.5)		
Intercept		
$A1$	$L2$	$f = l_2\alpha_1 - l_2\alpha_2 - 0.2381 l_2\beta_1 + 0.0952 l_2\beta_2 + 0.1429 l_2\beta_3$
$A2$	$-L2$	
$B1$	$-.2381\ L2$	
$B2$	$.0952\ L2$	
$B3$	$.1429\ L2$	

c. Estimable function obtained from output

In all cases the interpretation of these two columns of output is obtained by multiplying them together to get a function that we choose to call f:

$$f = \Sigma(\text{each parameter label}) \times (\text{corresponding function of } L1, L2, \ldots)\ .$$

Using parameter symbols μ, α_i, β_j and so on, and $l_1, l_2, \ldots$ in place of $L1, L2, \ldots$, the f for each example of Table 12.4 is:

(a) $$f = l_2\alpha_1 + l_3\alpha_2 - (l_2 + l_3)\alpha_3$$

and

(b) $$f = l_2\alpha_1 - l_2\alpha_2 - 0.2381 l_2\beta_1 + 0.0952 l_2\beta_2 + 0.1429 l_2\beta_3\ .$$

TABLE 12.5. NUMBERS OF OBSERVATIONS IN A 2-WAY CLASSIFICATION FOR PART (b) OF TABLE 12.4

	n_{ij}			$n_{i\cdot}$	
	3	3	1	7	
	2	1	—	3	
$n_{\cdot j}$:	5	4	1	10	$= n_{\cdot\cdot}$

Every such f has two important characteristics:

(I) Each f is an estimable function (of the parameters of the model being used) no matter what values are given to $l_1, l_2, \ldots,$ in f. In Table 12.4, using $l_1 = l_2 = -1$ and then $l_1 = 1 = l_2$ in (a) gives

$$\text{(a)} \qquad f_1 = \alpha_1 - \alpha_2 \quad \text{and} \quad f_2 = \alpha_1 - \alpha_3; \tag{3}$$

and using $l_1 = 1$ in (b) gives

$$\text{(b)} \qquad f_1 = \alpha_1 - \alpha_2 - 0.2381\beta_1 + 0.0952\beta_2 + 0.1429\beta_3\ . \tag{4}$$

(II) The number of different symbols $l_1, l_2, \ldots$ in each f equals the degrees of freedom, $r_{\mathbf{A}}$, of the sum of squares, call it $\mathbf{y}'\mathbf{A}\mathbf{y}$, that the f is associated with. (Take $r_{\mathbf{A}} = r$, say.) In Table 12.4 both sums of squares are $R(\boldsymbol{\alpha} \mid \mu)$; for part (a)

$$R(\boldsymbol{\alpha} \mid \mu) = \sum_{i=1}^{3} n_i(\bar{y}_{i\cdot} - \bar{y}_{\cdot\cdot})^2 \quad \text{and} \quad r_{\mathbf{A}} = 2,$$

and for part (b)

$$R(\boldsymbol{\alpha} \mid \mu) = \sum_{i=1}^{2} n_{i\cdot}(\bar{y}_{i\cdot\cdot} - \bar{y}_{\cdot\cdot\cdot})^2 \quad \text{and} \quad r_{\mathbf{A}} = 1\ .$$

d. Estimable function provides the hypothesis

Having expressed the output as f, it is used as follows. Denote the $r_{\mathbf{A}}$ different symbols $l_1, l_2, \ldots, l_r$ in f by the vector $\mathbf{l}' = [l_1 \quad l_2 \quad \cdots \quad l_r]$. Use *any* $r_{\mathbf{A}}$ linearly independent values for $\mathbf{l}$ in f to yield $r_{\mathbf{A}}$ specific estimable functions $f_1, f_2, \ldots, f_r$. Then the associated hypothesis for the sum of squares corresponding to f is

$$\text{H:}\ f_i = 0 \quad \textit{for } i = 1, \ldots, r_{\mathbf{A}}\ . \tag{5}$$

For the examples of Table 12.4:

Part (a) has $r_{\mathbf{A}} = 2$. The two values of $[l_1 \quad l_2]$ used to derive (3) are linearly independent. Therefore (5) applied to (3) yields the associated hypothesis as

$$\text{H:} \begin{cases} \alpha_1 - \alpha_2 = 0 \\ \alpha_1 - \alpha_3 = 0 \end{cases}, \quad \text{i.e.,} \quad \text{H:}\ \alpha_1 = \alpha_2 = \alpha_3; \tag{6}$$

and this, we know, is indeed the case.

Part (b) has $r_A = 1$. Choosing $l_1 = 1$ gives f_1 of (4), and using that in (5) gives the associated hypothesis for part (b) of Table 12.4 as

$$H: \ \alpha_1 - \alpha_2 - 0.2381\beta_1 + 0.0952\beta_2 + 0.1429\beta_3 = 0 \ . \tag{7}$$

This does not, in general, look very useful (and indeed it is not). Nor does it look familiar; and yet it is just an alternative formulation (as it should be) of the general form of hypothesis tested by $R(\alpha \mid \mu)$ in the no-interaction 2-way classification model. This is, from Table 9.2 in Section 9.2g,

$$H: \ \alpha_i + \Sigma_j n_{ij}\beta_j / n_{i\cdot} \ \textit{equal} \ \ \forall \ i \ . \tag{8}$$

And for the example of Table 8.8, which is the same as Table 12.5 used in Table 12.4, the hypothesis (8) is

$$H: \ \alpha_1 + (3\beta_1 + 3\beta_2 + \beta_3)/7 = \alpha_2 + (2\beta_1 + \beta_3)/3, \tag{9}$$

as shown in Table 9.2E. This is

$$H: \ \alpha_1 - \alpha_2 + (\tfrac{3}{7} - \tfrac{2}{3})\beta_1 + (\tfrac{3}{7} - \tfrac{1}{3})\beta_2 + \tfrac{1}{7}\beta_3 = 0 \tag{10}$$

which, after decimalizing the fractions, is (7). Thus (7) is indeed a correct hypothesis for part (b) of Table 12.4.

e. Summary

The SAS GLM output labeled "estimable function" corresponding to a sum of squares Q consists, in each case, of a column of parameter labels and a column of functions of Ls. To obtain the associated hypothesis for Q, form

f = sum of products of each parameter label multiplied
by its corresponding function of Ls .

For r being the degrees of freedom of Q, there will be r different Ls in f; thus giving them (the Ls) r different sets of linearly independent values yields r different forms of f, call them $f_1, f_2, \ldots, f_r$. Then the associated hypothesis for Q is, as in (5), H: $f_i = 0$ for $i = 1, \ldots, r_A$.

f. Comment

The examples of Table 12.4 illustrate an important feature of a SAS GLM estimable function. In many cases the coefficients of the Ls in the output are (positive or negative) integers—indeed often $+1$ or -1, as in part (a) of Table 12.4. But this is not always so—as in part (b) of Table

12.4. Moreover, those coefficients may bear no easy nor "obvious" relationship to the model or the n_{ij}s. But relationship there is, through the eigenvectors of **A**, as is now shown.

g. Verification

-i. A general result. Reasons are given in Section 7.6b-iv as to why, for $\mathbf{y} \sim \mathcal{N}(\mathbf{X}\boldsymbol{\beta}, \sigma^2\mathbf{I})$, an F-statistic of the form $\mathbf{y}'\mathbf{A}\mathbf{y}/r_{\mathbf{A}}\hat{\sigma}^2$ tests

$$\text{H: } \mathbf{T}'\mathbf{X}\boldsymbol{\beta} = \mathbf{0} \tag{11}$$

where, with **A** being symmetric and idempotent, $\mathbf{T}'$ has full row rank $r_{\mathbf{A}}$, its rows being eigenvectors of **A** corresponding to the unity eigenvalue of **A**. We show how (11) can be expressed in the form of (5).

Consider not just (11), but any testable hypothesis H: $\mathbf{K}'\boldsymbol{\beta} = \mathbf{0}$ where $\mathbf{K}'$ has full row rank s. For any arbitrary nonsingular matrix **L** of order s, that hypothesis is equivalent to

$$\text{H: } \mathbf{L}\mathbf{K}'\boldsymbol{\beta} = \mathbf{0}\ . \tag{12}$$

But, since arbitrary **L** can be viewed as any s linearly independent (LIN) vectors $\mathbf{l}'$ of order s, the hypothesis (12) can also be stated as

$$\text{H: } \mathbf{l}'\mathbf{K}'\boldsymbol{\beta} = \mathbf{0} \text{ for any } s \text{ LIN vectors } \mathbf{l}' \text{ of order } s\ . \tag{13}$$

The expression $\mathbf{l}'\mathbf{K}'\boldsymbol{\beta}$ in (13) is an estimable function (because $\mathbf{K}'\boldsymbol{\beta}$ is estimable). Define it as f:

$$f = \mathbf{l}'\mathbf{K}'\boldsymbol{\beta}\ . \tag{14}$$

Hence, on using s LIN numerical vectors $\mathbf{l}'_1, \mathbf{l}'_2, \ldots, \mathbf{l}'_s$ to calculate s values of f as $f_i = \mathbf{l}'_i\mathbf{K}'\boldsymbol{\beta}$ for $i = 1, \ldots, s$, the hypothesis in (13) can be written as

$$\text{H: } \begin{cases} f_1 = 0 \\ f_2 = 0 \\ \vdots \\ f_s = 0 \end{cases} \tag{15}$$

This is exactly the form in (5).

The estimable function f, in (14), is precisely the form of estimable function that appears in SAS GLM output associated with each sum of squares $\mathbf{y}'\mathbf{A}\mathbf{y}$ for which $\mathbf{y}'\mathbf{A}\mathbf{y}/r_{\mathbf{A}}\hat{\sigma}^2$ is an F-statistic. Hence, from (11), that f can be derived as

$$f = \mathbf{l}'\mathbf{T}'\mathbf{X}\boldsymbol{\beta} \tag{16}$$

for $\mathbf{T}'$ as described following (11). The SAS output lists the elements of $\boldsymbol{\beta}$, the parameters of the model, and alongside each is its coefficient appropriate to $\mathbf{l}'\mathbf{T}'\mathbf{X}$ where the elements of $\mathbf{l}' = \{\mathbf{l}_i\}$ are printed in the output as L1, L2, L3, ..., for a total of s different Ls.

Hypothesis (12) can also be written as $H{:}\mathbf{L}\mathbf{S}^{-1}\mathbf{K}'\boldsymbol{\beta} = \mathbf{0}$ for any matrices $\mathbf{L}$ and $\mathbf{S}$ that are nonsingular. Therefore the form of f customarily used in place of (16) is

$$f = \mathbf{l}'\mathbf{S}^{-1}\mathbf{T}'\mathbf{X}\boldsymbol{\beta}, \tag{17}$$

where $\mathbf{S}$ is chosen so that the coefficient of each of the first s elements of $\boldsymbol{\beta}$ that occur in f is just a single element of $\mathbf{l}' = \{l_i\}$ for $i = 1, \ldots, s$.

-ii. ***Examples.*** The sum of squares in part (a) of Table 12.4 is, with $n_1 = 3$, $n_2 = 2$ and $n_3 = 1$,

$$R(\boldsymbol{\alpha}\,|\,\mu) = \sum_{i=1}^{3} n_i(\bar{y}_{i\cdot} - \bar{y}_{\cdot\cdot})^2$$

$$= \mathbf{y}'\mathbf{A}\mathbf{y} \quad \text{for } \mathbf{A} = \begin{bmatrix} \bar{\mathbf{J}}_3 & \mathbf{0} & \mathbf{0} \\ \mathbf{0} & \bar{\mathbf{J}}_2 & \mathbf{0} \\ \mathbf{0} & \mathbf{0} & \bar{\mathbf{J}}_1 \end{bmatrix} - \bar{\mathbf{J}}_7 \text{ and } r_{\mathbf{A}} = 2\ .$$

It can be shown that a suitable value of $\mathbf{T}'$ of $r_{\mathbf{A}}$ rows, each being an eigenvector of $\mathbf{A}$ corresponding to the unity eigenvalue of $\mathbf{A}$, is

$$\mathbf{T}' = \begin{bmatrix} 2 & 2 & 2 & 0 & 0 & -3 & -3 \\ 0 & 0 & 0 & 2 & 2 & -2 & -2 \end{bmatrix}.$$

Then with

$$\mathbf{X}\boldsymbol{\beta} = \begin{bmatrix} 1 & 1 & \cdot & \cdot \\ 1 & 1 & \cdot & \cdot \\ 1 & 1 & \cdot & \cdot \\ 1 & \cdot & 1 & \cdot \\ 1 & \cdot & 1 & \cdot \\ 1 & \cdot & \cdot & 1 \\ 1 & \cdot & \cdot & 1 \end{bmatrix} \begin{bmatrix} \mu \\ \alpha_1 \\ \alpha_2 \\ \alpha_3 \end{bmatrix}, \quad \mathbf{T}'\mathbf{X}\boldsymbol{\beta} = \begin{bmatrix} 0 & 6 & 0 & -6 \\ 0 & 0 & 4 & -4 \end{bmatrix} \begin{bmatrix} \mu \\ \alpha_1 \\ \alpha_2 \\ \alpha_3 \end{bmatrix}$$

so that f of (17) is taken as

$$f = [l_2 \quad l_3] \begin{bmatrix} 6 & 0 \\ 0 & 4 \end{bmatrix}^{-1} \begin{bmatrix} 0 & 6 & 0 & -6 \\ 0 & 0 & 4 & -4 \end{bmatrix} \begin{bmatrix} \mu \\ \alpha_1 \\ \alpha_2 \\ \alpha_3 \end{bmatrix}$$

$$= l_2\alpha_1 + l_3\alpha_2 - (l_2 + l_3)\alpha_3\ .$$

This is the estimable function (or at least the interpretation of it) produced by SAS GLM, as shown in part (a) of Table 12.4; and we have shown in (6) that it is appropriate for displaying the associated hypothesis for (12).

For example (b) of Table 12.4 with $n_1 = 7$ and $n_2 = 3$,

$$R(\boldsymbol{\alpha} \mid \mu) = \sum_{i=1}^{2} n_{i\cdot}(\bar{y}_{i\cdot\cdot} - \bar{y}_{\cdot\cdot\cdot})^2 = \mathbf{y}'\mathbf{A}\mathbf{y} \quad \text{for} \quad \mathbf{A} = \begin{bmatrix} \bar{\mathbf{J}}_7 & \mathbf{0} \\ \mathbf{0} & \bar{\mathbf{J}}_3 \end{bmatrix} - \bar{\mathbf{J}}_{10} .$$

A suitable value for the single ($r_{\mathbf{A}} = 1$) eigenvector corresponding to the unity eigenvalue of $\mathbf{A}$ is

$$t' = [3 \quad 3 \quad 3 \quad 3 \quad 3 \quad 3 \quad 3 \quad -7 \quad -7 \quad -7] .$$

Then, for the no-interaction model $E(y_{ijk}) = \mu + \alpha_i + \beta_j$, we have for the n_{ij}-values of Table 12.5,

$$\mathbf{X}\boldsymbol{\beta} = \begin{bmatrix} 1 & 1 & \cdot & 1 & \cdot & \cdot \\ 1 & 1 & \cdot & 1 & \cdot & \cdot \\ 1 & 1 & \cdot & 1 & \cdot & \cdot \\ 1 & 1 & \cdot & \cdot & 1 & \cdot \\ 1 & 1 & \cdot & \cdot & 1 & \cdot \\ 1 & 1 & \cdot & \cdot & 1 & \cdot \\ 1 & 1 & \cdot & \cdot & \cdot & 1 \\ 1 & \cdot & 1 & 1 & \cdot & \cdot \\ 1 & \cdot & 1 & 1 & \cdot & \cdot \\ 1 & \cdot & 1 & \cdot & 1 & \cdot \end{bmatrix} \begin{bmatrix} \mu \\ \alpha_1 \\ \alpha_2 \\ \beta_1 \\ \beta_2 \\ \beta_3 \end{bmatrix}, \tag{18}$$

$$\mathbf{t}'\mathbf{X}\boldsymbol{\beta} = [0 \quad 21 \quad -21 \quad -5 \quad 2 \quad 3][\mu \quad \alpha_1 \quad \alpha_2 \quad \beta_1 \quad \beta_2 \quad \beta_3] .$$

Then f of (18) is taken as

$$f = l_2(21)^{-1}[0 \quad 21 \quad -21 \quad -5 \quad 2 \quad 3][\mu \quad \alpha_1 \quad \alpha_2 \quad \beta_1 \quad \beta_2 \quad \beta_3]'$$

$$= l_2(\alpha_1 - \alpha_2) - \tfrac{5}{21}\beta_1 + \tfrac{2}{21}\beta_2 + \tfrac{3}{21}\beta_3,$$

which, on decimalizing the fractions and using (5), is (7).

12.4. SOLUTION VECTOR OUTPUT

The BLUE of an estimable function $\mathbf{q}'\boldsymbol{\beta}$ is $\mathbf{q}'\boldsymbol{\beta}^\circ$ for $\boldsymbol{\beta}^\circ$ being a solution vector of the normal equations. We therefore concentrate on identifying a

$\boldsymbol{\beta}^\circ$ whenever it occurs in the output of the five routines of Table 12.1. Attention is also directed to availability (if any) of output for calculating the estimated variance of $\mathbf{q}'\boldsymbol{\beta}^\circ$, namely $\mathbf{q}'\mathbf{G}\mathbf{q}\hat{\sigma}^2$ [see (132) in Section 8.7c].

a. BMDP2V

For models with all interactions, and with all-cells-filled data, the observed cell means (e.g., $\bar{y}_{ij\cdot}$ in the 2-way classification) in the output provide estimators of population cell means. But the output's STANDARD DEVIATIONS are not values of s.e.$(\bar{y}_{ij\cdot}) = \sqrt{\hat{\sigma}^2/n_{ij}}$ but are within-cell standard deviations; e.g., $[\Sigma_k(y_{ijk} - \bar{y}_{ij\cdot})^2/(n_{ij} - 1)]^{\frac{1}{2}}$. However, $\hat{\sigma}^2$- and n_{ij}-values are given, so that $\hat{\sigma}^2/n_{ij}$ can easily be obtained. Thus any linear function of cell means, $\Sigma\lambda_{ij}\mu_{ij}$, can be estimated, as $\Sigma\lambda_{ij}\bar{y}_{ij\cdot}$, and so can its standard error, as $[\hat{\sigma}^2\Sigma\lambda_{ij}^2/n_{ij}]^{\frac{1}{2}}$.

No output is provided for some-cells-empty data from with-interaction models. And for without-interaction models (for which the absence of interactions must be specified with data input), no solution of the normal equations is given.

b. GENSTAT ANOVA

This routine uses the "estimated missing values" technique for unbalanced data and has no solution of the normal equations amongst its output. The means in the output include the estimated missing values.

c. SAS GLM

-i. Using a generalized inverse. Available output from SAS GLM includes a solution vector and a generalized inverse $\mathbf{G}$, of $\mathbf{X}'\mathbf{X}$, corresponding to the solution vector. The $\mathbf{G}$ is always symmetric and reflexive (see Section 7.1h) and is labeled $\mathbf{G}_2$ in the output, to indicate its satisfying the first two Penrose conditions [equation (6), Section 7.1h]. Furthermore, the $\mathbf{G}$ always corresponds to the constraints on the solution elements that some of them are zero (Section 9.5a). Both $\mathbf{X}'\mathbf{X}$ and $\mathbf{G}$ can be obtained in the output. Having $\mathbf{G}$ is essential for knowing what the solution vector is estimating and for being able to derive standard errors of estimates.

-ii. The solution vector. Output containing the solution vector and ancillary information appears under the titles PARAMETER, ESTIMATE, T FOR HO PARAMETER = 0, PR > |T|, and STD ERROR OF ESTIMATE.

PARAMETER heads a list of a parameter symbol for each effect, using the same labels as indicated in Section 12.3b. The corresponding solution elements are listed alongside, in the column headed ESTIMATE. Unfortunately, in terms of the overparameterized models that SAS GLM is geared to, the heading "ESTIMATE" is misleading. The numerical values listed there are only a solution vector $\boldsymbol{\beta}^\circ = \mathbf{GX'y}$ of the normal equations $\mathbf{X'X}\boldsymbol{\beta}^\circ = \mathbf{X'y}$, and they are not the BLUE of $\boldsymbol{\beta}$ (which, in overparameterized models, does not exist). This state of affairs is suggested by the B that follows many (or all) solution elements in the output. That B refers the reader to a footnote in the output indicating that the values so labeled in the ESTIMATE column are biased estimates of the corresponding parameters.

That is true, but since $\boldsymbol{\beta}^\circ$ is not the BLUE of $\boldsymbol{\beta}$, and should not be thought of as an estimator of $\boldsymbol{\beta}$, it is more important to use

$$E(\boldsymbol{\beta}^\circ) = \mathbf{H}\boldsymbol{\beta} \quad \text{for } \mathbf{H} = \mathbf{GX'X}$$

(of Section 8.3a) and concentrate attention on

$$\text{BLUE}(\mathbf{H}\boldsymbol{\beta}) = \boldsymbol{\beta}^\circ; \tag{19}$$

i.e., $\boldsymbol{\beta}^\circ$ is the BLUE of $\mathbf{H}\boldsymbol{\beta}$. Users of $\boldsymbol{\beta}^\circ$ would be well advised, therefore, to always have $\mathbf{G}$ in their output, and use it to calculate $\mathbf{H}\boldsymbol{\beta}$ in order to derive the parametric function for which $\boldsymbol{\beta}^\circ$ is the BLUE.

The manner in which $\mathbf{G}$ corresponds to certain elements of $\boldsymbol{\beta}^\circ$ being zero means, in many instances, that the $\mathbf{H}\boldsymbol{\beta}$ estimated by $\boldsymbol{\beta}^\circ$ will have as its elements certain sums and differences of elements of $\boldsymbol{\beta}$.

Example. In a 1-way classification with 3 classes where

$$\boldsymbol{\beta}' = [\mu \quad \alpha_1 \quad \alpha_2 \quad \alpha_3],$$

$\boldsymbol{\beta}^\circ$ from SAS GLM usually has $\alpha_3^\circ = 0$ and is the BLUE of

$$\mathbf{H}\boldsymbol{\beta} = [\mu + \alpha_3 \quad \alpha_1 - \alpha_3 \quad \alpha_2 - \alpha_3 \quad 0]\ .$$

-iii. Standard errors and t-statistics output. The variance–covariance matrix of $\boldsymbol{\beta}^\circ$ is, from (52) of Section 8.3a, $\text{var}(\boldsymbol{\beta}^\circ) = \mathbf{GX'XG'}\sigma^2 = \mathbf{G}\sigma^2$, this last equality because a symmetric reflexive $\mathbf{G}$ is being used. Thus

$$\widehat{\text{var}}(\boldsymbol{\beta}^\circ) = \mathbf{G}\hat{\sigma}^2\ . \tag{20}$$

Each entry in the output headed STD ERROR OF ESTIMATE is the square root of a diagonal element of (20); for the kth element of $\boldsymbol{\beta}^\circ$ it is

$$\widehat{\text{s.e.}}(\beta_k^\circ) = \sqrt{g_{kk}\hat{\sigma}^2}, \tag{21}$$

where g_{kk} is the kth diagonal element of $\mathbf{G}$. An accurate description of (21) is "estimated standard error of solution element."

Using (20), values of the t-statistics

$$t_k = \frac{\beta^\circ}{\widehat{\text{s.e.}}(\beta_k^\circ)} = \frac{\beta^\circ}{\sqrt{g_{tt}\hat{\sigma}^2}} \tag{22}$$

are listed under the heading T FOR HO PARAMETER = 0. This heading appears to indicate that t_k is testing the hypothesis H: $\beta_k = 0$. But is it not; because of (19) it is testing

$$\text{H: } \mathbf{h}_k'\boldsymbol{\beta} = 0, \tag{23}$$

where $\mathbf{h}_k'$ is the kth row of $\mathbf{H} = \mathbf{GX'X}$. Hypothesis (23) is radically different from H: $\beta_k = 0$, a fact that must be recognized should one be tempted to use the output t-statistics for making inferences. Finally, PR > $|T|$ in the output is the probability that a random variable which has the t-distribution will be outside the interval $-T$ to $+T$, for T being the calculated t-statistic for testing (23).

-iv. ***Estimating estimable functions.*** The BLUE of an estimable $\mathbf{q}'\boldsymbol{\beta}$ is $\mathbf{q}'\boldsymbol{\beta}^\circ$ and is readily calculated from $\boldsymbol{\beta}^\circ$ given in the ESTIMATE output—the same function of $\boldsymbol{\beta}^\circ$ as $\mathbf{q}'\boldsymbol{\beta}$ is of $\boldsymbol{\beta}$. But the standard error of $\mathbf{q}'\boldsymbol{\beta}^\circ$ cannot be obtained from the STD ERROR OF ESTIMATE output. That output only provides values of $\sqrt{g_{tt}\hat{\sigma}^2}$, as in (21).

The variance of $\mathbf{q}'\boldsymbol{\beta}^\circ$ is

$$v(\mathbf{q}'\boldsymbol{\beta}^\circ) = \mathbf{q}'\mathbf{G}\mathbf{q}\sigma^2, \tag{24}$$

as in (132) of Section 8.7b, and this requires more than what is provided by the values of (21). For example, suppose β_k and β_m are the kth and mth elements of $\boldsymbol{\beta}$ and that $\beta_k - \beta_m$ is estimable. Then, using (24), and similar to (86) of Section 9.2l,

$$\hat{v}(\beta_k^\circ - \beta_m^\circ) = (g_{kk} + g_{mm} - 2g_{km})\hat{\sigma}^2 . \tag{25}$$

The first two terms of (25) are provided by (21) but the last is not. It must be obtained from $\mathbf{G}$. That is why having $\mathbf{G}$ as part of one's output is essential for calculating estimated variances and (hence estimated standard errors) of estimated estimable functions.

-v. ***Other features.*** SAS GLM has numerous other features that allow the user to have a variety of calculations done beyond the standard calculations already described. For example, procedure CONTRAST will estimate contrasts [see (104) of Section 8.12e] and test hypotheses about them. This and the other features pertaining to fixed effects models are quite straightforward.

d. SAS HARVEY

The output column labeled CONSTANT ESTIMATES is a solution of the normal equations that satisfies the Σ-restrictions.

The output labeled LISTING OF INVERSE ELEMENTS provides $(\mathbf{X}_r'\mathbf{X}_r)^{-1}$, where $\mathbf{X}_r$ is $\mathbf{X}$ after amending the model to satisfy the Σ-restrictions, through replacing in the model the last effect of each factor by minus the sum of all other effects in that factor.

Example. Consider the n_{ij}-values of Table 12.5. In the no-interaction, unrestricted, overparameterized model, $\mathbf{X}\boldsymbol{\beta}$ is as in (15) and

$$\mathbf{X}'\mathbf{X} = \begin{bmatrix} 10 & 7 & 3 & 5 & 4 & 1 \\ 7 & 7 & \cdot & 3 & 3 & 1 \\ 3 & \cdot & 3 & 2 & 1 & 0 \\ 5 & 3 & 2 & 5 & \cdot & \cdot \\ 4 & 3 & 1 & \cdot & 4 & \cdot \\ 1 & 1 & 0 & \cdot & \cdot & \cdot \end{bmatrix}$$

with

$$\mathbf{G} = \frac{1}{39}\begin{bmatrix} \cdot & \cdot & \cdot & \cdot & \cdot & \cdot \\ \cdot & 39 & -39 & -39 & -39 & \cdot \\ \cdot & -39 & 59 & -47 & -44 & \cdot \\ \cdot & -39 & -47 & 50 & 41 & \cdot \\ \cdot & -39 & -44 & 41 & 50 & \cdot \\ \cdot & \cdot & \cdot & \cdot & \cdot & \cdot \end{bmatrix} . \tag{26}$$

The Σ-restricted form of this model is $E(y_{ijk}) = \dot{\mu} + \dot{\alpha}_i + \dot{\beta}_j$ with $\dot{\alpha}_1 + \dot{\alpha}_2 = 0$ and $\dot{\beta}_1 + \dot{\beta}_2 + \dot{\beta}_3 = 0$. Suppose we use these restrictions to make replacements in the model equations in the form

$$\dot{\alpha}_2 = -\dot{\alpha}_1 \quad \text{and} \quad \dot{\beta}_3 = -\left(\dot{\beta}_1 + \dot{\beta}_2\right) . \tag{27}$$

We call this the *replace the last effect* algorithm. (It occurs again in Section 12.5.) On using it in the form of (27) applied to (18), we get what shall be called $\mathbf{X}_r\boldsymbol{\beta}_r$ as

$$\mathbf{X}_r\boldsymbol{\beta}_r = \begin{bmatrix} 1 & 1 & 1 & \cdot \\ 1 & 1 & 1 & \cdot \\ 1 & 1 & 1 & \cdot \\ 1 & 1 & \cdot & 1 \\ 1 & 1 & \cdot & 1 \\ 1 & 1 & \cdot & 1 \\ 1 & 1 & -1 & -1 \\ 1 & -1 & 1 & \cdot \\ 1 & -1 & 1 & \cdot \\ 1 & -1 & \cdot & 1 \end{bmatrix} \begin{bmatrix} \dot{\mu} \\ \dot{\alpha}_1 \\ \dot{\beta}_1 \\ \dot{\beta}_2 \end{bmatrix} . \tag{28}$$

SAS HARVEY yields $\boldsymbol{\beta}^\circ$ using

$$\boldsymbol{\beta}_r' = \left[\hat{\mu} \quad \hat{\alpha}_1 \quad \hat{\beta}_1 \quad \hat{\beta}_2\right]' = (\mathbf{X}_r'\mathbf{X}_r)^{-1}\mathbf{X}_r'\mathbf{y},$$

and we have $\boldsymbol{\beta}^{\circ\prime} =$

$$\left[\mu^\circ = \hat{\mu} \;\; \alpha_1^\circ = \hat{\alpha}_1 \;\; \alpha_2^\circ = -\hat{\alpha}_1 \;\; \beta_1^\circ = \hat{\beta}_1 \;\; \beta_2^\circ = \hat{\beta}_2 \;\; \beta_3^\circ = -\left(\hat{\beta}_1 + \hat{\beta}_2\right)\right]. \quad (29)$$

Then any estimable function $\mathbf{q}'\boldsymbol{\beta}$ does, as usual, have BLUE($\mathbf{q}'\boldsymbol{\beta}$) = $\mathbf{q}'\boldsymbol{\beta}^\circ$, which can be calculated from $\boldsymbol{\beta}^\circ$ obtained from $\hat{\boldsymbol{\beta}}_r$ as just illustrated in (29).

The sampling variance of $\mathbf{q}'\boldsymbol{\beta}^\circ$ can also be obtained. It comes from $(\mathbf{X}_r'\mathbf{X}_r)^{-1}$ that is provided by the routine, but which first has to be expanded from

$$\text{var}\left(\hat{\boldsymbol{\beta}}_r\right) = (\mathbf{X}_r'\mathbf{X}_r)^{-1}\sigma^2 \quad (30)$$

to be var($\boldsymbol{\beta}^\circ$) using, from (29),

$$\alpha_2^\circ = -\hat{\alpha}_1 = -\alpha_1^\circ \quad \text{and} \quad \beta_3^\circ = -\left(\hat{\beta}_1 + \hat{\beta}_2\right) = -\left(\beta_1^\circ + \beta_2^\circ\right). \quad (31)$$

Thus with $\mathbf{X}_r'\mathbf{X}_r = \begin{bmatrix} 10 & 4 & 4 & 3 \\ 4 & 10 & 0 & 1 \\ 4 & 0 & 6 & 1 \\ 3 & 1 & 1 & 5 \end{bmatrix}$, var$\left(\hat{\boldsymbol{\beta}}_r\right)$ is

$$\text{var}\begin{bmatrix} \hat{\mu} \\ \hat{\alpha}_1 \\ \hat{\beta}_1 \\ \hat{\beta}_2 \end{bmatrix} = (\mathbf{X}_r'\mathbf{X}_r)^{-1}\sigma^2 = \frac{1}{702}\begin{bmatrix} 142 & -51 & -85 & -58 \\ -51 & 90 & 33 & 6 \\ -85 & 33 & 172 & 10 \\ -58 & 6 & 10 & 172 \end{bmatrix}. \quad (32)$$

To obtain var($\boldsymbol{\beta}^\circ$) for $\boldsymbol{\beta}^\circ$ of (29) one now uses (32) to derive from (31) variances and covariances involving α_2° and β_3°; e.g.,

$$v(\alpha_2^\circ) = v\left(\hat{\alpha}_1\right) = 90/702; \; \text{cov}(\alpha_1^\circ, \alpha_2^\circ) = -v\left(\hat{\alpha}_1\right) = -90/702;$$

$$v(\beta_3^\circ) = v\left(\hat{\beta}_1 + \hat{\beta}_2\right) = (172 + 172 + 20)/702\,. \quad (33)$$

In this way $\text{var}(\boldsymbol{\beta}^\circ) = \text{var}[\mu^\circ \quad \alpha_1^\circ \quad \alpha_2^\circ \quad \beta_1^\circ \quad \beta_2^\circ \quad \beta_3^\circ]'$ is

$$\frac{1}{702}\begin{bmatrix} \begin{pmatrix} 142 & -51 \\ -51 & 90 \end{pmatrix} & \begin{matrix} 51 \\ -90 \end{matrix} & \begin{pmatrix} -85 & -58 \\ 33 & 6 \end{pmatrix} & \begin{matrix} 85 + 58 = & 143 \\ -33 - 6 = & -39 \end{matrix} \\ \begin{matrix} 51 & -90 \end{matrix} & 90 & \begin{matrix} -33 & -6 \end{matrix} & \begin{matrix} 33 + 6 = & 39 \end{matrix} \\ \begin{pmatrix} -85 & 33 \\ -58 & 6 \end{pmatrix} & \begin{matrix} -33 \\ -6 \end{matrix} & \begin{pmatrix} 172 & 10 \\ 10 & 172 \end{pmatrix} & \begin{matrix} -172 - 10 = & -182 \\ -10 - 172 = & -182 \end{matrix} \\ \begin{matrix} 143 & -39 \end{matrix} & 39 & \begin{matrix} -182 & -182 \end{matrix} & \begin{matrix} 172 + 172 + 20 = & 364 \end{matrix} \end{bmatrix} . \tag{34}$$

For clarity, the submatrices in (34) that came directly from (32) are shown in parentheses; and derivation of elements of the last column (and, therefore, of the last row) is shown in terms of expressions such as (33). There would be similar expressions for elements of the third row and column were there to be more than two αs in the model.

In this way, then, we can use elements of $(\mathbf{X}_r'\mathbf{X}_r)^{-1}$ to obtain $\text{var}(\boldsymbol{\beta}^\circ)$ for the corresponding $\boldsymbol{\beta}^\circ$; and, of course, not only for the SAS HARVEY routine, but for any routine that uses Σ-restrictions and includes $(\mathbf{X}_r'\mathbf{X}_r)^{-1}$ in its output. In passing, it is interesting to illustrate that for the BLUE of any estimable function both (26) and (34) do yield the same value. For example, from $\text{var}(\boldsymbol{\beta}^\circ) = \mathbf{G}\sigma^2$ using (26) we get

$$v(\alpha_1^\circ - \alpha_2^\circ) = (39 + 59 - 78)/39 \qquad = 20/39$$

and

$$v(\beta_1^\circ - \beta_3^\circ) = (50 + 0 + 0)/39 \qquad = 50/39;$$

and (34) gives

$$v(\alpha_1^\circ - \alpha_2^\circ) = (90 + 90 + 180)/702 \quad = 20/39$$

and

$$v(\beta_1^\circ - \beta_3^\circ) = (172 + 364 + 364)/702 = 50/39 \ .$$

e. SPSS ANOVA

No solution of the normal equations is given in the analysis of variance section of the output. But a section originally labeled "Multiple Classification Analysis" under the heading ADJUSTED FOR INDEPENDENTS sub-headed DEV'N, contains a solution vector of the normal equations after dropping interactions from the model. As well as being for this no-interaction situation, the solution elements also satisfy the weighted restrictions, such as $\Sigma_i n_{i\cdot}\alpha_i^\circ = 0$. It is to be noted that even when the analysis of variance part of the output is for a model with interactions they are omitted from the calculations of the output headed DEV'N as just described. No information relevant to standard errors is given.

12.5. A FAULTY COMPUTING ALGORITHM

Equations (142) of Section 9.4b-i are the model equations for the example of Table 9.4 for an overparameterized, unrestricted, with-interaction, 2-way classification model. One can observe in those equations that in any row of the **X**-matrix the coefficient of each ϕ_{ij} is the product of the coefficients of α_i and β_j. The same is evident in (144) of Chapter 9, for the Σ-restricted form of that model. This *product rule*, as we shall call it, is well known and dates back many years. It was especially useful before the advent of computing packages that could handle non full rank models as they stand, and when, by the use of Σ-restrictions and this product rule, such models had to be adapted to packages designed for (full rank) regression analysis. Although using Σ-restrictions and the product rule can be successfully employed in many situations to convert a non full rank model to a full rank model, it cannot be so used in all cases of some-cells-empty data. We illustrate its breakdown in terms of an example.

$\mathbf{X\beta}$ for the no-interaction model for the example of Table 12.5 is shown in (18). For the with-interaction model it is

$$\mathbf{X\beta} = \left[\begin{array}{c|cc|ccc|ccccc} 1 & 1 & \cdot & 1 & \cdot & \cdot & 1 & \cdot & \cdot & \cdot & \cdot \\ 1 & 1 & \cdot & 1 & \cdot & \cdot & 1 & \cdot & \cdot & \cdot & \cdot \\ 1 & 1 & \cdot & 1 & \cdot & \cdot & 1 & \cdot & \cdot & \cdot & \cdot \\ 1 & 1 & \cdot & \cdot & 1 & \cdot & \cdot & 1 & \cdot & \cdot & \cdot \\ 1 & 1 & \cdot & \cdot & 1 & \cdot & \cdot & 1 & \cdot & \cdot & \cdot \\ 1 & 1 & \cdot & \cdot & 1 & \cdot & \cdot & 1 & \cdot & \cdot & \cdot \\ 1 & 1 & \cdot & \cdot & \cdot & 1 & \cdot & \cdot & 1 & \cdot & \cdot \\ 1 & \cdot & 1 & 1 & \cdot & \cdot & \cdot & \cdot & \cdot & 1 & \cdot \\ 1 & \cdot & 1 & 1 & \cdot & \cdot & \cdot & \cdot & \cdot & 1 & \cdot \\ 1 & \cdot & 1 & \cdot & 1 & \cdot & \cdot & \cdot & \cdot & \cdot & 1 \end{array}\right] \begin{bmatrix} \mu \\ \alpha_1 \\ \alpha_2 \\ \beta_1 \\ \beta_2 \\ \beta_3 \\ \phi_{11} \\ \phi_{12} \\ \phi_{13} \\ \phi_{21} \\ \phi_{22} \end{bmatrix}, \qquad (35)$$

where, for ease of observation, dashed lines partition **X** conformably with the different parameter symbols in $\boldsymbol{\beta}$. Again, the product rule is evident: e.g., elements of **X** corresponding to ϕ_{11} are products of elements corresponding to α_1 and β_1.

We now investigate what can happen to this rule when using Σ-restrictions. The Σ-restricted form of the no-interaction model equations (18) is shown in (28), obtained by using what we called in (27) the "replace the last effect" algorithm of $\dot{\alpha}_2 = -\dot{\alpha}_1$ and $\dot{\beta}_3 = -(\dot{\beta}_1 + \dot{\beta}_2)$. Doing the same for the interaction effects in (35) leads to

$$\left.\begin{array}{ll} \dot{\phi}_{11} + \dot{\phi}_{12} + \dot{\phi}_{13} & = 0 \\ \dot{\phi}_{21} + \dot{\phi}_{22} & = 0 \\ \dot{\phi}_{11} + \dot{\phi}_{21} & = 0 \\ \dot{\phi}_{12} + \dot{\phi}_{22} & = 0 \\ \dot{\phi}_{13} & = 0 \end{array}\right\} \text{implying} \left\{\begin{array}{l} \dot{\phi}_{22} = \dot{\phi}_{11} \\ \dot{\phi}_{21} = -\dot{\phi}_{11} \\ \dot{\phi}_{12} = -\dot{\phi}_{11} \end{array}\right. . \tag{36}$$

This gives $\mathbf{X}_r\boldsymbol{\beta}_r$ from (35) being (28) with the addition of providing for ϕ_{11}:

$$\mathbf{X}_r\boldsymbol{\beta}_r = \begin{bmatrix} 1 & 1 & 1 & \cdot & 1 \\ 1 & 1 & 1 & \cdot & 1 \\ 1 & 1 & 1 & \cdot & 1 \\ 1 & 1 & \cdot & 1 & -1 \\ 1 & 1 & \cdot & 1 & -1 \\ 1 & 1 & \cdot & 1 & -1 \\ 1 & 1 & -1 & -1 & 0 \\ 1 & -1 & 1 & \cdot & -1 \\ 1 & -1 & 1 & \cdot & -1 \\ 1 & -1 & \cdot & 1 & 1 \end{bmatrix} \begin{bmatrix} \dot{\mu} \\ \dot{\alpha}_1 \\ \dot{\beta}_1 \\ \dot{\beta}_2 \\ \dot{\phi}_{11} \end{bmatrix} . \tag{37}$$

It can be seen that the product rule is not upheld in (37): if it were, the elements of the last column in $\mathbf{X}_r$ of (37) would be the products of the corresponding elements in the second and third columns. And they are not: e.g., in rows 4, 5 and 6 of (37), $1 \times 0 = 0 \neq -1$; and in row 7, $1 \times (-1) = -1 \neq 0$. Hence, using the "replace the last effect algorithm" on the main effect parameters, and then using the product rule for interactions would not produce the correct $\mathbf{X}_r$ shown in (37). Therefore any computing package that does this will be wrong in this example, and can be wrong for other cases. Evidence of this, for a larger example, is given for the SAS HARVEY package in Searle and Henderson (1983).

The product rule in combination with the "replace the last effect" algorithm for main effects does not work in (37) because it works only if the last row and column of the data grid have all cells filled. This is so because having these cells filled ensures that in each Σ-restriction for $\dot{\phi}$s there is in that last row and column a $\dot{\phi}$ (in the model for the data) that can be replaced by other $\dot{\phi}$s. For example, for Table 12.5 the Σ-restriction $\dot{\phi}_{21} + \dot{\phi}_{22} = 0$ contains no $\dot{\phi}_{23}$ corresponding to the last column. Of course, an appropriate replacement is readily ascertained in easy cases like this one, but a general computing algorithm using this procedure is usually tied to making all replacements from one row and one column, and so requires this row and column to have all cells filled. This explains why the SAS HARVEY computing package requires that the data have (or be re-sequenced to have) the last row and column having all cells filled. And, based on a "replace the first effect" algorithm, this also explains why one option of the SPSS ANOVA package requires data to have (or be re-sequenced to have) the first row and column having all cells filled. Further description of this situation is given in Searle and Henderson (1983).

12.6. OUTPUT FOR ANALYSIS WITH COVARIATES

Analysis in the presence of covariates can have many variations on the same theme, even for the simplest case of the 1-way classification with but one covariate. Chapter 6 shows evidence of that. Combine this with the multitude of computations that can be done by the array of packages available today, and one has a complicated setting in which to answer the simple-sounding question "What is the output for analysis with covariates?". The extent of the annotated computer output for covariance of Searle *et al.* (1982) bears evidence to this.

We therefore make no attempt whatever to be complete, or even nearly so, in providing information about what the different routines provide as output for analysis with covariates. The reader is referred to the aforementioned annotated computer output for details. Attention is confined to the 1-way classification, overparameterized models, with single slope and with multiple slopes.

Dealing first with the single slope model, all routines calculate the estimate slope as $\hat{b} = W_{yz}/W_{zz}$ [equation (33), Chapter 6]. In an adaptation from Searle and Hudson (1982), certain other characteristics of the packages are summarized in Table 12.6: whether the covariate is handled in the form of z_{ij} or $z_{ij} - \bar{z}_{..}$, what restrictions (if any) are used, what output values are given (if any) as solutions to the normal equations, and whether or not adjusted class means $A_i = \bar{y}_{i\cdot} + \hat{b}(\bar{z}_{i\cdot} - \bar{z})$ are calculated. Not all options of the packages are shown; only those involved in the standard use of a package are indicated. Astute users can invoke options available in some packages to do things not shown in Table 12.6, e.g., fitting a no-intercept model.

An immediate reaction to Table 12.6 is that no two of the computing packages are exactly the same. Other features of the packages could be listed, which would perhaps make some packages look more alike, such as the output of means, standard deviations, analyses of variance and so on.

But the object of Table 12.6 is to show features where the packages differ in important ways.

Tables 12.7 and 12.8 summarize for the single slope and multiple slopes models, respectively, what sums of squares are delivered by what routines. The presentation of this information is greatly complicated by the several options that are available in the different routines, e.g., in BMDP4V, the WEIGHT option; in SAS GLM, the sequencing of classes before or after the covariate (A, Z or Z, A) and the Types I–IV sums of squares; in SPSS ANOVA, the options identified as 7, 8, 9 or 10; and in SPSS MANOVA, the two kinds of contrasts, 1 and 2, and the two sets of sums of squares, SEQ and UNIQ. Tables 12.7 and 12.8 attempt to accommodate these options. Readers wanting more complete details are referred to Chapter 6 and to Searle *et al.* (1982).

TABLE 12.6. DISTINCTIVE FEATURES OF COMPUTING ROUTINES FOR ANALYSIS OF A 1-WAY CLASSIFICATION WITH ONE COVARIATE AND THE SINGLE SLOPE MODEL

Computing Package	Form of Covariate[1]	Restrictions[2]	Solutions[3] to Normal Equations: For μ	For α_i	Adjusted means[3,4]
BMD:					
P1V	z_{ij}	Σ	—	—	A_i
P2V	z_{ij}	Σ	—	—	A_i
P4V	z_{ij}	Σ or W[5]	—	—	No
GENSTAT:					
ANOVA	$z_{ij} - \bar{z}_{..}$	W	$\bar{y}_{..}$	$A_i - \bar{y}_{..}$	A_i
GLM[6]	z_{ij}	None	$\bar{y}_{1\cdot} - \hat{b}\bar{z}_{1\cdot}$	$A_i - A_1$	No
SAS:					
GLM	z_{ij}	None	$\bar{y}_{a\cdot} - \hat{b}\bar{z}_{a\cdot}$	$A_i - A_a$	A_i
HARVEY	$z_{ij} - \bar{z}_{..}$	Σ	$\bar{A}$	$A_i - \bar{A}$	A_i
SPSS:					
ANOVA	$z_{ij} - \bar{z}_{..}$	W	$\bar{y}_{..}$	$A_i - \bar{y}_{..}$	No
MANOVA[7]	z_{ij}	W[8]	$\bar{A} - \hat{b}\bar{z}_{..}$	$A_i - \bar{A}$	A_i + d[9]

[1]All routines calculate $\hat{b} = W_{yz}/W_{zz}$; those using z_{ij} can be made to use $z_{ij} - \bar{z}_{..}$.
[2]Σ denotes Σ-restrictions: $\Sigma\alpha_i = 0$; W denotes W-restrictions: $\Sigma n_i\alpha_i = 0$.
[3]Labeled with various names.
[4]Adjusted mean for class i is $A_i = \bar{y}_{i\cdot} - \hat{b}(\bar{z}_{i\cdot} - \bar{z}_{..})$; and $\bar{A} = \Sigma A_i/a$.
[5]User specifies cell weights: unity gives Σ, n_i gives W.
[6]Also available as GLIM.
[7]Used in univariate mode.
[8]Specifying METHOD = SSTYPE(UNIQUE) uses Σ; default is W.
[9]$d = \hat{b}(\Sigma\bar{z}_{i\cdot}/a - \bar{z})$.

TABLE 12.7. SUMS OF SQUARES CALCULATED BY DIFFERENT COMPUTING ROUTINES FOR THE SINGLE SLOPE 1-WAY CLASSIFICATION COVARIANCE MODEL

Computing Package	Symbols Indicate Similar Output	Sums of Squares: Mean				Classes		Covar.		Notes
		$R(\mu)$	$R(\dot{\mu}\|\boldsymbol{\alpha}, b_z)_\Sigma$	$R(\dot{\mu}\|\boldsymbol{\alpha}, b_z)_W$	$R(\dot{\mu}\|\boldsymbol{\alpha}, b_\delta)_\Sigma$	$R(\boldsymbol{\alpha}\|\mu)$	$R(\boldsymbol{\alpha}\|\mu, b)$	$R(b\|\mu)$	$R(b\|\mu, \boldsymbol{\alpha})$	
BMD P4V										
WGT = EQUAL	+		√				√		√	WGT = n_i
WGT = SIZE	□			√			√		√	
GENSTAT ANOVA	•						√		√	
SAS GLM										
Type I: A, Z	†					√			√	Input order: A, Z
Type I: Z, A							√	√		Z, A
Type II, III, IV	•						√		√	
SAS HARVEY					√		√		√	
SPSS ANOVA										User options are 7, 8, 9, 10 and default
Default or 10							√	√		
7 or 9	•						√		√	
8, or 7/10, or 8/10	†					√			√	
SPSS MANOVA (in univariate mode)										
1 SEQ	□			√			√		√	1 and 2 represent alternative controls SEQ and UNIQ are available options
1 UNIQ	+						√		√	
2 SEQ A, Z		√				√			√	
2 SEQ Z, A		√					√	√		
2 UNIQ	+		√				√		√	

TABLE 12.8. SUMS OF SQUARES CALCULATED BY DIFFERENT COMPUTING ROUTINES FOR THE INTRA-CLASS REGRESSION MODEL OF THE 1-WAY CLASSIFICATION

Computing Package[1]	Sums of Squares: Mean	Classes		Covariate					Notes
	$R(\mu)$	$R(\boldsymbol{\alpha}\mid\mu)$	$R(\boldsymbol{\alpha}\mid\mu, \mathbf{b})$	$R(\mathbf{b}\mid\mu)$	$R(\mathbf{b}\mid\mu, \boldsymbol{\alpha})$	$R(b\mid\mu, \boldsymbol{\alpha})$	$R(\mathbf{b}\mid\mu, \boldsymbol{\alpha}) - R(b\mid\mu, \boldsymbol{\alpha})$	$R(\dot{b}\mid\dot{\mu}, \dot{\boldsymbol{\alpha}}, c)_\Sigma$	
SAS GLM									(1) Model: $E(y_{ij}) = \mu + \alpha_i + b_i z_{ij}$
Type I: A, Z		√			√				
(1) Type I: Z, A			√	√					
Type II, III, IV			√		√				
Type I: $A, Z, A*Z$		√				√	√		(2) Model: $E(y_{ij}) = \mu + \alpha_i + (b + c_i) z_{ij}$
(2) Type I: $Z, A, A*Z$			√	√			√		
Type II		√				√	√		$R(\boldsymbol{\alpha}\mid\mu, \mathbf{b}) \equiv R(\boldsymbol{\alpha}\mid\mu, b, \mathbf{c})$
Type III, IV			√				√	√	$R(\mathbf{b}\mid\mu) \equiv R(b, \mathbf{c}\mid\mu)$; $R(\mathbf{b}\mid\mu, \boldsymbol{\alpha}) - R(b\mid\mu, \boldsymbol{\alpha}) \equiv R(\mathbf{c}\mid\mu, \boldsymbol{\alpha}, b)$
SPSS MANOVA (in univariate mode) 2 SEQ	√		√	√				√	Sequence $Z, A\ A*Z$

[1]Routines BMD P4V, GENSTAT ANOVA, SAS HARVEY, and SPSS ANOVA do not handle the intra-class regression model.

CHAPTER 13

MIXED MODELS: A THUMBNAIL SURVEY

13.1. INTRODUCTION

Section 1.2a contains description of fixed effects and of random effects, of the corresponding fixed effects models and random (effects) models, and of mixed models which consist of a mixture of fixed effects and random effects. Although fixed effects models are the prime topic of this book (and they are the only models dealt with up to this point), we here give a brief review of the analysis of mixed models, beginning with a simple example. Since the literature of mixed models is very extensive we give nothing more here than a survey of some of the main results available for the estimation of fixed effects and of variance components in mixed models

a. An example

Consider a randomized complete blocks experiment of baking layer cakes at different pre-assigned temperatures. Let y_{ij} be the height of cake from the jth batch of dough cooked at the ith temperature. We use the model

$$y_{ij} = \mu + \alpha_i + \delta_j + e_{ij} \tag{1}$$

where μ is a general mean, α_i is a treatment (temperature) effect, δ_j is a batch (or block) effect and e_{ij} is an error term. Presumably the different temperatures used in the experiment were carefully pre-assigned. They are temperatures of interest. Consequently the α_i effects are thought of as fixed effects. In contrast, the batches of dough are each just a sample of dough taken from one large cauldron of dough. Therefore the batch effects, the δ_j, are random effects. To them we attribute the properties of zero mean, homoscedastic variances and zero covariances:

$$E(\delta_j) = 0 \quad \forall\ j, \tag{2}$$

$$E\left[\delta_j - E(\delta_j)\right]^2 = E(\delta_j)^2 = \sigma_\delta^2 \quad \forall\ j \tag{3}$$

$$E\left[\delta_j - E(\delta_j)\right]\left[\delta_{j'} - E(\delta_{j'})\right] = E(\delta_j\delta_{j'}) = 0 \quad \forall\ j \neq j'\ . \tag{4}$$

With α_i being a fixed effect and δ_j being a random effect, (1) is the model equation for what we call a mixed model having, as it does, a mixture of fixed and random effects. Having δ_j as a random effect, with properties (2)–(4), then necessitates more careful definition of e_{ij} than is used in the fixed effects counterpart of (1), wherein e_{ij} is defined as $y_{ij} - E(y_{ij})$. Now, in the mixed model, it is defined as

$$e_{ij} = y_{ij} - E(y_{ij} \mid \delta_j), \tag{5}$$

where $E(y_{ij} \mid \delta_j)$ means expected value of y_{ij}, given δ_j. Then

$$E(e_{ij}) = 0 \tag{6}$$

and to the e_{ij}s we attribute uniform variance

$$E\left[e_{ij} - E(e_{ij})\right]^2 = E\left(e_{ij}^2\right) = \sigma_e^2 \quad \forall\ i,\ j, \tag{7}$$

and zero covariance between every pair of e_{ij}s,

$$E\left[e_{ij} - E(e_{ij})\right]\left[e_{i'j'} - E(e_{i'j'})\right] = E(e_{ij}e_{i'j'}) = 0,$$

except for $i = i'$ and $j = j'$ together, and zero covariance between every δ_j and every e_{ij},

$$E\left[e_{ij} - E(e_{ij})\right]\left[\delta_{j'} - E(\delta_{j'})\right] = E\left(e_{ij}\delta_{j'}\right) = 0 \quad \forall\ i,\ j \text{ and } j'\ . \tag{8}$$

These assumptions are typical of those made about random effects and error terms in all mixed models.

The expected values in (2), (5) and (6) mean that whereas

$$E(y_{ij} \mid \delta_j) = \mu + \alpha_i + \delta_j, \tag{9}$$

$$E(y_{ij}) = \mu + \alpha_i\ . \tag{10}$$

And the variances in (3) and (7) give

$$\sigma^2_{y_{ij}} = \sigma^2_\delta + \sigma^2_e \; . \tag{11}$$

It is this relationship which gives to the variances σ^2_δ and σ^2_e the name *variance components*; they are the (variance) components of the variance of any y_{ij}.

All of the models considered previously in this book have involved but a single variance, σ^2_e. And that did not affect estimation of the other parameters of those models, namely the fixed effects. Nor was estimation of σ^2_e difficult. But with mixed models there is σ^2_e and other variances, namely the variances of the random effects, e.g., σ^2_δ of (11). And in almost all cases of unbalanced data those variances do affect estimation of the fixed effects; also, those variances are often difficult to estimate. These are the difficulties of the mixed model: for unbalanced data, estimating fixed effects in the presence of variance components, and estimating the variance components themselves. The "mixed model problem" is what some writers call those difficulties. (In the presence of balanced data most of the difficulties evaporate.)

b. A general description

To describe a general mixed model we expand the $\mathbf{y} = \mathbf{X}\boldsymbol{\beta} + \mathbf{e}$ of Chapter 8 to

$$\mathbf{y} = \mathbf{X}\boldsymbol{\beta} + \mathbf{Z}\mathbf{u} + \mathbf{e}, \tag{12}$$

where $\mathbf{y}$ and $\mathbf{X}\boldsymbol{\beta}$ are as previously, and $\mathbf{Zu}$ represents the occurrence of random effects. Just as $\boldsymbol{\beta}$ is the vector of fixed effects with corresponding model matrix $\mathbf{X}$, so is $\mathbf{u}$ the vector of random effects with corresponding model matrix $\mathbf{Z}$, and $\mathbf{e}$ is a vector of error terms defined as $\mathbf{e} = \mathbf{y} - E(\mathbf{y} \mid \mathbf{u})$, where $E(\mathbf{y} \mid \mathbf{u}) = \mathbf{X}\boldsymbol{\beta} + \mathbf{Zu}$ and

$$E(\mathbf{y}) = \mathbf{X}\boldsymbol{\beta} \; . \tag{13}$$

The symbol $\mathbf{V}$ is used for the dispersion matrix of $\mathbf{y}$:

$$\mathbf{V} = \text{var}(\mathbf{y}) \; . \tag{14}$$

Then, from (12),
$$\mathbf{V} = \text{var}(\mathbf{Zu} + \mathbf{e}) \; . \tag{15}$$

Two properties customarily incorporated in (15) are that

$$\text{cov}(\mathbf{u}, \mathbf{e}') = \mathbf{0} \quad \text{and} \quad \text{var}(\mathbf{e}) = \sigma^2_e \mathbf{I} \; . \tag{16}$$

Then
$$\mathbf{V} = \mathbf{Z}\,\text{var}(\mathbf{u})\mathbf{Z}' + \sigma^2_e \mathbf{I} \; . \tag{17}$$

Now partition $\mathbf{u}$ into r sub-vectors

$$\mathbf{u}' = [\mathbf{u}_1' \quad \mathbf{u}_2' \quad \cdots \quad \mathbf{u}_i' \quad \cdots \quad \mathbf{u}_r'], \tag{18}$$

where each sub-vector $\mathbf{u}_i$ has as elements the effects corresponding to all levels (in the data) of a single, random effects factor. For example, $\mathbf{u}_1$ for (1) would be the δs corresponding to the batches (of dough) in the data. And if a random effect for a treatment-by-batch interaction were to be included in the model, that could be represented by $\mathbf{u}_2$. To be conformable with $\mathbf{u}$ of (18), $\mathbf{Z}$ is partitioned as

$$\mathbf{Z} = [\mathbf{Z}_1 \quad \mathbf{Z}_2 \quad \cdots \quad \mathbf{Z}_i \quad \cdots \quad \mathbf{Z}_r] . \tag{19}$$

This gives

$$\mathbf{y} = \mathbf{X}\boldsymbol{\beta} + \sum_{i=1}^{r} Z_i\mathbf{u}_i + \mathbf{e}; \tag{20}$$

$$\mathbf{V} = \text{var}(\mathbf{y}) = \text{var}\left(\sum_{i=1}^{r} \mathbf{Z}_i\mathbf{u}_i + \mathbf{e}\right) . \tag{21}$$

Then to the $\mathbf{u}_i$s we attribute mean and variance properties that are direct extensions of (2)–(4):

$$E(\mathbf{u}_i) = \mathbf{0}, \quad \forall\ i, \tag{22}$$

$$\text{var}(\mathbf{u}_i) = \sigma_i^2\mathbf{I}_{q_i}, \quad \forall\ i \tag{23}$$

and

$$\text{cov}(\mathbf{u}_i, \mathbf{u}_{i'}') = \mathbf{0}, \quad \forall\ i \neq i'; \tag{24}$$

and, like (8),

$$\text{cov}(\mathbf{u}_i, \mathbf{e}') = \mathbf{0} . \tag{25}$$

q_i in (23) is the order of $\mathbf{u}_i$, the number of levels in the data of the random effects factor represented by $\mathbf{u}_i$. Using (23)–(25) gives (21) as

$$\mathbf{V} = \sum_{i=1}^{r} \mathbf{Z}_i\mathbf{Z}_i'\sigma_i^2 + \sigma_e^2\mathbf{I} . \tag{26}$$

Finally, to be even more concise we can define

$$\mathbf{u}_0 = \mathbf{e}, \qquad \sigma_0^2 = \sigma_e^2 \quad \text{and} \quad \mathbf{Z}_0 = \mathbf{I} \tag{27}$$

and have

$$\mathbf{y} = \mathbf{X}\boldsymbol{\beta} + \sum_{i=0}^{r} \mathbf{Z}_i\mathbf{u}_i \tag{28}$$

and

$$\mathbf{V} = \sum_{i=0}^{r} \mathbf{Z}_i\mathbf{Z}_i'\sigma_i^2 . \tag{29}$$

This is the description of the most frequently-used general form of mixed model. It can, of course, be made more complicated insofar as variance–covariance structure is concerned, by allowing $\text{var}(\mathbf{u}_i)$ to be something other than the simple $\sigma_i^2\mathbf{I}$ of (23); e.g., by allowing variances of the elements of $\mathbf{u}_i$ to be something other than all equal; and by allowing those elements to have nonzero covariances. Further complication can be included by allowing $\text{cov}(\mathbf{u}_i, \mathbf{u}'_{i'})$ of (24) to be something other than zero. In permitting these more general properties one could write (28) and (29) as

$$\mathbf{y} = \mathbf{X}\boldsymbol{\beta} + \boldsymbol{\varepsilon} \tag{30}$$

with

$$\text{var}(\mathbf{y}) = \text{var}(\boldsymbol{\varepsilon}) = \mathbf{W} \tag{31}$$

for $\mathbf{W}$ being any positive definite matrix (or even positive semi-definite). The form $\mathbf{W} = \mathbf{V}$ for $\mathbf{V}$ of (29) is then just one special case—albeit the case that is widely used and to which we shall confine ourselves, with $\mathbf{V}$ assumed positive definite.

A simple form of (28) for $\boldsymbol{\beta}$ being the vector of cell means, $\boldsymbol{\mu}$, is

$$\mathbf{y} = \mathbf{X}\boldsymbol{\mu} + \sum_{i=0} \mathbf{Z}_i\mathbf{u}_i, \tag{32}$$

with

$$\mathbf{X} = \mathbf{D}\{\mathbf{1}_{n_t}\} \tag{33}$$

or, as in (30) and (31),

$$\mathbf{y} = \mathbf{X}\boldsymbol{\mu} + \boldsymbol{\varepsilon} \quad \text{with} \quad \text{var}(\boldsymbol{\varepsilon}) = \mathbf{W}\ . \tag{34}$$

Then (32) and (34) are cell means formulations of mixed models with $\mathbf{X}$ of (33) being of full column rank. We use these formulations in the sequel.

Example (continued). In (1) and (10) write

$$\mu_i = \mu + \alpha_i$$

and the model equation (1) is

$$y_{ij} = \mu_i + \delta_j + e_{ij}\ . \tag{35}$$

Suppose there are a temperatures and b batches of dough, so that $i = 1, \ldots, a$ and $j = 1, \ldots, b$. Then with $\mathbf{y}$ and $\mathbf{e}$ being, respectively, the vectors of y_{ij} and e_{ij} in lexicon order, and with

$$\boldsymbol{\mu}' = [\mu_1 \quad \cdots \quad \mu_a] \quad \text{and} \quad \mathbf{u}' = [\delta_1 \quad \cdots \quad \delta_b],$$

the model equation (35) is

$$\mathbf{y} = \mathbf{X}\boldsymbol{\mu} + \mathbf{Z}\mathbf{u} + \mathbf{e} \tag{36}$$

with

$$\mathbf{X} = \{_d\, \mathbf{1}_b\}_{i=1}^{i=a} \quad \text{and} \quad \mathbf{Z} = [\mathbf{I}_b \ldots \mathbf{I}_b]', \tag{37}$$

with the matrix $\mathbf{I}_b$ occurring a times in $\mathbf{Z}$.

c. Estimation

In the fixed effects models of preceding chapters, where $\mathbf{y} = \mathbf{X}\boldsymbol{\beta} + \mathbf{e}$ with $\mathbf{e} \sim (0, \sigma^2\mathbf{I})$, the method of estimation used was least squares. It led to normal equations $\mathbf{X}'\mathbf{X}\boldsymbol{\beta}^\circ = \mathbf{X}'\mathbf{y}$ of (14) in Section 8.2a. Those equations can still be used for the mixed model formulated either as (28) and (29), or as (30) and (31), or as (34). They still provide estimators; but they are estimators that are obtained without regard for the variance components; indeed, in mixed models they are not always BLU estimators. To obtain BLUEs, the method given in Section 8.10 must be used: for the model (28) and (29) this is

$$\text{BLUE}(\boldsymbol{\lambda}'\mathbf{X}\boldsymbol{\beta}) = \boldsymbol{\lambda}'\mathbf{X}(\mathbf{X}'\mathbf{V}^{-1}\mathbf{X})^{-}\mathbf{X}'\mathbf{V}^{-1}\mathbf{y} \tag{38}$$

In this way, the variance components get taken into account in estimating estimable functions $\boldsymbol{\lambda}'\mathbf{X}\boldsymbol{\beta}$ of the fixed effects: the variance of (38) is

$$\text{var}[\text{BLUE}(\boldsymbol{\lambda}'\mathbf{X}\boldsymbol{\beta})] = \boldsymbol{\lambda}'\mathbf{X}(\mathbf{X}'\mathbf{V}^{-1}\mathbf{X})^{-}\mathbf{X}'\boldsymbol{\lambda}\,. \tag{39}$$

This method of estimation is preferred over using ordinary least squares, $\mathbf{X}'\mathbf{X}\boldsymbol{\beta}^\circ_{\text{OLSE}} = \mathbf{X}'\mathbf{y}$, and is the method illustrated at equation (103) that follows.

A difficulty with (38) and (39) is that they require $\mathbf{V} = \text{var}(\mathbf{y})$, and the variance of any observation involves the variance components; as in (11), for example. Thus (38) and (39) involve the variance components. When those are known, there is no trouble in using (38) and (39). But in many situations the variance components are not known, and one is then faced with two questions: (i) How are the σ^2s estimated? (ii) How are the resulting estimates used for estimating $\boldsymbol{\lambda}'\mathbf{X}\boldsymbol{\beta}$? Fortunately, although the first of these questions is full of difficulties, the second is less so. Given a set of estimated variance components, use those estimates in place of the components themselves in $\mathbf{V}$. This gives what shall be called $\hat{\mathbf{V}}$. Then calculate (38) with $\hat{\mathbf{V}}$ in place of $\mathbf{V}$:

$$\text{BLUE}_{\hat{\mathbf{V}}}(\boldsymbol{\lambda}'\mathbf{X}\boldsymbol{\beta}) = \boldsymbol{\lambda}'\mathbf{X}(\mathbf{X}'\hat{\mathbf{V}}^{-1}\mathbf{X})^{-}\mathbf{X}'\hat{\mathbf{V}}^{-1}\mathbf{y}\,. \tag{40}$$

Kackar and Harville (1981) show for a broad class of estimators of the variance components that (40) is an unbiased estimator of $\boldsymbol{\lambda}'\mathbf{X}\boldsymbol{\beta}$; and in Kackar and Harville (1984) they show how to calculate an approximate variance estimate of (40).

Simplification of (38) occurs when the cell means formulation, (32) and (33), of the mixed model is used. Then $\mathbf{X}$ of (33) is of full column rank, and so $\boldsymbol{\mu}$ is estimable and $\mathbf{X}'\mathbf{V}^{-1}\mathbf{X}$ is nonsingular. Hence, instead of (38), we have the BLUE of $\boldsymbol{\mu}$ as

$$\hat{\boldsymbol{\mu}} = (\mathbf{X}'\mathbf{V}^{-1}\mathbf{X})^{-1}\mathbf{X}'\mathbf{V}^{-1}\mathbf{y} \quad \text{with} \quad \text{var}(\hat{\boldsymbol{\mu}}) = (\mathbf{X}'\mathbf{V}^{-1}\mathbf{X})^{-1} . \tag{41}$$

Again, as in (40), $\hat{\mathbf{V}}$ can be used in (41) to overcome the difficulty of not knowing $\mathbf{V}$.

One important situation when neither (38) nor (41) is needed is that of balanced data— or at least a very broad class of balanced data, as formally defined through a notation based on direct products of matrices, as used by Smith and Hocking (1978), Searle and Henderson (1979) and Seifert (1979). With the aid of that notation and (192) of Section 8.10, Searle (1987) has shown that for such balanced data

$$\text{OLSE}(\boldsymbol{\lambda}'\mathbf{X}\boldsymbol{\beta}) = \boldsymbol{\lambda}'\mathbf{X}(\mathbf{X}'\mathbf{X})^{-}\mathbf{X}'\mathbf{y} \tag{42}$$

$$= \boldsymbol{\lambda}'\mathbf{X}(\mathbf{X}'\mathbf{V}^{-1}\mathbf{X})^{-}\mathbf{X}'\mathbf{V}^{-1}\mathbf{y} = \text{BLUE}(\boldsymbol{\lambda}'\mathbf{X}\boldsymbol{\beta}) . \tag{43}$$

This is a useful result because, for balanced data, $\text{OLSE}(\boldsymbol{\lambda}'\mathbf{X}\boldsymbol{\beta})$ is usually a simple function of observed means. Therefore, by (43), so is $\text{BLUE}(\boldsymbol{\lambda}'\mathbf{X}\boldsymbol{\beta})$.

Example (continued). $\text{OLSE}(\alpha_i - \alpha_{i'})$ for the model (1) ignores the random δ_j effects and is the same as it is for (10); and the latter is simply a 1-way classification and so

$$\text{OLSE}(\alpha_i - \alpha_{i'}) = \bar{y}_{i\cdot} - \bar{y}_{i'\cdot} .$$

Therefore, by (43), for balanced data, the BLUE of $\alpha_i - \alpha_{i'}$ is

$$\text{BLUE}(\alpha_i - \alpha_{i'}) = \text{OLSE}(\alpha_i - \alpha_{i'}) = \bar{y}_{i\cdot} - \bar{y}_{i'\cdot} .$$

13.2. ESTIMATING VARIANCE COMPONENTS FROM BALANCED DATA

The estimation of variance components is an extensive topic, whether one wants estimates for their own sake, or for use in (40) in order to estimate the BLUE of an estimable function that otherwise demands knowing the variance components. Indeed, the whole subject of estimating

TABLE 13.1. DATA FOR A RANDOMIZED COMPLETE BLOCKS EXPERIMENT OF 3 TREATMENTS AND 5 BLOCKS

	y_{ij}					$y_{i\cdot}$	$\bar{y}_{i\cdot}$
	2	1	1	2	4	10	2
	8	6	4	3	9	30	6
	5	5	1	4	5	20	4
$y_{\cdot j}$	15	12	6	9	18	$60 = y_{..}$	
$\bar{y}_{\cdot j}$	5	4	2	3	6		$4 = \bar{y}_{..}$

variance components is too extensive for a full detailed account here, but a brief survey of available methods is in order. We begin with balanced data, the easiest case, and the origin of much methodology for unbalanced data.

Estimation of variance components from balanced data is almost always done by what is known as the analysis of variance (ANOVA) method. This method obtains estimators by equating the sums of squares (or mean squares) of an analysis of variance table to their expected values. The latter are linear combinations of the variance components. Hence the method yields equations linear in the variance components that can be solved; and the solutions are taken as the estimators.

a. Example (continued)

Suppose for the example in (1) the data are as in Table 13.1, with the corresponding analysis of variance being as shown in Table 13.2. As a result

TABLE 13.2. ANALYSIS OF VARIANCE FOR TABLE 13.1

Source of Variation	Degrees of Freedom	Sum of Squares	
Mean	$1 = 1$	$ab\bar{y}_{..}^2$	$= 240$
Rows	$a - 1 = 2$	$b\sum_{i=1}^{a} (\bar{y}_{i\cdot} - \bar{y}_{..})^2$	$= 40$
Columns	$b - 1 = 4$	$a\sum_{j=1}^{b} (\bar{y}_{\cdot j} - \bar{y}_{..})^2$	$= 30$
Error	$(a-1)(b-1) = 8$	$\sum_{i=1}^{a}\sum_{j=1}^{b} (y_{ij} - \bar{y}_{i\cdot} - \bar{y}_{\cdot j} + \bar{y}_{..})^2$	$= 14$
Total	$ab = 15$	$\sum_{i=1}^{a}\sum_{j=1}^{b} y_{ij}^2$	$= 324$

of properties (2)–(4), it will be found that expected values of the columns and error sums of squares of Table 13.2 are:

$$E\,a\sum_{j=1}^{b}(\bar{y}_{\cdot j}-\bar{y}_{\cdot\cdot})^2=(b-1)\left(a\sigma_\delta^2+\sigma_e^2\right) \tag{44}$$

$$E\sum_{i=1}^{a}\sum_{j=1}^{b}(y_{ij}-\bar{y}_{i\cdot}-\bar{y}_{\cdot j}+\bar{y}_{\cdot\cdot})^2=(a-1)(b-1)\sigma_e^2\,. \tag{45}$$

Then, with these sums of squares having the computed values in Table 13.2 of 30 and 14, respectively, the ANOVA method of estimating the variance components is to equate these computed values to the left-hand sides of (44) and (45), but writing each σ^2 therein as $\hat{\sigma}^2$ to denote the corresponding estimate. This gives

$$30=4\left(3\hat{\sigma}_\delta^2+\hat{\sigma}_e^2\right)\quad\text{and}\quad 14=2(4)\hat{\sigma}_e^2\,. \tag{46}$$

Thus

$$\hat{\sigma}_e^2=\tfrac{14}{8}\quad\text{and}\quad\hat{\sigma}_\delta^2=\left(\tfrac{30}{4}-\tfrac{14}{8}\right)/3=\tfrac{23}{12}\,. \tag{47}$$

Properties of these estimators are, of course, derived by writing the estimators, using (44) and (45), as

$$\hat{\sigma}_e^2=\sum_{i=1}^{a}\sum_{j=1}^{b}(y_{ij}-\bar{y}_{i\cdot}-\bar{y}_{\cdot j}+\bar{y}_{\cdot\cdot})^2/(a-1)(b-1) \tag{48}$$

and

$$\hat{\sigma}_\delta^2=\frac{1}{a}\left[\frac{a}{b-1}\Sigma_j(\bar{y}_{\cdot j}-\bar{y}_{\cdot\cdot})^2-\hat{\sigma}_e^2\right]. \tag{49}$$

Some of the merits and demerits of estimating variance components from balanced data in this manner are as follows.

b. Merits

-i. ***Broad applicability.*** Application of the ANOVA method to balanced data from any mixed model is straightforward, and details for many particular cases are available in several texts. In almost all cases the required computation is easy. Furthermore, no distributional assumptions are required of the data, other than variance and covariance assumptions of the nature illustrated in (2)–(4) and generalized in (22)–(25).

-ii. ***Unbiasedness.*** Estimators are always unbiased—indeed they are best quadratic unbiased; i.e., of all quadratic functions of the observations

that are unbiased for each variance component, the ANOVA estimator has minimum variance. Moreover, under normality assumptions on the error terms and the random effects, the estimators are best unbiased; i.e., of all unbiased estimators, the ANOVA estimators have minimum variance.

-iii. Sampling variances. Under normality assumptions, sampling variances of the estimators are readily available—because the mean squares in an analysis of variance table are independent and have distributions that are proportional to χ^2-distributions. Thus, if

$$\hat{\sigma}^2 = \Sigma k_i M_i, \tag{50}$$

where M_i is a mean square with degrees of freedom f_i such that

$$f_i M_i / E(M_i) \sim \chi^2_{f_i}$$

and where the M_is are independent, then the sampling variance of $\hat{\sigma}^2$ is

$$v(\hat{\sigma}^2) = 2\Sigma_i \frac{k_i^2 [E(M_i)]^2}{f_i} \, . \tag{51}$$

The difficulty with (51) as a usable variance is that it is, through $E(M_i)$, a function of the unknown variance components. To use it on data one therefore needs to estimate it. This is done unbiasedly as follows.

The normality assumptions give

$$v(M_i) = 2[E(M_i)]^2/f_i \, .$$

But, by definition

$$v(M_i) = E(M_i^2) - [E(M_i)]^2 \, . \tag{52}$$

Equating these two expressions gives

$$E(M_i^2) = \frac{f_i + 2}{f_i}[E(M_i)]^2, \tag{53}$$

which suggests that an unbiased estimator of $[E(M_i)]^2/f_i$ is $M_i^2/(f_i + 2)$. Hence from (51)

$$\hat{v}(\hat{\sigma}^2) = 2\Sigma_i \frac{k_i^2 M_i^2}{f_i + 2} \tag{54}$$

is an unbiased estimator of $v(\hat{\sigma}^2)$.

Example (continued). Rewrite (49) and (50) as

$$\hat{\sigma}_e^2 = \text{MSE} \quad \text{and} \quad \hat{\sigma}_\delta^2 = \frac{\text{MSB} - \text{MSE}}{a}, \tag{55}$$

where MSB and MSE are the mean squares for "Columns" and "Error" in Table 13.2 with, from (44) and (45),

$$E(\text{MSB}) = a\sigma_\delta^2 + \sigma_e^2 \quad \text{and} \quad E(\text{MSE}) = \sigma_e^2 \,. \tag{56}$$

Then, from (51)

$$v(\hat{\sigma}_\delta^2) = \frac{2}{a^2}\left\{\frac{[E(\text{MSB})]^2}{b-1} + \frac{[E(\text{MSE})]^2}{(a-1)(b-1)}\right\}$$

$$= \frac{2}{a^2}\left[\frac{(a\sigma_\delta^2 + \sigma_e^2)^2}{b-1} + \frac{\sigma_e^4}{(a-1)(b-1)}\right]$$

and from (54) an unbiased estimator of this variance is

$$\hat{v}(\hat{\sigma}_\delta^2) = \frac{2}{a^2}\left[\frac{\text{MSB}^2}{b-1+2} + \frac{\text{MSE}^2}{(a-1)(b-1)+2}\right]. \tag{57}$$

For the data in Table 13.2, with MSB $= \frac{30}{4}$ and MSE $= \frac{14}{8}$, the estimate from (57) is that for $\hat{\sigma}_\delta^2 = \frac{23}{12}$ of (47):

$$\hat{v}(\hat{\sigma}_\delta^2) = \frac{2}{9}\left[\frac{(\frac{30}{4})^2}{5-1+2} + \frac{(\frac{14}{8})^2}{(3-1)(5-1)+2}\right] = \frac{1549}{720} = 2.12 \,.$$

c. Demerits

-i. Negative estimates. The method does not preclude occurrence of negative estimates. For example, suppose the experiment of Table 13.1 had been for just two rows and two columns, with observations 4 and 14 in the first row and 6 and 2 in the second. Then it will be found that the equations in place of (46) and (47) are, from (49) and (50), $9 = 2\hat{\sigma}_\delta^2 + \hat{\sigma}_e^2$ and $49 = \hat{\sigma}_e^2$, so that $\hat{\sigma}_e^2 = 49$ and $\hat{\sigma}_\delta^2 = \frac{1}{2}(9 - 49) = -20$. Clearly, a negative estimate of a parameter (a variance) that by definition is positive is an embarrassment. Nevertheless it can happen, and sometimes with real data it does happen; indeed, in some circumstances there is a positive probability that it will happen [e.g., Searle (1971), p. 415]. Despite this, the attractive properties listed as (iv) and (v) under "Merits" still apply.

*-ii. **Distributional properties.*** Even under normality assumptions, the distribution of ANOVA estimators is unknown. This is because such estimators are not just sums of independent χ^2-variables (which, themselves are χ^2-variables) but are linear combinations of χ^2-variables that usually include some negative coefficients. So, although sampling variances can be obtained, distributional properties cannot, except in some cases as infinite sums of weighted χ^2-distributions (see Robinson, 1965). Exact confidence intervals are, therefore, not available.

13.3. ESTIMATING VARIANCE COMPONENTS FROM UNBALANCED DATA

The overriding problem with estimating variance components from unbalanced data is that many methods of estimation are available and deciding between them can be a matter of some difficulty. As computers are developed that are larger and faster, and cheaper to use per arithmetic operation, the practical choice may settle down to one or other of two maximum-likelihood based methods, but until such broad acceptance of but one methodology occurs (if it ever does), the reader needs to be aware of the several methods currently available. We therefore briefly summarize nine methods: a general ANOVA method, Henderson's Methods I, II and III; and ML, REML, MINQUE, MIVQUE and I-MINQUE.

a. General ANOVA methodology

The principle of the ANOVA method used with balanced data of equating sums of squares to their expected values, as just illustrated, can be generalized. The generalization is to use any of a broad class of quadratic forms in place of the sums of squares.

Let $\boldsymbol{\sigma}^2$ be the vector of variance components that are to be estimated. And let $\mathbf{q}$ be a vector, the same order as $\boldsymbol{\sigma}^2$, of any linearly independent quadratic forms of the observations. Then $E(\mathbf{q})$ can be obtained from Theorem 7.2a of Section 7.5a. Suppose $\mathbf{q}$ is such that

$$E(\mathbf{q}) = \mathbf{C}\boldsymbol{\sigma}^2 \tag{58}$$

for some matrix $\mathbf{C}$. Then, if $\mathbf{C}$ is nonsingular,

$$\hat{\boldsymbol{\sigma}}^2 = \mathbf{C}^{-1}\mathbf{q} \tag{59}$$

is an unbiased estimator of $\boldsymbol{\sigma}^2$. The elements of $\mathbf{q}$ can be any quadratic forms in the observations such that (58) holds. Thus there is little limit as to how many different ways (59) can be used. The salient question is "can (59) be used in some optimal manner, i.e., is there some optimal set of quadratic forms that can be used as elements of **q**?" For unbalanced data there seems to be no answer to this question, whereas for balanced data the sums of squares of the analysis of variance do provide estimators that have certain minimum variance and unbiased properties as described in Section 13.2b-ii.

From (58) the dispersion matrix of $\hat{\boldsymbol{\sigma}}^2$ is

$$\text{var}(\hat{\boldsymbol{\sigma}}^2) = \mathbf{C}^{-1}\text{var}(\mathbf{q})\mathbf{C}^{-1\prime}, \tag{60}$$

where elements of var(**q**) are variances of and covariances between the quadratic forms used as elements of **q**; and under normality assumptions, these can be derived from (58) of Section 7.5a.

Notation. A useful operator [Searle (1982), Section 12.9] on symmetric matrices is vech: it creates a vector of order $\frac{1}{2}n(n+1)$ from a symmetric matrix of order n, consisting of all the elements on and above the diagonal. For example,

$$\text{vech}\begin{bmatrix} a & b & c \\ b & x & y \\ c & y & p \end{bmatrix} = \begin{bmatrix} a \\ b \\ c \\ x \\ y \\ p \end{bmatrix}.$$

We use this operator to derive an unbiased estimator of $\text{var}(\hat{\boldsymbol{\sigma}}^2)$.

Elements of $\text{var}(\hat{\boldsymbol{\sigma}}^2)$ are variances of, and covariances between, elements $\hat{\sigma}_i^2$ and $\hat{\sigma}_j^2$, say, of $\hat{\boldsymbol{\sigma}}^2$. And, by the definition of variance and covariance, with $\hat{\sigma}_i^2$ being an unbiased estimator of σ_i^2,

$$E\left(\hat{\sigma}_i^4\right) = v\left(\hat{\sigma}_i^2\right) + \sigma_i^4 \tag{61}$$

and

$$E\left(\hat{\sigma}_i^2\hat{\sigma}_j^2\right) = \text{cov}\left(\hat{\sigma}_i^2, \hat{\sigma}_j^2\right) + \sigma_i^2\sigma_j^2 . \tag{62}$$

Now define

$$\mathbf{v} = \text{vech}\left[\text{var}(\hat{\boldsymbol{\sigma}}^2)\right] \quad \text{and} \quad \boldsymbol{\gamma} = \text{vech}(\boldsymbol{\sigma}^2\boldsymbol{\sigma}^{2\prime}) . \tag{63}$$

Then **v** is a vector of all the variances of, and covariances between, the $\hat{\sigma}^2$s; and $\boldsymbol{\gamma}$ is a vector of the squares and products of the σ^2s. Now consider var(**q**) in (60). In many situations, and certainly under normality assumptions, elements of var(**q**) are quadratic functions of the σ^2s; and hence by

(60), so are elements of $\text{var}(\hat{\boldsymbol{\sigma}}^2)$. Under these circumstances, using $\mathbf{v}$ and $\boldsymbol{\gamma}$ defined in (63), equation (60) can be rewritten as

$$\mathbf{v} = \mathbf{A}\boldsymbol{\gamma} \tag{64}$$

for some $\mathbf{A}$. Further define $\hat{\boldsymbol{\gamma}} = \text{vech}(\hat{\boldsymbol{\sigma}}^2\hat{\boldsymbol{\sigma}}^{2\prime})$. Then from using (61) and (62) in (63), $E(\hat{\boldsymbol{\gamma}}) = \mathbf{v} + \boldsymbol{\gamma}$. Therefore

$$E(\mathbf{A}\hat{\boldsymbol{\gamma}}) = \mathbf{A}\mathbf{v} + \mathbf{A}\boldsymbol{\gamma} = (\mathbf{A} + \mathbf{I})\mathbf{v}$$

using (64). Hence an unbiased estimator of $\mathbf{v}$ is

$$\hat{\mathbf{v}} = (\mathbf{A} + \mathbf{I})^{-1}\mathbf{A}\hat{\boldsymbol{\gamma}}\ . \tag{65}$$

This provides the facility for obtaining unbiased estimators of variances of, and covariances between, estimated variance components.

Example. To have minimal algebra, consider the balanced data case of (55): $\hat{\sigma}_\delta^2 = (\text{MSB} - \text{MSE})/a$ and $\hat{\sigma}_e^2 = \text{MSE}$. Under normality, cov(MSB, MSE) = 0 and

$$v(\text{MSB}) = \frac{2(a\sigma_\delta^2 + \sigma_e^2)^2}{b-1} \quad \text{and} \quad v(\text{MSE}) = \frac{2\sigma_e^4}{(a-1)(b-1)}\ . \tag{66}$$

Therefore (60) is, for $f = (a-1)(b-1)$

$$\text{var}\begin{bmatrix} \hat{\sigma}_\delta^2 \\ \hat{\sigma}_e^2 \end{bmatrix} = \begin{bmatrix} \dfrac{2(a\sigma_\delta^2 + \sigma_e^2)^2}{a^2(b-1)} + \dfrac{2\sigma_e^4}{a^2 f} & \dfrac{-2\sigma_e^4}{af} \\ \dfrac{-2\sigma_e^4}{af} & \dfrac{2\sigma_e^4}{f} \end{bmatrix}. \tag{67}$$

Hence

$$\begin{bmatrix} v(\hat{\sigma}_\delta^2) \\ v(\hat{\sigma}_e^2) \\ \text{cov}(\hat{\sigma}_\delta^2, \hat{\sigma}_e^2) \end{bmatrix} = \begin{bmatrix} \dfrac{2}{b-1} & \dfrac{2}{a^2(b-1)} + \dfrac{2}{a^2 f} & \dfrac{4}{a(b-1)} \\ 0 & \dfrac{2}{f} & 0 \\ 0 & \dfrac{-2}{af} & 0 \end{bmatrix} \begin{bmatrix} \sigma_\delta^4 \\ \sigma_e^4 \\ \sigma_\delta^2\sigma_e^2 \end{bmatrix}. \tag{68}$$

This is $\mathbf{v} = \mathbf{A}\boldsymbol{\gamma}$ of (64); it yields, (65) as

$$\left[\hat{v}(\hat{\sigma}_\delta^2) \quad \hat{v}(\hat{\sigma}_e^2) \quad \widehat{\text{cov}}(\hat{\sigma}_\delta^2, \hat{\sigma}_e^2)\right]'$$

$$= \begin{bmatrix} \dfrac{2}{(b-1)+2} & \dfrac{2}{a^2}\left[\dfrac{1}{(b-1)+2} + \dfrac{1}{f+2}\right] & \dfrac{4/a}{(b-1)+2} \\ 0 & \dfrac{2}{f+2} & 0 \\ 0 & \dfrac{-2/a}{f+2} & 0 \end{bmatrix} \begin{bmatrix} \hat{\sigma}_\delta^4 \\ \hat{\sigma}_e^4 \\ \hat{\sigma}_\delta^2\hat{\sigma}_e^2 \end{bmatrix}. \quad (69)$$

The results in (69) are the same as obtained directly (for this balanced data example) from (54) and (55); e.g., $\hat{v}(\hat{\sigma}_\delta^2)$ is the same as (57). Even though (65) has been illustrated in (69) for balanced data, it is to be emphasized that (65) does apply for unbalanced data.

-i. *Merits.* Whenever some form of general ANOVA method is applicable, i.e., whenever (58) and (59) exist, the method is easy to use, it yields unbiased estimators, and unbiased estimates of sampling variances of these estimators can be calculated.

-ii. *Demerits.* Using any ANOVA method incurs a number of disadvantages.

(A) There is no unique, nor optimal way of choosing the quadratic forms that can be used as elements of $\mathbf{q}$ for $E(\mathbf{q}) = \mathbf{C}\boldsymbol{\sigma}^2$.

(B) Many quadratic forms will, for mixed models, contain quadratic functions of the fixed effects—in which case $E(\mathbf{q}) = \mathbf{C}\boldsymbol{\sigma}^2$ of (58), which is the basis of the method, will not apply.

(C) Every set of quadratics for which (58) does apply will yield a set of unbiased estimates—but there is no way of deciding, for unbalanced data, which set of quadratics is better or worse than some other set; and there is an infinite number of sets. For the example they can be as trite as $(y_1 - y_2)^2$ and $(y_1 + y_2 - 2y_3)^2$, where y_1, y_2 and y_3 are any three observations.

(D) Negative estimates can occur.

(E) Distributional properties of the estimators are unknown, even under normality assumptions except, often, for $\hat{\boldsymbol{\sigma}}_e^2$.

b. Henderson's Method I

Henderson (1953) described three methods of estimating variance components that are just three different ways of using the general ANOVA method of (58) and (59). They differ only in the different quadratics (not always sums of squares) used for $\mathbf{q}$. All three also suffer from demerits of the general ANOVA method—that for unbalanced data no optimal application of the method is known, the methods can yield negative estimates, and distributional properties of the estimators are not known. Nevertheless, since the methods have been in constant use over the last thirty years, they merit description.

In Method I the quadratics used are analogous to the sums of squares used for balanced data, the analogy being such that certain sums of squares in balanced data become, for unbalanced data, quadratic forms that are not non-negative definite, and so they are not sums of squares. Thus, with a 2-way crossed classification with n observations per cell, the sum of squares

$$\Sigma_i bn(\bar{y}_{i\cdot\cdot} - \bar{y}_{\cdot\cdot\cdot})^2 = \Sigma_i bn\bar{y}_{i\cdot\cdot}^2 - abn\bar{y}_{\cdot\cdot\cdot}^2$$

becomes, for unbalanced data,

$$\Sigma_i n_{i\cdot}(\bar{y}_{i\cdot\cdot} - \bar{y}_{\cdot\cdot\cdot})^2 = \Sigma_i n_{i\cdot}\bar{y}_{i\cdot\cdot}^2 - n_{\cdot\cdot}\bar{y}_{\cdot\cdot\cdot}^2 \ .$$

It is the right-hand side of this which is used in Henderson's Method I. But not all such analogies with balanced data expressions are sums of squares. For example, with n observations per cell, the interaction sum of squares (the counterpart of the error sum of squares in Table 13.2) is

$$\Sigma_i\Sigma_j n(\bar{y}_{ij\cdot} - \bar{y}_{i\cdot\cdot} - \bar{y}_{\cdot j\cdot} + \bar{y}_{\cdot\cdot\cdot})^2$$
$$= n\Sigma_i\Sigma_j\bar{y}_{ij\cdot}^2 - bn\Sigma_i\bar{y}_{i\cdot\cdot}^2 - an\Sigma_j\bar{y}_{\cdot j\cdot}^2 + abn\bar{y}_{\cdot\cdot\cdot}^2 \ .$$

Analogous to the right-hand side of this, the expression used for unbalanced data in Henderson's Method I is

$$\Sigma_i\Sigma_j n_{ij}\bar{y}_{ij\cdot}^2 - \Sigma_i n_{i\cdot}\bar{y}_{i\cdot\cdot}^2 - \Sigma_j n_{\cdot j}\bar{y}_{\cdot j\cdot}^2 + n_{\cdot\cdot}\bar{y}_{\cdot\cdot\cdot}^2; \qquad (70)$$

and this, it can be demonstrated, is not a sum of squares: it is not non-negative definite. For example, with the data

6	4	10
6, 27	12	45
39	16	55

(70) has the negative value $-14\frac{1}{2}$. But, of course, this possible negativity of (70) does not negate using it as an element of $\mathbf{q}$ in (58) and (59).

Properties. Method I is easy to compute, even for very large data sets; and, for random models, it yields estimators that are unbiased; and analytic expressions for, or leading to, var($\hat{\boldsymbol{\sigma}}^2$) are available for several specific models [Searle (1971), Chapter 11].

But Method I cannot be used for mixed models. It can be adapted to a mixed model by altering that model and treating the fixed effects either as non-existent or as random—in which case the estimated variance components for the true random effects will be biased.

c. Henderson's Method II

This is designed to capitalize on the easy computability of Method I and to broaden its use by removing the limitation of Method I that it cannot be used for mixed models. The method has two parts. First, make the temporary assumption that the random effects are fixed, and for the model $\mathbf{y} = \mathbf{X}\boldsymbol{\beta} + \mathbf{Z}\mathbf{u} + \mathbf{e}$ of (12) solve the normal equations

$$\begin{bmatrix} \mathbf{X}'\mathbf{X} & \mathbf{X}'\mathbf{Z} \\ \mathbf{Z}'\mathbf{X} & \mathbf{Z}'\mathbf{Z} \end{bmatrix} \begin{bmatrix} \boldsymbol{\beta}^\circ \\ \mathbf{u}^\circ \end{bmatrix} = \begin{bmatrix} \mathbf{X}'\mathbf{y} \\ \mathbf{Z}'\mathbf{y} \end{bmatrix} \tag{71}$$

for $\boldsymbol{\beta}^\circ$. Then consider the vector of data adjusted for $\boldsymbol{\beta}^\circ$, namely $\mathbf{z} = \mathbf{y} - \mathbf{X}\boldsymbol{\beta}^\circ$. If the solution for $\boldsymbol{\beta}^\circ$ has been obtained in a very particular manner [see Searle (1968) and Henderson *et al.* (1974)] the model for $\mathbf{z}$ will (provided $\mathbf{X}\boldsymbol{\beta}$ includes a general mean term $\mathbf{1}\mu$) be

$$\mathbf{z} = \mathbf{1}\mu^* + \mathbf{Z}\mathbf{u} + \mathbf{K}\mathbf{e}, \tag{72}$$

where μ^* differs from μ and where $\mathbf{K}$ is known. Apart from $\mathbf{K}$, (72) is then the same form as the μ and random effects parts of the model equation (12) for $\mathbf{y}$. Hence Method I is easily applied to $\mathbf{z}$, except for adaptations arising from having $\mathbf{K}\mathbf{e}$ as the error term rather than $\mathbf{e}$, and these are known [Henderson *et al.* (1974), pp. 586–588].

Properties. Method II is relatively easy to compute, especially when the number of fixed effects is not too large. And although it can be used for a wide variety of mixed models, it cannot be used for those mixed models that have interactions between fixed and random factors, whether those interactions are defined as random effects (the usual case) or as fixed effects. Searle (1968) has proof of this. Also, analytic forms of sampling variances are unknown.

d. Henderson's Method III

This uses sums of squares that arise in fitting an overparameterized model and sub-models thereof—as described in Chapters 4, 5 and 9.

Properties. Method III can be used for any mixed model and yields estimators that are unbiased. Although the method uses sums of squares that are known (at least in some cases) to be useful in certain fixed effects models, no analytic evidence is available that these sums of squares have optimal properties for estimating variances—but their origin in the fixed model context is comforting to some users.

The disadvantage of Method III is that through its confinement to sums of squares for fitting overparameterized models, there is a problem of too many sums of squares being available, albeit in limited form. For example, consider the 2-way crossed classification overparameterized model with equation

$$y_{ijk} = \mu + \alpha_i + \beta_j + \phi_{ij} + e_{ijk} \tag{73}$$

as in Chapter 9. Suppose all effects in (73) are random. There are then four variance components to estimate: σ_α^2, σ_β^2, σ_γ^2 and σ_e^2. But for that model, there are five different sums of squares in Table 9.2 that can be used for Method III: $R(\boldsymbol{\alpha}\,|\,\mu)$, $R(\boldsymbol{\beta}\,|\,\mu, \boldsymbol{\alpha})$, $R(\boldsymbol{\beta}\,|\,\mu)$, $R(\boldsymbol{\alpha}\,|\,\mu, \boldsymbol{\beta})$ and $R(\boldsymbol{\alpha}\,|\,\mu, \boldsymbol{\alpha}, \boldsymbol{\beta})$ as well as SSE. From these, at least three sets suggest themselves as possible candidates for Method III estimation:

(a)	(b)	(c)
$R(\boldsymbol{\alpha}\,|\,\mu)$	$R(\boldsymbol{\beta}\,|\,\mu)$	$R(\boldsymbol{\alpha}\,|\,\mu, \boldsymbol{\beta})$
$R(\boldsymbol{\beta}\,|\,\mu, \boldsymbol{\alpha})$	$R(\boldsymbol{\alpha}\,|\,\mu, \boldsymbol{\beta})$	$R(\boldsymbol{\beta}\,|\,\mu, \boldsymbol{\alpha})$
$R(\boldsymbol{\gamma}\,|\,\mu, \boldsymbol{\alpha}, \boldsymbol{\beta})$	$R(\boldsymbol{\gamma}\,|\,\mu, \boldsymbol{\alpha}, \boldsymbol{\beta})$	$R(\boldsymbol{\gamma}\,|\,\mu, \boldsymbol{\alpha}, \boldsymbol{\beta})$
SSE	SSE	SSE

All three sets yield the same estimators of σ_γ^2 and σ_e^2. Two different estimators of σ_α^2 and σ_β^2 arise as follows:

(a)	(b)	(c)
$\hat{\sigma}^2_{\alpha,1}$	$\hat{\sigma}^2_{\alpha,2}$	$\hat{\sigma}^2_{\alpha,2}$
$\hat{\sigma}^2_{\beta,2}$	$\hat{\sigma}^2_{\beta,1}$	$\hat{\sigma}^2_{\beta,2}$

No procedure that I know of can tell us which (if any) of these sets of estimators is to be preferred over the others.

Another disadvantage of Method III is that, in some situations, dimensions of the matrices involved in calculating some of the sums of squares are so large as to preclude doing the calculations.

e. ML (Maximum Likelihood)

Estimation by maximum likelihood is a well-established method of estimation, originating with Fisher (1925). It was first applied to the general mixed model by Hartley and Rao (1967). On assuming that the error terms and random effects are all normally distributed and using the variance–covariance specifications of (12)–(29) the starting point is

$$\mathbf{y} = \mathbf{X}\boldsymbol{\beta} + \mathbf{Z}\mathbf{u} + \mathbf{e} \sim N(\mathbf{X}\boldsymbol{\beta}, \mathbf{V})$$

with
$$\mathbf{V} = \sum_{i=1}^{r} \sigma_i^2 \mathbf{Z}_i\mathbf{Z}_i' + \sigma_e^2 \mathbf{I}_N = \sum_{i=0}^{r} \sigma_i^2 \mathbf{Z}_i\mathbf{Z}_i',$$

of (26) and (29). The likelihood function is then

$$L = (2\pi)^{-\frac{1}{2}N}|\mathbf{V}|^{-\frac{1}{2}}\exp\left[-\tfrac{1}{2}(\mathbf{y}-\mathbf{X}\boldsymbol{\beta})'\mathbf{V}^{-1}(\mathbf{y}-\mathbf{X}\boldsymbol{\beta})\right].$$

Maximizing L with respect to elements of $\boldsymbol{\beta}$ and the variance components (the σ_i^2s that occur in $\mathbf{V}$) leads to equations that have to be solved to yield the ML estimators of $\boldsymbol{\beta}$ and of $\boldsymbol{\sigma}^2 = \{\sigma_i^2\}_{i=0}^{i=r}$. These equations can be written in a variety of ways with numerous attendant details that are available in a number of places [and extensively, for example, in Searle (1979)]. Hence only the equations are given here. They are

$$\mathbf{X}'\tilde{\mathbf{V}}^{-1}\mathbf{X}\tilde{\boldsymbol{\beta}} = \mathbf{X}'\tilde{\mathbf{V}}^{-1}\mathbf{y} \tag{74}$$

and the $r + 1$ equations

$$\operatorname{tr}\left(\tilde{\mathbf{V}}^{-1}\mathbf{Z}_i\mathbf{Z}_i'\right) = \left(\mathbf{y}-\mathbf{X}\tilde{\boldsymbol{\beta}}\right)\tilde{\mathbf{V}}^{-1}\mathbf{Z}_i\mathbf{Z}_i'\tilde{\mathbf{V}}^{-1}\left(\mathbf{y}-\mathbf{X}\tilde{\boldsymbol{\beta}}\right) \tag{75}$$

for $i = 0, 1, \ldots, r$.

Equations (74) and (75) have to be solved for $\tilde{\boldsymbol{\beta}}$ and $\tilde{\boldsymbol{\sigma}}^2$, the elements of $\tilde{\boldsymbol{\sigma}}^2$ being implicit in $\tilde{\mathbf{V}}$. Clearly, these equations are not linear in those elements $\tilde{\sigma}_i^2$, although once the $\tilde{\sigma}_i^2$-values are obtained they can be used in (74) to obtain $\tilde{\boldsymbol{\beta}}$; and it can be noted that (74) is just like the BLUE equations of Section 8.10, only with $\tilde{\mathbf{V}}$ in place of $\mathbf{V}$.

The non-linearity of (74) and (75), especially insofar as the variance components are concerned, means that those equations cannot be solved analytically (except in a few simple, balanced-data situations). So they have to be solved numerically, by iteration. For convenience, write

$$\mathbf{P} = \mathbf{V}^{-1} - \mathbf{V}^{-1}\mathbf{X}(\mathbf{X}'\mathbf{V}^{-1}\mathbf{X})^{-}\mathbf{X}'\mathbf{V}^{-1}. \tag{76}$$

And with
$$\mathbf{I} = \mathbf{V}^{-1}\mathbf{V} = \mathbf{V}^{-1}\sum_{i=0}^{r}\mathbf{Z}_i\mathbf{Z}_i'\sigma_i^2$$

used inside the trace operation on the left-hand side of (75), that set of $r + 1$ equations can be written as

$$\left\{\mathrm{tr}\left(\tilde{\mathbf{V}}^{-1}\mathbf{Z}_i\mathbf{Z}_i'\tilde{\mathbf{V}}^{-1}\mathbf{Z}_j\mathbf{Z}_j'\right)\right\}\left\{\tilde{\sigma}_i^2\right\} = \left\{\mathbf{y}'\tilde{\mathbf{P}}\mathbf{Z}_i\mathbf{Z}_i'\tilde{\mathbf{P}}\mathbf{y}\right\} \tag{77}$$

for $i, j = 0, 1, \ldots, r$. These provide easier visualization of an iterative process than do (74) and (75); in (77) we can use a starting value for $\tilde{\boldsymbol{\sigma}}^2$ in $\tilde{\mathbf{V}}$ and $\tilde{\mathbf{P}}$, solve (77) and repeat the process.

There are several numerical problems associated with solving either (74) and (75), or (77), which shall not be addressed here (because they belong mainly under the rubric of the numerical analyst) but which are important to recognize. First is the problem of whether the choice of a starting value for $\tilde{\boldsymbol{\sigma}}^2$ affects the final value obtained when (if?) the iterative process has converged. Second is the allied problem of whether the final value obtained for $\tilde{\boldsymbol{\sigma}}^2$ is giving a global maximum of L or only a local maximum. Third is the problem of what to do if some intermediate (or even the final) value of $\tilde{\boldsymbol{\sigma}}^2$ is such that the corresponding $\tilde{\mathbf{V}}$ is singular; or if it is nonsingular, that it not be positive definite. [Harville (1977) addresses this problem by invoking some of the extensive matrix manipulations that can be made on $\mathbf{V}^{-1}$ arising from the form of $\mathbf{V}$ in (26) and (29).] Fourth is the sheer volume of calculations that can be involved, arising essentially from the inversion of $\mathbf{V}$, the order of which is the number of observations—since $\mathbf{V} = \mathrm{var}(\mathbf{y})$. The Hemmerle and Hartley (1973) W-transformation is an assistance in this regard, but real advances in speeding up the inversion of matrices of orders of tens of thousands will come only as computer sizes and speeds increase considerably. And this is already occurring. Finally (so far as we are concerned here), is the fact that the solution for $\tilde{\boldsymbol{\sigma}}^2$ to (77) is not necessarily the ML estimator. This is so because, by the definition of variance, the parameter space for $\boldsymbol{\sigma}^2$ is defined by $\sigma_0^2 > 0$ and $\sigma_i^2 \geq 0$ for $i = 1, \ldots, r$. And, by definition, ML estimators must be in their parameter space. Therefore if some $\tilde{\sigma}_i^2 < 0$, one must put $\tilde{\sigma}_i^2 = 0$ and re-estimate the other σ^2s. The effect of this is, of course, to change the model, because implicit in having a variance component zero is the consequence of having all levels of the corresponding factor being the same; i.e., that factor must be deleted from the model. That done, we have a new model and so must re-estimate the variance components for it.

What has just been described in terms of the final value of some $\tilde{\sigma}_i^2$ being negative does, in fact, apply not just to final values of the iterative process but to intermediate values also. There is then the following additional difficulty: if $\tilde{\sigma}_t^2 < 0$ after the kth round of iteration and we replace it with $\tilde{\sigma}_t^2 = 0$, then for the $(k + 1)$th round of iteration there is no tth random factor in the model and no value of $\tilde{\sigma}_t^2$, and this will be the case for all rounds after the $(k + 1)$th too. This may not be the correct thing to do: suppose the negative value for $\tilde{\sigma}_t^2$ after the kth round was just a numerical idiosyncrasy which, if it had been retained for the start of the $(k + 1)$th round, would have yielded a positive value at the end of that round. Then putting $\tilde{\sigma}_t^2 = 0$ at the first occurrence of a negative in a manner that leaves it that way for all remaining iterations is clearly not appropriate.

These are just some of the simpler difficulties that beset the numerical analyst who plans to computerize a methodology for solving (77) and deriving an ML estimator of $\boldsymbol{\sigma}^2$. It is no task for amateurs; and especially so since computing packages are available that do these calculations (e.g., SAS VARCOMP and BMDP3V) and others are known to be in varying stages of development (e.g., Giesbrecht, 1983, 1985).

Despite the numerical difficulties involved in solving (77) and obtaining ML estimators of variance components, I think this is the method of estimation that is to be preferred or, even more preferably, restricted maximum likelihood (REML) as described in the next section. One reason for thinking this way is that the maximum likelihood method of estimation is well defined and the resulting estimators have attractive, well-known large-sample properties: they are normally distributed and their sampling variances are known (e.g., Searle, 1970). True it is that these properties are valid only under large-sample conditions; and what is meant by this in variance components models is not necessarily easy to define (e.g., Hartley and Rao, 1967; Miller, 1977). Nevertheless, it seems reasonable to prefer these properties over those, for example, of ANOVA estimators from unbalanced data. Such estimators can be negative (which ML precludes); and their distribution is unknown and in many cases their variances are unknown. Furthermore, ANOVA estimation of variance components gives no lead on how to then estimate the fixed effects in a mixed model. ML does: use the BLUE equations with the ML estimator $\tilde{\mathbf{V}}$ in place of $\mathbf{V}$, as in (167).

f. REML (Restricted Maximum Likelihood)

A variant of maximum likelihood estimation in the mixed model is restricted maximum likelihood, or REML. Used by Patterson and Thompson (1971) for block designs, REML estimators are obtained from maximizing that part of the likelihood which is invariant to the location parameter; i.e., in terms of the mixed model $\mathbf{y} = \mathbf{X}\boldsymbol{\beta} + \mathbf{Zu} + \mathbf{e}$, invariant to $\mathbf{X}\boldsymbol{\beta}$. Another way of looking at it is that REML maximizes the likelihood of a vector of linear combinations of the observations that are invariant to $\mathbf{X}\boldsymbol{\beta}$. Suppose $\mathbf{Ly}$ is such a vector. Then $\mathbf{Ly} = \mathbf{LX}\boldsymbol{\beta} + \mathbf{LZu} + \mathbf{Le}$ is invariant to $\mathbf{X}\boldsymbol{\beta}$ if and only if $\mathbf{LX} = \mathbf{0}$. But $\mathbf{LX} = \mathbf{0}$ if and only if $\mathbf{L} = \mathbf{TM}$ for

$\mathbf{M} = \mathbf{I} - \mathbf{X}(\mathbf{X}'\mathbf{X})^-\mathbf{X}'$ and some $\mathbf{T}$ (see E10.7). Clearly, $\mathbf{L}$ must be of full row rank: so therefore $\mathbf{T}$ is also. Hence $r_{\mathbf{L}} = r_{\mathbf{T}}$ and $r_{\mathbf{L}} \le r_{\mathbf{M}}$ with $r_{\mathbf{M}} = N - r_{\mathbf{X}}$; and so $r_{\mathbf{L}} = N - r_{\mathbf{X}}$.

For $\mathbf{L}$ as just described it can be shown (e.g., Searle, 1979) that under normality the likelihood function of $\mathbf{Ly}$ is invariant to $\mathbf{L}$. Thus it does not matter what functions $\mathbf{Ly}$ are used as long as $\mathbf{L}$ has full row rank $N - r_{\mathbf{X}}$ and has the form $\mathbf{TM}$ for any $\mathbf{T}$ of full row rank $N - r_{\mathbf{X}}$. Harville (1977) has termed an element of any such $\mathbf{Ly}$ as an "error contrast": it has expected value that is zero:

$$E(\mathbf{Ly}) = \mathbf{L}E(\mathbf{X}\boldsymbol{\beta} + \mathbf{Zu} + \mathbf{e}) = \mathbf{LX}\boldsymbol{\beta} = \mathbf{0} \ .$$

With the preceding prescription for $\mathbf{L}$ the resulting equations for REML estimation of $\boldsymbol{\sigma}^2$ are, akin to (77), for $i, j = 0, 1, \ldots, r$,

$$\left\{\mathrm{tr}\left(\tilde{\mathbf{P}}\mathbf{Z}_i\mathbf{Z}_i'\tilde{\mathbf{P}}\mathbf{Z}_j\mathbf{Z}_j'\right)\right\}\left\{\sigma_i^2\right\} = \left\{\mathbf{y}'\tilde{\mathbf{P}}\mathbf{Z}_i\mathbf{Z}_i'\tilde{\mathbf{P}}\mathbf{y}\right\} \ . \tag{78}$$

These are, it will be noted, the same as the ML equations (77), except for having $\tilde{\mathbf{P}}$ in place of $\tilde{\mathbf{V}}^{-1}$ on the left-hand side.

Computational problems for obtaining the iterative solution of (78), the only way of solving (78) for unbalanced data, are the same as those for solving the ML equations (77). Nevertheless, being a form of maximum likelihood estimators they have the same large-sample properties as already discussed for ML estimators. The REML estimation procedure does not, however, include estimating $\boldsymbol{\beta}$. Nevertheless, having obtained $\tilde{\boldsymbol{\sigma}}^2$ using REML one would undoubtedly use it to derive $\tilde{\mathbf{V}}$ from $\mathbf{V}$ and thence solve $\mathbf{X}'\tilde{\mathbf{V}}^{-1}\mathbf{X}\boldsymbol{\beta} = \mathbf{X}'\tilde{\mathbf{V}}^{-1}\mathbf{y}$ as in (74) of ML estimation.

The question will be asked, very naturally, "Which of the ML and REML methods is to be preferred?" There seems to be no categorical answer. The relative advantage of ML is that it provides estimation of $\mathbf{X}\boldsymbol{\beta}$, whereas REML does not. On the other hand the REML equations with balanced data provide solutions that are identical to ANOVA estimators which are unbiased and have attractive minimum variance properties. In this sense REML is said to take account of the degrees of freedom involved in estimating the fixed effects, whereas ML estimators do not. The easiest example of this is the case of a simple sample of n observations from a $N(\mu, \sigma^2)$-distribution. The two estimators of σ^2 are

$$\hat{\sigma}^2_{\mathrm{ML}} = \frac{\Sigma_i(x_i - \bar{x})^2}{n} \quad \text{and} \quad \hat{\sigma}^2_{\mathrm{REML}} = \frac{\Sigma_i(x_i - \bar{x})^2}{n - 1} \ . \tag{79}$$

The REML estimator has taken account of the one degree of freedom required for estimating μ, whereas the ML estimator has not.

In (79) the REML estimator is also unbiased, but the ML estimator is not. In the general case of unbalanced data neither the ML estimators from (77) nor the REML estimators from (78) are unbiased.

g. MINQUE (Minimum Norm Quadratic Unbiased Estimation)

LaMotte (1970, 1971, 1973) and Rao (1970, 1971a, b, 1972) describe an estimation method which is derived by requiring that the estimator minimize a (Euclidean) norm, be a quadratic form of the observations and be unbiased. Its development involves extensive algebra. More importantly, its concept demands the use of some pre-assigned weights that effectively play the part of *à priori* values for the unknown variance components. Denoting these weights by $\mathbf{w}$, the same order as $\boldsymbol{\sigma}^2$, denote by $\mathbf{V}_w$ the matrix $\mathbf{V}$ with $\mathbf{w}$ in place of $\boldsymbol{\sigma}^2$; and similarly $\mathbf{P}_w$ is $\mathbf{P}$ of (76) with $\mathbf{V}_w$ in place of $\mathbf{V}$. Then the equations for the corresponding MINQUE estimator of $\boldsymbol{\sigma}^2$ are, for $i, j = 0, 1, \ldots, r$,

$$\left\{\mathrm{tr}\left(\mathbf{P}_w\mathbf{Z}_i\mathbf{Z}_i'\mathbf{P}_w\mathbf{Z}_j\mathbf{Z}_j'\right)\right\}\{\hat{\sigma}_i\} = \{\mathbf{y}'\mathbf{P}_w\mathbf{Z}_i\mathbf{Z}_i'\mathbf{P}_w\mathbf{y}\} . \tag{80}$$

This method has two advantages: it involves no normality assumptions as do ML and REML as previously described. And the equations that yield the estimator, namely (80), do not have to be solved iteratively. This is so because $\mathbf{P}_w$ is a numerical matrix and does not involve $\boldsymbol{\sigma}^2$. Therefore (80) is just a set of straightforward linear equations in the elements of $\boldsymbol{\sigma}^2$ and so presents no difficulty insofar as calculating a solution is concerned. On the other hand, that solution does depend on the pre-assigned values used in $\mathbf{w}$; different values of $\mathbf{w}$ can give different estimators from the same data set. One must therefore talk about "a" MINQUE estimator and not "the" MINQUE estimator. This appears to be a troublesome feature of the MINQUE procedure. Also, its estimators can be negative and they are only unbiased if indeed the true, unknown value of $\boldsymbol{\sigma}^2$ is $\mathbf{w}$.

One particular value of $\mathbf{w}$ that had some appeal for a number of years was $\mathbf{w}$ being a null vector except for the element 1.0 corresponding to σ_e^2. This reduces $\mathbf{V}_w$ to being $\mathbf{I}$ and thence $\mathbf{P}_w$ to being $\mathbf{M} = \mathbf{I} - \mathbf{X}(\mathbf{X}'\mathbf{X})^-\mathbf{X}'$. Computationally this was very advantageous [see, for example, Goodnight and Hemmerle (1978)], but simulation studies of Quaas and Bolgiano (1979) and of Swallow and Monahan (1984) indicate that it yields estimators of dubious quality. They have large mean square error values.

A relationship that exists between REML and MINQUE is important. If in the REML equations (78) one replaces $\tilde{\mathbf{P}}$ by $\mathbf{P}_w$ then the MINQUE equations (80) are obtained. Yet this is precisely the first step when solving the REML equations iteratively: one uses a starting value for $\tilde{\boldsymbol{\sigma}}^2$ in $\tilde{\mathbf{P}}$ of (76) and solves for $\tilde{\boldsymbol{\sigma}}^2$. If that starting value is $\mathbf{w}$, then the first solution is the MINQUE estimator based on $\mathbf{w}$. Thus we have the connection

$$\text{a MINQUE solution} = \text{a first iterate of REML} . \tag{81}$$

h. MIVQUE (Minimum Variance Quadratic Unbiased Estimation)

MINQUE demands no assumptions about the form of the distribution of $\mathbf{y}$. But if the usual normality assumptions are invoked, the MINQUE solution has the properties of being that unbiased quadratic form of the observations which has minimum variance; i.e., it is a *mi*nimum *v*ariance *q*uadratic *u*nbiased *e*stimator, MIVQUE.

i. I-MINQUE (Iterative MINQUE)

As already pointed out, the MINQUE procedure demands using a weight vector $\mathbf{w}$ as if it were a pre-assigned value of $\boldsymbol{\sigma}^2$ and then solving (80). No iteration is involved. But having obtained a solution, $\tilde{\boldsymbol{\sigma}}_1^2$ say, its existence prompts the idea of using it as a new $\mathbf{w}$ from which a new set of equations (80) can be established and solved, yielding $\tilde{\boldsymbol{\sigma}}_2^2$ say. This leads to using the MINQUE equations iteratively to yield iterative MINQUE, or I-MINQUE estimators. They are, of course, if one iterates to convergence, the same as REML estimators—by virtue of the similarity of the MINQUE equations (80) and the REML equations (77) when viewed in the light of iterating. Hence

$$\text{I-MINQUE} = \text{REML} . \tag{82}$$

Furthermore, Brown (1976) has shown that even without normality assumptions on $\mathbf{y}$, the I-MINQUE solutions do have large-sample normality properties. In view of (81) and (82) it seems reasonable to concentrate attention on REML rather than MINQUE.

13.4. PREDICTION OF RANDOM VARIABLES

In the mixed model $\mathbf{y} = \mathbf{X}\boldsymbol{\beta} + \mathbf{Z}\mathbf{u} + \mathbf{e}$, we have taken $\mathbf{u}$ as a vector of random variables. In fact, for a given data vector $\mathbf{y}$, the vector $\mathbf{u}$ is a set of realized values of random variables, albeit values that usually cannot be measured. A common example of this is where $\mathbf{y}$ represents animal or plant production records and $\mathbf{u}$ is a vector of genotypes that occur in the genetic heritage of those animals or plants. For instance, if $\mathbf{y}$ consisted of milk yields from dairy cows, $\mathbf{u}$ might be the genetic values of the sires of those cows (with different groups of cows having the same sire); or $\mathbf{y}$ might be annual fleece weights over several years from a group of ewes and **u** might be the genotypic values of those ewes. In situations such as these a question that is often of interest is: "Given the data vector **y**, how can we predict the value of the vector **u** that could be associated with such a **y**?" Or, put another way, "Of all the possible realizations **u** that could be associated with the particular value of **y** that has been observed, what is an estimator of the average, or expected value, of those values of **u**?" In other words, "What is an estimator of the conditional mean $E(\mathbf{u}\,|\,\mathbf{y})$?"

Interest in $E(\mathbf{u}\,|\,\mathbf{y})$ can also be motivated more formally. **u** and **y** are (realizations of) random variables. Whatever their joint distribution function is, we can show [e.g., Cochran (1951), Rao (1965, pp. 79 and 220–222) and Searle (1974)] that the predictor of **u** which minimizes the mean square error of prediction is $E(\mathbf{u}\,|\,\mathbf{y})$. Thus $E(\mathbf{u}\,|\,\mathbf{y})$ is called the best predictor of **u**. Under normality it takes the form

$$\tilde{\mathbf{u}} = E(\mathbf{u}) + \text{cov}(\mathbf{u}, \mathbf{y}')[\text{var}(\mathbf{y})]^{-1}[\mathbf{y} - E(\mathbf{y})] \ . \tag{83}$$

The best predictor, $E(\mathbf{u}\,|\,\mathbf{y})$, is not necessarily linear in **y** as it is in (83) under normality. Nevertheless, one can also show that no matter what the joint distribution of **u** and **y** is, the best linear predictor (linear in **y**) is $\tilde{\mathbf{u}}$. Thus $\tilde{\mathbf{u}}$ can be thought of either as the best linear predictor or as the best predictor under normality.

In the case of the mixed model $\mathbf{y} = \mathbf{X\beta} + \mathbf{Zu} + \mathbf{e}$, with

$$\text{var}(\mathbf{u}) = \mathbf{D}\left\{\sigma_i^2\mathbf{I}_{q_i}\right\}, \quad = \mathbf{D} \text{ say,} \tag{84}$$

we have
$$\text{cov}(\mathbf{u}, \mathbf{y}') = \text{cov}(\mathbf{u}, \mathbf{u}'\mathbf{Z}') = \mathbf{DZ}' \tag{85}$$

and
$$\text{var}(\mathbf{y}) = \mathbf{ZDZ}' + \sigma_e^2\mathbf{I} = \mathbf{V},$$

so that (84) is
$$\tilde{\mathbf{u}} = \mathbf{DZ}'\mathbf{V}^{-1}(\mathbf{y} - \mathbf{X\beta}) \ .$$

Thought of as a function of $\mathbf{X\beta}$, which is estimable, we then have with

$$\text{BLUE}(\tilde{\mathbf{u}}) = \mathbf{DZ}'\mathbf{V}^{-1}(\mathbf{y} - \mathbf{X\beta}^\circ), \tag{86}$$

$$\text{BLUE}(\mathbf{X\beta}) = \mathbf{X\beta}^\circ = \mathbf{X}(\mathbf{X}'\mathbf{V}^{-1}\mathbf{X})^-\mathbf{X}'\mathbf{V}^{-1}\mathbf{y}, \tag{87}$$

where $\mathbf{X\beta}^\circ$ is, of course, the same as $\mathbf{X\beta}^\circ_{\text{BLUE}}$ of Section 8.10. With the two concepts of prediction and best linear unbiased estimation being combined in (86), that result is often referred to (especially in animal breeding literature) as a best linear unbiased prediction: BLUP. Thus (86) is

$$\text{BLUP}(\mathbf{u}) = \mathbf{DZ}'\mathbf{V}^{-1}(\mathbf{y} - \mathbf{X\beta}^\circ) \ . \tag{88}$$

When $\mathbf{D}$ and σ^2 are known, i.e., when $\sigma_1^2, \ldots, \sigma_a^2$ and σ_e^2 are known, calculation of (88) presents no difficulties, other than the numerical problems of inverting matrices of large order—which may well be the case with $\mathbf{V}$ since its order is the number of observations. This difficulty, of a matrix of large order, can be reduced somewhat by the use of what have come to be called *mixed model equations*. They can be described as follows. Make the temporary assumption that $\mathbf{u}$ represents fixed and not random effects. Then $\text{var}(\mathbf{y}) = \sigma_e^2\mathbf{I}$ and the OLS equations for $\boldsymbol{\beta}$ and $\mathbf{u}$ can be written as

$$\begin{bmatrix} \mathbf{X'X} & \mathbf{X'Z} \\ \mathbf{Z'X} & \mathbf{Z'Z} \end{bmatrix} \begin{bmatrix} \hat{\boldsymbol{\beta}} \\ \hat{\mathbf{u}} \end{bmatrix} = \begin{bmatrix} \mathbf{X'y} \\ \mathbf{Z'y} \end{bmatrix} .$$

Amend these equations by adding $\sigma_e^2\mathbf{D}^{-1}$ to $\mathbf{Z'Z}$ to get

$$\begin{bmatrix} \mathbf{X'X} & \mathbf{X'Z} \\ \mathbf{Z'X} & \mathbf{Z'Z} + \sigma_e^2\mathbf{D}^{-1} \end{bmatrix} \begin{bmatrix} \boldsymbol{\beta}^\circ \\ \tilde{\mathbf{u}} \end{bmatrix} = \begin{bmatrix} \mathbf{X'y} \\ \mathbf{Z'y} \end{bmatrix} . \tag{89}$$

These equations are what are known as the "mixed model equations". Proof that $\boldsymbol{\beta}^\circ$ and $\tilde{\mathbf{u}}$ from (89) are indeed $\boldsymbol{\beta}^\circ$ of (87) and BLUP($\mathbf{u}$) of (88) was first given in Henderson *et al.* (1959) for $\mathbf{X}$ of full column rank and can be derived for $\mathbf{X}$ not of full rank using the generalized inverse of a partitioned matrix given in (38) of Section 7.3—see E13.1. Compared to (87) and (88) for $\mathbf{X}\boldsymbol{\beta}^\circ$ and BLUP($\mathbf{u}$), the advantage of (89) is that the matrix to be inverted in (87), namely $\mathbf{V}$, has order N, whereas the matrix on the left-hand side of (89) has order $p + q$, where p and q are the numbers of fixed and random effects, respectively: and $p + q$ is often much less than N, the number of observations.

The preceding development assumes that $\mathbf{V}$ is known, i.e., that $\sigma_1^2, \ldots, \sigma_r^2$ and σ_e^2 are all known. When this is not the case, then these variances must be estimated, using one or other of the methods discussed in Section 13.3. Then, just as in (40), $\mathbf{V}$ in (87) must be replaced by $\hat{\mathbf{V}}$ to give

$$\mathbf{X}\boldsymbol{\beta}^\circ_{\hat{\mathbf{V}}} = \mathbf{X}(\mathbf{X}'\hat{\mathbf{V}}^{-1}\mathbf{X})^-\mathbf{X}'\hat{\mathbf{V}}^{-1}\mathbf{y};$$

and this and $\hat{\sigma}_e^2$ and $\hat{\sigma}_1^2, \ldots, \hat{\sigma}_r^2$ must be used in (86) to give

$$\text{BLUP}(\mathbf{u})_{\hat{\mathbf{V}}} = \hat{\mathbf{D}}\mathbf{Z}'\hat{\mathbf{V}}^{-1}(\mathbf{y} - \mathbf{X}\boldsymbol{\beta}^\circ_{\hat{\mathbf{V}}}), \tag{90}$$

where $\hat{\mathbf{D}}$ is $\mathbf{D}$ of (85) with $\hat{\sigma}_i^2$ replacing σ_i^2 for $i = 0, \ldots, r$.

13.5. AN EXAMPLE: THE 1-WAY CLASSIFICATION

We show here some (but not all) of the details of applying the preceding methodology to the 1-way classification.

a. The model

In Chapter 2 we used $y_{ij} = \mu_i + e_{ij}$ as the model equation for the 1-way classification, with μ_i being a fixed effect to be estimated. Two adaptations of this to the mixed model are $y_{ij} = \mu + \varepsilon_{ij}$,

$$\text{with} \qquad \operatorname{cov}(\varepsilon_{ij}, \varepsilon_{i'j'}) = \begin{cases} \sigma_\alpha^2 + \sigma_e^2 & \text{for} \quad i = i' \text{ and } j = j' \\ \sigma_\alpha^2 & \text{for} \quad i = i' \text{ and } j \neq j' \\ 0 & \text{for} \quad i \neq i' \text{ and } j \neq j' \end{cases} .$$

The equivalent model that is more customarily used is $y_{ij} = \mu + \alpha_i + e_{ij}$, with μ being fixed and $\boldsymbol{\alpha} = \{\alpha_i\}_{i=1}^{i=a}$ being random with $E(\boldsymbol{\alpha}) = \mathbf{0}$ and $E(\mathbf{e}) = 0$ and

$$\operatorname{cov}(e_{ij}, e_{i'j'}) = \sigma_e^2 \text{ for } i = i' \text{ and } j = j', \text{ and 0 otherwise;}$$

$$\operatorname{cov}(\alpha_i, \alpha_{i'}) = \sigma_\alpha^2 \text{ for } i = i' \text{ and 0 otherwise;}$$

$$\operatorname{cov}(\alpha_i, e_{i'j'}) = 0 \text{ for all } i,\ i' \text{ and } j\ .$$

This is just a specific case of the general mixed model described in Section 13.1b. In terms of that general model formulation $\mathbf{y} = \mathbf{X}\boldsymbol{\beta} + \mathbf{Z}\mathbf{u} + \mathbf{e}$, we here have

$$\mathbf{X} = \mathbf{1}_N, \qquad \boldsymbol{\beta} = \mu, \qquad \mathbf{Z} = \mathbf{D}\{\mathbf{1}_{n_i}\} \quad \text{and} \quad \mathbf{u} = \boldsymbol{\alpha}\ . \tag{91}$$

Although this model is known as the random model, in truth it is a mixed model since μ is fixed. The variance–covariance structure defined above is

$$\operatorname{var}(\mathbf{e}) = \sigma_e^2\mathbf{I}_N, \qquad \operatorname{var}(\boldsymbol{\alpha}) = \mathbf{D} = \sigma_\alpha^2\mathbf{I}_a \quad \text{and} \quad \operatorname{cov}(\boldsymbol{\alpha}, \mathbf{e}') = \mathbf{0}_{a\times N}, \tag{92}$$

so that with (92) and $E(\mathbf{y}) = \mu\mathbf{1}_N$ we have $\mathbf{y} \sim (\mu\mathbf{1}, \mathbf{V})$ for

$$\mathbf{V} = \sigma_\alpha^2\mathbf{D}\{\mathbf{J}_{n_i}\} + \sigma_e^2\mathbf{I}_N = \mathbf{D}\{\sigma_e^2\mathbf{I}_{n_i} + \sigma_\alpha^2\mathbf{J}_{n_i}\}\ . \tag{93}$$

TABLE 13.3. EXPECTED VALUES OF SUMS OF SQUARES FOR THE 1-WAY CLASSIFICATION, RANDOM EFFECTS MODEL

Source of Variation	d.f.	Sum of Squares[1]	Expected Sum of Squares[1,2,3]
Mean	1	$R(\mu) \quad = N\bar{y}_{..}^2$	$N\mu^2 + (\Sigma_i n_i^2/N)\sigma_\alpha^2 + \sigma_e^2$
Classes	$a-1$	$R(\alpha \mid \mu) = \Sigma_i n_i(\bar{y}_{i\cdot} - \bar{y}_{..})^2$	$(N - \Sigma_i n_i^2/N)\sigma_\alpha^2 + (a-1)\sigma_e^2$
Residual	$N-a$	$\text{SSE} \quad = \Sigma_i\Sigma_j(y_{ij} - \bar{y}_{i\cdot})^2$	$(N-a)\sigma_e^2$
Total	N	$\text{SST} \quad = \Sigma_i\Sigma_j y_{ij}^2$	

[1]Summations are for $i = 1, \ldots, a$ and for $j = 1, \ldots, n_i$.
[2]For unbalanced data, $R(\mu)$ and $R(\alpha \mid \mu)$ do not have χ^2-distributions.
[3]For balanced data, $R(\mu)$ and $R(\alpha \mid \mu)$ do have χ^2-distributions, with expected values $ER(\mu) = an\mu^2 + n\sigma_\alpha^2 + \sigma_e^2$, $ER(\alpha \mid \mu) = (a-1)(n\sigma_\alpha^2 + \sigma_e^2)$ and $E(\text{SSE}) = a(n-1)\sigma_e^2$.

b. Analysis of variance

Using Theorem 7.2a it is easily established that expected values of the sums of squares of Table 13.3 are as shown in that table. Also, under normality assumptions, i.e., $\mathbf{y} \sim \mathcal{N}(\mu\mathbf{1}, \mathbf{V})$ for $\mathbf{V}$ of (97), using Theorem 7.3 shows that the sums of squares $R(\mu)$ and $R(\alpha \mid \mu)$ do *not*, with unbalanced data, have χ^2-distributions. They do in the fixed effects model, where $\mathbf{V} = \sigma_e^2\mathbf{I}$, and also in the mixed model with balanced data, but not in the mixed model with unbalanced data. Thus, although in the mixed model

$$F(\alpha \mid \mu) = \frac{R(\alpha \mid \mu)/(a-1)}{\text{SSE}/(N-a)} \quad \text{tests H: } \sigma_\alpha^2 = 0 \text{ with balanced data,}$$

it does not with unbalanced data, because $R(\alpha \mid \mu)$ does not then have a χ^2-distribution.

c. Testing the mean

$$F(\mu) = \frac{N\bar{y}_{..}^2}{\text{SSE}/(N-a)} \text{ tests } \quad \text{H: } \Sigma n_i\mu_i/N = 0 \text{ in the fixed effects model}$$

in Table 2.7. In contrast, in Table 13.3, under normality assumptions,

$$\frac{N\bar{y}_{..}^2}{R(\alpha \mid \mu)/(a-1)} \quad \text{tests} \quad \text{H: } \mu = 0 \text{ with balanced data;}$$

it does not do so with unbalanced data. A test comparable to this for unbalanced data could be based on the standardized normal variable

$$z = \frac{\bar{y}_{..}}{\sqrt{\sigma_\alpha^2 \sum_{i=1}^{a} n_i^2 \Big/ N^2 + \sigma_e^2/N}} = \sqrt{\frac{N\bar{y}_{..}^2}{\sigma_\alpha^2(\Sigma n_i^2/N) + \sigma_e^2}}$$

if σ_α^2 and σ_e^2 were known. But usually they are not known. In this case, were there to be a mean square, MS say, that was an unbiased estimator of $\sigma_\alpha^2(\Sigma n_i^2/N) + \sigma_e^2$, independent of $N\bar{y}_{..}^2$ and distributed proportionally to a χ^2, then $N\bar{y}_{..}^2/\text{MS}$ would provide a test of H: $\mu = 0$ for the mixed model, with unbalanced data.

Another test is provided by BLUE($\mathbf{X\beta}$) of (87), which for (91) and (93) gives $\hat{\mu} = \mathbf{1'V^{-1}y/1'V^{-1}1}$. This becomes (E13.2), with $\mathbf{V}$ of (93) and

$$\mathbf{V}^{-1} = \frac{1}{\sigma_e^2}\mathbf{d}\left\{\mathbf{I}_{n_i} - \frac{\sigma_\alpha^2}{\sigma_e^2 + n_i\sigma_\alpha^2}\mathbf{J}_{n_i}\right\},$$

$$\hat{\mu} = \sum_{i=1}^{a} \frac{n_i\bar{y}_{i\cdot}}{\sigma_e^2 + n_i\sigma_\alpha^2} \Bigg/ \sum_{i=1}^{a} \frac{n_i}{\sigma_e^2 + n_i\sigma_\alpha^2} = \sum_{i=1}^{a} w_i\bar{y}_{i\cdot} \Bigg/ \sum_{i=1}^{a} w_i, \tag{94}$$

where $1/w_i = \sigma_\alpha^2 + \sigma_e^2/n_i$ is the variance of $\bar{y}_{i\cdot}$. Again, if σ_e^2 and σ_α^2 are known and normality is assumed,

$$\frac{\hat{\mu}}{\sqrt{v(\hat{\mu})}} = \sum_{i=1}^{a} w_i\bar{y}_{i\cdot} \Bigg/ \sqrt{\sum_{i=1}^{a} w_i}$$

is a standardized normal variable for testing H: $\mu = 0$. In that case $\hat{\mu}$ is the maximum likelihood estimator of μ and (94) would be the preferred statistic. These results are summarized in Table 13.4.

Note, in passing that $\hat{\mu}$ of (94), the BLUE of μ, reduces for balanced data (i.e., all n_is equal to n) to $\hat{\mu} = \bar{y}_{..}$. This result would also arise were there to be no random effects in the model. Hence $\bar{y}_{..}$ is the ordinary least squares estimator (OLSE) of μ:

$$(\mathbf{X'X})^{-}\mathbf{X'y} = (\mathbf{1}'_N\mathbf{1}_N)^{-1}\mathbf{1}'_N\mathbf{y} = \bar{y}_{..}\ .$$

Thus for balanced data in the mixed model, the BLUE of μ is the same as the OLSE. This is the simplest example of a more general result in (43), namely, that for a broad class of balanced data the BLUE of estimable functions of fixed effects in mixed models is the same as the OLSE.

TABLE 13.4. STATISTICS FOR "TESTING THE MEAN" IN THE 1-WAY CLASSIFICATION MODEL, ASSUMING NORMALITY[1]

Statistic	Null Hypothesis
Fixed Model	
A. $N\bar{y}_{..}^2/\sigma^2 \sim F_{1,N-a}$	H: $\sum_{i=1}^{a} n_i\mu_i/N = 0$
B. $(\Sigma\bar{y}_{i\cdot})^2\left(\sum_{i=1}^{a} n_i^{-1}\right) \sim F_{1,N-a}$	H: $\sum_{i=1}^{a} \mu_i/a = 0$
Mixed Model	
C. $N\bar{y}_{..}^2/[R(\alpha\mid\mu)/(a-1)] \sim F_{1,a-1}$	H: $\mu = 0$, for balanced data
D. $z = \sqrt{\dfrac{N\bar{y}_{..}^2}{\sigma_\alpha^2\left(\sum_{i=1}^{a} n_i^2/N\right) + \sigma_e^2}} \sim N(0,1)$	H: $\mu = 0$, for unbalanced data
E. $u = \sum_{i=1}^{a} w_i\bar{y}_{i\cdot}\Big/\sqrt{\sum_{i=1}^{a} w_i} \sim N(0,1)$ for $w_i = n_i/(n_i\sigma_\alpha^2 + \sigma_e^2)$	H: $\mu = 0$, for unbalanced data

[1] Lines D and E require knowing σ_α^2 and σ_e^2. When $n_i = n \ \forall\ i$, A = B and D = E; and D and E are equivalent to C because their denominators are estimated by $R(\alpha\mid\mu)/(a-1)$.

d. Prediction

Substituting from (91)–(94) into (86) gives the BLUP of $\boldsymbol{\alpha}$ as

$$\text{BLUP}(\boldsymbol{\alpha}) = [\sigma_\alpha^2\mathbf{I}_a]\{{}_d\,\mathbf{1}'_{n_i}\}\left\{\frac{1}{{}_d\,\sigma_e^2}\left(\mathbf{I}_{n_i} - \frac{\sigma_\alpha^2}{\sigma_e^2 + n_i\sigma_\alpha^2}\mathbf{J}_{n_i}\right)\right\}(\mathbf{y} - \hat{\mu}\mathbf{1}_N)$$

$$= \frac{\sigma_\alpha^2}{\sigma_e^2}\left\{{}_d\,\mathbf{1}'_{n_i} - \frac{\sigma_\alpha^2}{\sigma_e^2 + n_i\sigma_\alpha^2}n_i\mathbf{1}'_{n_i}\right\}(\mathbf{y} - \hat{\mu}\mathbf{1}_N)$$

$$= \frac{\sigma_\alpha^2}{\sigma_e^2}\left\{{}_d\left(1 - \frac{n_i\sigma_\alpha^2}{\sigma_e^2 + n_i\sigma_\alpha^2}\right)\mathbf{1}'_{n_i}\right\}(\mathbf{y} - \hat{\mu}\mathbf{1}_N) .$$

Hence the ith element is

$$\text{BLUP}(\alpha_i) = \frac{\sigma_\alpha^2}{\sigma_e^2 + n_i\sigma_\alpha^2}(y_{i\cdot} - n_i\hat{\mu}) = \frac{n_i\sigma_\alpha^2}{\sigma_e^2 + n_i\sigma_\alpha^2}(\bar{y}_{i\cdot} - \hat{\mu}) .$$

Animal breeders, who use the heritability ratio $h = 4\sigma_\alpha^2/(\sigma_\alpha^2 + \sigma_e^2)$, will recognize the equivalent form

$$\text{BLUP}(\alpha_i) = \frac{nh}{4 + (n-1)h}(\bar{y}_{i\cdot} - \hat{\mu}) \ .$$

e. Estimating variance components

The analysis of variance method of estimating variance components equates the expected mean squares (or sums of squares) of Table 13.3 to their observed values. This gives the equations

$$\left(N - \Sigma n_i^2/N\right)\hat{\sigma}_\alpha^2 + (a-1)\hat{\sigma}_e^2 = R(\boldsymbol{\alpha}\,|\,\mu) \equiv (a-1)\text{MSA}$$

and

$$(N-a)\hat{\sigma}_e^2 = \text{SSE} \equiv (N-a)\text{MSE},$$

where MSA and MSE are the mean squares between and within classes, respectively. Solving these equations gives estimators

$$\hat{\sigma}_e^2 = \text{MSE} \quad \text{and} \quad \hat{\sigma}_\alpha^2 = \frac{\text{MSA} - \text{MSE}}{\left(N - \Sigma_i n_i^2/N\right)/(a-1)} \ .$$

For balanced data these have the simpler form

$$\hat{\sigma}_e^2 = \text{MSE} \quad \text{and} \quad \hat{\sigma}_\alpha^2 = (\text{MSA} - \text{MSE})/n \ .$$

These estimators are, of course, unbiased; and under normality their sampling variances are known. The distribution of $\hat{\sigma}_e^2$ is proportional to a χ^2 but that of $\hat{\sigma}_\alpha^2$ has no closed form. Full details of these and other properties of the estimators are available in Searle (1971, Chapter 9).

13.6. EXERCISES

E13.1. Show that $\mathbf{X}\boldsymbol{\beta}^\circ$ from equation (89) is the same as $\mathbf{X}\boldsymbol{\beta}^\circ$ in (87); and $\tilde{\mathbf{u}}$ of (89) is BLUP($\mathbf{u}$) of (87).

E13.2. In the 1-way classification of Table 13.3, show that under normality $R(\mu)$ and $R(\boldsymbol{\alpha}\,|\,\mu)$ have distributions that are multiples of χ^2-distributions

(a) in the fixed effects model, with either balanced or unbalanced data;

(b) in the mixed model with balanced data, but

(c) not in the mixed model with unbalanced data.

E13.3. **(Randomized blocks, unbalanced data)** The model equation for randomized blocks can be taken as

$$y_{ijk} = \tau_i + \beta_j + e_{ijk},$$

where τ_i is a (fixed) treatment effect, β_j is a (random) block effect and e_{ijk} is a customary error term. Take $i = 1, \ldots, a$, $j = 1, \ldots, b$ and, for unbalanced data, $k = 1, 2, \ldots, n_{ij} \geq 0$ so that the numbers of observations on the treatments is not necessarily the same within a block, nor from block to block, including the possibility of some cases of no observations. Then do the following.

(a) Show that

$$\text{BLUE}\{\tau_i\}_{i=1}^{i=a} = \mathbf{L}^{-1}\left\{y_{i\cdot\cdot} - \sigma_\beta^2 \sum_{j=1}^{b} \frac{n_{ij}y_{\cdot j\cdot}}{\sigma_e^2 + n_{\cdot j}\sigma_\beta^2}\right\}_{i=1}^{i=a}$$

for

$$\mathbf{L}^{-1} = \left[\mathbf{D}\{n_{i\cdot}\}_{i=1}^{i=a} - \sigma_\beta^2 \sum_{j=1}^{b} \frac{1}{\sigma_e^2 + n_{\cdot j}\sigma_\beta^2}\mathbf{c}_j\mathbf{c}_j'\right]^{-1},$$

where $\mathbf{c}_j = \{n_{ij}\}_{i=1}^{i=a}$, the column vector of numbers of observations (including any zeros) on the treatments in block j. Note that this refutes the argument of Steinhorst (1982) that cell means models are not adaptable to randomized block designs.

(b) Show also that the sampling variance–covariance matrix of the BLUE of $\boldsymbol{\tau}$ is $\mathbf{L}^{-1}\sigma_e^2$.

Note: Having the y_{ij}s ordered by k within i within j (rather than within j within i) greatly simplifies the derivation of the result in (a).

E13.4. **(Balanced incomplete blocks)** Balanced incomplete blocks experiments usually have one observation on k treatments in each block, with each of the t treatments occurring in r different blocks and with each treatment occurring with each other treatment in exactly λ blocks.

(a) Show that the BLUE in E13.3(a) simplifies to give

$$\hat{\tau}_i = \frac{r(\sigma_e^2 + k\sigma_\beta^2)}{r\sigma_e^2 + \lambda t\sigma_\beta^2}\left[\bar{y}_{i\cdot} - \frac{k\sigma_\beta^2}{\sigma_e^2 + k\sigma_\beta^2}\bar{y}_i^* + \frac{\lambda t\sigma_\beta^2}{r(\sigma_e^2 + k\sigma_\beta^2)}\bar{y}_{\cdot\cdot}\right],$$

where $\bar{y}_i^*$ is the average of the block averages of the r blocks where treatment i occurs.

(b) For $\rho = \sigma_\beta^2/\sigma_e^2$, show that

$$v(\hat{\tau}_i) = \frac{\sigma_e^2(1 + k\rho)(r + \lambda\rho)}{r + \lambda t\rho}$$

and
$$\text{cov}(\tilde{\tau}_i, \tilde{\tau}_{i'}) = \frac{\sigma_e^2\lambda\rho(1 + r\lambda\rho)}{r(r + \lambda t\rho)}$$

(c) Show that $\hat{\tau}_i$ is the same as the inter- and intra-class estimator given, for example, in Kempthorne (1952, Section 26.4), Federer (1955, p. 418), Cochran and Cox (1957, p. 444), Scheffé (1959, pp. 165–178) and John (1971, pp. 224–234).

REFERENCES

Alalouf, I. S. and Styan, G. P. H. (1973). Characterizations of estimability in the general linear model. *Ann. Stat.* **7**, 194–200.

Anderson, R. D., Henderson, H. V., Pukelsheim, F. and Searle, S. R. (1984). Best estimation of variance components from balanced data with arbitrary kurtosis. *Math. Operationsforsch. Stat. Ser. Stat.* **15**, 163–176.

Arnold, S. F. (1981). *The Theory of Linear Models and Multivariate Analysis*. John Wiley & Sons, New York.

Baksalary, J. K. and Kala, R. (1983). On equalities between BLUEs, WLSEs and SLSEs. *Can. J. Stat.* **11**, 119–123.

Bittner, A. C., Jr. (1974). Exact linear restrictions on parameters in a linear regression model. *The Amer. Stat.* **28**, 36.

Brown, K. G. (1976). Asymptotic behavior of MINQUE-type estimators of variance components. *Ann. Stat.* **4**, 746–754.

Burdick, D. S. (1979). Rows adjusted for columns versus simple row means in the two-way ANOVA: When are the two hypotheses equivalent? *J. Amer. Stat. Assoc.* **74**, 457–458.

Cochran, W. G. (1951). Improvement by means of selection. *Proc. 2nd Berkeley Symp.*, 449–470.

Cochran, W. G. and Cox, G. M. (1957). *Experimental Designs*, 2nd ed. John Wiley & Sons, New York.

Cox, P. (1980). The all-elements-unit matrix. *The Amer. Stat.* **34**, 191–192.

Driscoll, M. F. and Gundberg, W. R. (1986). A history of the development of Craig's theorem. *The Amer. Stat.* **40**, 65–71.

Emerson, J. D., Hoaglin, D. C. and Kempthorne, P. C. (1984). Leverage in least squares additive-plus-multiplicative fits for two-way tables. *J. Amer. Stat. Assoc.* **79**, 329–335.

Federer, W. T. (1955). *Experimental Design*. Macmillan, New York.

Fisher, R. A. (1925). Theory of statistical estimation. *Proc. Cambridge Philos. Soc.* **22**, 700–725.

Giesbrecht, F. G. (1983). An efficient procedure for computing MINQUE of variance components and generalized least squares estimates of fixed effects. *Commun. Stat. A. Theory and Methods* **12**, 2, 2169–2177.

Giesbrecht, F. G. (1985). MIXMOD, a SAS procedure for analysing mixed models. Technical Report No. 1659, Institute of Statistics, North Carolina State University, Raleigh, North Carolina.

Goodnight, J. H. and Hemmerle, W. J. (1978). A simplified algorithm for the W-transformation in variance component estimation. SAS Technical Report R-104, SAS Institute, Cary, North Carolina.

Graybill, F. A. (1976). *Theory and Applications of the Linear Model*. Duxbury, North Scituate, Massachusetts.

Guttman, I. (1982). *Linear Models, An Introduction*. John Wiley & Sons, New York.

Hartley, H. O. and Rao, J. N. K. (1967). Maximum likelihood estimation for the mixed analysis of variance model. *Biometrika* **54**, 93–108.

Harville, D. A. (1977). Maximum likelihood approaches to variance component estimation and to related problems. *J. Amer. Stat. Assoc.* **72**, 320–340.

Hemmerle, W. J. and Hartley, H. O. (1973). Computing maximum likelihood estimates for the mixed A. O. V. model using the W-transformation. *Technometrics* **15**, 819–831.

Henderson, C. R. (1953). Estimation of variance and covariance components. *Biometrics* **9**, 226–252.

Henderson, C. R., Kempthorne, O., Searle, S. R. and Von Krosigk, C. N. (1959). Estimation of environmental and genetic trends from records subject to culling. *Biometrics* **15**, 192–218.

Henderson, C. R., Searle, S. R. and Schafer, L. R. (1974). The invariance and calculation of Method 2 for estimating variance components. *Biometrics* **30**, 583–588.

Hoaglin, D. C. and Welsch, R. E. (1978). The hat matrix in regression and ANOVA. *The Amer. Stat.* **32**, 17–22 (and 146).

Hocking, R. R. (1985). *The Analysis of Linear Models*. Brooks/Coles, Monterey, CA.

Hudson, G. F. S. and Searle, S. R. (1982). Hypothesis testing with Type IV sums of squares of the computer routine SAS GLM. *Proceedings, 7th Annu. SAS User's Group Int. Conf.*, 521–527.

Hurlburt, R. T. and Spiegel, D. K. (1976). Dependence of F ratios sharing a common denominator mean square. *The Amer. Stat.* **30**, 74–75, 78.

John, P. W. M. (1971). *Statistical Design and Analysis of Experiments*. Macmillan, New York.

Johnson, N. L. and Kotz, S. (1970). *Continuous Univariate Distributions*, 1 and 2. John Wiley & Sons, New York.

Kackar, R. N. and Harville, D. A. (1981). Unbiasedness of two-stage estimation and prediction procedures for mixed linear models. *Commun. Stat. A. Theory and Methods* **10**, 1249–1261.

Kackar, R. N. and Harville, D. A. (1984). Approximations for standard errors of estimators of fixed and random effects in mixed linear models. *J. Amer. Stat. Assoc.* **79**, 853–861.

Kempthorne, O. (1952). *The Design and Analysis of Experiments*. John Wiley & Sons, New York.

Kempthorne, O. (1975). Fixed and mixed models in analysis of variance. *Biometrics* **31**, 473–486.

Kempthorne, O. (1980). The term design matrix. *The Amer. Stat.* **34**, 249.

LaMotte, L. R. (1970). A class of estimators of variance components. Technical Report No. 10, Department of Statistics, University of Kentucky, Lexington, Kentucky.

LaMotte, L. R. (1971). Locally best quadratic estimators of variance components. Technical Report No. 22, Department of Statistics, University of Kentucky, Lexington, Kentucky.

LaMotte, L. R. (1973). Quadratic estimation of variance components. *Biometrics* **29**, 311–330.

Li, C. C. (1964). *Introduction to Experimental Statistics*. McGraw-Hill, New York.

Mantell, E. H. (1973). Exact linear restrictions on parameters in the classical linear regression model. *The Amer. Stat.* **27**, 86–87.

Marasinghe, M. G. and Johnson, D. E. (1981). Testing subhypotheses in the multiplicative interaction model. *Technometrics* **23**, 385–393.

Marasinghe, M. G. and Johnson, D. E. (1982). A test of incomplete additivity in the multiplicative interaction model. *J. Amer. Stat. Assoc.* **77**, 869–877.

Marsaglia, G. and Styan, G. P. H. (1974). Rank conditions for generalized inverses of partitioned matrices. *Sankhyā* **36**, 437–442.

Martin, K. J. (1980). A partition of a 2-factor interaction, with an agricultural example. *Appl. Stat.* **29**, 149–155.

Mathew, T. and Bhimasankaram, P. (1983). Optimality of BLUEs in a general linear model. *J. Stat. Planning and Inference* **8**, 315–319.

Meredith, M. P. and Cady, F. B. (1984). Methodology analysis of treatment means from factorial experiments with unequal replication. Technical Report BU-859-M, Biometrics Unit, Cornell University, Ithaca, New York.

Miller, J. J. (1977). Asymptotic properties of maximum likelihood estimates in the mixed model of the analysis of variance. *Ann. Stat.* **5**, 746–762.

Moore, E. H. (1920). On the reciprocal of the general algebraic matrix. *Amer. Math. Soc. Bull.* **26**, 394–395.

Nelder, J. A. (1976). Hypothesis testing in linear models. *The Amer. Stat.* **30**, 103.

Nelder, J. A. (1977). A reformulation of linear models. *J. R. Stat. Soc. Ser. A* **140**, 48–77.

Patterson, H. D. and Thompson, R. (1971). Recovery of inter-block information when block sizes are unequal. *Biometrika* **58**, 545–554.

Penrose, R. A. (1955). A generalized inverse for matrices. *Proc. Cambridge Philos. Soc.* **51**, 406–413.

Quaas, R. L. and Bolgiano, D. C. (1979). Sampling variances of the MIVQUE and Method 3 estimators of sire component of variance. *Proceedings "Variance Components and Animal Breeding, a Conference in Honor of C. R. Henderson*," L. D. Van Vleck and S. R. Searle, Editors, Cornell University, Ithaca, New York, 99–106.

Rao, C. R. (1962). A note on a generalized inverse of a matrix with applications to problems in mathematical statistics. *J. R. Stat. Soc. Ser. B* **24**, 152–158.

Rao, C. R. (1965). *Linear Statistical Inference and Its Applications*, 1st. ed. John Wiley & Sons, New York.

Rao, C. R. (1970). Estimation of heteroscedastic variances in linear models. *J. Amer. Stat. Assoc.* **65**, 161–172.

Rao, C. R. (1971a). Estimation of variance and covariance components—MINQUE theory. *J. Multivar. Anal.* **1**, 257–275.

Rao, C. R. (1971b). Minimum variance quadratic unbiased estimation of variance components. *J. Multivar. Anal.* **1**, 445–456.

Rao, C. R. (1972). Estimation of variance and covariance components in linear models. *J. Amer. Stat. Assoc.* **67**, 112–115.

Robinson, J. (1965). The distribution of a general quadratic form in normal variables. *Aust. J. Stat.* **7**, 110–114.

Scheffé, H. (1959). *The Analysis of Variance*. John Wiley & Sons, New York.

Searle, S. R. (1968). Another look at Henderson's methods of estimating variance components. *Biometrics* **24**, 749–788.

Searle, S. R. (1970). Large sample variances of maximum likelihood estimators of variance components. *Biometrics* **26**, 505–524.

Searle, S. R. (1971). *Linear Models*. John Wiley & Sons, New York.

Searle, S. R. (1974). Prediction, mixed models and variance components. *Reliability and Biometry*, F. Proschan and R. J. Serfling, Editors, Society of Industrial and Applied Mathematics, Philadelphia, pp. 229–266.

Searle, S. R. (1979). Alternative covariance models for the 2-way crossed classification. *Commun. Stat. A. Theory and Methods* **8**, 799–818.

Searle, S. R. (1982). *Matrix Algebra Useful for Statistics*. John Wiley & Sons, New York.

Searle, S. R. (1984a). Detailed proofs (class notes) of necessary and sufficient conditions for independence and chi-square distributions properties of quadratic forms of normal variables. Technical Report BU-834-M, Biometrics Unit, Cornell University, Ithaca, New York.

Searle, S. R. (1984b). Restrictions and generalized inverses in linear models. *The Amer. Stat.* **38**, 53–54.

Searle, S. R. (1987). Best linear unbiased estimation in mixed models of the analysis of variance. *Felicitation Volume in Honor of Franklin A. Graybill*, J. Srivastava, Editor (in press).

Searle, S. R. and Henderson, H. V. (1979). Dispersion matrices for variance components models. *J. Amer. Stat. Assoc.* **74**, 465–470.

Searle, S. R. and Henderson, H. V. (1983). Faults in a computing algorithm for reparameterizing linear models. *Commun. Stat. B. Simulation and Computation* **12**, 67–76.

Searle, S. R. and Hudson, G. F. S. (1982). Some distinctive features of output from statistical computing packages for analysis of covariance. *Biometrics* **38**, 337–345.

Searle, S. R., Hudson, G. F. S. and Federer, W. T. (1982). Annotated computer output for analysis of covariance. *The Amer. Stat.* **37**, 172–173.

Searle, S. R. and Pukelsheim, F. (1985). Establishing χ^2 properties of sums of squares using induction. *The Amer. Stat.* **39**, 301–303.

Searle, S. R. and Speed, F. M. (1982). Estimability in the cell means general linear model. Technical Report BU-730-M, Biometrics Unit, Cornell University, Ithaca, New York.

Searle, S. R., Speed, F. M. and Henderson, H. V. (1981). Some computational and model equivalences in analysis of variance of unequal-subclass-numbers data. *The Amer. Stat.* **35**, 16–33.

Searle, S. R., Speed, F. M. and Milliken, G. A. (1980). Population marginal means in the linear model: an alternative to least squares means. *The Amer. Stat.* **34**, 216–221.

Searle, S. R., Swallow, W. T. and McCulloch, C. E. (1984). Non-testable hypotheses in linear models. *SIAM J. Algebraic and Discrete Methods* **5**, 486–496.

Seber, G. A. F. (1977). *Linear Regression Analysis*. John Wiley & Sons, New York.

Seely, J. (1977). Estimability and linear hypotheses. *The Amer. Stat.* **31**, 121–123.

Seifert, B. (1979). Optimal testing for fixed effects in general balanced mixed classifications models. *Math. Operationsforsch. Stat. Ser. Stat.* **10**, 237–256.

Smith, D. and Hocking, R. R. (1978). Maximum likelihood analyses of the mixed model: the balanced case. *Commun. Stat. A. Theory and Methods* **7**, 1253–1266.

Snedecor, G. W. and Cochran, W. G. (1980). *Statistical Methods*, 7th ed. Iowa State University Press, Ames, Iowa.

Snee, R. D. (1982). Nonadditivity in a two-way classification: is it interaction or nonhomogeneous variance? *J. Amer. Stat. Assoc.* **77**, 515–519.

Steel, R. G. D. and Torrie, J. H. (1980). *Principles and Procedures of Statistics: A Biometrical Approach*, 2nd ed. McGraw-Hill, New York.

Steinhorst, R. K. (1982). Resolving current controversies in analysis of variance. *The Amer. Stat.* **36**, 138–139.

Stigler, S. M. (1984). Kruskal's proof of the joint distribution of $\bar{X}$ and S^2. *The Amer. Stat.* **38**, 134–135.

Swallow, W. H. and Monahan, J. F. (1984). Monte Carlo comparison of ANOVA, MIVQUE, REML and ML estimators of variance components. *Technometrics* **26**, 47–58.

Timm, N. H. and Carlson, J. E. (1973). Analysis of variance through full rank models. Working Paper No. 21, Department of Educational Research, University of Pittsburgh.

Turnbull, H. W. (1947). *Theory of Equations*. Oliver & Boyd, Edinburgh.

Urquhart, N. S. (1982). Adjustment in covariance when one factor affects the covariable. *Biometrics* **38**, 651–660.

Urquhart, N. S. and Weeks, D. L. (1978). Linear models in messy data: some problems and alternatives. *Biometrics* **34**, 696–705.

Weeks, D. L. and Williams, D. R. (1964). A note on the determination of connectedness in an *N*-way cross classification. *Technometrics* **6**, 319–324.

Winer, B. J. (1971). *Statistical Principles in Experimental Design*, 2nd ed. McGraw-Hill, New York.

Yates, F. (1934). The analysis of multiple classifications with unequal numbers in the different classes. *J. Amer. Stat. Assoc.* **29**, 51–66.

Zyskind, G. (1967). On canonical forms, nonnegative covariance matrices and best and simple least squares linear estimators in linear models. *Ann. Math. Stat.* **38**, 1092–1109.

Zyskind, G. and Johnson, P. A. (1973). On a zero residual sum in regression. *The Amer. Stat.* **27**, 43–44.

STATISTICAL TABLES

The following abridged tables of the normal distributions and central t-, χ^2- and F-distributions are given solely for convenience. They in no way represent detailed coverage of these distributions.

TABLE 1. VALUES P_x AND x ON THE NORMAL DISTRIBUTION

$N(0, 1)$-distribution

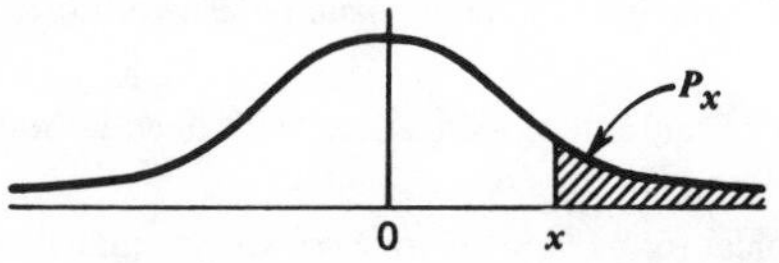

$P_x = \Pr\{N(0, 1)\text{-variable} \geq x\}$

P_x	x	P_x	x	P_x	x	P_x	x	P_x	x
.50	0.00	**.050**	**1.64**	.030	1.88	.020	2.05	**.010**	**2.33**
.45	0.13	.048	1.66	.029	1.90	.019	2.07	.009	2.37
.40	0.25	.046	1.68	.028	1.91	.018	2.10	.008	2.41
.35	0.39	.044	1.71	.027	1.93	.017	2.12	.007	2.46
.30	0.52	.042	1.73	.026	1.94	.016	2.14	.006	2.51
.25	0.67	.040	1.75	**.025**	**1.96**	.015	2.17	**.005**	**2.58**
.20	0.84	.038	1.77	.024	1.98	.014	2.20	.004	2.65
.15	1.04	.036	1.80	.023	2.00	.013	2.23	.003	2.75
.10	**1.28**	.034	1.83	.022	2.01	.012	2.26	.002	2.88
.05	**1.64**	.032	1.85	.021	2.03	.011	2.29	**.001**	**3.09**
								0.000	∞

Bold-face values are those often (but not exclusively) used when testing hypotheses and/or establishing confidence intervals.

Source: Table 2 of Lindley and Miller (1958), *Cambridge Elementary Statistical Tables*, published by Cambridge University Press, with kind permission of the authors and publishers.

TABLE 2. VALUES OF $t_{n,\alpha}$ ON THE $t(n)$-DISTRIBUTION

$t(n)$-distribution

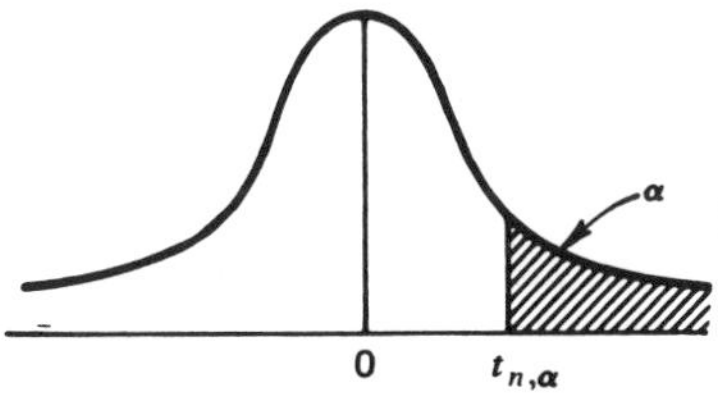

Note: $\Pr\{t(n)\text{-variable} \geq t_{n,\alpha}\} = \alpha$
$\Pr\{t(n)\text{-variable} \leq -t_{n,\alpha}\} = \alpha$
$\Pr\{|t(n)\text{-variable}| \geq t_{n,\alpha}\} = 2\alpha$

n (d.f.)	α .10	.05	.025	.010	.005
1	3.08	6.31	12.71	31.82	63.66
2	1.89	2.92	4.30	6.97	9.92
3	1.64	2.35	3.18	4.54	5.84
4	1.53	2.13	2.78	3.75	4.60
5	1.48	2.02	2.57	3.36	4.03
6	1.44	1.94	2.45	3.14	3.71
7	1.42	1.89	2.36	3.00	3.50
8	1.40	1.86	2.31	2.90	3.36
9	1.38	1.83	2.26	2.82	3.25
10	1.37	1.81	2.23	2.76	3.17
12	1.36	1.78	2.18	2.68	3.06
14	1.34	1.76	2.14	2.62	2.98
16	1.34	1.75	2.12	2.58	2.92
18	1.33	1.73	2.10	2.55	2.88
20	1.32	1.72	2.09	2.53	2.84
30	1.31	1.70	2.04	2.46	2.75
40	1.30	1.68	2.02	2.42	2.70
60	1.30	1.67	2.00	2.39	2.66
120	1.29	1.66	1.98	2.36	2.62
∞ [$N(0, 1)$]	1.28	1.64	1.96	2.33	2.58

Source: Table 2 is adapted from Table III of Fisher and Yates (1963), *Statistical Tables for Biological, Agricultural and Medical Research*, 6th Ed., published by Oliver and Boyd, Edinburgh, with kind permission of the authors and publishers.

TABLE 3. VALUES OF $\chi^2_{n,\alpha}$ ON THE $\chi^2(n)$-DISTRIBUTION

$$\Pr\{\chi^2(n)\text{-variable} \geq \chi^2_{n,\alpha}\} = \alpha$$

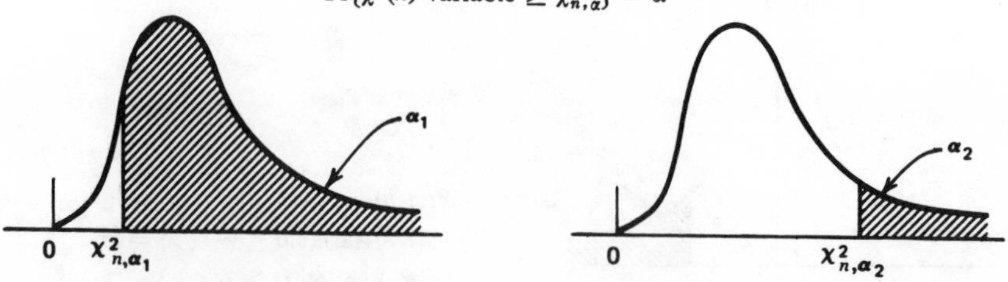

n	α_1			α_2		
(d.f.)	.99	.975	.95	.05	.025	.01
			Values of $\chi^2_{n,\alpha}$			
1	.000157	.000982	.00393	3.84	5.02	6.63
2	.020	.051	.103	5.99	7.38	9.21
3	.115	.216	.352	7.81	9.35	11.34
4	.297	.484	.711	9.49	11.14	13.28
5	.554	.831	1.145	11.07	12.83	15.09
6	.872	1.24	1.64	12.59	14.45	16.81
7	1.24	1.69	2.17	14.07	16.01	18.48
8	1.65	2.18	2.73	15.51	17.53	20.09
9	2.09	2.70	3.33	16.92	19.02	21.67
10	2.56	3.25	3.94	18.31	20.48	23.21
11	3.05	3.82	4.57	19.68	21.92	24.73
12	3.57	4.40	5.23	21.03	23.34	26.22
13	4.11	5.01	5.89	22.36	24.74	27.69
14	4.66	5.63	6.57	23.68	26.12	29.14
15	5.23	6.26	7.26	25.00	27.49	30.58
20	8.26	9.59	10.85	31.41	34.17	37.57
25	11.52	13.12	14.61	37.65	40.65	44.31
30	14.95	16.79	18.49	43.77	46.98	50.89
40	22.16	24.43	26.51	55.76	59.34	63.69
50	29.71	32.36	34.76	67.50	71.42	76.15
60	37.48	40.48	43.19	79.08	83.30	88.38
70	45.44	48.76	51.74	90.53	95.02	100.4
80	53.54	57.15	60.39	101.9	106.6	112.3
90	61.75	65.65	69.13	113.1	118.1	124.1
100	70.06	74.22	77.93	124.3	129.6	135.8

Source: Abridged from Table 8 of Pearson and Hartley (1954), *Biometrika Tables for Statisticians, Volume I,* published at the Cambridge University Press for *Biometrika* Trustees, with kind permission of the authors and publishers.

TABLE 4A. VALUES OF $F_{n_1, n_2, \alpha}$ ON THE $F(n_1, n_2)$-DISTRIBUTION

$F(n_1, n_2)$-distribution

$\Pr\{F(n_1, n_2)\text{-variable} \geq F_{n_1,n_2,\alpha}\} = \alpha = .05$

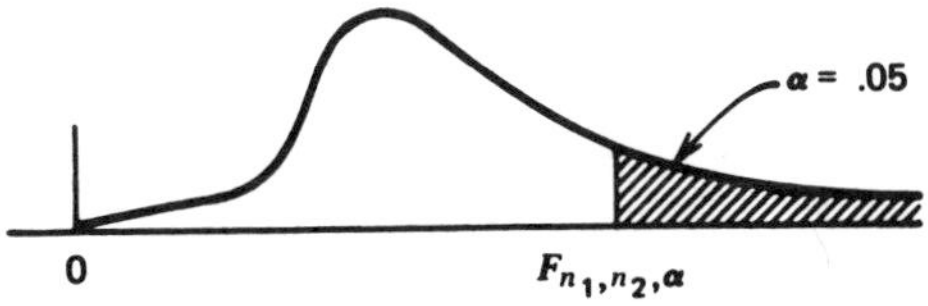

$\alpha = .05$

n_2 (denom. d.f.)	1	2	4	6	8	10	12	24	∞
	$[t_{n_2,.025}]^2$				Values of $F_{n_1,n_2,\alpha}$				
1	161.4	199.5	224.6	234.0	238.9	241.9	243.9	249.1	254.3
2	18.51	19.00	19.25	19.33	19.37	19.40	19.41	19.45	19.50
3	10.13	9.55	9.12	8.94	8.85	8.79	8.74	8.64	8.53
4	7.71	6.94	6.39	6.16	6.04	5.96	5.91	5.77	5.63
5	6.61	5.79	5.19	4.95	4.82	4.74	4.68	4.53	4.36
6	5.99	5.14	4.53	4.28	4.15	4.06	4.00	3.84	3.67
7	5.59	4.74	4.12	3.87	3.73	3.64	3.57	3.41	3.23
8	5.32	4.46	3.84	3.58	3.44	3.35	3.28	3.12	2.93
9	5.12	4.26	3.63	3.37	3.23	3.14	3.07	2.90	2.71
10	4.96	4.10	3.48	3.22	3.07	2.98	2.91	2.74	2.54
11	4.84	3.98	3.36	3.09	2.95	2.85	2.79	2.61	2.40
12	4.75	3.89	3.26	3.00	2.85	2.75	2.69	2.51	2.30
13	4.67	3.81	3.18	2.92	2.77	2.67	2.60	2.42	2.21
14	4.60	3.74	3.11	2.85	2.70	2.60	2.53	2.35	2.13
15	4.54	3.68	3.06	2.79	2.64	2.54	2.48	2.29	2.07
20	4.35	3.49	2.87	2.60	2.45	2.35	2.28	2.08	1.84
25	4.24	3.39	2.76	2.49	2.34	2.24	2.16	1.96	1.71
30	4.17	3.32	2.69	2.42	2.27	2.16	2.09	1.89	1.62
40	4.08	3.23	2.61	2.34	2.18	2.08	2.00	1.79	1.51
60	4.00	3.15	2.53	2.25	2.10	1.99	1.92	1.70	1.39
120	3.92	3.07	2.45	2.17	2.02	1.91	1.83	1.61	1.25
∞	3.84	3.00	2.37	2.10	1.94	1.83	1.75	1.52	1.00

Source: Abridged from Table 18 of Pearson and Hartley (1954), *Biometrika Tables for Statisticians, Volume I*, published at the Cambridge University Press for *Biometrika* Trustees, with kind permission of the authors and publishers.

TABLE 4B. VALUES OF $F_{n_1, n_2, \alpha}$ ON THE $F(n_1, n_2)$-DISTRIBUTION

$\Pr\{F(n_1, n_2)\text{-variable} \geq F_{n_1,n_2,\alpha}\} = \alpha = .01$

$F(n_1, n_2)$-distribution

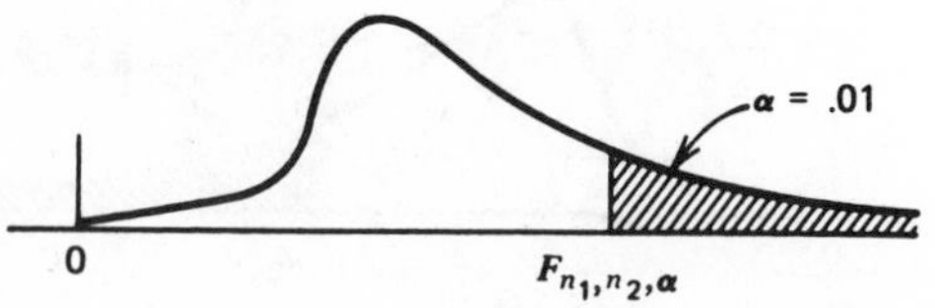

$\alpha = .01$

n_2 (denom. d.f.)	1	2	4	6	8	10	12	24	∞
	$[t_{n_2,.005}]^2$	Values of $F_{n_1,n_2,\alpha}$							
1	4052	5000	5625	5859	5982	6056	6106	6235	6366
2	98.50	99.00	99.25	99.33	99.37	99.40	99.42	99.46	99.50
3	34.12	30.82	28.71	27.91	27.49	27.23	27.05	26.60	26.13
4	21.20	18.00	15.98	15.21	14.80	14.55	14.37	13.93	13.46
5	16.26	13.27	11.39	10.67	10.29	10.05	9.89	9.47	9.02
6	13.75	10.92	9.15	8.47	8.10	7.87	7.72	7.31	6.88
7	12.25	9.55	7.85	7.19	6.84	6.62	6.47	6.07	5.65
8	11.26	8.65	7.01	6.37	6.03	5.81	5.67	5.28	4.86
9	10.56	8.02	6.42	5.80	5.47	5.26	5.11	4.73	4.31
10	10.04	7.56	5.99	5.39	5.06	4.85	4.71	4.33	3.91
11	9.65	7.21	5.67	5.07	4.74	4.54	4.40	4.02	3.60
12	9.33	6.93	5.41	4.82	4.50	4.30	4.16	3.78	3.36
13	9.07	6.70	5.21	4.62	4.30	4.10	3.96	3.59	3.17
14	8.86	6.51	5.04	4.46	4.14	3.94	3.80	3.43	3.00
15	8.68	6.36	4.89	4.32	4.00	3.80	3.67	3.29	2.87
20	8.10	5.85	4.43	3.87	3.56	3.37	3.23	2.86	2.42
25	7.77	5.57	4.18	3.63	3.32	3.13	2.99	2.62	2.17
30	7.56	5.39	4.02	3.47	3.17	2.98	2.84	2.47	2.01
40	7.31	5.18	3.83	3.29	2.99	2.80	2.66	2.29	1.80
60	7.08	4.98	3.65	3.12	2.82	2.63	2.50	2.12	1.60
120	6.85	4.79	3.48	2.96	2.66	2.47	2.34	1.95	1.38
∞	6.63	4.61	3.32	2.80	2.51	2.32	2.18	1.79	1.00

Source: Abridged from Table 18 of Pearson and Hartley (1954), *Biometrika Tables for Statisticians*, *Volume I*, published at the Cambridge University Press for the *Biometrika* Trustees, with kind permission of the authors and publishers.

LIST OF TABLES AND FIGURES

CHAPTER 4:

CHAPTER 5:

CHAPTER 6:

CHAPTER 11:

Table

CHAPTER 12:

Table

CHAPTER 13:

INDEX

Note: The index should be used in conjunction with the Table of Contents, which is very detailed. For index entries that refer to chapters, section titles in the Table of Contents are relevant sub-entries for the index.

Applied Probability and Statistics (Continued)

JUDGE, HILL, GRIFFITHS, LÜTKEPOHL and LEE • Introduction to the Theory and Practice of Econometrics
JUDGE, GRIFFITHS, HILL, LÜTKEPOHL and LEE • The Theory and Practice of Econometrics, *Second Edition*
KALBFLEISCH and PRENTICE • The Statistical Analysis of Failure Time Data
KISH • Statistical Design for Research
KISH • Survey Sampling
KUH, NEESE, and HOLLINGER • Structural Sensitivity in Econometric Models
KEENEY and RAIFFA • Decisions with Multiple Objectives
LAWLESS • Statistical Models and Methods for Lifetime Data
LEAMER • Specification Searches: Ad Hoc Inference with Nonexperimental Data
LEBART, MORINEAU, and WARWICK • Multivariate Descriptive Statistical Analysis: Correspondence Analysis and Related Techniques for Large Matrices
LINHART and ZUCCHINI • Model Selection
LITTLE and RUBIN • Statistical Analysis with Missing Data
McNEIL • Interactive Data Analysis
MAINDONALD • Statistical Computation
MALLOWS • Design, Data, and Analysis by Some Friends of Cuthbert Daniel
MANN, SCHAFER and SINGPURWALLA • Methods for Statistical Analysis of Reliability and Life Data
MARTZ and WALLER • Bayesian Reliability Analysis
MIKÉ and STANLEY • Statistics in Medical Research: Methods and Issues with Applications in Cancer Research
MILLER • Beyond ANOVA, Basics of Applied Statistics
MILLER • Survival Analysis
MILLER, EFRON, BROWN, and MOSES • Biostatistics Casebook
MONTGOMERY and PECK • Introduction to Linear Regression Analysis
NELSON • Applied Life Data Analysis
OSBORNE • Finite Algorithms in Optimization and Data Analysis
OTNES and ENOCHSON • Applied Time Series Analysis: Volume I, Basic Techniques
OTNES and ENOCHSON • Digital Time Series Analysis
PANKRATZ • Forecasting with Univariate Box-Jenkins Models: Concepts and Cases
PIELOU • Interpretation of Ecological Data: A Primer on Classification and Ordination
PLATEK, RAO, SARNDAL and SINGH • Small Area Statistics: An International Symposium
POLLOCK • The Algebra of Econometrics
RAO and MITRA • Generalized Inverse of Matrices and Its Applications
RÉNYI • A Diary on Information Theory
RIPLEY • Spatial Statistics
RIPLEY • Stochastic Simulation
ROSS • Introduction to Probability and Statistics for Engineers and Scientists
ROUSSEEUW and LEROY • Robust Regression and Outlier Detection
RUBIN • Multiple Imputation for Nonresponse in Surveys
RUBINSTEIN • Monte Carlo Optimization, Simulation, and Sensitivity of Queueing Networks
SCHUSS • Theory and Applications of Stochastic Differential Equations
SEARLE • Linear Models
SEARLE • Linear Models for Unbalanced Data

(*continued from front*)